Quantum Mechanics in Nanoscience and Engineering

Quantum Mechanics in Nanoscience and Engineering covers both elementary and advanced quantum mechanics within a coherent and self-contained framework. Undergraduate students of physics, chemistry, and engineering will find comprehensive coverage of their introductory quantum mechanics courses, and graduate students will gain an understanding of additional tools and concepts necessary to describe real-world phenomena. Each topic presented is first motivated by an experimental technique, phenomenon, or concept derived directly from the realm of nanoscience and technology. The machinery of quantum mechanics is described and reinforced through the perspective of nanoscale phenomena, and in this manner practical and fundamental questions are raised and answered. The main text remains fluent and accessible by leaving technical details and mathematical proofs to guided exercises. Introductory readers may overlook these exercises, while rigorous students can benefit from reading the guidance or solving the exercises in full to strengthen and consolidate their understanding of the material.

Uri Peskin is Professor of Chemistry and a member of the Russell Berrie Nanotechnology Institute and the Helen Diller quantum center at Technion – Israel Institute of Technology. His expertise lies in atomic and molecular physics and scientific computing, with emphasis on quantum dynamics on the nanoscale. He was a postdoctoral fellow at the University of California at Berkeley, and a visiting professor at Harvard and Freiburg universities. His research and teaching of quantum mechanics has won him several awards, including the Yanai Prize for Excellence in Academic Education.

My deepest gratitude to my science teachers

Meira and Igal Peskin, Naphtali Shoham, Menachem Fisch, Avinoam Ben-Shaul, Ronnie Kosloff, Raphael D. Levine, Maurice Cohen, Robert B. Gerber, Nimrod Moiseyev, Roland Lefebvre, Gabriel Kventsel, Ruben Pauncz, Jacob Katriel, Tsofar Maniv, Yitzhak Apeloig, Itzhak Oref, Claude Leforestier, Ofir E. Alon, Nir Ben-Tal, Naomi Rom, Rami Rom, William H. Miller, Hanna Reisler, Ron Naaman, Ilya Vorobeichick, Rob D. Coalson, David J. Tannor, Amnon Stanger, Lorenz. S. Cederbaum, Åke Edlund, Yoav Eichen, Soliman Khatib, Musa Abu-Hilu, Tamar Seidman, Frank A. Weinhold, Wolfgang Domcke, Eli Pollak, Shmuel Gurvitz, Hans-Dieter Meyer, Daniel Neuhauser, Ilan Bar-On, Michal Steinberg, Oded Godsi, Asher Schmidt, Abraham Nitzan, Michael Galperin, Mark A. Ratner, Shammai Speiser, Joshua Jortner, Lihu Berman, Maytal Caspary Toroker, Alon Malka, Daria Brisker-Klaiman, Roie Volkovich, Shachar Klaiman, Shira Weissman, Sabre Kais, Daly Davis, Michael A. Collins, Edvardas Narevicius, Vitali Averbukh, Michael Thoss, Rainer Härtle, Yossi Levy, Roman Vaxenburg, Tamar Goldzak, Yoram Selzer, Tal Simon, Yehudit J. Dori, Vered Dangur, Roni Pozner, Efrat Lifshitz, Maayan Kuperman, Ariel D. Levine, Anat Kira, Michael Iv, Lev Chuntonov, Semion Saikin, Christoph Kreisbeck, Alán Aspuru-Guzik, Doran I. G. Bennett, Hossein R. Sadeghpour, Yossi Paltiel, Andre Erpenbeck, Avner Fleischer, Yaling Ke, Roi Baer, Leeor Kronik, Yonatan Dubi, Gilad Haran, Danny Porath, Ferdinand Evers, Oren Tal, Ioan Baldea, Spiros Skourtis, Eberhard K. U. Gross, Todd Martinez, Eitan Geva, Eran Rabani, Oded Hod, Milan Šindelka, Arik Landau, Ido Gilary, Eli Kolodney, Alon Hoffman, Zohar Amitay, Saar Rahav, David Gelbwaser-Klimovsky, E-Dean Fung, Latha Venkataraman, Yoni Eshel, Yuval Agam, Nadav Amdursky, Amir Ilan, Linoy Nagar, Yonathan Langbeheim.

To Hani, Noa and Sagi, Shiri and Ori,
I cannot thank you enough. I can only apologize for not being there as much as I should while writing and putting this book together.

To my parents Meira and Igal, and to Noam, Niv, Yali and their families,
thank you for always being there. My special thanks to Noam Peskin for legal advice.

Quantum Mechanics in Nanoscience and Engineering

URI PESKIN

Technion – Israel Institute of Technology

CAMBRIDGE
UNIVERSITY PRESS

Shaftesbury Road, Cambridge CB2 8BS, United Kingdom

One Liberty Plaza, 20th Floor, New York, NY 10006, USA

477 Williamstown Road, Port Melbourne, VIC 3207, Australia

314–321, 3rd Floor, Plot 3, Splendor Forum, Jasola District Centre, New Delhi – 110025, India

103 Penang Road, #05–06/07, Visioncrest Commercial, Singapore 238467

Cambridge University Press is part of Cambridge University Press & Assessment, a department of the University of Cambridge.

We share the University's mission to contribute to society through the pursuit of education, learning and research at the highest international levels of excellence.

www.cambridge.org
Information on this title: www.cambridge.org/9781108834902
DOI: 10.1017/9781108877787

First published 2023

Printed in the United Kingdom by TJ Books Limited, Padstow Cornwall

A catalogue record for this publication is available from the British Library.

ISBN 978-1-108-83490-2 Hardback

Additional resources for this publication at cambridge.org/peskin.

Contents

Preface: Who Can Benefit from Reading This Book?

Nanoscience and nano-engineering are rapidly growing fields that break the boundaries between classical disciplines such as Physics, Chemistry, Biology, and Engineering. The potential for utilizing nanoscale devices in very different applications, including, for example, medicine, electronics, or information and data processing, drives research efforts toward unraveling the way nature works on the nanoscale. One thing that immediately becomes apparent is that the rules of Quantum Mechanics (QM) are necessary for understanding nature on this scale. In fact, it was the development of experimental techniques with nanoscale resolution that revealed the wave–particle duality and contributed to the formulation of QM in the early twentieth century (for example, the X-ray crystallography contributed to the discovery of the wave nature of electrons in the Davisson–Germer experiment, discussed in Chapter 1).

Along with the potential for advancement and the growing interest in nanoscience and nano-engineering comes the challenge of making the principles of QM accessible and useful to researchers from different fields. Scientists and engineers turning to nanotechnology from diverse disciplines such as biology, chemical engineering, or mechanical engineering often lack formal education in QM, and sometimes lack even the mathematical experience or technical skills needed to learn QM from a standard introductory textbook. Moreover, the research in nanoscience and engineering is motivated by "real-world" problems that require rather advanced topics not usually addressed in introductory textbooks. For example, most textbooks would address the exact solution of the Schrödinger equation for an isolated hydrogen atom, but not the "real-world" phenomena of electron tunneling into an atom on a surface, or electron transfer between impurities in a condensed phase environment. While the mathematics needed to describe (at least approximately) the latter phenomenon is less cumbersome in comparison to the analytic solution of the hydrogen atom problem, advanced applications are excluded from introductory textbooks since they require the introduction of advanced QM topics relevant to open quantum systems (Chapters 17–19) or nonequilibrium states (Chapter 20).

In this book we propose a pedagogical approach aimed at making advanced QM theory accessible for readers interested in QM and its applications in general, for researchers entering the fields of nanoscience and engineering, as well as for deep thinkers who wish to master the field. Three features make our approach unique and different from standard approaches to teaching QM. The first feature is the close relation to "real-world" phenomena and applications from nanoscience and technology, which often motivate the theoretical discussions and/or summarize each topic. In many of the chapters, the study of the theory will be motivated by problems from the realm of nanotechnology, mapping them onto fundamental questions and answering these questions via acquaintance with the principles of QM. The manifestation of the postulates and the mathematical structure of the theory in experimental observations will be emphasized. The second feature is the inclusion of advanced topics (commonly met in applications to "real-world" problems) within an introductory-level textbook. The third feature is the "layered" structure of the text, which should be appealing to both "rigorous" and "easy" readers. For the benefit of the latter, the reading is kept relatively fluent by defining the most technical details and mathematical proofs as exercises appearing next to the main text. Some of the exercises are standard, but most of them are guided instructions for proving equations or justifying claims made in the main text. Just reading these guided exercises may be convincing enough for some of the more technically oriented readers, while the most rigorous readers would probably want to actively succeed in solving the exercises, thus gaining a complete hold on any claim and equation mentioned in the book.

It is emphasized that the postulates and the mathematical structure of nonrelativistic QM are presented at a fairly rigorous level, along with mathematical technicalities required, for example, for solution of ordinary differential equations or dealing with tensor product spaces. In this sense, the book can also serve the more rigorous readers who wish to master the field, including undergraduate and graduate Physics students.

In Chapter 1 we introduce the reader to the necessity for using QM to understand phenomena on the nanoscale, where the wave–particle duality naturally appears. This chapter leads to a basic introduction to wave functions and probability densities (Chapter 2), observables as operators (Chapter 3), and the Schrödinger equation (Chapter 4). While discussing the solutions of the stationary Schrödinger equation for different model systems in the following chapters (Chapters 5–10), we emphasize relations to applications in nanotechnology. For example, the ability to control the observed color (light absorption or emission) associated with nanoparticles of different size is relevant in numerous applications. This phenomenon motivates our fundamental discussion of energy quantization in QM in Chapter 5. Indeed, relating "color" to energy level spacing, we provide a rigorous discussion of energy quantization and of the "quantum size effect." Similarly, the discussion of experimental methods for characterization of nanoscale objects, such as Scanning Tunneling Microscopy (STM), motivates our theoretical discussion of the quantum tunneling phenomena, first for bound particles (quantum wells) (Chapter 6), and then for free particles (quantum barriers) (Chapter 7). The latter motivates the introduction of theoretical concepts such as probability current density and the first encounter with quantum scattering theory. The investigation of mechanical motions in nanoscale systems by infrared and microwave spectroscopies motivates our introduction to quantization of vibrations and rotations in many-atom systems (molecules, solids) and to the fundamental models of the quantum harmonic oscillator and the quantum rigid rotor in Chapters 8–9. In Chapter 10 we address the structure of atoms as the building blocks of matter on the nanoscale by starting from the solution of the Schrödinger equation for the single-electron (hydrogen like) atom. After extensive exposure to solutions of the Schrödinger equation for simple systems in Chapters 5–10, the mathematical formulation and the postulates of QM are rigorously given in Chapter 11. To establish a basis for the more advanced discussions, approximation methods based on perturbation theory and on the variation principle are introduced in Chapter 12. The variation of electronic properties between different elements (the periodic table) motivates our discussion of the electronic structure of many-electron atoms in Chapter 13, within the mean-field (orbital) approximation. In Chapter 14 the structure of many-atom systems is addressed in molecules and in periodic crystals, where the nature of chemical bonds is analyzed. The relation between the chemical composition of a material and its electric conductivity is thus also revealed. The more advanced themes start with a rigorous introduction to quantum dynamics in Chapter 15 and proceed to quantum thermodynamic systems (mixed ensembles) in Chapter 16. Elementary rate processes in nanoscale systems are most relevant for "real-world" applications such as photovoltaic cells or electro-optical devices. Their study motivates our discussion of transport processes and quantum kinetics. In Chapter 17 the emergence of unidirectional rate processes in QM is introduced within the framework of time-dependent perturbation theory, where Fermi's golden rule for the rate constant is derived. Applications to processes such as light absorption and emission, charge, and energy (exciton) transfer between impurities in bulk materials and between molecules in solution are discussed in Chapter 18. In Chapter 19 we address the dynamics in open quantum systems, the validity of the Markovian approximation, and the emergence of irreversible dynamics and relaxation to equilibrium. Chapter 20 provides the theoretical basis needed for the description of quantum charge transport in nanoscale devices under nonequilibrium conditions, motivated by the applications of molecular electronic devices.

In conclusion, this book exposes the reader to the foundations of quantum mechanics, to the richness of quantum phenomena on the nanoscale, and to the theoretical understanding of the physics underlying these phenomena. The reader gains a solid basis in the theory of nonrelativistic QM, its fundamental postulates, and its practical implementation. The book provides the language and the concepts needed for addressing more specified QM texts such as advanced textbooks and research articles dealing with nanoscale phenomena and their applications. Moreover, the reading and the active learning associated with solving the many guided exercises in this book aim to provide our readers also with skills that should enhance their own creativity in addressing "real-world" problems encountered on the nanoscale.

Motivation

1.1 The Wave–Particle Duality and de Broglie Wavelength

It so happens that nature does not always follow the strict rules set by classical mechanics. When attempting to describe phenomena occurring on the nanoscale, this reality becomes most apparent, as we shall see in numerous examples throughout this book. The structure of atoms and molecules, the organization of matter at the atomistic level, the dynamics of charge and energy in different materials, diffraction and scattering of electromagnetic radiation or elementary particles from nanoscale structures, and chemical reactivity are but a few phenomena that cannot be explained within the realm of classical mechanics.

As a concrete example of the breakdown of classical mechanics on the subnanometer length scale, let us consider the scattering of an electron beam from a surface of a metal, as studied by Davisson and Germer in the year 1927 [1.1]. The electron beam was initially regarded as a flux of particles (the electrons) in classical terms. Considering the mass and kinetic energy, E, each electron is associated with a linear momentum $\mathbf{p} = (p_x, p_y, p_z)$, where (neglecting relativistic corrections)

$$\mathbf{p}^2 = 2mE. \tag{1.1.1}$$

In the simplest terms, the metal acts as a perfect mirror when reflecting the electron beam. In the scattering process, the momentum flips its sign in the direction perpendicular to the surface (e.g., $(p_x, p_y, p_z) \to (p_x, p_y, -p_z)$), and consequently, the angle between the direction of the reflected beam and the surface plane (α_{out}) equals the impact angle (α_{in}), as depicted in Fig. 1.1.1.

According to this classical interpretation, the reflected flux at $\alpha_{out}(= \alpha_{in})$ should depend on the incoming particle flux, but not on the energy of each particle, E. Yet,

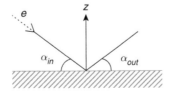

Figure 1.1.1 The classical description of electron beam scattering from a metal surface.

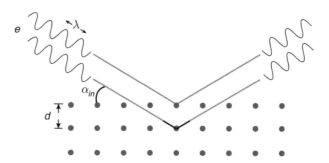

Figure 1.1.2 The wave description of electron beam scattering from a metallic crystal. The pathway of wave scattering from the internal atomic layer is longer by $2\,d\sin(\alpha)$ in comparison to wave scattering from the outer layer, resulting in wave interference.

in the famous Davisson–Germer experiment a clear dependence of the reflected flux on the energy of the incoming electrons was observed. Moreover, *the intensity of the reflected beam showed distinctive maxima at specific E values, which was reminiscent of the result of electromagnetic wave reflection from ordered atomic crystals* (X-rays, in particular). This result couldn't be interpreted within classical mechanics.

In wave theory, modulations in the reflected flux can be readily explained in terms of interference of waves originating from different sources, or incoming through different pathways. When waves are scattered from an atomic crystal, the first and second atomic layers can be regarded as two reflecting mirrors displaced from each other by the interatomic distance, d. (Scattering from deeper layers may be neglected for moderate beam energies when the penetration of the beam into the crystal is low.) For particle scattering, the reflection from the two mirrors is additive and therefore equivalent to an effective single mirror. In contrast, for wave scattering, the two mirrors define two scattering pathways with remarkable consequences for the reflected flux owing to interference. In particular, when the two pathways differ in length by an integer multiplicity of the wavelength, λ, the interference is constructive, and the reflected beam intensity should obtain a maximum. Given the scattering angle, $\alpha_{out}(=\alpha_{in})$, the two pathways differ in length by $2d\sin(\alpha_{in})$ (see Fig. 1.1.2). In this case, constructive interference should occur for a set of discrete wavelengths, satisfying the (Bragg) condition:

$$\lambda_n = \frac{2\,d\sin(\alpha_{in})}{n} \quad ; \quad n = 1,2,3,\ldots. \tag{1.1.2}$$

According to the wave theory, the reflected flux should therefore depend on the scattering angle, obtaining a set of maximal values as a function of the wavelength. *But how could such a theory apply to a beam of particles? How could one relate the modulation in the particle's energy to modulation in some respective wavelength?* The experimentally observed similarities between scattering of electrons and of electromagnetic waves by a metal crystal led Davisson and Germer to conclude that [1.1] "a description of the occurrence and behavior of the electron diffraction beams in terms of the scattering of an equivalent wave radiation by the atoms of the crystal, and its subsequent interference, is not only possible, but most simple and natural." Indeed, it was already proposed by Louis de Brogliede-Broglie in 1924 [1.2] that *a beam of particles (electrons, included) can be associated with a wavelength, according to the following postulate*:

$$\lambda \equiv \frac{h}{p}, \tag{1.1.3}$$

where h is Planck's constant (already known by then, from the theory of electromagnetic radiation [1.3]), and p is the particle's momentum in the direction of the beam propagation. *The Davisson–Germer experiment reinforced a dual description of an electron beam in terms of both particles and waves.* The maxima in the reflected beam flux as a function of the electron energy (E) could be correlated with wave properties not only qualitatively, but also quantitatively. By implementing the conditions for constructive wave interference (such as Eq. (1.1.2)), the de Broglie relation between momentum and wavelength (Eq. (1.1.3)), and the classical relation between the momentum and energy (Eq. (1.1.1)), that is, $\lambda_n = \frac{h}{\sqrt{2mE_n}}$, the maxima in the reflected flux as a function of the kinetic energy of the electrons could be explained.

Notice that while the wave properties of the electron beam turn out to be inherent to its nature, their manifestation in the experiment required special conditions. For the different interference peaks in the reflected flux to be distinctively resolved, the ratio $(\lambda_n - \lambda_{n+1})/\lambda_{n+1}$ should not be too small. From Eq. (1.1.2) it follows that this ratio equals $1/n$, which means that the wave properties are most clearly pronounced for $n = 1$, or in physical terms, when the de Broglie wavelength of the electrons is of the order of the interatomic distance, $\lambda_1 = 2d\sin(\alpha_{in})$. As the de Broglie wavelength of the electrons becomes smaller ($\lambda_n/d << 1$), for example, at higher kinetic energies, n becomes too large to enable resolving specific interference peaks. Therefore, the discovery of the wave nature of the electron required that the de Broglie wavelength (set by the kinetic energy) would be of the order of the (subnanometer) interatomic distances in the atomic crystal. Moreover, the quantitative verification of de Broglie's relation (Eq. (1.1.3)) was based on prior knowledge of the interatomic distance (d), which was already inferred from X-ray diffraction measurements. Remarkably, it took more than two centuries after the establishment of classical mechanics for the revelation of the electron's wave–particle duality. The characterization of the structure of matter on the subnanometer scale was an essential precondition for this discovery.

Exercise 1.1.1 *The distance between adjacent atomic layers in a nickel crystal is $d = 2.03 \times 10^{-10}$ meter. An electron beam is scattered from the face of the crystal at an angle, $\alpha_{in} = 45^o$. Given the electron mass, $m = 9 \cdot 10^{-31}$ kg, calculate the kinetic energy for the three most resolved maxima in the reflected flux at $\alpha_{out} = 45^o$ (see Figs. 1.1.1 and 1.1.2).*

Exercise 1.1.2 *In the "classical world," the de Broglie wavelength is typically much smaller than the characteristic length scale of the system under consideration. Consider, for example, a tennis ball at a mass $m = 0.058$ kg, flying at a typical serve-velocity, for example, 50 meter/sec. What is the associated de Broglie wavelength? How does it compare with the length of a tennis court (24 meter), or with the diameter of the ball itself (6.7 cm)?*

1.2 Quantum Mechanics for Nanoscience and Engineering

Quantum mechanics, as detailed in the following chapters of this book, provides a theoretical framework that accounts consistently and comprehensively for the wave–particle duality. Indeed, any matter is associated with wave properties, which are expressed most profoundly when the associated de Broglie wavelength is of the order of the characteristic length scale of the system under consideration,

$$\lambda \approx d. \tag{1.2.1}$$

In the realm of nanoscience and engineering, the characteristic length scales are associated with the typical dimensions of atoms (10^{-10}m), molecules (10^{-9}m), or nanocrystals (10^{-8} m). The relevant masses are those of electrons ($\sim 10^{-30}$ kg) and small atoms ($\sim 10^{-28} - 10^{-27}$ kg), and the relevant kinetic energy scales are typically $\sim 10^{-2} - 10^{1}$ eV. Translating the masses and energy values to wavelength according to de Broglie (Eq. (1.1.3)), it immediately follows that the condition for manifestation of wave properties, Eq. (1.2.1), is often fulfilled. Quantum mechanics is therefore essential for a proper description of phenomena associated with the nanoscale. These include, for example, the colors of matter (absorption spectra of atoms, molecules, nanoparticles, etc.), electrical conductivity, interatomic forces, the structure of matter, chemical reactivity, heat and energy transport, and many more.

Bibliography

[1.1] C. Davisson and L. H. Germer, "Diffraction of electrons by a crystal of nickel," Physical Review 30, 705 (1927).

[1.2] L. de Broglie, "Recherches sur la théorie des quanta" (doctoral dissertation, Migration-université en cours d'affectation, 1924).

[1.3] M. Planck, "Zur Theorie des Gesetzes der Energieverteilung im Normalspectrum," Verhandlungen der Deutschen Physikalischen Gesellschaft 2, 237 (1900).

2 The State of a System

2.1 Probability Densities

In this chapter we shall get some informal familiarity with fundamental assumptions (postulates) of quantum mechanics, which deal with the perception of the state of a closed quantum system (a formal and complete presentation of the postulates is given in Chapter 11 of this book). Imagine a collection of well-resolved particles in the sense that to characterize their collective physical state, we only need to know the properties of the particles themselves (e.g., position, momentum), their mutual interactions, and any conservative external forces acting on them. These particles constitute a closed system. *According to the postulates of quantum mechanics, the state of a closed system is associated with a proper, complex-valued wave function, $\psi(t)$, which contains all measurable information on the system at time, t.* The function, $\psi(t)$ depends on time as well as on variables that represent the system, that is, $\psi(x_1, x_2, x_3, \ldots, t)$. The number of these variables reflects the number of particles and the dimensions of the relevant physical space.

As a concrete example, let us consider a single point particle, whose motion is restricted to a single dimension and confined to some region in space. Denoting by x the position of the particle in this one-dimensional space ($-\infty < x < \infty$), any measurable property of the particle is contained in a proper wave function, $\psi(x,t)$. A property of natural interest is the position of the particle at time t; namely, we may wish to know, *Where is the particle?* However, according to another postulate of quantum mechanics, this question does not have a unique answer. Instead, we can obtain an answer to the question, *What is the probability of finding the particle in a given region of space?* This information is encoded in the wave function, $\psi(x,t)$, which defines the probability density, $\rho(x,t)$, for locating the particle near a point, x, at time t. According to the postulate,

$$\rho(x,t) \equiv \psi^*(x,t)\psi(x,t) = |\psi(x,t)|^2. \tag{2.1.1}$$

Here ψ^* denotes the complex conjugate of the complex-valued wave function, ψ. *The wave function therefore obtains the meaning of a probability amplitude.* Notice that the probability of locating the particle at any specific point must vanish, since a point has a zero measure. (This is true for the probability of measuring any physical quantity with a continuous set of measurable values.) Nevertheless, the probability of locating the particle anywhere between two points is finite. In particular, the probability

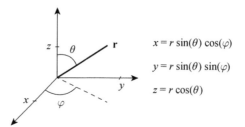

$x = r \sin(\theta) \cos(\varphi)$

$y = r \sin(\theta) \sin(\varphi)$

$z = r \cos(\theta)$

Figure 2.1.1 Cartesian and spherical coordinate representations of a position vector.

of locating the particle within an infinitesimal interval dx at the vicinity of the point x equals $\rho(x,t)dx$. For a finite interval, one needs to integrate; namely, the probability of locating the particle between points a and b at time t is given as

$$P_{[a,b]}(t) = \int_a^b \rho(x,t)dx = \int_a^b |\psi(x,t)|^2 dx. \tag{2.1.2}$$

Similarly, the probability of locating the particle *anywhere* in space should be a unity at any time, namely,

$$P_{total} = \int_{-\infty}^{\infty} \rho(x,t)dx = \int_{-\infty}^{\infty} |\psi(x,t)|^2 dx = 1. \tag{2.1.3}$$

(The fact that P_{total} is indeed time-independent is assured by additional postulates, to be discussed in the following chapters.)

Generalizations for multidimensional coordinate spaces are straightforward. For a single particle in a three-dimensional space, the probability density for locating the particle in an infinitesimal volume element in the vicinity of a point in space reads $|\psi(x,y,z,t)|^2 dxdydz$ in a Cartesian coordinate system, or $|\psi(r,\theta,\varphi,t)|^2 r^2 \sin(\theta)drd\theta d\varphi$ in spherical coordinates (see Fig. 2.1.1). For a system of N particles in a D-dimensional coordinate space, the wave function depends on time t and on $N \times D$ spatial coordinates. The probability of locating simultaneously each of the particles in the vicinity of a certain point in space reads (e.g., in Cartesian coordinates) $|\psi(x_1,y_1,z_1;x_2,y_2,z_2;\ldots,t)|^2 dx_1 dy_1 dz_1 dx_2 dy_2 dz_2 \ldots$.

The association of the wave function with a probability density implies that the wave function obtains physical dimensions, which depend on the system. For N particles in a D-dimensional space, $\psi(t)$ obtains the dimensions:

$$[\psi(t)] = [length]^{-N \times D/2}. \tag{2.1.4}$$

2.2 Proper Wave Functions

The properties of a closed system are therefore encoded in a proper wave function. *But what does "proper" mean?* The association of $|\psi(t)|^2$ with a probability density in the physical space imposes strict limitations on the properties of $\psi(t)$ itself. Particularly, to

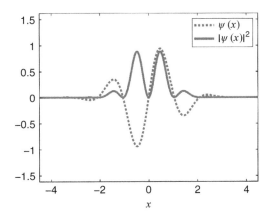

Figure 2.2.1 A proper one-dimensional wave function and the respective probability density for locating the particle in space.

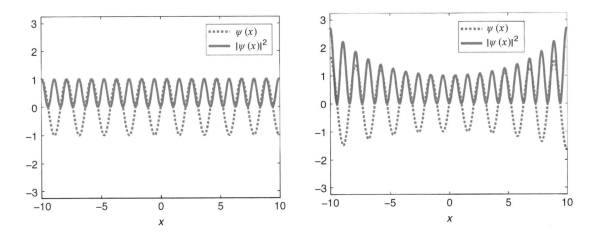

Figure 2.2.2 Improper one-dimensional wave functions and their respective absolute squares. The left and right plots represent periodic and diverging functions, respectively.

fulfil the condition in Eq. (2.1.3), the wave function must not be identically zero, and at the same time, the integral of $|\psi(t)|^2$ over the entire space must obtain a finite positive value. ***In mathematical terms, a proper wave function must be square-integrable.***

For example, let us refer to a one-dimensional Cartesian space. Since the probability density $(|\psi(x,t)|^2)$ is nonnegative, a proper wave function must satisfy the following boundary conditions:

$$\psi(x,t) \underset{x\to\pm\infty}{\longrightarrow} 0. \tag{2.2.1}$$

In Fig. 2.2.1, a specific proper one-dimensional wave function and the respective probability density are plotted for illustration. Examples of improper functions and their respective absolute value squares are shown in Fig. 2.2.2.

Generalization of the one-dimensional condition (Eq. (2.2.1)) to multidimensional systems is straightforward: a wave function is proper only if the corresponding probability for locating any part of the system in an infinitesimal volume vanishes at the boundaries of the (multidimensional) coordinate space.

Note that the boundary conditions that are necessary for "properness" are but the first constraint on wave functions that can be associated with the state of a physical system. In the following chapters we shall get familiar with the Schrödinger equation, which imposes additional physical as well as mathematical constraints on $\psi(t)$ for any given system.

2.3 Normalization

Notice that for any proper wave function, $\psi(t)$, the integral of $|\psi(t)|^2$ over the entire space obtains a finite positive value. Nevertheless, to associate $\psi(t)$ with a probability density, as in Eqs. (2.1.2) and (2.1.3), the integral should equal unity. A proper wave function that satisfies the latter condition is termed "normalized."

Any given proper wave function can become normalized by multiplying it by a constant number. For example, given a proper wave function, $\psi(x,t)$, we have $\int\limits_{-\infty}^{\infty} |\psi(x,t)|^2\,dx = N$, where N is some finite positive number. The corresponding normalized wave function is given as $\tilde{\psi}(x,t) = \frac{1}{\sqrt{N}}\psi(x,t)$, where the condition for probability conservation (Eq. (2.1.3)) immediately follows: $\int\limits_{-\infty}^{\infty} dx|\tilde{\psi}(x,t)|^2 = \int\limits_{-\infty}^{\infty} dx\frac{1}{N}|\psi(x,t)|^2 = \frac{N}{N} = 1$.

The normalization procedure can be readily generalized for different coordinate systems and to higher dimensions. For example, for a single point particle moving on a perfect ring, the particle's position is defined by an angular variable, $0 \leq \varphi < 2\pi$. Given any proper wave function, $\psi(\varphi,t)$, the normalized wave function reads $\tilde{\psi}(\varphi,t) = \sqrt{\dfrac{1}{\int\limits_{0}^{2\pi} d\varphi|\psi(\varphi,t)|^2}}\,\psi(\varphi,t)$. Or, for a particle moving in three dimensions, the normalized wave function would be $\tilde{\psi}(x,y,z,t) = \sqrt{\dfrac{1}{\int\limits_{-\infty}^{\infty} dx \int\limits_{-\infty}^{\infty} dy \int\limits_{-\infty}^{\infty} dz|\psi(x,y,z,t)|^2}}\,\psi(x,y,z,t)$.

Exercise 2.3.1 *The probability density of finding a point particle along the x-axis at a certain time is given by $\rho(x) = \alpha e^{-\frac{(x-x_0)^2}{2\sigma^2}}$. (a) What is the most probable position for this particle? (Does it depend on the value of α?). (b) Determine the value of α for which the probability density is normalized, recalling that $\int\limits_{-\infty}^{\infty} e^{-\beta y^2}\,dy = \sqrt{\frac{\pi}{\beta}}$. (c) The average position associated with a probability density, $\rho(x)$, is defined as $\langle x \rangle \equiv \int\limits_{-\infty}^{\infty} x\rho(x)dx$. Calculate the average position of the given point particle. (d) The corresponding standard deviation in the position probability distribution is defined as $\Delta x \equiv \sqrt{\langle x^2 \rangle - \langle x \rangle^2}$. Calculate the standard deviation in the position of the given point particle.*

Exercise 2.3.2   *The state of a particle is described as a "superposition of wave functions," namely, $\psi(x) = \sum\limits_{n=1}^{\infty} c_n \phi_n(x)$, where $\{c_n\}$ are given scalar expansion coefficients, and $\{\phi_n(x)\}$ is a given set of "orthonormal wave functions," namely, $\int\limits_{-\infty}^{\infty} \phi_m^*(x) \phi_n(x) dx = \begin{cases} 0 & m \neq n \\ 1 & m = n \end{cases}$. Normalize $\psi(x)$. (Express the normalized wave function in terms of the given expansion coefficients.)*

Exercise 2.3.3   *An isolated hydrogen-like atom is composed of a nucleus with Z protons and a single electron. At the minimal energy state of the atom, the probability density for finding the electron at a given position reads $\rho(\mathbf{r}) = \alpha e^{-2Zr/a_0}$; $r = |\mathbf{r}|$, where $\mathbf{r}$ is the three-dimensional vector of the relative position between the nucleus and the electron, and $a_0 = 0.0529$ nm is the Bohr radius. Show that the most probable distance r between the electron and the nucleus in this state is a_0/Z. What is the probability density of finding the electron at the most probable distance, r, in this state?*

Observables and Operators

3.1 Physical Observables and Operators

According to quantum mechanics, the information with respect to the physical properties of a system is contained in a mathematical object, the wave function, $\psi(t)$. Here we elaborate on the mathematical representation of the physical properties themselves. Particularly, we shall start from the most elementary dynamical variables needed to describe a system of particles, namely, the particles' positions in space and their momenta (or velocities, given their masses). Then, we shall move on to properties that are derived from the particles' positions (e.g., potential energy), momenta (e.g., kinetic energies), or both (e.g., angular momenta).

The postulates of quantum mechanics associate each measurable property with a specific operator, namely, a unique mathematical operation that maps one proper function on another. For example, let us consider the operator that maps any function on its derivative. The equation $\frac{d}{dx}f(x) = g(x)$ can be written as $\hat{D}_x f(x) = g(x)$, where $\hat{D}_x$ marks the derivative operator with respect to x. (Operators are denoted by the hat superscript). It follows that: $\hat{D}_x x^2 = 2x$, $\hat{D}_x \sin(x) = \cos(x)$, $\hat{D}_x c = 0$, and so on. Other common operators are multiplication operators. For example, multiplying any function $f(x)$ by the function $g(x)$ can be written as $g(x) \cdot f(x) = \hat{G}f(x)$, where $\hat{G}$ is the corresponding operator.

A critical point to notice is that operators do not necessarily commute. Namely, the result of two successive operations on a function may lead to different results, depending on the order in which the operators are applied. For example, considering the operators $\hat{D}_x$ and $\hat{G}$ just defined, we readily obtain $\hat{D}_x \hat{G} f(x) = [\frac{d}{dx}g(x)]f(x) + g(x)\frac{d}{dx}f(x)$, whereas changing the order yields $\hat{G}\hat{D}_x f(x) = g(x)\frac{d}{dx}f(x)$. It follows that $\hat{G}\hat{D}_x \neq \hat{D}_x \hat{G}$. This property is expressed in terms of the commutator (or the commutation relation). For any two operators, $\hat{A}$ and $\hat{B}$, the commutator is defined as

$$[\hat{A}, \hat{B}] \equiv \hat{A}\hat{B} - \hat{B}\hat{A}. \qquad (3.1.1)$$

If $\hat{A}$ and $\hat{B}$ commute with each other, then $[\hat{A}, \hat{B}] = 0$, and the order of operations on any (proper) function does not matter. Otherwise, the operators do not commute with each other, and the order of operations must be carefully specified.

An important feature of the operators that correspond to measurable properties (we shall often term such operators hereafter as "quantum mechanical operators") is their

linearity. An operator $\hat{A}$ is linear if for any two functions, $f_1(x)$ and $f_2(x)$ (in the space of proper functions), and for any (complex valued) scalars a_1, a_2, we have

$$\hat{A}[a_1 f_1(x) + a_2 f_2(x)] = a_1 \hat{A} f_1(x) + a_2 \hat{A} f_2(x). \tag{3.1.2}$$

Exercise 3.1.1   *Write explicit expressions for the following commutators:* $[\sin x, \frac{d}{dx}]$, $[\frac{1}{x}, \frac{d}{dx}], [\frac{d}{dx}, \frac{d^2}{dx^2}]$.

Exercise 3.1.2   *Verify the following identities for any linear operators ($\hat{A}$, $\hat{B}$, $\hat{C}$):*

$$[\hat{A}, \hat{B}] = -[\hat{B}, \hat{A}] ; [\hat{A}, \hat{B} + \hat{C}] = [\hat{A}, \hat{B}] + [\hat{A}, \hat{C}] ; [\hat{A}, \hat{B}\hat{C}] = [\hat{A}, \hat{B}]\hat{C} + \hat{B}[\hat{A}, \hat{C}].$$

Exercise 3.1.3   *The operators: $\hat{A}$; $\hat{B}$; $\hat{C}$; $\hat{D}$ are defined by the results of their operation on any function $f(x)$:* $\hat{A}f(x) = x^n f(x)$; $\hat{B}f(x) = \frac{d}{dx}f(x)$; $\hat{C}f(x) = \sin(f(x))$; $\hat{D}f(x) = \sqrt{f(x)}$. *Which ones are linear?*

Exercise 3.1.4   *Given that $\hat{A}$ and $\hat{B}$ are linear operators, show that $\hat{C} = \hat{A}\hat{B}$ and $\hat{D} = \hat{A} + \hat{B}$ are also linear operators.*

Exercise 3.1.5   *$\hat{D}$ is a linear differential operator, and ϕ and φ are two solutions of the homogeneous linear equation defined by $\hat{D}$,*

$$\hat{D}\phi = 0 \quad ; \quad \hat{D}\varphi = 0.$$

Show that any linear combination of ϕ and φ (i.e., $a\phi + b\varphi$ with constant scalars a and b) is also a solution of the homogeneous equation (the "superposition principle").

3.2 The Canonical Operators

The postulates of quantum mechanics identify the operators corresponding to the position and the momentum of a particle restricted to a given Cartesian axis (x) as

$$\hat{x} = x \cdot \tag{3.2.1}$$

$$\hat{p}_x = -i\hbar \frac{d}{dx} \quad ; \quad (i = \sqrt{-1}, \hbar = \frac{h}{2\pi}). \tag{3.2.2}$$

The position is a simple multiplication operator, namely, $\hat{x}\psi(x) = x\psi(x)$ for any $\psi(x)$. It is a local operator in the sense that the value of $\hat{x}\psi(x)$ at a certain position along the x-axis depends on $\psi(x)$ only at that position. The momentum involves a derivative with respect to the position and a multiplication by $-i\hbar$, where $h = 6.62607004 \times 10^{-34}$ m^{2}kg/s is Planck's constant. The momentum is a nonlocal operator in the sense that the value of $\hat{p}_x\psi(x)$ at a point depends on the function at the vicinity of this point. The commutator of these two canonical operators reads

$$[\hat{x}, \hat{p}_x] = i\hbar. \tag{3.2.3}$$

In a three-dimensional coordinate space, the position and momentum of a particle correspond to vector operators. In a Cartesian coordinate system,

$$\hat{\mathbf{r}} = (\hat{x}, \hat{y}, \hat{z}), \tag{3.2.4}$$

$$\hat{\mathbf{p}} = (\hat{p}_x, \hat{p}_y, \hat{p}_z) = -i\hbar\left(\frac{\partial}{\partial x}, \frac{\partial}{\partial y}, \frac{\partial}{\partial z}\right) = -i\hbar\nabla, \tag{3.2.5}$$

where, for example, $\frac{\partial}{\partial x}$ is the partial derivative with respect to the coordinate x, and ∇ is the gradient vector, $\nabla\psi(x,y,z) = \frac{\partial\psi(x,y,z)}{\partial x}\mathbf{i} + \frac{\partial\psi(x,y,z)}{\partial y}\mathbf{j} + \frac{\partial\psi(x,y,z)}{\partial z}\mathbf{k}$, (where $\mathbf{i}$, $\mathbf{j}$, and $\mathbf{k}$ are the unit vectors defining the Cartesian system axes). Representations of these operators in different coordinate systems are obtained straightforwardly by the proper transformation of variables.

Exercise 3.2.1   *Show that,* $[\hat{\mathbf{r}}, \hat{\mathbf{p}}] = 3i\hbar$.

3.3 The Angular Momentum Operators

The definitions of the canonical operators enable us to associate other dynamical variables with operators in a consistent manner, based on their relations to the position and momentum variables as known in classical mechanics. Particularly, we can derive an explicit form for the angular momentum operators for a single particle, using the classical relation, $\mathbf{L} = \mathbf{r} \times \mathbf{p}$. Associating the position and momentum vectors with the respective operators (Eqs. (3.2.4, 3.2.5)), the angular momentum vector operator reads

$$\hat{\mathbf{L}} = \hat{\mathbf{r}} \times \hat{\mathbf{p}} = (\hat{L}_x, \hat{L}_y, \hat{L}_z), \tag{3.3.1}$$

where

$$\begin{aligned} \hat{L}_x &= \hat{y}\hat{p}_z - \hat{z}\hat{p}_y \\ \hat{L}_y &= \hat{z}\hat{p}_x - \hat{x}\hat{p}_z. \\ \hat{L}_z &= \hat{x}\hat{p}_y - \hat{y}\hat{p}_x \end{aligned} \tag{3.3.2}$$

Using the commutator of the canonical operators, we readily obtain the set of commutation relations between the different angular momentum operators,

$$\begin{aligned} [\hat{L}_x, \hat{L}_y] &= i\hbar\hat{L}_z, \\ [\hat{L}_y, \hat{L}_z] &= i\hbar\hat{L}_x, \\ [\hat{L}_z, \hat{L}_x] &= i\hbar\hat{L}_y. \end{aligned} \tag{3.3.3}$$

Defining the total angular momentum operator,

$$\hat{L}^2 \equiv \hat{L}_x^2 + \hat{L}_y^2 + \hat{L}_z^2, \tag{3.3.4}$$

we obtain

$$[\hat{L}^2, \hat{L}_x] = [\hat{L}^2, \hat{L}_y] = [\hat{L}^2, \hat{L}_z] = 0. \tag{3.3.5}$$

Exercise 3.3.1   *Prove the relations in Eqs. (3.3.3, 3.3.5).*

3.4 Functions of Operators

Physical properties that depend on a particle's position and/or momentum in classical mechanics are associated with quantum mechanical operators, which are the respective functions of the canonical operators.

A function f of an operator $\hat{A}$ is defined as its power series, $f(\hat{A}) \equiv \sum_{n=0}^{\infty} f_n \hat{A}^n$ (Taylor's expansion, for analytic functions). $\{f_n\}$ are the expansion coefficients of the function $f(\alpha)$, namely, $f(\alpha) \equiv \sum_{n=0}^{\infty} f_n \alpha^n$, for any scalar variable α.

Exercise 3.4.1 *Show that the following function of the linear momentum operator, $\hat{M}_\alpha = e^{\frac{i\alpha}{\hbar}\hat{p}_x}$, is a displacement operator, namely, $\hat{M}_\alpha \psi(x) = \psi(x+\alpha)$, where α is a constant.*

As an example, consider a particle of mass m moving along the x-axis. In classical mechanics, the particle's potential energy is a function of its position in space, $V = V(x)$, whereas the particle's kinetic energy is a function of its linear momentum, $T_x = p_x^2/(2m)$.

The corresponding quantum mechanical operator associated with the particle's potential energy therefore reads

$$\hat{V}_x = V(\hat{x}). \tag{3.4.1}$$

Using the formal power expansion of the function V, we obtain

$$\hat{V}_x \psi(x) = \sum_{n=0}^{\infty} V_n \hat{x}^n \psi(x) = \sum_{n=0}^{\infty} V_n x^n \psi(x) = V(x)\psi(x), \tag{3.4.2}$$

where we have used the identification of $\hat{x}$ as a multiplication operator (Eq. (3.2.1)). Similarly, the quantum mechanical operator associated with the particle's kinetic energy obtains the form

$$\hat{T}_x = \frac{1}{2m}(\hat{p}_x)^2. \tag{3.4.3}$$

Here we use the identification of the momentum operator, $\hat{p}_x$, Eq. (3.2.2), to obtain

$$\hat{T}_x \psi(x) = \frac{1}{2m}(-i\hbar\frac{d}{dx})^2 \psi(x) = \frac{-\hbar^2}{2m}\frac{d^2}{dx^2}\psi(x). \tag{3.4.4}$$

As we can see, the potential energy corresponds to a local (multiplication) operator, whereas the kinetic energy is a nonlocal (second derivative) operator, in accordance with the properties of the canonical position and momentum operators.

In three dimensions, the position and momentum operators are vectors (Eqs. (3.2.4, 3.2.5)), and the corresponding potential and kinetic energy operators read

$$\hat{V}_r \psi(\mathbf{r}) = V(\mathbf{r})\psi(\mathbf{r}), \tag{3.4.5}$$

$$\hat{T}_r \psi(\mathbf{r}) = \frac{-\hbar^2}{2m}\nabla \cdot \nabla \psi(\mathbf{r}) = \frac{-\hbar^2}{2m}\Delta\psi(\mathbf{r}). \tag{3.4.6}$$

Here, Δ is the Laplacian operator, for example, using Cartesian coordinates, $\Delta\psi(x,y,z) = (\frac{\partial^2}{\partial x^2} + \frac{\partial^2}{\partial y^2} + \frac{\partial^2}{\partial z^2})\psi(x,y,z)$.

The generalization to many-particle systems is straightforward. Considering a system of N particles with masses $m_1, m_2, \ldots, m_N$ and respective coordinates $\mathbf{r}_1, \mathbf{r}_2, \ldots, \mathbf{r}_N$, the potential and kinetic energy operators read

$$\hat{V}_{\mathbf{r}_1,\mathbf{r}_2,\ldots,\mathbf{r}_N}\psi(\mathbf{r}_1,\mathbf{r}_2,\ldots,\mathbf{r}_N) = V(\mathbf{r}_1,\mathbf{r}_2,\ldots,r_N)\psi(\mathbf{r}_1,\mathbf{r}_2,\ldots,\mathbf{r}_N), \tag{3.4.7}$$

$$\hat{T}_{\mathbf{r}_1,\mathbf{r}_2,\ldots,\mathbf{r}_N}\psi(\mathbf{r}_1,\mathbf{r}_2,\ldots,\mathbf{r}_N) = \sum_{i=1}^{N}\frac{-\hbar^2}{2m_i}\Delta_{\mathbf{r}_i}\psi(\mathbf{r}_1,\mathbf{r}_2,\ldots,\mathbf{r}_N). \tag{3.4.8}$$

Finally, we can associate the Hamilton function of classical mechanics [3.1], $H = T + V$, for any system of particles, with a quantum mechanical Hamiltonian operator,

$$\hat{H} \equiv \hat{T} + \hat{V}. \tag{3.4.9}$$

The Hamiltonian has a special role in quantum mechanics. In addition to the representation of the total (kinetic and potential) energy of the system, the Hamiltonian of a system uniquely defines the time evolution of the wave function, $\psi(t)$, which contains the entire information on the physical properties of a system. This is formulated in terms of the time-dependent Schrödinger equation, to be discussed in the following chapters.

It is instructive at this point to give explicit expressions for the Hamiltonians for some systems of interest. Let us consider first a single, isolated atom, containing a nucleus with Z protons and N electrons (for a neutral atom, $N = Z$). The (negative) electron charge is denoted e, and the nucleus charge is, accordingly, $Z|e|$. Without loss of generality, we consider the nucleus at rest, and we set its position to the origin of the coordinate system, as illustrated in Fig. 3.4.1. The kinetic energy, $\hat{T}_e$, is therefore a sum over the kinetic energies of the electrons (each of mass m_e). The potential energy is the sum of electrostatic pair interactions between all the charges. It amounts to the attraction of each electron to the nucleus, $\hat{V}_{en}$, and the repulsion between the electrons, $\hat{V}_{ee}$. **The (nonrelativistic) Hamiltonian for an isolated atom, $\hat{H} = \hat{T}_e + \hat{V}_{en} + \hat{V}_{ee}$, therefore reads**

$$\hat{H} = \sum_{i=1}^{N}\frac{-\hbar^2}{2m_e}\Delta_{\mathbf{r}_i} + \sum_{i=1}^{N}\frac{-KZe^2}{|\mathbf{r}_i|} + \sum_{j>i}^{N}\sum_{i=1}^{N}\frac{Ke^2}{|\mathbf{r}_i - \mathbf{r}_j|}, \tag{3.4.10}$$

where the electron rest mass is $m_e = 9.1094 \times 10^{-31}$ kg, and $K = 9.988 \times 10^9$ N × m² × C^{-2} is the Coulomb constant. (Notice that an exact account for the physical state of an isolated atom must also include its interaction with the vacuum state of the electromagnetic field, as well as relativistic corrections, including spin-orbit coupling. The effect of these corrections on the energy is comparably small with respect to the electrostatic interactions, whose contributions are accounted for explicitly in Eq. (3.4.10)).

Now let us consider a molecule composed of n nuclei and a total number of N electrons. Each nucleus has a mass, m_α ($\alpha = 1, 2, \ldots, n$), according to its number of nucleons, and an electric charge, $Z_\alpha|e|$, according to the number of protons, Z_α. The total kinetic energy is a sum over the kinetic energies of the electrons (each of mass m_e) and the nuclei, $\hat{T} = \hat{T}_n + \hat{T}_e$. The potential energy is the sum of electrostatic

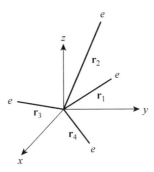

Figure 3.4.1 Electrons surrounding a nucleus located at the origin. The electron position vectors are denoted $\mathbf{r}_i$.

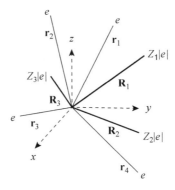

Figure 3.4.2 Electrons and nuclei charges and positions (denoted as $\mathbf{r}_i$ and $\mathbf{R}_\alpha$, respectively) in a generic molecule.

pair interactions between all the charges, including the repulsion between the negatively charged electrons, and between the positively charged nuclei, and the attraction between any electron and all the nuclei, $\hat{V} = \hat{V}_{ee} + \hat{V}_{nn} + \hat{V}_{en}$. Denoting the position vectors of the ith electron and the αth nucleus in a reference coordinate system as $\mathbf{r}_i$ and $\mathbf{R}_\alpha$, respectively (see Fig. 3.4.2), *we can write down the (nonrelativistic) quantum Hamiltonian of any molecule as follows:*

$$\hat{H} = \sum_{\alpha=1}^{n} \frac{-\hbar^2}{2m_\alpha}\Delta_{\mathbf{R}_\alpha} + \sum_{i=1}^{N} \frac{-\hbar^2}{2m_e}\Delta_{\mathbf{r}_i}$$

$$+ \sum_{\beta>\alpha}^{n}\sum_{\alpha=1}^{n} \frac{Ke^2 Z_\alpha Z_\beta}{|\mathbf{R}_\alpha - \mathbf{R}_\beta|} + \sum_{\alpha=1}^{n}\sum_{i=1}^{N} \frac{-KZ_\alpha e^2}{|\mathbf{R}_\alpha - \mathbf{r}_i|} + \sum_{j>i}^{N}\sum_{i=1}^{N} \frac{Ke^2}{|\mathbf{r}_i - \mathbf{r}_j|}. \qquad (3.4.11)$$

(As in Eq. (3.4.10), interaction with the electromagnetic vacuum and relativistic corrections are ignored.)

Bibliography

[3.1] H. Goldstein, "Classical Mechanics," 2nd ed. (Addison Wesley, 1981).

4 The Schrödinger Equation

4.1 The Time-Dependent Schrödinger Equation

In the previous chapters we became familiar with the association of the state of a system with a proper wave function, $\psi(t)$ and the identification of dynamical observables with specific operators. Particularly, we became familiar with the Hamiltonian operator $(\hat{H})$, which corresponds to the total energy (the sum of kinetic and potential energy) of a system of particles. In this chapter we shall become familiar with another postulate of quantum mechanics, the one that states that *the time-evolution of a given system is uniquely defined by its Hamiltonian operator. This relation is formulated explicitly in terms of the time-dependent Schrödinger equation*,

$$i\hbar\frac{\partial}{\partial t}\psi(t) = \hat{H}\psi(t). \tag{4.1.1}$$

Notice that the equation imposes additional constraints on proper wave functions representing a given physical system. First, for $\psi(t)$ to be physically relevant, the operation of $\hat{H}$ on $\psi(t)$ must be well defined and must yield a proper wave function, at all times. Second, being the solution of a first-order differential equation in time, *the wave function representing a system is uniquely defined at any time, given that it is known at a certain point in time (an initial condition).* From an ideal perspective, the experimental preparation of the system in a given state at a given time, t_0, is associated with a corresponding specific wave function at that time. The wave function then evolves in time according to the Schrödinger equation, namely under the influence of the system's Hamiltonian, and the information regarding a measurement to be performed at a future time, $t > t_0$, is encoded in the time-evolving wave function. Notice that for a given initial condition (i.e., a given state of a set of particles) there can be, in principle, an infinite number of different Schrödinger equations, accounting for different Hamiltonian operators, that represent, for example, different external forces experienced by the given set of particles.

For a given system (i.e., a given Hamiltonian) there is a unique Schrödinger equation, but there is an infinite number of different solutions to this equation. By different, we mean that the solutions are linearly independent, namely, they differ by more than a multiplication by a constant (which is equivalent to the normalization of the probability density function). Mathematically, the different solutions correspond to different initial conditions; and physically, they correspond to different preparations of the system, initiating different processes within the system.

The operator that maps $\psi(t_0)$ on $\psi(t)$ is termed the time-evolution operator (or the propagator,

$$\psi(t) = \hat{U}(t,t_0)\psi(t_0).$$ (4.1.2)

One can readily see that the Schrödinger equation (Eq. (4.1.1)) means that, $\hat{U}(t,t_0) = e^{\frac{-i}{\hbar}\hat{H}(t-t_0)}$. However, in more advanced formulation of quantum mechanics effective Hamiltonians are introduced, which may depend explicitly on time ($\hat{H} \mapsto \hat{H}(t)$). In such cases the expression for the time-evolution operator is more involved. (For details, see the discussion of quantum dynamics and Dyson's expansion in Chapter 15.) Notice that time itself is not associated with a quantum mechanical operator, but rather with a scalar parameter that identifies the system's evolution. (Advanced formulations in which time is associated with an operator were proposed, but these are beyond the scope of the present discussion [4.1].)

4.2 Time-Dependent Solutions

In this section we discuss some examples of time-dependent solutions to Schrödinger equations. For concreteness we shall focus on systems composed of a single point particle of mass m, moving along a one-dimensional coordinate x. The state of the particle is associated with a proper wave function, $\psi(x,t)$. As an initial condition we chose a normalized "Gaussian wave packet" [4.2],

$$\psi(x,0) = \left(\frac{1}{2\pi\sigma^2}\right)^{1/4} e^{\frac{-(x-x_0)^2}{4\sigma^2}} e^{ip_0x/\hbar}.$$ (4.2.1)

This wave function is characterized by three parameters, σ, x_0, and p_0. One can readily show (see Ex. 2.3.1) that x_0 and σ correspond, respectively, to the particle's averaged position in space, $<x>=x_0$, and to the standard deviation in the particle's position distribution, $\sigma = \sqrt{\langle x^2 \rangle - \langle x \rangle^2}$. The parameter p_0 has no effect on the initial position probability density, $|\psi(x,0)|^2$ (see Eq. (4.2.1)), but has a dramatic effect on the wave function evolution, $\psi(x,t)$. Indeed, this parameter corresponds to the averaged initial momentum of the particle, as will be discussed in detail in Chapter 15. Here we only make a brief intuitive comment in this context: for $\sigma \to \infty$, one can see that $\psi(x,0)$ approaches the form of a periodic plane wave in any finite region of space, $\psi(x,0) \tilde{\propto} e^{ip_0x/\hbar} \equiv e^{i2\pi x/\lambda}$. Using the de Broglie relation (Eq. (1.1.3)), p_0, therefore obtains the meaning of the particle's momentum, at least in this limit.

The general Hamiltonian operator for a point particle of mass m, moving along a one-dimensional coordinate, x, takes the form (see Eqs. (3.4.2, 3.4.4, 3.4.9))

$$\hat{H} = \frac{-\hbar^2}{2m}\frac{\partial^2}{\partial x^2} + V(x),$$ (4.2.2)

and the corresponding time-dependent Schrödinger equation therefore reads

$$i\hbar\frac{\partial}{\partial t}\psi(x,t) = \frac{-\hbar^2}{2m}\frac{\partial^2}{\partial x^2}\psi(x,t) + V(x)\psi(x,t). \tag{4.2.3}$$

$V(x)$ is the potential energy function of the particle, as known from classical mechanics. The details with respect to a particular system, namely, the conservative forces experienced by the particle at each point in space, are derived from this function. In what follows we consider three different choices of $V(x)$, corresponding to different systems, each associated with a different Schrödinger equation.

Let us consider first the case $V(x) = const$. This corresponds to a zero net force on the particle anywhere in space, $F(x) = -dV(x)/dx = 0$ (a free particle). The time evolution of the initial Gaussian wave packet for a free particle can be obtained by solving the time-dependent Schrödinger equation analytically (see, e.g. [4.3] and also Chapter 15 in this book). Here we skip the details of the quantitative analysis, focusing on some qualitative features of the wave function evolution, as represented in Fig. 4.2.1. In particular, the probability density ($\rho(x,t) = |\psi(x,t)|^2$) and the underlying probability amplitude ($Re[\psi(x,t)]$) are plotted at different times. As one can see, the average position of the particle along the x-axis changes in time, as expected for a freely moving particle. (A quantitative analysis shows that it increases linearly with time and in fact follows a classical trajectory for a free particle, $<x(t)> =x_0 + p_0t/m$.) In parallel, the position uncertainty as expressed in the standard deviation of the probability density function ("the width") increases in time. This broadening effect suggests that the particle associated with this solution of Schrödinger's equation does not have a well-defined momentum. Envisioning, instead, a distribution of momenta, larger and smaller momentum values should be associated, respectively, with faster and slower changes of the particle's position, resulting in broadening of the position probability density function. This dispersion effect is indeed supported by analyzing the particle's momentum distribution (see Chapter 15). It is noticed qualitatively already in Fig. 4.2.1, where the "local wavelength" in $Re[\psi(x,t)]$ as a function of x becomes shorter with increasing x, namely, toward the wave packet front along the propagation direction. (This becomes more apparent at the later times.) Invoking the de Broglie relation (Eq. (1.1.3)), the shortening of the wavelength translates to increasing momentum toward the front of the wave packet, in accord with the wave packet broadening effect.

As a second example, we refer to the case where the particle is scattered against a potential energy barrier,

$$V(x) = \begin{cases} 0 & ; & x < 0 \\ V_0 & ; & x \geq 0 \end{cases}.$$

The net force on the particle vanishes anywhere in space, except for at the origin, where it becomes infinite in the direction of decreasing x (a "reflecting" force, toward negative x). Again, we focus on some qualitative features of the wave function evolution, as represented in Fig. 4.2.2. The Gaussian wave packet is shown to propagate initially toward the barrier position, similarly to a free particle (compare to Fig. 4.2.1). In the vicinity

(a)   **Probability Density**

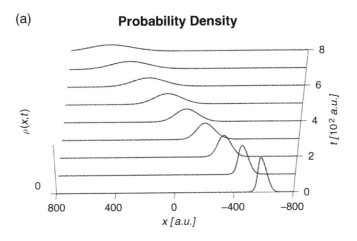

(b)   **Probability Amplitude**

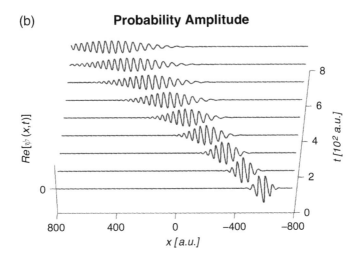

Figure 4.2.1   The time evolution of a free particle represented as a Gaussian wave packet. (a) the probability density, $\rho(x,t)$. (b) the real part of the respective probability amplitude, $\text{Re}\left[\psi(x,t)\right]$ (the wave function normalization is arbitrary). The particle's mass was taken as $m = 0.1$, and the initial wave packet parameters are $x_0 = -600$, $\sigma = 40/\sqrt{2}$, and $p_0 = 0.1414$, all in atomic units (where $\hbar$, the Bohr radius, and the electron rest mass all equal unity).

of the potential energy step the wave is scattered, where the probability density deviates significantly from the initial Gaussian form. The consecutive dynamics depends on the barrier energy: for a relatively high barrier ((a) and (b) in Fig. 4.2.2), the wave packet is scattered primarily backward, resulting in a free particle-like evolution, but in the direction opposite to the original one. This is reminiscent of the scattering of a classical particle whose total energy is below the potential energy barrier. For a lower barrier energy ((c) and (d) in Fig. 4.2.2), the wave packet is shown to split into two parts, one moving forward in the barrier region, and the other reflected backward. This implies

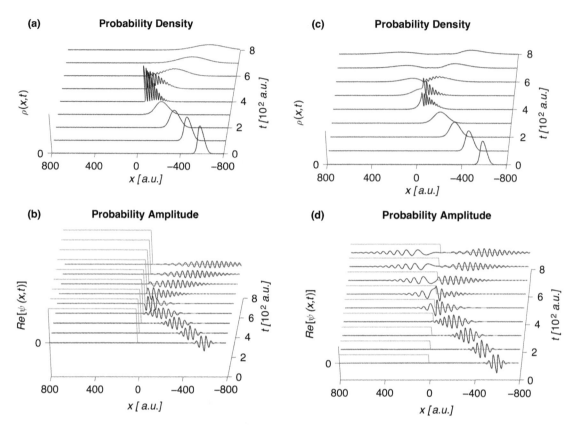

Figure 4.2.2   The time evolution of a Gaussian wave packet scattered from a potential energy barrier. Top (a and c): the probability density, $\rho(x,t)$. Bottom (b and d): the real part of the respective probability amplitude, $\mathrm{Re}[\psi(x,t)]$ (the normalization is arbitrary). The potential energy barriers are also marked for each snapshot. The left column (a and b) corresponds to a relatively high energy barrier ($V_0 = 0.5$ a.u.), where the probability density appears to be completely reflected. The right column (c and d) corresponds to a lower energy barrier ($V_0 = 0.1$ a.u.), where the probability density splits into transmitted and reflected parts. The particle's mass was taken as $m = 0.1$, and the initial wave packet parameters are $x_0 = -600$, $\sigma = 40/\sqrt{2}$, and $p_0 = 0.1414$, all in atomic units.

that there is a significant probability for the particle to either pass the potential energy step or be reflected backward. This result has no analogue in the classical description of a particle's trajectory. Again, it suggests that the particle associated with such a solution to the relevant Schrödinger equation does not have a well-defined momentum, as has already been hinted, but rather some combination of negative and positive momenta.

Finally, let us consider the case of a particle trapped in a potential energy well. Specifically, we consider the potential of a harmonic oscillator centered at the origin, $V(x) = \frac{1}{2}m\omega^2 x^2$, where the particle experiences a position-dependent reflecting force, $F(x) = -m\omega^2 x$. The change in the particle's position according to classical mechanics is oscillatory at a frequency, ω. A thorough analysis of the solution to the

(a)   **Probability Density**

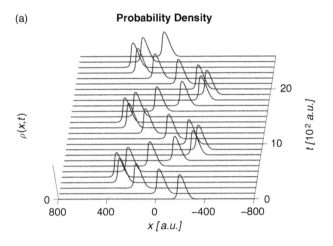

(b)   **Probability Amplitude**

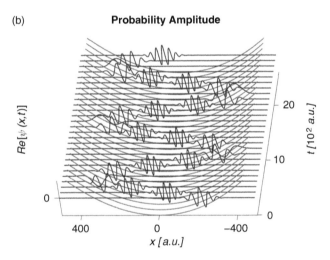

Figure 4.2.3   The time evolution of a Gaussian wave packet in a harmonic potential energy trap. (a) the probability density, $\rho(x,t)$. (b) the real part of the respective probability amplitude, $\mathrm{Re}[\psi(x,t)]$ (the wave function normalization is arbitrary). The potential energy is also marked for each snapshot. The harmonic potential frequency was set to $\omega = 1/160$, the particle's mass was taken as $m = 0.1$, and the initial wave packet parameters are $x_0 = -226$, $\sigma = 40/\sqrt{2}$, and $p_0 = 0.1414$, all in atomic units.

Schrödinger equation for Gaussian wave packets [4.2] in harmonic potentials will be given in Chapter 15, where, as in the preceding discussion, here we are interested only in some qualitative considerations. In Fig. 4.2.3 we present snapshots from the time evolution of a specific initial Gaussian wave packet (Eq. (4.2.1)) in the potential energy well. In accord with the classical results, the probability density is confined to the region of the well. Moreover, both the probability density and the probability amplitude demonstrate periodic time evolution, which are reminiscent of the classical trajectory for a particle in a harmonic potential well.

The same Schrödinger equation can have different solutions, depending on the initial conditions. Let us consider again the harmonic potential well, but this time set the parameters of the initial Gaussian wave packet, x_0 and p_0, to zero. The results, as presented in Fig. 4.2.4, are remarkably different. Particularly, while the probability amplitude keeps changing periodically in time, *the probability density is time-independent, implying that the probability of locating the particle anywhere in space does not change in time. This is a specific example to an important set of solutions to the Schrödinger equation, termed the stationary solutions.* In the following sections, we shall focus on the physical meaning of these stationary solutions and on their special role in quantum mechanics.

Probability Density

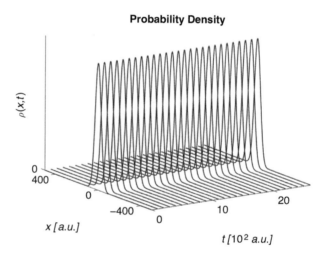

Probability Amplitude

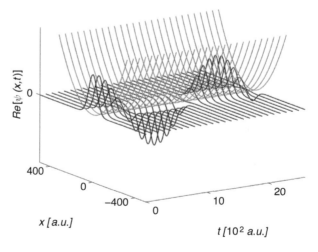

Figure 4.2.4   The same as Fig. 4.2.3, but for different initial wave packet parameters, with $x_0 = p_0 = 0$.

4.3 Stationary Solutions and the Time-Independent Schrödinger Equation

The time evolution of solutions to the Schrödinger equation ($\psi(t)$) is not necessarily associated with a change of the probability density function ($|\psi(t)|^2$). Indeed, the time-dependent Schrödinger equation has solutions for which the probability density is time-independent. These solutions are termed "stationary solutions" and are analogues to "standing wave" solutions of similar wave equations (e.g., the scalar wave equation in electrodynamics). Let us consider a proper time-dependent wave function in the form

$$\psi(t) \equiv e^{-i\alpha t}\psi(0), \tag{4.3.1}$$

where α is any real-valued scalar. It immediately follows that although the probability amplitude, $\psi(t)$, depends explicitly on time, the corresponding probability density is time-independent in this case:

$$\rho(t) = |\psi(t)|^2 = |\psi(0)|^2. \tag{4.3.2}$$

$\psi(t)$ is therefore termed a stationary wave function. However, for $\psi(t)$ to be associated with a state of a physical system, it must be a solution of the time-dependent Schrödinger equation, Eq. (4.1.1); namely, it must satisfy $\hat{H}e^{-i\alpha t}\psi(0) = i\hbar\frac{\partial}{\partial t}e^{-i\alpha t}\psi(0)$. This imposes an implicit relation between the parameter α and the function $\psi(0)$,

$$\hat{H}\psi(0) = \alpha\hbar\psi(0). \tag{4.3.3}$$

The relation is formulated as an eigenvalue equation. A function f is termed an eigenfunction of an operator, $\hat{O}$, if $\hat{O}f$ equals a scalar times f, namely, $\hat{O}f = \lambda f$. The scalar (λ) is termed the eigenvalue of the operator, $\hat{O}$, that corresponds to the eigenfunction, f. Eq. (4.3.3) means that *a function, $\psi(t) \equiv e^{-i\alpha t}\psi(0)$, can be a (stationary) solution to the time-dependent Schrödinger equation if and only if $\psi(0)$ is an eigenfunction of the system Hamiltonian*.

Exercise 4.3.1 *Show that if $f_1(x)$ is an eigenfunction of a linear operator $\hat{A}$ with an eigenvalue α, so is $f_2(x) = cf_1(x)$, where c is a constant scalar.*

Exercise 4.3.2 *(a) Determine whether the functions $f_1(x) = x^2$, $f_2(x) = e^{iax}$, and $f_3(x) = \sin(ax)$ are (independently) eigenfunctions of the operator $\frac{d}{dx}$. In cases where the answer is positive, determine the respective eigenvalue. (b) Repeat the exercise for the operator $\frac{d^2}{dx^2}$.*

Exercise 4.3.3 *In Section 4.2 we encountered two different solutions to the Schrödinger equation with a harmonic potential energy well, $V(x) = \frac{1}{2}m\omega^2x^2$. The respective probability density was time-dependent only in one of these cases (presented in Fig. 4.2.3), whereas the other solution (presented in Fig. 4.2.4) was stationary. These two solutions correspond to different choices of the parameters (x_0, p_0) in the initial wave function (see*

Eq. (4.2.1)). Check whether $\psi(x,0)$ is an eigenfunction of the system Hamiltonian for these two choices and explain the observed difference between the two solutions.

Each eigenfunction of the Hamiltonian (energy) operator $\hat{H}$ is associated with an eigenvalue, which is real-valued (see what follows). The eigenvalues are conventionally marked by "E" (for "energy"), and the respective eigenfunctions are denoted as $\psi_E(0)$. Noticing that $\psi_E(0)$ itself is time-independent, we can therefore omit the zero-time notation and rewrite Eq. (4.3.3) as follows:

$$\hat{H}\psi_E = E\psi_E. \tag{4.3.4}$$

This equation is known as the time-independent (or the stationary) Schrödinger equation. Each solution to this equation (each Hamiltonian eigenfunction) corresponds to a stationary solution of the time-dependent Schrödinger equation, which reads

$$\psi_E(t) = e^{-iEt/\hbar}\psi_E. \tag{4.3.5}$$

Exercise 4.3.4   *An N-dimensional system is associated with the spatial coordinates $x_1, x_2, \ldots, x_N$. Let us consider the case where the system Hamiltonian is separable, namely, it can be written as a sum, $\hat{H} = \sum_{j=1}^{N} \hat{H}_{x_j}$, where $\hat{H}_{x_j}$ is a Hamiltonian of a system associated only with the coordinate x_j. The eigenfunctions and eigenvalues of the jth Hamiltonians are defined by the set of eigenvalue equations, $\hat{H}_{x_j}\varphi_{n_j}(x_j) = E_{n_j}\varphi_{n_j}(x_j)$. Show that any product function, $\psi_{n_1,n_2,\ldots,n_N}(x_1, x_2, \ldots, x_N) = \varphi_{n_1}(x_1)\varphi_{n_2}(x_2)\cdots\varphi_{n_N}(x_N)$, is an eigenfunction of the full Hamiltonian, $\hat{H}$, with the corresponding eigenvalue, $E_{n_1,n_2,\ldots,n_N} = E_{n_1} + E_{n_2} + \ldots + E_{n_N}$.*

4.4 Interpretation of The Hamiltonian Eigenvalues

The stationary solutions encountered in the preceding discussion have a special meaning. They correspond to the physical states in which the total energy of a system is well defined. To realize this, we must become familiar with another postulate of quantum mechanics. We have already learned that the state of a system is associated with a proper wave function and that observables are associated with specific operators. As elaborated later in Chapter 11, *the postulates of quantum mechanics state additionally that in a single measurement on a system the measured value can only be an eigenvalue of the operator that represents the measured physical variable.* Namely, if the wave function that represents the state of the system, $\psi(t)$, happens to be an eigenfunction of an operator $\hat{A}$, that is, $\hat{A}\psi(t) = \lambda\psi(t)$, then the result of the measurement of the quantity associated with the operator $\hat{A}$ can only be the respective eigenvalue λ. Particularly, if the system is found in a stationary solution, that is, it is an eigenfunction of the energy operator, $\hat{H}\psi_E(t) = E\psi_E(t)$ (see Eqs. (4.3.4, 4.3.5)), a measurement of its energy will yield the eigenvalue, E. It follows that when a system is prepared in a state that corresponds to a stationary solution of the Schrödinger equation, its energy is well defined.

It also follows that when a system has a well-defined energy, the value of that energy cannot obtain just any number. It must coincide with one of the eigenvalues of the system Hamiltonian. *The set of Hamiltonian eigenvalues therefore corresponds to the energy levels of the system.*

Notice that probability density associated with any stationary $\psi_E(t)$ is time-independent (for any real-valued E; see the following discussion). The "time-independence" associated with the stationary solutions is not restricted, however, to the probability density. When discussing the postulates of quantum mechanics in Chapter 11 we shall see that *when a system is in a state that corresponds to a stationary solution (Eq. (4.3.5)), any measurable information on the system is in fact time-independent.*

4.5 The Hamiltonian as a Hermitian Operator

The postulates of quantum mechanics associate measurable quantities with operators, as we discovered in Chapter 3, and measured values with eigenvalues of these operators, as indicated in the previous section. The latter means that the eigenvalues of the operators that represent measurable quantities must be real-valued. This is indeed the case, owing to the facts that *in quantum mechanics the measurable quantities are represented by Hermitian operators in the space of proper functions* and that *the eigenvalues of Hermitian operators in this space are real-valued.* In this section we introduce the definition of a Hermitian operator and show that the operators that were associated with measurable quantities in Chapter 3 are indeed Hermitian. We then discuss several general properties of the eigenvalues and eigenfunctions of Hermitian operators.

An operator $\hat{A}$ is termed Hermitian in a space of functions if, for any two functions in the space (e.g., $f(x)$ and $g(x)$ in a one-dimensional Cartesian space), the following identity is satisfied:

$$\int_{-\infty}^{\infty} f(x)\hat{A}g(x)dx = \left[\int_{-\infty}^{\infty} g^*(x)\hat{A}f^*(x)dx\right]^*. \tag{4.5.1}$$

Here z^* denotes the complex conjugate of the complex number z. The generalization of this definition to operators in a multidimensional coordinate space is straightforward, replacing the one-dimensional integral by the corresponding multidimensional integral. For simplicity, we shall stick to a one-dimensional coordinate space in the discussion that follows.

A Hermitian operator is self-adjoint; namely, it is equal to its Hermitian conjugate. The Hermitian conjugate of any operator $\hat{A}$ is denoted as $\hat{A}^\dagger$, and it must satisfy the following identity for any proper functions, $f(x)$ and $g(x)$:

$$\int_{-\infty}^{\infty} f(x)\hat{A}^\dagger g(x)dx = \left[\int_{-\infty}^{\infty} g^*(x)\hat{A}f^*(x)dx\right]^*. \tag{4.5.2}$$

For $\hat{A}$, which is Hermitian, one therefore has

$$\hat{A}^\dagger = \hat{A}. \tag{4.5.3}$$

We can readily verify that the quantum mechanical operators discussed in Chapter 3 are Hermitian in the space of proper functions.

Let us start with the canonical operators, $\hat{x}$ and $\hat{p}_x$. Considering first the position operator ($\hat{x}g(x) = xg(x)$), one obtains for any proper $f(x)$ and $g(x)$

$$\int_{-\infty}^{\infty} f(x)\hat{x}g(x)dx = \int_{-\infty}^{\infty} f(x)xg(x)dx = \left[\int_{-\infty}^{\infty} g^*(x)x^*f^*(x)dx\right]^*$$

$$= \left[\int_{-\infty}^{\infty} g^*(x)xf^*(x)dx\right]^* = \left[\int_{-\infty}^{\infty} g^*(x)\hat{x}f^*(x)dx\right]^*, \tag{4.5.4}$$

where the fact that x itself is real-valued was used. For the momentum operator, $\hat{p}_xg(x) = -i\hbar\frac{dg(x)}{dx}$, we can use integration by parts and the asymptotic vanishing of proper wave functions (Eq. (2.2.1)) to validate its Hermiticity:

$$\int_{-\infty}^{\infty} f(x)\hat{p}_xg(x)dx = -i\hbar\int_{-\infty}^{\infty} f(x)\frac{dg(x)}{dx}dx = -i\hbar f(x)g(x)|_{-\infty}^{\infty} + i\hbar\int_{-\infty}^{\infty} g(x)\frac{df(x)}{dx}dx$$

$$= \left[-i\hbar\int_{-\infty}^{\infty} g^*(x)\frac{df^*(x)}{dx}dx\right]^* = \left[\int_{-\infty}^{\infty} g^*(x)\hat{p}_xf^*(x)dx\right]^*. \tag{4.5.5}$$

Exercise 4.5.1 *Given a Hermitian operators, $\hat{A}$, prove that (a) $(\hat{A})^n$ is also Hermitian (for any natural n); (b) $\alpha\hat{A}$ is also Hermitian (for any real-valued α); (c) $i\hat{A}$ is anti-Hermitian (namely $(i\hat{A})^\dagger = -(i\hat{A})$).*

Exercise 4.5.2 *Given two Hermitian operators, $\hat{A}$ and $\hat{B}$, prove that (a) $\hat{A} + \hat{B}$ is also Hermitian; (b) if $\hat{A}$ and $\hat{B}$ commute, $\hat{A}\hat{B}$ is also Hermitian.*

Exercise 4.5.3 *Using the results of Exs. 4.5.1 and 4.5.2, prove, independently, that the potential energy operator ($V(\hat{x})$), the kinetic energy operator ($\hat{p}_x^2/(2m)$), the Hamiltonian, ($\hat{H} = \hat{p}_x^2/(2m) + V(\hat{x})$), and the angular momentum operators ($\hat{L}_x, \hat{L}_y,$ and $\hat{L}_z$, as defined in Eq. (3.3.2)) are all Hermitian.*

As we discussed in Section 3.4, other quantum mechanical operators are derived from the canonical operators. Using the results in Exs. 4.5.1 and 4.5.2, it follows that these operators are also Hermitian (see Ex. 4.5.3). In particular, the components of the angular momentum of a particle in a three-dimensional space, $\hat{L}_x, \hat{L}_y,$ and $\hat{L}_z$ (Eq. (3.3.2)), the potential energy, $\hat{V} = V(\hat{x})$ (Eq. (3.4.7)), and the kinetic energy, $\hat{T} = \frac{\hat{p}_x^2}{2m}$ (Eq. (3.4.8)) are Hermitian operators, and so is the Hamiltonian $\hat{H} = \hat{T} + \hat{V}$, for *any closed system of particles*. It is worthwhile to emphasize that non-Hermitian operators are also useful and are commonly met in quantum mechanics (see, e.g., Ex. 4.5.4, as well as the following chapters of this book), but these do not represent directly measurable physical quantities.

Exercise 4.5.4 *An operator $\hat{U}$ is termed unitary if its Hermitian conjugate equals its inverse, namely, $\hat{U} = (\hat{U}^{-1})^{\dagger}$, where $\hat{U}^{-1}\hat{U} = \hat{I}$. Use the Hermiticity of the Hamiltonian to prove that the time evolution operator for a time-independent Hamiltonian, $e^{-i(t-t_0)\hat{H}/\hbar}$ (Eq. (4.1.2)), is unitary.*

The eigenvalues and eigenfunctions of Hermitian operators have unique properties with important consequences. The first property has already been mentioned, namely, the eigenvalues of a Hermitian operator in the space of proper wave functions are real-valued. Let $\hat{A}$ be a Hermitian operator, and let α be any one of its eigenvalues, associated with the eigenfunction, $\varphi_\alpha(x)$. Since $\hat{A}$ is Hermitian, Eq. (4.5.1) applies for any proper functions, and particularly for the choices, $f(x) \equiv \varphi_\alpha^*(x)$ and $g(x) \equiv \varphi_\alpha(x)$, such that $\int_{-\infty}^{\infty} \varphi_\alpha^*(x)\hat{A}\varphi_\alpha(x)dx = \left[\int_{-\infty}^{\infty} \varphi_\alpha^*(x)\hat{A}\varphi_\alpha(x)dx\right]^*$. Since $\hat{A}\varphi_\alpha(x) = \alpha\varphi_\alpha(x)$, one obtains $\alpha \int_{-\infty}^{\infty} \varphi_\alpha^*(x)\varphi_\alpha(x)dx = \alpha^* \int_{-\infty}^{\infty} \varphi_\alpha^*(x)\varphi_\alpha(x)dx$. For any proper function, $\int_{-\infty}^{\infty} \varphi_\alpha^*(x)\varphi_\alpha(x)dx \neq 0$ (see Eq. (2.1.3)), and therefore we conclude that any eigenvalue must be real-valued ($\alpha = \alpha^*$). This property is consistent with the interpretation of the eigenvalues of quantum mechanical operators as directly measurable quantities.

4.6 Properties of the Stationary Solutions

The stationary solutions of the Schrödinger equation have additional special properties. Particularly, the entire set of stationary solutions for a given system can be mapped on a set of orthonormal functions. As we shall discuss in Chapters 11 and 15, such a set spans the space of physically meaningful solutions to the time-dependent Schrödinger equation for the system; namely, any time-dependent solution can be expanded as a linear combination of the stationary solutions. In this section we focus on the orthogonality property of different stationary solutions, which is derived directly from their being solutions of the time-independent Schrödinger equation (Eq. (4.3.4)), namely, being eigenfunctions of the Hermitian (Hamiltonian) operator.

Let us start by defining the overlap integral between two normalized functions, $f(x)$ and $g(x)$, as follows:

$$S = \int_{-\infty}^{\infty} f^*(x)g(x)dx, \tag{4.6.1}$$

where $\int_{-\infty}^{\infty} |f(x)|^2 dx = \int_{-\infty}^{\infty} |g(x)|^2 dx = 1$. Notice that S is generally complex-valued. However, one can show that its absolute value satisfies $0 \leq |S|^2 \leq 1$, and it can therefore provide a measure for the "similarity" between the two functions. (For a detailed discussion in the context of an "inner product" definition, see Chapter 11.) The maximal value corresponds to $f(x)$ and $g(x)$ being equal, whereas the minimal value corresponds to the functions being inherently different (see Ex. 4.6.1), that is, orthogonal to each other. *Two functions, $f(x)$ and $g(x)$, are termed orthogonal to each other if their overlap integral vanishes, namely,*

$$\int_{-\infty}^{\infty} f^*(x)g(x)dx = 0. \tag{4.6.2}$$

It is easy to see that *any two eigenfunctions of a Hermitian operator, $\hat{A}$, that is, $\varphi_\alpha(x)$ and $\varphi_\beta(x)$, associated respectively with two different eigenvalues, $\alpha \neq \beta$, are orthogonal.* Since $\hat{A}$ is Hermitian, Eq. (4.5.1) applies for any proper functions, and particularly for the choices, $f(x) \equiv \varphi_\alpha^*(x)$ and $g(x) \equiv \varphi_\beta(x)$, such that $\int_{-\infty}^{\infty} \varphi_\alpha^*(x)\hat{A}\varphi_\beta(x)dx = \left[\int_{-\infty}^{\infty} \varphi_\beta^*(x)\hat{A}\varphi_\alpha(x)dx\right]^*$. Since $\hat{A}\varphi_\alpha(x) = \alpha\varphi_\alpha(x)$ and $\hat{A}\varphi_\beta(x) = \beta\varphi_\beta(x)$, with real-valued α and β one obtains $\beta \int_{-\infty}^{\infty} \varphi_\alpha^*(x)\varphi_\beta(x)dx = \alpha \int_{-\infty}^{\infty} \varphi_\alpha^*(x)\varphi_\beta(x)dx$. Finally, since $\alpha \neq \beta$, one has $\int_{-\infty}^{\infty} \varphi_\alpha^*(x)\varphi_\beta(x)dx = 0$, namely, $\varphi_\alpha(x)$ and $\varphi_\beta(x)$ must be orthogonal (see Eq. (4.6.2)). Since the Hamiltonian is Hermitian, it therefore follows that any two solutions to the time-independent Schrödinger equation, associated with different eigenvalues (different energy values), are orthogonal. *But what if two solutions are associated with the same eigenvalue?*

It often happens that two (or more) different (namely, linearly independent) eigenfunctions of an operator are associated with the same eigenvalue. Such functions are termed degenerate. Notice that different solutions of the Schrödinger equation (Eq. (4.1.1)) correspond to different physical states of a given system. However, *when two solutions are degenerate with respect to an operator, $\hat{A}$, the corresponding physical states cannot be distinguished by a measurement of the property represented by $\hat{A}$*, since the measured value would be the same for both. In particular, an energy measurement cannot distinguish between different degenerate stationary states (Hamiltonian eigenfunctions) of a system. To do so, one needs to perform a different measurement, associated with another operator for which the two functions have different eigenvalues. Examples are discussed in the following chapters.

Exercise 4.6.1 *Let $f_e(x)$ and $f_o(x)$ be any even and odd functions respectively, namely, $f_e(-x) = f_e(x)$ and $f_o(-x) = -f_o(x)$. Prove that $f_e(x)$ and $f_o(x)$ are orthogonal according to the definition in Eq. (4.6.2).*

Exercise 4.6.2 *The parity operator is defined by its operation: $\hat{P}f(x) = f(-x)$. (a) Prove that $\hat{P}$ is Hermitian. (b) Show that even or odd functions of x are eigenfunctions of $\hat{P}$. What are the corresponding eigenvalues? (c) Explain the result of Ex. 4.6.1, using (a) and (b).*

Exercise 4.6.3 *Prove that if $\hat{A}$ is a linear operator, and if $\varphi_1(x), \varphi_2(x), \ldots, \varphi_N(x)$ are degenerate eigenfunctions of $\hat{A}$, namely, $\hat{A}\varphi_n(x) = \alpha\varphi_n(x)$, for $n \in 1, 2, \ldots, N$, then any linear combination of these functions is an eigenfunction of $\hat{A}$, with the eigenvalue α.*

Exercise 4.6.4 *Given two degenerate normalized eigenfunctions of a linear operator $\hat{A}$, $\hat{A}\varphi_1(x) = \alpha\varphi_1(x)$ and $\hat{A}\varphi_2(x) = \alpha\varphi_2(x)$, show that the following functions,*

$$\psi_1(x) = \varphi_1(x) \quad ; \quad \psi_2(x) = \varphi_2(x) - \varphi_1(x)\int\limits_{-\infty}^{\infty} dx'\, \varphi_1^*(x')\varphi_2(x'),$$

are two orthogonal eigenfunctions of $\hat{A}$ corresponding to the same eigenvalue α.

Recalling that the quantum mechanical operators are linear, it follows that any linear combination of degenerate eigenfunctions of a quantum mechanical operator is itself an eigenfunction of the same operator, associated with the same eigenvalue (see Ex. 4.6.3)). When a set of N linearly independent degenerate functions spans the subspace of degenerate functions, the corresponding eigenvalue is termed N-fold degenerate. Clearly, degenerate eigenfunctions of an operator are not defined uniquely. In particular, two degenerate eigenfunctions of a given Hermitian operator are not necessarily orthogonal to each other. Nevertheless, *for an N-fold degenerate eigenvalue of a given operator, it is always possible to construct N orthogonal eigenfunctions to the operator that correspond to the same eigenvalue*. An example for $N = 2$, given explicitly in Ex. 4.6.4, can be generalized to any N [4.4]. It therefore follows that *for any (Hermitian and linear) quantum mechanical operator there exists a set of eigenfunctions that are orthonormal*, namely a set in which each function is normalized, and every two functions are orthogonal to each other. Denoting the functions in the orthonormal set as, $\varphi_1(x), \varphi_2(x), \ldots$, one has

$$\int\limits_{-\infty}^{\infty} \varphi_n^*(x), \varphi_{n'}(x)dx = \delta_{n,n'}, \tag{4.6.3}$$

where the Kronecker delta is defined as

$$\delta_{n,n'} = \begin{cases} 0 & ; \quad n \neq n' \\ 1 & ; \quad n = n' \end{cases}. \tag{4.6.4}$$

Notice that Eq. (4.6.3) applies in particular to the set of the Hamiltonian eigenfunctions (stationary solutions), defined by Eq. (4.3.4).

Exercise 4.6.5 *$\varphi_1(x)$, $\varphi_2(x)$, and $\varphi_3(x)$ are eigenfunctions of a Hamiltonian $\hat{H}_x$, that is, $\hat{H}_x\varphi_n(x) = E_n\varphi_n(x)$, where $E_1 \neq E_2 = E_3$. Each $\varphi_n(x)$ corresponds to a stationary solution of the time-dependent Schrödinger equation, $\psi_n(x,t) = \varphi_n(x)e^{\frac{-iE_n t}{\hbar}}$. Show that the following linear combinations of stationary solutions (a_1, a_2, and a_3 are non-zero scalars) are solutions to the time-dependent Schrödinger equation. Which of these solutions is stationary (namely, associated with a time-independent probability density function)? What is the conclusion?*

a. $\Psi(x,t) = a_1\psi_1(x,t) + a_2\psi_2(x,t)$
b. $\Psi(x,t) = a_3\psi_3(x,t) + a_2\psi_2(x,t)$
c. $\Psi(x,t) = a_3\psi_3(x,t) + a_1\psi_1(x,t)$

Exercise 4.6.6 *Show that the two Hamiltonians $\hat{H}_x$ and $\hat{H}_x + \alpha$, where α is a scalar constant, have the same eigenfunctions. What is the relation between the corresponding eigenvalues?*

Bibliography

[4.1] U. Peskin and N. Moiseyev, "The solution of the time-dependent Schrödinger equation by the (t, t') method: theory, computational algorithm and applications," The Journal of Chemical Physics 99, 4590 (1993).

[4.2] E. J. Heller, "Time-dependent approach to semiclassical dynamics," The Journal of Chemical Physics 62, 1544 (1975).

[4.3] C. Cohen-Tannoudji, B. Diu and F. Laloë, "Quantum Mechanics," vols. 1–2: (John Wiley & Sons, 2020).

[4.4] W. E. Arnoldi, "The principle of minimized iterations in the solution of the matrix eigenvalue problem," Quarterly of Applied Mathematics 9, 17 (1951).

5 Energy Quantization

5.1 The Energy Spectrum

In Chapter 4 we became familiar with the existence of stationary (standing wave) solutions to the time-dependent Schrödinger equation. These solutions were associated with eigenfunctions of the Hamiltonian operator (Eq. (4.3.4)), where the corresponding eigenvalues were identified with the real-valued energy levels of the system.

Typically, the Hamiltonian of a system has numerous eigenfunctions (different standing wave solutions), which correspond to a set of energy levels. This set is termed the energy spectrum of the system. Since the Hamiltonian eigenvalues are real-valued, they can be ordered: $E_n, n = 0, 1, 2, 3, \ldots$. However, the set of energy levels of a given system within a finite energy interval can be either discrete ("quantized") or continuous. For particles in extended (or "open") systems (such as solid lattices or in free space) the spectrum is continuous, but for particles bounded in (or confined to) a restricted region of space, for example, within atoms, molecules, or nanoparticles, the energy spectrum becomes discrete (see Fig. 5.1.1). To emphasize this, we rewrite Eq. (4.3.4) for a discrete spectrum as follows:

$$\hat{H}_x \psi_n(x) = E_n \psi_n(x). \tag{5.1.1}$$

As we discuss in detail here and also in Chapters 6–10, the discrete spectrum emerges naturally following mathematical constraints imposed on the solutions to Eq. (5.1.1).

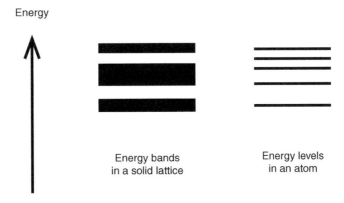

Energy

Energy bands in a solid lattice

Energy levels in an atom

Figure 5.1.1 The emergence of energy level quantization for particles confined in space.

These constraints assure that the wave functions are proper and are therefore suitable for representing what we can consider to be a physical reality (probability conservation, continuity in space, etc.). The mathematical constraints typically include specific boundary conditions imposed on the wave functions (e.g., $\psi_n(x) \xrightarrow[x \to \pm\infty]{} 0$, Eq. (2.2.1)) which can only be satisfied for a discrete set of energy values. An important manifestation of this "quantization" of the energy for particles confined in space is the "quantum size effect" discussed in what follows.

5.2 The "Quantum Size Effect"

Nanocrystals are relatively small crystals of semiconductors (containing typically 10^3–10^4 atoms), with a diameter of several nanometers. These crystals are useful in numerous applications including Light Emitting Diodes (LEDs), display screens, solar cells, medical treatment of diseases, and many others. The reason for their usefulness is that their electronic properties are sensitively determined by particle shape and size, which can be relatively easily controlled during their preparation process. This is reflected in size-controlled properties such as absorption and emission of electromagnetic radiation in the visible spectrum (the observed "color"), interparticle charge and energy transport, chemical catalytic activity, and so on. Consider, for example, light emission from nanocrystals made of the same semiconductor material, but at different particle sizes, as illustrated in Fig. 5.2.1. The red light is emitted from particles of 6 nm in diameter, and the blue is emitted from particles of 2 nm in diameter. *How can the*

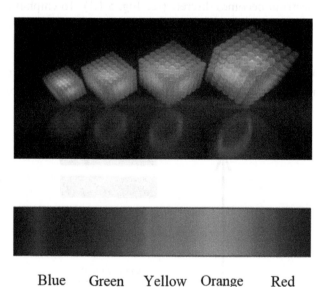

Blue   Green   Yellow   Orange   Red

Figure 5.2.1 (Greyscale) illustration of the effect of the size of colloidal nanoparticles (quantum dots) on the color of their emitted radiation (Image by Serg Myshkovsky/Photodisc/Getty Images).

size of a particle affect so dramatically (and usefully!) the color of the emitted light? Why is the emission shifted toward the red color as the particle diameter increases? Why is it necessary to have particles of just a few nanometers in diameter to observe the size-dependence of the emitted light?

A rigorous answer to these questions requires familiarity with the electronic structure of materials (and semiconductor crystals in particular), discussed in Chapter 14, as well as with the phenomenon of radiation emission, discussed in Chapter 18. Nevertheless, to explain the essence of the phenomenon we provide here some necessary facts that will be elaborated when we address light–matter interactions in Chapters 15 and 18. Considering the colors associated with the visible spectrum of the electromagnetic radiation, the change of color from red to blue corresponds to a change in the wavelength (λ) of the emitted radiation from $\sim$700 nm to $\sim$400 nm, where the emitted wavelength is inversely proportional to the radiation frequency (v),

$$v = \frac{c}{\lambda}. \tag{5.2.1}$$

(Here c is the speed of light.) The frequency is proportional to the energy lost by electrons in the material during the radiation emission (Planck's formula [1.3]):

$$E_f - E_i = hv = \frac{hc}{\lambda}. \tag{5.2.2}$$

E_i and E_f are respectively the "initial" and "final" electronic energy levels within the nanoparticle. It follows that a shorter emission wavelength indicates a larger energy gap between the relevant initial and final electronic energy levels. The observation in Fig. 5.2.1 therefore suggests that the energy gap between E_i and E_f increases with decreasing

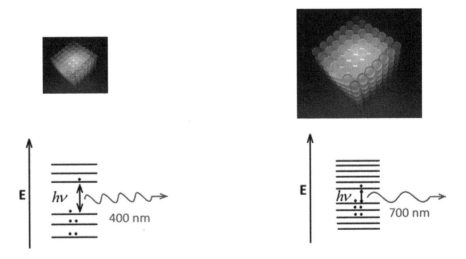

Figure 5.2.2 The correlation between the size of a nanoparticle and the color of emission. The smaller (left) nanoparticle is associated with a relatively larger spacing between the discrete energy levels, and the emission wavelength is thus shorter. Electrons occupying the energy levels are indicated by dots. (Image by Serg Myshkovsky/Photodisc/Getty Images.)

particle diameter. This phenomenon is a manifestation of ***"the quantum size effect,"*** ***which means that the spectrum of a quantum system becomes less dense (namely, the*** ***energy gap between two given states increases) as the size of the system decreases.*** This is demonstrated qualitatively in Fig. 5.2.2.

The quantum size effect is ubiquitous in quantum mechanics. In what follows we demonstrate its emergence by solving the time-independent Schrödinger equation for the simplest textbook model of a confined particle: the "particle-in-a-box."

5.3 Energy Quantization for a "Particle-in-a-Box"

The model assumes a single point particle of mass m and a one-dimensional Cartesian coordinate, x. In classical mechanics, the particle would be free to move inside a "square box" in the range $0 \le x \le L$ (see Fig. 5.3.1), and its motion would be confined to this box by the following potential energy function:

$$V(x) = \begin{cases} \infty & 0 > x \\ 0 & 0 \le x \le L \\ \infty & L < x \end{cases}. \tag{5.3.1}$$

Moving to the quantum mechanical description, we seek the solutions to the time-independent Schrödinger equation for this system. Outside the box, the probability density for locating the particle must vanish on physical grounds. (This result can be derived rigorously by considering a finite external potential energy [see Chapter 6], and taking the limit to infinity in the resulting solutions.) This condition reads

$$|\psi(x)|^2 = 0; \quad x < 0; \ L < x. \tag{5.3.2}$$

It follows that the wave function itself must also vanish ($\psi(x) = 0$) anywhere outside the box. Inside the box, the time-independent Schrödinger equation (Eq. (5.1.1)) reads

$$\frac{-\hbar^2}{2m} \frac{\partial^2}{\partial x^2} \psi(x) = E \psi(x). \tag{5.3.3}$$

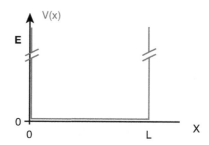

Figure 5.3.1   The potential energy for a particle in an infinite one-dimensional "box."

Defining

$$k \equiv \sqrt{\frac{2mE}{\hbar^2}}, \tag{5.3.4}$$

the general solution of Eq. (5.3.3) for any k reads

$$\psi(x) = ae^{ikx} + be^{-ikx}. \tag{5.3.5}$$

On physical grounds, we require that the probability density for locating the particle in space is a continuous function of x. Particularly, since the probability density vanishes anywhere outside the box, it must vanish also at the box boundaries, in accordance with Eq. (5.3.2). This requirement translates to boundary conditions on the wave function itself, namely,

$$\psi(0) = 0, \tag{5.3.6a}$$

$$\psi(L) = 0. \tag{5.3.6b}$$

Using Eq. (5.3.6a) in Eq. (5.3.5) imposes a dependency between a and b, and the space of solutions is restricted accordingly:

$$\psi(x) = 2ia \sin(kx), \tag{5.3.7}$$

which holds for any value of k. However, imposing the second boundary condition, (5.3.6b), the allowed values of k are shown to be restricted to a discrete set, $k_n = \frac{n\pi}{L}$:

$$\psi_n(x) = 2ia \sin(\frac{n\pi}{L}x); \quad n = 0, \pm 1, \pm 2, \dots. \tag{5.3.8}$$

We note that for $n = 0$ the wave function ($\psi_0(x)$) is identically zero in the entire coordinate space, and therefore it is improper (see Section 2.2). Additionally, the solutions $\psi_{-n}(x)$ and $\psi_n(x)$ are linearly dependent. (They differ only by sign; namely, they are the same up to a multiplication by a constant.) Normalizing the solutions in Eq. (5.3.7), we obtain the following set of proper, linearly independent solutions to the Schrödinger equation for the particle in a box:

$$\psi_n(x) = \begin{cases} 0 & 0 > x \\ \sqrt{\frac{2}{L}} \sin\left(\frac{n\pi}{L}x\right) & 0 \le x \le L \\ 0 & L < x \end{cases} \quad ; \quad n = 1, 2, 3, \dots. \tag{5.3.9}$$

We emphasize that this discrete set of solutions was "filtered" out from the full space of solutions (Eq. (5.3.5)) by requirements that were based on our physical interpretation of the solutions, such as the continuity and properness of the wave functions.

The energy spectrum corresponding to the physically meaningful solutions is, therefore, also discrete. Using $k_n = \frac{n\pi}{L}$ and Eq. (5.3.4), we obtain the expression for the discrete energy levels,

$$E_n = \frac{\hbar^2 k_n^2}{2m} = \frac{\hbar^2 \pi^2 n^2}{2mL^2} \quad ; \quad n = 1, 2, 3, \dots. \tag{5.3.10}$$

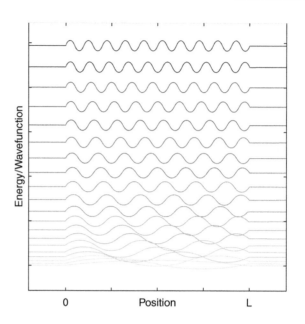

Figure 5.3.2   Energy levels and stationary wave functions for the "particle-in-a-box." The different plots correspond to different energy levels. The plots are displaced from each other by the respective energy differences.

This result provides a first demonstration of the "quantum size effect." We can see that the spacing between any two levels (denoted by the indexes, n and n′) increases, as the box becomes smaller:

$$E_n - E_{n'} = \frac{\hbar^2 \pi^2 (n^2 - n'^2)}{2mL^2} \propto \frac{1}{L^2}. \qquad (5.3.11)$$

Indeed, for an infinitely large box, the energy spectrum becomes continuous. The quantization of the spectrum emerges due to the confinement of the particle to a finite region in coordinate space, and becomes more apparent as the confinement becomes stronger, namely, as the box becomes smaller. Notice that this qualitative result is universal for confined particles, although the quantitative dependence, $\Delta E \propto \frac{1}{L^2}$, is specific for the one-dimensional box with an infinite external potential energy.

Fig. 5.3.2 illustrates the energy spectrum and the corresponding wave functions for the particle in a box model. The set of stationary solutions for this model demonstrates some general features of stationary solutions to Schrödinger equations, which we briefly review here. It is worthwhile to notice the following points:

I. The emergence of a "zero-point energy": as it turns out, the lowest energy level is above the zero of the potential energy in the system. This means that the particle in the box has some minimal energy that cannot be "drained" out. This phenomenon is typical to all confining potentials. It is related to the uncertainty principle (discussed in Chapter 11), which shows that for a proper state it is impossible to simultaneously have a strictly zero momentum (zero kinetic energy) and some certainty with respect to the position of the particle (namely, its confinement to the box region).

II. For small n values, the probability density is not uniformly distributed along the box and the wave nature of the stationary solutions is pronounced.

III. For large n values, the probability tends to distribute uniformly over the entire box, which is reminiscent of the classical behavior for a particle in a box at a given energy.

IV. The number of oscillations (the number of zero values, or "nodes") in the wave function increases with energy. This is somewhat reminiscent of the de Broglie wavelength formula (Eq. (1.1.3)), but notice that the particle in the box is not a free particle since its momentum vector (the direction of motion) is not well-defined.

Exercise 5.3.1   *Prove that the solutions of the time-independent Schrödinger equation for the particle-in-a-box model (denoted as, $\psi_n(x)$), associated with different energy levels, are orthogonal to each other, as required by the Hermiticity of the Hamiltonian (Eq. (4.6.3)).*

5.4 Energy Quantization for a "Particle-on-a-Ring"

Our second example for the emergence of quantization for spatially confined particles is the model of a "particle on a ring," illustrated in Fig. 5.4.1. This model has similarities to the particle-in-a-box model, but also some important differences, that demonstrate the effect of different boundary conditions on the energy level quantization. The model assumes a single point particle of mass μ, restricted to a uniform ring at a constant radius r in the (x, y) plane. The time-independent Schrödinger equation in the plane reads

$$\frac{-\hbar^2}{2\mu} \left[\frac{\partial^2}{\partial x^2} + \frac{\partial^2}{\partial y^2} \right] \psi(x, y) + V(x, y) \psi(x, y) = E \psi(x, y). \tag{5.4.1}$$

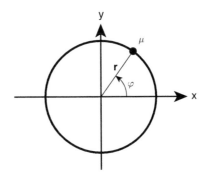

Figure 5.4.1   The "particle-on-a-ring"

However, since r is restricted to a constant value, it is useful to switch from Cartesian to polar coordinates, $(x, y) \rightarrow (r, \varphi)$, where

$$
\begin{aligned}
x &= r\cos(\varphi), \\
y &= r\sin(\varphi), \\
r &= \sqrt{x^2 + y^2}, \\
\varphi &= \arccos(x/\sqrt{x^2 + y^2}) = \arcsin(y/\sqrt{x^2 + y^2}).
\end{aligned}
\tag{5.4.2}
$$

For a given r, the only dynamical variable is the polar angle, φ, in the range $0 \leq \varphi < 2\pi$. The potential energy on the uniform ring assumes a constant value, which is taken to be zero. Equation (5.4.1) therefore leads to a one-dimensional Schrödinger equation for the particle on the ring,

$$
\frac{-\hbar^2}{2\mu r^2} \frac{\partial^2 \Phi(\varphi)}{\partial \varphi^2} = E\Phi(\varphi) \quad ; \quad 0 \leq \varphi < 2\pi.
\tag{5.4.3}
$$

Exercise 5.4.1 *(a) Use Eq. (5.4.2) and the chain rule to derive Eq. (5.4.3) from Eq. (5.4.1). (b) Show that the z-component of the angular momentum reads $\hat{L}_z = -i\hbar\frac{\partial}{\partial \varphi}$. (c) Show that the kinetic energy operator for the particle on the ring reads $\hat{H} = \frac{\hat{L}_z^2}{2\mu r^2}$.*

The Schrödinger equation (Eq. (5.4.3)) is perfectly analogous to Eq. (5.3.3), and so are its general solutions. Defining,

$$
\alpha \equiv \sqrt{\frac{2\mu r^2 E}{\hbar^2}},
\tag{5.4.4}
$$

we have (for any choice of constants, a and b)

$$
\Phi(\varphi) = ae^{i\alpha\varphi} + be^{-i\alpha\varphi}.
\tag{5.4.5}
$$

However, the condition for probability continuity reads differently for the angular variable φ. Instead of vanishing at the box boundaries (as in Eq. (5.3.6)), the probability density (and thus the wave function $\Phi(\varphi)$), must satisfy periodic boundary conditions, as φ and $\varphi + 2\pi$ correspond to the same geometry,

$$
\Phi(0) = \Phi(2\pi).
\tag{5.4.6}
$$

This condition imposes a restriction on the space of solutions (Eq. (5.4.5)), since α must obtain only integer values to satisfy Eq. (5.4.6). These integer values are commonly denoted by a quantum number, $m = 0, \pm 1, \pm 2, \ldots$. Without loss of generality, we can set b in Eq. (5.4.5) to zero, and set $a = \frac{1}{\sqrt{2\pi}}$ such that the wave functions are normalized on the ring. The discrete set of solutions is therefore

$$
\Phi_m(\varphi) = \frac{1}{\sqrt{2\pi}} e^{im\varphi} \quad ; \quad m = 0, \pm 1, \pm 2, \pm 3, \ldots.
\tag{5.4.7}
$$

Using Eq. (5.4.4), we obtain the discrete energy spectrum for the particle on a ring,

$$
E_m \equiv \frac{\hbar^2 m^2}{2\mu r^2} \quad ; \quad m = 0, \pm 1, \pm 2, \pm 3, \ldots.
\tag{5.4.8}
$$

This result is our second demonstration of the "quantum size effect." Indeed, the energy spacing between any two different energy levels of the particle on the ring increases as it becomes more confined in space. In the present case, the confinement is associated with the finite radius of the ring, where the level spacing increases as the radius of the ring becomes smaller, $E_m - E_{m'} \propto \frac{1}{r^2}$, by analogy to the case of a box.

In spite of the analogy between the box and the ring, there is an important difference between the two cases. Notice that the energy associated with the two quantum numbers, m and $-m$, is the same. Nevertheless, the two wave functions, $\Phi_m(\varphi)$ and $\Phi_{-m}(\varphi)$, are generally different (except for the case, $m = 0$). They are linearly independent (orthogonal, in fact (see Ex. 5.4.2)) and correspond to degenerate states of the particle on the ring. Although corresponding to different physical states of the particle, the two states cannot be distinguished by an energy measurement. (See Section 4.6 for the discussion of degenerate states.)

To understand the origin of the degeneracy in the particle's energy spectrum, we notice that the stationary solutions $\Phi_m(\varphi)$ and $\Phi_{-m}(\varphi)$, are eigenfunctions not only of the Hamiltonian, but also of the angular momentum operator, associated with the in-plane motion, $\hat{L}_z = -i\hbar \frac{\partial}{\partial \varphi}$ (see Ex. 5.4.1):

$$\hat{L}_z \Phi_{\pm m}(\varphi) = \pm m\hbar \Phi_{\pm m}(\varphi). \tag{5.4.9}$$

However, the two eigenfunctions are nondegenerate with respect to $\hat{L}_z$, as they are associated with two different eigenvalues of this operator, namely $\pm m\hbar$. Recalling the quantum mechanical postulate that associates a measurement with specific eigenvalues of an operator (Section 4.4), the physical meaning of the last result is as follows: measuring the angular momentum on a state $\Phi_{\pm m}(\varphi)$ would yield different results, depending on whether the state is associated with the "plus" or the "minus" signs. While the absolute value of the angular momentum is the same in the two cases (which explains why the two states are associated with the same kinetic energy), the sign of the angular momentum, namely, the direction of the rotation, is opposite in the two states. Indeed, the energy of rotation should be the same if the particle rotates clockwise or anticlockwise, but the physical states are distinguishable by a measurement of the angular momentum vector.

Notice that $\Phi_m(\varphi)$ and $\Phi_{-m}(\varphi)$ are orthogonal to each other, as expected for nondegenerate eigenfunctions of the Hermitian operator, $\hat{L}_z$ (see Section 4.6 and Ex. 5.4.2). Also notice that the Hamiltonian of the particle on the ring and the angular momentum operator $\hat{L}_z$ share a common set of eigenfunctions. *This implies that the quantities that are represented by these operators can be defined simultaneously.* This property does not hold for any two operators, since to have a common set of eigenfunctions, the operators need to commute. We shall return to this point in Chapter 11 in relation to Heisenberg's uncertainty principle.

Exercise 5.4.2 *Prove that the solutions to the time-independent Schrödinger equation for the particle on a ring model, defined in Eq. (5.4.7), are orthogonal to each other. Recall that the relevant coordinate space is $0 \le \varphi < 2\pi$.*

5.5 Particles in Three-Dimensional Boxes: Quantum Wells, Wires, and Dots

The solutions of the one-dimensional particle in a box model enable us to gain useful insight into the effect of spatial confinement and the associated quantum size effect also in higher dimensions. Particularly, it can provide estimates for the density of quantum states for a particle at an energy E in different structures that confine the particle's motion in different dimensions.

Let us consider a particle in a three-dimensional square box, as defined by the following potential energy function:

$$V(x,y,z) = \begin{cases} 0 & ; \quad 0 \le x \le L_x, 0 \le y \le L_y, 0 \le z \le L_z \\ \infty & ; \quad \text{elsewhere} \end{cases}. \qquad (5.5.1)$$

As in the one-dimensional case, the probability for locating the particle anywhere outside the box must vanish, and it is sufficient to solve the time-dependent Schrödinger equation inside the box, subject to the condition $\psi(x,y,z) = 0$, anywhere on the box boundary. Inside the box, the potential energy is zero, such that the Hamiltonian includes only the kinetic energy,

$$\hat{H} = \frac{-\hbar^2}{2m} \left(\frac{\partial^2}{\partial x^2} + \frac{\partial^2}{\partial y^2} + \frac{\partial^2}{\partial z^2} \right). \qquad (5.5.2)$$

We notice that the box Hamiltonian is separable in the three Cartesian coordinates:

$$\hat{H} \equiv \hat{H}_x + \hat{H}_y + \hat{H}_z. \qquad (5.5.3)$$

Therefore (using the conclusion of Ex. 4.3.4), the solutions to the corresponding three-dimensional, time-independent Schrödinger equation can be expressed in terms of multiplications of solutions to the respective one-dimensional boxes, namely,

$$\psi_{n_x,n_y,n_z}(x,y,z) = \sqrt{\frac{2}{L_x}} \sqrt{\frac{2}{L_y}} \sqrt{\frac{2}{L_z}} \sin\left(\frac{n_x \pi x}{L_x}\right) \sin\left(\frac{n_y \pi y}{L_y}\right) \sin\left(\frac{n_z \pi z}{L_z}\right). \qquad (5.5.4)$$

The corresponding energy levels are summations,

$$E_{n_x,n_y,n_z} = \frac{\hbar^2 \pi^2}{2m} \left[\left(\frac{n_x}{L_x}\right)^2 + \left(\frac{n_y}{L_y}\right)^2 + \left(\frac{n_z}{L_z}\right)^2 \right], \qquad (5.5.5)$$

where each solution is associated with a set of three independent quantum numbers,

$$n_x = 1,2,3,\dots \quad ; \quad n_y = 1,2,3,\dots \quad ; \quad n_z = 1,2,3,\dots. \qquad (5.5.6)$$

It is interesting to examine the energy spectrum of the box in several limits:

Case I : $L_z = L_x = L_y = L$ under weak confinement (Characteristic of bulk material):

For a large L of a macroscopic size, the three-dimensional box can be regarded as a naïve model for the available electronic states in an idealized bulk material. The model enables us to estimate the number of quantum states per single particle of mass m, available up to an energy, E, namely, $N(E)$. Using Eq. (5.5.5), this translates to counting the number of different sets of quantum numbers (n_x, n_y, n_z), such that

$$\frac{\hbar^2 \pi^2}{2mL^2}(n_x^2 + n_y^2 + n_z^2) \leq E. \tag{5.5.6}$$

In a system of macroscopic size it is justified to consider the classical limit and to replace the quantum numbers by continuous variables x, y, z such that the problem of counting the states is replaced by calculating a three-dimensional volume integral, defined by $x^2 + y^2 + z^2 \leq \frac{2mL^2E}{\hbar^2\pi^2}$, for positive x, y, and z. This translates to a volume of a sphere of radius $R = \sqrt{\frac{2mL^2E}{\hbar^2\pi^2}}$, divided by 8, which yields the following estimate for the number of states:

$$N(E) = \frac{2^{1/2}}{3}\frac{m^{3/2}L^3}{\hbar^3\pi^2}E^{3/2}. \tag{5.5.7}$$

The number of states per infinitesimal energy interval is the density of states (which plays an important role in quantum dynamics and kinetics, to be discussed in Chapters 17–20):

$$\rho(E) \equiv \frac{dN(E)}{dE} = \frac{m^{3/2}L^3}{2^{1/2}\hbar^3\pi^2}E^{1/2}. \tag{5.5.8}$$

Important qualitative results are that the density of single-particle states in a macroscopic three-dimensional box increases for a heavier particle ($\rho(E) \propto m^{3/2}$) and for a larger spatial volume ($\rho(E) \propto L^3$), in accordance with the quantum size effect. Additionally, the density of states in three dimensions increases monotonically with the energy and scales as $\rho(E) \propto E^{1/2}$.

Case II : $L_z << L_x = L_y = L$ (a quantum well):

In this case the particle is much more strongly confined in one of the three dimensions. Let us consider first the limit in which the energy levels in the box along the z direction are so far apart that only levels associated with $n_z = 1$ are relevant for describing the lowest energy states of the system (up to some relevant energy). Therefore, within this interval, the energy depends only on n_x and n_y (see Eq. (5.5.5)):

$$E_{n_x, n_y} = \frac{\hbar^2\pi^2}{2mL^2}[n_x^2 + n_y^2] + E_1. \tag{5.5.9}$$

This is a typical situation for "quantum wells," which are effective two-dimensional systems, owing to strong confinement in the third dimension. Examples are semiconductor heterostructures made of AlGaAs-GaAs-AlGaAs, as presented in Fig. 5.5.1. Electrons are trapped in a narrow potential energy well between the layers along the z direction, where the energy levels are quantized. In contrast, the box dimensions along the x and y directions are of macroscopic size, with an essentially continuous energy spectrum. Electrons in such structures are often referred to as a "two-dimensional electron gas" [5.1].

(a)

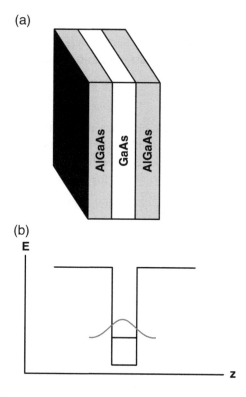

(b)

E

Figure 5.5.1   A schematic representation of a "quantum well." (a) A heterostructure of doped semiconductor layers (AlGaAs-GaAs-AlGaAs). (b) The formation of an effective potential energy well (a quantum well) between the confining layers along the direction perpendicular to the layers. A confined wave function is illustrated inside the potential energy well.

In this case, counting the number of quantum states per particle up to energy E using Eq. (5.5.9) translates (in the continuum limit) to calculating the area of a circle of radius $R = \sqrt{\frac{2mL^2(E-E_1)}{\hbar^2 \pi^2}}$, divided by 4, which yields the following estimate for the number of states:

$$N(E) = \begin{cases} 0 & ; E \leq E_1 \\ \frac{mL^2(E-E_1)}{2\hbar^2 \pi} & ; E > E_1 \end{cases}, \tag{5.5.10}$$

and the corresponding density of states (for $E > E_1$),

$$\rho(E) = \frac{mL^2}{2\hbar^2 \pi}. \tag{5.5.11}$$

Interestingly, the density of states in a quantum well (a two-dimensional box) does not depend on energy. Notice, however, that if the confinement along the z direction is weaker (L_z gets larger) such that the levels corresponding to $n_z = 2, 3, \ldots$ enter the relevant energy interval, $N(E)$ increases accordingly, and $\rho(E)$ increases by steps

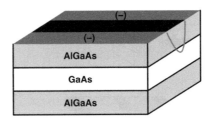

Figure 5.5.2 From a quantum well to a quantum wire: charging selected regions on top of a quantum well, made of a semiconductor heterostructure, confines the motion of electrons into an effective one-dimensional wire. The charged plates on top of the structure are marked by (-).

whose positions reflect the quantized energy levels along the z direction of the quantum well.

Case III : $L_z, L_y << L_x = L$ (a quantum wire):

In this case the particle is strongly confined along two of the three dimensions. In the limiting case, the quantum numbers that refer to the z and y directions are both restricted to their lowest values, $n_z = n_y = 1$, resulting in the following formula for the energy (see Eq. (5.5.5)):

$$E_{n_x} = \frac{\hbar^2 \pi^2 n_x^2}{2mL^2} + E_{1,1}. \tag{5.5.12}$$

This situation defines a "quantum wire" [5.2], where the particle is effectively restricted to a one-dimensional system. Quantum wires are naturally obtained in nanoscale systems including atomic chains and molecules [5.3] [5.4], or they can be fabricated by artificially confining an electron gas in a two-dimensional heterostructure into an effective one-dimensional structure (a "point contact"), by structural design [5.5], or by using appropriate static fields (see Fig. 5.5.2).

Also here, we can count the number of quantum states per particle up to an energy, E, by considering Eq. (5.5.12) in the continuum limit, which yields

$$N(E) = \begin{cases} 0 & ; \quad E \leq E_{1,1} \\ \frac{L\sqrt{2m(E-E_{11})}}{\hbar \pi} & ; \quad E > E_{1,1} \end{cases}, \tag{5.5.13}$$

where the corresponding density of states for $E > E_{1,1}$ reads

$$\rho(E) = \frac{L\sqrt{m}}{\sqrt{2}\pi\hbar} (E - E_{11})^{-1/2}. \tag{5.5.14}$$

Notice that the density of states in a one-dimensional box decreases as a function of energy. However, in typical realizations of quantum wires each combination of the lateral quantum numbers (e.g., n_y, n_z) would contribute to the total number of states as long as $E > E_{n_y, n_z}$. The result is a sharp increase in the density of state as a function of energy, whenever $E \gtrsim E_{n_y, n_z}$, followed by the decrease characteristic of the one-dimensional box (Eq. 5.5.14).

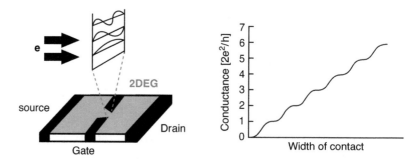

Conductance through a point contact. The two-dimensional electron gas is manipulated into a narrow quantum wire with a "point contact" at its bottleneck. The electric conductance through the point contact is an integer number of conductance quanta $(G_0 = 2e^2/h)$, which correspond to the quantum number associated with the quantization in the lateral direction, perpendicular to the current flow. As the width of the contact increases, more such lateral states become accessible, and the conductance increases in steps.

The jumps in the density of states in a quantum wire can be directly measured in electronic transport through quantum point contacts [5.4] (see Fig. 5.5.3). For a rigorous derivation based on quantum scattering theory, we refer to Ref. [5.6], whereas here we give only a brief qualitative explanation: the electric conductance is proportional to the charge flux through the contact, which is proportional to the density of states and to the particle velocities. Since the density of states scales as $(E - E_{n_x,n_y})^{-1/2}$ (by Eq. (5.5.14)), while the particle velocity scales as, $(E - E_{n_x,n_y})^{1/2}$, the conductance is independent of the energy as long as the number of energetically accessible lateral states (e.g., the number of n_y and n_z values, for which $E > E_{n_y,n_z}$) does not change. Each set of lateral quantum numbers therefore contributes a universal factor termed the conductance quantum, $G_0 = \frac{2e^2}{h}$, to the overall conductance. Increasing the lateral box dimensions (L_x and/or L_y) increases the number of energetically accessible lateral states, which leads to a stepwise increase of the conductance, characteristic of the lateral size of the quantum point contact.

Case IV : $L_z = L_x = L_y = L$ under strong confinement (a quantum dot):

Finally, when the particle is strongly confined in all three dimensions, the system is often termed "a quantum dot." In Section 5.2, we have already encountered the unique properties of nanocrystals, which are a type of quantum dots and reveal the quantum size effect. Indeed, the discrete spectrum of a quantum dot is apparent, and its density of state is discontinuous. For the three-dimensional box model, each combination of the three quantum numbers (n_x, n_y, n_z) contributes an isolated peak to the density of states, $\rho(E) = \sum_{n_x,n_y,n_z} \delta(E - E_{n_x,n_y,n_z})$, where E_{n_x,n_y,n_z} is given by Eq. (5.5.5), and δ is Dirac's delta function (discussed in detail in Chapter 11). Notice that different combinations of quantum numbers may be associated with the same energy. Such accidental degeneracies are demonstrated in Fig. 5.5.4 for a particle in a cubical box, where $L_x = L_y = L_z$.

Energy

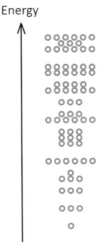

Figure 5.5.4 Energy levels and degeneracies in a symmetric three-dimensional box model. Each circle corresponds to a different set of quantum numbers, (n_x, n_y, n_z), where the lowest energy corresponds to $(1,1,1)$, the next level corresponds to the three degenerate combinations, $(2,1,1)$, $(1,2,1)$, $(1,1,2)$, and so on.

Bibliography

[5.1] R. L. Anderson, "Germanium-gallium arsenide heterojunctions," IBM Journal of Research and Development 4, 283 (1960).

[5.2] S. Datta, "Electronic Transport in Mesoscopic Systems" (Cambridge University Press, 1997).

[5.3] C. Dekker, "Carbon nanotubes as molecular quantum wires," Physics Today 52, 22 (1999).

[5.4] B. J. Van Wees, H. Van Houten, C. W. J. Beenakker et al., "Quantized conductance of point contacts in a two-dimensional electron gas," Physical Review Letters 60, 848 (1988).

[5.5] R. Schuster, E. Buks, M. Heiblum et al., "Phase measurement in a quantum dot via a double-slit interference experiment," Nature 385, 417 (1997).

[5.6] U. Peskin, "An introduction to the formulation of steady-state transport through molecular junctions," Journal of Physics B: Atomic, Molecular and Optical Physics 43, 153001 (2010).

6 Wave Function Penetration, Tunneling, and Quantum Wells

6.1 The Scanning Tunneling Microscopy

In Chapter 1 we learned how the wave nature of the electrons was revealed for the first time through their interaction with ordered matter on the subnanometer scale. We start this chapter by discussing how the wave nature of electrons can be utilized for unravelling the structure of matter on this length scale. Indeed, the Scanning Tunneling Microscopy (STM), based on the wave phenomenon of electron tunneling, has become one of the prominent experimental techniques for characterizing the detailed structure of nanoscale objects on surfaces.

The STM apparatus (see Fig. 6.1.1) contains a sharp metal tip positioned above a surface. The lateral position of the tip and its distance from the surface can be sensitively controlled piezoelectrically with an atomic-scale resolution. The tip and the surface are connected to a voltage source, which applies a controlled potential bias between them. During operation, the tip and the surface are not in contact, and the bias potential is insufficient for detaching an electron from the tip and/or the surface. According to classical mechanics, in this situation electric current cannot flow between the surface and the tip, since the electron's potential energy at the spatial gap between them is larger than the electron's total energy (the gap region is said to be "classically forbidden"). Nevertheless, a residual electric current does flow, which is attributed to the quantum tunneling phenomenon. This current is exponentially sensitive to the distance between the tip and the surface, which makes it a sensitive probe for the surface structure. Structural changes on the surface, or the presence of adsorbed atoms/molecules, affect the spatial gap and are reflected in changes in the tunneling current. Changing the lateral position of the tip, while adjusting the height of the tip such that the tunneling current remains constant (a constant current topography mode), or keeping a fixed height and monitoring the changes in the current (constant height topography mode), can be translated into detailed images of the surface plane with atomic-scale resolution (see Fig. 6.1.2). The apparatus can also be used for Scanning Tunneling Spectroscopy (STS). Changing the bias voltage at a fixed lateral tip position and measuring the current at steady state, the local electrical conductivity can be measured, and from it the local electronic density of states of the surface, or of adsorbates on the surface, can be inferred (see Chapter 20). Changing the bias voltage and

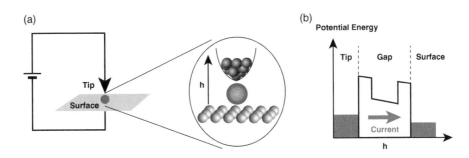

(a) A schematic illustration of an STM apparatus. (b) A schematic representation of the effective potential energy for electrons at the vicinity of the tunneling gap. Electrons are marked in gray.

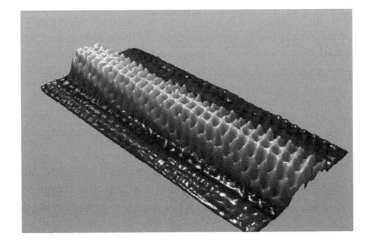

STM image of a carbon nanotube (image by Cees Dekker group at TU Delft, The Netherlands).

monitoring the transient response of the current can be used for following dynamical processes on the surfaces [6.1], [6.2].

In Chapter 5 we discussed models of potential wells, where the external potential was infinite, and the particle was perfectly confined to a finite region in space. However, realistic models should account for finite-valued potential energy wells and/or barriers. As we shall see in what follows, in these cases, wave functions can penetrate "classically forbidden" regions, which facilitate transfer of particles "through" (in contrast to "above") potential energy barriers (see Fig. 6.1.1). This is the origin of the tunneling phenomenon, and the basis for the STM operation. We emphasize that tunneling is facilitated only for finite potential energy barriers, whereas for infinite barriers, the wave function penetration vanishes and the phenomenon is suppressed. In the next sections of this chapter, we shall become familiar with the solutions of the Schrödinger equation for simple models of finite potential energy wells and barriers, which will provide us with some knowledge about the wave function penetration, and the quantum tunneling phenomena.

6.2 A Particle in a Finite Square Well Potential and Wave Function Penetration

The "particle in a box" model discussed in Section. 5.3 provides important insights into the energy levels and stationary solutions of bound particles. Nevertheless, the model is idealized in many important ways. Primarily, it assumes that the confining external potential is infinite, resulting in perfect confinement of the particle within a finite region in space. A more realistic description must account for a finite potential energy, where quantum mechanics allows particles to 'leak out' of their potential energy wells into regions that are 'forbidden' according to the laws of classical mechanics. In this section we will become familiar with the phenomena of wave function penetration into classically forbidden regions, by solving the time-independent Schrödinger equation for a particle trapped in a one-dimensional finite square well potential. The model potential reads

$$V(x) = \begin{cases} V_0 & 0 > x \\ 0 & 0 \leq x < L. \\ V_0 & L \leq x \end{cases}$$

In classical mechanics, the particle would be free to move anywhere in space, provided that its total energy is above the threshold energy of the potential well, namely, $E \geq V_0$. In this energy range the quantum system is in fact open, and the physically meaningful solutions of the Schrödinger equation need not be proper. We shall postpone the discussion of free particles and improper states to Chapters 7 and 11, and focus here on the bound states of the particle, namely, on the energy range, $E < V_0$. Classical mechanics implies that in this energy range the particle's position would be strictly limited to the "square box," $0 \leq x \leq L$ (marked as region '(2)' in Fig. 6.2.1). In quantum mechanics the bound particle defines a closed system, where the physically meaningful solutions of the Schrödinger equation must be proper. However, given that V_0 is finite, there is no reason to assume a priori that the probability density strictly vanishes outside the square box. ***We should therefore look for proper solutions to the time-independent Schrödinger equation in the entire coordinate space.*** Since the potential is piecewise constant, it is convenient to first solve the Schrödinger equation in each segment separately, and then assure that the solutions are proper also at the boundary points between the segments.

Numbering the three segments, $n = 1, 2, 3$, as in Fig. 6.2.1, the potential energy obtains the respective constant values, $\{V_n\}$, and the respective Schrödinger equation (Eq. (4.3.4)) in each segment reads

$$\frac{-\hbar^2}{2m} \frac{\partial^2}{\partial x^2} \psi_n(x) + V_n \psi_n(x) = E \psi_n(x). \tag{6.2.1}$$

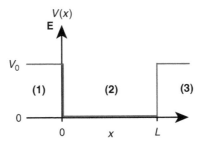

Figure 6.2.1 The potential energy for a particle in a finite one-dimensional square potential well. The regions of constant potential energy are numbered as (1), (2), and (3).

Defining,

$$k_n \equiv \sqrt{\frac{2m(E - V_n)}{\hbar^2}}, \tag{6.2.2}$$

a general solution for the nth segment reads

$$\psi_n(x) = a_n e^{ik_n x} + b_n e^{-ik_n x}. \tag{6.2.3}$$

On physical grounds, we must restrict ourselves to proper solutions. First, the wave function must decay to zero at the asymptotes, $x \to \pm\infty$ (see Eq. (2.2.1)). Second, as a solution of a second-order differential equation, the proper wave function must be continuously differentiable anywhere in space. The latter is readily satisfied within each segment by Eq. (6.2.3), but must be imposed at the boundary points between nearby segments; namely, both the function and its derivative must be continuous at the boundary points ($x = 0$ and $x = L$).

Considering first the asymptote at segment (3), the following condition must be fulfilled, $\psi_3(x) = a_3 e^{ik_3 x} + b_3 e^{-ik_3 x} \xrightarrow[x \to +\infty]{} 0$. For $E < V_0$, k_3 obtains an imaginary value, $k_3 = \sqrt{\frac{2m(E - V_0)}{\hbar^2}} \equiv \frac{i}{\gamma}$ (where $i = \sqrt{-1}$). It follows that the first exponent indeed converges to zero as $x \to \infty$, but the second exponent diverges. The only way to assure a proper function $\psi_3(x)$ is therefore to set the coefficient b_3 to zero. Consequently,

$$\psi_3(x) = a_3 e^{ik_3 x} = a_3 e^{-x/\gamma}. \tag{6.2.4}$$

The result has several important characteristics. First of all, *there is a finite probability density to find the particle in a "classically forbidden" region, namely in a region of space, where the total energy, E, is smaller than the potential energy, V_0.* This phenomenon is referred to as the wave function penetration into a classically forbidden region. Second, within this region, the probability is shown to decay exponentially, as the distance from the classically allowed region (the potential well) increases, with a "penetration length," $\gamma = \frac{\hbar}{\sqrt{2m(V_0 - E)}}$. *The penetration length is shown to be larger for lighter particles (smaller mass, m), and at higher particle energies, E*, namely, as the energy approaches the threshold, V_0. Notice that this result is in line with the classical result, which states that above the threshold, the particle should not be confined to the well region at all.

Similarly, the asymptotic condition at segment (1) reads, $\psi_1(x) = a_1 e^{ik_1 x} + b_1 e^{-ik_1 x} \xrightarrow[x \to -\infty]{} 0$. Since k_1 is also imaginary, $k_1 = \sqrt{\frac{2m(E-V_0)}{\hbar^2}} = k_3$, the only way to assure a proper $\psi_1(x)$ is to set the coefficient a_1 to zero. Consequently,

$$\psi_1(x) = b_1 e^{-ik_1 x} = b_1 e^{x/\gamma}. \tag{6.2.5}$$

Also here, the wave function is shown to penetrate exponentially into the classically forbidden region of space (in the direction of negative x).

In the intermediate segment, (2), the wave function is not restricted by the asymptotic conditions and generally reads (Eq. (6.2.3))

$$\psi_2(x) = a_2 e^{ik_2 x} + b_2 e^{-ik_2 x}. \tag{6.2.6}$$

However, the requirement on the solutions of the Schrödinger equation to be continuously differentiable anywhere in space imposes additional constraints on the proper solutions. Particularly, the wave function and its first derivative must be continuous also at the boundary points between nearby segments ($x = 0$ and $x = L$):

$$\psi_1(0) = \psi_2(0)$$

$$\frac{d}{dx}\psi_1(x)\Big|_{x=0} = \frac{d}{dx}\psi_2(x)\Big|_{x=0}$$

$$\psi_2(L) = \psi_3(L)$$

$$\frac{d}{dx}\psi_2(x)\Big|_{x=L} = \frac{d}{dx}\psi_3(x)\Big|_{x=L}. \tag{6.2.7}$$

Any proper solution is therefore defined by a set of variables, (b_1, a_2, b_2, a_3), which must satisfy Eq. (6.2.7). The four continuity conditions translate into a homogeneous system of linear equations for these variables:

$$\begin{bmatrix} 1 & 1 & -1 & 0 \\ ik_2 & -ik_2 & ik_1 & 0 \\ e^{ik_2 L} & e^{-ik_2 L} & 0 & -e^{+ik_3 L} \\ ik_2 e^{ik_2 L} & -ik_2 e^{-ik_2 L} & 0 & -ik_3 e^{ik_3 L} \end{bmatrix} \begin{bmatrix} a_2 \\ b_2 \\ b_1 \\ a_3 \end{bmatrix} = \begin{bmatrix} 0 \\ 0 \\ 0 \\ 0 \end{bmatrix}. \tag{6.2.8}$$

Notice that the trivial solution to this matrix equation, $b_1 = a_2 = b_2 = a_3 = 0$, corresponds to an improper wave function, $\psi(x) = 0$. Nontrivial solutions exist, for which the determinant of the coefficient matrix vanishes. Since the determinant depends on the energy via $k_1 = k_3 = \sqrt{\frac{2m(E-V_0)}{\hbar^2}}$ and $k_2 = \sqrt{\frac{2mE}{\hbar^2}}$, it vanishes only for specific values of the particle's energy E (for a discrete set). Indeed, *the energy spectrum for a bound particle in a finite square potential well is quantized*. This is demonstrated in Fig. 6.2.2 for a model of an electron bound in a 1-nm-long square potential well. The plot of the determinant as a function of energy reveals a set of local minima that point to

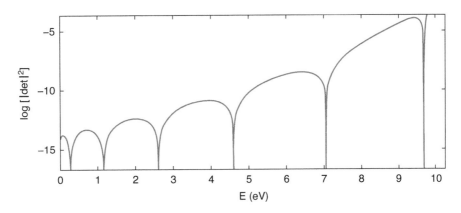

Figure 6.2.2   The emergence of energy quantization in a finite square well potential. The minima point to a discrete set of zeros, which correspond to the energy level of the bound particle. The finite square potential well parameters are $m = 1$ a.u., $L = 1$ nm, and $V_0 = 10$ eV.

the energy levels of the particle in the finite potential well, for which the determinant vanishes.

For each energy level, the corresponding stationary solution (the vector (b_1, a_2, b_2, a_3)) can be determined. Notice that the solutions to the system of equations, Eq. (6.2.8), are defined only up to a multiplication by a constant, which can be set by the normalization requirement (see Section 2.3). Probability densities associated with the stationary solutions of the bound particle in the finite potential well are plotted in Fig. 6.2.3 (solid lines). It is instructive to compare the energy spectrum and the stationary solutions to those of a particle in the corresponding infinite box of the same length (the dashed lines in Fig. 6.2.3). The results have apparent similarities, where the energy levels are quantized in both cases, and the stationary solutions corresponding to the same quantum number (the energy-level index) are qualitatively similar in terms of the shape of the probability density distribution and the number of nodes. There are, however, several important differences. In the case of a finite potential well: first, the probability density is shown to penetrate the classically forbidden region, as we have discussed (Eqs. (6.2.4, 6.2.5)), which becomes more pronounced as the energy increases; and second, the density of states is larger, namely, the gap between nearby energy levels is smaller. Notice that this is yet another manifestation of the quantum size effect: the effectively lower spatial confinement of the particle in the finite potential well (due to the wave function penetration) is correlated with a denser energy spectrum.

Exercise 6.2.1   *Derive Eq. (6.2.8).*

Exercise 6.2.2   *Show that, when $V_0 \to \infty$, the energy levels and the stationary solutions of the Schrödinger equation for the symmetric finite square well potential converge to the results for an infinite box, obtained in Chapter 5.*

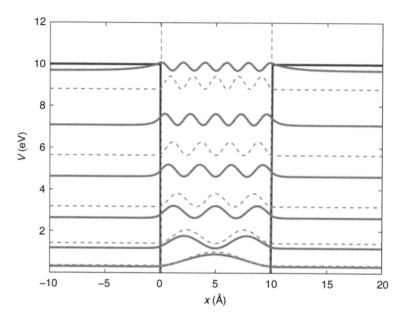

Figure 6.2.3 Energy levels and probability densities for a particle in a finite square well potential. The different plots correspond to different energy levels. The plots are displaced from each other by the respective energy differences. Solid lines correspond to the finite square well potential (see Fig. 6.2.2). Dashed lines correspond to an infinite external potential $V_0 \to \infty$, discussed in Chapter 5.

6.3 The Schrödinger Equation for a Piecewise Constant Potential Energy

The method implemented in the previous section is readily generalizable for solving the time-independent Schrödinger equation in cases where the potential energy is piecewise constant, as defined in Eq. (6.3.1):

$$V(x) = \begin{cases} x < x_1 & ; & V_1 \\ x_1 \le x < x_2 & ; & V_2 \\ & \vdots & \\ x_{N-2} \le x < x_{N-1} & ; & V_{N-1} \\ x_{N-1} \le x & ; & V_N \end{cases} \tag{6.3.1}$$

The coordinate space is divided into N segments, in which the potential energy obtains a constant value, V_n (see Fig. 6.3.1). The Schrödinger equation within each

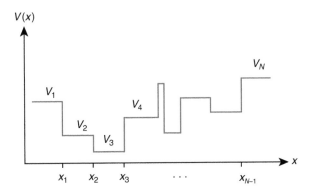

Figure 6.3.1   A schematic representation of a piecewise constant potential energy.

segment, $x_{n-1} \leq x < x_n$, obtains the form of Eq. (6.2.3), where the general solution is given by Eqs. (6.2.2, 6.2.3):

$$\psi(x) = \begin{cases} x < x_1 & ; & a_1 e^{ik_1 x} + b_2 e^{-ik_1 x} \\ x_1 \leq x < x_2 & ; & a_2 e^{ik_2 x} + b_2 e^{-ik_2 x} \\ \vdots & & \vdots \\ x_{N-2} \leq x < x_{N-1} & ; & a_{N-1} e^{ik_{N-1} x} + b_{N-1} e^{-ik_{N-1} x} \\ x_{N-1} \leq x & ; & a_N e^{ik_N x} + b_N e^{-ik_N x} \end{cases} \tag{6.3.2}$$

A specific solution is defined by the set of $2N$ coefficients, $(a_1, a_2, \ldots, a_N, b_1, b_2, \ldots, b_N)$. Again, we shall restrict the discussion to the energy range in which a classical particle would be trapped in a finite range in space, namely, a bound particle (free particles will be considered in the next chapter). This means

$$E < \min(V_1, V_N). \tag{6.3.3}$$

In this energy range, the quantum system is closed and the physically meaningful solutions to the Schrödinger equation must be proper. It follows that the wave functions must vanish asymptotically,

$$\psi_1(x) \xrightarrow[x \to -\infty]{} 0 \quad ; \quad \psi_N(x) \xrightarrow[x \to \infty]{} 0, \tag{6.3.4}$$

as well as be continuously differentiable at each of the $N - 1$ boundary points between the segments:

$$\psi_n(x_n) = \psi_{n+1}(x_n)$$
$$\left. \frac{d}{dx} \psi_n(x) \right|_{x=x_n} = \left. \frac{d}{dx} \psi_{n+1}(x) \right|_{x=x_n}. \tag{6.3.5}$$

The requirement for proper solutions therefore imposes $2N$ equations for the $2N$ wave function coefficients. Eq. (6.3.4) sets two of the coefficients in the asymptotic segments to zero (see the discussion of Eqs. (6.2.4) and (6.2.5) in the previous section for the asymptotic solutions),

$$a_1 = 0 \quad ; \quad b_N = 0, \tag{6.3.6}$$

and Eq. (6.3.5) leads to $N - 1$ pairs of linear equations for the remaining $2N - 2$ coefficients,

$$a_n e^{ik_n x_n} + b_n e^{-ik_n x_n} - a_{n+1} e^{ik_{n+1} x_n} - b_{n+1} e^{-ik_{n+1} x_n} = 0$$

$$ik_n e^{ik_n x_n} a_n - ik_n e^{-ik_n x_n} b_n - ik_{n+1} e^{ik_{n+1} x_n} a_{n+1} + ik_{n+1} e^{-ik_{n+1} x_n} b_{n+1} = 0, \tag{6.3.7}$$

where, $n = 1, 2, \ldots, N - 1$. The result is a homogeneous matrix equation for the coefficients,

$$\begin{bmatrix} e^{-ik_1 x_1} & -e^{ik_2 x_1} & -e^{-ik_2 x_1} \\ -ik_1 e^{-ik_1 x_1} & -ik_2 e^{ik_2 x_1} & ik_2 e^{-ik_2 x_1} \\ & e^{ik_2 x_2} & e^{-ik_2 x_2} & -e^{ik_3 x_2} & -e^{-ik_3 x_2} \\ & ik_2 e^{ik_2 x_2} & -ik_2 e^{-ik_2 x_2} & -ik_3 e^{ik_3 x_2} & ik_3 e^{-ik_3 x_2} \\ & & & \ddots \\ & & & & e^{ik_{N-1} x_{N-1}} & e^{-ik_{N-1} x_{N-1}} & -e^{ik_N x_{N-1}} \\ & & & & ik_{N-1} e^{ik_{N-1} x_{N-1}} & -ik_{N-1} e^{-ik_{N-1} x_{N-1}} & -ik_N e^{ik_N x_{N-1}} \end{bmatrix} \begin{bmatrix} b_1 \\ a_2 \\ b_2 \\ a_3 \\ b_3 \\ \vdots \\ a_{N-1} \\ b_{N-1} \\ a_N \end{bmatrix} = \begin{bmatrix} 0 \\ 0 \\ 0 \\ 0 \\ 0 \\ \vdots \\ 0 \\ 0 \\ 0 \end{bmatrix}, \tag{6.3.8}$$

which obtains a nontrivial solution only when the matrix determinant vanishes. Since the matrix entries depend on the energy via $k_n = \sqrt{\frac{2m(E - V_n)}{\hbar^2}}$, the determinant vanishes only for a set of E values, which constitute the energy levels of the particle in the general potential well. Given E, the corresponding coefficient vector can be determined (up to a constant). Imposing the normalization condition, a specific solution to the time-independent Schrödinger equation (in the form of Eq. (6.3.2)) is obtained for each energy level.

6.4 The Symmetric Double Well Potential and Quantum Tunneling

In this section we consider the Schrödinger equation for a particle bound in a symmetric double well potential, whose solutions demonstrate the phenomenon of quantum tunneling in bound systems. The potential energy for this model is defined as follows:

$$V(x) = \begin{cases} x < -(D/2 + L) & ; & V_0 \\ -(D/2 + L) \leq x < -D/2 & ; & 0 \\ -D/2 \leq x < D/2 & ; & V_0 \\ D/2 \leq x < D/2 + L & ; & 0 \\ D/2 + L \leq x & ; & V_0 \end{cases} . \tag{6.4.1}$$

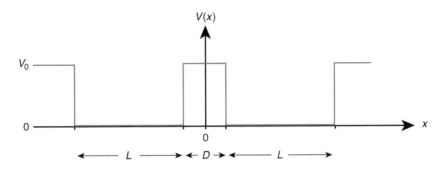

Figure 6.4.1 A symmetric double square well potential

The length of each well is set to L, and the energy "depth" is set to V_0, just as for the single square potential well discussed in Section 6.2 (see Fig. 6.2.1). In this model system, however, there are two wells, separated by a distance D (Fig. 6.4.1). Here also, we are interested in the bound states of the particle, namely, in the energy range where the total energy is below the potential energy threshold, $E < V_0$. Notice that according to classical mechanics, a bound particle in this energy range can be confined to one, and only one, of the two wells, since the region between the wells is classically forbidden. In contrast, we shall see that quantum mechanics allows for delocalization of a bound particle between the wells, which manifests the quantum tunneling phenomenon in bound systems. In precise terms, the bound state solutions to the Schrödinger equation for this system correspond to probability densities that are distributed between the two wells at the same time. As we shall see, the tunneling phenomenon is closely related to the penetration of the wave functions into the classically forbidden region.

Let us consider the stationary solutions for the double well system, as calculated according to the general methodology described in Section 6.3. In Fig. 6.4.3 the solutions corresponding to the two lowest energy levels are presented for different distances (D) between the two wells. It is instructive to analyze the results in comparison to the solutions for the respective single well potential. When the two wells are far apart (Fig. 6.4.3a), the two lowest energy levels of the double well potential become nearly degenerate and converge to the value of the lowest energy of the respective single well potential ($E_1 = 2$ eV, for the given parameters). The corresponding stationary wave functions in each well seem to coincide (up to a sign, in each well) with the lowest energy solutions of two isolated potential wells. This behavior is characteristic of inter-well distances, which are much larger than the penetration length of the single well solution into the classically forbidden region, namely, $D >> \gamma = \frac{\hbar}{\sqrt{2m(V_0-E_1)}}$. In the other extreme (Fig. 6.4.3c), the distance between the wells is much smaller than the penetration length, $D << \gamma$. In this range the two lowest energy solutions of the double well potential appear to resemble the solutions of a single well potential, whose length is the sum of the two well lengths ($\sim 2L$). In the intermediate regime (Fig. 6.4.3b), the distance between the wells is of the order of the penetration length of the single well solution, $D \sim \gamma$. The two lowest energy levels as well as the corresponding stationary

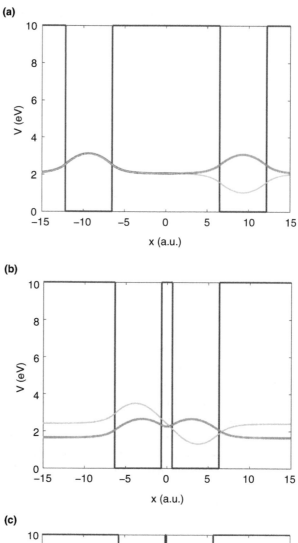

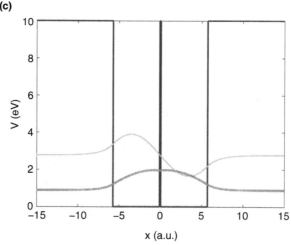

Figure 6.4.2 The two lowest energy levels and the respective wave functions for a particle in a double square well potential. The different plots correspond to different energy levels. The plots are displaced from each other by the respective energy difference. (a), (b), and (c) correspond to decreasing inter-well distance, where $D = 10\gamma, \gamma, 0.1\gamma$, respectively. The model parameters are $m = 1$ electron mass, $L = 0.3$ nm, and $V_0 = 10$ eV. The lowest energy level of the isolated single well is $E_1 = 2$ eV, and the penetration length of the corresponding stationary state is $\gamma = 0.069$ nm.

solutions in this regime cannot be approximated by any single well solution, and reflect the unique characteristics of the double well potential.

Let us reemphasize that the solutions depicted in Fig. 6.4.3 were all obtained for the double well potential, and their analysis in terms of solutions to a single well potential in the limits, $D << \gamma$, and $D >> \gamma$ is merely a convenient interpretation. The first apparent characteristic common to all cases is that each stationary solution in the double well system is delocalized over the two wells. This means that there is a probability for the particle to be located at the two wells in a stationary state (namely, "simultaneously"), in contrast to the results of classical mechanics. Moreover, the probabilities for locating the particle in the left and right wells are equal. This is a result of the right–left symmetry of the model Hamiltonian. Specifically, the Hamiltonian for the symmetric double well system commutes with the parity operator (see Ex. 4.6.2). Consequently, the Hamiltonian eigenfunctions are also eigenfunctions of the parity operator (see Ex. 6.4.1); namely, they are either even or odd functions of the position, x (see Ex. 4.6.2). It follows that the absolute squares of the stationary solutions that correspond to a bound electron are even function of x.

Exercise 6.4.1   *Show that if the two linear operators, denoted $\hat{H}$ and $\hat{P}$, commute ($[\hat{H},\hat{P}] = 0$), and if ψ is an eigenfunction of $\hat{H}$ that corresponds to a nondegenerate eigenvalue, then ψ is also an eigenfunction of $\hat{P}$.*

We now turn to discussing the dynamics of a bound particle in the double well potential. Recalling that any linear combination of stationary solutions to the time-dependent Schrödinger equation is also a solution (see Ex. 4.6.5), let us consider a specific (normalized) linear combination of the stationary solutions corresponding to the two lowest energy levels, E_1 and E_2:

$$\psi(x,t) = \frac{1}{\sqrt{2}} \psi_{E_1}(x)e^{-iE_1 t/\hbar} + \frac{1}{\sqrt{2}} \psi_{E_2}(x)e^{-iE_2 t/\hbar}. \tag{6.4.2}$$

Since the solutions are nondegenerate (with $E_2 > E_1$), $\psi(x,t)$ is not a stationary solution, and the probability density depends on time:

$$\rho(x,t) = \frac{1}{2}|\psi_{E_1}(x) + \psi_{E_2}(x)e^{-i(E_2-E_1)t/\hbar}|^2. \tag{6.4.3}$$

The functions $\psi_{E_1}(x)$ and $\psi_{E_2}(x)$ are even and odd functions of x, respectively (see Fig. 6.4.3 and Ex. 6.4.1), which have the same sign for $x < 0$, and opposite signs for $x > 0$. For $t_n = \frac{2n\pi\hbar}{E_2-E_1}$ ($n = 0,\pm1,\pm2,\ldots$), the probability density reads $\rho(x,t_n) \propto |\psi_{E_1}(x) + \psi_{E_2}(x)|^2$. At these times, $\psi_{E_1}(x)$ and $\psi_{E_2}(x)$ interfere constructively for $x < 0$, and destructively for $x > 0$, such that the probability density favors the population of the left potential well. In contrast, at different times, $t_n = \frac{(2n+1)\pi\hbar}{E_2-E_1}$, the probability density reads $\rho(x,t_n) \propto |\psi_{E_1}(x) - \psi_{E_2}(x)|^2$, and the population of the right potential is favored. It follows that the probability for locating the bound particle in space oscillates periodically between the two wells, with a time period,

$$\tau = \frac{h}{E_2 - E_1}. \tag{6.4.4}$$

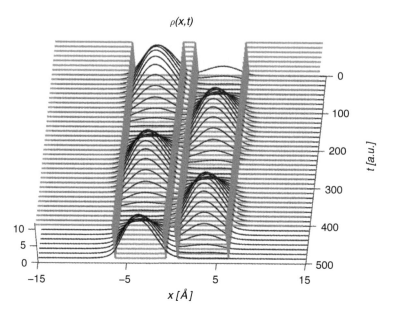

$\rho(x,t)$

Figure 6.4.3 Time-dependent probability density for a particle in a symmetric double square well potential. The model parameters are as for Fig. 6.4.2b.

This is demonstrated in Fig. 6.4.3. Notice that a bound classical particle located at one of the wells would have zero probability to be found in the other well at any time. *Quantum mechanics allows the bound particle to tunnel through the potential barrier between the wells and transfer to the other well.* Nevertheless, the probability of finding the particle at the barrier region remains relatively low at any time. This is the case when the inter-well distance is similar to or larger than the penetration length of the stationary wave functions associated with the two independent wells.

Notice that as $E_2 \rightarrow E_1$, the time for transfer between the wells approaches infinity (Eq. (6.4.4)). This is the limiting case of the situation illustrated in Fig. 6.4.3a, where the distance between the wells is much larger than the penetration length ($D \gg \gamma$). Indeed, in this limit the two lowest-energy (even and odd) eigenfunctions of the double well Hamiltonian become degenerate, such that their linear combinations (Eq. (6.4.2)) are also eigenfunctions of the Hamiltonian (see Ex. 4.6.3). In particular, the wave functions, $\psi_{E_1}(x) \pm \psi_{E_2}(x)$, which confine the probability density to only one of the wells, become stationary solutions. This limit is in line with the classical results. However, as the distance (D) between the wells becomes finite, the energy splitting ($E_2 - E_1$) increases, and the time period for quantum tunneling between the wells decreases accordingly.

We end this section by noting that while the tunneling phenomenon facilitates powerful experimental analysis tools, such as STM and STS, it is a major problem in the context of miniaturization of nanoscale electronics devices. The tendency of electrons to delocalize between nearby quantum wells hinders their confinement into specific, independent conductance channels, resulting in leakage currents, when the conductors

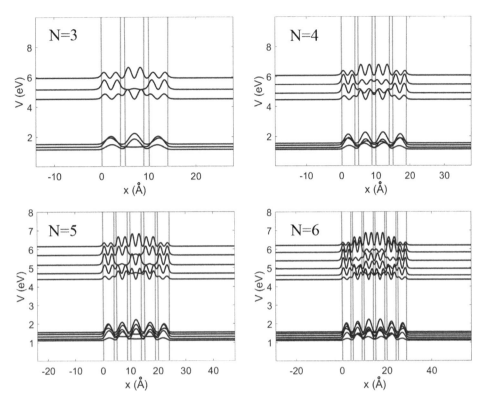

Figure 6.5.1   Energy levels and probability densities for a particle ($m = 1$ a.u.) in several multiple square well potentials, demonstrating the formation of energy bands and gaps with increasing number of wells. The different plots in each frame correspond to different energy levels. The plots are displaced from each other by the respective energy differences.

are too closely packed. This poses a physical lower bound for miniaturization of electronic devices. The quest for denser information processing devices must therefore account for nonclassical architectures, designed to operate in the quantum tunneling regime.

6.5 A Particle in Multiple Quantum Wells and Energy Bands

In this section we briefly discuss the solution of the Schrödinger equation for a particle in multiple (N) potential wells. The stationary solutions of the time-independent Schrödinger equation for such systems can be readily obtained as described in Section 6.3. In Fig. 6.5.1 we present the corresponding energy levels and probability densities for systems of different N. Each stationary solution is shown to be delocalized over all the potential wells, which implies that the bound particle is not confined to any particular potential well.

The characteristics of the solutions can be rationalized by referring again to the case of the double well potential ($N = 2$). In Fig. 6.4.3, we notice that when the two potential wells are sufficiently far apart, $D \gtrsim \gamma$, the energy levels appear in nearby pairs, separated by larger energy gaps, where each pair can be correlated to a single energy level of an independent single potential well. (This becomes most apparent in the limit, $D >> \gamma$.) A similar trend is observed also for $N > 2$. The energy spectrum is composed of relatively dense groups of N levels, separated by larger energy gaps, where each group can be correlated to one of the energy levels of the independent single potential well. As N becomes larger, each group becomes denser, such that *in the limit, $N \to \infty$, the energy spectrum consists of continuous energy bands separated by "forbidden" energy gaps*. The formation of energy bands and gaps is essential for understanding the properties of extended materials. We postpone a more rigorous discussion of the emergence of energy bands and gaps in the spectrum of different materials and its consequences to Chapter 14. Here we only comment that energy bands are characteristic of spatially periodic structures, and that as the number of potential energy wells becomes large, the system of quantum wells can be well approximated as a periodic structure [6.3]. This is most relevant for understanding the mobility of valence electrons in perfect atomic lattices, and the phenomenon of conductance, as will be discussed in Chapter 14.

Bibliography

[6.1] H. J. Lee and Wilson Ho. "Single-bond formation and characterization with a scanning tunneling microscope," Science 286, 1719 (1999).

[6.2] T. L. Cocker, D. Peller, P. Yu, J. Repp, and R. Huber, "Tracking the ultrafast motion of a single molecule by femtosecond orbital imaging," Nature 539, 263 (2016).

[6.3] R. de L. Kronig and W. G. Penney, "Quantum mechanics of electrons in crystal lattices," Proceedings of the Royal Society of London. Series A 130, 499 (1931).

7 The Continuous Spectrum and Scattering States

7.1 The Continuity Equation and the Probability Current Density

In previous chapters we focused on closed systems, in which the particles are confined to some finite region in coordinate space. It is of interest, however, to consider also open systems, in which particles (or energy) may leak out, of enter. Postponing a more formal discussion of open quantum systems to Chapter 19, here we focus on a particular kind of open system, in which there is a stationary flow of particles through the entire space. This description corresponds to scattering experiments, where a well-defined stationary current (i.e., a flux) of particles is directed toward a target. The resulting currents of particles scattered from the target are continuously monitored, and their spatial distribution in space provides useful information with respect to the target. Indeed, scattering of particles such as electrons, neutrons, and other light particles, at energies in which their de Broglie wavelength is of the order of interatomic distances, provided most valuable information with respect to the structure of matter on the nanoscale. In this chapter we give a modest introduction to the solutions of the Schrödinger equation in the scattering regime, emphasizing fundamental differences between scattering and bound states.

We already learned that particle positions in space are associated with a probability density functions. *But what is a corresponding suitable description of particle currents?* To answer this, we notice that the time-dependent Schrödinger equation can be rewritten as a universal continuity equation. Without loss of generality, let us consider a system of N particles of masses $(m_1, m_2, \ldots, m_N)$, in a three-dimensional space, where the nth particle is associated with the Cartesian coordinates, $(x_{1,n}, x_{2,n}, x_{3,n})$. The multi-dimensional wave function reads $\psi(x_{1,1}, x_{2,1}, x_{3,1}, x_{1,2}, \ldots, x_{3,N}, t) \equiv \psi(\mathbf{x}, t)$. Given that $\psi(\mathbf{x}, t)$ is a solution to the time-dependent Schrödinger equation (Eqs. (3.4.7, 3.4.8, 4.1.1)), it follows that (see Ex. 7.1.1)

$$\frac{\partial}{\partial t}|\psi(\mathbf{x}, t)|^2 = -\nabla \cdot \mathbf{J}(\mathbf{x}, t), \tag{7.1.1}$$

where the gradient vector reads $\nabla = (\frac{\partial}{\partial x_{1,1}}, \frac{\partial}{\partial x_{2,1}}, \frac{\partial}{\partial x_{3,1}}, \frac{\partial}{\partial x_{1,2}}, \ldots, \frac{\partial}{\partial x_{3,N}})$, and the corresponding components of the vector $\mathbf{J}(\mathbf{x}, t)$ read

$$J_{j,n}(\mathbf{x}, t) \equiv \frac{\hbar}{m_n} \text{Im}[\psi^*(\mathbf{x}, t) \frac{\partial}{\partial x_{j,n}} \psi(\mathbf{x}, t)]. \tag{7.1.2}$$

Recalling the association of $|\psi(\mathbf{x},t)|^2$ as a probability density, **Eq. (7.1.1) has the universal form of a continuity equation. The vector $\mathbf{J}(\mathbf{x},t)$ is therefore identified as the probability current density (or "probability flux") corresponding to the wave function** $\psi(\mathbf{x},t)$. In its integral form, Eq. (7.1.1) means that the loss of probability from any volume in the multidimensional space equals the total probability current exiting that volume through the surface surrounding it,

$$-\frac{\partial}{\partial t}\int_V |\psi(\mathbf{x},t)|^2 d\mathbf{x} = \oint_S \mathbf{J}(\mathbf{x},t)\cdot d\mathbf{S}. \tag{7.1.3}$$

Exercise 7.1.1   *A system of N particles is associated with the Hamiltonian $\hat{H} = \sum_{n=1}^{N}\frac{-\hbar^2}{2m_n}\left[\frac{\partial^2}{\partial x_{1,n}^2} + \frac{\partial^2}{\partial x_{2,n}^2} + \frac{\partial^2}{\partial x_{3,n}^2}\right] + V(x_{1,1},x_{2,1},x_{3,1},x_{1,2},\ldots,x_{3,N})$, where the nth particle is associated with the Cartesian coordinates, $(x_{1,n},x_{2,n},x_{3,n})$. The state of the system is represented by a solution to the time-dependent Schrödinger equation, $\psi(x_{1,1},x_{2,1},x_{3,1},x_{1,2},\ldots,x_{3,N},t) \equiv \psi(\mathbf{x},t)$. Prove the following identity: $\frac{\partial}{\partial t}|\psi(\mathbf{x},t)|^2 = -\sum_{n=1}^{N}\sum_{j=1}^{3}\frac{\partial}{\partial x_{j,n}}J_{j,n}(\mathbf{x},t)$, where $J_{j,n}(\mathbf{x},t) \equiv \frac{\hbar}{m_n}\text{Im}[\psi^*(\mathbf{x},t)\frac{\partial}{\partial x_{j,n}}\psi(\mathbf{x},t)]$.*

We now focus specifically on the stationary solutions to the time-dependent Schrödinger equation, which are associated with well-defined energies, E, and obtain the form, $\psi(\mathbf{x},t) = \psi_E(\mathbf{x})e^{-iEt/\hbar}$. Both the probability density, $|\psi(\mathbf{x},t)|^2 = |\psi_E(\mathbf{x})|^2$, and the corresponding probability flux, $\mathbf{J}_E(\mathbf{x})$ (see Eq. (7.1.2)), are time-independent in this case, which means that the continuity equation (Eq. (7.1.1)) takes a simpler form:

$$\nabla\mathbf{J}_E(\mathbf{x}) = 0 \quad ; \quad \oint_S \mathbf{J}_E(\mathbf{x})\cdot d\mathbf{S} = 0. \tag{7.1.4}$$

The vanishing divergence $\nabla\mathbf{J}_E$ means that the probability flux exiting from the surface of any volume element (infinitesimal, or finite) in coordinate space must vanish. Expanding the volume element to the entire coordinate space, one can distinguish between two cases: if the system is a closed one, the probability flux itself, $\mathbf{J}_E(\mathbf{x})$, vanishes at the boundary surface, and the integral in Eq. (7.1.4) is trivially satisfied. This is consistent with $\psi_E(\mathbf{x})$ being a proper wave function, as discussed in Chapter 2, which vanishes at the boundary surface (see, e.g., Eq. (2.2.1) for the one-dimensional case). If the system is open, however, where particles can flow in and out of its space, $\mathbf{J}_E(\mathbf{x})$ does not have to vanish at the boundary surface. Probability conservation, as expressed in Eq. (7.1.4), requires only that any probability fluxes entering the entire space must equal the probability fluxes exiting from it. Importantly, nonvanishing fluxes at the space boundaries are associated with nonvanishing wave functions. *It follows that stationary wave functions that satisfy Eq. (7.1.4) need not be proper, and that in this case, $|\psi_E(\mathbf{x})|^2$ cannot be interpreted as a probability density.* (Indeed, the formulation of probability conservation for stationary states (Eq. (7.1.4)) does not depend explicitly on $|\psi_E(\mathbf{x})|^2$, only on $\mathbf{J}_E(\mathbf{x})$.) Importantly, improper stationary solutions, $\psi(\mathbf{x},t) = \psi_E(\mathbf{x})e^{-iEt/\hbar}$, satisfying Eqs. (7.1.2, 7.1.4), when they exist, are still associated with a well-defined physical meaning. ***These are the "scattering wave functions" of*** *the system [7.1], corresponding to stationary flows of particles at a given energy.* Information on the scattering process is contained in the relative distribution of probability fluxes at the space boundaries, where fluxes of particles can be measured by proper experiments. As we have discussed, the boundary conditions imposed on the scattering wave functions are less strict than the boundary conditions imposed on proper solutions (as the wave function does not need to vanish asymptotically). Consequently, as we shall see in the following section, *the scattering states are not quantized, and their energy spectrum is continuous.*

7.2 Scattering in One Dimension

In this section we shall discuss in some detail scattering wave functions and their corresponding probability fluxes in the case of stationary scattering of a single particle of mass m in a one-dimensional coordinate space (x). In this case the continuity equation, Eq. (7.1.1), simplifies to

$$\frac{\partial}{\partial t}|\psi(x,t)|^2 \equiv -\frac{\partial}{\partial x}J(x,t), \tag{7.2.1}$$

where the probability current density reads

$$J(x,t) \equiv \frac{\hbar}{m}\text{Im}[\psi^*(x,t)\frac{\partial}{\partial x}\psi(x,t)]. \tag{7.2.2}$$

For a stationary solution, $\psi(x,t) = e^{-iEt/\hbar}\psi_E(x)$, the probability flux is readily shown to be time-independent,

$$J_E(x) = \frac{\hbar}{m}\text{Im}[\psi_E^*(x)\frac{\partial}{\partial x}\psi_E(x)], \tag{7.2.3}$$

where Eq. (7.2.1) yields in this case a constant probability flux,

$$J_E(x) = Const, \tag{7.2.4}$$

which is analogous to Eq. (7.1.4) in the more general case.

Exercise 7.2.1   *A single particle in a one-dimensional coordinate space (x) is associated with a proper stationary wave function. Show that the probability flux vanishes anywhere in space in this case.*

First, let us focus on a stationary flow of free particles, in a system in which the potential energy is constant, $V(x) = V_0$. The corresponding time-independent Schrödinger equation reads

$$\frac{-\hbar^2}{2m}\frac{\partial^2}{\partial x^2}\psi_E(x) + V_0\psi_E(x) = E\psi_E(x), \tag{7.2.5}$$

and the general solutions (see Eqs. (6.2.1–6.2.3)) obtain the form

$$\psi_E(x) = ae^{ikx} + be^{-ikx} \quad ; \quad k = \sqrt{\frac{2m(E - V_0)}{\hbar^2}}. \tag{7.2.6}$$

Specific solutions are the "plane waves" associated with $E > V_0$, defined as

$$\psi_{E_\pm}(x) \equiv e^{\pm ikx}. \tag{7.2.7}$$

Each plane wave is an eigenfunction of the linear momentum operator, $\hat{p}_x = -i\hbar \frac{\partial}{\partial x}$,

$$\hat{p}_x e^{\pm ikx} = \pm k\hbar e^{\pm ikx}, \tag{7.2.8}$$

with a real eigenvalue, $p_x = \pm k\hbar$. According to the postulates of quantum mechanics (see Section 4.4, and Chapter 11), this implies that these solutions represent a flow of particles with well-defined momentum, where the different waves, e^{+ikx} and e^{-ikx}, correspond to particles moving in the left-to-right and right-to-left directions, respectively. Substitution in Eq. (7.2.3) yields the corresponding particle fluxes:

$$J_{E_\pm}(x) = \frac{\hbar}{m} \text{Im}[e^{\mp ikx} \frac{\partial}{\partial x} e^{\pm ikx}] = \pm \frac{k\hbar}{m}. \tag{7.2.9}$$

The probability fluxes associated with the plane wave solutions can therefore be identified with the particle's velocities, p_x/m. A general solution to Eq. (7.2.5) is therefore a superposition of plane waves propagating in opposite directions, where the momentum is not necessarily defined. (Indeed, $\psi_E(x)$ in Eq. (7.2.6) is generally not an eigenfunction of $\hat{p}_x$.) Nevertheless, the probability flux obtains a well-defined value for any general solution according to Eq. (7.2.3),

$$J_E = |a|^2 J_{E_+} - |b|^2 J_{E_-} = \frac{k\hbar}{m}(|a|^2 - |b|^2), \tag{7.2.10}$$

which is the weighted difference between the left-to-right and the right-to-left fluxes.

We now consider the scattering of particles through a one-dimensional space with an alternating potential energy, which is piecewise constant, as defined in Eq. (6.3.1) (see Fig. 7.2.1). The formal solution of the time-independent Schrödinger equation in this case is given by Eq. (6.3.2). In Section 6.3 we discussed the bound states corresponding to particles trapped in space, whose energy is smaller than the asymptotic potential energy thresholds (V_1 on the left, and V_N on the right, see Fig. 7.2.1). Here we discuss the scattering state for the same system, at particle energies for which a classical particle could propagate to $x \to \infty$ and/or $x \to -\infty$, namely,

$$E > \min(V_1, V_N). \tag{7.2.11}$$

In this energy range the quantum system is open. Consequently, the physically meaningful stationary solutions must satisfy Eq. (7.2.4), but they don't need to be proper. Therefore, the boundary conditions imposed at the asymptotes are different from the ones applied in the case of bound states.

Without loss of generality, let us consider a source of particles located infinitely far on the left, ($x \to -\infty$), in the segment $n = 1$. The corresponding asymptotic wave function is $\psi_1(x) = a_1 e^{ik_1 x} + b_1 e^{-ik_1 x}$, where $k_1 = \sqrt{\frac{2m(E-V_1)}{\hbar^2}}$. The source transmits a flux of

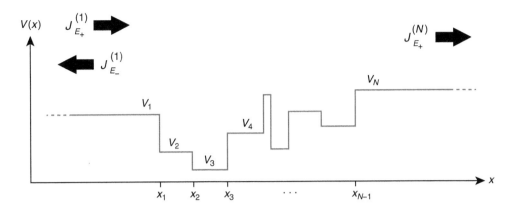

Figure 7.2.1   A schematic representation of a piecewise constant potential energy. The incoming probability flux ($J_{E_+}^{(1)}$) as well as the transmitted ($J_{E_+}^{(N)}$) and reflected ($J_{E_-}^{(1)}$) fluxes in the asymptotic regions are marked by arrows.

free particles from left to right, which corresponds to the incoming plane wave, $a_1 e^{ik_1 x}$. Using Eq. (7.2.3), this flux corresponds to

$$J_{E_+}^{(1)} = |a_1|^2 \frac{\hbar k_1}{m}. \tag{7.2.12}$$

It follows that a_1 must obtain a finite value in this case. Since the solution to the Schrödinger equation is defined up to a multiplication by a constant, it is conventional to set it to unity,

$$a_1 = 1. \tag{7.2.13}$$

The same consideration applies to the asymptotic region on the right ($x \to +\infty$, in the segment $n = N$), where the asymptotic wave function is $\psi_N(x) = a_N e^{ik_N x} + b_N e^{-ik_N x}$. Having no source of particles in that region, the incoming right-to-left flux must vanish, which means

$$b_N = 0. \tag{7.2.14}$$

Equations (7.2.13, 7.2.14) constitute the selected scattering boundary conditions. (Notice that similar boundary conditions would correspond to a flux source located on the right, setting $a_1 = 0$ and $b_N = 1$.) In any event, these boundary conditions differ from the asymptotic boundary conditions for bound states (which read $a_1 = b_N = 0$, see Eq. (6.3.6)) in that one of the incoming asymptotic waves has a nonvanishing contribution.

Recalling that a solution to the Schrödinger equation must be continuously differentiable, the scattering wave function and its derivative must be continuous at each one of the $N - 1$ boundary points between nearby segments (Eq. (6.3.5)). The resulting $2(N - 1)$ equations plus the two boundary conditions (Eqs. (7.2.13, 7.2.14)) can be cast into a system of linear equations for the wave function coefficients in all segments,

$$
\begin{bmatrix}
1 \\
e^{ik_1 x_1} & e^{-ik_1 x_1} & -e^{ik_2 x_1} & -e^{-ik_2 x_1} \\
ik_1 e^{ik_1 x_1} & -ik_1 e^{-ik_1 x_1} & -ik_2 e^{ik_2 x_1} & ik_2 e^{-ik_2 x_1} \\
& & e^{ik_2 x_2} & e^{-ik_2 x_2} & -e^{ik_3 x_2} & -e^{-ik_3 x_2} \\
& & ik_2 e^{ik_2 x_2} & -ik_2 e^{-ik_2 x_2} & -ik_3 e^{ik_3 x_2} & ik_3 e^{-ik_3 x_2} \\
& & & & & & \ddots \\
& & & & & & & e^{ik_{N-1} x_{N-1}} & e^{-ik_{N-1} x_{N-1}} & -e^{ik_N x_{N-1}} \\
& & & & & & & ik_{N-1} e^{ik_{N-1} x_{N-1}} & -ik_{N-1} e^{-ik_{N-1} x_{N-1}} & -ik_N e^{ik_N x_{N-1}}
\end{bmatrix}
$$

$$
\times
\begin{bmatrix}
a_1 \\
b_1 \\
a_2 \\
b_2 \\
a_3 \\
b_3 \\
\vdots \\
b_{N-1} \\
a_N
\end{bmatrix}
=
\begin{bmatrix}
1 \\
0 \\
0 \\
0 \\
0 \\
0 \\
\vdots \\
0
\end{bmatrix} .
\tag{7.2.15}
$$

This equation is similar to the one obtained for the bound states (Eq. (6.3.8)), except for the constraint, $a_1 = 1$, which introduces a source term, thus converting the equation into an inhomogeneous one. ***The fact that the scattering wave function coefficients are defined by an inhomogeneous equation has a remarkable consequence: a nontrivial solution exists at any energy.*** This is different than in the case of bound states, where the homogeneous equation has a nontrivial solution only for specific energies. It follows that the energy spectrum of the open system ($E > \min(V_1, V_N)$) is continuous.

First, let us consider the scattering energies, which are larger than both of the asymptotic potential energy thresholds, namely, $E > V_1, V_N$. The information of prime interest is the probability of scattering from the target into the different possible directions in space. In the one-dimensional case, the asymptotic incoming flux can be either transmitted forward, or reflected backward. For the piecewise constant potential energy, an incoming particles flux from, for example, the left asymptote (segment '1'), $J_{E+}^{(1)}$, can be either transmitted to the Nth segment or reflected to the 1st segment. The corresponding transmitted and reflected fluxes, $J_{E+}^{(N)}$ and $J_{E-}^{(1)}$, are associated with the plane waves, $a_N e^{ik_N x}$, and $b_1 e^{-ik_1 x}$, respectively. Using Eq. (7.2.3), one obtains

$$
J_{E+}^{(N)} = |a_N|^2 \frac{\hbar k_N}{m} .
\tag{7.2.16}
$$

$$
J_{E-}^{(1)} = -|b_1|^2 \frac{\hbar k_1}{m} .
\tag{7.2.17}
$$

The ratios between these scattered fluxes and the incoming flux (in absolute values) defines the transmission and reflection probabilities, $T(E)$ and $R(E)$,

$$
T(E) = \frac{|J_{E+}^{(N)}|}{|J_{E+}^{(1)}|} = \frac{|a_N|^2 k_N}{|a_1|^2 k_1} \quad ; \quad R(E) = \frac{|J_{E-}^{(1)}|}{|J_{E+}^{(1)}|} = \frac{|b_1|^2}{|a_1|^2} .
\tag{7.2.18}
$$

Recalling that for stationary solutions the flux must be constant anywhere in space (Eq. (7.2.4)), namely, $J_{E+}^{(N)} = J_{E+}^{(1)} + J_{E-}^{(1)}$, it follows that the transmission and reflection probabilities must sum to unity (this result is a manifestation of the unitarity of the scattering matrix [7.1]):

$$R(E) + T(E) = 1. \tag{7.2.19}$$

The reflection probability is expressed in terms of the ratio of wave function coefficients, $R(E) \equiv |r_1|^2 \equiv |\frac{b_1}{a_1}|^2$. For the piecewise constant potential (Eq. (6.3.1)) this ratio can be calculated for any energy by reformulating Eq. (7.2.15) as a recursion relation,

$$r_n = e^{2ik_n x_n} \frac{(k_{n+1} - k_n)e^{ik_{n+1}x_n} - (k_{n+1} + k_n)e^{-ik_{n+1}x_n}r_{n+1}}{-(k_{n+1} + k_n)e^{ik_{n+1}x_n} + (k_{n+1} - k_n)e^{-ik_{n+1}x_n}r_{n+1}}. \tag{7.2.20}$$

Starting from the Nth segment, where $r_N = 0$, and progressing backward, one can obtain an explicit expression for $r_{N-1}, r_{N-2}, \ldots, r_2, r_1$, in terms of the potential energy parameters ($V_1, V_2, \ldots, V_N$ and ($x_1, x_2, \ldots, x_{N-1}$)) and the particle mass. Consequently, $R(E) = |r_1|^2$ and $T(E) = 1 - R(E)$ are obtained.

7.3 Degeneracy of Scattering States

Notice that for any energy in this range ($E > V_1, V_N$) there are two degenerate scattering wave functions. The first corresponds to an incoming flux from the left, as has been discussed in Eqs. (7.2.12–7.2.20), with the asymptotic wave functions,

$$\psi_{L \to R}(x) = \begin{cases} e^{ik_1 x} + b_1 e^{-ik_1 x} & ; \quad x \leq x_1 \\ \vdots \\ a_N e^{ik_N x} & ; \quad x \geq x_{N-1} \end{cases} \tag{7.3.1}$$

The second corresponds to an incoming flux from the right, associated with different asymptotic wave functions:

$$\psi_{R \to L}(x) = \begin{cases} \bar{a}_1 e^{-ik_1 x} & ; \quad x \leq x_1 \\ \vdots \\ e^{-ik_N x} + \bar{b}_N e^{ik_N x} & ; \quad x \geq x_{N-1} \end{cases} \tag{7.3.2}$$

Interestingly, regardless of any symmetry requirement on the potential energy, the transmission and reflection probabilities associated with the two scattering directions are identical, namely,

$$R(E)_{L \to R} = |b_1|^2 = |\bar{b}_N|^2 = R(E)_{R \to L}$$

$$T(E)_{L \to R} = |a_N|^2 \frac{k_N}{k_1} = |\bar{a}_1|^2 \frac{k_1}{k_N} = T(E)_{R \to L}. \tag{7.3.3}$$

This result means that the probability of an incoming flux from the left, associated with an initial momentum, $+\hbar k_1$, to be scattered through space into a final momentum $+\hbar k_N$, equals the probability of an incoming flux from the right with an initial momentum, $-\hbar k_N$, to be scattered into an asymptotic state with a final momentum, $-\hbar k_1$. It follows directly from the time-reversal symmetry of the corresponding time-dependent Schrödinger equation, which means that both $\psi(x,t)$ and $\psi^*(x,-t)$ are solutions to the time-dependent Schrödinger equation, as can be readily verified. For stationary solutions, this means that $\psi_E(x)$ and $\psi_E^*(x)$ are degenerate. To prove Eq.(7.3.3), we can show that $\psi_{R \to L}(x)$ is in fact a superposition of the two degenerate solutions, $\psi_{L \to R}(x)$ and its complex conjugate, such that scattering probabilities associated with $\psi_{R \to L}(x)$ are expressed in terms of the parameters of $\psi_{L \to R}(x)$ (Ex. 7.3.1).

Exercise 7.3.1 *Given the scattering states $\psi_{L \to R}(x)$ and $\psi_{R \to L}(x)$ in Eqs. (7.3.1, 7.3.2), show that $\psi_{R \to L}(x)$ can be expressed as a linear combination of $\psi_{L \to R}(x)$ and $\psi_{L \to R}^*(x)$, and prove the result in Eq. (7.3.3).*

We now consider scattering energies that are larger than one of the asymptotic potentials but smaller than the other one, for example, without loss of generality, $V_1 < E < V_N$. In this case, the asymptotic scattering wave functions obtain the form

$$\psi_{L \to R}(x) = \begin{cases} a_1 e^{ik_1 x} + b_1 e^{-ik_1 x} & ; \quad x \le x_1 \\ \quad \vdots & \\ 0 & ; \quad x \to \infty \end{cases} \tag{7.3.4}$$

Flux conservation (Eq. (7.2.4)) leads in this case to $|a_1|^2 - |b_1|^2 = 0$. This means that in this energy range, an incoming particles flux from the left asymptote (segment '1') must be reflected with a unit probability, namely, $R(E) = |b_1|^2/|a_1|^2 = 1$ (see also Ex. 7.3.2). Time reversal symmetry means that the complex conjugate of $\psi_{L \to R}(x)$ is also a solution associated with the same energy. However, since $|a_1|^2 = |b_1|^2$, one can readily verify that, $\frac{a_1}{b_1^*} \psi_{L \to R}^*(x) = \psi_{L \to R}(x)$, namely $\psi_{L \to R}^*(x)$ and $\psi_{L \to R}(x)$ are linearly dependent. It follows that in this energy range the solutions are nondegenerate. Indeed, scattering from a right flux source is excluded since the wave function vanishes asymptotically in the classically forbidden region.

Exercise 7.3.2 *Given a single potential energy step, $V(x) = \begin{cases} x \le 0 & ; V_1 \\ x > 0 & ; V_2 \end{cases}$, where $V_2 > V_1$, use Eq. (7.2.20) to show that the reflection probability equals unity in the scattering energy range, $V_1 < E < V_2$.*

7.4 Scattering from a Single Potential Barrier or Well

We now consider explicitly the case of scattering from a symmetric potential energy barrier (or well) of a finite length, where the potential energy reads

$$V(x) = \begin{cases} V_0 & ; \quad x < 0 \\ V_1 & ; \quad 0 \le x < L. \\ V_0 & ; \quad L \le x \end{cases} \tag{7.4.1}$$

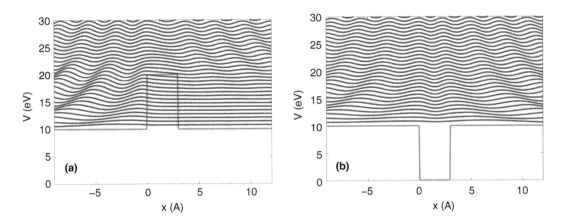

Figure 7.4.1 Potential energy and scattering wave functions for a potential energy barrier (a), and well (b). The wave functions are represented by their real part, where the different plots correspond to different scattering energies. The plots are displaced from each other by the respective energy differences. The particle's mass was set to an electron mass, and the barrier (well) width was set to $L = 0.3$ nm. The barrier (well) energy was shifted by $+10$ (-10) eV with respect to the asymptotic potential energy.

The scattering wave functions are the stationary solutions of the time-dependent Schrödinger equation, for $E > V_0$. Considering an incoming flux source from the left, the scattering states obtain the form

$$\psi(x) = \begin{cases} a_1 e^{ik_1 x} + b_1 e^{-ik_1 x} & ; \quad x < 0 \\ a_2 e^{ik_2 x} + b_2 e^{-ik_2 x} & ; \quad 0 \le x < L, \\ a_3 e^{ik_1 x} & ; \quad L \le x \end{cases} \tag{7.4.2}$$

where $k_1 = k_3 \equiv \sqrt{\frac{2m(E-V_0)}{\hbar^2}}$ and $k_2 \equiv \sqrt{\frac{2m(E-V_1)}{\hbar^2}}$. Some of the scattering wave functions, corresponding to selected energy values (out of the continuous energy spectrum), are plotted in Fig. 7.4.1. Implementing the recursion relation (Eq. (7.2.20)) for this case, one has $r_3 = 0$, $r_2 = e^{2ik_2 L} \frac{k_2 - k_1}{k_2 + k_1}$, and, $r_1 = \frac{k_1^2 - k_2^2}{k_2^2 + k_1^2 + 2ik_1 k_2 \cot(k_2 L)}$, which yields the following explicit expressions for the reflection and the transmission probabilities,

$$R(E) = \frac{(k_1^2 - k_2^2)^2}{(k_2^2 + k_1^2)^2 + 4k_1^2 k_2^2 \cot^2(k_2 L)} \quad ; \quad T(E) = 1 - R(E). \tag{7.4.3}$$

Exercise 7.4.1 *Derive the result Eq. (7.4.3) for the transmission and reflection probabilities for particles scattering from a square potential energy barrier (or well).*

The probability for transmission through a potential barrier ($V_1 > V_0$) in plotted in Fig. 7.4.2. For energies smaller than the potential barrier height, $E < V_1$, a classical particle would not be able to penetrate a forbidden region in space to pass through the barrier. It follows that according to classical mechanics the transmission probability should be strictly zero. In quantum mechanics, however, as apparent in Fig. 7.4.2,

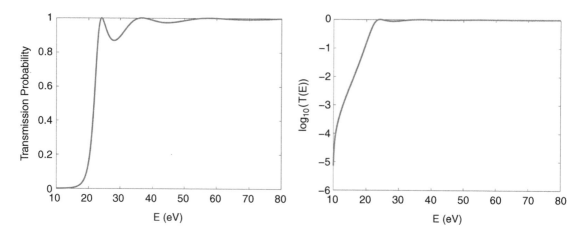

Transmission probability, $T(E)$, and its logarithm, $\log_{10}(T(E))$, as a function of the energy for scattering from a potential energy barrier ($V_1 = 20$ eV; see Fig. 7.4.1 for all the parameters).

there is a finite transmission probability even in this low-energy regime. This phenomenon is attributed to the penetration of the scattering wave function into the classical forbidden region (recall that k_2 is purely imaginary in the range $E < V_1$). See also Fig. 7.4.1a), which allows for quantum tunneling through the potential energy barrier. Indeed, we encountered already the possibility for quantum tunneling when discussing bound particles in quantum wells (see Chapter 6). For energies above the potential barrier height, $E > V_1$, the transmission probability should be unity according to classical mechanics, since a classical particle would have sufficient kinetic energy to cross the barrier region. Nevertheless, the corresponding quantum mechanical transmission probability is generally less than unity (see Fig. 7.4.2). This phenomenon is familiar from wave scattering at boundaries between different media. Equation (7.4.2) shows that the reflection strictly vanishes (and therefore the transmission probability is unity) only when $\cot^2(k_2L) \to \infty$, namely, when, $\sin(k_2L) = 0$. This strictly happens only at specific energies, for which the barrier length equals an integer multiple of half de Broglie's wavelengths, $\lambda = \frac{2\pi\hbar}{p}$,

$$L = \frac{\pi n}{k_2} = \frac{\hbar\pi n}{\sqrt{2m(E - V_1)}} = \frac{n\lambda}{2}. \tag{7.4.4}$$

Changing the potential energy such that $V_1 < V_0$, the barrier turns into a potential energy well (see Fig. 7.4.1b). The transmission probability for scattering above the potential well is plotted in Fig. 7.4.3. Again, the classical transmission probability for $E > V_0 > V_1$ is unity, but the corresponding quantum mechanical transmission probability is generally smaller. As for the case of a potential energy barrier, the reflection strictly vanishes (and therefore the transmission probability is unity) only when $\sin(k_2L) = 0$ namely, when the length of the potential energy well equals an integer multiplication of half de Broglie's wavelengths (Eq. (7.4.4)).

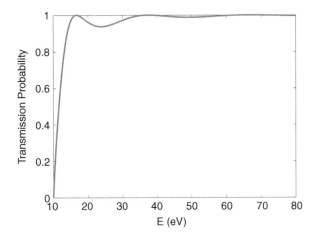

Figure 7.4.3 Transmission probability, $T(E)$, as a function of the energy for scattering above a potential energy well ($V_0 = 10$ eV, $V_1 = 0$; see Fig. 7.4.1 for the other parameter values).

7.5 The Resonant Tunneling Phenomenon

We now return to scattering through potential barriers ($V_1 > V_0$) in the quantum tunneling regime ($E < V_1$). As we have seen, the transmission probability for scattering through a single barrier increases monotonically with the scattering energy in this regime. Here we ask: *what would be the effect of a consecutive barrier, located after the first one, along the propagation direction?* In classical mechanics, as long as the second barrier height is smaller than or equal to the first one, there should be no effect at all. In quantum mechanics, the result turns out to be very different. Consider, for example, the symmetric double barrier potential presented in Eq. (7.5.1) and Fig. 7.5.1,

$$V(x) = \begin{cases} V_0 & ; \quad x < x_1 \\ V_1 & ; \quad x_1 \leq x < x_1 + L \\ V_0 & ; \quad x_1 + L \leq x < x_1 + L + d \\ V_1 & ; \quad x_1 + L + d \leq x < x_1 + 2L + d \\ V_0 & ; \quad x_1 + 2L + d \leq x \end{cases} \qquad (7.5.1)$$

The transmission probability, calculated by implementing the general formulation presented in Eq. (7.2.20), is plotted in Fig. 7.5.2. A remarkable feature is revealed: for some energy values within the tunneling range, $E < V_1$, the presence of an additional consecutive barrier increases the transmission probability through the target. The effect is known as the resonant tunneling phenomenon. It is known in scattering of electromagnetic waves through resonators, and in optics it is named after Fabry and Pérot, the developers of the optical resonator [7.2]. In essence, the two barriers define an effective finite potential energy well between them. If the barriers' widths extended to infinity ($L \to \infty$), this potential well would support a finite number of bound states (see

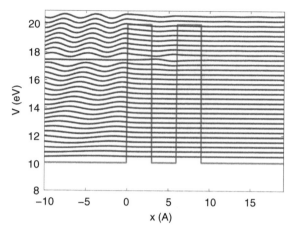

Figure 7.5.1 Potential energy and scattering wave functions for a symmetric double barrier potential. The wave functions are represented by their real part, where the different plots correspond to different scattering energies. The plots are displaced from each other by the respective energy differences. The particle's mass was set to an electron mass, and the double barrier parameters are $L = d = 0.3$ nm, $V_0 = 10$ eV, and $V_1 = 20$ eV.

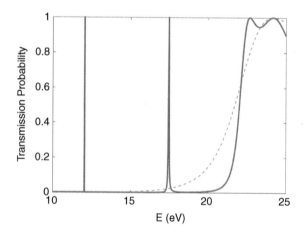

Figure 7.5.2 Transmission probability, $T(E)$, as a function of the energy for scattering from the symmetric double barrier potential presented in Fig. 7.5.1. The result for a single barrier is plotted for comparison (dashed line).

Section 6.2). For a finite L, however, owing to wave function penetration, there are no bound states. Instead, for sufficiently large L, each bound state would correspond to a "quasi-bound state" (or "resonance state"). This means that if the bound state is selected as an initial condition for the time-dependent Schrödinger equation, the probability for populating the intermediate potential well would decay exponentially in time. (See Chapter 19 for a general analysis of exponential decays.) The decay rate is denoted $\Gamma/\hbar$, where Γ is termed the "resonance energy width" (formally associated with twice the imaginary part of a complex "resonance energy"). In a scattering experiment, when the energy of the incoming stationary flux matches the quasi-bound state energy (within the energy-width Γ), the tunneling probability is significantly enhanced. Remarkably, for a symmetric double barrier, the transmission can reach unity; namely, the consecutive barrier turns the double barrier structure into a transparent one for selected energies (see Ex. 7.5.1). Resonant tunneling phenomena play a crucial role in transport through nanoscale conductors (see [5.6] and Chapter 20). The interested reader is directed to complementary literature focusing on quantum scattering theory in general [7.1] and resonance states in particular [7.3].

Exercise 7.5.1   *Given a symmetric double barrier potential,*

$$V(x) = \begin{cases} V_0 & ; & x < -(L+d/2) \\ V_1 & ; & -(L+d/2) \leq x < -d/2 \\ V_0 & ; & -d/2 \leq x < d/2 \\ V_1 & ; & d/2 \leq x < d/2+L \\ V_0 & ; & d/2+L \leq x \end{cases} \quad .$$

obtain an equation for the scattering energies in which the transmission probability is 100%.

Bibliography

[7.1] J. R. Taylor, "Scattering Theory: The Quantum Theory of Nonrelativistic Collisions" (Dover, 2006).

[7.2] G Hernández, "Fabry-Perot Interferometers" (Cambridge University Press, 1988).

[7.3] N. Moiseyev, "Non-Hermitian Quantum Mechanics" (Cambridge University Press, 2011).

Mechanical Vibrations and the Harmonic Oscillator Model

8.1 Molecular Vibrations

Let us consider a group of atoms composing a molecule in one of its stable geometrical arrangements. In classical mechanics, given some finite kinetic energy, the atoms will perform small amplitude motions, namely vibrations, around the geometry of minimal energy. In what follows, we address the questions: *What is the corresponding quantum mechanical description of vibrational motion? What are the energy levels and wave functions that account for internal molecular vibrations?*

A glimpse into the nature of molecular vibrations is obtained through their interaction with electromagnetic radiation. The relative motion of partially charged atoms against each other is associated with dipole oscillations, which can exchange energy with the electromagnetic field. In Fig. 8.1.1 absorption spectra of electromagnetic radiation are plotted for two organic molecules as a function of the radiation wave number, $\frac{1}{\lambda} = \frac{\nu}{c}$, in the infrared regime. The spectrum is shown to be sensitive to the specific molecule, which is useful as a "molecular fingerprint" for identification of the molecular structure and composition. Indeed, each molecule is characterized by a unique combination of dipole oscillations with characteristic frequencies, which reflect the interatomic forces and the atomic masses.

While a classical mechanical description may qualitatively explain the differences between different molecules, a quantitative analysis of the absorption spectrum necessitates a quantum mechanical description. As we shall see, the absorption peak positions correspond to the quantized vibrational energy levels in the molecule (which turn out to be related to the frequencies of the classical vibrations). Moreover, the intensity of each peak is subject to quantum mechanical "selection rules" that are derived from the corresponding wave functions, namely the stationary solutions to the Schrödinger equation, and the widths of the peaks are associated with the interaction between each molecular vibration and other degrees of freedom within the molecule as well as with its surroundings. In this chapter we shall focus on the relevant vibrational energy levels and the corresponding stationary

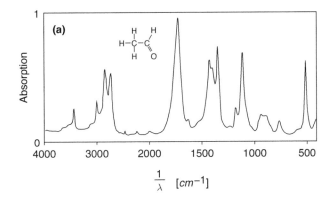

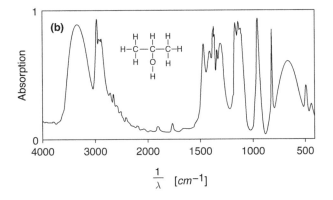

Figure 8.1.1 Infrared absorption spectra of two molecules: (a) C_2H_4O, and (b) C_3H_8O. The absorption is defined here as, $1 - I/I_0$, where I_0 and I are the incoming and transmitted radiation fluxes.

wave functions. A more detailed discussion of spectral line shapes will be given in Chapter 18.

8.2 The Normal Modes of a Many-Particle System and the Harmonic Approximation

In the previous chapters, our solutions of the Schrödinger equation referred to systems in which a single particle (or a flux of noninteracting particles) was subject to an "external" potential energy. In this chapter we address systems in which a given number of particles are bounded to each other by interparticle interactions, associated with a many-particle potential energy function of a general form. Particularly, we shall address the geometrical arrangement of the particles relative to each other in a stable state of the system, namely near a minimum of the potential energy function. As explained below, this scenario has a universal form, which can be approximately mapped onto a set of harmonic oscillators. Indeed, vibrations of atoms in molecules, nanoparticles, and solid lattices are often referred to in terms of the harmonic approximation. The energy levels and the stationary probability density distributions of the relative atomic locations in these systems are therefore inferred from the solution of the time-independent Schrödinger equation for the harmonic oscillator model, as detailed in the following sections. Notice that historically, the notion that the radiation emerging from oscillating dipoles must be quantized in energy preceded the formulation of quantum mechanics. It was the basis for Max Planck's successful explanation of the "black body" radiation in the year 1900, which was a major driving force for pursuing the origin of energy quantization.

Let us consider a system of N particles with masses, $m_1, m_2, \ldots, m_N$ (e.g., atoms in a polyatomic molecule) in a Cartesian coordinate system. The energy of the system depends on the vector of $3N$ Cartesian positions of all the particles, denoted here as $\mathbf{x}$, and the corresponding momenta. We are interested in the potential energy as a function of the coordinates at the vicinity of a minimum, $\mathbf{x}^{(eq)}$, which corresponds to a stable ("equilibrium") geometry of the particles. At a minimum point, $\left. \frac{\partial V(\mathbf{x})}{\partial x_j} \right|_{\mathbf{x}=\mathbf{x}^{(eq)}} = 0$, for $j = 1, 2, \ldots, 3N$. For small deviations around the minimum, the Taylor expansion of the potential function can be truncated in the second order. Setting the zero of the potential energy to, $V(\mathbf{x}^{(eq)}) = 0$, one therefore obtains

$$V(\mathbf{x}) \approx \sum_{j,j'=1}^{3N} \frac{1}{2} \left. \frac{\partial^2 V(\mathbf{x})}{\partial x_j \partial x_{j'}} \right|_{\mathbf{x}=\mathbf{x}^{(eq)}} (x_j - x_j^{(eq)})(x_{j'} - x_{j'}^{(eq)}). \tag{8.2.1}$$

Defining the Hessian matrix, $\mathcal{H}_{j,j'} = \left. \frac{\partial^2 V(\mathbf{x})}{\partial x_j \partial x_{j'}} \right|_{\mathbf{x}=\mathbf{x}^{(eq)}}$, the mass matrix, $\mathcal{M}_{j,j'} = m_j \delta_{j,j'}$ (where m_j is the mass of the particle associated with the coordinate x_j), and the displacement vector, $\mathbf{d} = \mathbf{x} - \mathbf{x}^{(eq)}$, the classical system Hamiltonian can be expressed as follows:

$$H = \frac{1}{2}\mathbf{d}^t \mathcal{H} \mathbf{d} + \frac{1}{2}\dot{\mathbf{d}}^t \mathcal{M} \dot{\mathbf{d}}. \tag{8.2.2}$$

This bilinear form can be simplified by diagonalizing the (symmetric) mass-weighted Hessian matrix, $\mathcal{M}^{-1/2} \mathcal{H} \mathcal{M}^{-1/2}$, namely, using $\left[\mathcal{M}^{-1/2} \mathcal{H} \mathcal{M}^{-1/2} \right] \mathbf{U} = \mathbf{U} \mathbf{K}$, where $\mathbf{K}$ is a diagonal matrix. Using the eigenvectors matrix $\mathbf{U}$ to define a new set of coordinates (linear combinations of the particles Cartesian coordinates),

$$\mathbf{q} \equiv \mathbf{U}^t \mathcal{M}^{1/2} \mathbf{d}, \tag{8.2.3}$$

the Hamiltonian obtains the form

$$H = \frac{1}{2}\mathbf{q}^t \mathbf{K} \mathbf{q} + \frac{1}{2}\dot{\mathbf{q}}^t \dot{\mathbf{q}} = \sum_{j=1}^{3N} \frac{1}{2}(K_{j,j}q_j^2 + \dot{q}_j^2) \equiv \sum_{j=1}^{3N} H_j. \tag{8.2.4}$$

The Hamiltonian is therefore expressed as sum of independent terms, each corresponding to a single "global" coordinate, $q_j \equiv \sum_{j'=1}^{3N} \mathcal{U}_{j',j} \mathcal{M}_{j',j'}^{1/2} (x_{j'} - x_{j'}^{(eq)})$. These global coordinates are termed the "normal modes" of the system. Notice that generally only $3N - 6$ degrees of freedom (out of the $3N$) correspond to interparticle vibrations. Three out of the remaining six coordinates correspond to the position of the center of mass of the system, and the other three correspond to the global orientation of the particles in space as a rigid body. Indeed, when the potential energy function depends only on the interparticle distances (namely, in the absence of external forces), there are only $3N - 6$ nonzero K_j values, associated with interparticle vibrations. (For a linear equilibrium arrangement $\mathbf{x}^{(eq)}$ of the particles, the global orientation is defined by only two variables instead of three, and the number of interparticle vibrations increases to $3N - 5$.)

Since the Hamiltonian, Eq. (8.2.4), is separable in the different modes ($H = \sum_{j=1}^{3N} H_j$), *the analysis of the many-particle system near its stable geometry can be performed independently for each mode (see Ex. 4.3.4). Moreover, this result is universal for any collection of particles of different types and numbers.* For example, when the particles under consideration are atoms, the normal mode analysis applies to small molecules (see Section 8.1), nanoparticles, quantum dots containing thousands of atoms, and also to solid lattices, extended to include macroscopic numbers of atoms. (In the latter case, the excitations of the normal modes is referred to as the lattice phonons.)

Each single mode is characterized by H_j in Eq. (8.2.4). This Hamiltonian corresponds to a kinetic energy of a particle of (unit) mass and a potential energy term, which is quadratic in the respective coordinate. Before turning to the quantum mechanical treatment of this Hamiltonian, let us recall the classical mechanics description of the corresponding motion. For generality, we shall consider a particle of mass m. The particle's coordinate and conjugate momentum will be denoted as q and p, respectively. The classical Hamiltonian reads

$$H = \frac{1}{2}\frac{p^2}{m} + \frac{1}{2}Kq^2, \tag{8.2.5}$$

where the force experienced by the particle satisfies Hooke's law,

$$f = -\frac{\partial H}{\partial q} = -Kq. \tag{8.2.6}$$

Newton's equation of motion for this model reads $m\ddot{q} = -Kq$, and its general solution for the coordinate reads

$$q(t) = a\cos(\omega t) + b\sin(\omega t) \quad ; \quad \omega \equiv \sqrt{\frac{K}{m}}, \tag{8.2.7}$$

where a and b are scalars, set by the initial conditions. It follows that the particle performs a simple harmonic motion in time. The Hamiltonian in Eq. (8.2.5) is therefore referred to as the harmonic oscillator model, where the oscillation frequency, ω, is set by the force constant, K, and the mass m. Notice that the harmonic oscillator model appears naturally in the normal mode analysis of atomic vibrations as we have outlined, but its applications extend to many areas of physics, including nuclear structure theory, quantum electrodynamics and quantum optics, mechanical engineering, and so forth.

8.3 The Solutions of the Schrödinger Equation for the Harmonic Oscillator

We now turn to a quantum mechanical description of a harmonic oscillator, associated with a mass m, a force constant, $K = m\omega^2$, a Cartesian position coordinate, q, $-\infty < q < \infty$, and a corresponding classical momentum, $p = m\dot{q}$. The quantum mechanical Hamiltonian operator is obtained by using the quantization rule for the canonical position and momentum operators, Eqs. (3.2.1, 3.2.2), namely, $\hat{H} = \frac{-\hbar^2}{2m}\frac{\partial^2}{\partial q^2} + \frac{1}{2}m\omega^2 q^2$. The time-independent Schrödinger equation (Eq. (4.3.4)) therefore reads

$$\frac{-\hbar^2}{2m}\frac{\partial^2}{\partial q^2}\psi(q) + \frac{1}{2}m\omega^2 q^2\psi(q) = E\psi(q), \tag{8.3.1}$$

where we seek for its normalized solutions, for which

$$\int_{-\infty}^{\infty}|\psi(q)|^2 dq = 1. \tag{8.3.2}$$

It is convenient to introduce at this point dimensionless variables, $y = \sqrt{\frac{m\omega}{\hbar}}q$ and $\lambda = E/(\hbar\omega)$, for the oscillator's position and energy, respectively. Defining $\psi(q) = \varphi(y)$, the Schrödinger equation Eq. (8.3.1) for $\varphi(y)$ obtains the form

$$\frac{1}{2}[y^2 - \frac{\partial^2}{\partial y^2}]\varphi(y) = \lambda\varphi(y). \tag{8.3.3}$$

Eq. (8.3.3) is an ordinary second-order differential equation, regular in the entire space. It therefore has power series solutions for any y, namely, $\varphi(y) = \sum_{k=0}^{\infty}a_k y^k$. Since the potential energy diverges asymptotically (at $y \to \pm\infty$), all the stationary solutions correspond to bound states, and therefore must be proper; namely, they must vanish asymptotically, $\varphi(y) \xrightarrow{y\to\pm\infty} 0$ (see Eq. (2.2.1)). Focusing on the asymptotic regime, Eq. (8.3.3) means that for any finite λ (finite energy), λ becomes negligible next to y^2, and therefore, $\frac{\partial^2}{\partial y^2}\varphi(y) \xrightarrow{y\to\pm\infty} y^2\varphi(y)$. One can readily verify that a power series that satisfies this equation must take the form $\varphi_n(y) = e^{\pm y^2/2}P_n(y)$, where $P_n(y) = \sum_{k=0}^{n}a_k y^k$ is a polynomial of a **finite** degree, n (see Ex. 8.3.1). Excluding the asymptotically diverging solutions associated with, $e^{+y^2/2}$, the proper solutions obtain the form

$$\varphi_n(y) = e^{-y^2/2}P_n(y). \tag{8.3.4}$$

The lowest-degree polynomial corresponds to $n = 0$. The solution to Eq. (8.3.3), subject to normalization (Eq. (8.3.2)), and the respective eigenvalue, read in this case

$$\varphi_0(y) = \left(\frac{m\omega}{\hbar\pi}\right)^{1/4}e^{-y^2/2} \quad ; \quad \lambda_0 = \frac{1}{2}. \tag{8.3.5}$$

Exercise 8.3.1 *Show that the function $\chi(y) = e^{\pm y^2/2}y^n$ satisfies the asymptotic equation for the harmonic oscillator; namely, $\frac{\partial^2}{\partial y^2}\varphi(y) \xrightarrow{y\to\pm\infty} y^2\varphi(y)$, for any finite n.*

$$H_0(y) = 1$$
$$H_1(y) = 2y$$
$$H_2(y) = 4y^2 - 2$$
$$H_3(y) = 8y^3 - 12y$$
$$H_4(y) = 16y^4 - 48y^2 + 12$$
$$H_5(y) = 32y^5 - 160y^3 + 120y$$
$$H_6(y) = 64y^6 - 480y^4 + 720y^2 - 120$$
$$\vdots$$

Figure 8.3.1   A list of the first seven Hermit polynomials

Exercise 8.3.2   *Verify that $\varphi_0(y)$ and λ_0, as defined in Eq. (8.3.5), are indeed an eigenfunction and its corresponding eigenvalue of Eq. (8.3.3).*

To obtain the polynomials of higher degrees we can make use of the observation that if $\varphi_n(y)$ is an eigenfunction of Eq. (8.3.3), associated with the eigenvalue, λ_n, then $\left[y - \frac{\partial}{\partial y}\right]\varphi_n(y)$ is also an eigenfunction of the same equation, associated with the eigenvalue, $\lambda_n + 1$ (see Ex. 8.3.3). Using $\lambda_0 = \frac{1}{2}$ and $\varphi_0(y) \propto e^{-y^2/2}$, it follows that the eigenfunctions obtain the form $\varphi_n(y) = \left[y - \frac{\partial}{\partial y}\right]^n e^{-y^2/2}$, which is indeed of the form of Eq. (8.3.4), $\left(\left[y - \frac{\partial}{\partial y}\right]^n e^{-y^2/2} \equiv e^{-y^2/2}P_n(y)\right)$, where the polynomial degree is $n = 0,1,2,\ldots$, and the corresponding eigenvalues are $\lambda_n = n + \frac{1}{2}$. Furthermore, if $\varphi_n(y) = e^{-y^2/2}P_n(y)$ is normalized, so is $\varphi_{n+1}(y) = \frac{1}{\sqrt{2(n+1)}}\left[y - \frac{\partial}{\partial y}\right]\varphi_n(y)$ (see Ex. 8.3.4). Using Eq. (8.3.5), it follows that the normalized eigenfunctions are $\varphi_n(y) = \left(\frac{m\omega}{\hbar\pi}\right)^{1/4}\sqrt{\frac{1}{2^n n!}}[y - \frac{\partial}{\partial y}]^n e^{-y^2/2}$. Introducing the definition of the known Hermit polynomials [8.1], $H_n(y) \equiv e^{y^2/2}(y - \frac{\partial}{\partial y})^n e^{-y^2/2}$, we therefore obtain

$$\varphi_n(y) = \left(\frac{m\omega}{\hbar\pi}\right)^{1/4}\sqrt{\frac{1}{2^n n!}}e^{-y^2/2}H_n(y). \tag{8.3.6}$$

A few of the lower-order Hermit polynomials are presented in Fig. 8.3.1.

Exercise 8.3.3   *Show that if $\varphi_n(y)$ is a solution to the eigenvalue equation for a harmonic oscillator, that is, $\frac{1}{2}[y^2 - \frac{\partial^2}{\partial y^2}]\varphi_n(y) = \lambda_n \varphi_n(y)$, then:*

(a) $[y - \frac{\partial}{\partial y}]\varphi_n(y)$ is also an eigenstate solution, with the respective eigenvalue, $(\lambda_n + 1)$, and

(b) $[y + \frac{\partial}{\partial y}]\varphi_n(y)$ is also an eigenstate solution, with the respective eigenvalue, $(\lambda_n - 1)$.

You can use the commutators, $[(y^2 - \frac{\partial^2}{\partial y^2}), (y \mp \frac{\partial}{\partial y})] = \pm 2(y \mp \frac{\partial}{\partial y})$.

Exercise 8.3.4   *Let $\varphi_n(y)$ be a normalized solution to the Schrödinger equation for the harmonic oscillator: $\frac{1}{2}[y^2 - \frac{\partial^2}{\partial y^2}]\varphi_n(y) = (n + \frac{1}{2})\varphi_n(y)$. Show that:*

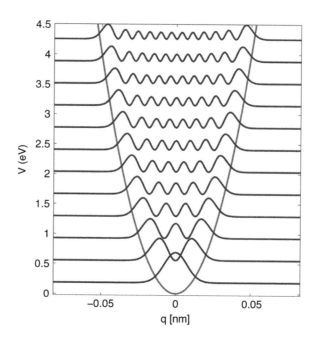

Figure 8.3.2 Energy levels and probability densities ($|\psi_n(q)|^2$) for the harmonic oscillator model. The different plots correspond to different energy levels. The plots are displaced from each other by the respective energy differences. The model parameters are $m = 1789$ a.u., and $\hbar\omega = 0.358$ eV, corresponding to the harmonic approximation for the vibration in the $H^{35}Cl$ molecule.

(a) $\varphi_{n+1}(y) = \frac{1}{\sqrt{2(n+1)}}\left[y - \frac{\partial}{\partial y}\right]\varphi_n(y)$ is also normalized.

(b) $\varphi_{n-1}(y) = \frac{1}{\sqrt{2n}}\left[y + \frac{\partial}{\partial y}\right]\varphi_n(y)$ is also normalized.

Returning to the original variables, $E_n = \hbar\omega\lambda_n$ and $q = \sqrt{\frac{\hbar}{m\omega}}y$, the energy levels and stationary wave functions of the harmonic oscillator read

$$E_n = \hbar\omega\left(n + \frac{1}{2}\right) \quad ; \quad n = 0, 1, 2, \ldots \tag{8.3.7}$$

$$\psi_n(q) = \left(\frac{m\omega}{\hbar\pi}\right)^{1/4}\sqrt{\frac{1}{n!2^n}}H_n\left[\sqrt{\frac{m\omega}{\hbar}}q\right]e^{-\frac{m\omega}{2\hbar}q^2}. \tag{8.3.8}$$

Probability densities corresponding to the lowest-energy solutions are plotted in Fig. 8.3.2. The stationary solutions of the harmonic oscillator reveal several important characteristics. Some of these are generic to solutions of the one-dimensional Schrödinger equation, as discussed in the previous chapters (Chapters 5–7), and others are unique for the harmonic oscillator model:

I. **Energy quantization**: the energy spectrum is discrete (which is general for a bound system), with a uniform spacing between nearest levels (this uniformity is specific to this model), $E_{n+1} - E_n = \hbar\omega = \hbar\sqrt{K/m}$. The level spacing decreases with increasing mass (the classical limit).

II. The number of oscillations in the stationary wave functions, namely, the number of zero values ("nodes") in the respective probability density functions, increases with increasing energy.

III. The probability densities are shown to penetrate the classically forbidden regions, for which $\sqrt{\frac{2E_n}{m\omega^2}} < |q|$ (namely, $E_n < \frac{1}{2}m\omega^2 q^2$).

IV. For small n values, the probability density is distributed nonuniformly in space and the wavelike nature of the stationary solutions is pronounced.

V. For large n values, the probability density tends to peak near the classical turning points, q_n, in which the total energy equals the potential energy, $E_n = \frac{1}{2}m\omega^2 q_n^2$. This observation is in accord with the motion of a classical oscillator, which spends longer times at the classical turning points, where the kinetic energy vanishes.

VI. *The quantum size effect*: Comparing between oscillators with different force constants, the level spacing near a given energy, E_n, decreases when the distance between the classical turning points increases (namely, when the oscillator wave function is less localized); see Ex. 8.3.5.

VII. *The "zero-point energy"*: as in other examples of bound systems that we encountered before, the energy of the oscillator at its lowest energy state does not vanish, and equals

$$E_0 = \frac{1}{2}\hbar\omega = \frac{1}{2}\hbar\sqrt{\frac{K}{m}}. \qquad (8.3.9)$$

This phenomenon is related to the uncertainty principle (to be discussed in Chapter 11), which prohibits simultaneous determination of a strictly zero momentum (zero kinetic energy) and a strict position at the minimum of the potential energy function. Let us emphasize that the zero-point energy associated with a chemical bond within a molecule has a remarkable effect on the bond mechanical stability: the lower is the zero point energy, the more energy is needed to dissociate the bond (see the next section). Since the zero-point energy decreases with increasing oscillator mass (see Eq. (8.3.9)), replacing an atom with its heavier isotope increases the bond stability and therefore slows down chemical reactions involving the relevant bond. (Notice that the bond force-constant does not change upon isotope substitution, since the electric charges are unchanged.) The effect is especially pronounced for bonds involving light atoms, such as hydrogen. Indeed, replacing hydrogen by deuterium can lead to a remarkable reduction in reaction rates and serves as a standard tool in chemical kinetics investigations.

Exercise 8.3.5 *Let us denote the classical amplitude of motion for a harmonic oscillator at energy E_n as Δ_n. Show that the level spacing near E_n, namely, $E_{n+1} - E_n$, is inversely proportional to Δ_n, namely, a larger amplitude of motion corresponds to a more dense energy spectrum (the quantum size effect).*

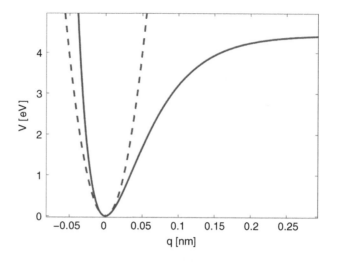

Figure 8.4.1 Solid: potential energy function for a diatomic molecule at its electronic ground state, as a function of the deviation from the classical equilibrium geometry. Dashed: the harmonic approximation for the potential energy function. The plots correspond to the $H^{35}Cl$ molecule with a dissociation energy, $V(\infty) - V(R_0) = 4.47$ eV [8.2]. The harmonic frequency is $8.66 \cdot 10^{13}$ Hz, and the equilibrium distance is, $R_0 = 0.127$ nm.

8.4 The Infrared Absorption Spectrum of Diatomic Molecules

In Section 8.2 we discussed the normal modes of a polyatomic system. Here we focus on an isolated diatomic molecule (e.g., in the gas phase), in which two atoms with masses m_1 and m_2 are connected via a chemical bond. The positions of the two atom centers in space are defined by six degrees of freedom, for example, the Cartesian coordinates of each atom (denoted by the vectors $\mathbf{R}_1$ and $\mathbf{R}_2$). It is useful to transform into linear combinations of these coordinates, separating the motion of the molecular center of mass, associated with the coordinates, $\mathbf{R}_{cm} = \frac{m_1}{m_1+m_2}\mathbf{R}_1 + \frac{m_2}{m_1+m_2}\mathbf{R}_2$, and the total mass, $M = m_1 + m_2$, from the relative motion of the atoms, associated with the coordinates, $\mathbf{R} = \mathbf{R}_1 - \mathbf{R}_2 \equiv (R, \theta, \varphi)$, and a reduced mass, $m = \frac{m_1 m_2}{m_1+m_2}$ (see Chapter 9 for a detailed discussion). Since the interaction between the atoms is due to the electrically charged electrons and atomic nuclei (see Chapter 14 for a detailed account for the nature of the chemical bond), the potential energy depends only on the absolute relative distance between the atom centers, namely, $V(\mathbf{R}) = V(R)$ where the rotation of the vector $\mathbf{R}$ induces an additional centrifugal force that depends on the distance, R. However, for discussing the vibrations, this force can be neglected to a good approximation for typical molecules and rotational energies (see Chapter 9), such that the vibrational motion can be approximated as being one-dimensional, in the coordinate R.

A typical potential energy function, $V(R)$, is plotted in Fig. 8.4.1. For a large interatomic distance, the potential energy obtains an asymptotic constant value, implying that the interatomic forces (potential energy derivatives) vanish at large distances.

Indeed, for sufficiently high vibrational energy, the chemical bond can dissociate, and the two atoms detach from each other and behave as nearly free particles. In contrast, for small interatomic distances the potential energy keeps increasing as the distance gets shorter. This manifests the increased repulsion between the nuclear charges, as well as the quantum mechanical exclusion principle (see Chapter 13), which are opposing the fusion of the two atoms into the same point in space. At some finite interatomic distance, R_0, the potential energy obtains a minimum. This is the classical "bond-length" or the "equilibrium distance" of the molecule. For small deviations around the minimum, $|R - R_0| << |R_0|$, one can replace R by a Cartesian variable (see Section 9.4 for a more detailed discussion), $q \equiv R - R_0$, and write an approximated one-dimensional quantum mechanical Hamiltonian for the molecular vibration,

$$\hat{H} = \frac{-\hbar^2}{2m} \frac{\partial^2}{\partial q^2} + V(R_0 + q). \tag{8.4.1}$$

Additionally, for small deviations from R_0 the potential energy function can be approximated to second order in a Taylor series. Setting the zero of potential energy to $V(R_0) = 0$, one obtains in the harmonic approximation,

$$V(R_0 + q) \cong \frac{1}{2} \left. \frac{\partial^2 V(R_0 + q)}{\partial q^2} \right|_{q=0} q^2 \equiv \frac{1}{2} K q^2. \tag{8.4.2}$$

Clearly, this approximation breaks down for large deviations from the equilibrium distance (see Fig. 8.4.1). Nevertheless, for a stable molecule at its lowest energy state the approximation becomes accurate. An example is given in Fig. 8.4.2 for the diatomic molecule, $H^{35}Cl$, where the energy levels and the respective stationary probability densities are plotted. The results obtained with the anharmonic potential are compared to those obtained using its quadratic (harmonic) approximation (see Eq. (8.4.2)). At the lowest quantum numbers, the energy levels and stationary wave functions are well approximated by the harmonic model; in particular, the value of the zero-point energy, $E_0 = \frac{1}{2}\hbar\omega$, and the first excitation energy, $E_1 - E_0 = \hbar\omega$.

   As it turns out, for most chemical bonds at standard experimental conditions the low-energy vibrational states are the most relevant ones. This holds due to the fact that the typical excitation energies ($\hbar\omega \approx 0.1 - 0.5\ eV$) are much larger than the thermal energy, which for standard room temperature equals $k_B T \approx 0.026\ eV$. Thermal ensembles will be dealt with only in Chapters 16–20; but let us mention already here that in thermal equilibrium the condition, $k_B T << \hbar\omega$, implies that the bond oscillator is found primarily at its ground state, where the probability of populating higher lying vibrational states decays exponentially with the excitation number. *It follows that the harmonic approximation provides a qualitatively correct description for most of the stable chemical bonds in equilibrium conditions*.

   An important consequence of the validity of the harmonic approximation for low vibrational energies is the observed infrared absorption spectrum of molecules interacting with electromagnetic field (see Section 8.1). Indeed, excitation of a harmonic oscillator from its ground state to higher vibrational levels is subject to a selection rule

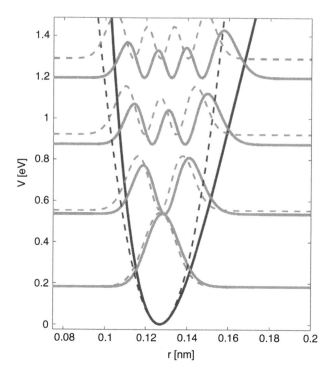

Figure 8.4.2 Energy levels and probability densities as functions of the interatomic distance in the $H^{35}Cl$ molecule. The plots are displaced from each other by the respective energy differences. The solid and dashed lines correspond to solutions obtained for the anharmonic potential (see Fig. 8.4.1) and its harmonic approximation (see Fig. 8.3.2), respectively.

derived from the symmetry of the stationary wave functions. Particularly, excitation from the $n = 0$ level via a (weak) electromagnetic field is restricted to the first energy level, $n = 1$ (see Ex. 8.5.3). This rule applies approximately also for real bond oscillators, since they are typically found at their ground vibrational state, and therefore, they are well approximated by the harmonic model. It follows that a specific bond vibration is associated with a distinctive transition energy, $E_1 - E_0 = \hbar \omega$. Using Planck's formula (Eq. (5.2.2), and see Chapter 18 for discussion of radiation absorption and emission), the transition is associated with a distinctive radiation wavelength,

$$\frac{hc}{\lambda} = E_1 - E_0 = \hbar \omega = \hbar \sqrt{\frac{K}{m}}. \tag{8.4.3}$$

First, we notice that given the typical vibrational excitation energies in chemical bonds ($\hbar \omega \approx 0.1 - 0.5$ eV), the corresponding absorption spectra are indeed in the infrared regime ($\frac{1}{\lambda} \sim 1000 - 4000\,\mathrm{cm}^{-1}$). Second, each bond is associated with a specific wavelength, determined by the force constant, and the reduced mass of the atoms. Finally, let us recall that in a polyatomic molecule, the infrared absorption spectrum (see Fig. 8.1.1) corresponds to the normal modes of the molecule rather than to local chemical bonds between neighboring atoms. The specific (global) normal mode frequencies are a unique characteristic of the molecule, which can be inferred from its infrared absorption spectrum.

8.5 Dirac's Ladder Operators

As we saw (Exs. 8.3.3, 8.3.4), the normalized nth and $(n \pm 1)$th stationary solutions of the Schrödinger equation for the harmonic oscillator are related as follows:

$$\sqrt{(n+1)}\varphi_{n+1}(y) = \frac{1}{\sqrt{2}}\left[y - \frac{\partial}{\partial y}\right]\varphi_n(y), \tag{8.5.1}$$

$$\sqrt{n}\varphi_{n-1}(y) = \frac{1}{\sqrt{2}}\left[y + \frac{\partial}{\partial y}\right]\varphi_n(y). \tag{8.5.2}$$

Following Dirac, it is useful to identify ladder operators, which map the nth solution onto the $(n \pm 1)$th ones. Since the corresponding change in the oscillator energy is $\pm\hbar\omega$ (see Eq. (8.3.7)), which corresponds to a creation or annihilation of a single energy quantum, the operators are termed the creation ($\hat{b}^\dagger$) and annihilation ($\hat{b}$) operators,

$$\hat{b}\varphi_n(y) = \sqrt{n}\varphi_{n-1}(y), \tag{8.5.3}$$

$$\hat{b}^\dagger \varphi_n(y) = \sqrt{n+1}\varphi_{n+1}(y). \tag{8.5.4}$$

The operators are defined according to Eqs. (8.5.1, 8.5.2):

$$\hat{b} \equiv \frac{1}{\sqrt{2}}\left(y + \frac{\partial}{\partial y}\right), \tag{8.5.5}$$

$$\hat{b}^\dagger \equiv \frac{1}{\sqrt{2}}\left(y - \frac{\partial}{\partial y}\right). \tag{8.5.6}$$

These operators are non-Hermitian, where $\hat{b}^\dagger$ is the Hermitian conjugate of $\hat{b}$ (see Ex. 8.5.1), and they satisfy the commutation relation (Ex. 8.5.2),

$$[\hat{b}, \hat{b}^\dagger] = 1. \tag{8.5.7}$$

Using the relations,

$$y = \frac{1}{\sqrt{2}}(\hat{b} + \hat{b}^\dagger) \tag{8.5.8}$$

$$\frac{\partial}{\partial y} = \frac{1}{\sqrt{2}}(\hat{b} - \hat{b}^\dagger), \tag{8.5.9}$$

and recalling that $y = \sqrt{\frac{m\omega}{\hbar}}q$, the oscillator position and momentum operators are related to the ladder operators as follows:

$$\hat{q} = \sqrt{\frac{\hbar}{2m\omega}}(\hat{b} + \hat{b}^\dagger) \quad ; \quad \hat{b} = \sqrt{\frac{m\omega}{2\hbar}}\hat{q} + i\sqrt{\frac{1}{2m\omega\hbar}}\hat{p} \tag{8.5.10}$$

$$\hat{p} = -i\sqrt{\frac{m\omega\hbar}{2}}(\hat{b} - \hat{b}^\dagger) \quad ; \quad \hat{b}^\dagger = \sqrt{\frac{m\omega}{2\hbar}}\hat{q} - i\sqrt{\frac{1}{2m\omega\hbar}}\hat{p}. \tag{8.5.11}$$

Finally, the Hamiltonian of the harmonic oscillator can be expressed in terms of the ladder operators:

$$\hat{H} = \frac{1}{2m}\hat{p}^2 + \frac{1}{2}m\omega^2\hat{q}^2 = \hbar\omega\left[\hat{b}^\dagger\hat{b} + \frac{1}{2}\right]. \tag{8.5.12}$$

Exercise 8.5.1   *Use the definition of a Hermitian conjugate, Eq. (4.5.2), and show that $\hat{b}^\dagger$ is the Hermitian conjugate of $\hat{b}$, using their definitions in Eqs. (8.5.5, 8.5.6).*

Exercise 8.5.2   *Use the definition of the creation and annihilation operators (Eqs. (8.5.5, 8.5.6)), and show that $[\hat{b},\hat{b}^\dagger] = 1$.*

Exercise 8.5.3   *The rate of transitions between stationary states of a system via a "weak" external perturbation is proportional to the "perturbation matrix element" squared (see Chapters 17–20). In the case of a molecular vibration interacting with an electromagnetic field, the perturbation operator is the molecular dipole, which is proportional to the interatomic distance, y, and the stationary states are approximated as the harmonic oscillator eigenfunctions. The transition rate between two stationary states, $\varphi_n(y)$ and $\varphi_{n'}(y)$, is therefore given by $k_{n\to n'} \propto \left|\int_{-\infty}^{\infty} \varphi_{n'}^*(y)y\varphi_n(y)dy\right|^2$. (a) Use Eqs. (8.5.3, 8.5.4, 8.5.8) and the orthonormality of the stationary states to show that the transition is subject to a "selection rule": $k_{n\to n'} \propto \left[n\delta_{n',n-1} + (n+1)\delta_{n',n+1}\right]$. (b) Use this result to show that the transition from the ground state is restricted to the first excited state.*

Exercise 8.5.4   *Prove Eq. (8.5.12) using Eqs. (8.5.7, 8.5.10, 8.5.11).*

Since the stationary solutions are eigenfunctions of the Hamiltonian, $\hat{H}\psi_n(q) = \lambda_n\psi_n(q)$, with the eigenvalues, $\lambda_n = \hbar\omega(n+\frac{1}{2})$ (see Eqs. (8.3.7, 8.3.8)) it follows that

$$\hat{b}^\dagger\hat{b}\psi_n(q) = n\psi_n(q) \quad ; \quad n = 0,1,2,\ldots. \tag{8.5.13}$$

The stationary solutions are therefore eigenfunctions of the operator $\hat{b}^\dagger\hat{b}$, with integer eigenvalues that count the corresponding number of energy quanta. This operator is therefore termed the number operator,

$$\hat{b}^\dagger\hat{b} = \hat{N}, \tag{8.5.14}$$

where $\hat{H} = \hbar\omega[\hat{N} + \frac{1}{2}]$. In applications of the harmonic oscillator model to quantum field theory the single excitation quantum of the oscillator is regarded as an energy-carrying particle (a boson). Examples are phonons representing solid lattice vibrations, or photons corresponding to elementary excitations of the electromagnetic radiation field. We shall return to the discussion of boson fields in Chapter 18.

Bibliography

[8.1] M. Abramowitz and A. S. Irene, eds., "Handbook of Mathematical Functions with Formulas, Graphs, and Mathematical Tables." National Bureau of Standards Applied Mathematics Series, vol. 55. (US Government Printing Office, 1964).

[8.2] J. D. D. Martin and J. W. Hepburn, "Determination of bond dissociation energies by threshold ion-pair production spectroscopy: An improved D_0 (HCl)." The Journal of Chemical Physics 109, 8139 (1998).

Two-Body Rotation and Angular Momentum

9.1 The Two-Body Problem with a Central Potential

In this chapter we discuss some universal aspects of interparticle interactions, focusing on an interacting pair of particles ("the two-body problem"). For this purpose, let us consider a closed system composed of two point-particles, associated with masses, m_1, m_2, and the position vectors, $\mathbf{r}_1 = (x_1, y_1, z_1)$ and $\mathbf{r}_2 = (x_2, y_2, z_2)$, in a Cartesian coordinates system. In the most general case, the potential energy depends on the absolute positions of the particles, namely, $V(\mathbf{r}_1, \mathbf{r}_2)$; but we shall limit the discussion to the absence of external fields, where the potential energy depends only on the relative position between the particles, namely, $V(\mathbf{r}_1, \mathbf{r}_2) = V(\mathbf{r}_2 - \mathbf{r}_1)$. It is therefore useful to introduce a new set of coordinates, corresponding to the relative position, $\mathbf{r}$, and to the center-of-mass position, $\mathbf{r}_{cm}$ (see Fig. 9.1.1):

$$\mathbf{r} \equiv \mathbf{r}_2 - \mathbf{r}_1$$
$$\mathbf{r}_{cm} \equiv \frac{m_1}{m_1+m_2}\mathbf{r}_1 + \frac{m_2}{m_1+m_2}\mathbf{r}_2 \tag{9.1.1}$$

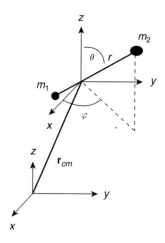

Figure 9.1.1
Representation of two particle positions using their center-of-mass coordinates ($\mathbf{r}_{cm}$) and the relative position coordinates ($\mathbf{r} = (r, \theta, \varphi)$).

Using these variables, the classical Hamiltonian for the "two-body problem" obtains the form

$$H = \frac{m_1 \dot{\mathbf{r}}_1^2}{2} + \frac{m_2 \dot{\mathbf{r}}_2^2}{2} + V(\mathbf{r}_2 - \mathbf{r}_1) = \frac{(m_1 + m_2)\dot{\mathbf{r}}_{cm}^2}{2} + \frac{m_1 m_2 \dot{\mathbf{r}}^2}{2(m_1 + m_2)} + V(\mathbf{r}). \quad (9.1.2)$$

The apparent advantage of changing the coordinates from $\mathbf{r}_1$ and $\mathbf{r}_2$ to $\mathbf{r}$ and $\mathbf{r}_{cm}$ is that the system Hamiltonian becomes separable in the latter, $H_{\mathbf{r}_{cm},\mathbf{r}} \equiv H_{\mathbf{r}_{cm}} + H_{\mathbf{r}}$. It immediately follows that the Schrödinger equation can be solved independently for the relative and for the center-of-mass motions (see Ex. 4.3.4). The quantum mechanical Hamiltonian for the center-of-mass motion is trivial (see Eq. 3.4.6), $\hat{H}_{\mathbf{r}_{cm}} = \frac{-\hbar^2}{2(m_1 + m_2)} \hat{\Delta}_{\mathbf{r}_{cm}}$, and corresponds to the motion of a single free particle, whose mass is the sum of the two particle masses. Our focus will naturally be on $\hat{H}_{\mathbf{r}}$, which accounts for the interparticle interaction,

$$\hat{H}_{\mathbf{r}} = \frac{-\hbar^2}{2\mu} \hat{\Delta}_{\mathbf{r}} + V(\hat{\mathbf{r}}). \quad (9.1.3)$$

This Hamiltonian corresponds to an effective single particle with a mass, $\mu \equiv \frac{m_1 m_2}{m_1 + m_2}$ (the "reduced mass" of the two particles), in the presence of the interaction.

We now restrict the discussion further, focusing on an important class of pair interactions, associated with "a central potential." In this case the potential energy is invariant to the relative orientation between the particles and depends only on the absolute distance between them, namely on $r \equiv |\mathbf{r}|$. It is therefore most natural to represent the three-dimensional vector of relative distance in spherical coordinates (see Fig. 2.1.1), using $\mathbf{r} = (r, \theta, \varphi)$, where

$$V(\hat{\mathbf{r}}) = V(\hat{r}). \quad (9.1.4)$$

In spherical coordinates the kinetic energy operator takes the form (see Ex. 9.1.1)

$$\frac{-\hbar^2}{2\mu}\hat{\Delta}_{\mathbf{r}} = \frac{-\hbar^2}{2\mu}\left[\frac{\partial^2}{\partial r^2} + \frac{2}{r}\frac{\partial}{\partial r}\right] + \frac{-\hbar^2}{2\mu r^2}\left[\frac{\partial^2}{\partial \theta^2} + \frac{1}{tg(\theta)}\frac{\partial}{\partial \theta} + \frac{1}{\sin^2(\theta)}\frac{\partial^2}{\partial \varphi^2}\right]. \quad (9.1.5)$$

Exercise 9.1.1 *Use the transformation from Cartesian to spherical coordinates, $x = r\sin(\theta)\cos(\varphi)$; $y = r\sin(\theta)\sin(\varphi)$; $z = r\cos(\theta)$, to derive Eq. (9.1.5) from the kinetic energy in Cartesian coordinates, $\frac{-\hbar^2}{2\mu}\hat{\Delta}_{\mathbf{r}} = \frac{-\hbar^2}{2\mu}\left(\frac{\partial^2}{\partial x^2} + \frac{\partial^2}{\partial y^2} + \frac{\partial^2}{\partial z^2}\right)$.*

The term containing derivatives with respect to r is associated with the *radial* kinetic energy. In classical mechanics, it would correspond to dynamical changes in the absolute interparticle distance. The term containing derivatives with respect to the angles is associated with a *rotational* kinetic energy; namely, in classical mechanics, it would correspond to a motion in which the orientation of the particle in space (θ and/or φ) changes in time. Recalling the kinetic energy of a classical rotor, that is, $\frac{|\mathbf{L}|^2}{2\mu r^2}$, where $\mathbf{L}$ is the vector of angular momentum, the rotational kinetic energy can be expressed in terms of the angular momentum operator $|\hat{\mathbf{L}}|^2 \equiv \hat{L}^2$:

$$\hat{L}^2 = -\hbar^2 \left(\frac{\partial^2}{\partial \theta^2} + \frac{1}{tg(\theta)}\frac{\partial}{\partial \theta} + \frac{1}{\sin^2(\theta)}\frac{\partial^2}{\partial \varphi^2}\right). \quad (9.1.6)$$

Indeed, recalling the definitions of the three components of the angular momentum vector operator in Cartesian coordinates (Eq. (3.3.2)), and transforming to spherical coordinates (Ex. 9.1.2), we obtain

$$
\hat{L}_x = i\hbar \left[\sin(\varphi) \frac{\partial}{\partial \theta} + \cot(\theta) \cos(\varphi) \frac{\partial}{\partial \varphi} \right]
$$

$$
\hat{L}_y = -i\hbar \left[\cos(\varphi) \frac{\partial}{\partial \theta} - \cot(\theta) \sin(\varphi) \frac{\partial}{\partial \varphi} \right]
$$

$$
\hat{L}_z = -i\hbar \frac{\partial}{\partial \varphi}
$$ (9.1.7)

$$
\hat{L}^2 = \hat{L}_x^2 + \hat{L}_y^2 + \hat{L}_z^2
$$

where the expression for the total angular momentum (Eq. (9.1.6)) is reproduced.

Exercise 9.1.2   *Using the transformation from Cartesian to spherical coordinates, $x = r\sin(\theta)\cos(\varphi)$; $y = r\sin(\theta)\sin(\varphi)$; $z = r\cos(\theta)$, derive (a) the explicit expressions for $\hat{L}_x, \hat{L}_y, \hat{L}_z$ in Eq. (9.1.7); (b) Eq. (9.1.6) by summing over the component of the angular momentum vector, $\hat{L}^2 = \hat{L}_x^2 + \hat{L}_y^2 + \hat{L}_z^2$.*

The information regarding the interaction within the pair of particles is contained in the solutions to the time-independent Schrödinger equation for their relative position. For any central potential, $V(r)$, this equation obtains the form (Eqs. (9.1.3–9.1.6))

$$
\left[\frac{-\hbar^2}{2\mu} \left[\frac{\partial^2}{\partial r^2} + \frac{2}{r} \frac{\partial}{\partial r} \right] + \frac{\hat{L}^2}{2\mu r^2} + V(r) \right] \psi(r, \theta, \varphi) = E \psi(r, \theta, \varphi).
$$ (9.1.9)

The equation has solutions that are products of functions of the radial and angular variables, namely, $\psi(r, \theta, \varphi) \equiv Y(\theta, \varphi)R(r)$. It can be readily verified (Ex. 9.1.3) that these solutions (based on separation of variables) must simultaneously satisfy an angular equation,

$$
\hat{L}^2 Y(\theta, \varphi) = \lambda Y(\theta, \varphi),
$$ (9.1.10)

and a radial equation,

$$
\frac{-\hbar^2}{2\mu} \left[\frac{\partial^2}{\partial r^2} + \frac{2}{r} \frac{\partial}{\partial r} \right] R(r) + [V(r) + \frac{\lambda}{2\mu r^2}] R(r) = E R(r).
$$ (9.1.11)

The angular equation (Eq. (9.1.10)) is the eigenvalue equation for the angular momentum operator, $\hat{L}^2$ (as defined in Eq. (9.1.6)). Its solutions constitute the angular part of the stationary solutions of the Schrödinger equation with any central potential and are discussed in the following section (Section 9.2). The radial equation depends on the specific interaction term, $V(r)$. In Chapter 10 we discuss in detail an important example of the radial equation, which corresponds to the hydrogen atom.

Exercise 9.1.3   *Show that any product function $\psi(r, \theta, \varphi) \equiv Y(\theta, \varphi)R(r)$, where $Y(\theta, \varphi)$ and $R(r)$ are defined solutions to Eq. (9.1.10) and Eq. (9.1.11), respectively, is a solution to Eq. (9.1.9).*

9.2 Angular Momentum Eigenstates

Focusing on the angular dependence of the solutions to the time-independent Schrödinger equation, we seek for proper solutions to the angular equation, $\hat{L}^2 Y(\theta, \varphi) = \lambda Y(\theta, \varphi)$ (Eq. (9.1.10)). We can readily identify this equation as the eigenvalue equation for the angular momentum operator, $\hat{L}^2$; namely, we seek for the eigenfunctions of this operator, and their corresponding eigenvalues, denoted here by λ.

We first notice that the operators $\hat{L}^2$ and $\hat{L}_z$ commute, that is, $[\hat{L}^2, \hat{L}_z] = 0$ (see Eq. (3.3.5)). It follows (see Ex. 6.4.1) that the eigenfunctions of $\hat{L}_z$ are also eigenfunctions of $\hat{L}^2$. Recalling the explicit form of the $\hat{L}_z$-eigenfunctions (encountered already in Chapter 5 in the context of the "particle on a ring"; see Eqs. (5.4.7–5.4.9)),

$$\hat{L}_z \frac{1}{\sqrt{2\pi}} e^{im\varphi} = m\hbar \frac{1}{\sqrt{2\pi}} e^{im\varphi} \quad ; \quad m = 0, \pm 1, \pm 2, \pm 3, \ldots, \tag{9.2.1}$$

we can conclude that $\hat{L}_z$ and $\hat{L}^2$ have a common set of eigenfunctions, each associated with a quantum number m, which we denote as $Y_{\lambda,m}(\theta, \varphi)$. The quantum number m defines the eigenvalue of $\hat{L}_z$, and λ is the respective $\hat{L}^2$-eigenvalue, namely,

$$\hat{L}_z Y_{\lambda,m}(\theta, \varphi) = m\hbar Y_{\lambda,m}(\theta, \varphi), \tag{9.2.2}$$

$$\hat{L}^2 Y_{\lambda,m}(\theta, \varphi) = \lambda Y_{\lambda,m}(\theta, \varphi). \tag{9.2.3}$$

Eqs. (9.2.1, 9.2.2) imply that $Y_{\lambda,m}(\theta, \varphi)$ must obtain the following form:

$$Y_{\lambda,m}(\theta, \varphi) \equiv \Theta_{\lambda,m}(\theta) \frac{1}{\sqrt{2\pi}} e^{im\varphi}. \tag{9.2.4}$$

Seeking for the unknown function, $\Theta_{\lambda,m}(\theta)$, we can substitute $Y_{\lambda,m}(\theta, \varphi)$ in the $\hat{L}^2$-eigenvalue equation (Eq. (9.1.10)). Using Eq. (9.1.6), we have

$$-\hbar^2 \left(\frac{\partial^2}{\partial \theta^2} + \frac{1}{tg(\theta)} \frac{\partial}{\partial \theta} + \frac{1}{\sin^2(\theta)} \frac{\partial^2}{\partial \varphi^2} \right) Y_{\lambda,m}(\theta, \varphi) = \lambda Y_{\lambda,m}(\theta, \varphi), \tag{9.2.5}$$

which yields an explicit equation for $\Theta_{\lambda,m}(\theta)$:

$$-\hbar^2 \left(\frac{\partial^2}{\partial \theta^2} + \frac{1}{tg(\theta)} \frac{\partial}{\partial \theta} + \frac{-m^2}{\sin^2(\theta)} \right) \Theta_{\lambda,m}(\theta) = \lambda \Theta_{\lambda,m}(\theta). \tag{9.2.6}$$

Notice that the quantum number m, which determines the dependence of $Y_{\lambda,m}(\theta, \varphi)$ on the azimuthal angle, φ, determines also the dependence of $Y_{\lambda,m}(\theta, \varphi)$ on the polar angle θ, through the function $\Theta_{\lambda,m}(\theta)$. This correlation between the two angular dependencies is due to the nonseparable form of the operator $\hat{L}^2$ in the two angular variables (see Eq. (9.2.5)).

To identify the proper solutions to Eq. (9.2.6), we recall that the polar angle is defined on the finite interval, $0 \leq \theta \leq \pi$, where the wave function normalization condition reads $\int_0^\pi |\Theta_{\lambda,m}(\theta)|^2 \sin(\theta)d\theta = 1$. It is convenient to define a new variable and a corresponding function:

$$\xi = \cos(\theta) \quad ; \quad \Theta_{\lambda,m}(\theta) = f_{\lambda,m}(\xi). \tag{9.2.7}$$

The normalization condition translates to $\int_{-1}^{1} |f_{\lambda,m}(\xi)|^2 d\xi = 1$, and Eq. (9.2.6) turns into the following equation for $f_{\lambda,m}(\xi)$ on the interval $-1 \leq \xi \leq 1$:

$$\left[(\xi^2 - 1)\frac{d^2}{d\xi^2} + 2\xi\frac{d}{d\xi} + \frac{m^2}{1-\xi^2} \right] f_{\lambda,m}(\xi) = \frac{\lambda}{\hbar^2} f_{\lambda,m}(\xi). \tag{9.2.8}$$

Let us consider first the case, $m = 0$. In this case [9.1] the corresponding ordinary differential equation is guaranteed to have solutions that are polynomials of a *finite* degree in ξ:

$$f_{\lambda,0}(\xi) = \sum_{k=0}^{l} a_k^{(\lambda,0)} \xi^k. \tag{9.2.9}$$

Indeed, substituting $f_{\lambda,0}(\xi)$ in Eq. (9.2.8) for $m = 0$ and shifting the series index, we obtain the following closed equation for the coefficient of the finite (lth) degree polynomials:

$$\sum_{k=0}^{l-2} (k+1)(k+2)a_{k+2}^{(\lambda,0)} \xi^k = \sum_{k=0}^{l} \left[k(k+1) - \frac{\lambda}{\hbar^2} \right] a_k^{(\lambda,0)} \xi^k. \tag{9.2.10}$$

Comparing the coefficients of ξ^k in both sides of the equation for any k, we have

$$[l(l+1) - \tfrac{\lambda}{\hbar^2}]a_l^{(\lambda,0)} = 0 \quad ; \qquad k = l$$
$$[l(l-1) - \tfrac{\lambda}{\hbar^2}]a_{l-1}^{(\lambda,0)} = 0 \quad ; \qquad k = l-1 \tag{9.2.11}$$
$$a_{k+2}^{(\lambda,0)} = \tfrac{k(k+1)-\lambda/\hbar^2}{(k+1)(k+2)} a_k^{(\lambda,0)} \quad ; \quad 0 \leq k \leq l-2.$$

Since $a_l^{(\lambda,0)} \neq 0$ for an lth-degree polynomial it follows that $\left[l(l+1) - \frac{\lambda}{\hbar^2} \right] = 0$. We can therefore identify a set of discrete values for λ, namely for the eigenvalues of $\hat{L}^2$:

$$\lambda_l = \hbar^2 l(l+1) \quad ; \quad l = 0, 1, 2, \ldots. \tag{9.2.12}$$

According to Eqs. (9.2.11, 9.2.12), the coefficient $a_{l-1}^{(\lambda,0)}$ must vanish, and the rest of the coefficients for the lth-degree polynomial can be obtained recursively. Notice that the nonzero coefficients in each polynomial correspond to either even or odd powers of ξ (see also Ex. 9.2.1).

Exercise 9.2.1 *Recalling the definition of the parity operator: $\hat{P}f(\xi) = f(-\xi)$, do the following. (a) Prove that $\hat{P}$ commutes with the differential operator on the left-hand side of Eq. (9.2.8). (b) Use the result of Ex. 6.4.1 and Eq. (9.2.12) to show that the eigenfunctions of Eq. (9.2.8) must be either even or odd functions of ξ.*

The polynomials satisfying Eq. (9.2.11) can be readily identified as the associated Legendre polynomials of zero order, $\{P_l^0(\xi)\}$ (also termed the lth-degree Legendre polynomials [8.1]). These polynomials are defined by the function, $P_l^0(\xi) = \frac{1}{2^l l!} \frac{d^l}{d\xi^l}$ $(\xi^2 - 1)^l$, and they satisfy an orthogonality relation, $\int_{-1}^{1} P_{l'}^0(\xi) P_l^0(\xi) = \frac{2}{2l+1} \delta_{l,l'}$. It follows that the normalized proper solutions to Eqs. (9.2.6, 9.2.7) with $m = 0$ are

$$f_{\lambda,0}(\xi) \equiv \sqrt{\frac{2l+1}{2}} P_l^0(\xi). \qquad (9.2.13)$$

Using Eqs. (9.2.4, 9.2.7), the corresponding $\hat{L}^2$-eigenfunctions read

$$Y_{\lambda,0}(\theta, \varphi) = \sqrt{\frac{2l+1}{4\pi}} P_l^0(\cos(\theta)). \qquad (9.2.14)$$

To obtain the eigenfunctions associated with $m \neq 0$, we can solve an ordinary differential equation (Eq. (9.2.8)) directly for $f_{\lambda,m}(\xi)$. Alternatively, we can make use of angular momentum "ladder operators," defined as

$$\hat{L}_+ \equiv \hat{L}_x + i\hat{L}_y$$
$$\hat{L}_- \equiv \hat{L}_x - i\hat{L}_y. \qquad (9.2.15)$$

These operators are Hermitian conjugates of each other, $\hat{L}_- = \hat{L}_+^\dagger$, and satisfy the following commutation relations (Ex. 9.2.2):

$$[\hat{L}_z, \hat{L}_\pm] = \pm \hbar \hat{L}_\pm, \qquad (9.2.16)$$
$$[\hat{L}^2, \hat{L}_\pm] = 0. \qquad (9.2.17)$$

Exercise 9.2.2 *The angular momentum ladder operators, $\hat{L}_+$ and $\hat{L}_-$, are defined in Eq. (9.2.15). Show the following:*

(a) $\hat{L}_+$ and $\hat{L}_-$ are Hermitian conjugates of each other.
(b) $[\hat{L}_+, \hat{L}_-] = 2\hbar \hat{L}_z$, $[\hat{L}_z, \hat{L}_\pm] = \pm \hbar \hat{L}_\pm$, and $[\hat{L}^2, \hat{L}_\pm] = 0$.
(c) $\hat{L}^2 - \hat{L}_z^2 = \frac{1}{2}(\hat{L}_+ \hat{L}_- + \hat{L}_- \hat{L}_+).$

We can readily verify that if $Y_{\lambda,m}(\theta, \varphi)$ is a simultaneous eigenfunction of $\hat{L}^2$ and $\hat{L}_z$ (associated with the eigenvalues λ and $\hbar m$, respectively), so are the functions $\hat{L}_\pm Y_{\lambda,m}(\theta, \varphi)$. Indeed, using the commutation relations in Eqs. (9.2.16, 9.2.17) we have

$$\hat{L}_z[\hat{L}_\pm Y_{\lambda,m}(\theta, \varphi)] = ([\hat{L}_z, \hat{L}_\pm] + \hat{L}_\pm \hat{L}_z) Y_{\lambda,m}(\theta, \varphi) = (m \pm 1)\hbar [\hat{L}_\pm Y_{\lambda,m}(\theta, \varphi)]. \quad (9.2.18)$$
$$\hat{L}^2[\hat{L}_\pm Y_{\lambda,m}(\theta, \varphi)] = \hat{L}_\pm \hat{L}^2 Y_{\lambda,m}(\theta, \varphi) = \lambda [\hat{L}_\pm Y_{\lambda,m}(\theta, \varphi)]. \qquad (9.2.19)$$

Consequently, we can identify

$$Y_{\lambda,m\pm1}(\theta, \varphi) \propto \hat{L}_\pm Y_{\lambda,m}(\theta, \varphi). \qquad (9.2.20)$$

It follows that ***any common eigenfunction of $\hat{L}^2$ and $\hat{L}_z$, namely any $Y_{\lambda,m}(\theta, \varphi)$, can be related to $Y_{\lambda,0}(\theta, \varphi)$ as follows:***

$$Y_{\lambda,\pm|m|}(\theta,\varphi) \propto (\hat{L}_\pm)^{|m|} Y_{\lambda,0}(\theta,\varphi), \tag{9.2.21}$$

where the $\hat{L}^2$-eigenvalues, $\{\lambda\}$, are independent on m. These eigenvalues were already obtained explicitly for the case $m = 0$ (see Eq. (9.2.12)), that is, $\lambda_l = \hbar^2 l(l+1)$. Since λ_l is uniquely defined by the quantum number l, *we shall hereafter denote the common eigenfunctions of $\hat{L}^2$ and $\hat{L}_z$ by two quantum numbers, l and m:*

$$Y_{\lambda,m}(\theta,\varphi) \equiv Y_{l,m}(\theta,\varphi). \tag{9.2.22}$$

The explicit expressions for any $Y_{l,m}(\theta,\varphi)$ can be derived, recalling that $Y_{l,0}(\theta,\varphi) \propto P_l^0(\cos(\theta))$ (see Eq. (9.2.14)) and that the ladder operators take the explicit form (see Ex. 9.2.3):

$$\hat{L}_\pm = \hbar e^{\pm i\varphi}[\pm \frac{\partial}{\partial\theta} + i\cot(\theta)\frac{\partial}{\partial\varphi}]. \tag{9.2.23}$$

Using Eq. (9.2.21), it follows that $Y_{l,m}(\theta,\varphi)$ is identified as

$$Y_{l,m}(\theta,\varphi) \propto P_l^{|m|}(\cos(\theta))e^{im\varphi}, \tag{9.2.24}$$

where $\{P_l^{|m|}(\xi)\}$ are the associated Legendre polynomials, defined as (for $m \geq 0$)

$$P_l^m(\xi) \equiv (-1)^m (1-\xi)^{\frac{m}{2}}\frac{d^m}{d\xi^m}P_l^0(\xi) \quad ; \quad m \geq 0. \tag{9.2.25}$$

Indeed, using this definition, we can readily see (Ex. 9.2.4) that

$$\begin{aligned} m \geq 0 &\quad ; \quad \hat{L}_+ P_l^m(\xi)e^{im\varphi} = \hbar P_l^{m+1}(\xi)e^{i(m+1)\varphi} \\ m \leq 0 &\quad ; \quad \hat{L}_- P_l^{|m|}(\xi)e^{im\varphi} = -\hbar P_l^{|m|+1}(\xi)e^{i(m-1)\varphi}, \end{aligned} \tag{9.2.26}$$

where it follows that the function $P_l^{|m|}(\cos(\theta))e^{im\varphi}$ is obtained for any m (up to normalization) by $|m|$ successive operations of the ladder operators on $P_l^0(\xi)$, as required for $Y_{l,m}(\theta,\varphi)$ in Eq. (9.2.21). Notice, however, that for $|m| > l$, $P_l^{|m|}(\xi)$ must vanish identically, since $P_l^0(\xi)$ is a polynomial of degree l in ξ. A proper (i.e., a nontrivial) eigenfunction $Y_{l,m}(\theta,\varphi)$ is therefore limited to

$$|m| \leq l, \tag{9.2.27}$$

where it follows that for any eigenvalue of $\hat{L}^2$ (namely, for any value of the quantum number, $l = 0,1,2,\ldots$), there are $2l+1$ degenerate eigenfuctions, $Y_{l,m}(\theta,\varphi)$, corresponding to the values $m = -l, -l+1, \ldots, 0, \ldots, l-1, l$.

Exercise 9.2.3 *(a) Derive Eq. (9.2.23) using Eqs. (9.1.7, 9.2.15). (b) Changing variable, $\xi = \cos(\theta)$, show that $\hat{L}_\pm = \hbar e^{\pm i\varphi}\left[\mp\sqrt{1-\xi^2}\frac{d}{d\xi} + i\frac{\xi}{\sqrt{1-\xi^2}}\frac{\partial}{\partial\varphi}\right]$.*

Exercise 9.2.4 *The associated Legendre polynomials are defined in Eq. (9.2.25). (a) Show that (for nonnegative m) we have $P_l^{m+1}(\xi) = (-1)\left[\sqrt{1-\xi^2}\frac{d}{d\xi} + \frac{m\xi}{\sqrt{1-\xi^2}}\right]P_l^m(\xi)$. (b) Derive the relations in Eq. (9.2.26).*

Finally, we address the eigenfunctions normalization. We start by recalling the normalized function for $m = 0$, namely, $Y_{l,0}(\theta, \varphi) = \sqrt{\frac{2l+1}{4\pi}} P_l^0(\cos(\theta))$ (see Eq. (9.2.14)). Then, we identify the following relations between the normalized functions, $Y_{l,m}(\theta, \varphi)$ and $Y_{l,m\pm1}(\theta, \varphi)$ for $|m| < l$ (see Ex. 9.2.5):

$$Y_{l,m+1}(\theta, \varphi) = \frac{1}{\hbar\sqrt{(l-m)(l+m+1)}} \hat{L}_+ Y_{l,m}(\theta, \varphi)$$

$$Y_{l,m-1}(\theta, \varphi) = \frac{1}{\hbar\sqrt{(l+m)(l-m+1)}} \hat{L}_- Y_{l,m}(\theta, \varphi). \tag{9.2.28}$$

It follows that any normalized $Y_{l,m}(\theta, \varphi)$ is obtained by $|m|$ successive operations of $\hat{L}_+$, or $\hat{L}_-$ (for positive or negative m, respectively) on $Y_{l,0}(\theta, \varphi)$. Together with Eq. (9.2.26), this yields (Ex. 9.2.6)

$$Y_{l,m}(\theta, \varphi) = (-1)^m \sqrt{\frac{2l+1}{4\pi}} \sqrt{\frac{(l-|m|)!}{(l+|m|)!}} P_l^{|m|}(\cos(\theta)) e^{im\varphi}, \tag{9.2.29}$$

where the $(-1)^m$ is conventional.

Exercise 9.2.5 *Let $Y_{l,m}(\theta, \varphi)$ and $Y_{l,m\pm1}(\theta, \varphi)$ be normalized functions: $\int\limits_0^{2\pi} d\varphi \int\limits_0^\pi \sin(\theta)$*
$d\theta Y_{l,m}^*(\theta, \varphi) Y_{l,m}(\theta, \varphi) = \int\limits_0^{2\pi} d\varphi \int\limits_0^\pi \sin(\theta) d\theta Y_{l,m\pm1}^*(\theta, \varphi) Y_{l,m\pm1}(\theta, \varphi) = 1$, and let $Y_{l,m\pm1}$
$(\theta, \varphi) = c_\pm \hat{L}_\pm Y_{l,m}(\theta, \varphi)$. Use the results of Ex. 9.2.2 to show that (for $|m| < l$) $c_\pm = \frac{1}{\hbar\sqrt{(l\pm m+1)(l\mp m)}}$.

Exercise 9.2.6 *Use Eqs. (9.2.26, 9.2.28) and $Y_{l,0}(\theta, \varphi) = \sqrt{\frac{2l+1}{4\pi}} P_l^0(\cos(\theta))$ to derive Eq. (9.2.29).*

The normalized functions, $Y_{l,m}(\theta, \varphi)$, are known in the literature as the spherical harmonics [9.2]. Some examples, corresponding to the angular quantum numbers, $l = 0, 1, 2$, are presented explicitly in Fig. 9.2.1. A summary of important properties of the spherical harmonics follows:

I. The set $\{Y_{l,m}(\theta, \varphi)\}$ consists of proper simultaneous solutions to the following eigenvalue equations:

$$\hat{L}^2 Y_{l,m}(\theta, \varphi) = \hbar^2 l(l+1) Y_{l,m}(\theta, \varphi) \quad ; \quad l = 0, 1, 2, \ldots, \tag{9.2.30}$$

$$\hat{L}_z Y_{l,m}(\theta, \varphi) = m\hbar Y_{l,m}(\theta, \varphi) \quad ; \quad |m| \leq l. \tag{9.2.31}$$

II. The set $\{Y_{l,m}(\theta, \varphi)\}$ is an orthonormal set of angular functions in the spherical coordinate space ($0 \leq \theta \leq \pi$, and $0 \leq \varphi < 2\pi$):

$$\int\limits_0^{2\pi} d\varphi \int\limits_0^\pi \sin(\theta) d\theta Y_{l,m}^*(\theta, \varphi) Y_{l',m'}(\theta, \varphi) = \delta_{l,l'} \delta_{m,m'}. \tag{9.2.32}$$

$$Y_{0,0}(\theta,\varphi) = \sqrt{\frac{1}{4\pi}}$$

$$Y_{1,0}(\theta,\varphi) = \sqrt{\frac{3}{4\pi}}\cos(\theta)$$

$$Y_{1,1}(\theta,\varphi) = \sqrt{\frac{3}{8\pi}}\sin(\theta)e^{i\varphi}$$

$$Y_{1,-1}(\theta,\varphi) = -\sqrt{\frac{3}{8\pi}}\sin(\theta)e^{-i\varphi}$$

$$Y_{2,0}(\theta,\varphi) = \sqrt{\frac{5}{16\pi}}(3\cos^2(\theta)-1)$$

$$Y_{2,1}(\theta,\varphi) = \sqrt{\frac{5}{24\pi}}\,3\cos(\theta)\sin(\theta)e^{i\varphi}$$

$$Y_{2,-1}(\theta,\varphi) = \sqrt{\frac{5}{24\pi}}\,3\cos(\theta)\sin(\theta)e^{-i\varphi}$$

$$Y_{2,2}(\theta,\varphi) = \sqrt{\frac{5}{96\pi}}\,3\sin^2(\theta)e^{2i\varphi}$$

$$Y_{2,-2}(\theta,\varphi) = \sqrt{\frac{5}{96\pi}}\,3\sin^2(\theta)e^{-2i\varphi}$$

Figure 9.2.1   A list of the spherical harmonics associated with the angular quantum numbers, $l = 0, 1, 2$.

III. For any quantum number, l, there is a set of $2l+1$ degenerate orthogonal functions, corresponding to $m = -l,\ldots,0,\ldots,l$. In atomic physics and chemistry (see Chapter 10), it is customary to replace the set of spherical harmonics associated with a given l by a set of real-valued linear combinations of spherical harmonics (see Ex. 4.6.3). For example, the set $\{Y_{1,-1}(\theta,\varphi), Y_{1,0}(\theta,\varphi), Y_{1,1}(\theta,\varphi)\}$ is replaced by

$$Y_{1,0}(\theta,\varphi),$$
$$\frac{1}{\sqrt{2}}[Y_{1,1}(\theta,\varphi)+Y_{1,-1}(\theta,\varphi)],$$
$$\frac{1}{\sqrt{2}}[Y_{1,1}(\theta,\varphi)-Y_{1,-1}(\theta,\varphi)],$$

and the set $\{Y_{2,-2}(\theta,\varphi), Y_{2,-1}(\theta,\varphi), Y_{2,0}(\theta,\varphi), Y_{2,1}(\theta,\varphi), Y_{2,2}(\theta,\varphi)\}$ is replaced by

$$Y_{2,0}(\theta,\varphi)$$
$$\frac{1}{\sqrt{2}}[Y_{2,1}(\theta,\varphi)+Y_{2,-1}(\theta,\varphi)]$$
$$\frac{1}{\sqrt{2}}[Y_{2,1}(\theta,\varphi)-Y_{2,-1}(\theta,\varphi)]$$
$$\frac{1}{\sqrt{2}}[Y_{2,2}(\theta,\varphi)+Y_{2,-2}(\theta,\varphi)]$$
$$\frac{1}{\sqrt{2}}[Y_{2,2}(\theta,\varphi)-Y_{2,-2}(\theta,\varphi)].$$

IV. *The spherical harmonics (or linear combinations of degenerate spherical harmonics) constitute the angular part of the stationary states of the Schrödinger equation in any central potential problem* (see Eq. (9.1.9)).

9.3 The Rigid Rotor Model and Rotational Spectrum of Diatomic Molecules

When the radial motion is highly constrained or suppressed, a model of a "rigid rotor" becomes instrumental. Referring to the time-independent Schrödinger equation for a general central potential (Eq. (9.1.9)), the rigid rotor corresponds to the case where the radial distance obtains a fixed value, $r = r_0$. The Schrödinger equation therefore obtains the form

$$\frac{\hat{L}^2}{2\mu r_0^2} \psi(r_0, \theta, \varphi) = E\psi(r_0, \theta, \varphi), \tag{9.3.1}$$

where the Hamiltonian reduces to the angular kinetic energy operator, with the relative distance coordinate, r, replaced by a parameter, r_0:

$$\hat{H} = \frac{\hat{L}^2}{2\mu r_0^2}. \tag{9.3.2}$$

Since the Hamiltonian is proportional to the angular momentum operator, $\hat{L}^2$, the stationary solutions to the Schrödinger equation coincide with the $\hat{L}^2$-eigenfunctions (see Eqs. (9.2.29–9.2.31)), and the corresponding energy levels are proportional to the $\hat{L}^2$-eigenvalues, namely,

$$\hat{H}Y_{l,m}(\theta, \varphi) = E_l Y_{l,m}(\theta, \varphi) \tag{9.3.3}$$

$$E_l = \frac{\hbar^2 l(l+1)}{2\mu r_0^2} \quad ; \quad l = 0, 1, 2, \ldots. \tag{9.3.4}$$

The rigid-rotor model can be successfully applied in the analysis of the relative rotational motion of atoms in molecules. Particularly, let us consider here a diatomic molecule, as discussed in Section 8.4, where two atoms with masses, m_1 and m_2, are connected via a chemical bond. The relative motion of the two atoms is associated with the relative position vector, $\mathbf{r} = \mathbf{r}_1 - \mathbf{r}_2 \equiv (r, \theta, \varphi)$, and a reduced mass, $m = \frac{m_1 m_2}{m_1 + m_2}$ (see Fig. 9.1.1). Since the interaction between the atoms is due to the electrically charged electrons and atomic nuclei (see Chapter 14), the potential energy depends only on the absolute relative distance between the atom centers, namely, $V(\mathbf{r}) = V(r)$. Consequently, the relative motion is associated with a Schrödinger equation for a central potential (Eq. (9.1.9)). In Section 8.4, we focused on the radial part, associated with the interatomic vibrations, where the coupling between the radial and the angular motions via the centrifugal term, $\frac{\hat{L}^2}{2\mu r^2}$, was neglected. Indeed, for typical rotational and vibrational energies in diatomic molecules, rotations can be ignored to a good approximation for the purpose of describing the vibrations. Similarly, for an approximate qualitative discussion of the rotational motion, the vibrational motion can be ignored. In classical mechanical terms, the justification for the separate treatment is derived from a timescale separation, where rotational motion is typically much slower than the vibrations in molecules. This implies that the rotation is nearly frozen on the vibration timescale, while the fast vibrational motion can be averaged in time on the rotational timescale. In quantum mechanical terms (see an extended discussion of quantum dynamics in Chapter 15), the timescale separation corresponds to the fact that a slower (rotational, in this case) motion is associated with a denser spectrum of energy levels. Indeed, rotational excitation energies are typically much smaller than vibrational excitation energies, such that the rotational dynamics can be well approximated by assuming a constant vibrational quantum number. (This argument is often referred to as "an adiabatic approximation," to be discussed in more detail in Chapter 14 in the context of separating the treatment of the "fast" electrons from the "slow" nuclei in molecules.) The timescale separation argument implies that the diatomic molecule can be regarded approximately as a rigid rotor, with a constant moment of inertia, $I = \mu r_0^2$, set by the reduced mass and by the typical (e.g., averaged) interatomic distance, associated with its vibrational ground state (see Section 8.4). In Section 9.4, the validity of this approximation is discussed in quantitative terms.

The relevance of the rigid rotor model to the rotational energy spectrum of a diatomic molecule can be revealed in experiments when the molecule interacts with an electromagnetic radiation. Specifically, molecules having a finite permanent electric dipole moment (namely, molecules composed of two different atoms) change their energy upon interaction with the external electric field. This change is given as, $-\mathbf{\mu} \cdot \mathbf{E}$, where $\mathbf{\mu}$ is the molecular dipole, and $\mathbf{E}$ is the electric field vector. Since the projection $\mathbf{\mu} \cdot \mathbf{E} = |\mu||E|\cos(\chi)$ depends on an angle (χ) between the molecular dipole (directed along the axis connecting the two atomic centers) and the electric field vector, the energy of the molecule in the field changes upon rotation. The molecular rotational energy can therefore increase (via radiation absorption) or decrease (via radiation emission) in the presence of the field.

Transitions between stationary states, induced by a weak interaction with the field, are subject to selection rules derived from the properties of the respective stationary wave functions (see Chapter 18 for a detailed discussion of field induced transitions). For a perfectly rigid rotor, the stationary states are the spherical harmonics (Eq. (9.3.3)), and the corresponding selection rules imply that transitions are restricted to changes of the angular momentum quantum number, l, by ± 1 (see Ex. 9.3.1). Focusing on absorption of rotational energy (namely, $l \rightarrow l+1$) and using Planck's formula (Eq. (5.2.2)), it follows that each transition is associated with a distinctive radiation wavelength, $\lambda_l^{(ab)}$, given by $\frac{hc}{\lambda_l^{(ab)}} = E_{l+1} - E_l$. Using Eq. (9.3.4) for the rotational energy levels and defining a rotational constant, B, the set of absorption wavelengths for any specific molecule is given as

$$\frac{1}{\lambda_l} = 2B(l+1) \quad ; \quad B \equiv \frac{\hbar}{4\pi c \mu r_0^2} \quad ; \quad l = 0,1,2,\ldots. \tag{9.3.5}$$

As we can readily perceive (see also Fig. 9.4.2), the inverse wavelengths are evenly spaced, $\frac{1}{\lambda} \in 2B, 4B, 6B, \ldots.,$. Notice that while the sequence is formally infinite, in practice, transitions corresponding to large l values are not observed in experiments at typical thermal equilibrium conditions (see, e.g., [9.3]). The main reason is that rotational energies that exceed the thermal energy $E_l \gg K_B T$ are rarely found at thermal equilibrium (see Chapter 16 for a detailed discussion of thermal ensembles), and rare events are missed at a finite detection resolution.

Exercise 9.3.1 *The rate of transitions between stationary states of a system via a "weak" external perturbation is proportional to the "perturbation matrix element" squared (see Chapters 17–20). In the case of rotation of a diatomic molecule interacting with an electromagnetic field, the perturbation operator is the projection of the molecular dipole on the electric field vector ($-\mu \cdot \mathbf{E}$). When the direction of the electric field vector is fixed in the lab reference frame, it is convenient to identify it with the z-axis direction of the molecular reference frame. The perturbation operator is then proportional to $\cos(\theta)$, where θ is the polar angle of the spherical coordinate system. The rate of field-induced transitions between two rotational eigenfunctions, $Y_{l,m}(\theta,\varphi)$ and $Y_{l',m'}(\theta,\varphi)$, reads $k_{l,m \to l',m'} \propto$*

$$\left| \int_0^{2\pi} d\varphi \int_0^{\pi} \sin(\theta) d\theta\, Y_{l',m'}^*(\theta,\varphi) \cos(\theta) Y_{l,m}(\theta,\varphi) \right|^2 .$$ *Use the relation of the spherical harmonics to the associated Legendre polynomials (Eq. (9.2.29)), a known recursive relation for associated Legendre polynomials, $(2l+1)\cos(\theta)P_l^m(\cos(\theta)) = (l-m+1)P_{l+1}^m(\cos(\theta)) + (l+m)P_{l-1}^m(\cos(\theta))$ [9.4], and the orthonormality of the spherical harmonics, (Eq. (9.2.32)) to derive the "selection rule" for rotational transitions induced by a (weak) electromagnetic field, $k_{l,m \to l',m'} \propto \delta_{m',m} \delta_{l',l \pm 1}$.*

The interpretation of the rotational absorption spectra of diatonic molecules in terms of the rigid rotor model immediately leads to several practical conclusions. First, the typical absorption wave lengths are within the microwave regime (defined as $\frac{1}{\lambda} \sim 0.01 - 10$ cm^{-1}). Second, knowing the atomic masses, one can extract the interatomic distance, r_0, with remarkable precision. Let us consider, for example, the molecule $H^{35}Cl$ (discussed in Section 8.4), where the measured rotational constant corresponds to $B \cong 10.6$ cm^{-1}. Considering the reduced mass, $\mu = 1.6 \times 10^{-27}$ kg, this readily yields the interatomic distance, $r_0 \equiv \sqrt{\frac{\hbar}{4\pi c \mu B}} \approx 0.128$ nm. In other cases, the interatomic distance, r_0, is already known, and the reduced mass can be deduced by measuring the rotational constant. Particularly, while different isotopes of the same chemical element are associated with different masses, they have the same electronic structure. Therefore, diatomic molecules containing different isotopes experience the same interatomic forces and have nearly the same interatomic distance r_0 (strictly the same, within the harmonic approximation for the vibrations), but a different reduced mass, μ. Consequently, diatomic molecules containing different isotopes have different rotational constants. For example, the rotational constant of $H^{35}Cl$ is larger by 0.15% than that of $H^{37}Cl$, and by 95% than that of $D^{35}Cl$ (see Ex. 9.3.2). These differences are readily detectable in experiments.

Exercise 9.3.2 *Use the definition of the rotational constant to show that the rotational constant of $H^{35}Cl$ is larger by 0.15% than that of $H^{37}Cl$, and by 95% than that of $D^{35}Cl$.*

9.4 Beyond the Rigid Rotor Model: Vibration–Rotation Coupling

In this section we address the coupling between vibration and rotation, beyond the rigid rotor approximation. Specifically, we shall analyze the manifestation of this coupling in the spectrum of diatomic molecules, and the validity of the timescale separation argument, which justifies the separate treatment of vibration and rotation. Our discussion would be restricted to small deviations of the interatomic distance from r_0, and to the lowest vibrational state, where the potential energy can be approximated within the harmonic approximation (see Section 8.4):

$$V(r) = \frac{1}{2}\mu\omega^2(r - r_0)^2. \tag{9.4.1}$$

The stationary Schrödinger equation for the coupled vibration and rotation degrees of freedom of a diatomic molecule therefore reads

$$\left[\frac{-\hbar^2}{2\mu}\left[\frac{\partial^2}{\partial r^2} + \frac{2}{r}\frac{\partial}{\partial r}\right] + \frac{\hat{L}^2}{2\mu r^2} + \frac{1}{2}\mu\omega^2(r - r_0)^2\right]\psi(r,\theta,\varphi) = E\psi(r,\theta,\varphi). \tag{9.4.2}$$

As for any central potential problem, the corresponding stationary solutions are of the form $\psi(r,\theta,\varphi) = Y_{l,m}(\theta,\varphi)R(r)$, where $Y_{l,m}(\theta,\varphi)$ are the angular momentum eigenfunctions (see Eqs. (9.1.9–9.1.11)). Substitution in Eq. (9.4.2) leads to the corresponding radial equation for the function $R(r)$,

$$\frac{-\hbar^2}{2\mu}\left[\frac{\partial^2}{\partial r^2} + \frac{2}{r}\frac{\partial}{\partial r}\right]R(r) + \left[\frac{\mu\omega^2}{2}(r - r_0)^2 + \frac{\hbar^2 l(l+1)}{2\mu r^2}\right]R(r) = ER(r), \tag{9.4.3}$$

where an l-dependent centrifugal potential energy term, $\frac{\hbar^2 l(l+1)}{2\mu r^2}$, is added to the radial potential energy. The equation can be simplified by defining $\chi(r) \equiv rR(r)$. Substitution in Eq. (9.4.3) yields (see Ex. 9.4.1)

$$-\frac{\hbar^2}{2\mu}\frac{\partial^2}{\partial r^2}\chi(r) + \left[\frac{\mu\omega^2}{2}(r - r_0)^2 + \frac{\hbar^2 l(l+1)}{2\mu r^2}\right]\chi(r) = E\chi(r). \tag{9.4.4}$$

Exercise 9.4.1 *Use the definition $\chi(r) \equiv rR(r)$, and derive Eq. (9.4.4) from Eq. (9.4.3).*

Focusing on the deviations from r_0, we define a vibration coordinate, $q \equiv r - r_0$, and change variables, $\chi(r) = \phi(q)$. The equation for $\phi(q)$ (in the range, $q \geq -r_0$) reads

$$-\frac{\hbar^2}{2\mu}\frac{\partial^2}{\partial q^2}\phi(q) + \left[\frac{\mu\omega^2}{2}q^2 + \frac{\hbar^2 l(l+1)}{2\mu(r_0 + q)^2}\right]\phi(q) = E\phi(q), \tag{9.4.5}$$

where the vibrational coordinate is coupled to the angular momentum, $\hbar^2 l(l+1)$, via the q-dependent moment of inertia. It is instructive to associate the centrifugal potential energy with a "rotational frequency." Motivated by the classical relation between the rotation frequency, the angular momentum, and the moment of inertia, $\omega_{rot} = \frac{L}{I}$, we define

$$\omega_l \equiv \frac{\sqrt{\hbar^2 l(l+1)}}{\mu r_0^2}, \tag{9.4.6}$$

and rewrite Eq. (9.4.5) as follows:

$$-\frac{\hbar^2}{2\mu}\frac{\partial^2}{\partial q^2}\phi(q) + \left[\frac{\mu\omega^2}{2}q^2 + \frac{\mu\omega_l^2}{2}r_0^2\frac{1}{(1+q/r_0)^2}\right]\phi(q) = E\phi(q). \tag{9.4.7}$$

This equation can be solved analytically in a most important parameter regime, which corresponds to small vibrational amplitudes. Indeed, in typical diatomic molecules, the distribution of interatomic distances is much narrower than the average bond length, namely,

$$|q/r_0| \ll 1. \tag{9.4.8}$$

In this regime it is useful to expand, $\frac{1}{(1+q/r_0)^2} = (1 - 2q/r_0 + 3(q/r_0)^2 + \cdots)$, and to keep only the linear term in q. Notice, however, that terms of order $o(q^2)$ in the centrifugal potential can be neglected in Eq. (9.4.7) only if they are small with respect to the harmonic potential energy, $\frac{\mu\omega^2}{2}q^2$, which amounts to

$$\frac{\omega_l}{\omega} \ll 1. \tag{9.4.9}$$

This condition limits the validity of the linear approximation to small rotational quantum numbers, for which the rotational frequency is small with respect to the bond harmonic frequency. Under the restrictions of Eqs. (9.4.8, 9.4.9), the equation for the vibrational functions becomes a Schrödinger equation for a linearly displaced harmonic oscillator,

$$-\frac{\hbar^2}{2\mu}\frac{\partial^2}{\partial q^2}\phi(q) + \left[\frac{\mu\omega^2}{2}q^2 - \mu\omega_l^2 r_0 q + \frac{\mu\omega_l^2 r_0^2}{2}\right]\phi(q) = E\phi(q). \tag{9.4.10}$$

Defining a displaced coordinate, q_l, and a corresponding energy, ε_l, Eq. (9.4.10) transforms into (see Ex. 9.4.2)

$$-\frac{\hbar^2}{2\mu}\frac{\partial^2}{\partial q_l^2}\tilde{\phi}(q_l) + \frac{\mu\omega^2}{2}q_l^2\tilde{\phi}(q_l) = (E - \varepsilon_l)\tilde{\phi}(q_l), \tag{9.4.11}$$

where $\tilde{\phi}(q_l) = \phi(q)$, and

$$q_l = q - r_0\frac{\omega_l^2}{\omega^2} = r - r_0\left(1 + \frac{\omega_l^2}{\omega^2}\right), \tag{9.4.12}$$

$$\varepsilon_l = \frac{\mu\omega_l^2 r_0^2}{2}\left(1 - \frac{\omega_l^2}{\omega^2}\right). \tag{9.4.13}$$

Exercise 9.4.2 *Use Eqs. (9.4.12–9.4.13) to derive Eq. (9.4.11) from Eq. (9.4.10).*

Eq. (9.4.11) is the time-independent Schrödinger equation for a harmonic oscillator (Eq. (8.3.1)). For the standard boundary conditions, $\tilde{\phi}(q_l) \xrightarrow{q_l \to \pm \infty} 0$, the exact eigenvalues and eigenfunctions are given by Eqs. (8.3.7, 8.3.8). Notice that the variable q_l is related to the radial variable, $r \geq 0$ (Eq. (9.4.12)), and therefore obtains physically meaningful values only for $q_l \geq -r_0(1 + \frac{\omega_l^2}{\omega^2})$. From a mathematical point of view, however, extending q_l to the range, $-\infty \leq q_l \leq \infty$, has a negligible effect on eigenfunctions for which $|\tilde{\phi}(q_l)|^2$ is confined in a narrow range, $|q_l| << r_0(1 + \frac{\omega_l^2}{\omega^2})$. Since the latter condition holds within our restriction to small amplitude vibrations (see Eqs. (9.4.8, 9.4.12)), the relevant vibration eigenfunctions are well approximated by the standard harmonic oscillator model. Using these solutions in the full Schrödinger equation (Eq. (9.4.2)) for the nonrigid diatomic molecule (for $\omega_l \ll \omega$), the eigenvalues are given as

$$E_{n,l} = \hbar\omega \left(n + \frac{1}{2}\right) + \frac{\hbar^2 l(l+1)}{2\mu r_0^2}\left(1 - \frac{\omega_l^2}{\omega^2}\right). \tag{9.4.14}$$

The corresponding eigenfunctions are, $\psi_{n,l,m}(r,\theta,\varphi) = Y_{l,m}(\theta,\varphi)R_{n,l}(r)$, where the radial functions, confined in a range $|r - r_0| \ll r_0$, obtain the form (Ex. 9.4.3)

$$R_{n,l}(r) = \frac{1}{r}\left(\frac{\mu\omega}{\hbar\pi}\right)^{1/4}\sqrt{\frac{1}{n!2^n}}H_n\left[\sqrt{\frac{\mu\omega}{\hbar}}\left(r - r_0(1 + \frac{\omega_l^2}{\omega^2})\right)\right]e^{-\frac{\mu\omega}{2\hbar}(r-r_0(1+\frac{\omega_l^2}{\omega^2}))^2}. \tag{9.4.15}$$

The coupling between vibration and rotation is apparent from the dependence of the radial functions (Eq. (9.4.15)) on the angular quantum number. In particular, the typical interatomic distance is shown to shift to larger values with increasing angular momentum, $r_0 \to r_0(1 + \frac{\omega_l^2}{\omega^2})$ (see Ex. 9.4.4), which means that *the interatomic bond distance "stretches" with increasing angular momentum*. This is indeed expected, owing to the centrifugal force.

Exercise 9.4.3   *Use the solutions of the Schrödinger equation for the harmonic oscillator (Eqs. (8.3.1, 8.3.7, 8.3.8)), and change variables, to obtain Eqs. (9.4.14, 9.4.15).*

Exercise 9.4.4   *A nonrigid diatomic molecule is associated with a (normalized) stationary state, $\psi_{n,l,m}(r,\theta,\varphi) = Y_{l,m}(\theta,\varphi)R_{n,l}(r)$, where $R_{n,l}(r)$ is given by Eq. (9.4.15). Show that the average interatomic distance becomes larger with increasing angular momentum quantum number, $<r_{n,l}> = \int_0^\pi d\theta \int_0^{2\pi} d\varphi \int_0^\infty dr \cdot r \cdot r^2 \sin(\theta)|\psi_{n,l,m}(r,\theta,\varphi)|^2 = r_0 (1 + \frac{\omega_l^2}{\omega^2})$. (Notice that when the probability density is confined to a range where $|r| << r_0(1 + \frac{\omega_l^2}{\omega^2})$, the boundaries of the radial integral can be changed: $\int_0^\infty dr \to \int_{-\infty}^\infty dr$.)*

Another remarkable effect of the vibration–rotation coupling is on the rotational absorption spectrum. From Eq. (9.4.14) it follows that for $l > 0$, the energy level for the nonrigid rotor is lower than the energy level for a rigid rotor, corresponding to the same quantum number. Moreover, the decrease in energy is larger for a larger l (see Fig. 9.4.1). This means that the energy spectrum of a nonrigid rotor is denser than

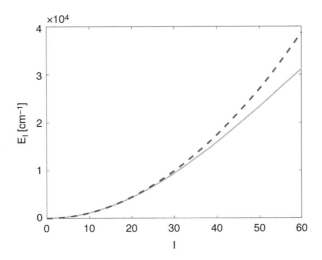

Figure 9.4.1 Calculated rotational energy ($E_l = E_{0,l} - \hbar\omega/2$) as a function of the rotation quantum number, l. The solid and dashed lines correspond to a nonrigid and rigid rotor models, respectively. The parameters correspond to the diatomic molecule $H^{35}Cl$ ($\mu = 1.6 \times 10^{-27}$ kg, $r_0 = 0.128$ nm, $\omega = 5.4412 \times 10^{14}$ rad/sec. The maximal quantum number in the plot, $l = 60$, corresponds to $\omega_l^2/\omega^2 = 0.2$).

that of a rigid rotor, which is closely related to the effect of "bond stretching" (yet another manifestation of a "quantum size effect"). Indeed, a larger radius means a larger moment of inertia, and therefore a lower rotational excitation energy. In quantitative terms, using Eq. (9.4.14) for the energy levels, the absorption wavelengths change from that of a rigid rotor, Eq. (9.3.5), into (see Fig. 9.4.2)

$$\frac{1}{\lambda_l} = 2B(1 - \frac{\omega_l^2}{\omega^2})(l+1) \quad ; \quad l = 0, 1, 2, \dots. \tag{9.4.16}$$

It follows that unlike for a rigid rotor, where the peaks are evenly spaced, for a nonrigid rotor the distance between two successive absorption peaks decreases with increasing l (see Ex. 9.4.5 and Fig. 9.4.2),

$$\frac{1}{\lambda_{l+1}} - \frac{1}{\lambda_l} = 2B(1 - \frac{\omega_l^2 + 2\omega_{l+1}^2}{\omega^2}). \tag{9.4.17}$$

Exercise 9.4.5 *Use the definition of the rotational frequency (Eq. (9.4.6)) and Eq. (9.4.16) to derive Eq. (9.4.17).*

The manifestations of the vibration–rotation coupling are shown to depend on the ratio between the rotational frequency, ω_l (defined in Eq. (9.4.6)) and the vibrational frequency, ω. As $\frac{\omega_l^2}{\omega^2} \to 0$, these manifestations are expected to vanish. First, the rotational energy spectrum coincides with that of a rigid rotor, $\frac{\hbar^2 l(l+1)}{2\mu r_0^2}(1 - \frac{\omega_l^2}{\omega^2}) \to \frac{\hbar^2 l(l+1)}{2\mu r_0^2}$

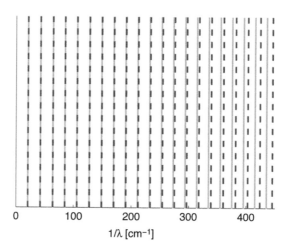

Figure 9.4.2 Calculated rotational absorption wavelengths for the diatomic molecule, $H^{35}Cl$ (see Fig. 9.4.1 for parameters). The solid and dashed lines correspond to nonrigid and rigid rotor models, respectively.

(see Eq. (9.4.14). Second, $R_{n,l}(r)$ becomes independent of the angular quantum number, as $r_0(1 + \frac{\omega_l^2}{\omega^2}) \to r_0$ (see Eq. (9.4.15)), which justifies a separable treatment of the vibration and rotation degrees of freedom, as in the rigid rotor model (Section 9.3).

Notice that in the classical mechanical sense, $\frac{\omega_l^2}{\omega^2} \ll 1$ corresponds to a timescale separation between the (slow) rotation and the (fast) vibration, where the rotation time period $(2\pi/\omega_l)$ exceeds by far the vibration time period $(2\pi/\omega)$. Using the definitions of the rotational frequency (Eq. (9.4.6)), this condition corresponds to $\frac{\hbar\sqrt{l(l+1)}}{\mu r_0^2} \ll \omega$. Recalling the definition of the typical vibrational absorption wave number, $\frac{1}{\lambda_{vib}} = \frac{\omega}{2\pi c}$ (see Eq. (8.4.3)), and the definition of a rotational constant, $B \equiv \frac{\hbar}{4\pi c \mu r_0^2}$ (Eq. (9.3.5)), this translates to

$$2B\sqrt{l(l+1)} \ll \frac{1}{\lambda_{vib}}. \tag{9.4.18}$$

Considering that for typical diatomic molecules the values of $\frac{1}{\lambda_{vib}}$ are in the range of $\sim 1000 - 4000\,\mathrm{cm}^{-1}$ (see Section 8.4), whereas the values for B are in the range of $\sim 1 - 10\,\mathrm{cm}^{-1}$ (see Section 9.3), the limit in Eq. (9.4.18) is a reasonable approximation as long as the rotational quantum number l is not too large. This is also apparent in Figs. 9.4.1 and 9.4.2. As already mentioned, the observed values of l are typically limited by the thermal energy, such that the condition (Eq. (9.4.18)) is relevant for most diatomic molecules in typical thermal conditions. While the rigid rotor model provides a good approximation, effects of vibration–rotation coupling are also apparent in experiments. These include, for example, the deviations from equally spaced absorption spectral lines, as we have discussed, as well as other effects associated with the anharmonicity of the interatomic potential energy function in real molecules. These are left for further reading elsewhere.

Bibliography

[9.1] V. I. Arnold, "Ordinary Differential Equations" (Springer-Verlag, 2006).

[9.2] G. B. Arfken, H. J. Weber and F. E. Harris "Mathematical Methods for Physicists" (Elsevier, 2013).

[9.3] J. M. Hollas, "Modern Spectroscopy," 4th ed. (John Wiley & Sons, 2004).

[9.4] E. W. Weisstein, "Associated Legendre Polynomial." *From MathWorld –* A Wolfram Web Resource. Accessed at https://mathworld.wolfram.com/ AssociatedLegendrePolynomial.html

10 The Hydrogen-Like Atom

10.1 Rydberg's Formula

Atoms, as originally conceived by the atomistic philosophers in ancient Greece, are the "indivisible" building blocks of any known material. At the center of each atom there is a positively charged nucleus, composed of protons and neutrons, where the number of protons in the nucleus defines the chemical identity of each element. The nucleus is surrounded by electrons, whose number (for an electrically neutral atom) equals the number of protons. Recalling that the typical length scale (e.g., an effective radius) of a single atom is of the order of 1 Å (0.1 nanometer), atomic resolution is necessary for a full description of the structure and properties of matter on the nanoscale. Indeed, a detailed account of the internal structure of atoms is essential for understanding interatomic forces as well as binding between atoms to form molecules and extended lattices, which further influence structure and reactivity of materials on the macroscopic length scale. In this chapter we start our discussion of the internal structure of atoms, and particularly of the binding of electrons to the nucleus.

It turns out that the way electrons are bounded to atomic nuclei is strongly influenced by the laws of quantum mechanics. ***The mere existence of an atom as a stable entity and of its most basic properties, such as size (volume) and internal particle arrangement, contradicts the laws of classical mechanics.*** In fact, the accumulation of experimental evidence with respect to the internal structure of atoms was a cornerstone for formulating quantum mechanics at the beginning of the twentieth century. Particularly, let us focus here on a remarkable observation associated with the emission of electromagnetic radiation by atoms in a gas tube, following their electronic excitation by an electric discharge. For hydrogen, for example, the emitted radiation appears in a set of specific wavelengths (see Fig. 10.1.1), which fits an empirical formula proposed by Johannes Rydberg [10.1]:

$$\frac{hc}{\lambda} = R_H \left(\frac{1}{n_1^2} - \frac{1}{n_2^2} \right) \quad ; \quad n_1 = 1, 2, 3, \ldots \quad ; \quad n_2 = n_1 + 1, n_1 + 2, \ldots . \quad (10.1.1)$$

This formula, with the constant (named after Rydberg) $R_H = 13.6$ eV, strictly applies to hydrogen atoms, but similar patterns emerge in the emission spectra of other excited atoms [10.1].

The quest for the origin of these empirical findings in the late nineteenth century led to an important breakthrough toward the formulation of quantum mechanics in the

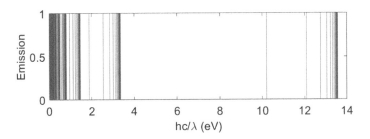

Figure 10.1.1   The emission spectral lines for hydrogen atoms according to the Rydberg formula.

year 1913, when Niels Bohr proposed his model for the atom. While the Bohr model relied on classical mechanical equations of motion for the electron, emission of electromagnetic radiation by an accelerated charged particle was ignored, and additionally, discretization (quantization) of angular momentum (resulting in energy quantization) was assumed. The success of these additional assumptions in reproducing Rydberg's formula for the emission spectrum of hydrogen indeed accelerated the formulation of quantum mechanics. The interested reader may learn more about Bohr's model elsewhere; here, we turn to the fully quantum mechanical treatment.

10.2 The Stationary States of a Hydrogen-Like Atom

The hydrogen-like atom is a two-particle system, composed of a single electron and a nucleus containing Z protons. This strictly holds for the series of single-electron atoms (or ions) $H, He^+, Li^{2+}, \ldots$ and so on. However, many-electron atoms and molecules can sometimes be approximated (especially in electronically excited states) as hydrogen-like atoms, when one of the electrons is located far from a positively charged core containing the nucleus (or nuclei), and from the rest of the electrons. Atoms or molecules of the latter type are often referred to as "Rydberg atoms" or "Rydberg molecules" [10.2].

The electron and nucleus are associated with point charges $-|e|$ and $Z|e|$, respectively, where the two oppositely charged particles are coupled via the classical electrostatic (Coulomb) force. Denoting the nucleus and the electron position vectors as $\mathbf{R}_n$ and $\mathbf{r}_e$, respectively, the potential energy is given by Coulomb's law (where $K = 8.99 \times 10^9 \text{ N} \cdot \text{m}^2 \cdot \text{C}^{-2}$ is Coulomb's constant in a vacuum):

$$V(\mathbf{R}_n, \mathbf{r}_e) = \frac{-KZe^2}{|\mathbf{R}_n - \mathbf{r}_e|}. \tag{10.2.1}$$

Notice that relativistic spin-orbit coupling as well as the interaction of the electric dipole with the vacuum state of the electromagnetic field (the Lamb shift [10.3], [10.4]) are ignored here. Inclusion of these effects would indeed lead to deviations from Rydberg's formula for the emission spectrum. Nevertheless, the latter are typically small on the energy scale set by the electrostatic interactions and therefore observed only at high resolution. Accordingly, a theory that accounts only for the electrostatic interactions is sufficient for reproducing the emission spectra as captured by the simple Rydberg formula.

Since the potential energy in Eq. (10.2.1) depends only on the absolute distance between the two particles and not on their relative orientation, *the hydrogen-like atom is a two-body problem with a central potential*, as discussed in Chapter 9 (see Fig. 9.1.1). As in any two-body problem, we can focus on the relative electron–nucleus Hamiltonian independently from the atom center-of-mass Hamiltonian. The relative distance and the center-of-mass coordinates are defined as

$$\mathbf{r} \equiv \mathbf{r}_e - \mathbf{R}_n$$
$$\mathbf{R} \equiv \frac{m_e}{m_e + m_n}\mathbf{r}_e + \frac{m_n}{m_e + m_n}\mathbf{R}_n , \tag{10.2.2}$$

with the corresponding reduced mass and total mass,

$$\mu \equiv \frac{m_e m_n}{m_e + m_n}. $$
$$M \equiv m_e + m_n. \tag{10.2.3}$$

Notice that the ratio between the nucleus and electron masses is considerably large. Indeed, even for the smallest possible nucleus (corresponding to a single proton, in the case of the hydrogen atom) this ratio is $m_n/m_e \approx 1.8 \times 10^3$. This means that the reduced mass associated with the relative electron–nucleus Hamiltonian approximately equals the (light) electron mass, $\mu \approx m_e$, whereas the center-of-mass Hamiltonian corresponds to the (heavy) nucleus mass, $M \approx m_n$. It follows that the relative motion can be identified (approximately) with the motion of the electron around a fixed nucleus.

Using spherical coordinates for the relative position vector, $\mathbf{r} = (r, \theta, \varphi)$, the quantum Hamiltonian that corresponds to the relative electron–nucleus system in a hydrogen-like atom reads (see Eqs. (9.1.3–9.1.6)) for any central potential)

$$\hat{H} = \frac{-\hbar^2}{2\mu}\left(\frac{\partial^2}{\partial r^2} + \frac{2}{r}\frac{\partial}{\partial r}\right) + \frac{-KZe^2}{r} + \frac{\hat{L}^2}{2\mu r^2}, \tag{10.2.4}$$

where $\hat{L}^2$ is the angular momentum operator. The properties of the atom are captured in the set of energy levels and corresponding stationary solutions to the Schrödinger equation

$$\hat{H}\psi(r, \theta, \varphi) = E\psi(r, \theta, \varphi). \tag{10.2.5}$$

As for any central potential, the solutions ($\psi(r, \theta, \varphi)$) obtain a universal angular dependence, corresponding to the spherical harmonics ($Y_{l,m}(\theta, \varphi)$; see Eq. (9.2.29)):

$$\psi(r, \theta, \varphi) = R(r)Y_{l,m}(\theta, \varphi). \tag{10.2.6}$$

The quantum numbers l and m are associated, respectively, with the relative angular momentum ($\hat{L}^2$) and its projection of the z-axis, ($\hat{L}_z$), and are subject to the following constraints (see Eqs. (9.2.30, 9.2.31)):

$$l = 0, 1, 2, \ldots . \quad ; \quad m = -l, -l+1, \ldots, 0, \ldots, l-1, l. \tag{10.2.7}$$

The radial equation for $R(r)$ (Eq. (9.1.11)) is obtained from the full Schrödinger equation (Eq. (10.2.5)) by using explicitly Eq. (10.2.4) for the Hamiltonian, and Eq. (10.2.6) for the wave functions. This yields for any angular quantum number, l, the following radial equation:

$$\left[\frac{-\hbar^2}{2\mu}\left(\frac{\partial^2}{\partial r^2} + \frac{2}{r}\frac{\partial}{\partial r}\right) + \frac{-KZe^2}{r} + \frac{\hbar^2 l(l+1)}{2\mu r^2}\right] R(r) = ER(r), \qquad (10.2.8)$$

where the angular quantum number appears in the differential operator through the radial centrifugal potential, $\frac{\hbar^2 l(l+1)}{2\mu r^2}$. The radial equation can be simplified by defining

$$\chi(r) \equiv rR(r). \qquad (10.2.9)$$

Substitution in Eq. (10.2.8) readily yields an equation for $\chi(r)$ (see Ex. 9.4.1 for a similar transformation):

$$\frac{-\hbar^2}{2\mu}\frac{\partial^2}{\partial r^2}\chi(r) + \left[\frac{-KZe^2}{r} + \frac{\hbar^2 l(l+1)}{2\mu r^2}\right]\chi(\mathbf{r}) = E\chi(\mathbf{r}). \qquad (10.2.10)$$

It is convenient at this point to change variables, $\chi(r) = \Phi(\rho)$, where ρ is a dimensionless position variable,

$$\rho = \frac{\mu e^2 K}{\hbar^2}r. \qquad (10.2.11)$$

(Approximating the reduced electron–nucleus mass by the electron mass, Eq. (10.2.11) reads $\rho = r/a_0$, where $a_0 = \frac{\hbar^2}{m_e e^2 K} = 0.05292$ nm is the known Bohr radius). In the new variable, the radial equation reads (see Ex. 10.2.1)

$$\frac{\partial^2}{\partial\rho^2}\Phi(\rho) + [\frac{2Z}{\rho} - \frac{l(l+1)}{\rho^2}]\Phi(\rho) = \lambda^2\Phi(\rho), \qquad (10.2.12)$$

where the energy is associated with a dimensionless variable,

$$-\lambda^2 \equiv \frac{2\hbar^2}{\mu e^4 K^2}E. \qquad (10.2.13)$$

Notice that by associating the energy, E, with a negative value $(-\lambda^2)$ we restrict the discussion below to the bound states of the electron–nucleus system. These states are associated with a total energy smaller than the asymptotic potential energy, such that the distribution of the relative electron–nucleus distances is confined to finite values. Since the Coulomb potential energy vanishes asymptotically, $V(\mathbf{R}_n, \mathbf{r}_e) \to 0$ as $r \to \infty$ (see Eq. (10.2.1)); the bound states are therefore associated with negative energy levels.

Exercise 10.2.1   *Use the definitions of the dimensionless variables (Eqs. (10.2.11, 10.2.13)) to derive Eq. (10.2.12) from Eq. (10.2.10).*

We seek proper eigenfunctions of Eq. (10.2.12) and their corresponding eigenvalues. The proper solutions are subject to a normalization condition,

$$\int\limits_0^{2\pi} d\varphi \int\limits_0^{\pi} d\theta \int\limits_0^{\infty} dr r^2 \sin(\theta)|\psi(r, \theta, \varphi)|^2 = 1. \qquad (10.2.14)$$

Using Eqs. (10.2.6, 10.2.9, 10.2.11), this translates to the following condition on the radial function, $\Phi(\rho)$:

$$\int_0^\infty d\rho |\Phi(\rho)|^2 = \frac{1}{a_0}, \tag{10.2.15}$$

which means that $\Phi(\rho)$ must vanish asymptotically,

$$\Phi(\rho) \xrightarrow[\rho \to \infty]{} 0. \tag{10.2.16}$$

At the origin it is sufficient that $\Phi(\rho)$ does not diverge, fulfilling the normalization condition, Eq. (10.2.15). However, a more stringent restriction is needed to assure continuity of the radial probability density function (namely, $|R(r)|^2 r^2$) toward vanishing in the nonphysical coordinate regime, $\rho < 0$ (namely, $r < 0$). We therefore seek for solutions that vanish also at the origin,

$$\Phi(\rho) \xrightarrow[\rho \to 0]{} 0. \tag{10.2.17}$$

We now notice that Eq. (10.2.12) is an ordinary second-order differential equation [9.1] with a regular singular point at the origin. Therefore, it has power series solutions in the form $\Phi(\rho) = \rho^s \sum_{k=0}^\infty a_k \rho^k$, where Eq. (10.2.17) holds only for $s > 0$. The solution must also satisfy the asymptotic condition at $\rho \to \infty$, where Eq. (10.2.12) takes the form $\frac{\partial^2}{\partial \rho^2} \Phi(\rho) \to \lambda^2 \Phi(\rho)$. It can be readily verified that this restricts the power series solutions to the form $\Phi(\rho) = \rho^s e^{\pm \lambda \rho} P_q(\rho)$, where s *is finite*, and $P_q(\rho) = \sum_{k=0}^q a_k^{(q)} \rho^k$ is a polynomial of a *finite degree*, q (see Ex. 10.2.2). Excluding the asymptotically diverging solutions associated with $e^{+\lambda \rho}$, the proper solutions, which satisfy the two boundary conditions (Eqs. (10.2.16, 10.2.17)), are of the form

$$\Phi(\rho) = e^{-\lambda \rho} \rho^s \sum_{k=0}^q a_k^{(q)} \rho^k. \tag{10.2.18}$$

Substituting Eq. (10.2.18) in Eq. (10.2.12) and shifting the series index, we obtain closed equations for the coefficients $\{a_k^{(q)}\}$ of each q-degree polynomial (Ex. 10.2.3),

$$\sum_{k=0}^q [(s+k)(s+k-1) - l(l+1)] a_k^{(q)} \rho^{s+k-2} = \sum_{k=1}^{q+1} 2[\lambda(s+k-1) - Z] a_{k-1}^{(q)} \rho^{s+k-2}. \tag{10.2.19}$$

Exercise 10.2.2 *Show that the functions* $\chi(y) = e^{\pm \lambda \rho} \rho^p$ *satisfy the asymptotic radial equation for the hydrogen-like atom, namely,* $\frac{\partial^2}{\partial \rho^2} \Phi(\rho) \xrightarrow[\rho \to \infty]{} \lambda^2 \Phi(\rho)$*, for any nonnegative finite power, p, and use it to show that* $\Phi(\rho)$*, defined in Eq. (10.2.18), is a solution to this equation.*

Exercise 10.2.3 *Derive Eq. (10.2.19), using Eqs. (10.2.12, 10.2.18).*

Comparing the coefficients of ρ^k for any k in Eq. (10.2.19), we obtain

$$[s(s-1) - l(l+1)]a_0^{(q)} = 0 \quad ; \quad k = 0,$$

$$a_k^{(q)} = \frac{2[\lambda(s+k-1)-Z]}{(s+k)(s+k-1)-l(l+1)}a_{k-1}^{(q)} \quad ; \quad 1 \le k \le q, \qquad (10.2.20)$$

$$[\lambda(s+q) - z]a_q^{(q)} = 0 \quad ; \quad k = q+1.$$

Notice that a proper solution, $\Phi(\rho)$, must be associated with $a_0^{(q)} \ne 0$. (Otherwise, the recursion relation means that all the coefficients, $a_0^{(q)}, a_1^{(q)}, a_2^{(q)}, \ldots, a_q^{(q)}$, vanish, and $\Phi_q(\rho) = 0$.) It therefore follows from the first condition that $s(s-1) = l(l+1)$. The (only) positive value of s (that assures consistency with Eq. (10.2.17)) is

$$s = l + 1. \qquad (10.2.21)$$

Additionally, $a_q^{(q)} \ne 0$ for a qth-degree polynomial. Therefore, the third condition in Eq. (10.2.20) means that $\lambda(q+s) = Z$. Recalling that $q = 0, 1, 2, \ldots$ stands for the polynomial degree, and using Eq. (10.2.21), it is useful to define a new integer index,

$$n \equiv q + s = q + l + 1, \qquad (10.2.22)$$

where

$$Z = \lambda n. \qquad (10.2.23)$$

Using the relation of λ to the energy (Eq. (10.2.13)), we can readily identify the set of eigenvalues of the Hamiltonian Eq. (10.2.4) for the hydrogen-like atom:

$$E_n = -\frac{\mu e^4 K^2}{2\hbar^2}\frac{Z^2}{n^2} \quad ; \quad n = l+1, l+2, \ldots. \qquad (10.2.24)$$

Approximating the reduced electron–nucleus mass by the electron mass m_e (see Eq. (10.2.3)), the energy levels of a hydrogen-like atom are given as $E_n = -R_H\frac{Z^2}{n^2}$, where the constant

$$R_H = \frac{m_e e^4 K^2}{2\hbar^2} = 13.6\,\text{eV} \qquad (10.2.25)$$

is identified as the experimentally known Rydberg's constant.

According to this quantum mechanical result, the specific lines (wavelengths) appearing in the emission spectrum of a hydrogen-like atom (Eq. (10.1.1)) are associated with transitions between stationary states of the atom, induced by (weak) coupling to the radiation field. Using Planck's formula (Eq. (5.2.2); also, see Chapters 18 for a detailed discussion of field-induced transitions), each transition $n_i \to n_f$ corresponds to a particular radiation wavelength,

$$\frac{hc}{\lambda_{n_i,n_f}} = E_{n_i} - E_{n_f} = -R_H Z^2 \left(\frac{1}{n_i^2} - \frac{1}{n_f^2}\right) \quad ; \quad n_f < n_i \quad ; \quad n_f = 1, 2, \ldots. \qquad (10.2.26)$$

In Fig. 10.2.1, two series of emission spectral lines, corresponding to transitions from excited electronic states into the ground ($n_f = 1$) and into the first excited ($n_f = 2$) states, are illustrated.

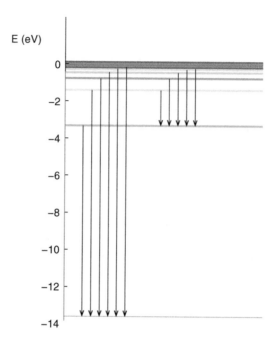

Figure 10.2.1 Emission spectral lines for a hydrogen-like atom, according to the quantum model.

The two quantum numbers, n and l, therefore define a proper solution to the radial equation. Using Eqs. (10.2.21–10.2.23) in Eq. (10.2.18), the solutions obtain the form

$$\Phi_{n,l}(\rho) = e^{-Z\rho/n}\rho^{l+1}\sum_{k=0}^{n-l-1} a_k^{(n,l)}\rho^k, \tag{10.2.27}$$

where the polynomial coefficients, $\{a_k^{(n,l)}\}$, defined by the recursion relation, Eq. (10.2.20), can be calculated explicitly using Eqs. (10.2.21–10.2.23) (see Ex. 10.2.4):

$$a_k^{(n,l)} = \left(-\frac{2Z}{n}\right)^k \frac{(n-l-1)!(2l+1)!}{(n-l-1-k)!(2l+1+k)!k!}a_0^{(n,l)}. \tag{10.2.28}$$

The radial solutions can be expressed compactly in terms of the known associated Laguerre polynomials [10.5], defined as

$$L_q^p(\xi) \equiv \sum_{k=0}^{q}(-1)^k\frac{(p+q)!}{(q-k)!(p+k)!k!}\xi^k. \tag{10.2.29}$$

Identifying $q = n-l-1$ and $p = 2l+1$ in Eqs. (10.2.27, 10.2.28), it follows that (see Ex. 10.2.5)

$$\Phi_{n,l}(\rho) = c_{n,l}e^{-Z\rho/n}\rho^{l+1}L_{n-l-1}^{2l+1}\left(\frac{2Z}{n}\rho\right), \tag{10.2.30}$$

where $c_{n,l} = a_0^{(n,l)}(n-l-1)!(2l+1)!/(l+n)!$. Notice that the constant $a_0^{(n,l)}$ can be selected to assure the normalization condition, Eq. (10.2.15). Using the property of the associated Laguerre polynomials [10.5], $\int_0^\infty e^{-x}x^{p+1}[L_q^p(x)]^2dx = (2q+p+1)(q+p)!/q!$,

$$\psi_{1,0,0}(r,\theta,\varphi) = \sqrt{\frac{Z^3}{\pi a_0^3}} e^{-Zr/a_0}$$

$$\psi_{2,0,0}(r,\theta,\varphi) = \frac{1}{8}\sqrt{\frac{2Z^3}{\pi a_0^3}}\left(2 - \frac{Z}{a_0}r\right)e^{-rZ/2a_0}$$

$$\psi_{2,1,0}(r,\theta,\varphi) = \frac{1}{8}\sqrt{\frac{2Z^5}{\pi a_0^5}}\,r e^{-rZ/2a_0}\cos(\theta)$$

$$\psi_{2,1,1}(r,\theta,\varphi) = \frac{1}{8}\sqrt{\frac{Z^5}{\pi a_0^5}}\,r e^{-rZ/2a_0}\sin(\theta)e^{i\varphi}$$

$$\psi_{2,1,-1}(r,\theta,\varphi) = \frac{1}{8}\sqrt{\frac{Z^5}{\pi a_0^5}}\,r e^{-rZ/2a_0}\sin(\theta)e^{-i\varphi}$$

$$\psi_{3,0,0}(r,\theta,\varphi) = \frac{1}{81}\sqrt{\frac{Z^3}{3\pi a_0^3}}\left(27 - 18\frac{Z}{a_0}r + 2\frac{Z^2}{a_0^2}r^2\right)e^{-Zr/3a_0}$$

$$\psi_{3,1,0}(r,\theta,\varphi) = \frac{1}{81}\sqrt{\frac{2Z^5}{\pi a_0^5}}\,r\left(6 - \frac{Z}{a_0}r\right)e^{-Zr/3a_0}\cos(\theta)$$

$$\psi_{3,1,1}(r,\theta,\varphi) = \frac{1}{81}\sqrt{\frac{Z^5}{\pi a_0^5}}\,r\left(6 - \frac{Z}{a_0}r\right)e^{-Zr/3a_0}\sin(\theta)e^{i\varphi}$$

$$\psi_{3,1,-1}(r,\theta,\varphi) = \frac{1}{81}\sqrt{\frac{Z^5}{\pi a_0^5}}\,r\left(6 - \frac{Z}{a_0}r\right)e^{-Zr/3a_0}\sin(\theta)e^{-i\varphi}$$

$$\psi_{3,2,0}(r,\theta,\varphi) = \frac{1}{81}\sqrt{\frac{Z^7}{6\pi a_0^7}}\,r^2 e^{-Zr/3a_0}\left(3\cos^2(\theta) - 1\right)$$

$$\psi_{3,2,1}(r,\theta,\varphi) = \frac{1}{81}\sqrt{\frac{Z^7}{\pi a_0^7}}\,r^2 e^{-Zr/3a_0}\sin(\theta)\cos(\theta)e^{i\varphi}$$

$$\psi_{3,2,-1}(r,\theta,\varphi) = \frac{1}{81}\sqrt{\frac{Z^7}{\pi a_0^7}}\,r^2 e^{-Zr/3a_0}\sin(\theta)\cos(\theta)e^{-i\varphi}$$

$$\psi_{3,2,2}(r,\theta,\varphi) = \frac{1}{162}\sqrt{\frac{Z^7}{\pi a_0^7}}\,r^2 e^{-Zr/3a_0}\sin^2(\theta)e^{2i\varphi}$$

$$\psi_{3,2,-2}(r,\theta,\varphi) = \frac{1}{162}\sqrt{\frac{Z^7}{\pi a_0^7}}\,r^2 e^{-Zr/3a_0}\sin^2(\theta)e^{-2i\varphi}$$

Figure 10.2.2   A list of some low-energy eigenfunctions of the hydrogen-like atom Hamiltonian

this yields $c_{n,l} = \sqrt{\frac{Z(n-l-1)!}{a_0 n^2 (n+l)!}}\left(\frac{2Z}{n}\right)^{l+1}$. Returning to the original radial-coordinate variable, $\Phi_{n,l}(\rho) = \chi_{n,l}(r) = rR_{n,l}(r)$, we obtain (Ex. 10.2.6)

$$R_{n,l}(r) = \sqrt{\frac{4Z^3(n-l-1)!}{a_0^3 n^4 (n+l)!}}\left(\frac{2Zr}{na_0}\right)^l e^{\frac{-Zr}{na_0}} L_{n-l-1}^{2l+1}\left(\frac{2Zr}{na_0}\right). \tag{10.2.31}$$

Exercise 10.2.4 *The radial wave functions for a hydrogen-like atom are of the form* $\Phi_{n,l}(\rho) = e^{-Z\rho/n}\rho^{l+1}\sum_{k=0}^{n-l-1} a_k^{(n,l)}\rho^k$. *The polynomial coefficients,* $\{a_k^{(n,l)}\}$, *are defined by the recursion relation, Eq. (10.2.20). Use Eqs. (10.2.21–10.2.23) and derive the explicit expression for these coefficients, Eq. (10.2.28).*

Exercise 10.2.5 *Use Eqs. (10.2.27, 10.2.28) for the radial wave functions of the hydrogen-like atom, and Eq. (10.2.29) for the associated Laguerre polynomials, to derive Eq. (10.2.30).*

Exercise 10.2.6 *Use the normalization condition (Eq. (10.2.15)) to normalize the radial function given by Eq. (10.2.30), and change variables to obtain* $R_{n,l}(r)$ *in Eq. (10.2.31).*

Finally, the three-dimensional stationary wave functions corresponding to the hydrogen-like atom, as defined in Eqs. (10.2.5, 10.2.6), are given as

$$\psi_{n,l,m}(r,\theta,\varphi) = R_{n,l}(r)Y_{l,m}(\theta,\varphi). \tag{10.2.32}$$

In Fig. 10.2.2, the wave functions corresponding to some of the smallest quantum numbers are given explicitly, and in Fig. 10.2.3 some functions are plotted for illustration along a one-dimensional cut through the three-dimensional space, on top of the underlying coulomb potential energy.

Let us highlight here some general properties of the set of hydrogen-like atom eigenfunctions $\{\psi_{n,l,m}\}$:

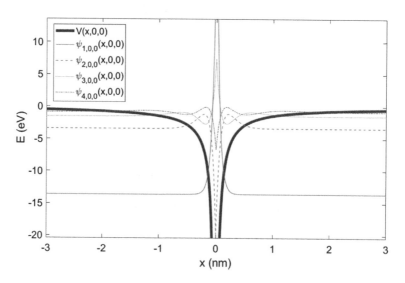

Figure 10.2.3 Stationary wave functions, $\psi_{n,l,m}(x,y,z)$, for the hydrogen-like atom (for $Z = 1$), represented along a one-dimensional cut through the three-dimensional space, $\psi_{n,l,m}(x,0,0)$. The four functions associated with the lowest energy levels are displaced from each other by the respective energy differences. The corresponding Coulomb potential energy curve, $V(x,0,0)$, is plotted (thick line) for reference.

I. Each function, $\psi_{n,l,m}(r,\theta,\varphi)$, is characterized by three different quantum numbers. These quantum numbers are subject to the following restrictions (see Eqs. (9.2.27, 10.2.22, 10.2.24)), derived from the requirement for proper wave functions:

$$n = 1, 2, 3 \ldots,$$
$$l = 0, 1, 2, \ldots, n-1, \tag{10.2.33}$$
$$m = -l, -l+1, \ldots, 0, \ldots, l-1, l.$$

II. Each function, $\psi_{n,l,m}(r,\theta,\varphi)$, is a common eigenfunction of three commuting operators (Ex. 10.2.7),

$$\hat{H}\psi_{n,l,m}(r,\theta,\varphi) = \frac{-R_H Z^2}{n^2}\psi_{n,l,m}(r,\theta,\varphi),$$
$$\hat{L}^2\psi_{n,l,m}(r,\theta,\varphi) = \hbar^2 l(l+1)\psi_{n,l,m}(r,\theta,\varphi), \tag{10.2.34}$$
$$\hat{L}_z\psi_{n,l,m}(r,\theta,\varphi) = m\hbar\psi_{n,l,m}(r,\theta,\varphi),$$

where $\hat{H} = [\frac{-\hbar^2}{2\mu}(\frac{\partial^2}{\partial r^2} + \frac{2}{r}\frac{\partial}{\partial r}) + \frac{-KZe^2}{r} + \frac{\hat{L}^2}{2\mu r^2}]$ is the hydrogen-like atom Hamiltonian, and $\hat{L}^2$, $\hat{L}_z$ are the corresponding angular momentum operators. The quantum numbers n, l, and m define uniquely the eigenvalues of $\hat{H}$, $\hat{L}^2$, and $\hat{L}_z$, namely, $-R_H Z^2/n^2$, $\hbar^2 l(l+1)$, and $m\hbar$, respectively. They are commonly referred to as the main (or principal) quantum number (n), the angular momentum quantum number (l), and the magnetic quantum number (m).

III. The set of functions $\{\psi_{n,l,m}(r,\theta,\varphi)\}$ is an orthonormal set:

$$\int_0^{2\pi} d\varphi \int_0^{\pi} \sin(\theta)d\theta \int_0^{\infty} r^2 dr \psi_{n',l',m'}^*(r,\theta,\varphi)\psi_{n,l,m}(r,\theta,\varphi) = \delta_{n,n'}\delta_{l,l'}\delta_{m,m'}. \tag{10.2.35}$$

IV. Degeneracy: For any quantum number, n, there are n^2 degenerate eigenfunctions of $\hat{H}$, corresponding to the same eigenvalue, $E_n = -R_H Z^2/n^2$. The different functions correspond to all the different combinations of the quantum numbers, l and m, associated with the given n, where l takes on the values, $0, 1, \ldots, n-1$, and the number of m-values for each l is $2l+1$ (see Ex. 10.2.8).

V. In atomic physics and chemistry, the set $\{\psi_{n,l,m}\}$ is often replaced by an alternative orthonormal set of real-valued functions, which are linear combinations of degenerate $\psi_{n,l,m}$, associated with the same n and l (see Figs. 10.2.4, 10.3.2). Each real-valued function is denoted by the principal quantum number, n, followed by a letter that corresponds to the angular quantum number ($s, p, d, f, g, h, \ldots$ mark, respectively, $l = 0, 1, 2, 3, 4, 5, \ldots$), and a function in Cartesian variables that represents the angular dependence of the function. For example, $\psi_{3d_{xz}}$ corresponds to $n = 3$, $l = 2$, and an angular function, proportional to $x \cdot z = \cos(\theta)\sin(\theta)\cos(\varphi)$. The letters, s, p, d are reminiscent of the historical names given to the spectral lines for the corresponding electronic transitions: Sharp, Principal, and Diffused.

Exercise 10.2.7 *Show that the three operators, $\hat{L}_z$, $\hat{L}^2$, and the hydrogen-like atom Hamiltonian commute with each other. Prove that these three operators have a set of joint eigenfunctions (use Eq. (10.2.32)).*

$$\psi_{1s} \equiv \psi_{1,0,0} = R_{1,0}(r)Y_{0,0}(\theta,\varphi)$$

$$\psi_{2s} \equiv \psi_{2,0,0} = R_{2,0}(r)Y_{0,0}(\theta,\varphi)$$

$$\psi_{2p_z} \equiv \psi_{2,1,0} = R_{2,1}(r)Y_{1,0}(\theta,\varphi)$$

$$\psi_{2p_x} \equiv \frac{1}{\sqrt{2}}(\psi_{2,1,1}+\psi_{2,1,-1}) = \frac{1}{\sqrt{2}}R_{2,1}(r)[Y_{1,1}(\theta,\varphi)+Y_{1,-1}(\theta,\varphi)]$$

$$\psi_{2p_y} \equiv \frac{1}{\sqrt{2}}(\psi_{2,1,1}-\psi_{2,1,-1}) = \frac{1}{\sqrt{2}}R_{2,1}(r)[Y_{1,1}(\theta,\varphi)-Y_{1,-1}(\theta,\varphi)]$$

$$\psi_{3s} \equiv \psi_{3,0,0} = R_{3,0}(r)Y_{0,0}(\theta,\varphi)$$

$$\psi_{3p_z} \equiv \psi_{2,1,0} = R_{2,1}(r)Y_{1,0}(\theta,\varphi)$$

$$\psi_{3p_x} \equiv \frac{1}{\sqrt{2}}(\psi_{3,1,1}+\psi_{3,1,-1}) = \frac{1}{\sqrt{2}}R_{3,1}(r)[Y_{1,1}(\theta,\varphi)+Y_{1,-1}(\theta,\varphi)]$$

$$\psi_{3p_y} \equiv \frac{1}{\sqrt{2}}(\psi_{3,1,1}-\psi_{3,1,-1}) = \frac{1}{\sqrt{2}}R_{3,1}(r)[Y_{1,1}(\theta,\varphi)-Y_{1,-1}(\theta,\varphi)]$$

$$\psi_{3d_{z^2}} = \psi_{3,2,0} = R_{3,2}(r)Y_{2,0}(\theta,\varphi)$$

$$\psi_{3d_{xz}} = \frac{1}{\sqrt{2}}(\psi_{3,2,1}+\psi_{3,2,-1}) = \frac{1}{\sqrt{2}}R_{3,2}(r)[Y_{2,1}(\theta,\varphi)+Y_{2,-1}(\theta,\varphi)]$$

$$\psi_{3d_{yz}} = \frac{1}{\sqrt{2}}(\psi_{3,2,1}-\psi_{3,2,-1}) = \frac{1}{\sqrt{2}}R_{3,2}(r)[Y_{2,1}(\theta,\varphi)-Y_{2,-1}(\theta,\varphi)]$$

$$\psi_{3d_{x^2-y^2}} = \frac{1}{\sqrt{2}}(\psi_{3,2,2}+\psi_{3,2,-2}) = \frac{1}{\sqrt{2}}R_{3,2}(r)[Y_{2,2}(\theta,\varphi)+Y_{2,-2}(\theta,\varphi)]$$

$$\psi_{3d_{xy}} = \frac{1}{\sqrt{2}}(\psi_{3,2,2}-\psi_{3,2,-2}) = \frac{1}{\sqrt{2}}R_{3,2}(r)[Y_{2,2}(\theta,\varphi)-Y_{2,-2}(\theta,\varphi)]$$

Figure 10.2.4   A list of real-valued eigenfunctions of the hydrogen-like atom Hamiltonian

Exercise 10.2.8   *Show that the energy levels of a hydrogen-like atom, $E_n = -R_H Z^2/n^2$, are n^2-fold degenerate. (Notice that, $\sum_{l=0}^{n-1}(2l+1) = n^2$.)*

The probability distribution functions for the vector of relative position between the electron and the nucleus in a hydrogen-like atom in its stationary states are discussed in the following section.

10.3 Probability Density Distributions and Atomic Orbitals

The stationary solutions to the Schrödinger equation for a hydrogen-like atom ena-ble us to address directly a fundamental question relating to atoms in general and to hydrogen-like atoms in particular: *What is the size (volume) of an atom?* This ques-tion does not have a unique answer, to the best of our current knowledge, since the laws of quantum mechanics state that the relative positions of the electron and the nucleus cannot be defined deterministically. Nevertheless, a closely related question can be answered quite accurately (at least for a hydrogen-like atom) on the basis of solutions to the Schrödinger equation: *Given an atom in a stationary state, what is the*

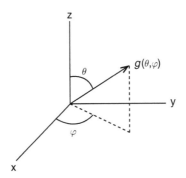

Representing an angular distribution function, $g(\theta, \varphi)$, by a vector of length, $g(\theta, \varphi)$, pointing in the direction set by the angles, θ and φ.

probability density for finding the electron and the nucleus at a relative position vector **r** $= (r, \theta, \varphi)$? It is customary to represent a probability density function, $\rho(\mathbf{r}) = \rho(r, \theta, \varphi)$, by considering separately its angular and radial dependencies.

Angular Distributions

For representation of an angular probability density distribution, each pair of angles (θ, φ) is associated with a vector in the direction defined by θ and φ (see Fig. 10.3.1), whose length is proportional to $\rho(r, \theta, \varphi)$ for some fixed value of the radial coordinate, $r = r_0$. Notice that the stationary atomic wave functions discussed in Section 10.2 are products of radial and angular functions (see Fig. 10.2.4), $\rho(r_0, \theta, \varphi) = |R_{n,l}(r_0)|^2 g_l(\theta, \varphi)$, such that the choices of r_0 as well as the quantum number n affect only the prefactor, $|R_{n,l}(r_0)|^2$, which does not bear any information regarding the angular distribution. The shape of the angular distribution function is therefore encoded in the function, $g_l(\theta, \varphi)$. Each angular quantum number, l, corresponds to $2l + 1$ different angular distributions, corresponding to degenerate spherical harmonics, $g_l(\theta, \varphi) = |Y_{l,m}(\theta, \varphi)|^2$, or their linear combinations, $g_l(\theta, \varphi) = |\sum_{m=-l}^{l} c_m Y_{l,m}(\theta, \varphi)|^2$.

In Fig. 10.3.2 angular distributions ($g_l(\theta, \varphi)$) are plotted for some real-valued stationary solutions, associated with $l = 0, 1, 2$ (see Fig. 10.2.4). As we can see, the shape of the angular distribution function changes with the angular quantum number, l. Particularly, as l increases, the number of nodal surfaces in the angular distribution increases. The orbitals of type $s, p,$ and d are associated, respectively, with zero, one, and two nodal surfaces on which the probability density vanishes, as reflected in the changes of the sign of the wave function.

Radial Distributions

We now change focus and discuss the radial probability distributions. Specifically, we wish to answer the following question: *Given an atom in a stationary state, what is the probability density for finding the electron and the nucleus at a relative*

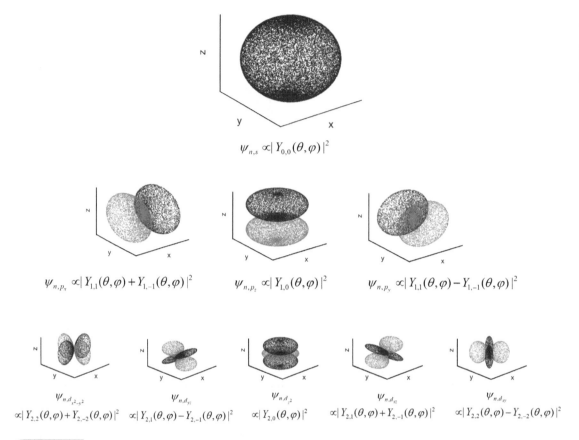

$$\psi_{n,s} \propto |Y_{0,0}(\theta,\varphi)|^2$$

$$\psi_{n,p_x} \propto |Y_{1,1}(\theta,\varphi) + Y_{1,-1}(\theta,\varphi)|^2 \qquad \psi_{n,p_z} \propto |Y_{1,0}(\theta,\varphi)|^2 \qquad \psi_{n,p_y} \propto |Y_{1,1}(\theta,\varphi) - Y_{1,-1}(\theta,\varphi)|^2$$

$$\psi_{n,d_{x^2-y^2}} \qquad \psi_{n,d_{yz}} \qquad \psi_{n,d_{z^2}} \qquad \psi_{n,d_{xz}} \qquad \psi_{n,d_{xy}}$$
$$\propto |Y_{2,2}(\theta,\varphi) + Y_{2,-2}(\theta,\varphi)|^2 \quad \propto |Y_{2,1}(\theta,\varphi) - Y_{2,-1}(\theta,\varphi)|^2 \quad \propto |Y_{2,0}(\theta,\varphi)|^2 \quad \propto |Y_{2,1}(\theta,\varphi) + Y_{2,-1}(\theta,\varphi)|^2 \quad \propto |Y_{2,2}(\theta,\varphi) - Y_{2,-2}(\theta,\varphi)|^2$$

Figure 10.3.2 Angular distribution functions for a hydrogen-like atom, corresponding to $l = 0, 1, 2$. The different shades in each plot add information corresponding to changes in the sign of the wave function, for which the probability density is presented.

distance, r, regardless of their relative spatial orientation (θ, φ)? The answer is given by recalling the normalization condition for any proper solutions, Eq. (10.2.14), $\int_0^\infty dr \int_0^{2\pi} d\varphi \int_0^\pi d\theta\, r^2 \sin(\theta)|\psi(r,\theta,\varphi)|^2 = 1$. We can readily rewrite this equation as

$$\int_0^\infty \rho(r)dr = 1, \qquad (10.3.1)$$

where the radial distribution function, $\rho(r)$, is defined as

$$\rho(r) \equiv r^2 \int_0^{2\pi} d\varphi \int_0^\pi d\theta \sin(\theta)|\psi(r,\theta,\varphi)|^2. \qquad (10.3.2)$$

Since the stationary atomic wave functions of a hydrogen-like atom are products of (normalized) angular and radial functions (see Fig. 10.2.4), $|\psi(r,\theta,\varphi)|^2 = |R_{n,l}(r)|^2 g_l(\theta,\varphi)$, the radial probability distribution depends only on the principal and angular quantum numbers (n and l) and is of the form

$$\rho_{n,l}(r) = r^2 |R_{n,l}(r)|^2. \tag{10.3.3}$$

In Fig. 10.3.3 radial distributions are plotted for different stationary states, $\{\psi_{n,l,m}\}$. The different functions correspond to different values of the principal quantum number, $n = 1, 2, 3$, where the angular and magnetic numbers, $l = 0$ and $m = 0$, are the same. The most apparent observation is that the radial probability density spreads toward larger values of the electron–nucleus relative distance as n increases (and accordingly the energy, $E_n = -R_H Z^2 / n^2$, increases). Another observation is that the number of nodes in the radial probability density increases with increasing quantum number, n, and therefore with increasing energy, E_n. This behavior is reminiscent of the trend observed in solutions of the Schrödinger equation for one-dimensional model Hamiltonians (see Chapters 5 and 8). Here the number of nodes equals $n - 1$, which reflects the degree of the associated Laguerre polynomial, $L_{n-1}^1\left(\frac{2Zr}{na_0}\right)$ (see Eq. 10.2.31).

The correlation between increasing energy and increasing electron–nucleus distance is expected on classical mechanics grounds, where the Coulomb law associates a larger relative distance between the oppositely charged electron and nucleus with a higher energy. This correlation was also the basis for the historical Bohr model for the hydrogen atom. In that model the set of discrete energy levels of the electron in the atom, $E_n = -R_H Z^2 / n^2$, were associated with circulating orbits of the electron around the nucleus, each having a radius, $r_n = a_0 n^2 / Z$. Remarkably, the same qualitative trend is obtained in quantum mechanics for the *average* electron–nucleus distance, which reads $\langle r \rangle_n = \frac{3a_0 n^2}{2Z}$ (for zero angular momentum); but the important difference is that in Bohr's model each energy level corresponds to a deterministic electron–nucleus distance, which is the radius of a circular orbit, whereas in quantum mechanics, the distance is defined only probabilistically. *The state of the electron is associated with an "orbital" rather than with a classical orbit. An orbital in the hydrogen-like atom is the quantum mechanical wave function, which defines the probability density for finding the electron and nucleus in any relative distance. In general, the term orbital is often used in atomic, molecular, and solid-state physics for describing any single-electron quantum mechanical wave function* (see Chapters 13 and 14).

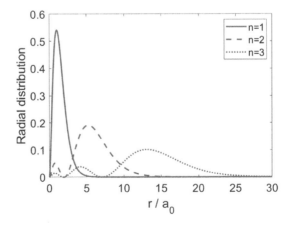

Radial distribution functions, $\rho_{n,l}(r)$, for a hydrogen-like atom ($Z = 1$), corresponding to $l = 0$ and $n = 1, 2, 3$.

Exercise 10.3.1   *Use Eqs. (10.3.3) and (10.2.31) for the radial probability distribution to show that (a) the most probable relative distance between the electron and the nucleus in the ground state of a hydrogen-like atom is* $r = a_0/Z$; *(b) the probability for finding the electron and the nucleus at any distance in the range,* $0 < r < \gamma\frac{a_0}{Z}$, *equals* $P(\gamma) =$ $1 - e^{-2\gamma}(2\gamma^2 + 2\gamma + 1)$ *(you can use the identity* $\int_x^\infty dy\, y^2 e^{-\alpha y} = \frac{d^2}{d\alpha^2}\int_x^\infty dy\, e^{-\alpha y}$*).*

The "effective volume" of the hydrogen-like atom can be estimated by considering the radial probability density associated with the ground state, $\psi_{1,0,0}(r, \theta, \varphi)$. The most probable distance between the electron and the nucleus in this state is $r = a_0/Z$ (see Ex. 10.3.1), which defines a sphere with a volume, $V = 4\pi a_0^3/(3Z^3)$. For $Z = 1$, which corresponds, for example, to the hydrogen atom itself, $r = 0.5292$ Å, and the corresponding volume equals 0.62 cubic Å. Notice, however, that this sphere is too small to account for the volume of the atom, since there is a substantial probability of finding the relative electron–nucleus distance outside this sphere. It turns out that this sphere accounts only for $\sim 32\%$ of all possible electron–nucleus distances (see Ex. 10.3.1). Clearly, the sphere volume needs to be extended to infinity to include *any accessible* relative distance. However, a substantial portion of the possible distances ($\sim 94\%$) is contained in a sphere of a radius that is three times larger, $r = 3a_0 \approx 1.6$ Å (see Ex. 10.3.1), defining an effective volume of 16.8 cubic Å.

Let us turn now to discussing the effect of the angular quantum number, l, on the radial distributions. Figure 10.3.4 demonstrates that the probability density distribution of the relative electron–nucleus distances does depend on l; but the changes in the distribution are mild in comparison to the changes associated with the principal quantum number, n. In particular, the changes in the most probable electron–nucleus distance as l changes are relatively mild (compare to Fig. 10.3.3). Nevertheless, an important effect is the decrease in the number of nodes in the radial distribution with increasing angular quantum number, l. Notice that the number of nodes equals $n - l - 1$, whereas the number of angular nodal surfaces is equal to l. Since each node

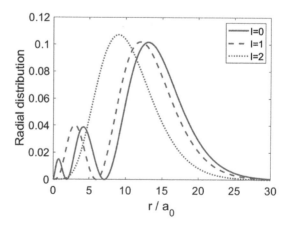

Figure 10.3.4   Radial distribution functions, $\rho_{n,l}(r)$, for a hydrogen-like atom ($Z = 1$), corresponding to $n = 3$ and $l = 0, 1, 2$.

in the radial distribution defines a surfaces of a sphere on which the wave function vanishes (and changes sign), the overall number of nodal surfaces equals $n-1$ and is therefore correlated with the energy, E_n, as we have already mentioned in discussing other quantum systems (see Chapters 5 and 8).

Bibliography

[10.1] J. R. Rydberg, "On the structure of the line-spectra of the chemical elements," The London, Edinburgh, and Dublin Philosophical Magazine and Journal of Science 29, 331 (1890).

[10.2] C. H. Greene, A. S. Dickinson and H. R. Sadeghpour, "Creation of polar and nonpolar ultra-long-range Rydberg molecules," Physical Review Letters 85, 2458 (2000).

[10.3] W. E. Lamb, Jr and R. C. Retherford, "Fine structure of the hydrogen atom by a microwave method," Physical Review 72, 241 (1947).

[10.4] H. A. Bethe, "The electromagnetic shift of energy levels," Physical Review 72, 339 (1947).

[10.5] E. W. Weisstein, "Associated Laguerre Polynomial." From MathWorld – A Wolfram Web Resource. Accessed at https://mathworld.wolfram.com/AssociatedLaguerrePolynomial.html

11 The Postulates of Quantum Mechanics

11.1 A Summary of the Postulates

In the previous chapters we already encountered some of the postulates of quantum mechanics. These postulates provide the needed guidelines for associating physically measurable quantities of interest with mathematical objects. In Chapter 2, the state of a closed physical system was associated with a proper, complex-valued, wave function, which contains all measurable information on the system, where the extension of the wave functions description to open systems was given in Chapter 7. In Chapter 3 we discussed the representation of measurable quantities in terms of mathematical operators that are linear and Hermitian, and we saw how relations between physically measurable quantities (energy, angular momentum, position, momentum, etc.) are associated with mathematical relations between the corresponding operators. In Chapter 4 we discussed the Schrödinger equation, which associates the Hamiltonian operator of a system with the time evolution of any physical state within this system. We learned that the measured values in a single measurement on a single system can only be eigenvalues of the operator that represents the measured physical quantity, and therefore, when two operators share a common set of eigenfunctions (e.g., commuting operators), the quantities that are represented by these operators can be defined simultaneously.

In Chapters 6–10 we discussed solutions to the Schrödinger equation for simple and yet most informative model Hamiltonians, which enable us to understand some basic phenomena on the nanoscale. Many other phenomena, however, are beyond the scope of exact solutions of the Schrödinger equation. To address these, the full mathematical structure of quantum mechanics theory must be exploited. The present chapter will be devoted to a rigorous formulation of the postulates of quantum mechanics, and in the next chapter we will introduce some of the basic approximation methods in quantum mechanics that are necessary for treating complex, many-particle systems.

For the sake of continuity with the previous chapters, we shall first introduce the postulates in the realm of wave functions, namely focusing on the position representation. Then we shall generalize the formulation to the abstract Hilbert space of quantum states, introducing Dirac's notations. We chose to group the postulates into four groups and to introduce them according to the following order: Postulate 1 deals with the mathematical representation of the physical state of a system in terms of a wave function, and postulate 2 deals with the representation of measurable quantities in terms of operators. Postulate 3 deals with the dynamics of the physical states, and postulate 4 deals with the consequences of a measurement of an observable (represented by an operator) on a system (represented by a wave function).

Postulate 1: ***The state of a physical system is associated with a complex-valued wave function***, which contains all the measurable information on the system. This function depends on the coordinates of all the particles composing the system. Denoting here the vector of all the particle-coordinates as $\mathbf{r} \equiv \mathbf{r}_1, \mathbf{r}_2, \mathbf{r}_3, \ldots$, the wave function is denoted, $\psi(\mathbf{r})$.

In a closed system, namely, where all the particles are bound within a finite region in space, ***the wave function must be proper (square integrable), and its absolute square value has the meaning of a probability density in the position space:***

$$\psi^*(\mathbf{r})\psi(\mathbf{r}) = |\psi(\mathbf{r})|^2 \equiv \rho(\mathbf{r}). \tag{11.1.1}$$

Here $\rho(\mathbf{r})d\mathbf{r}$ is the probability for locating the particles in the system in the spatial arrangement defined by the position vector $\mathbf{r}$, within an infinitesimal volume, $d\mathbf{r}$. ***For a proper wave function, the probability of finding the particles in <u>any</u> position in space can be normalized to unity, such that***

$$\int_{V_{ps}} \rho(\mathbf{r})d\mathbf{r} = 1. \tag{11.1.2}$$

Here V_{ps} is the volume of the entire physical space. The association of $\rho(\mathbf{r})$ with a probability density means that the wave function itself is interpreted as a probability amplitude and obtains physical dimensions, $[\psi] = [\mathbf{r}]^{-1/2}$. The information content in the wave function with respect to the physical state of the system (e.g., the ratio between probabilities of finding the system at different spatial configurations) is unchanged when $\psi(\mathbf{r})$ is multiplied by a finite (complex) number. This means that any proper wave function with a finite (nonzero) integral, $\int_{V_{ps}} \rho(\mathbf{r})d\mathbf{r} \equiv c$, can be

normalized by its multiplication with $1/\sqrt{c}$. Detailed discussions as well as exercises concerning the definition of probability densities, proper wave functions, and wave function normalization in spaces of different dimensions can be found in Sections 2.1, 2.2, and 2.3.

When the system is open, for example, when particles are scattered through (enter and/or leave) the physical space, the state of the physical system is still associated with a wave function, $\psi(\mathbf{r})$, but the latter is not necessarily proper. When the system is associated with an improper scattering wave function, probability conservation is formulated in terms of probability fluxes, $\mathbf{J}(\mathbf{r})$, rather than densities, and reads

$$\oint_{S_{ps}} \mathbf{J}(\mathbf{r}) \cdot d\mathbf{S} = 0. \tag{11.1.3}$$

Here, S_{ps} is a surface surrounding the boundaries of the entire relevant physical space (e.g., the interaction region in a scattering experiment [7.1]). The improper wave functions are conventionally normalized either to a specified incoming flux (see Chapter 7) or to Dirac's delta function, as discussed in Section 11.5. Examples of scattering wave functions and their implementations were discussed in Chapter 7 of this book.

Notice that while the probabilistic interpretation of the wave function imposes some necessary condition, such as Eqs. (11.1.2, 11.1.3), on $\psi(\mathbf{r})$, these conditions are not sufficient. The full set of constraints on "physically meaningful" wave functions accounts also for their being solutions of the Schrödinger equation, to be discussed in what follows.

Postulate 2: **Dynamical variables are associated with operators. In particular, the position (x) the momentum (p_x) of a particle along a given Cartesian axis are associated with the operators, $\hat{x}$ and $\hat{p}_x$, where**

$$\hat{x}\psi \equiv x\psi(x)$$

$$\hat{p}_x\psi \equiv -i\hbar\frac{d}{dx}\psi(x). \tag{11.1.4}$$

In the three-dimensional space, the position and momentum are associated with the vector operators,

$$\hat{\mathbf{r}} \equiv (\hat{x}, \hat{y}, \hat{z})$$

$$\hat{\mathbf{p}} \equiv (\hat{p}_x, \hat{p}_y, \hat{p}_z). \tag{11.1.5}$$

The position operator, $\hat{\mathbf{r}}$, is a local operator in the position representation, namely, its operation is well defined at each point in the coordinate space, $\hat{\mathbf{r}}\psi \equiv (x \cdot \psi(\mathbf{r}), y \cdot \psi(\mathbf{r}), z \cdot \psi(\mathbf{r}))$. The momentum operator, $\hat{\mathbf{p}}$, is a nonlocal (differential) operator in the position representation. Its operation requires knowledge of the wave function also at the vicinity of each point, $\hat{\mathbf{p}}\psi = -i\hbar\left(\frac{\partial\psi(\mathbf{r})}{\partial x}, \frac{\partial\psi(\mathbf{r})}{\partial y}, \frac{\partial\psi(\mathbf{r})}{\partial z}\right) = -i\hbar\nabla\psi(\mathbf{r})$.

Other observables are represented as operators according to their relations to the canonical position and momentum operators, as known from classical mechanics. Examples are:

The angular momentum of a particle, which is a vector operator,

$$\hat{\mathbf{L}} = \hat{\mathbf{r}} \times \hat{\mathbf{p}} = (\hat{y}\hat{p}_z - \hat{z}\hat{p}_y, \hat{z}\hat{p}_x - \hat{x}\hat{p}_z, \hat{x}\hat{p}_y - \hat{y}\hat{p}_x). \tag{11.1.6}$$

Scalar or vector potentials are represented in terms of the corresponding functions of the canonical operators. For example, the operator that corresponds to the scalar potential energy function of a particle, $V(\mathbf{r})$, is a function of the local position operator,

$$\hat{V} = V(\hat{x}, \hat{y}, \hat{z}). \tag{11.1.7}$$

The kinetic energy of a particle of mass m is a scalar function of its momentum, $T = \frac{|\mathbf{p}|^2}{2m}$, represented in terms of the nonlocal momentum operator,

$$\hat{T} = \frac{1}{2m}\hat{\mathbf{p}}\cdot\hat{\mathbf{p}} = \frac{1}{2m}(\hat{p}_x^2 + \hat{p}_y^2 + \hat{p}_z^2), \tag{11.1.8}$$

where $\hat{T}\psi = \frac{-\hbar^2}{2m}\left(\frac{\partial^2\psi(\mathbf{r})}{\partial x^2} + \frac{\partial^2\psi(\mathbf{r})}{\partial y^2} + \frac{\partial^2\psi(\mathbf{r})}{\partial z^2}\right) = \frac{-\hbar^2}{2m}\Delta\psi(\mathbf{r})$.

The Hamiltonian of classical mechanics [3.1], $H = T + V$, is represented as the sum of the corresponding quantum mechanical operators,

$$\hat{H} \equiv \hat{T} + \hat{V}. \tag{11.1.9}$$

The generalization to many-particle systems is straightforward. Considering a system of N particles with masses, $m_1, m_2, \ldots, m_N$, and respective coordinates, $\mathbf{r}_1, \mathbf{r}_2, \ldots, \mathbf{r}_N$, the potential and kinetic energy operators read $\hat{V}_{\mathbf{r}_1, \mathbf{r}_2, \ldots, \mathbf{r}_N} \psi = V(\mathbf{r}_1, \mathbf{r}_2, \ldots, \mathbf{r}_N) \psi(\mathbf{r}_1, \mathbf{r}_2, \ldots, \mathbf{r}_N)$ and $\hat{T}_{\mathbf{r}_1, \mathbf{r}_2, \ldots, \mathbf{r}_N} \psi = \sum_{i=1}^{N} \frac{-\hbar^2}{2m_i} \Delta_{\mathbf{r}_i} \psi(\mathbf{r}_1, \mathbf{r}_2, \ldots, \mathbf{r}_N)$, respectively. Detailed discussions as well as exercises concerning the definition of quantum mechanical operators, functions of operators, and their properties can be found in Chapter 3.

The operators representing dynamical variables share common properties. The two of the most important ones are:

The operators are linear in the space of wave functions (see Section 3.1 for relevant exercises). An operator $\hat{A}$ is linear in the space $\{\psi(\mathbf{r})\}$, if the following identity holds for any two functions in the space, $\psi_1(\mathbf{r})$ and $\psi_2(\mathbf{r})$, and for any scalars, a_1 and a_2,

$$\hat{A}(a_1 \psi_1(\mathbf{r}) + a_2 \psi_2(\mathbf{r})) = a_1 \hat{A} \psi_1(\mathbf{r}) + a_2 \hat{A} \psi_2(\mathbf{r}). \tag{11.1.10}$$

The operators are Hermitian (self-adjoint) in the space of wave functions (see Sections 4.5, 4.6 for relevant exercises). An operator $\hat{A}$ is Hermitian in the space $\{\psi(\mathbf{r})\}$, if the following identity holds for any two functions in the space, $\psi_1(\mathbf{r})$ and $\psi_2(\mathbf{r})$:

$$\int_{V_{ps}} \psi_1^*(\mathbf{r}) \hat{A} \psi_2(\mathbf{r}) d\mathbf{r} = \left[\int_{V_{ps}} \psi_2^*(\mathbf{r}) \hat{A} \psi_1(\mathbf{r}) d\mathbf{r} \right]^* \tag{11.1.11}$$

Hermitian operators are characterized by a set of real eigenvalues, and orthonormal eigenfunctions. (See Section 4.6 for a detailed discussion.)

Postulate 3: ***The time evolution of a system is uniquely defined by its initial state and its Hamiltonian operator. The equation of motion for the state of a system is the time-dependent Schrödinger equation.*** Let $\psi(\mathbf{r}, t)$ represent the state of a system at any time, t; then,

$$i\hbar \frac{\partial}{\partial t} \psi(\mathbf{r}, t) = \hat{H} \psi(\mathbf{r}, t). \tag{11.1.12}$$

The time-dependent Schrödinger equation is a first-order differential equation in t, which means that ***the wave function representing a system is uniquely defined at all times, given that it is known at a certain point in time*** (i.e., an initial time, $\psi(\mathbf{r}, t)|_{t=0} = \psi_0(\mathbf{r})$). The time-dependent Schrödinger equation is a homogeneous linear differential equation, $\left[\hat{H} - i\hbar \frac{\partial}{\partial t} \right] \psi(\mathbf{r}, t) = 0$, from which follows ***the superposition principle: Any linear combination of solutions is also a solution.*** Specifically, a solution to the Schrödinger equation is defined only up to multiplication by a scalar constant. The constant is conventionally chosen to impose normalization on the solution (Eq. (11.1.2) or

Eq. (11.1.3)). Since the Hamiltonian is a Hermitian operator, the normalization is conserved at all times. (See Chapter 15 for a general proof.):

$$\int_{V_{ps}} \psi^*(\mathbf{r},t)\psi(\mathbf{r},t)d\mathbf{r} = 1 \quad \text{or} \quad \oint_{S_{ps}} \mathbf{J}(\mathbf{r},t)\cdot d\mathbf{S} = 0. \tag{11.1.13}$$

Notice that the Schrödinger equation imposes mathematical constraints on $\psi(\mathbf{r}, t)$, in addition to the ones associated with its interpretation as a probability amplitude (e.g., the normalization requirements, Eqs. (11.1.2, 11.1.3)). Specifically, the wave function $\psi(\mathbf{r},t)$ must be continuously differentiable with respect to any of the system position variables. The time-dependent Schrödinger equation was already discussed in Chapter 4 of this book. An extended discussion of quantum dynamics is given also in Chapter 15.

Postulate 4: **In a single measurement of an observable on a single system, the measured value can only be an eigenvalue of the operator that represents the observable. The probability for measuring a specific eigenvalue is defined by the wave function representing the system and by the operator representing the observable.**

Let a system be in a state associated with the wave function, $\psi(\mathbf{r},t)$, prior to a measurement at time t', namely at $t \leq t'$, and let the measured observable be represented by the operator, $\hat{A}$. If $\psi(\mathbf{r},t')$ is an eigenfunction of $\hat{A}$, namely, if $\hat{A}\psi(\mathbf{r},t') = \alpha\psi(\mathbf{r},t')$, with an eigenvalue, α, then the result of the measurement would yield the value α, with perfect certainty. Notice that α is real-valued, since observables are represented by Hermitian operators (postulate 2).

In the more general case, where $\psi(\mathbf{r},t')$ is not necessarily an eigenfunction of $\hat{A}$, the following holds:

I. The result of a single measurement of the observable associated with $\hat{A}$ can only be one of the eigenvalues of $\hat{A}$, as defined by its eigenvalue equation,

$$\hat{A}\varphi_n(\mathbf{r}) = \alpha_n\varphi_n(\mathbf{r}). \tag{11.1.14}$$

II. For a proper wave function, $\psi(\mathbf{r},t')$, the probability of measuring an isolated eigenvalue (from a discrete spectrum) at the measurement time, $P(\alpha_n,t')$, is given by the overlap integral between the (normalized) wave function at the time of measurement, $\psi(\mathbf{r},t')$, and the corresponding (normalized) eigenfunction, $\varphi_n(\mathbf{r})$, namely

$$P(\alpha_n,t') \equiv |\int_{V_{ps}} \varphi_n^*(\mathbf{r})\psi(\mathbf{r},t')d\mathbf{r}|^2, \tag{11.1.15}$$

where $\int_{V_{ps}} |\psi(\mathbf{r},t')|^2 d\mathbf{r} = 1$ and $\int_{V_{ps}} |\varphi_n(\mathbf{r})|^2 d\mathbf{r} = 1$. If the eigenvalue, α_n, is N-fold degenerate, namely $\hat{A}\varphi_k(\mathbf{r}) = \alpha_n\varphi_k(\mathbf{r})$, where $\varphi_1, \varphi_2,\ldots,\varphi_N$ are a subset of ortho-normal functions, $\int_{V_{ps}} \varphi_{k'}^*(\mathbf{r})\varphi_k(\mathbf{r})d\mathbf{r} = \delta_{k,k'}$, the probability for measuring α_n is given

by the summation, $P(\alpha_n,t') \equiv \sum_{k=1}^{N} |\int_{V_{ps}} \varphi_k^*(\mathbf{r})\psi(\mathbf{r},t')d\mathbf{r}|^2.$

When the operator $\hat{A}$ has a continuous spectrum, namely $\hat{A}\varphi_\alpha(\mathbf{r}) = \alpha\varphi_\alpha(\mathbf{r})$, with α being a continuous variable, the probability for measuring the eigenvalue α, within an interval $d\alpha$, is $\rho(\alpha,t')d\alpha$, where the probability density reads

$$\rho(\alpha,t') \equiv |\int_{V_{ps}} \varphi_\alpha^*(\mathbf{r})\psi(\mathbf{r},t')d\mathbf{r}|^2. \tag{11.1.16}$$

Also in this case, if α is a degenerate eigenvalue, the right-hand side must be extended to include the sum over all degenerate states.

Notice that if the entire eigenvalue spectrum of $\hat{A}$ is discrete, probability conservation means that

$$\sum_n P(\alpha_n,t') = 1, \tag{11.1.17}$$

where the sum is over the entire set of eigenvalues of the operator (which can also be infinite). If the entire eigenvalue spectrum of $\hat{A}$ is continuous, probability conservation means that

$$\int \rho(\alpha,t')d\alpha = 1. \tag{11.1.18}$$

As we shall discuss in Section 11.4, Eqs. (11.1.17, 11.1.18) indeed hold from a mathematical point of view, since the corresponding set of eigenfunctions of any linear Hermitian operator is a complete orthonormal system that spans the space of wave functions.

III. Given a measurement on the system at time t', which yields a specific eigenvalue of $\hat{A}$, for example, α_n, the state of the system collapses to the respective eigenfunction of $\hat{A}$,

$$\psi(\mathbf{r},t \leq t') \mapsto \varphi_n(\mathbf{r},t'). \tag{11.1.19}$$

Notice that the Schrödinger equation dictates the evolution of the physical state of the system before the measurement ($\psi(\mathbf{r},t \leq t')$) and also after the measurement time (where $\varphi_n(\mathbf{r},t')$ is an initial condition for the future evolution at $t > t'$), but not during the measurement. Notice also that this postulate refers to a full collapse of the wave function, which refers to a perfect precision in the determination of the measured eigenvalue. This is an idealized limit, and in practice the finite precision of the measurement also needs to be accounted for [11.1].

11.2 The Hilbert Space of Proper Quantum States and Dirac's Notations

In this section we turn to an abstract formulation of the postulates of quantum mechanics, which is fully consistent with, but not limited to wave functions in the position space, as introduced so far. We start by noticing that the solutions to the Schrödinger equation for a given system define a vector space over the field of complex numbers.

Particularly, this space is closed with respect to addition of functions and to their multiplications by scalar complex numbers (i.e., a sum of wave functions and/or a wave function multiplied by a complex number is a wave function), where the following set of properties, defining a vector space, hold trivially. Indeed, addition of wave function satisfies ***commutativity*** (for any wave functions, $\psi_1(\mathbf{r})$ and $\psi_2(\mathbf{r})$, we have $\psi_1(\mathbf{r}) + \psi_2(\mathbf{r}) = \psi_2(\mathbf{r}) + \psi_1(\mathbf{r})$), ***associativity*** (for any wave functions $\psi_1(\mathbf{r})$, $\psi_2(\mathbf{r})$, and $\psi_3(\mathbf{r})$, we have $\psi_1(\mathbf{r}) + [\psi_2(\mathbf{r}) + \psi_3(\mathbf{r})] = [\psi_1(\mathbf{r}) + \psi_2(\mathbf{r})] + \psi_3(\mathbf{r})$), the existence of an ***additive identity*** (namely the trivial solution, $\psi(\mathbf{r}) = \psi(\mathbf{r}) + 0$), and the existence of an ***additive inverse*** for any $\psi(\mathbf{r})$ (namely $\psi(\mathbf{r}) - \psi(\mathbf{r}) = 0$). Additionally, multiplication of a wave function by a complex valued scalar satisfies ***distributivity*** (for any scalars a and b, and any wave functions, $\psi_1(\mathbf{r})$ and $\psi_2(\mathbf{r})$, we have $a[\psi_1(\mathbf{r}) + \psi_2(\mathbf{r})] = a\psi_1(\mathbf{r}) + a\psi_2(\mathbf{r})$ and $(a+b)\psi_1(\mathbf{r}) = a\psi_1(\mathbf{r}) + b\psi_1(\mathbf{r})$), ***associativity*** (for any scalars a and b, $(ab)\psi(\mathbf{r}) = a[b\psi(\mathbf{r})]$), and the existence of ***an identity*** ($1 \cdot \psi(\mathbf{r}) = \psi(\mathbf{r})$). We can accordingly associate the physical states of a given system with abstract vectors, each corresponding to a realization in terms of a wave function, $\psi(\mathbf{r})$. Following Dirac [11.2], an abstract vector representing a physical state is termed a "ket," where (as elaborated in Section 11.5) there is a unique correspondence between a ket, denoted as $|\psi\rangle$, and its realization in terms of a wave function,

$$\psi(\mathbf{r}) \leftrightarrow |\psi\rangle. \tag{11.2.1}$$

Let us consider first the vector space of all proper (square-integrable) complex-valued wave functions representing the possible physical states of a given closed system. According to the postulates (Eqs. (11.1.1, 11.1.2)) the information with respect to any measurable quantity in a system characterized by a ket vector, $|\psi\rangle$, is contained in this vector. This information is expressed in terms of probabilities (or probability densities), which must be real-valued and nonnegative. In particular, the state vector, $|\psi\rangle$, must be normalizable, namely, must have a unique norm, expressed in terms of a real-valued (orderable), positive scalar. Moreover, the definition of the norm must account for the fact that the vector space is defined over the field of complex numbers. A natural choice is to associate the norm of a vector with its "length," namely the positive square root of its inner product with itself (a Euclidian norm). For this purpose, a suitable inner product between vectors in the space needs to be defined. Following Dirac, the inner product between the vectors $|\psi\rangle$ and $|\phi\rangle$ is termed a "braket," and is denoted as

$$\langle \varphi(\mathbf{r}), \psi(\mathbf{r}) \rangle \leftrightarrow \langle \varphi | \psi \rangle, \tag{11.2.2}$$

where, in order for $\langle \psi | \psi \rangle$ to be real-valued, the following must hold:

$$\langle \varphi | \psi \rangle = \langle \psi | \varphi \rangle^*. \tag{11.2.3}$$

Notice that the inner product between two different ket vectors can be complex in general. Nevertheless, the condition, Eq. (11.2.3), assures that each physically meaningful (nonzero) solution can be associated with a positive norm, $\sqrt{\langle \psi | \psi \rangle} > 0$. It is customary to multiply the vector $|\psi\rangle$ by a scalar in order to obtain the normalized ket,

$|\tilde{\psi}\rangle = \frac{1}{\sqrt{\langle\psi|\psi\rangle}}|\psi\rangle$, which corresponds to a normalized wave function (see Eqs. (11.1.1, 11.1.2)):

$$\int_{V_{ps}} \tilde{\psi}^*(\mathbf{r})\tilde{\psi}(\mathbf{r})d\mathbf{r} = 1 \leftrightarrow \langle\tilde{\psi}|\tilde{\psi}\rangle = 1. \tag{11.2.4}$$

The vector appearing in the left place within the bracket is termed the "bra." Eq. (11.2.3) implies that a bra, denoted as $\langle\varphi|$, is related uniquely to a ket vector, $|\varphi\rangle$, in terms of Hermitian conjugation. Namely, for any $|\chi\rangle$ in the vector space, the scalars, $\langle\varphi|\chi\rangle$ and $\langle\chi|\varphi\rangle$, are complex conjugates. In linear algebra terms (namely, in finite-dimensional spaces), a ket would be analogous to a column vector of N complex-valued entries, and the corresponding bra would be associated with its **Hermitian conjugate** (complex-transposed) vector, namely the row vector consisting of the complex conjugates of these entries. The bra-ket inner product is therefore analogous to the scalar (dot) product between two vectors of a finite dimension. Nevertheless, the dimension of the ket space can be infinite, in general, that is, $N \to \infty$. This can be readily seen by inspecting, for example, Eq.(11.2.4). Replacing the normalization integral by a Riemann sum over finite volume elements, the inner product can be *approximately* associated with the dot product of a complex vector with itself, namely $\langle\tilde{\psi}|\tilde{\psi}\rangle \leftrightarrow \lim_{d\mathbf{r}\to 0} d\mathbf{r} \sum_{n=1}^{V_{ps}/d\mathbf{r}} \tilde{\psi}_n^*(\mathbf{r})\tilde{\psi}_n(\mathbf{r})$, where the nth vector element is $\tilde{\psi}_n(\mathbf{r})\sqrt{d\mathbf{r}}$. Nevertheless, the exact result of the normalization integral would be reproduced only when the volume element becomes infinitesimally small, and the vector dimension therefore becomes infinite, as $V_{ps}/d\mathbf{r} \to \infty$. *The infinite dimension of the ket space reflects the infinity of the physical space.* Indeed, even for a closed system composed of a fixed number of particles, the idea that the particle's position changes continuously in space (with resolution limited only by the measurement precision) implies that the position can obtain an infinite number of entries, and a similar argument is true for other physical properties such as momentum, energy, and so on. Notice in this context that even when the spectrum of an observable is discrete (rather than continuous) the number of eigenvalues, and therefore the different outcomes of a measurement, can still be infinite. We already encountered this when discussing the discrete energy spectrum of several bound systems, including the one-dimensional harmonic oscillator, a particle in a one-dimensional infinite box, or the hydrogen atom.

It is therefore natural that the postulates of quantum mechanics associate the space of proper physical states of a system with a Hilbert space. *The Hilbert space is an infinite vector space over the field of complex numbers, which admits an inner product (e.g., Eq. (11.2.3)) and a finite norm (e.g. Eq. (11.2.4)) for any nonzero vector. Moreover, a Hilbert space is the span of an infinite (countable) set of orthonormal vectors (a complete orthonormal set).* The latter property extends the existence of a finite orthonormal basis for finite-dimensional inner-product vector spaces to infinite dimensions. While a mathematical proof of existence is beyond the scope of this book, in the following section we shall show how such sets can be constructed, and we shall discuss their essential role in the interpretation of the postulates of quantum mechanics.

Given a complete orthonormal set for spanning a Hilbert space of *proper physical states*, each vector within the set, denoted as $|\varphi_n\rangle$, can be associated with a natural index, $n = 1, 2, 3, \ldots$. The orthonormality of the set is expressed in terms of the Kronecker delta (Eq. (4.6.4)),

$$\langle \varphi_{n'} | \varphi_n \rangle = \delta_{n',n} \tag{11.2.5}$$

where any ket $|\psi\rangle$ in the Hilbert space can be expanded as

$$|\psi\rangle = \sum_{n=1}^{\infty} a_n |\varphi_n\rangle. \tag{11.2.6}$$

The series of scalars $\{a_n\}$ are the elements of the "vector representation" of the state in the selected set. These can be readily identified by taking the inner product of $|\psi\rangle$ with any specific ket, $\langle \varphi_{n'} | \psi \rangle$. Using the orthogonality condition (Eq. (11.2.5)), this yields

$$a_n = \langle \varphi_n | \psi \rangle, \tag{11.2.7}$$

where the expansion, Eq. (11.2.6), obtains the form,

$$|\psi\rangle = \sum_{n=1}^{\infty} |\varphi_n\rangle \langle \varphi_n | \psi \rangle. \tag{11.2.8}$$

Since for any two kets, $|\psi\rangle$ and $|\chi\rangle$, we have $\langle \psi | \chi \rangle = \langle \chi | \psi \rangle^*$, and since $\langle \chi | \psi \rangle^* = \sum_{n=1}^{\infty} \langle \chi | \varphi_n \rangle^* \langle \varphi_n | \psi \rangle^* = \sum_{n=1}^{\infty} \langle \psi | \varphi_n \rangle \langle \varphi_n | \chi \rangle$, it follows that the bra state, $\langle \psi |$, can be expanded similarly as follows:

$$\langle \psi | = \sum_{n=1}^{\infty} \langle \psi | \varphi_n \rangle \langle \varphi_n |. \tag{11.2.9}$$

In analogy with a finite-dimensional space, the ket corresponds to a column vector, ψ, whose entries, $[\psi]_n = \langle \varphi_n | \psi \rangle$, are the representation of the vector $|\psi\rangle$ in the selected orthonormal set. The corresponding bra is the complex conjugate transposed vector, namely $\psi^\dagger$.

Since the expansions (Eqs. (11.2.8, 11.2.9)) hold for any $|\psi\rangle$, we can identify an expression of the identity operator in the Hilbert space, which reads

$$\sum_{n=1}^{\infty} |\varphi_n\rangle \langle \varphi_n | = \hat{I}. \tag{11.2.10}$$

The "ket-bra" structure, $|\psi_1\rangle \langle \psi_2 |$, associated with any two states, $|\psi_1\rangle$ and $|\psi_2\rangle$, indeed defines an operator, since it maps any ket on a ket within the space, that is, $|\psi_1\rangle \langle \psi_2 | \cdot |\psi\rangle = \langle \psi_2 | \psi \rangle |\psi_1\rangle$. In a finite-dimensional space this structure is analogous to a matrix, formed by an outer product of two vectors, $\psi_1 \otimes \psi_2^\dagger$, where the ket corresponds to a column vector, ψ_1, and the bra corresponds to the complex conjugate transposed vector, $\psi_2^\dagger$.

The identity operator is shown to be a sum over all outer products of the basis vectors with themselves, which manifests the completeness of the set. Moreover, any linear

operator in the Hilbert space can be represented using a complete orthonormal set.
Indeed, for any linear operator, $\hat{A}$, we obtain

$$\hat{A} = \hat{I}\hat{A}\hat{I} = \sum_{n=1}^{\infty} |\varphi_n\rangle\langle\varphi_n|\hat{A} \sum_{n'=1}^{\infty} |\varphi_{n'}\rangle\langle\varphi_{n'}| = \sum_{n,n'=1}^{\infty} A_{n,n'} |\varphi_n\rangle\langle\varphi_{n'}|. \tag{11.2.11}$$

In analogy with finite vector spaces, the scalar $A_{n,n'} \equiv \langle\varphi_n|\hat{A}|\varphi_{n'}\rangle$ is termed "the matrix element" of the operator $\hat{A}$, between the n and n' basis vectors, where the indexes, n and n', can be associated, respectively, with the row and column index of an infinite "matrix representation" of the operator $\hat{A}$ in the selected basis,

$$A_{n,n'}, \equiv [\mathbf{A}]_{n,n'}. \tag{11.2.12}$$

Notice that a matrix representation of an operator $\hat{A}$ depends on the choice of basis, and therefore it is not unique. Let $\mathbf{A}$ and $\tilde{\mathbf{A}}$ be the matrix representations of $\hat{A}$ obtained using two different orthonormal sets, $\{|\varphi_n\rangle\}$, and $\{|\chi_m\rangle\}$, respectively, as defined according to Eqs. (11.2.11, 11.2.12). It then follows that $\tilde{A}_{m,m'} = \sum_{n,n'=1}^{\infty} \langle\chi_m|\varphi_n\rangle A_{n,n'}\langle\varphi_{n'}|\chi_{m'}\rangle$. Associating the inner products between the different basis vectors with the elements of a matrix, $[\mathbf{S}]_{m,n} = \langle\chi_m|\varphi_n\rangle$, the last equation can be formally written as

$$\tilde{\mathbf{A}} = \mathbf{S}\mathbf{A}\mathbf{S}^{\dagger}, \tag{11.2.13}$$

where the Hermitian conjugate matrix, $\mathbf{S}^{\dagger}$, is defined by $[\mathbf{S}^{\dagger}]_{n,m} = [\mathbf{S}]_{m,n}^{*} = \langle\varphi_n|\chi_m\rangle$. Since each of the two sets $\{|\varphi_n\rangle\}$ and $\{|\chi_m\rangle\}$ spans the space, it immediately follows (see Ex. 11.2.1), that the matrix $\mathbf{S}$ is unitary, namely

$$\mathbf{S}\mathbf{S}^{\dagger} = \mathbf{S}^{\dagger}\mathbf{S} = \mathbf{I}. \tag{11.2.14}$$

Changing the matrix representation of an operator, $\hat{A}$, from one complete orthonormal set to another is therefore associated with a unitary transformation, Eq. (11.2.13).

Similarly, the vector representation of a state $|\psi\rangle$ depends on the choice of basis. Expanding the ket $|\psi\rangle$ in two different orthonormal basis sets, $|\psi\rangle = \sum_{n=1}^{\infty} |\varphi_n\rangle\langle\varphi_n|\psi\rangle = \sum_{m=1}^{\infty} |\chi_m\rangle\langle\chi_m|\psi\rangle$, we readily obtain $\langle\chi_m|\psi\rangle = \sum_{n=1}^{\infty} \langle\chi_m|\varphi_n\rangle\langle\varphi_n|\psi\rangle$. Denoting the elements of the two different vector representations as $\psi_n = \langle\varphi_n|\psi\rangle$ and, $\tilde{\psi}_m = \langle\chi_m|\psi\rangle$, we formally obtain

$$\tilde{\boldsymbol{\psi}} = \mathbf{S}\boldsymbol{\psi}. \tag{11.2.15}$$

Changing the vector representation of a state $|\psi\rangle$ from one complete orthonormal set to another is therefore formally equivalent to a unitary matrix operation.

Exercise 11.2.1 *(a) $\mathbf{S}$ is a matrix whose elements are given as $[\mathbf{S}]_{m,n} = \langle\chi_m|\varphi_n\rangle$, where the sets $\{|\varphi_1\rangle, |\varphi_2\rangle, |\varphi_3\rangle\ldots\}$ and $\{|\chi_1\rangle, |\chi_2\rangle, |\chi_3\rangle,\ldots\}$ are two different complete orthonormal sets of vectors, which span the Hilbert space. Use Eq. (11.2.10) to show that the matrix $\mathbf{S}$ is unitary, namely $[\mathbf{S}^{\dagger}\mathbf{S}] = \mathbf{I}$ (or, $[\mathbf{S}^{\dagger}\mathbf{S}]_{m,n} = \delta_{m,n}$).*

It is instructive to define a Hermitian conjugate to a linear operator $\hat{A}$ as $\hat{A}^{\dagger}$, whose matrix representation in any complete orthonormal set is identified with the complex conjugate transposed matrix representation of $\hat{A}$, namely

$$[\mathbf{A}^{\dagger}]_{n,n'} \equiv [\mathbf{A}]^{*}_{n',n}. \tag{11.2.16}$$

Using the definition of the matrix elements and the completeness relation (Eq. (11.2.11)), the following identity holds for any states, $|\psi\rangle$ and $|\chi\rangle$, in the Hilbert space (see Ex. 11.2.2):

$$\langle\psi|\hat{A}^{\dagger}|\chi\rangle = \langle\chi|\hat{A}|\psi\rangle^{*}. \tag{11.2.17}$$

Exercise 11.2.2   *Use Eqs. (11.2.11, 11.2.12, 11.2.16) and show that Eq. (11.2.17) holds for any states, $|\psi\rangle$ and $|\chi\rangle$, in the Hilbert space.*

The Hermitian conjugate operator is instrumental for defining the bra associated with $\hat{A}|\psi\rangle \equiv |\hat{A}\psi\rangle$. Since for any $|\chi\rangle$ in the Hilbert space we have $\langle\chi|\hat{A}\psi\rangle = \langle\hat{A}\psi|\chi\rangle^{*}$, and using Eq. (11.2.17), it follows that

$$\langle\hat{A}\psi| = \langle\psi|\hat{A}^{\dagger}. \tag{11.2.18}$$

Exercise 11.2.3   *The Schrödinger equation for a ket reads $\frac{\partial}{\partial t}|\psi(t)\rangle = \frac{1}{i\hbar}\hat{H}|\psi(t)\rangle$, where $\hat{H}$ is the Hamiltonian operator. Use Eq. (11.2.18) and show that the equation of motion for the corresponding bra reads $\frac{\partial}{\partial t}\langle\psi(t)| = \frac{-1}{i\hbar}\langle\psi(t)|\hat{H}$.*

Notice that for a Hermitian operator ($\hat{A} = \hat{A}^{\dagger}$), the matrix representation is Hermitian (see Eq. (11.2.16)),

$$\hat{A} = \hat{A}^{\dagger} \Leftrightarrow \mathbf{A} = \mathbf{A}^{\dagger}, \tag{11.2.19}$$

where

$$\hat{A} = \hat{A}^{\dagger} \Leftrightarrow \langle\chi|\hat{A}|\psi\rangle = \langle\psi|\hat{A}|\chi\rangle^{*}. \tag{11.2.20}$$

As discussed in Section 11.5, when the state vectors and the operators are presented in the continuous position representation, Eq. (11.2.20) coincides with the definition of Hermiticity in terms of integrals over the coordinate space for proper wave functions, as introduced in Section 4.5 (Eq. (4.5.1)).

11.3 Extending the Vector Space to Include Improper States

We now turn to consider improper states, which are outside the Hilbert space of square integrable functions. Following Dirac [11.2], it is most useful to extend the vector space of physical states beyond that Hilbert space, such that certain improper states can be included. We have already encountered improper solutions to the Schrödinger equation, which are not subject to the normalization requirement, Eqs. (11.1.1, 11.1.2), and yet conserve probability fluxes (Eq. (11.1.3)). These states obtain their physical meaning as stationary scattering states in open systems (see Chapter 7). Moreover, although scattering wave functions are not square integrable, a continuous set of scattering wave functions can replace a complete orthonormal set for expanding square integrable functions. A well-known example (to be discussed in detail in what follows) is the representation of square integrable functions as Fourier transforms, which are integral expansions in terms of a continuous set of non–square-integrable plane waves (encountered in Chapter 7). Such integral expansions exist also for other continuous sets of improper functions, as discussed in what follows.

In order to treat (physically meaningful) sets of improper functions within an extended ket space, we utilize the existence of integral transforms, namely representations of proper states within the Hilbert space in terms of a specified uncountable set of improper states. Denoting the set of improper functions as ket vectors, $\{|\varphi_\alpha\rangle\}$, each vector is marked by the continuous index, α (which can be a scalar or a vector, depending on dimensions of the physical system). *The existence of an integral transform implies that each ket in the Hilbert space can be expanded as follows,*

$$|\psi\rangle \equiv \int d\alpha\, \psi(\alpha)|\varphi_\alpha\rangle, \tag{11.3.1}$$

where the expansion coefficients $\{\psi(\alpha)\}$ are the inner products between the proper ket, $|\psi\rangle$, and the improper kets, $|\varphi_\alpha\rangle$ (subject to Eq. (11.2.3)),

$$\langle\varphi_\alpha|\psi\rangle \equiv \psi(\alpha). \tag{11.3.2}$$

The function, $\psi(\alpha)$, is often termed "the continuous vector representation" of the state vector $|\psi\rangle$. Substitution of Eq. (11.3.2) into Eq. (11.3.1), we obtain $|\psi\rangle = \int d\alpha |\varphi_\alpha\rangle\langle\varphi_\alpha|\psi\rangle$, which implies that *the identity operator in the Hilbert space can be expressed in terms of the specified set of improper states,*

$$\hat{I} = \int d\alpha |\varphi_\alpha\rangle\langle\varphi_\alpha|. \tag{11.3.3}$$

Using the normalization condition for any proper state, $\langle\psi|\psi\rangle = 1$, and the identity operator representation, it follows that

$$\int d\alpha\langle\psi|\varphi_a\rangle\langle\varphi_\alpha|\psi\rangle = \int d\alpha|\psi(\alpha)|^2 = 1, \tag{11.3.4}$$

which reassures that $\psi(\alpha)$ *is a square integrable function of* α.

Eqs. (11.3.1–11.3.3) can be regarded as continuous analogues to Eqs. (11.2.6, 11.2.7, 11.2.10), which refer to spanning the Hilbert space of proper states by a complete orthonormal set of proper vectors. However, let us reemphasize that the orthonormality condition, Eq. (11.2.5), does not apply to the set $\{|\varphi_\alpha\rangle\}$. In particular, the improper states are, by definition, not normalizable, namely, $\langle\varphi_a|\varphi_a\rangle$ does not strictly exist. Nevertheless, the existence of the expansions, Eqs. (11.3.1, 11.3.2), enables us to extend the definition of the orthonormality condition beyond Eq. (11.2.5) and to provide a formal basis for an analogy between the expansion of proper states in terms of a complete orthonormal system of proper states within the Hilbert space and in terms of a specified set of improper states. Using Eqs. (11.3.1, 11.3.2), we readily obtain

$$\psi(\alpha) = \int d\alpha' \langle \varphi_\alpha | \varphi_{\alpha'} \rangle \psi(\alpha'). \tag{11.3.5}$$

This can hold for any α, only if the inner product between $|\varphi_\alpha\rangle$ and $|\varphi_{\alpha'}\rangle$ is identified with "Dirac's delta function" [11.2],

$$\langle \varphi_\alpha | \varphi_{\alpha'} \rangle = \delta(\alpha' - \alpha), \tag{11.3.6}$$

which (for a one-dimensionl Cartesian variable) is defined as follows,

$$\int_{-\infty}^{\infty} \delta(\alpha' - \alpha) d\alpha = 1 \quad ; \quad \delta(\alpha' - \alpha) = \begin{cases} 0 & ; \quad \alpha \neq \alpha' \\ \infty & ; \quad \alpha = \alpha' \end{cases}, \tag{11.3.7}$$

such that

$$\psi(\alpha) = \int_{-\infty}^{\infty} \delta(\alpha' - \alpha) \psi(\alpha') d\alpha'. \tag{11.3.8}$$

Strictly speaking, Dirac's delta is not a well-defined function of its arguments. Nevertheless, it is a limit of well-defined functions, where Eqs. (11.3.7, 11.3.8) strictly hold under a limit (see, e.g., Ex. 11.3.1). Moreover, Eq. (11.3.8) is an integral transform, which expresses a square integrable function, $\psi(\alpha)$, in terms of a set of (improper) delta functions. As we shall discuss in Section 11.5, sets of Dirac's delta functions not only are useful for conveniently representing square integrable functions, but also obtain important physical meaning.

Exercise 11.3.1 *One of the definitions of Dirac's delta is the limit of an infinitely narrow normalized Gaussian distribution,* $\delta(\alpha) = \lim_{\varepsilon \to 0} \sqrt{\frac{1}{4\pi\varepsilon}} e^{\frac{-\alpha^2}{4\varepsilon}}$. *Calculate the integral explicitly and show that indeed* $\lim_{\varepsilon \to 0} \int_{-\infty}^{\infty} \sqrt{\frac{1}{4\pi\varepsilon}} e^{\frac{-(\alpha-\alpha')^2}{4\varepsilon}} \psi(\alpha) d\alpha = \psi(\alpha')$, *where* $\psi(\alpha)$ *is a Gaussian distribution of a fimite width,* $\psi(\alpha) = \sqrt{\frac{1}{2\pi\sigma^2}} e^{\frac{-(\alpha-\alpha_0)^2}{2\sigma^2}}$.

Eqs. (11.3.3, 11.3.6) are the analogues to Eqs. (11.2.10, 11.2.5), which enable one to extend the vector space of physical states beyond the Hilbert space of square-integrable functions. Thereby, proper and improper states can be regarded as kets (and bras) using the same formal terms. For this purpose, it is instructive to review the expressions of the inner product between the different types of vectors in the extended space, in terms of an uncountable set of improper states, $\{|\varphi_\alpha\rangle\}$.

I. The inner product between two proper states, $|\psi\rangle$ and $|\chi\rangle$, can be expanded as

$$\langle \chi | \psi \rangle = \int d\alpha \langle \chi | \varphi_\alpha \rangle \langle \varphi_\alpha | \psi \rangle \tag{11.3.9}$$

where both $\chi(\alpha) \equiv \langle \varphi_a | \chi \rangle$ and $\psi(\alpha) \equiv \langle \varphi_\alpha | \psi \rangle$ are proper functions of α.

II. The inner product between a proper vector, $|\psi\rangle$, and an improper vector, $|\chi_\beta\rangle$, can be expanded as

$$\langle\chi_\beta|\psi\rangle = \int d\alpha\,\langle\chi_\beta|\varphi_\alpha\rangle\langle\varphi_\alpha|\psi\rangle, \qquad (11.3.10)$$

where $\psi(\beta) \equiv \langle\chi_\beta|\psi\rangle$ and $\psi(\alpha) \equiv \langle\varphi_\alpha|\psi\rangle$ are proper function of α, and $\langle\chi_\beta|\varphi_\alpha\rangle$ is an improper function of α and β.

III. The inner product between two improper vectors, $|\psi_\gamma\rangle$, and $|\chi_\beta\rangle$, can be expanded as

$$\langle\chi_\beta|\psi_\gamma\rangle = \int d\alpha\,\langle\chi_\beta|\varphi_\alpha\rangle\langle\varphi_\alpha|\psi_\gamma\rangle, \qquad (11.3.11)$$

where $\langle\chi_\beta|\psi_\gamma\rangle$, $\langle\chi_\beta|\varphi_\alpha\rangle$, and $\langle\varphi_\alpha|\psi_\gamma\rangle$ are improper functions of their respective arguments (α, β and γ).

For the completeness of the discussion, we notice that the expansion of an improper state, for example, $|\varphi_\alpha\rangle$, in terms of a complete set of proper orthonormal functions does not strictly hold, namely $|\varphi_\alpha\rangle \neq \sum_{n=0}^{\infty} |\varphi_n\rangle\langle\varphi_n|\varphi_\alpha\rangle$. Nevertheless, an improper function that is a limit of a proper functions can be expanded using proper states. An important example is the formal representation of Dirac's delta in term of a complete orthogonal set, which reads

$$\delta(\alpha - \alpha') = \sum_{n=0}^{\infty} \langle\varphi_{\alpha'}|\varphi_n\rangle\langle\varphi_n|\varphi_\alpha\rangle = \sum_{n=0}^{\infty} \varphi_n(\alpha')\varphi_n^*(\alpha). \qquad (11.3.12)$$

Thereby, proper and improper states can be regarded as kets and bras using the same formal terms.

11.4 Rationalizing The Postulates

According to the postulates of quantum mechanics: (i) The measurable information on a system in a given state is entirely contained in a wave function (Section 11.1), where, as discussed in Section 11.2, the wave function is associated with an abstract vector in the space of physical states, (ii) measurable quantities are represented by linear operators in that vector space, and (iii) the results of a single measurement are expressed in terms of the eigenvalues and eigenvectors of these operators. In this section we discuss again these seemingly unrelated postulates in the context of the abstract vector space, attempting to unravel the rationale that links them.

So far, we used the fact that the vector space of physical states is spanned by a complete orthonormal set of vectors, but we did not attempt to specify how such orthonormal sets can be identified or constructed. For this purpose, we first recall that observables are represented by operators that are linear and Hermitian in the space of wave functions. We already noticed that the eigenvalues of Hermitian operators in the space of proper (square integrable) functions are real-valued, and therefore there is no fundamental problem in associating them with results of physical measurements (see Section 4.5). We also noticed that the eigenvalue spectrum of a linear Hermitian operator can be associated with an orthonormal set of eigenfunctions in the position representation (see Section 4.6). In what follows we reformulate these results within the abstract vector space of physical states, as introduced in Sections 11.2 and 11.3.

Using Dirac's notations, the eigenvalue equation for a linear Hermitian operator, $\hat{A}$, reads

$$\hat{A}|\varphi_\alpha\rangle = \alpha|\varphi_\alpha\rangle. \tag{11.4.1}$$

Here $|\varphi_\alpha\rangle$ is the eigenvector, and α is the respective scalar eigenvalue. Since $\hat{A}$ is Hermitian, the following identity holds for any two vectors in the space (Eq. (11.2.20)), and particularly for any two eigenvectors of $\hat{A}$, $|\varphi_\alpha\rangle$, and $|\varphi_\beta\rangle$,

$$\langle\varphi_\beta|\hat{A}|\varphi_\alpha\rangle = \langle\varphi_\alpha|\hat{A}|\varphi_\beta\rangle^*. \tag{11.4.2}$$

Using Eq. (11.4.1) in Eq. (11.4.2), we have

$$(\alpha - \beta^*)\langle\varphi_\beta|\varphi_\alpha\rangle = 0. \tag{11.4.3}$$

Setting $|\varphi_\beta\rangle = |\varphi_\alpha\rangle$ and noticing that $\langle\varphi_\alpha|\varphi_\alpha\rangle \neq 0$ (whether $|\varphi_\alpha\rangle$ is proper or improper, where, in the latter case, $\langle\varphi_{\alpha'}|\varphi_\alpha\rangle = \delta(\alpha - \alpha')$; see Eqs. (11.3.6, 11.3.7)), Eq. (11.4.3) is shown to hold only if the eigenvalue is real-valued,

$$\alpha = \alpha^*. \tag{11.4.4}$$

Considering now two different eigenvalues, $\alpha \neq \beta$, Eq. (11.4.3) can hold only if the inner product between the two corresponding eigenvectors vanishes; namely the two vectors corresponding to different eigenvalues, must be orthogonal:

$$\langle\varphi_\beta|\varphi_\alpha\rangle|_{\beta \neq a} = 0. \tag{11.4.5}$$

Notice that this result holds regardless of whether the two eigenvectors are proper or not. Also notice that when $|\varphi_\alpha\rangle$ and $|\varphi_\beta\rangle$ are degenerate vectors (associated with the same eigenvalue), their orthogonality is not imposed by Eq. (11.4.5). Nevertheless, it is always possible to construct linear combinations of degenerate eigenvectors that are mutually orthonormal by using a Gram Schmidt orthonormalization process (see, e.g., Exs. 4.6.3, 4.6.4, for proper functions) [4.4]. Consequently, the set of eigenvectors of any linear Hermitian operator $\hat{A}$ can be mapped onto an orthogonal set. Hereafter we denote this set as $\{|\varphi_\gamma,\rangle\}$, where γ is a parameter that identifies uniquely each vector within the set. (For example, γ can denote the eigenvalue itself, and/or an identifying index within a degenerate subspace, or a quantum number, etc.) Notice that in manyparticle and/or multidimensional systems, a unique definition of the eigenvector requires a group of parameters (e.g., a vector, $\boldsymbol{\gamma} = (\gamma_1, \gamma_2, \ldots)$). Without loss of generality, and for simplicity of notations, we shall restrict the discussion to a single parameter, where the generalization is straightforward. The eigenvalue equation therefore reads

$$\hat{A}|\varphi_\gamma\rangle = \alpha_\gamma|\varphi_\gamma\rangle, \tag{11.4.6}$$

where α_γ is the eigenvalue associated with $|\varphi_\gamma\rangle$. Eigenvectors corresponding to proper states (as, e.g., the Hamiltonian eigenstates in bound physical systems) are associated with a discrete spectrum, which means that γ is countable and can be denoted γ_n, with $n = 1, 2, 3 \ldots$. In this case, the orthonormality condition reads

$$\langle \varphi_{\gamma_{n'}} | \varphi_{\gamma_n} \rangle = \delta_{n,n'} \tag{11.4.7}$$

Eigenvectors corresponding to improper states (as, e.g., scattering states in open systems, as discussed in Chapter 7, or position eigenstates, as discussed in what follows), are associated with a continuous spectrum, which means that γ is uncountable. In this case (see Eq. (11.3.6)), the orthonormality condition corresponds to

$$\langle \varphi_\gamma | \varphi_{\gamma'} \rangle = \delta(\gamma - \gamma'). \tag{11.4.8}$$

In many cases, the spectrum of a linear Hermitian operator contains both continuous and discrete parts, which correspond, respectively to improper and proper wave functions (see, e.g., the discussion of the Hamiltonian eigenstates for a particle in a finite square well potential in Chapter 7). Eq. (11.4.5) means that eigenvectors associated with different parts of the spectrum are necessarily orthogonal, and therefore when the set of eigenvectors includes both countable $\{|\varphi_{\gamma_n}\rangle\}$ and uncountable $\{|\varphi_\gamma\rangle\}$ parts we have

$$\langle \varphi_{\gamma_n} | \varphi_\gamma \rangle = 0. \tag{11.4.9}$$

In finite, N-dimensional inner product vector spaces, a linear Hermitian operator (namely a Hermitian matrix) defines a series of N orthonormal eigenvectors (a basis), which spans the N-dimensional vector space [11.3]. In analogy, *an infinite set of orthonormal eigenvectors of a linear Hermitian operator is a complete orthonormal system, which spans the vector space of physical states*, in the sense that the only vector in the space that is orthogonal to the entire system is the zero vector. A rigorous proof of this statement in beyond our scope here, and the reader is directed to complementary literature on functional analysis.

As discussed in the previous section, the identity operator in the vector space can be expanded in terms of a complete orthonormal system. Choosing an orthonormal set of eigenvectors of a linear Hermitian operator, $\hat{A}$, we have (in the most general case)

$$\hat{I} = \sum_n |\varphi_{\gamma_n}\rangle\langle\varphi_{\gamma_n}| + \int d\gamma |\varphi_\gamma\rangle\langle\varphi_\gamma|, \tag{11.4.10}$$

where we recall that γ_n and γ correspond, respectively, to the discrete (proper) and continuous (improper) parts of the spectrum of $\hat{A}$. In many cases of interest the spectrum is either discrete (finite or infinite) or continuous, such that the identity is expanded in either the countable summation or the integral, in Eq. (11.4.10).

Since the postulates of quantum mechanics associate each observable with a linear Hermitian operator, and since a set of eigenvectors of such an operator spans the space of physical states, it is natural to analyze the results of the measurement of a particular observable in terms of the eigenvectors of the corresponding operator. Notice that the state of the system at the measurement time (e.g., $|\psi\rangle$) can be expanded as a linear combination of any complete orthonormal system. Nevertheless, *the postulates associate the results of a measurement represented by $\hat{A}$, with the expansion of $|\psi\rangle$ in a specific orthonormal system, corresponding to the set of $\hat{A}$ eigenvectors, as defined by $\hat{A}|\varphi_\gamma\rangle = \alpha_\gamma|\varphi_\gamma\rangle$.*

Let us consider first the case where the eigenvectors of $\hat{A}$ correspond to proper states, and the eigenvalues are countable. The vector that represents the state of the system, $|\psi\rangle$, can then be expanded as follows:

$$|\psi\rangle = \sum_{n=0}^{\infty} |\varphi_{\gamma_n}\rangle\langle\varphi_{\gamma_n}|\psi\rangle = \sum_{n=0}^{\infty} \psi_n |\varphi_{\gamma_n}\rangle. \qquad (11.4.11)$$

The scalar expansion coefficients are the inner products, $\psi_n \equiv \langle\varphi_{\gamma_n}|\psi\rangle$ (see Eqs. (11.2.6, 11.2.7)). Invoking again the analogy to a finite-dimensional (Euclidian) vector space, these inner products can be regarded as projections of the state vector $|\psi\rangle$ onto the different basis vectors, $|\varphi_{\gamma_n}\rangle$, just like the elements of the column vector, $[\psi]_n$, are projections (dot products) of the vector ψ on the elementary basis vectors, $\mathbf{e}_n$, namely $[\psi]_n = \mathbf{e}_n \cdot \psi$. In Euclidian geometrical terms, a larger absolute value of the inner product, implies a smaller angle between the vectors, namely a closer similarity between the vector directions. Moreover, for a normalized $|\psi\rangle$, Eq. (11.4.11) readily means

$$\sum_{n=0}^{\infty} |\langle\varphi_{\gamma_n}|\psi\rangle|^2 = \sum_{n=0}^{\infty} |\psi_n|^2 = 1. \qquad (11.4.12)$$

As we can see, a larger $|\langle\varphi_{\gamma_n}|\psi\rangle|^2$ implies a larger contribution of the basis vector, $|\varphi_{\gamma_n}\rangle$, to the norm of the state vector, $|\psi\rangle$, suggesting, again, that the "similarity" between $|\psi\rangle$ and the vector, $|\varphi_{\gamma_n}\rangle$, is larger in some sense.

A similar argument holds also when the observable represented by $\hat{A}$ is associated with a continuous eigenvalue spectrum. In this case, $|\psi\rangle$ can be expanded as an integral,

$$|\psi\rangle = \int d\gamma \langle\varphi_\gamma|\psi\rangle |\varphi_\gamma\rangle = \int d\gamma \psi(\gamma) |\varphi_\gamma\rangle. \qquad (11.4.13)$$

For a normalized $|\psi\rangle$, we have

$$\int d\gamma |\langle\varphi_\gamma|\psi\rangle|^2 = \int d\gamma |\psi(\gamma)|^2 = 1, \qquad (11.4.14)$$

where $\psi(\gamma)$ is a square integrable function of γ (see Eq. (11.3.4)).

According to the postulates, the results of a single measurement represented by an operator $\hat{A}$ on a single system represented by $|\psi\rangle$ are restricted to the eigenvalues of the operator $\hat{A}$, where the occurrence of any specific eigenvalue is inherently probabilistic. Given this framework, it is only natural to expect that the probability of measuring a particular eigenvalue, α_{γ_n} (or α_γ), and of collapsing the system into the corresponding eigenvector, $|\varphi_{\gamma_n}\rangle$ (or $|\varphi_\gamma\rangle$), should be larger with increasing similarity between the state vector $|\psi\rangle$ and that eigenvector. It is also natural to associate this similarity with the projection of $|\psi\rangle$ onto that eigenvector, namely, with the inner product, $\langle\varphi_{\gamma_n}|\psi\rangle$ (or $\langle\varphi_\gamma|\psi\rangle$) (see Ex. 11.4.1). The postulates indeed identify the probability for measuring α_γ with the nonnegative, bounded value, $|\langle\varphi_{\gamma_n}|\psi\rangle|^2$ (or $|\psi(\gamma)|^2 d\gamma$), where the completeness relations, Eqs. (11.4.12, 11.4.14), imply that *the probability of obtaining any eigenvalue of $\hat{A}$ in a measurement of the observable represented by $\hat{A}$ is unity*. Using Dirac's notations, the probability of measuring a specific eigenvalue in the case of a discrete spectrum is given as

$$P(\alpha_{\gamma_n}) = |\langle\varphi_{\gamma_n}|\psi\rangle|^2, \qquad (11.4.15)$$

where, in the case of a continuous spectrum, the probability density for measuring an eigenvalue α_γ, within an interval $d\gamma$, is expressed in terms of a probability density, $\rho(\gamma)d\gamma$, where

$$\rho(\gamma) \equiv |\langle \varphi_\gamma | \psi \rangle|^2. \tag{11.4.16}$$

Notice that in the original formulation of the postulates in terms of wave functions, the probabilities expressed in Eqs. (11.4.15, 11.4.16) were associated with overlap integrals. In the following section we shall show that an overlap integral is merely a particular representation of the inner product between two state vectors in their (continuous) position representation. Eqs. (11.1.15 and 11.1.16) therefore coincide with the "natural," more general definition of the probability of measurement as expressed in Eqs. (11.4.15, 11.4.16).

Exercise 11.4.1 *Let $S = \langle \chi | \psi \rangle$ be the inner product between two normalized proper vectors, $\langle \psi | \psi \rangle = 1$ and $\langle \chi | \chi \rangle = 1$. Show that the inner product satisfies the Cauchy-Schwarz inequality, namely $|S|^2 \leq 1$, which is consistent with the interpretation of $|S|^2$ as a probability. (You can use the identity, $\int d\gamma \int d\gamma' |\psi(\gamma)\chi(\gamma') - \chi(\gamma)\psi(\gamma')|^2 \geq 0$, which holds for any proper functions $\psi(\gamma)$ and $\chi(\gamma)$.)*

In summary, when attempting to measure a specific observable, the postulates restrict the outcome to a specific "ruler," composed of the set of all possible eigenvalues of the Hermitian linear operator that represents that observable. The probabilities for realization of a specific eigenvalue in the measurement process depend on the state of the system at the time of measurement. These probabilities are determined by expanding the corresponding state vector in a complete orthonormal system composed of the eigenvectors of the operator representing the observable. A larger expansion coefficient (in absolute value squared) of a specific eigenvector means that the probability of measuring the corresponding eigenvalue is larger. The expansion of the state vector is analogous to the representation of a vector in a Euclidian space in terms of its projections on a selected axis system, namely an orthonormal set of basis vectors. Indeed, each observable defines a unique, generally infinite, "axis system," associated with the eigenvectors of the corresponding operator. Each axis corresponds to a specific eigenvalue, and the magnitude of the projection of the state vector on any particular axis determines the probability of measuring that eigenvalue.

11.5 The Continuous Position and Momentum Representations

Let us return now to the postulates as formulated in Section 11.1. Postulate 1 associates the state of a system with a wave function, namely an explicit function of the particle positions, which (for a closed system) defines the probability density of measuring the particles in a specific position, $\rho(\mathbf{r}) = |\psi(\mathbf{r})|^2$ (Eq. (11.1.1)). For simplicity, let us consider a single particle in a one-dimensional coordinate space, where this identity takes the form

$$\rho(x) = |\psi(x)|^2, \tag{11.5.1}$$

and the particle's position is denoted by the real variable, x. According to postulate 3, each measurable particle position is associated with one of the eigenvalues of the (linear and Hermitian, Eq. (4.5.4)) position operator, $\hat{x}$, as defined by the following eigenvalue equation:

$$\hat{x}|\varphi_x\rangle = x|\varphi_x\rangle \tag{11.5.2}$$

First, we notice that the spectrum of $\hat{x}$ is continuous. Second, since the spectrum covers uniformly the entire axis, $-\infty < x < \infty$, the corresponding position eigenfunctions cannot be square integrable (proper) functions. Indeed, the position eigenvectors are extemal to the Hilbert space of square integrable functions (see the detailed discussion in Section 11.3). Nevertheless, as eigenvectors of a Hermitian operator, they constitute a set of vectors for expanding any physical state vector. Expressing the orthonormality condition for the improper position eigenvectors in terms of Dirac's delta (Eq. 11.3.6)),

$$\langle\varphi_{x'}|\varphi_x\rangle = \delta(x-x'), \tag{11.5.3}$$

and using the expansion of the identity in terms of the complete set (Eq. (11.3.3)),

$$\hat{I} = \int dx|\varphi_x\rangle\langle\varphi_x|, \tag{11.5.4}$$

any physical state vector, $|\psi\rangle$, can be expanded as an integral transform,

$$|\psi\rangle = \int_{-\infty}^{\infty} dx|\varphi_x\rangle\langle\varphi_x|\psi\rangle. \tag{11.5.5}$$

Similarly, any inner product between two state vectors, $|\psi\rangle$ and $|\chi\rangle$, can be expanded as

$$\langle\chi|\psi\rangle = \int_{-\infty}^{\infty} dx\langle\chi|\varphi_x\rangle\langle\varphi_x|\psi\rangle \tag{11.5.6}$$

where $\langle\varphi_x|\psi\rangle$ is the continuous representation of the state vector $|\psi\rangle$ in the set of position eigenstates. Therefore, according to postulate 3 (Eq. (11.4.16)), $|\langle\varphi_x|\psi\rangle|^2$ is the probability density for measuring the corresponding eigenvalue. Using the relation between the probability density and the wave function (Eq. (11.5.1)), *we can identify the wave function with the "position representation" of the abstract state, $|\psi\rangle$,*

$$\psi(x) = \langle\varphi_x|\psi\rangle, \tag{11.5.7}$$

where the inner product between $|\psi\rangle$ and $|\chi\rangle$ (Eq. (11.5.6) is identified with the overlap integral between the respective wave functions,

$$\langle\chi|\psi\rangle = \int_{\infty}^{\infty} dx\chi^*(x)\psi(x). \tag{11.5.8}$$

Particularly, projecting $|\psi\rangle$ on a specific position eigenstate, $|\varphi_{x'}\rangle$, we obtain

$$\langle\varphi_{x'}|\psi\rangle = \int_{-\infty}^{\infty} dx \langle\varphi_{x'}|\varphi_x\rangle\langle\varphi_x|\psi\rangle, \tag{11.5.9}$$

where, using Eq. (11.5.3), the integral transform of $\psi(x)$ obtains the familiar form,

$$\psi(x') = \int_{-\infty}^{\infty} dx\delta(x-x')\psi(x). \tag{11.5.10}$$

Eq. (11.5.10) can be regarded as an expansion of the wave function in terms of the complete orthonormal system of position eigenstates,

$$\psi(x') = \int_{-\infty}^{\infty} dx\psi(x)\varphi_x(x'), \tag{11.5.11}$$

where $\varphi_x(x') = \delta(x-x') = \langle\varphi_{x'}|\varphi_x\rangle$ is the "position representation" of the abstract position eigenstate, $|\varphi_x\rangle$. Indeed, considering an idealized measurement in absolute precision, in which the particle's position is determined to be, for example, x, the postulates imply that right after the measurement the wave function collapses to a delta-localized distribution at the position x.

Exercise 11.5.1 *A local operator is a function of the position operator (e.g., the scalar potential energy, $\hat{V}_x = V(\hat{x})$, Eq. (3.4.1)). Show that the matrix representation of a local operator in the position eigenstates is a diagonal matrix, namely $\langle\varphi_{x'}|\hat{V}|\varphi_x\rangle = V(x)\delta(x-x')$.*

Similar considerations apply to the particle's momentum. The measurable momentum values are the eigenvalues of the momentum operator, $\hat{p}_x$ (linear and Hermitian, Eq. (4.5.5)), as defined by the following eigenvalue equation:

$$\hat{p}_x|\varphi_{p_x}\rangle = p_x|\varphi_{p_x}\rangle. \tag{11.5.12}$$

As in the case of the position, the spectrum of $\hat{p}_x$ is continuous, and the eigenvalues cover uniformly the entire momentum axis, $-\infty < p_x < \infty$. The corresponding momentum eigenfunctions are improper and external to the Hilbert space of square integrable functions. Indeed, when discussing scattering processes in Chapter 7, we encountered the momentum eigenstates, as the improper "plane wave" solutions to the Schrödinger equation (see Eq. (7.2.8)). It is worthwhile to recall that this association of particles having a definite momentum with a plane wave by de Broglie (see Chapter 1) preceded the introduction of quantum mechanics, and later guided the elaborate formulation of its postulates. Associating the plane wave with the position representation of an abstract momentum eigenstate (see Eq. (11.5.7)), we have

$$\langle\varphi_x|\varphi_{p_x}\rangle = ce^{\frac{ip_x x}{\hbar}}, \tag{11.5.13}$$

where c is a complex-valued normalization constant. We recall that the improper plane waves are not normalizable in the sense of square integrable function. Nevertheless, as eigenvectors of a Hermitian linear operator, the momentum eigenstates constitute an orthonormal set. Expressing the orthonormality condition as Dirac's delta (Eq. 11.3.6),

$$\langle \varphi_{p'_x} | \varphi_{p_x} \rangle = \delta(p_x - p'_x), \tag{11.5.14}$$

and introducing the identity operator in the position representation (Eq. (11.5.4)), we obtain

$$\delta(p_x - p'_x) = \int dx \langle \varphi_{p'_x} | \varphi_x \rangle \langle \varphi_x | \varphi_{p_x} \rangle = |c|^2 \int_{-\infty}^{\infty} e^{\frac{-i(p'_x - p_x)x}{\hbar}} dx. \tag{11.5.15}$$

Using one of the standard representations of Dirac's delta, $\delta(k) = \frac{1}{2\pi} \int_{-\infty}^{\infty} e^{ikx} dx$, the normalization constant can be determined to be $c = 1/\sqrt{2\pi\hbar}$, and the normalized momentum eigenstate in the position representation reads

$$\varphi_{p_x}(x) = \langle \varphi_x | \varphi_{p_x} \rangle = \frac{1}{\sqrt{2\pi\hbar}} e^{\frac{ip_x x}{\hbar}}. \tag{11.5.16}$$

Expanding the identity in terms of a complete set of momentum eigenstates,

$$\hat{I} = \int dp_x | \varphi_{p_x} \rangle \langle \varphi_{p_x} |, \tag{11.5.17}$$

any physical state vector, $|\psi\rangle$, can be expanded as an integral transform,

$$|\psi\rangle = \int_{-\infty}^{\infty} dp_x | \varphi_{p_x} \rangle \langle \varphi_{p_x} | \psi \rangle, \tag{11.5.18}$$

where $\langle \varphi_{p_x} | \psi \rangle$ is the continuous representation of the state vector in the set of momentum eigenstates (the "momentum representation" of $|\psi\rangle$),

$$\overline{\psi}(p_x) = \langle \varphi_{p_x} | \psi \rangle. \tag{11.5.19}$$

Projecting $|\psi\rangle$ on a specific position eigenstate, $|\varphi_x\rangle$, Eq. (11.5.18) reads

$$\langle \varphi_x | \psi \rangle = \int_{-\infty}^{\infty} dp_x \langle \varphi_x | \varphi_{p_x} \rangle \langle \varphi_{p_x} | \psi \rangle. \tag{11.5.20}$$

Similarly (Ex. 11.5.1),

$$\langle \varphi_{p_x} | \psi \rangle = \int_{-\infty}^{\infty} dx \langle \varphi_{p_x} | \varphi_x \rangle \langle \varphi_x | \psi \rangle. \tag{11.5.21}$$

Using Eqs. (11.5.7, 11.5.16, 11.5.19), the last two equations can be written as follows:

$$\psi(x) = \frac{1}{\sqrt{2\pi\hbar}} \int_{-\infty}^{\infty} e^{\frac{ip_x x}{\hbar}} \overline{\psi}(p_x) dp_x, \tag{11.5.22}$$

$$\overline{\psi}(p_x) = \frac{1}{\sqrt{2\pi\hbar}} \int_{-\infty}^{\infty} e^{\frac{-ip_x x}{\hbar}} \psi(x) dx. \tag{11.5.23}$$

The momentum and position representations of any state vector are shown to be related by the Fourier transforms.

Exercise 11.5.2   *Given a state vector, $|\psi\rangle$, use Eq. (11.5.4) to derive Eq. (11.5.21).*

Exercise 11.5.3   *The kinetic energy of a particle is a function of the momentum operator, $\hat{T}_x = \frac{1}{2m}(\hat{p}_x)^2$ (Eq. (3.4.3)). Show that the matrix representation of the kinetic energy in the momentum eigenstates is a diagonal matrix, namely $\langle\varphi_{p'_x}|\hat{T}_x|\varphi_{p_x}\rangle = \frac{p_x^2}{2m}\delta(p_x - p'_x)$.*

Exercise 11.5.4   *Formulating the postulates in terms of wave functions, the linear momentum operator is defined as $\hat{p}_x = -i\hbar\frac{d}{dx}$. Use the expansions of the identity in terms of the position and momentum eigenstates, Eqs. (11.5.4 and 11.5.17), to show that $\langle\varphi_x|\hat{p}_x|\psi\rangle = -i\hbar\frac{d}{dx}\psi(x)$ and $\langle\varphi_x|\frac{\hat{p}_x^2}{2m}|\psi\rangle = \frac{-\hbar^2}{2m}\frac{d^2}{dx^2}\psi(x)$.*

Exercise 11.5.5   *Use Eqs. (11.5.12, 11.2.3, 11.5.17) to show that the momentum operator is Hermitian, namely $\langle\chi|\hat{p}_x|\psi\rangle = \langle\psi|\hat{p}_x|\chi\rangle^*$.*

Using the expansion of the identity in terms of eigenstates of an operator with a continuous spectrum, any operator can be formally associated with a "continuous matrix representation." For example, invoking the position representation (Eq. (11.5.4)), any operator takes the form

$$\hat{A} = \int dx \int dx' |\varphi_x\rangle\langle\varphi_x|\hat{A}|\varphi_{x'}\rangle\langle\varphi_{x'}| = \int dx \int dx' A(x,x')|\varphi_x\rangle\langle\varphi_{x'}|, \tag{11.5.24}$$

whereas in the momentum representation (Eq. (11.5.17)),

$$\hat{A} = \int dp_x \int dp'_x |\varphi_{p_x}\rangle\langle\varphi_{p_x}|\hat{A}|\varphi_{p'_x}\rangle\langle\varphi_{p'_x}| = \int dp_x \int dp'_x \overline{A}(p_x,p'_x)|\varphi_{p_x}\rangle\langle\varphi_{p'_x}|. \tag{11.5.25}$$

The functions $A(x,x')$ and $\overline{A}(p_x,p'_x)$ are different continuous (matrix) representations of the operator $\hat{A}$. In an appropriate basis, the matrix representation of an operator becomes diagonal. In the case of a continuous spectrum, the diagonal matrix corresponds to a function of the eigenvalues. For example, the scalar potential energy operator is diagonal in the coordinate representation (Ex. 11.5.1); hence, $V(x,x') = V(x)\delta(x - x')$, whereas the kinetic energy operator is diagonal in the momentum representation (Ex. 11.5.3), hence $T(p_x,p'_x) = \frac{p_x^2}{2m}\delta(p_x - p'_x)$.

11.6 The Vector Space of Multidimensional Systems

Let us consider, as examples, physical systems composed of single or groups of particles in a three-dimensional coordinate space. According to the first postulate, such systems are associated with a wave function of several independent variables, corresponding to the spatial coordinates of each particle composing the system. For concreteness, let us consider a system characterized by N independent Cartesian coordinate variables, x_1, $x_2, \ldots, x_N$. In this case the system position is an N-dimensional vector,

$$\mathbf{x} = (x_1, x_2, \ldots, x_N). \tag{11.6.1}$$

According to postulate 3, each specific N-dimensional position vector $\mathbf{x}$ is associated with one of the eigenvalues of the corresponding position vector operator (see Eq. (11.5.2) for the one-dimensional case),

$$\hat{\mathbf{x}}|\phi_\mathbf{x}\rangle = \mathbf{x}|\phi_\mathbf{x}\rangle, \tag{11.6.2}$$

where this form of writing means that the position eigenstate, $|\phi_\mathbf{x}\rangle$, is a simultaneous eigenvector of the N independent components of the position operator, namely

$$\hat{x}_j|\phi_\mathbf{x}\rangle = x_j|\phi_\mathbf{x}\rangle \quad ; \quad j = 1, 2, \ldots, N. \tag{11.6.3}$$

Defining the eigenstates of each one-dimensional position operator as (see Eq. (11.5.2))

$$\hat{x}_j|\varphi_{x_j}\rangle = x_j|\varphi_{x_j}\rangle \quad ; \quad j = 1, 2, \ldots, N, \tag{11.6.4}$$

the multidimensional position eigenstate ($|\phi_\mathbf{x}\rangle$, as defined in Eqs. (11.6.2, 11.6.3)) must therefore be proportional to each $|\varphi_{x_j}\rangle$, namely $|\phi_\mathbf{x}\rangle \propto |\varphi_{x_j}\rangle$ for every j. This means that $|\phi_\mathbf{x}\rangle$ must obtain a product form, that is, $|\phi_\mathbf{x}\rangle \propto |\varphi_{x_1}\rangle|\varphi_{x_2}\rangle \cdots |\varphi_{x_N}\rangle$, where the ordering is insignificant. Notice that each vector $|\varphi_{x_j}\rangle$ is associated with a different (extended) Hilbert space, corresponding to functions of the independent variable, x_j. The vector $|\phi_\mathbf{x}\rangle$ therefore belongs to a larger (extended) Hilbert space that corresponds to all the different combinations of products of the vectors $\{|\varphi_{x_j}\rangle\}$. The multiplication of vectors from different spaces is a tensor product (also known as a Kronecker product [11.4]), denoted as

$$|\phi_\mathbf{x}\rangle \equiv |\varphi_{x_1}\rangle \otimes |\varphi_{x_2}\rangle \otimes \cdots \otimes |\varphi_{x_N}\rangle. \tag{11.6.5}$$

(Please see the examples that follow of realizations of tensor products in standard linear algebra terms).

Notice that each complete set of position eigenvectors $\{|\varphi_{x_j}\rangle\}$ defines an identity operator in the space associated with functions of x_j only (see Eq. (11.5.4)), namely $\hat{I}_j \equiv \int dx_j |\varphi_{x_j}\rangle\langle\varphi_{x_j}|$. The corresponding identity in the space of multidimensional functions reads, accordingly,

$$\hat{I} = \int d\mathbf{x}|\phi_\mathbf{x}\rangle\langle\phi_\mathbf{x}|$$

$$= \int dx_1 \int dx_2 \ldots \int dx_N |\varphi_{x_1}\rangle \otimes |\varphi_{x_2}\rangle \otimes \cdots \otimes |\varphi_{x_N}\rangle\langle\varphi_{x_1}| \otimes \langle\varphi_{x_2}| \otimes \cdots \otimes \langle\varphi_{x_N}|, \tag{11.6.6}$$

$$= \int dx_1 \int dx_2 \ldots \int dx_N |\varphi_{x_1}\rangle\langle\varphi_{x_1}| \otimes |\varphi_{x_2}\rangle\langle\varphi_{x_2}| \otimes \cdots \otimes |\varphi_{x_N}\rangle\langle\varphi_{x_N}|$$

and the inner product between two multidimensional states reads

$$\langle\varphi|\psi\rangle = \int d\mathbf{x}\langle\varphi|\phi_\mathbf{x}\rangle\langle\phi_\mathbf{x}|\psi\rangle = \int dx_1 \int dx_2 \ldots \int dx_N \varphi^*(x_1, x_2, \ldots, x_N)\psi(x_1, x_2, \ldots, x_N), \tag{11.6.7}$$

where the multidimensional wave functions are identified as the projections (see Eq. (11.5.7)) of the state vector in the multidimensional space onto the multidimensional position eigenstate, $|\phi_{\mathbf{x}}\rangle$:

$$\psi(x_1, x_2, \ldots, x_N) = \langle \phi_{\mathbf{x}} | \psi \rangle$$
$$\varphi^*(x_1, x_2, \ldots, x_N) = \langle \varphi | \phi_{\mathbf{x}} \rangle. \tag{11.6.8}$$

Importantly, any $\psi(x_1, x_2, \ldots, x_N)$ is generally not a product function, namely

$$\psi(x_1, x_2, \ldots, x_N) \neq \psi_1(x_2) \psi_2(x_2) \ldots \psi_N(x_N). \tag{11.6.9}$$

However, it can be expanded in a complete orthonormal set of products, for example, using Eq. (11.6.6),

$$\psi(x_1, x_2, \ldots, x_N) = \int dx_1' \int dx_2' \ldots \int dx_N' \langle \varphi_{x_1} | \varphi_{x_1'} \rangle \langle \varphi_{x_2} | \varphi_{x_2'} \rangle \cdots \langle \varphi_{x_N} | \varphi_{x_N'} \rangle \psi(x_1', x_2', \ldots, x_N'). \tag{11.6.10}$$

Notice that the multidimensional identity operator, Eq. (11.6.6), can be readily reordered as a tensor product of identity operators (see the following examples for the definition in linear algebra terms) in the spaces of single coordinate functions,

$$\hat{I} = \left[\int dx_1 | \varphi_{x_1} \rangle \langle \varphi_{x_1} | \right] \otimes \left[\int dx_2 | \varphi_{x_2} \rangle \langle \varphi_{x_2} | \right] \otimes \cdots \otimes \left[\int dx_N | \varphi_{x_N} \rangle \langle \varphi_{x_N} | \right] = \hat{I}_1 \otimes \hat{I}_2 \otimes \cdots \otimes \hat{I}_N. \tag{11.6.11}$$

Since $\hat{I}_j$ can be expanded using any complete orthonormal set of eigenvectors of a Hermitian linear operator in the respective one-dimensional coordinate space, for example, $\hat{I}_j = \sum_{n_j} |\varphi_{n_j}\rangle \langle \varphi_{n_j}|$, where $\hat{A}_j |\varphi_{n_j}\rangle = \alpha_{n_j} |\varphi_{n_j}\rangle$, any multidimensional state vector, for example, $|\psi\rangle$, can be expanded in terms of tensor product states, which is the generalization of the expansion, Eq. (11.4.11), into the multidimensional coordinate space,

$$|\psi\rangle = \sum_{n_1, n_2 \ldots n_N} \psi_{n_1, n_2, \ldots, n_N} |\varphi_{n_1}\rangle \otimes |\varphi_{n_2}\rangle \otimes \cdots \otimes |\varphi_{n_N}\rangle, \tag{11.6.12}$$

and in the coordinate representation,

$$\psi(x_1, x_2, \ldots, x_N) = \sum_{n_1 n_2, \ldots, n_N} \psi_{n_1, n_2, \ldots, n_N} \varphi_{n_1}(x_1) \varphi_{n_2}(x_2) \cdots \varphi_{n_N}(x_N). \tag{11.6.13}$$

Similarly, any operator in the multidimensional space can be expanded as a sum of tensor products of operators in the one-dimensional coordinate spaces,

$$\hat{A} = \sum_{n_1, n_2, \ldots, n_N, n_1', n_2', \ldots, n_N'} A_{(n_1, n_2, \ldots, n_N), (n_1', n_2', \ldots, n_N')} |\varphi_{n_1}\rangle \langle \varphi_{n_1'}| \otimes |\varphi_{n_2}\rangle \langle \varphi_{n_2'}| \otimes \cdots \otimes |\varphi_{n_N}\rangle \langle \varphi_{n_N'}|. \tag{11.6.14}$$

It is instructive to discuss at this point tensor products of vector and operators in linear algebra terms, using the close analogy between the Hilbert space of physical states and finite-dimensional vector spaces, where each ket corresponds to a column vector, and a linear operator corresponds to a finite matrix. Let $\mathbf{a}$ and $\mathbf{b}$ be two vectors whose dimensions are M and N, respectively. The tensor product between these vectors is an $M \times N$ dimensional vector, defined as

$$\mathbf{a} \otimes \mathbf{b} = \begin{bmatrix} a_1 \\ a_2 \\ \vdots \\ a_M \end{bmatrix} \otimes \begin{bmatrix} b_1 \\ b_2 \\ \vdots \\ b_N \end{bmatrix} \equiv \begin{bmatrix} a_1 \begin{bmatrix} b_1 \\ b_2 \\ \vdots \\ b_N \end{bmatrix} \\ a_2 \begin{bmatrix} b_1 \\ b_2 \\ \vdots \\ b_N \end{bmatrix} \\ \vdots \\ a_M \begin{bmatrix} b_1 \\ b_2 \\ \vdots \\ b_N \end{bmatrix} \end{bmatrix} = \begin{bmatrix} a_1 b_1 \\ a_1 b_2 \\ \vdots \\ a_1 b_N \\ a_2 b_1 \\ a_2 b_2 \\ \vdots \\ a_2 b_N \\ \vdots \\ a_M b_1 \\ a_M b_2 \\ \vdots \\ a_M b_N \end{bmatrix}. \qquad (11.6.15)$$

Notice that each element in $\mathbf{a} \otimes \mathbf{b}$ is associated with two indexes,

$$[\mathbf{a} \otimes \mathbf{b}]_{k,l} \equiv a_k b_l. \qquad (11.6.16)$$

Similarly, let $\mathbf{A}$ and $\mathbf{B}$ be two matrices whose dimensions are M^2 and N^2, respectively. The tensor product between these matrices is an $(MN)^2$-dimensional matrix, defined as (e.g., for $M = 2$ and $N = 3$):

$$\mathbf{A} \otimes \mathbf{B} = \begin{bmatrix} a_{1,1} & a_{1,2} \\ a_{2,1} & a_{2,2} \end{bmatrix} \otimes \begin{bmatrix} b_{1,1} & b_{1,2} & b_{1,3} \\ b_{2,1} & b_{2,2} & b_{2,3} \\ b_{3,1} & b_{3,2} & b_{3,3} \end{bmatrix}$$

$$\equiv \begin{bmatrix} a_{1,1} \begin{bmatrix} b_{1,1} & b_{1,2} & b_{1,3} \\ b_{2,1} & b_{2,2} & b_{2,3} \\ b_{3,1} & b_{3,2} & b_{3,3} \end{bmatrix} & a_{1,2} \begin{bmatrix} b_{1,1} & b_{1,2} & b_{1,3} \\ b_{2,1} & b_{2,2} & b_{2,3} \\ b_{3,1} & b_{3,2} & b_{3,3} \end{bmatrix} \\ a_{2,1} \begin{bmatrix} b_{1,1} & b_{1,2} & b_{1,3} \\ b_{2,1} & b_{2,2} & b_{2,3} \\ b_{3,1} & b_{3,2} & b_{3,3} \end{bmatrix} & a_{2,2} \begin{bmatrix} b_{1,1} & b_{1,2} & b_{1,3} \\ b_{2,1} & b_{2,2} & b_{2,3} \\ b_{3,1} & b_{3,2} & b_{3,3} \end{bmatrix} \end{bmatrix}$$

$$= \begin{bmatrix} a_{1,1}b_{1,1} & a_{1,1}b_{1,2} & a_{1,1}b_{1,3} & a_{1,2}b_{1,1} & a_{1,2}b_{1,2} & a_{1,2}b_{1,3} \\ a_{1,1}b_{2,1} & a_{1,1}b_{2,2} & a_{1,1}b_{2,3} & a_{1,2}b_{2,1} & a_{1,2}b_{2,2} & a_{1,2}b_{2,3} \\ a_{1,1}b_{3,1} & a_{1,1}b_{3,2} & a_{1,1}b_{3,3} & a_{1,2}b_{3,1} & a_{1,2}b_{3,2} & a_{1,2}b_{3,3} \\ a_{2,1}b_{1,1} & a_{2,1}b_{1,2} & a_{2,1}b_{1,3} & a_{2,2}b_{1,1} & a_{2,2}b_{1,2} & a_{2,2}b_{1,3} \\ a_{2,1}b_{2,1} & a_{2,1}b_{2,2} & a_{2,1}b_{2,3} & a_{2,2}b_{2,1} & a_{2,2}b_{2,2} & a_{2,2}b_{2,3} \\ a_{2,1}b_{3,1} & a_{2,1}b_{3,2} & a_{2,1}b_{3,3} & a_{2,2}b_{3,1} & a_{2,2}b_{3,2} & a_{2,2}b_{3,3} \end{bmatrix} \qquad (11.6.17)$$

Each element in the matrix is associated with four indexes,

$$[\mathbf{A} \otimes \mathbf{B}]_{(k,l),(k',l')} \equiv a_{k,k'} b_{l,l'}. \qquad (11.6.18)$$

Notice that any operator in a specific subspace can be mapped onto the multidimensional space by a suitable tensor product with an identity operator. Consider, for example, the space defined by $\mathbf{A} \otimes \mathbf{B}$ (see, e.g., Eq. (11.6.17)). In this space the matrices $\mathbf{A}$ and $\mathbf{B}$ are mapped onto $\mathbf{A} \otimes \mathbf{I}$ and $\mathbf{I} \otimes \mathbf{B}$, respectively. Similarly, any vector in a specific subspace can be mapped onto the multidimensional space, where the representations of the two vectors, $\mathbf{a}$ and $\mathbf{b}$, in the space of $\mathbf{A} \otimes \mathbf{B}$ correspond to rectangular matrices, $\mathbf{a} \otimes \mathbf{I}$ and $\mathbf{I} \otimes \mathbf{b}$, respectively, where

$$[\mathbf{a} \otimes \mathbf{I}]_{(k,l),l'} \equiv a_k \delta_{l,l'} \quad ; \quad [\mathbf{I} \otimes \mathbf{b}]_{(k,l),k'} \equiv b_l \delta_{k,k'}. \qquad (11.6.19)$$

For example:

$$\mathbf{I} \otimes \mathbf{b} = \begin{bmatrix} 1 & 0 \\ 0 & 1 \end{bmatrix} \otimes \begin{bmatrix} b_1 \\ b_2 \\ b_3 \end{bmatrix} \equiv \begin{bmatrix} b_1 & 0 \\ b_2 & 0 \\ b_3 & 0 \\ 0 & b_1 \\ 0 & b_2 \\ 0 & b_3 \end{bmatrix} \quad \mathbf{a} \otimes \mathbf{I} = \begin{bmatrix} a_1 \\ a_2 \end{bmatrix} \otimes \begin{bmatrix} 1 & 0 & 0 \\ 0 & 1 & 0 \\ 0 & 0 & 1 \end{bmatrix} \equiv \begin{bmatrix} a_1 & 0 & 0 \\ 0 & a_1 & 0 \\ 0 & 0 & a_1 \\ a_2 & 0 & 0 \\ 0 & a_2 & 0 \\ 0 & 0 & a_2 \end{bmatrix}$$

$$(11.6.20)$$

The following are useful identities associated with tensor products (see Ex. 11.6.1). Let the matrices $\mathbf{A}$ and $\mathbf{C}$ be linear operators in a given vector space, where the matrices $\mathbf{B}$ and $\mathbf{D}$ are linear operators in another vector space. We have

$$(\mathbf{A} \otimes \mathbf{B})(\mathbf{C} \otimes \mathbf{D}) = \mathbf{AC} \otimes \mathbf{BD}. \qquad (11.6.21)$$

Let $\mathbf{c}$ and $\mathbf{d}$ be vectors in these two different vector spaces, respectively. We have

$$(\mathbf{A} \otimes \mathbf{B})(\mathbf{c} \otimes \mathbf{d}) = \mathbf{Ac} \otimes \mathbf{Bd} \qquad (11.6.22)$$

$$(\mathbf{A} \otimes \mathbf{B})(\mathbf{I} \otimes \mathbf{d}) = \mathbf{A} \otimes \mathbf{Bd} \quad ; \quad (\mathbf{A} \otimes \mathbf{B})(\mathbf{c} \otimes \mathbf{I}) = \mathbf{Ac} \otimes \mathbf{B}. \qquad (11.6.23)$$

Let the vectors $\mathbf{u}$ and $\mathbf{v}$ be eigenvectors of the matrices $\mathbf{A}$ and $\mathbf{B}$, associated with the eigenvalues $\lambda_{\mathbf{u}}$ and $\lambda_{\mathbf{v}}$, respectively; then,

$$(\mathbf{A} \otimes \mathbf{I} + \mathbf{I} \otimes \mathbf{B})\mathbf{u} \otimes \mathbf{v} = (\lambda_{\mathbf{u}} + \lambda_{\mathbf{v}})\mathbf{u} \otimes \mathbf{v}. \qquad (11.6.24)$$

It follows that an eigenvector of an operator, which is a sum over different operators restricted to different subspaces, is a tensor product of the corresponding eigenvectors of the subspace operators, where the eigenvalue is a sum over the eigenvalues associated with the eigenvectors composing the tensor product. Notice that this is a generalization of the property of separable Hamiltonians discussed in Ex. 4.3.4.

Exercise 11.6.1   *Use the definitions in Eqs. (11.6.15–11.6.20) to prove the identities given in Eqs. (11.6.21–11.6.24).*

An Example: The Vector Space of a Particle in a Three-Dimensional Coordinate Space

Let us consider a single point-particle of mass m in a three-dimensional physical space. In classical mechanics the particle's position is defined by the three Cartesian coordinates $\mathbf{r} = (x, y, z)$ (a specific case of Eq. (11.6.1)), where in quantum mechanics the particle's position operator corresponds to

$$\hat{\mathbf{r}} = (\hat{x}, \hat{y}, \hat{z}). \qquad (11.6.25)$$

Each of the position operators has a complete orthonormal system of eigenstates, defined by the independent eigenvalue equations (a specific case of Eq. (11.6.4)),

$$\hat{x}|\varphi_x\rangle = x|\varphi_x\rangle \quad ; \quad \hat{y}|\varphi_y\rangle = y|\varphi_y\rangle \quad ; \quad \hat{z}|\varphi_z\rangle = z|\varphi_z\rangle. \tag{11.6.26}$$

A unique particle position in the three-dimensional space corresponds, therefore, to a tensor product state (a specific case of Eq. (11.6.5)),

$$|\phi_{\mathbf{r}}\rangle \equiv |\varphi_x\rangle \otimes |\varphi_y\rangle \otimes |\varphi_z\rangle, \tag{11.6.27}$$

where the particle's state vector in the coordinate representation corresponds to the wave function of the three position variables (a specific case of Eq. (11.6.8)),

$$\psi(x,y,z) = \psi(\mathbf{r}) = \langle \phi_{\mathbf{r}}|\psi\rangle. \tag{11.6.28}$$

Introducing the identity operator in the three-dimensional coordinate space (a specific case of Eq. (11.6.11)),

$$\hat{I} = \int d\mathbf{r}|\phi_{\mathbf{r}}\rangle\langle\phi_{\mathbf{r}}| = \left[\int_{-\infty}^{\infty} dx|\varphi_x\rangle\langle\varphi_x|\right] \otimes \left[\int_{-\infty}^{\infty} dy|\varphi_y\rangle\langle\varphi_y|\right] \otimes \left[\int_{-\infty}^{\infty} dz|\varphi_z\rangle\langle\varphi_z|\right], \tag{11.6.29}$$

each state vector can be expanded as (see Ex. 11.6.2)

$$|\psi\rangle = \int d\mathbf{r}\psi(\mathbf{r})|\phi_{\mathbf{r}}\rangle = \int_{-\infty}^{\infty} dx \int_{-\infty}^{\infty} dy \int_{-\infty}^{\infty} dz\,\psi(x,y,z)|\varphi_x\rangle \otimes |\varphi_y\rangle \otimes |\varphi_z\rangle. \tag{11.6.30}$$

Exercise 11.6.2   *Use Eqs. (11.6.27–11.6.29) to derive Eq. (11.6.30).*

Similar considerations apply to the particle's momentum. In particular, the particle's momentum operator corresponds to

$$\hat{\mathbf{p}} = (\hat{p}_x, \hat{p}_y, \hat{p}_z). \tag{11.6.31}$$

Each component of the momentum operator defines a complete orthonormal system of eigenstates,

$$\hat{p}_x|\varphi_{p_x}\rangle = p_x|\varphi_{p_x}\rangle \quad ; \quad \hat{p}_y|\varphi_{p_y}\rangle = p_y|\varphi_{p_y}\rangle \quad ; \quad \hat{p}_z|\varphi_{p_z}\rangle = p_z|\varphi_{p_z}\rangle, \tag{11.6.32}$$

where a unique particle momentum in the three-dimensional space corresponds to a product of momentum eigenstates,

$$|\phi_{\mathbf{p}}\rangle \equiv |\varphi_{p_x}\rangle \otimes |\varphi_{p_y}\rangle \otimes |\varphi_{p_z}\rangle. \tag{11.6.33}$$

The particle's state vector in the momentum representation therefore corresponds to the wave function of the three momentum variables,

$$\overline{\psi}(p_x, p_y, p_z) = \overline{\psi}(\mathbf{p}) = \langle \phi_{\mathbf{p}}|\psi\rangle. \tag{11.6.34}$$

Introducing the identity operator in the three-dimensional momentum space,

$$\hat{I} = \int d\mathbf{p}|\phi_{\mathbf{p}}\rangle\langle\phi_{\mathbf{p}}| = \left[\int_{-\infty}^{\infty} dp_x|\varphi_{p_x}\rangle\langle\varphi_{p_x}|\right] \otimes \left[\int_{-\infty}^{\infty} dp_y|\varphi_{p_y}\rangle\langle\varphi_{p_y}|\right] \otimes \left[\int_{-\infty}^{\infty} dp_z|\varphi_{p_z}\rangle\langle\varphi_{p_z}|\right],$$
$$\tag{11.6.35}$$

any state vector can be expanded in terms of the momentum eigenstates,

$$|\psi\rangle = \int d\mathbf{p}\overline{\psi}(\mathbf{p})|\phi_\mathbf{p}\rangle = \int\limits_{-\infty}^{\infty} dp_x \int\limits_{-\infty}^{\infty} dp_y \int\limits_{-\infty}^{\infty} dp_z \overline{\psi}(p_x, p_y, p_z)|\varphi_{p_x}\rangle \otimes |\varphi_{p_y}\rangle \otimes |\varphi_{p_z}\rangle. \quad (11.6.36)$$

Notice that the position (Eq. (11.6.30)) and momentum (Eq. (11.6.36)) representations of the particle's state in the three-dimensional coordinate space can be transformed to any other representation by a resolution of the identity operator in terms of a complete orthonormal system of any Hermitian operator eigenstates (Eq. (11.6.12)).

Exercise 11.6.3 *Use the position and momentum representations of $|\psi\rangle$ in the three-dimensional space(Eqs. (11.6.30, 11.6.36)) and the explicit position representation of the momentum eigenstates (Eq. (11.5.16)) to show that the functions $\psi(\mathbf{r}) = \psi(x,y,z)$ and $\overline{\psi}(\mathbf{p}) = \overline{\psi}(p_x, p_y, p_z)$ are related to each other by the three-dimensional Fourier transforms,*

$$\overline{\psi}(\mathbf{p}) = \int d\mathbf{r}\psi(\mathbf{r}) \left(\frac{1}{\sqrt{2\pi\hbar}}\right)^3 e^{-i\mathbf{p}\mathbf{r}/\hbar}$$

$$\psi(\mathbf{r}) = \int d\mathbf{p}\overline{\psi}(\mathbf{p}) \left(\frac{1}{\sqrt{2\pi\hbar}}\right)^3 e^{i\mathbf{p}\mathbf{r}/\hbar}$$

(Compare to the one-dimensional case, Eqs. (11.5.22, 11.5.23)).

We now consider the representation of general operators in the three-dimensional coordinate space. Formally, any resolution of the identity can be invoked, and any operator can be represented according to Eq. (11.6.14). For example, in the coordinate representation,

$$\hat{A} = \int d\mathbf{r} \int d\mathbf{r}'|\phi_\mathbf{r}\rangle\langle\phi_\mathbf{r}|\hat{A}|\phi_{\mathbf{r}'}\rangle\langle\phi_{\mathbf{r}'}| = \int d\mathbf{r} \int d\mathbf{r}'A(\mathbf{r},\mathbf{r}')|\phi_\mathbf{r}\rangle\langle\phi_{\mathbf{r}'}|, \quad (11.6.37)$$

and in the momentum representation,

$$\hat{A} = \int d\mathbf{p} \int d\mathbf{p}'|\phi_\mathbf{p}\rangle\langle\phi_\mathbf{p}|\hat{A}|\phi_{\mathbf{p}'}\rangle\langle\phi_{\mathbf{p}'}| = \int d\mathbf{p} \int d\mathbf{p}'\overline{A}(\mathbf{p},\mathbf{p}')|\phi_\mathbf{p}\rangle\langle\phi_{\mathbf{p}'}|, \quad (11.6.38)$$

where $A(\mathbf{r},\mathbf{r}') = \langle\phi_\mathbf{r}|\hat{A}|\phi_{\mathbf{r}'}\rangle$ and $\overline{A}(\mathbf{p}, \mathbf{p}') = \langle\phi_\mathbf{p}|\hat{A}|\phi_{\mathbf{p}'}\rangle$ are continuous (matrix) representations of the operator $\hat{A}$ (see Eqs. (11.5.24, 11.5.25)).

Let us focus on local operators, such as the scalar potential energy:

$$\hat{V} = V(\hat{\mathbf{r}}) = V(\hat{x}, \hat{y}, \hat{z}) \quad (11.6.39)$$

When discussing a one-dimensional coordinate space, we noticed that the matrix representation of the potential energy is diagonal in the position representation (see Ex. (11.5.1)). To show that this holds also in the three-dimensional space, we restrict the discussion to analytic functions, where the classical potential energy function can be expanded in a power series, $y(x,y,z) = \sum\limits_{i,j,k} v_{i,j,k}x^iy^jz^k$. Since each product x^i, y^j, z^k translates to a tensor product of independent quantum mechanical operators, each associated with a different one-dimensional coordinate space, the potential energy operator obtains the form

$$\hat{V} = \sum_{i,j,k=0}^{\infty} v_{i,j,k} (\hat{x})^i \otimes (\hat{y})^j \otimes (\hat{z})^k. \tag{11.6.40}$$

Using this expansion and Eq. (11.6.37), we can readily see that the potential energy remains diagonal in the coordinate representation in the three-dimensional coordinate space as well (the generalization to higher dimensions is straightforward):

$$\begin{aligned} \langle \phi_{\mathbf{r}} | \hat{V} | \phi_{\mathbf{r}'} \rangle &= \langle \varphi_x | \otimes \langle \varphi_y | \otimes \langle \varphi_z | \sum_{i,j,k} v_{i,j,k} (\hat{x})^i \otimes (\hat{y})^j \otimes (\hat{z})^k | \varphi_{x'} \rangle \otimes | \varphi_{y'} \rangle \otimes | \varphi_{z'} \rangle \\ &= \sum_{i,j,k} v_{i,j,k} \langle \varphi_x | (\hat{x})^i | \varphi_{x'} \rangle \langle \varphi_y | (\hat{y})^j | \varphi_{y'} \rangle \langle \varphi_z | (\hat{z})^k | \varphi_{z'} \rangle \\ &= \sum_{i,j,k} v_{i,j,k} x^i y^j z^k \delta(x - x') \delta(y - y') \delta(z - z') \\ &= V(x,y,z) \delta(x - x') \delta(y - y') \delta(z - z') \\ &= V(\mathbf{r}) \delta(\mathbf{r} - \mathbf{r}'). \end{aligned} \tag{11.6.41}$$

Consequently, in the coordinate representation, the potential energy operation on any state vector in the three-dimensional coordinate space translates into a multiplication,

$$\langle \phi_{\mathbf{r}} | \hat{V} | \psi \rangle = \int d\mathbf{r}' \langle \phi_{\mathbf{r}} | \hat{V} | \phi_{\mathbf{r}'} \rangle \langle \phi_{\mathbf{r}'} | \psi \rangle = \int d\mathbf{p}' V(\mathbf{r}) \delta(\mathbf{r} - \mathbf{r}') \psi(\mathbf{r}') = V(\mathbf{r}) \psi(\mathbf{r}). \tag{11.6.42}$$

A parallel argument holds with respect to the nonlocal kinetic energy operator, when the momentum representation is considered:

$$\hat{T} = T(\hat{\mathbf{p}}) = T(\hat{p}_x, \hat{p}_y, \hat{p}_z). \tag{11.6.43}$$

In this case, the classical kinetic energy function reads $T(p_x, p_y, p_z) = \frac{1}{2m}(p_x^2 + p_y^2 + p_z^2)$, and the corresponding quantum mechanical operator in the three-dimensional coordinate space is a sum of tensor products of operators in the different one-dimensional coordinate spaces:

$$\hat{T} = \frac{1}{2m} \left((\hat{p}_x)^2 \otimes \hat{I}_y \otimes \hat{I}_z + \hat{I}_x \otimes (\hat{p}_y)^2 \otimes \hat{I}_z + \hat{I}_x \otimes \hat{I}_y \otimes (\hat{p}_z)^2 \right). \tag{11.6.44}$$

Using this expansion and Eq. (11.6.38), we can readily see that the kinetic energy is diagonal in the momentum representation, namely

$$\langle \phi_{\mathbf{p}} | \hat{T} | \phi_{\mathbf{p}'} \rangle = \frac{1}{2m} \cdot$$
$$\langle \varphi_{p_x} | \otimes \langle \varphi_{p_y} | \otimes \langle \varphi_{p_z} | ((\hat{p}_x)^2 \otimes \hat{I}_y \otimes \hat{I}_z + \hat{I}_x \otimes (\hat{p}_y)^2 \otimes \hat{I}_z + \hat{I}_x \otimes \hat{I}_y \otimes (\hat{p}_z)^2) | \varphi_{p_x'} \rangle \otimes | \varphi_{p_y'} \rangle \otimes | \varphi_{p_z'} \rangle$$
$$= \frac{1}{2m} (p_x^2 + p_y^2 + p_z^2) \delta(p_x - p_x') \delta(p_y - p_y') \delta(p_z - p_z'). \tag{11.6.45}$$

Consequently, in the momentum representation, the kinetic energy operation on any state vector in the three-dimensional coordinate space translates into a multiplication:

$$\langle \phi_{\mathbf{p}} | \hat{T} | \psi \rangle = \int d\mathbf{p}' \langle \phi_{\mathbf{p}} | \hat{T} | \phi_{\mathbf{p}'} \rangle \langle \phi_{\mathbf{p}'} | \psi \rangle = \int d\mathbf{p}' \frac{\mathbf{p}^2}{2m} \delta(\mathbf{p} - \mathbf{p}') \overline{\psi}(\mathbf{p}') = \frac{\mathbf{p}^2}{2m} \overline{\psi}(\mathbf{p}). \tag{11.6.46}$$

Finally, in the coordinate representation, the kinetic energy operation in the three-dimensional coordinate space obtains the familiar form,

$$\langle \phi_{\mathbf{r}} | \hat{T} | \psi \rangle = \frac{-\hbar^2}{2m} \hat{\Delta}_{\mathbf{r}} \psi(\mathbf{r}), \tag{11.6.47}$$

where $\hat{\Delta}_{\mathbf{r}} = \left(\frac{\partial^2}{\partial x^2} + \frac{\partial^2}{\partial y^2} + \frac{\partial^2}{\partial z^2} \right)$ is the Laplacian (see Ex. (11.6.5)).

Exercise 11.6.4 *Given that $\hat{V} = V(\hat{\mathbf{r}})$, and $T = T(\hat{\mathbf{p}})$, rederive Eqs. (11.6.42) and (11.6.46) by using the Hermiticity of the corresponding operators.*

Exercise 11.6.5 *(a) Use Eq. (11.6.46) and the explicit position representation of the momentum eigenstates (Eq. (11.5.16)) to show that $\langle \phi_{\mathbf{r}} | \hat{T} | \psi \rangle = \int d\mathbf{p} \frac{\mathbf{p}^2}{2m} \frac{e^{i\mathbf{p}\mathbf{r}/\hbar}}{(\sqrt{2\pi\hbar})^3} \overline{\psi}(\mathbf{p})$.*
(b) Show that $\mathbf{p}^2 e^{i\mathbf{p}\mathbf{r}/\hbar} = -\hbar^2 \left(\frac{\partial^2}{\partial x^2} + \frac{\partial^2}{\partial y^2} + \frac{\partial^2}{\partial z^2} \right) e^{i\mathbf{p}\mathbf{r}/\hbar}$.
(c) Use the results of (a) and (b), as well as the Fourier expansion of $\psi(\mathbf{r})$ in Ex. 11.6.3, to obtain Eq. (11.6.47).

11.7 Ensemble Measurements and the Uncertainty Principle

The postulates as previously formulated refer to idealized measurements in which a single measurement is performed on a single system, characterized by a single solution to the Schrödinger equation. In the vast majority of experiments, the situation in more involved. Typically, the limited accuracy, the large size, and the experimental noise associated with a measuring device exclude the isolation of a single system. Instead, the measurement records the result of a simultaneous measurement on a large number of systems. When the systems are identical and independent of each other, this is equivalent to a repeated experiment on a set of identically prepared systems. Either way, the outcome of the measurement corresponds to an ensemble of systems. In Chapter 16 we shall address the most general case in which different systems in the ensemble can be associated with different solutions to the Schrödinger equation. Here we focus on a pure ensemble, where all the systems prior to the measurement are identical and correspond to the same solution of the Schrödinger equation. Given the probabilistic nature of any single measurement on a single system, the outcome of an ensemble measurement is associated with a statistical distribution of measured values. It is customary to represent the distribution in terms of its first and second moments, namely the averaged measured value, and the measurement uncertainty, which is commonly associated with the standard deviation of the distribution about the average.

For an infinitely large number of identical systems, each associated with the same state, $|\psi(t)\rangle$, the ensemble averaged value of an observable represented by the operator $\hat{A}$, measured at time t, reads

$$\langle A(t) \rangle = \langle \psi(t) | \hat{A} | \psi(t) \rangle. \tag{11.7.1}$$

$\langle A(t) \rangle$ is often termed the "expectation value" (or "expected value") of $\hat{A}$ in the state $|\psi(t)\rangle$.

Exercise 11.7.1 *Prove that the expectation value of a Hermitian operator is real-valued.*

Exercise 11.7.2 *A system is found in a stationary state,* $|\psi_E(t)\rangle = e^{\frac{-iEt}{\hbar}}|\varphi_E\rangle$ *(Eq. (4.3.5)). Show that the expectation value of any operator is time-independent.*

Exercise 11.7.3 *Show that the expectation value of a local operator* $V(\hat{x})$ *(see Ex. 11.5.1) in a system associated with the wave function* $\psi(x,t)$ *reads* $\langle V(t) \rangle = \int_{-\infty}^{\infty} \psi^*(x,t)V(x)\psi(x,t)dx.$

Exercise 11.7.4 *Prove that the expectation value of the momentum operator,* $\hat{p}_x$*, vanishes for a system in a bound stationary state,* $\psi_E(x,t) = \varphi_E(x)e^{-iEt/\hbar}$*, where* $\text{Im}[\varphi_E^*(x)\frac{d}{dx}\varphi_E(x)] = 0.$

Exercise 11.7.5 *Use the time-dependent Schrödinger equation,* $i\hbar\frac{\partial}{\partial t}|\psi(t)\rangle = \left[\frac{\hat{p}_x^2}{2m} + V(\hat{x})\right]|\psi(t)\rangle$*, and prove the Ehrenfest theorem for a particle of mass m, in the presence of a one-dimensional potential energy,* $V(x)$: $\frac{\partial}{\partial t}\langle x(t) \rangle = \frac{\langle p_x(t) \rangle}{m}$ *and* $\frac{\partial}{\partial t}\langle p_x(t) \rangle = -\langle \frac{dV}{dx}(t) \rangle$*. Show that for a quadratic potential energy* $V(x) = \alpha + \beta x + \gamma x^2$*, the quantum mechanical expectation values follow the Hamilton equations of classical mechanics.*

Notice that Eq. (11.7.1) is not a new postulate, but rather a compact representation of the expectation value, which is consistent with the postulates. This can be readily verified by expanding the state vector of the system in the complete orthogonal set of eigenvectors of $\hat{A}$, the operator that represents the measured observable, $\hat{A}|\varphi_n\rangle = \gamma_n|\varphi_n\rangle$. (For simplicity, and without loss of generality, here $\hat{A}$ is assumed to have a discrete, nondegenerate spectrum.) Using $|\psi(t)\rangle = \sum_n a_n(t)|\varphi_n\rangle$ and $\langle\psi(t)| = \sum_n a_n^*(t)\langle\varphi_n|$, we obtain (Ex. 11.7.6)

$$\langle A(t) \rangle = \sum_n |a_n(t)|^2 \gamma_n. \tag{11.7.2}$$

Exercise 11.7.6 *Derive Eq. (11.7.2).*

According to the postulates, each single measurement yields one of $\hat{A}$'s eigenvalues, γ_n, at a probability, $P_{\gamma_n}(t) = |a_n(t)|^2 = |\langle\varphi_n|\psi(t)\rangle|^2$ Therefore, the expectation value as defined in Eq. (11.7.1) consistently equals the weighted average value over the distribution of single measurements.

Given the definition of the averaged measured value of any operator (Eq. (11.7.1)), the expression of the standard deviation in terms of the averaged square of the deviation from the average follows naturally,

$$\Delta A(t) = \sqrt{\langle\psi(t)|(\hat{A} - \langle A(t)\rangle)^2|\psi(t)\rangle} = \sqrt{\langle A^2(t) \rangle - \langle A(t) \rangle^2}. \tag{11.7.2}$$

Exercise 11.7.7   *A system is found in a stationary state,* $|\psi_E(t)\rangle = e^{\frac{-iEt}{\hbar}}|\varphi_E\rangle$ *(Eqs. (4.3.4, 4.3.5)). Show that: (a) The probability of measuring the energy E is 1. (b) The standard deviation in energy measurement, as defined in Eq. (11.7.2), vanishes.*

Notice that **when the state of the system happens to be an eigenstate of the operator representing the observable (i.e.,** $|\psi(t)\rangle = a(t)|\varphi_n\rangle$**), the uncertainty vanishes** (see Ex. 11.7.7). This means that the observable associated with $\hat{A}$ can be measured, in principle (excluding the limitations of the measuring device), with absolute precision. **However, even if the system was prepared to be in one of the eigenstates of** $\hat{A}$ **at the measurement time, such that the observable associated with** $\hat{A}$ **can be measured with absolute precision, other observables are likely to exhibit measurement uncertainties.** This holds since, in general, the eigenstate of $\hat{A}$ would not be an eigenstate of another operator. Specifically, if the state of the system, $|\psi(t)\rangle$, is an eigenstate of $\hat{A}$, but not of $\hat{B}$, the observable associated with $\hat{B}$ cannot be measured with absolute certainty. **Only if** $|\psi(t)\rangle$ **is a common eigenstate of both** $\hat{A}$ **and** $\hat{B}$ **can the two measurements, in principle, be performed with absolute certainty.** Moreover, the results would not depend on the order of measurements, since in each measurement (of either $\hat{A}$ or $\hat{B}$) the state of the system would be unaltered ($|\psi(t)\rangle$ would "collapse to itself").

Situations in which two operators share common eigenstates are not uncommon. In Ex. 6.4.1 we concluded that a nondegenerate eigenfunction of an operator $\hat{A}$ is also an eigenfunction of any operator $\hat{B}$ that commutes with $\hat{A}$ ($[\hat{A},\hat{B}] = 0$). This claim can be readily generalized to show that any commuting operators have a common set of eigenvectors [4.3]. Indeed, we already encountered examples of differential operators commuting with symmetry operators and having a common set of eigenfunctions (see, e.g., Ex. 9.2.1). For example, in Chapter 10, we saw that the hydrogen-like atomic orbitals, $\psi_{n,l,m}(r,\theta,\varphi)$, are common eigenfunctions of the commuting Hamiltonian and angular momentum operators, $\hat{H}$, $\hat{L}^2$, and $\hat{L}_z$.

In general, however, measurements are associated with uncertainties that reflect an inherent probabilistic nature of the outcome of single measurements. When two observables are associated with noncommuting operators, simultaneous certainty in their measurement is excluded. Even for commuting operators, the possibility for simultaneous certainty depends on the specific system state vector. The fundamental limitation on the certainty in the simultaneous measurement of two observables is commonly referred to as "Heisenberg's uncertainty principle." Formally, expressing the uncertainty as the standard deviation (Eq. (11.7.2)), the following inequality holds (see Ex. 11.7.8 for the proof) for any linear Hermitian operators, $\hat{A}$ and $\hat{B}$, and any normalizable (proper or improper) state vector $|\psi(t)\rangle$:

$$[\Delta A(t)]^2[\Delta B(t)]^2 \geq \frac{-\langle\psi(t)|[\hat{A},\hat{B}]|\psi(t)\rangle^2}{4} \qquad (11.7.3)$$

Exercise 11.7.8   *In order to prove Eq. (11.7.3): (a) Show that for any operator,* $\hat{O}$, *and its Hermitian conjugate,* $\hat{O}^\dagger$, *and for any state vector, the expectation value of* $\hat{O}^\dagger\hat{O}$ *is nonnegative, namely* $\langle\psi|\hat{O}^\dagger\hat{O}|\psi\rangle \geq 0$. *(b) Given two Hermitian linear operators,* $\hat{X}$ *and* $\hat{Y}$, *and a real-valued scalar,* α, *use (a) and the definition,* $\hat{O} \equiv \hat{X} - i\alpha\hat{Y}$, *to show that*

$\langle \psi | \hat{X}^2 | \psi \rangle + \alpha^2 \langle \psi | \hat{Y}^2 | \psi \rangle - i\alpha \langle \psi | [\hat{X}, \hat{Y}] | \psi \rangle \geq 0$. *(c) Since (b) holds for any real α, show that $\frac{-\langle \psi | [\hat{X}, \hat{Y}] | \psi \rangle^2}{4} \leq \langle \psi | \hat{X}^2 | \psi \rangle \langle \psi | \hat{Y}^2 | \psi \rangle$. (d) Let $\hat{X} \equiv \hat{A} - \langle \psi | \hat{A} | \psi \rangle$, and $\hat{Y} \equiv \hat{B} - \langle \psi | \hat{B} | \psi \rangle$, where $\hat{A}$ and $\hat{B}$ are any linear Hermitian operators. Use (c) and the definition of the standard deviation (Eq. (11.7.2)) to obtain the uncertainty inequality, Eq. (11.7.3).*

In line with the preceding discussion, simultaneous absolute precision in the measurement of two observables can only be reached, in principle, if the corresponding operators commute, and therefore $[\Delta A(t)]^2 [\Delta B(t)]^2 \geq 0$, where the equality corresponds to the case of $|\psi(t)\rangle$ being a common eigenfunction of both $\hat{A}$ and $\hat{B}$, which means $\Delta A(t) = \Delta B(t) = 0$. When $[\hat{A}, \hat{B}] \neq 0$, Eq. (11.7.3) means that $[\Delta A(t)]^2 [\Delta B(t)]^2 > 0$ for any nontrivial $|\psi(t)\rangle$ (Ex. 11.7.9). Namely, there is an intrinsic bound for simultaneous certainty. A well-known example is the Heisenberg relation between the position and the linear momentum uncertainties for a particle moving along a given direction (x). Using $[\hat{x}, \hat{p}_x] = i\hbar$, Eq. (11.7.3) yields

$$\Delta x(t) \cdot \Delta p(t) \geq \frac{\hbar}{2}. \tag{11.7.4}$$

Exercise 11.7.9 *Show that $-\langle \psi(t) | [\hat{A}, \hat{B}] | \psi(t) \rangle^2 > 0$ for any noncommuting Hermitian operators, $\hat{A}$ and $\hat{B}$, $[\hat{A}, \hat{B}] \neq 0$, and $|\psi(t)\rangle \neq 0$. (Show that $\langle \psi(t) | [\hat{A}, \hat{B}] | \psi(t) \rangle$ is an imaginary number.)*

Exercise 11.7.10 *A particle is associated at a certain time with a normalized Gaussian wave function, $\psi(x) = \left(\frac{1}{2\pi\sigma^2}\right)^{1/4} e^{\frac{-x^2}{4\sigma^2}}$. (a) Calculate the position and momentum standard deviations as defined by Eq. (11.7.2) and verify that their multiplication satisfies Eq. (11.7.4). (You can use the following integrals: $\int_{-\infty}^{\infty} e^{-\beta x^2} dx = \sqrt{\frac{\pi}{\beta}}$; $\int_{-\infty}^{\infty} x^2 e^{-\beta x^2} = \frac{\sqrt{\pi}}{2} \beta^{-3/2}$.) (b) **The Gaussian wave function is sometimes referred to as the minimal uncertainty state for the particle.** Explain this term in view of the result of (a).*

Bibliography

[11.1] Y. Aharonov, D. Z. Albert, and L. Vaidman, "How the result of a measurement of a component of the spin of a spin-1/2 particle can turn out to be 100," Physical Review Letters 60, 1351 (1988).

[11.2] P. A. M. Dirac, "The Principles of Quantum Mechanics" (Oxford University Press, 1958).

[11.3] L. Mirsky, "An Introduction to Linear Algebra" (Dover, 1990).

[11.4] C. F. Van Loan, "The ubiquitous Kronecker product," Journal of Computational and Applied Mathematics 123, 85 (2000).

12 Approximation Methods

12.1 Perturbation Theory for the Time-Independent Schrödinger Equation

So far, we have used the tools of quantum mechanics for explaining phenomena on the nanoscale, by associating different systems with simple model Hamiltonians and learning about them from the analytic solutions of the corresponding Schrödinger equation. This strategy of mapping a complex problem onto a simple, tractable model Hamiltonian is indeed essential for understanding the physics. Yet, in most cases, analytic solutions of the Schrödinger equation are not known even when the underlying model Hamiltonian is relatively simple, and numerical solutions to the Schrödinger equation are commonly practiced. In other cases, the complexity of the system is so large that minimal models are searched for, which capture only the essential ingredients of the system's Hamiltonian. In all these cases, approximation methods must be invoked for solving the relevant Schrödinger equation. The time-dependent Schrödinger equation will be discussed in this context in Chapters 15–20, while here we focus on the time-independent Schrödinger equation. One of the most important approximation methods is perturbation theory. The basic idea is to utilize the known solutions of a presumably simpler problem to approximate the solutions of a more complex one, which is too hard to solve "directly."

The perturbation theory due to Rayleigh and Schrödinger provides a formal framework for expressing the eigenvalues and eigenfunctions of the time-independent Schrödinger equation for a given Hamiltonian, $\hat{H}$, in terms of the known solutions of the equation for another ("zero-order") Hamiltonian, $\hat{H}_0$, in the same space. The two Hamiltonians can be formally connected via a continuous parameter, λ, which obtains the values 0 and 1 for $\hat{H}_0$ and $\hat{H}$, respectively. When the eigenvalues and eigenfunctions are analytic functions of λ on a contour, $0 \leq \lambda \leq 1$, they can be expanded as a power series in λ around zero. While, in general, the expansion may be infinite, perturbation theory becomes useful when the expansion converges at the lowest orders. In this case, $\hat{H}_0$ and $\hat{H}$ are sufficiently close in some sense (to be discussed in what follows), where $\hat{H}_1 = \hat{H} - \hat{H}_0$ is referred to as a "perturbation" to $\hat{H}_0$.

For the zero-order Hamiltonian, $\hat{H}_0 = \hat{H}(\lambda)|_{\lambda=0}$, the stationary solutions to the Schrödinger equation are assumed to be known,

$$\hat{H}_0 \left| \psi_n^{(0)} \right\rangle = E_n^{(0)} \left| \psi_n^{(0)} \right\rangle, \tag{12.1.1}$$

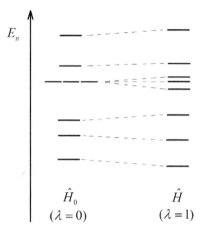

A schematic representation of the (known) energy levels of a zero-order Hamiltonian and the corresponding (unknown) energy levels of a full Hamiltonian. Roughly, perturbation theory is accurate when there is a one-to-one correspondence between the two ladders, and when the change at each of the energy levels is small in comparison to the energy spacing between nearby levels.

where

$$\left\langle \psi_m^{(0)} \middle| \psi_n^{(0)} \right\rangle = \delta_{m,n}. \tag{12.1.2}$$

We seek for the corresponding solutions to the Schrödinger equation with the full Hamiltonian, $\hat{H} = \hat{H}(\lambda)|_{\lambda=1}$ (see Fig. 12.1.1),

$$\hat{H}\middle|\psi_n\right\rangle = E_n|\psi_n\rangle. \tag{12.1.3}$$

By introducing explicitly a linear dependence in the parameter λ,

$$\hat{H}(\lambda) = \hat{H}_0 + \lambda \hat{H}_1, \tag{12.1.4}$$

Eq. (12.1.3) can be written as

$$[\hat{H}_0 + \lambda \hat{H}_1]|\psi_n(\lambda)\rangle = E_n(\lambda)|\psi_n(\lambda)\rangle, \tag{12.1.5}$$

where the solutions are expanded (up to normalization) in powers of λ,

$$E_n(\lambda) = \sum_{l=0}^{\infty} \lambda^l E_n^{(l)} \tag{12.1.6}$$

$$|\psi_n(\lambda)\rangle = \sum_{l=0}^{\infty} \lambda^l \middle|\psi_n^{(l)}\right\rangle. \tag{12.1.7}$$

The terms with $l > 0$ are corrections to $E_n^{(0)}$ and $\middle|\psi_n^{(0)}\right\rangle$, where $E_n^{(l)}$ and $\middle|\psi_n^{(l)}\right\rangle$ are referred to as the lth-order corrections to the energy and the wave function, respectively. Substitution of the expansions Eq. (12.1.6) and (12.1.7) in the Schrödinger equation, Eq. (12.1.5), and changing the summation index, we obtain (Ex. 12.1.1)

$$\sum_{l=0}^{\infty} \left[\hat{H}_0\middle|\psi_n^{(l)}\right\rangle + (1-\delta_{l,0})\hat{H}_1\middle|\psi_n^{(l-1)}\right\rangle - \sum_{l'=0}^{l} E_n^{(l')}\middle|\psi_n^{(l-l')}\right\rangle\right]\lambda^l = 0 \tag{12.1.8}$$

Since this result must hold for any $0 \leq \lambda \leq 1$, the coefficients of the different powers, λ^l (for $l = 0, 1, 2, \ldots$), should all vanish independently. For $l = 0$, Eq. (12.1.1) is reproduced, and for the higher powers, Eq. (12.1.8) yields

$$\hat{H}_0 \left| \psi_n^{(l)} \right\rangle = \sum_{l'=0}^{l} (E_n^{(l')} - \delta_{l',1} \hat{H}_1) \left| \psi_n^{(l-l')} \right\rangle \quad ; \quad l > 0. \tag{12.1.9}$$

Exercise 12.1.1 *Derive Eq. (12.1.8).*

Let us consider first the corrections to a nondegenerate eigenvalue of $\hat{H}_0$, corresponding to the energy, $E_n^{(0)}$, where the zero-order vector, $\left| \psi_n^{(0)} \right\rangle$, is defined uniquely (up to a constant multiplication). Noticing that the eigenstates of $\hat{H}_0$ are a complete orthonormal set (Eq. (12.1.2)), we can use them for expanding the (unnormalized) eigenfunctions of $\hat{H}$, that is, $\left| \psi_n(\lambda) \right\rangle = \left| \psi_n^{(0)} \right\rangle + \sum_{m \neq n} b_m(\lambda) \left| \psi_m^{(0)} \right\rangle$. Comparing to Eq. (12.1.7), it follows that $\sum_{l>0} \lambda^l \left| \psi_n^{(l)} \right\rangle \propto \sum_{m \neq n} b_m(\lambda) \left| \psi_m^{(0)} \right\rangle$, which means that for $l > 0$, the lth-order corrections to the wave function $\left| \psi_n(\lambda) \right\rangle$ are orthogonal to $\left| \psi_n^{(0)} \right\rangle$, namely

$$\left\langle \psi_n^{(0)} \middle| \psi_n^{(l)} \right\rangle = \delta_{l,0}. \tag{12.1.10}$$

Projecting Eq. (12.1.9) onto $\left| \psi_n^{(0)} \right\rangle$, using Eq. (12.1.10) and the hermiticity of $\hat{H}_0$, we obtain (Ex. 12.1.2)

$$E_n^{(l)} = \left\langle \psi_n^{(0)} \middle| \hat{H}_1 \middle| \psi_n^{(l-1)} \right\rangle. \tag{12.1.11}$$

The lth-order correction to the energy, is shown to depend on the wave function, corrected up to the order $l - 1$. The corrections, $\left\{ \left| \psi_n^{(l)} \right\rangle \right\}$, can be obtained by a recursive solution of Eq. (12.1.9). Using an orthonormal set of $\hat{H}_0$ eigenstates (excluding $\left| \psi_n^{(0)} \right\rangle$, in accordance with Eq. (12.1.10)), each correction can be expanded as a linear combination,

$$\left| \psi_n^{(l)} \right\rangle = \sum_{n' \neq n = 0}^{\infty} a_{n'}^{(l)} \left| \psi_{n'}^{(0)} \right\rangle \quad ; \quad a_{n'}^{(l)} = \left\langle \psi_{n'}^{(0)} \middle| \psi_n^{(l)} \right\rangle. \tag{12.1.12}$$

An explicit equation for the coefficients $\left\{ a_{n'}^{(l)} \right\}$ is obtained by substituting the expansion in Eq. (12.1.9) and projecting the two sides of the equation onto the corresponding eigenstates of $\hat{H}_0$, $\left| \psi_{n'}^{(0)} \right\rangle$, with $n \neq n'$. Since $E_n^{(0)}$ is nondegenerate, $E_n^{(0)} \neq E_{n'}^{(0)}$ for any n', and the result reads (Ex. 12.1.3)

$$a_{n'}^{(l)} = \frac{\left\langle \psi_{n'}^{(0)} \middle| \hat{H}_1 \middle| \psi_n^{(l-1)} \right\rangle}{E_n^{(0)} - E_{n'}^{(0)}} - \sum_{l'=1}^{l} \frac{\left\langle \psi_{n'}^{(0)} \middle| \psi_n^{(l-l')} \right\rangle E_n^{(l')}}{E_n^{(0)} - E_{n'}^{(0)}}. \tag{12.1.13}$$

Substitution of Eq. (12.1.12) in Eq. (12.1.13) yields explicit expressions for the corrections to the wave function. The results for the first and second orders read (Ex. 12.1.4)

$$\left|\psi_n^{(1)}\right\rangle = \sum_{n'\neq n} \frac{\left\langle \psi_{n'}^{(0)}\left|\hat{H}_1\right|\psi_n^{(0)}\right\rangle}{E_n^{(0)} - E_{n'}^{(0)}}\left|\psi_{n'}^{(0)}\right\rangle \tag{12.1.14}$$

$$\left|\psi_n^{(2)}\right\rangle = \sum_{n''\neq n}\sum_{n'\neq n} \frac{\left\langle \psi_{n''}^{(0)}\left|\hat{H}_1 - E_n^{(1)}\right|\psi_{n'}^{(0)}\right\rangle \left\langle \psi_{n'}^{(0)}\left|\hat{H}_1\right|\psi_n^{(0)}\right\rangle}{E_n^{(0)} - E_{n''}^{(0)}\quad E_n^{(0)} - E_{n'}^{(0)}}\left|\psi_{n''}^{(0)}\right\rangle. \tag{12.1.15}$$

Using Eq. (12.1.11), the corresponding corrections for the energy read

$$E_n^{(1)} = \left\langle \psi_n^{(0)}\left|\hat{H}_1\right|\psi_n^{(0)}\right\rangle \tag{12.1.16}$$

$$E_n^{(2)} = \sum_{n'\neq n} \frac{\left|\left\langle \psi_{n'}^{(0)}\left|\hat{H}_1\right|\psi_n^{(0)}\right\rangle\right|^2}{E_n^{(0)} - E_{n'}^{(0)}} \tag{12.1.17}$$

$$E_n^{(3)} = \sum_{n''\neq n}\sum_{n'\neq n} \frac{\left\langle \psi_n^{(0)}\left|\hat{H}_1\right|\psi_{n''}^{(0)}\right\rangle \left\langle \psi_{n''}^{(0)}\left|\hat{H}_1 - E_n^{(1)}\right|\psi_{n'}^{(0)}\right\rangle \left\langle \psi_{n'}^{(0)}\left|\hat{H}_1\right|\psi_n^{(0)}\right\rangle}{\left(E_n^{(0)} - E_{n'}^{(0)}\right)\left(E_n^{(0)} - E_{n''}^{(0)}\right)}. \tag{12.1.18}$$

Exercise 12.1.2 *(a) The projection of the vector* $\hat{H}_0\left|\psi_n^{(l)}\right\rangle$, *appearing in the left-hand side of Eq. (12.1.9), on the vector* $\left|\psi_n^{(0)}\right\rangle$ *is the inner product,* $\left\langle \psi_n^{(0)}\left|\hat{H}_0\right|\psi_n^{(l)}\right\rangle$. *Use the Hermiticity of* $\hat{H}_0$ *and Eqs. (12.1.1, 12.1.10) to show that this projection is zero. (b) Use the result obtained in (a) to show that projection of* $\left|\psi_n^{(0)}\right\rangle$ *on the vector appearing in the right-hand side of Eq. (12.1.9) leads to Eq. (12.1.11).*

Exercise 12.1.3 *Substitute the expansion of* $\left|\psi_n^{(l)}\right\rangle$ *(Eq. (12.1.12)) into Eq. (12.1.9), and project the two sides of the resulting equation on* $\left|\psi_{n'}^{(0)}\right\rangle$ *with* $n \neq n'$. *Derive Eq. (12.1.13) for the expansion coefficients, considering that* $E_n^{(0)}$ *is a nondegenerate eigenvalue of* $\hat{H}_0$.

Exercise 12.1.4 *(a) Use Eqs. (12.1.12, 12.1.13) to derive Eq. (12.1.14, 12.1.15). (b) Use the results of (a) and Eq. (12.1.11) to obtain Eqs. (12.1.16–12.1.18).*

The case where the eigenvalue of $\hat{H}_0$ **is degenerate (N-fold, with N > 1), requires a special consideration** (see the discussions of degenerate states in Sections 4.6 and 5.4). In this case, the zero-order wave functions corresponding to the energy $E_n^{(0)}$ are not uniquely defined in the sense that they can be taken as any linear combination of a finite set of N orthogonal vectors (a basis) in the degenerate subspace (see Ex. 4.6.3). Choosing such a basis, $\left\{\left|\psi_k^{(0)}\right\rangle\right\}$, where

$$\hat{H}_0\left|\psi_k^{(0)}\right\rangle = E_n^{(0)}\left|\psi_k^{(0)}\right\rangle \quad ; \quad \left\langle \psi_{k'}^{(0)}\middle|\psi_k^{(0)}\right\rangle = \delta_{k',k} \quad ; \quad k,k' \in 1,2,\ldots,N, \tag{12.1.19}$$

a general set of N orthonormal zero-order wave functions is therefore expressed as linear combinations:

$$\left|\psi_j^{(0)}\right\rangle = \sum_{k=1}^{N} a_{k,j}^{(0)}\left|\psi_k^{(0)}\right\rangle. \tag{12.1.20}$$

Our task is to derive equations for the corresponding lth-order corrections that approximate the eigenvalues and eigenfunctions of the full Hamiltonian, $\hat{H}$. Using a derivation identical to the one outlined in Eqs. (12.1.1–12.1.9), where $E_n^{(l)}$ and $\left| \psi_n^{(l)} \right\rangle$ are replaced by, $E_j^{(l)}$ and $\left| \psi_j^{(l)} \right\rangle$, respectively, we obtain the following equations for $E_j^{(l)}$ and $\left| \psi_j^{(l)} \right\rangle$:

$$\hat{H}_0 \left| \psi_j^{(l)} \right\rangle = \sum_{l'=0}^{l} (E_j^{(l')} - \delta_{l',1} \hat{H}_1) \left| \psi_j^{(l-l')} \right\rangle \quad ; \quad l > 0. \tag{12.1.21}$$

Recalling again that the eigenstates of $\hat{H}_0$ are a complete orthonormal set in the relevant space, any (unnormalized) eigenfunction of $\hat{H}$, $\left| \psi_j \right\rangle$ can be formally expanded as a linear combination, $\left| \psi_j(\lambda) \right\rangle = \sum_{k=1}^{N} a_{k,j}^{(0)} \left| \psi_k^{(0)} \right\rangle + \sum_{k' \neq \{1,2,...,N\}} b_{k'}(\lambda) \left| \psi_{k'}^{(0)} \right\rangle$. The set $\left\{ \left| \psi_k^{(0)} \right\rangle \right\}$ with $k = 1,2,...,N$ spans the degenerate subspace, whereas $\left\{ \left| \psi_{k'}^{(0)} \right\rangle \right\}$ with $k' \neq 1,2,...,N$ is the set of $\hat{H}_0$ eigenstates that are orthogonal to that space. Comparing this to the power expansion, $\left| \psi_j(\lambda) \right\rangle = \sum_{l=0}^{\infty} \lambda^l \left| \psi_j^{(l)} \right\rangle$, it follows that $\sum_{l>0} \lambda^l \left| \psi_j^{(l)} \right\rangle \propto \sum_{k' \neq \{1,2,...,N\}} b_{k'}(\lambda) \left| \psi_{k'}^{(0)} \right\rangle$, which means that for $l > 0$, the lth-order corrections to the wave function $\left| \psi_j(\lambda) \right\rangle$ are orthogonal to the degenerate subspace, namely

$$\left\langle \psi_k^{(0)} \middle| \psi_j^{(l)} \right\rangle = a_{k,j}^{(0)} \delta_{l,0} \quad ; \quad k \in 1,2,...,N. \tag{12.1.22}$$

Substitution of the expansion (Eq. 12.1.20) in Eq. (12.1.21) for $l = 1$, and projecting the two sides of the equation on the set which spans the degenerate subspace, $\left\{ \left| \psi_k^{(0)} \right\rangle \right\}$, we obtain the following set of equations (Ex. 12.1.5) for the expansion coefficients:

$$\sum_{k'=1}^{N} \left\langle \psi_k^{(0)} \middle| \hat{H}_1 \middle| \psi_{k'}^{(0)} \right\rangle a_{k',j}^{(0)} = E_j^{(1)} a_{k,j}^{(0)} \quad ; \quad k \in 1,2,...,N \tag{12.1.23}$$

Defining a square matrix, $[\mathbf{H}_1]_{k,k'} \equiv \left\langle \psi_k^{(0)} \middle| \hat{H}_1 \middle| \psi_{k'}^{(0)} \right\rangle$, and a vector, $\left[\mathbf{a}_j^{(0)} \right]_k = a_{k,j}^{(0)}$ in the N-dimensional degenerate subspace, Eq. (12.1.21) translates to an algebraic eigenvalue equation,

$$\mathbf{H}_1 \mathbf{a}_j^{(0)} = E_j^{(1)} \mathbf{a}_j^{(0)}, \tag{12.1.24}$$

where $\mathbf{H}_1$ is the matrix representation of the perturbation operator in a selected orthonormal basis for the degenerate subspace.

It follows that the first-order corrections to the energy are the eigenvalues of $\mathbf{H}_1$: $E_1^{(1)}, E_2^{(1)}, ... E_N^{(1)}$. The corresponding eigenvectors, $\mathbf{a}_1^{(0)}, \mathbf{a}_2^{(0)}, \mathbf{a}_N^{(0)}$, are specific sets of coefficients, definingspecific zero-order states,

$$E_j^{(1)} \leftrightarrow \left| \psi_j^{(0)} \right\rangle = \sum_{k=1}^{N} \left[\mathbf{a}_j^{(0)} \right]_k \left| \psi_k^{(0)} \right\rangle. \tag{12.1.25}$$

Notice that any linear combination of $\left\{ \left| \psi_k^{(0)} \right\rangle \right\}$ with $k = 1,2,...,N$ is an eigenvector of $\hat{H}_0$, but only specific linear combinations, satisfying Eq. (12.1.25), are consistent with *the expansions of the eigenvalues and eigenvectors of $\hat{H}$, in powers of λ.* While these specific zero-order states are degenerate with respect to $\hat{H}_0$, they are generally associated with different first-order corrections to the energy. In such a case the perturbation is said to "remove the degeneracy" between these states (see Fig. 12.1.1). Importantly, the degeneracy is not necessarily removed. It may happen (see some of the following examples) that some of the eigenvalues of $\mathbf{H}_1$ are zeros, which implies that the first-order corrections vanish, or it may happen that $\mathbf{H}_1$ has nonzero but still degenerate eigenvalues.

Exercise 12.1.5 *(a) Show that projecting Eq. (12.1.21) with $l = 1$ on the orthonormal set of degenerate vectors, defined in Eq. (12.1.19), yields $\langle \psi_k^{(0)} | (E_j^{(1)} - \hat{H}_1) | \psi_j^{(0)} \rangle = 0$. (b) Use the expansion, Eq. (12.1.20), to derive Eq. (12.1.23).*

Exercise 12.1.6 *(a) Use the orthonormality of the basis states (Eq. (12.1.19)) and recall that the eigenvectors of any Hermitian matrix (e.g., $\mathbf{H}_1$) can be chosen orthonormal,*

$$(\mathbf{a}_j^{(0)}, \mathbf{a}_{j'}^{(0)}) = \sum_{k=1}^{N} [\mathbf{a}_j^{(0)}]_k^* \cdot [\mathbf{a}_{j'}^{(0)}]_k = \delta_{j,j'}, \text{ to prove that the vectors defined in Eq. (12.1.25)}$$

can be chosen orthonormal, namely $\langle \psi_j^{(0)} | \psi_{j'}^{(0)} \rangle = \delta_{j,j'}$. (b) Show that for a normalized vector defined in Eq. (12.1.25), the first-order correction to the energy can be expressed as for a nondegenerate state (Eq. (12.1.16)), namely $E_j^{(1)} = \langle \psi_j^{(0)} | \hat{H}_1 | \psi_j^{(0)} \rangle$.

The higher-order corrections, namely $|\psi_j^{(l)}\rangle$ for $l > 0$ and $E_j^{(l)}$ for $l > 1$, can be obtained as in the case of nondegenerate zero-order states. Projecting the two sides of Eq. (12.1.21) on any of the zero-order solutions $\{|\psi_j^{(0)}\rangle\}$ defined in Eq. (12.1.25), we obtain the equation for the corrections to the energy (analogous to Eq. (12.1.11)),

$$E_j^{(l)} = \langle \psi_j^{(0)} | \hat{H}_1 | \psi_j^{(l-1)} \rangle. \tag{12.1.26}$$

The corrections to the wave functions for $l > 0$ can be obtained by expanding them in the set of $\hat{H}_0$ eigenstates, excluding the degenerate subspace, namely

$$|\psi_j^{(l)}\rangle = \sum_{k' \neq \{1,2,\ldots,N\}} a_{k',j}^{(l)} |\psi_{k'}^{(0)}\rangle \quad ; \quad a_{k',j}^{(l)} = \langle \psi_{k'}^{(0)} | \psi_j^{(l)} \rangle. \tag{12.1.27}$$

Substituting the expansions in Eq. (12.1.21) and projecting the two sides of the equation on the set $\{|\psi_{k'}^{(0)}\rangle\}$ with $k' \neq 1, 2, \ldots, N$, we obtain the following expression (analogous to Eq. (12.1.13)) for the expansion coefficients (see Ex. 12.1.3), recalling that $E_n^{(0)} \neq E_{k'}^{(0)}$, since all $\{|\psi_{k'}^{(0)}\rangle\}$ are external to the degenerate subspace:

$$a_{k',j}^{(l)} = \frac{\langle \psi_{k'}^{(0)} | \hat{H}_1 | \psi_j^{(l-1)} \rangle}{E_n^{(0)} - E_{k'}^{(0)}} - \sum_{l'=1}^{l} \frac{\langle \psi_{k'}^{(0)} | \psi_j^{(l-l')} \rangle E_n^{(l')}}{E_n^{(0)} - E_{k'}^{(0)}}. \tag{12.1.28}$$

Before turning to some practical examples of how perturbation theory is used, let us discuss the meaning of its "working equations," focusing on the lowest-order corrections. The first-order correction to the energy turns out to be the average of the perturbation (see Eq. (12.1.16) and Ex. 12.1.6). Indeed, given a system Hamiltonian, $\hat{H} = \hat{H}_0 + \lambda \hat{H}_1$, an energy can be associated with any state vector by means of the Hamiltonian expectation value (see Eq. (11.7.1)). Considering specifically the normalized zero-order state, $|\psi_n^{(0)}\rangle$, the associated energy reads

$$E_n \approx \langle \psi_n^{(0)} | \hat{H} | \psi_n^{(0)} \rangle = \langle \psi_n^{(0)} | \hat{H}_0 + \lambda \hat{H}_1 | \psi_n^{(0)} \rangle = E_n^{(0)} + \lambda \langle \psi_n^{(0)} | \hat{H}_1 | \psi_n^{(0)} \rangle, \qquad (12.1.29)$$

where the coefficient of λ is naturally associated with the expectation value of the perturbation. The first-order correction to the wave function, Eq. (12.1.14) extends beyond the space of the zero-order states associated with $E_n^{(0)}$. The relative contribution of each of the different eigenstates, $\left| \psi_{n'}^{(0)} \right\rangle$, is weighted by (in absolute magnitude)

$\dfrac{\left| \left\langle \psi_{n'}^{(0)} \middle| \hat{H}_1 \middle| \psi_n^{(0)} \right\rangle \right|}{\left| E_n^{(0)} - E_{n'}^{(0)} \right|}$. As we can see, as the energy gap between each eigenstate, $E_{n'}^{(0)}$, and $E_n^{(0)}$

increases, the contribution of the n' state becomes smaller. Moreover, when the energy gap, $|E_n^{(0)} - E_{n'}^{(0)}|$, is much larger than the corresponding matrix element of the perturbation, $|\langle \psi_{n'}^{(0)} | \hat{H}_1 | \psi_n^{(0)} \rangle|$, the contribution weight becomes much smaller than unity. Notice that the perturbation can indeed be regarded as "small," and the low-order corrections are indeed meaningful in the regime where

$$\frac{|\langle \psi_{n'}^{(0)} | \hat{H}_1 | \psi_n^{(0)} \rangle|}{|E_n^{(0)} - E_{n'}^{(0)}|} << 1 \qquad (12.1.29)$$

for all the eigenstates with $n' \neq n$. Setting $\lambda = 1$ to comply with the definition of the full Hamiltonian, the corrected vector, $|\psi_n^{(0)}\rangle + |\psi_n^{(1)}\rangle$, is dominated by $|\psi_n^{(0)}\rangle$ in this regime. In geometrical terms, the corrected vector still points (approximately) to the direction set by $|\psi_n^{(0)}\rangle$, in the space spanned by the set of all $\hat{H}_0$ eigenstates. A similar consideration applies to the next terms. Considering, for example, $|\psi_n^{(2)}\rangle$ (Eq. (12.1.15)), it is again expanded in the set of $\hat{H}_0$-eigenstates, beyond the space of $|\psi_n^{(0)}\rangle$. This time, however, the relative contributions of the different eigenstates are weighted by multiplications of the ratio between $\hat{H}_1$ matrix elements and corresponding $\hat{H}_0$ eigenvalue differences. Provided that all the ratios are smaller than unity, the second-order correction is small with respect to the previous terms, and so forth.

12.2 Perturbation Theory in Action

The Two-Level System

A commonly used model maps the physical space of a system onto a two-dimensional Hilbert space. In some cases, this description is exact, as, for example, the description of a single isolated spin in nonrelativistic quantum mechanics (see Chapter 13). In most cases, however, a two-level system (TLS) is a reduced approximate description of a multilevel system in some limit, useful for idealized descriptions of quantum systems with two distinctive states. In Chapters 17–20 we shall encounter several applications involving the generic TLS Hamiltonian.

Within a two-state model, any solution to the Schrödinger equation is a superposition of two orthonormal basis vectors,

$$|\psi\rangle = a_1|\chi_1\rangle + a_2|\chi_2\rangle \quad ; \quad \langle\chi_n|\chi_m\rangle = \delta_{n,m}, \tag{12.2.1}$$

where the corresponding Hamiltonian operator takes the generic form (Eq. (11.2.11)),

$$\hat{H} = H_{1,1}|\chi_1\rangle\langle\chi_1| + H_{2,2}|\chi_2\rangle\langle\chi_2| + H_{1,2}|\chi_1\rangle\langle\chi_2| + H_{2,1}|\chi_2\rangle\langle\chi_1|, \tag{12.2.2}$$

or in a matrix representation,

$$\mathbf{H} = \begin{bmatrix} H_{1,1} & H_{1,2} \\ H_{2,1} & H_{2,2} \end{bmatrix} \equiv \begin{bmatrix} \varepsilon_1 & \gamma \\ \gamma^* & \varepsilon_2 \end{bmatrix} \tag{12.2.3}$$

The basis states, $|\chi_1\rangle$ and $|\chi_2\rangle$, can be identified as the Hamiltonian eigenstates, for $\gamma = 0$. The entries, ε_1 and ε_2, therefore obtain the meaning of the energy levels associated with these states, The off-diagonal terms correspond to the interaction between the two basis states, which is expressed in terms of a single complex-valued parameter, $\gamma = H_{1,2} = H_{2,1}^*$, owing to the Hermiticity of $\hat{H}$.

The TLS model enables us to demonstrate fundamental aspects of perturbation theory, where exact solutions, readily available, enable us to test the accuracy of the approximation in the different parameter regimes. The exact time-independent Schrödinger equation for the TLS Hamiltonian (Eqs. (12.2.2, 12.2.3)) reads

$$\hat{H}|\psi_n\rangle = E_n|\psi_n\rangle. \tag{12.2.4}$$

It is convenient to define the average and the difference between the energies, ε_1 and ε_2, as $\bar{\varepsilon}$ and 2Δ, respectively:

$$\bar{\varepsilon} = \frac{\varepsilon_1 + \varepsilon_2}{2} \quad ; \quad \Delta = \frac{\varepsilon_1 - \varepsilon_2}{2}. \tag{12.2.5}$$

The exact eigenvalues of the TLS Hamiltonian are then readily obtained (see Ex. 12.2.1),

$$E_1 = \bar{\varepsilon} + \sqrt{\Delta^2 + |\gamma|^2} \quad ; \quad E_2 = \bar{\varepsilon} - \sqrt{\Delta^2 + |\gamma|^2}. \tag{12.2.6}$$

As for the eigenvectors, let us consider first the "nonsymmetric" case, $\Delta \neq 0$, introducing a dimensionless parameter,

$$\alpha \equiv \sqrt{\frac{\Delta^2 + |\gamma|^2}{\Delta^2}}. \tag{12.2.7}$$

Without loss of generality and for concreteness, we consider the case, $\gamma = -|\gamma|$, which yields the following expression for the orthonormal eigenvectors corresponding to E_1 and E_2 (see Ex. 12.2.1 for the general case),

$$|\psi_1\rangle = \sqrt{\frac{\alpha+1}{2\alpha}}|\chi_1\rangle - \sqrt{\frac{\alpha-1}{2\alpha}}|\chi_2\rangle \quad ; \quad |\psi_2\rangle = \sqrt{\frac{\alpha-1}{2\alpha}}|\chi_1\rangle + \sqrt{\frac{\alpha+1}{2\alpha}}|\chi_2\rangle, \tag{12.2.8}$$

where we notice that this expression applies also in the symmetric case ($\Delta = 0$), where $(\alpha \pm 1)/\alpha \xrightarrow[\Delta\to 0]{} 1$.

Exercise 12.2.1 *Derive the expressions for the eigenvalues (Eq. (12.2.6)) and the eigenvectors (Eq. (12.2.8)) of the TLS Hamiltonian, as defined in Eqs. (12.2.2, 12.2.3). In order to do this, express the eigenvectors as linear combinations of the basis vectors,*

$|\psi\rangle = a_1|\chi_1\rangle + a_2|\chi_2\rangle$, *project the corresponding eigenvalue equation* $\hat{H}|\psi\rangle = E|\psi\rangle$ *onto the basis vectors,* $|\chi_1\rangle$ *and* $|\chi_2\rangle$, *and obtain the algebraic eigenvalue equation,*
$\begin{bmatrix} \varepsilon_1 & \gamma \\ \gamma^* & \varepsilon_2 \end{bmatrix} \begin{bmatrix} a_1 \\ a_2 \end{bmatrix} = E \begin{bmatrix} a_1 \\ a_2 \end{bmatrix}$. *Then solve the equation: obtain the two eigenvalues,*
$E_\pm = \bar{\varepsilon} \pm \sqrt{\Delta^2 + |\gamma|^2}$, *and the corresponding eigenvector coefficients,* $a_1^{(\pm)} = \sqrt{\frac{\alpha\pm1}{2\alpha}}$,
$a_2^{(\pm)} = \pm\frac{|\gamma|}{\gamma}\sqrt{\frac{\alpha\mp1}{2\alpha}}$, *where* $\bar{\varepsilon} = (\varepsilon_1 + \varepsilon_2)/2, \Delta = (\varepsilon_1 - \varepsilon_2)/2$, *and* $\alpha \equiv \sqrt{1 + |\gamma|^2/\Delta^2}$.

We now turn to the implementation of the Rayleigh–Schrödinger perturbation theory for the TLS. Notice that the partitioning of the full Hamiltonian into zero-order and perturbation terms is not unique. Indeed, different partitions may be useful in different parameter regimes, where the more "correct" selection is the one yielding more accurate approximation to the exact solutions to Schrödinger's equation in a given parameter regime. Let us express the general TLS Hamiltonian (Eqs. (12.2.2, 12.2.3)) in terms of its three independent parameters, $\bar{\varepsilon}$, Δ, and γ:

$$\mathbf{H} = \bar{\varepsilon}\begin{bmatrix} 1 & 0 \\ 0 & 1 \end{bmatrix} + \Delta\begin{bmatrix} 1 & 0 \\ 0 & -1 \end{bmatrix} + \gamma\begin{bmatrix} 0 & 1 \\ 0 & 0 \end{bmatrix} + \gamma^*\begin{bmatrix} 0 & 0 \\ 1 & 0 \end{bmatrix} \qquad (12.2.9)$$

The first term amounts to an addition of a constant to the eigenvalues of $\mathbf{H}$ and has no effect on its eigenvectors (see Ex. 4.6.6). It can therefore be trivially included as part of the zero-order Hamiltonian. The effect of the other terms on the eigenvalues and eigenvectors depends on the ratio between the interstate interaction ($|\gamma|$) and the spacing (Δ) between the energy levels.

In Fig. 12.2.1, the two eigenvalues of the TLS (Eq. (12.2.6)) are plotted as functions of the ratio $|\gamma|/\Delta$. The dotted and dashed lines correspond to their approximations, based on second-order perturbation theory (derived in what follows) with different partitioning of the full Hamiltonian into zero-order and perturbation terms. As we can see, the exact values are remarkably well approximated by one (and only one) of these approximations, as long as $|\gamma|/\Delta \gg 1$ or $|\gamma|/\Delta \ll 1$.

When $|\gamma|$ is smaller than $|\Delta|$, a natural choice is to associate the perturbation operator with the interstate interaction term, proportional to $|\gamma|$. The zero-order Hamiltonian $\hat{H}_0$ takes the form

$$\hat{H}_0 = (\bar{\varepsilon} + \Delta)|\chi_1\rangle\langle\chi_1| + (\bar{\varepsilon} - \Delta)|\chi_2\rangle\langle\chi_2|, \qquad (12.2.10)$$

where its "preknown" eigenvalues and eigenvectors (Eq. (12.1.1)) are readily identified as

$$E_1^{(0)} = \bar{\varepsilon} + \Delta \quad ; \quad |\psi_1^{(0)}\rangle = |\chi_1\rangle$$
$$E_2^{(0)} = \bar{\varepsilon} - \Delta \quad ; \quad |\psi_2^{(0)}\rangle = |\chi_2\rangle \qquad (12.2.11)$$

The perturbation $\hat{H}_1$ is identified with the interstate interaction, where the full (λ-dependent) Hamiltonian obtains the form

$$\hat{H}(\lambda) = \hat{H}_0 + \lambda\hat{H}_1 \quad ; \quad \hat{H}_1 = \gamma|\chi_1\rangle\langle\chi_2| + \gamma^*|\chi_2\rangle\langle\chi_1|. \qquad (12.2.12)$$

Expanding the energy levels of $\hat{H}$ to the lowest powers in λ, using perturbation theory (Eqs. (12.1.16, 12.1.17)), we readily obtain (Ex. 12.2.2)

$$E_1 \cong E_1^{(0)} + \lambda E_1^{(1)} + \lambda^2 E_1^{(2)} = \bar{\varepsilon} + \Delta + \lambda^2 \frac{|\gamma|^2}{2\Delta}$$

$$E_2 \cong E_2^{(0)} + \lambda E_2^{(1)} + \lambda^2 E_2^{(2)} = \bar{\varepsilon} - \Delta - \lambda^2 \frac{|\gamma|^2}{2\Delta}. \qquad (12.2.13)$$

Notice that the exact energy levels can also be expressed as analytic functions of λ (replacing γ by $\lambda\gamma$ ($\lambda \in R$) in Eq. (12.2.6)),

$$E_{1,2} = \bar{\varepsilon} \pm \Delta \sqrt{1 + \frac{\lambda^2 |\gamma|^2}{\Delta^2}}. \qquad (12.2.14)$$

Expanding the square root function in a Taylor series, $\sqrt{1+x} = 1 + x/2 + \cdots$, we can readily identify the perturbation expansions (Eq. (12.2.13)) with the lowest-order terms in the corresponding expression of the exact energy levels (Eq. (12.2.14)). Indeed, for $\frac{\lambda^2 |\gamma|^2}{\Delta^2} \ll 1$, the truncated expansion is an accurate approximation to the exact eigenvalues (see the dashed lines in Fig. 12.2.1). Recalling that the "physical" TLS Hamiltonian is associated with $\lambda = 1$, the condition for validity of this low-order approximation reads

$$|\gamma| << |\Delta|. \qquad (12.2.15)$$

This result is consistent with our association of the interstate interaction with a "perturbation," when $|\gamma|/|\Delta| < 1$. Notice, however, that the accuracy of the low-order perturbation expansion depends on the interstate coupling energy, $|\gamma|$, and improves as it gets smaller in comparison to $|\Delta|$. Moreover, the result is accurate only when the resulting induced change to the level spacing is small.

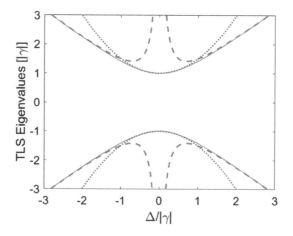

Figure 12.2.1   Solid lines: The exact two eigenvalues of a TLS defined by the diagonal matrix elements Δ and $-\Delta$ ($\bar{\varepsilon}$ was set to zero), and the off-diagonal, γ. The dashed and dotted lines represent approximated eigenvalues, calculated by second-order perturbation theory with different selections of the zero-order Hamiltonians: Dashed for diagonal and dotted for off-diagonal.

Exercise 12.2.2 *Implement perturbation theory (Eqs. (12.1.16, 12.1.17)) for the Hamiltonian defined in Eqs. (12.2.10–12.2.12). (a) Show that the first-order corrections to the energy vanish. (b) Calculate the second-order corrections to obtain Eq. (12.2.13).*

A similar consideration applies to the TLS eigenvectors. Using Eq. (12.1.14), the perturbation theory expansion yields the following (for concreteness, we set again, $\gamma = -|\gamma|$) (see Ex. 12.2.3):

$$|\psi_1\rangle \cong |\psi_1^{(0)}\rangle + \lambda|\psi_1^{(1)}\rangle = |\chi_1\rangle - \lambda\frac{|\gamma|}{2\Delta}|\chi_2\rangle$$

$$|\psi_2\rangle \cong |\psi_2^{(0)}\rangle + \lambda|\psi_2^{(1)}\rangle = |\chi_2\rangle + \lambda\frac{|\gamma|}{2\Delta}|\chi_1\rangle. \tag{12.2.16}$$

Renormalizing the exact eigenstates for (Eq. (12.2.7)) and replacing γ in Eq. (12.2.6) by $\lambda\gamma(\lambda \in R)$, the exact (unnormalized) TLS eigenvectors can be expressed as analytic functions of λ (Ex. 12.2.4):

$$|\psi_1\rangle \propto \left[|\chi_1\rangle + \frac{\Delta}{\lambda|\gamma|}\left(1 - \sqrt{1 + \frac{\lambda^2|\gamma|^2}{\Delta^2}}\right)|\chi_2\rangle\right]$$

$$|\psi_2\rangle \propto \left[|\chi_2\rangle + \frac{\Delta}{\lambda|\gamma|}\left(\sqrt{1 + \frac{\lambda^2|\gamma|^2}{\Delta^2}} - 1\right)|\chi_1\rangle\right]. \tag{12.2.17}$$

Again, the first-order approximations obtained using the perturbation expansion, Eq. (12.2.16), are shown to correspond to the first-order Taylor expansion of the square root function in Eq. (12.2.17). For $\lambda = 1$, the perturbation expansion (Eq. (12.2.16)) is shown to coincide with the exact wave function, as long as $|\gamma| << |\Delta|$, in line with Eq. (12.2.15). Notice that the exact eigenstates are dominated by the two eigenstates of $\hat{H}_0$ in this parameter regime, $|\psi_{1(2)}\rangle \approx |\chi_{1(2)}\rangle$, where the corrections are proportional to the small ratio, $|\gamma|/|\Delta|$.

Exercise 12.2.3 *Implement perturbation theory (Eq. (12.1.14)) for the Hamiltonian defined in Eqs. (12.2.10–12.2.12) to obtain the first-order corrections to the eigenstates, as given in Eq. (12.2.16).*

Exercise 12.2.4 *(a) Derive the expressions in Eq. (12.2.17) for the exact TLS eigenstates. (b) Show that the result obtained by first-order perturbation theory (Eq. (12.2.16)) is obtained by expanding the square root function in a first-order Taylor expansion.*

The identification of the perturbation with the interstate interaction (Eq. (12.2.12)) was shown to be useful only for $|\gamma| << |\Delta|$ (see Eq. (12.2.15)). However, perturbation theory is still applicable when the system parameters are in the complementary regime, namely, when $|\gamma| > |\Delta|$. In this case, however, the partition of $\hat{H}$ into zero-order and perturbation terms needs to be changed accordingly. Let us set the zero-order Hamiltonian to

$$\hat{H}_0 = \bar{\varepsilon}(|\chi_1\rangle\langle\chi_1| + |\chi_2\rangle\langle\chi_2|) + \gamma|\chi_1\rangle\langle\chi_2| + \gamma^*|\chi_2\rangle\langle\chi_1|, \tag{12.2.18}$$

where its "preknown" eigenvalues and eigenvectors (Eq. (12.1.1)) are readily identified as (again, we set $\gamma = -|\gamma|$ for concreteness)

$$E_1^{(0)} = \overline{\varepsilon} + |\gamma| \quad ; \quad |\psi_1^{(0)}\rangle = \sqrt{\frac{1}{2}}|\chi_1\rangle - \sqrt{\frac{1}{2}}|\chi_2\rangle$$

$$E_2^{(0)} = \overline{\varepsilon} - |\gamma| \quad ; \quad |\psi_2^{(0)}\rangle = \sqrt{\frac{1}{2}}|\chi_1\rangle + \sqrt{\frac{1}{2}}|\chi_2\rangle. \tag{12.2.19}$$

The perturbation $\hat{H}_1$ is now associated with the interaction-free Hamiltonian, where the full (λ-dependent) Hamiltonian obtains the form

$$\hat{H}(\lambda) = \hat{H}_0 + \lambda\hat{H}_1 \quad ; \quad \hat{H}_1 = \Delta|\chi_1\rangle\langle\chi_1| - \Delta|\chi_2\rangle\langle\chi_2| \tag{12.2.20}$$

The approximations to the energy levels of $\hat{H}$, using perturbation theory (Eqs. (12.1.16, 12.1.17)), read in this case (Ex. 12.2.5)

$$E_1 \simeq E_1^0 + \lambda E_1^{(1)} + \lambda^2 E_1^{(2)} = \overline{\varepsilon} + |\gamma| + \frac{\lambda^2\Delta^2}{2|\gamma|}$$

$$E_2 \simeq E_2^0 + \lambda E_2^{(1)} + \lambda^2 E_2^{(2)} = \overline{\varepsilon} - |\gamma| - \frac{\lambda^2\Delta^2}{2|\gamma|} \tag{12.2.21}$$

Expressing the exact energy levels as analytic functions of λ (replacing Δ by $\lambda\Delta$(where $\lambda \in R$) in Eq. (12.2.6)), we obtain

$$E_{1,2} = \overline{\varepsilon} \pm |\gamma|\sqrt{1 + \frac{\lambda^2\Delta^2}{|\gamma|^2}}, \tag{12.2.22}$$

where the perturbation expansions (Eq. (12.2.21)) are identified with the lowest-order expansion of the square root function, $\sqrt{1+x} = 1+x/2+\ldots$ Setting $\lambda = 1$, as in the full Hamiltonian, we conclude that the truncated expansion is an accurate approximation to the exact eigenvalues (see the dotted lines in Fig. 12.2.1), as long as

$$|\gamma| >> |\Delta|, \tag{12.2.23}$$

which is consistent with our choice of the interstate interaction as the zero-order Hamiltonian in the case, where $|\gamma|/|\Delta| > 1$. Again, the TLS eigenvectors can be approximated using perturbation theory (Ex. 12.2.6),

$$|\psi_1\rangle \cong |\psi_1^{(0)}\rangle + \frac{\lambda\Delta}{2|\gamma|}|\psi_2^{(0)}\rangle$$

$$|\psi_2\rangle = |\psi_2^{(0)}\rangle - \frac{\lambda\Delta}{2|\gamma|}|\psi_1^{(0)}\rangle. \tag{12.2.24}$$

Renormalizing the exact eigenstates for $\gamma = -|\gamma|$ (Eq. (12.2.7)) and replacing Δ in Eq. (12.2.7) by $\lambda\Delta$(where $\lambda \in R$), the exact (unnormalized) TLS eigenvectors can be expressed as analytic functions of λ (Ex. 12.2.7),

$$|\psi_1\rangle \propto \left(\sqrt{\sqrt{1 + \lambda^2\frac{\Delta^2}{|\gamma|^2}} + \lambda\frac{\Delta}{|\gamma|}}\sqrt{\frac{1}{2}}|\chi_1\rangle - \sqrt{\sqrt{1 + \lambda^2\frac{\Delta^2}{|\gamma|^2}} - \lambda\frac{\Delta}{|\gamma|}}\sqrt{\frac{1}{2}}|\chi_2\rangle \right)$$

$$|\psi_2\rangle \propto \left(\sqrt{\sqrt{1 + \lambda^2\frac{\Delta^2}{|\gamma|^2}} - \lambda\frac{\Delta}{|\gamma|}}\sqrt{\frac{1}{2}}|\chi_1\rangle + \sqrt{\sqrt{1 + \lambda^2\frac{\Delta^2}{|\gamma|^2}} + \lambda\frac{\Delta}{|\gamma|}}\sqrt{\frac{1}{2}}|\chi_2\rangle \right).$$

$$\tag{12.2.25}$$

Again, the approximations obtained using perturbation theory, Eq. (12.2.24), are shown to correspond to the first-order Taylor expansion of the square root functions in the exact expressions (Eq. (12.2.25)), which is valid for $|\Delta| \ll |\gamma|$. As we can see, in this regime the eigenstates of the full Hamiltonian are dominated by the two eigenstates of $\hat{H}_0$, $|\psi_{1(2)}\rangle \approx |\psi_{1(2)}^{(0)}\rangle$, where the corrections are proportional to the small ratio, $|\Delta|/|\gamma|$ (see Eqs. (12.2.19, 12.2.24)).

Exercise 12.2.5 *Implement perturbation theory (Eqs. (12.1.16, 12.1.17)) for the Hamiltonian defined in Eqs. (12.2.18–12.2.20). (a) Show that the first-order corrections to the energy vanish. (b) Calculate the second-order corrections to obtain Eq. (12.2.21).*

Exercise 12.2.6 *Implement perturbation theory (Eq. (12.1.14)) for the Hamiltonian defined in Eqs. (12.2.18–12.2.20) to obtain the first-order corrections to the eigenstates, as given in Eq. (12.2.24).*

Exercise 12.2.7 *(a) Derive the expressions in Eq. (12.2.25) for the exact TLS eigenstates. (b) Show that the result obtained by first-order perturbation theory (Eq. (12.2.24)) is obtained by expanding the square root function in a first-order Taylor expansion.*

Let us dwell on the specific case where the energy level spacing between the noninteracting states $|\chi_1\rangle$ and $|\chi_2\rangle$ strictly vanishes ($\Delta = 0$). On one hand, this is a limiting case of $|\Delta| \ll |\gamma|$, where the perturbation operator $\hat{H}_1 = \Delta|\chi_1\rangle\langle\chi_1| - \Delta|\chi_2\rangle\langle\chi_2|$ vanishes, and the result of perturbation theory with the proper partitioning, Eq. (12.2.20), trivially coincides with the exact results:

$$E_{1,2}|_{\Delta=0} = \bar{\varepsilon} \pm |\gamma| \quad ; \quad |\psi_{1,2}\rangle|_{\Delta=0} = \sqrt{\frac{1}{2}}|\chi_1\rangle \mp \sqrt{\frac{1}{2}}|\chi_2\rangle. \tag{12.2.26}$$

On the other hand, we may choose a partitioning of the Hamiltonian according to Eq. (12.2.12), and attempt to refer to the interaction operator as the perturbation and to the noninteracting states $|\chi_1\rangle$ and $|\chi_2\rangle$ as degenerate eigenstates of $\hat{H}_0$, associated with the interaction-free TLS Hamiltonian, Eq. (12.2.10). Recalling the procedure of implementing perturbation theory for degenerate states (Eq. (12.1.24)), our first task is to diagonalize the matrix representation of the perturbation operator in the degenerate subspace, which reads in this case

$$\mathbf{H}_1 = \begin{bmatrix} 0 & \gamma \\ \gamma^* & 0 \end{bmatrix}. \tag{12.2.27}$$

Since, for $\Delta = 0$, $\mathbf{H}_1$ differs from the full TLS Hamiltonian only by an addition of a constant $\bar{\varepsilon}$,

$$\mathbf{H}|_{\Delta=0} = \begin{bmatrix} \bar{\varepsilon} & \gamma \\ \gamma^* & \bar{\varepsilon} \end{bmatrix} = \mathbf{H}_1 + \bar{\varepsilon}\begin{bmatrix} 1 & 0 \\ 0 & 1 \end{bmatrix}, \tag{12.2.28}$$

the eigenvectors of the perturbation operator $\mathbf{H}_1$ coincide with the exact Hamiltonian eigenstates in this case, and the respective eigenvalues are the same up to the additional $\bar{\varepsilon}$ (Eq. (12.2.26)). Indeed, implementing perturbation theory for degenerate states is equivalent to a full solution of the Schrödinger equation, projected onto the degenerate subspace. In the case of the TLS model (as well as for other models, in which the degenerate space coincides with the full space), this is equivalent to a full solution of the Schrödinger equation, and therefore is generally not practical. In most cases of interest, however, the degenerate space is only a part of the full space, and perturbation theory for degenerate states is therefore useful as a first step for approximating the exact solutions to the Schrödinger equation in the entire space.

"Hybridization" of Atomic Orbitals

In Chapter 10 we became familiar with the states of an electron in an isolated hydrogen-like atom model. Here we use perturbation theory in order to study the effect of a small perturbation, associated with the presence of another point charge, on the energy levels and orbitals of the electron. As we shall see, even a remote, small charge can induce substantial changes to the electronic energy levels and stationary wave functions. Such changes are indeed meaningful in understanding the formation of much more complicated systems, including molecules and solid lattices (discussed in Chapter 14), where electrons are shared among several nuclei. Particularly, we shall see that the presence of external charges can induce a "hybridization" of atomic orbitals, which lowers the ground state energy.

Our zero-order model Hamiltonian would correspond to an isolated hydrogen-like atom where the nucleus is positioned at the origin, $\mathbf{R} = (0,0,0)$, and the electron coordinate is $\mathbf{r} = (x,y,z)$ (see Eq. (10.2.4), using Cartesian coordinates):

$$\hat{H}_0 = \frac{-\hbar^2}{2\mu} \left(\frac{\partial^2}{\partial x^2} + \frac{\partial^2}{\partial y^2} + \frac{\partial^2}{\partial z^2} \right) + \frac{-KZe^2}{\sqrt{x^2 + y^2 + z^2}}. \qquad (12.2.29)$$

The perturbation corresponds to the interaction of the atom with a remote point-charge, $q|e|$, located at a fixed position, $\mathbf{R}_q = (x_q, y_q, z_q)$, at a distance $R_q = |\mathbf{R}_q|$ from the origin. The addition to the Hamiltonian accounts for the electrostatic interaction of the remote charge with the electron and the nucleus,

$$\hat{H}_1 = \frac{KZqe^2}{|\mathbf{R}_q|} + \frac{-Kqe^2}{|\mathbf{r} - \mathbf{R}_q|} \qquad (12.2.30)$$

We shall focus here on perturbations whose effect on the ground state of the system becomes null as $R_q \to \infty$. This excludes, for example, cases in which $q = |Z|$, where the Hamiltonian $\hat{H}_0 + \hat{H}_1$ becomes symmetric with respect to exchanging the nucleus and the remote charge resulting in delocalization of the electronic wave functions between the two positive charges. Let us emphasize that orbital delocalization between atoms is considered as the "driving force" for chemical bond formation and for atomic orbital hybridizations, which is the basis for the valence bond theory [12.1]). Here we focus only on atomic orbital hybridization driven by weak electrostatic perturbations, where a detailed discussion of delocalization of atomic orbitals between different atoms and chemical bond formation will be given in Chapter 14. Cases in which $q > |Z|$ are also excluded here, since in that case the ground states of $\hat{H}_0 + \hat{H}_1$ will be associated with charge transfer to the remote point charge, whereas our focus here is on changes to the atomic orbitals near the atomic nucleus.

Restricting the following discussion to perturbations associated with $q < |Z|$, we shall focus on the first-order corrections to the ground and first excited energy states of $\hat{H}_0$. The zero-order atomic ground state is nondegenerate (see Fig. 10.2.2 and Eq. (10.2.32)) and corresponds to the wave function and energy level,

$$\langle \mathbf{r} | \psi_{1s}^{(0)} \rangle = \psi_{1s}(x,y,z) = \sqrt{\frac{Z^3}{\pi a_0^3}} e^{-Z\sqrt{x^2+y^2+z^2}/a_0} \quad ; \quad E_1^{(0)} = -R_H \frac{Z^2}{1^2} \tag{12.2.31}$$

The first-order correction to the **ground state energy** is therefore the matrix element of the perturbation operator (see Eq. (12.1.16)). For convenience, the line connecting the atomic nucleus (the origin) and the remote charge is identified here as the z-axis, namely $\mathbf{R}_q \equiv (0,0,R_q)$, which yields in this case (see Ex. 12.2.8)

$$E_1^{(1)} = \langle \psi_{1s}^{(0)} | \hat{H}_1 | \psi_{1s}^{(0)} \rangle = \frac{KZqe^2}{R_q} - Ke^2 q \int d\mathbf{r} \frac{|\psi_{1s}(\mathbf{r})|^2}{|\mathbf{r} - \mathbf{R}_q|}$$

$$= \frac{Kqe^2}{R_q} \left[Z - 1 + \left(\frac{ZR_q}{a_0} + 1 \right) e^{\frac{-2ZR_q}{a_0}} \right]. \tag{12.2.32}$$

Notice that at asymptotic distances $(R_q >> \frac{a_0}{Z})$ we obtain

$$E_1^{(1)} \xrightarrow{R_q \to \infty} \frac{Ke^2 q(Z-1)}{R_q}, \tag{12.2.33}$$

which has a clear physical meaning of the Coulomb interaction energy between the remote point charge $(q|e|)$ and the net atomic charge, obtained by subtracting the negative electron charge from the positive nucleus charge, $(Z-1)|e|$, as if the remote charge "sees" the atom as a point charge. Indeed, when the external charge is positioned far from the atom $\left(\frac{a_0}{Z} << R_q \right)$, its interaction energy with the electron is insensitive to the details of the electron's probability density distribution. The latter is confined to some radius of the order $\sim \frac{a_0}{Z}$ around the atomic nucleus, and therefore when integrating over the entire space in Eq. (12.2.32), the interaction energy $\frac{-Ke^2 q}{|\mathbf{r}-\mathbf{R}_q|}$ is projected into this (relatively small) region. Consequently, we can approximate the perturbation operator by its first-order Taylor expansion as a function of $\mathbf{r}$ near the origin,

$$\hat{H}_1 = \frac{Ke^2 Zq}{|\mathbf{R}_q|} + \frac{-Ke^2 q}{|\mathbf{r}-\mathbf{R}_q|} \cong \frac{Ke^2 Zq}{R_q} + \frac{-Ke^2 q}{R_q} \left(1 + \frac{x_q x}{R_q^2} + \frac{y_q y}{R_q^2} + \frac{z_q z}{R_q^2} \right). \tag{12.2.34}$$

Since the probability density, $|\psi_{1s}(\mathbf{r})|^2$, is an even function of $\mathbf{r}$, the integral over the linear terms in $\mathbf{r}$ vanishes, and the remaining result (see Ex. 12.2.9) amounts to $\frac{Ke^2 q(Z-1)}{R_q}$, multiplied by the normalization integral of $\psi_{1s}(\mathbf{r})$, which indeed coincides with the asymptotic value of $E_1^{(1)}$ in Eq. (12.2.33).

Exercise 12.2.8   *To calculate the integral* $-Ke^2 q \int d\mathbf{r} \frac{|\psi_{1s}(\mathbf{r})|^2}{|\mathbf{r}-\mathbf{R}_q|}$, *it is convenient to change variables to the elliptical coordinates, (λ,μ,φ), defined as*

$$\lambda \equiv \frac{r+r_q}{R_q} \quad ; \quad 1 < \lambda < \infty$$

$$\mu \equiv \frac{r_q-r}{R_q} \quad ; \quad -1 < \mu < 1,$$

where $r_q \equiv |\mathbf{r}-\mathbf{R}_q|$, $R_q \equiv |\mathbf{R}_q|$, $r \equiv |\mathbf{r}|$, *and* $\mathbf{R}_q \equiv (0,0,R_q)$.

(a) *Show that these definitions yield the following results:* $x = r\sin(\theta)\cos(\varphi) = \frac{R_q}{2}\sqrt{\lambda^2+\mu^2-1-\mu^2\lambda^2}\cos(\varphi)$, $y = r\sin(\theta)\sin(\varphi) = \frac{R_q}{2}\sqrt{\lambda^2+\mu^2-1-\mu^2\lambda^2}\sin(\varphi)$ *and* $z = r\cos(\theta) = \frac{R_q}{2}(1-\mu\lambda)$.

(b) *Let* $g(\lambda,\mu,\varphi) = f(x,y,z)$. *Calculate the corresponding Jacobian,* $\begin{vmatrix} \frac{\partial x}{\partial \lambda} & \frac{\partial x}{\partial \mu} & \frac{\partial x}{\partial \varphi} \\ \frac{\partial y}{\partial \lambda} & \frac{\partial y}{\partial \mu} & \frac{\partial y}{\partial \varphi} \\ \frac{\partial z}{\partial \lambda} & \frac{\partial z}{\partial \mu} & \frac{\partial z}{\partial \varphi} \end{vmatrix}$,

and show that $\int\limits_{-\infty}^{\infty} dx \int\limits_{-\infty}^{\infty} dy \int\limits_{-\infty}^{\infty} dz f(x,y,z) = \int\limits_{0}^{2\pi} d\varphi \int\limits_{1}^{\infty} d\lambda \int\limits_{-1}^{1} d\mu \frac{R_q^3}{8}(\lambda^2-\mu^2)g(\lambda,\mu,\varphi)$.

(c) *Let* $f(x,y,z) = -Ke^2 q \frac{|\psi_{1s}(\mathbf{r})|^2}{|\mathbf{r}-\mathbf{R}_q|}$, *with* $\psi_{1s}(\mathbf{r})$ *as defined in Eq. (12.2.31) and* $\mathbf{R}_q \equiv (0,0,R_q)$. *Derive the result for* $E_1^{(1)}$ *in Eq. (12.2.32).*

(d) *Show that as the distance to the point charge goes to infinity, the first-order correction to the energy approaches the Coulomb interaction energy between the remote charge and an effective point charge, in which the electron charge is subtracted from the nucleus charge, namely* $E_1^{(1)} \xrightarrow[R_q \to \infty]{} \frac{Ke^2 q}{R_q}(Z-1)$.

Exercise 12.2.9   *Calculate the first-order correction to the ground-state energy (Eq. (12.1.16)), owing to a remote point charge, using the approximation for the perturbation, Eq. (12.2.34). Compare the result to the exact calculation, Eq. (12.2.32), in the limit, $R_q \gg \frac{a_0}{Z}$.*

We now turn to the effect of a remote point charge on **the first excited state** of the hydrogen-like atom. In this case the energy level of $\hat{H}_0$ is 4-fold degenerate. For convenience, we shall choose the orthonormal set of real-valued functions $\{\psi_{2s}, \psi_{2p_x}, \psi_{2p_y}, \psi_{2p_z}\}$ (see Fig. 10.2.4) as a basis for the degenerate subspace (see Eq. (12.1.19)). The zero-order solutions therefore read

$$E_2^{(0)} = -R_H \frac{Z^2}{2^2},$$

$$\langle \mathbf{r} | \psi_{2s}^{(0)} \rangle = \psi_{2s}(x,y,z) = \frac{1}{8}\sqrt{\frac{2Z^3}{\pi a_0^3}}\left(2 - \frac{Z}{a_0}\sqrt{x^2+y^2+z^2}\right)e^{-\sqrt{x^2+y^2+z^2}\frac{Z}{2a_0}},$$

$$\langle \mathbf{r} | \psi_{2p_z}^{(0)} \rangle = \psi_{2p_z}(x,y,z) = \frac{1}{8}\sqrt{\frac{2Z^5}{\pi a_0^5}}z e^{-\sqrt{x^2+y^2+z^2}\frac{Z}{2a_0}},$$

$$\langle \mathbf{r} | \psi_{2p_y}^{(0)} \rangle = \psi_{2p_y}(x,y,z) = \frac{1}{8}\sqrt{\frac{2Z^5}{\pi a_0^5}} y e^{-\sqrt{x^2+y^2+z^2}\frac{Z}{2a_0}},$$

$$\langle \mathbf{r} | \psi_{2p_x}^{(0)} \rangle = \psi_{2p_x}(x,y,z) = \frac{1}{8}\sqrt{\frac{2Z^5}{\pi a_0^5}} x e^{-\sqrt{x^2+y^2+z^2}\frac{Z}{2a_0}}. \tag{12.2.35}$$

The first-order corrections to the energy in this case are the eigenvalues of the matrix representation of the perturbation operator in the degenerate subspace (Eq. (12.1.24)). In the present case (Eq. (12.2.30)), when $\mathbf{R}_q \equiv (0,0,R_q)$ in Cartesian coordinates), the matrix representation of the perturbation operator obtains the form

$$\mathbf{H}_1 = \begin{bmatrix} \langle \psi_{2s}^{(0)}|\hat{H}_1|\psi_{2s}^{(0)} \rangle & \langle \psi_{2s}^{(0)}|\hat{H}_1|\psi_{2p_z}^{(0)} \rangle & \langle \psi_{2s}^{(0)}|\hat{H}_1|\psi_{2p_y}^{(0)} \rangle & \langle \psi_{2s}^{(0)}|\hat{H}_1|\psi_{2p_x}^{(0)} \rangle \\ \langle \psi_{2p_z}^{(0)}|\hat{H}_1|\psi_{2s}^{(0)} \rangle & \langle \psi_{2p_z}^{(0)}|\hat{H}_1|\psi_{2p_z}^{(0)} \rangle & \langle \psi_{2p_z}^{(0)}|\hat{H}_1|\psi_{2p_y}^{(0)} \rangle & \langle \psi_{2p_z}^{(0)}|\hat{H}_1|\psi_{2p_x}^{(0)} \rangle \\ \langle \psi_{2p_y}^{(0)}|\hat{H}_1|\psi_{2s}^{(0)} \rangle & \langle \psi_{2p_y}^{(0)}|\hat{H}_1|\psi_{2p_z}^{(0)} \rangle & \langle \psi_{2p_y}^{(0)}|\hat{H}_1|\psi_{2p_y}^{(0)} \rangle & \langle \psi_{2p_y}^{(0)}|\hat{H}_1|\psi_{2p_x}^{(0)} \rangle \\ \langle \psi_{2p_x}^{(0)}|\hat{H}_1|\psi_{2s}^{(0)} \rangle & \langle \psi_{2p_x}^{(0)}|\hat{H}_1|\psi_{2p_z}^{(0)} \rangle & \langle \psi_{2p_x}^{(0)}|\hat{H}_1|\psi_{2p_y}^{(0)} \rangle & \langle \psi_{2p_x}^{(0)}|\hat{H}_1|\psi_{2p_x}^{(0)} \rangle \end{bmatrix}$$

$$= \begin{bmatrix} f_{2s}(\beta) & f_{s,p_z}(\beta) & 0 & 0 \\ f_{s,p_z}(\beta) & f_{2p_z}(\beta) & 0 & 0 \\ 0 & 0 & f_{2p_y}(\beta) & 0 \\ 0 & 0 & 0 & f_{2p_x}(\beta) \end{bmatrix}, \tag{12.2.36}$$

where the nonvanishing integrals are functions of the dimensionless distance parameter, $\beta \equiv ZR_q/(2a_0)$ (see Ex. 12.2.10 for the full calculation using elliptical coordinates):

$$f_{2s}(\beta) = \frac{Ke^2Z^2q}{2a_0\beta} + \frac{-ZKe^2q}{4a_0\beta^3}[2\beta^2 - (2\beta^5 + 2\beta^4 + 3\beta^3 + 2\beta^2)e^{-2\beta}],$$

$$f_{2p_z}(\beta) = \frac{Ke^2Z^2q}{2a_0\beta} + \frac{-ZKe^2q}{4a_0\beta^3}[2\beta^2 + 6 - (2\beta^5 + 6\beta^4 + 11\beta^3 + 14\beta^2 + 12\beta + 6)e^{-2\beta}],$$

$$f_{2p_x}(\beta) = f_{2p_y}(\beta) = \frac{Ke^2Z^2q}{2a_0\beta} + \frac{-ZKe^2q}{4a_0\beta^3}[2\beta^2 - 3 + (\beta^3 + 4\beta^2 + 6\beta + 3)e^{-2\beta}],$$

$$f_{s,p_z}(\beta) = \frac{-ZKe^2q}{4a_0\beta^3}[-3\beta + (2\beta^5 + 4\beta^4 + 6\beta^3 + 6\beta^2 + 3\beta)e^{-2\beta}]. \tag{12.2.37}$$

Exercise 12.2.10 *Using the explicit set of degenerate wave functions (Eq. (12.2.35)), calculate the matrix elements of the operator $\frac{-Ke^2q}{|\mathbf{r}-\mathbf{R}_q|}$, for $\mathbf{R}_q \equiv (0,0,R_q)$, and verify the results given in Eqs. (12.2.36, 12.2.37) (including the vanishing entries). For this purpose, it is recommended to change variables to the elliptical coordinates, (λ,μ,φ), following the practice of Ex. 12.2.8.*

The eigenvalues of $\mathbf{H}_1(\{E_2^{(1)}\})$, are the roots of the following determinant,

$$\begin{vmatrix} f_{2s}(\beta) - E_2^{(1)} & f_{s,p_z}(\beta) & 0 & 0 \\ f_{s,p_z}(\beta) & f_{2p_z}(\beta) - E_2^{(1)} & 0 & 0 \\ 0 & 0 & f_{2p_y}(\beta) - E_2^{(1)} & 0 \\ 0 & 0 & 0 & f_{2p_x}(\beta) - E_2^{(1)} \end{vmatrix} = 0, \tag{12.2.38}$$

where each root $(E_{2,j}^{(1)}, \ j = 1,2,3,4)$ corresponds to a specific eigenvector,

$$
\begin{bmatrix}
f_{2s}(\beta) - E_{2,j}^{(1)} & f_{s,p_z}(\beta) & 0 & 0 \\
f_{s,p_z}(\beta) & f_{2p_z}(\beta) - E_{2,j}^{(1)} & 0 & 0 \\
0 & 0 & f_{2p_y}(\beta) - E_{2,j}^{(1)} & 0 \\
0 & 0 & 0 & f_{2p_x}(\beta) - E_{2,j}^{(1)}
\end{bmatrix}
\begin{bmatrix}
a_{2s,j}^{(0)} \\
a_{2p_z,j}^{(0)} \\
a_{2p_y,j}^{(0)} \\
a_{2p_x,j}^{(0)}
\end{bmatrix}
=
\begin{bmatrix}
0 \\
0 \\
0 \\
0
\end{bmatrix}. \qquad (12.2.39)
$$

Each coefficient vector, $a_{2s,j}^{(0)}, a_{2p_z,j}^{(0)}, a_{2p_y,j}^{(0)}, a_{2p_x,j}^{(0)}$, corresponds to a zero-order eigenstate of $\hat{H}_0$,

$$
E_{2,j}^{(1)} \leftrightarrow |\psi_{2,j}^{(0)}\rangle = a_{2s,j}^{(0)}|\psi_{2s}^{(0)}\rangle + a_{2p_z,j}^{(0)}|\psi_{2p_z}^{(0)}\rangle + a_{2p_y,j}^{(0)}|\psi_{2p_y}^{(0)}\rangle + a_{2p_x,j}^{(0)}|\psi_{2p_x}^{(0)}\rangle, \qquad (12.2.40)
$$

which is consistent with the perturbation to the Hamiltonian (see Eq. (12.1.25)).

The corrections to the excited-state energy are readily obtained from Eq. (12.2.38),

$$
E_{2,1}^{(1)} = \frac{1}{2}(f_{2s}(\beta) + f_{2p_z}(\beta) + \sqrt{[f_{2s}(\beta) - f_{2p_z}(\beta)]^2 + 4[f_{s,p_z}(\beta)]^2})
$$
$$
E_{2,2}^{(1)} = \frac{1}{2}(f_{2s}(\beta) + f_{2p_z}(\beta) - \sqrt{[f_{2s}(\beta) - f_{2p_z}(\beta)]^2 + 4[f_{s,p_z}(\beta)]^2}),
$$
$$
E_{2,3}^{(1)} = f_{2p_y}(\beta)
$$
$$
E_{2,4}^{(1)} = E_{2,3}^{(1)} \qquad (12.2.41)
$$

As we can see, the degeneracy of the excited state orbitals is only partially removed in the presence of the remote point charge, where one of the corrected eigenvalues remains doubly degenerate. The block diagonal structure of Eq. (12.2.39) reveals that the two degenerate eigenvalues are (any) linear combinations of $|\psi_{2p_x}^{(0)}\rangle$ and $|\psi_{2p_y}^{(0)}\rangle$, whereas the nondegenerate states are specific linear combinations of $|\psi_{2s}^{(0)}\rangle$ and $|\psi_{2p_z}^{(0)}\rangle$. The remaining degeneracy is intuitive, considering that the point charge is positioned along the z-axis, and its interaction with the charge densities associated with the orbitals $|\psi_{2p_x}^{(0)}\rangle$ and $|\psi_{2p_y}^{(0)}\rangle$ is therefore invariant to a 90° rotation around the z-axis, which would map these two densities one on the other.

While an explicit calculation of the new orbitals obtained by mixing $|\psi_{2s}^{(0)}\rangle$ with $|\psi_{2p_z}^{(0)}\rangle$ is straightforward for any position of the point charge (solving explicitly Eq. (12.2.39)), it is insightful to consider again the asymptotic limit, in which the extra point charge is positioned far from the electronic density surrounding the atomic nucleus, namely at $R_q \gg a_0/Z$. The matrix representation of the perturbation obtains a simple form in this case (see Eq. (12.2.34)),

$$\mathbf{H}_1 = \frac{Ke^2Zq}{R_q}\begin{bmatrix} 1 & 0 & 0 & 0 \\ 0 & 1 & 0 & 0 \\ 0 & 0 & 1 & 0 \\ 0 & 0 & 0 & 1 \end{bmatrix} + \frac{-Ke^2q}{R_q}\begin{bmatrix} 1 & \frac{-3a_0}{ZR_q} & 0 & 0 \\ \frac{-3a_0}{ZR_q} & 1 & 0 & 0 \\ 0 & 0 & 1 & 0 \\ 0 & 0 & 0 & 1 \end{bmatrix}, \qquad (12.2.42)$$

with the following eigenvalues and eigenvectors,

$$E_{2,2}^{(1)} = \frac{Ke^2q}{R_q}\left(Z-1+\frac{3a_0}{ZR_q}\right) \quad ; \quad |\psi_{2,1}^{(0)}\rangle = \frac{1}{\sqrt{2}}\left(|\psi_{2s}^{(0)}\rangle + |\psi_{2p_z}^{(0)}\rangle\right)$$

$$E_{2,2}^{(1)} = \frac{Ke^2q}{R_q}\left(Z-1-\frac{3a_0}{ZR_q}\right) \quad ; \quad |\psi_{2,2}^{(0)}\rangle = \frac{1}{\sqrt{2}}\left(|\psi_{2s}^{(0)}\rangle - |\psi_{2p_z}^{(0)}\rangle\right)$$

$$E_{2,3}^{(1)} = \frac{Ke^2q}{R_q}(Z-1) \quad ; \quad |\psi_{2,3}^{(0)}\rangle = |\psi_{2p_y}^{(0)}\rangle$$

$$E_{2,4}^{(1)} = \frac{Ke^2q}{R_q}(Z-1) \quad ; \quad |\psi_{2,4}^{(0)}\rangle = |\psi_{2p_x}^{(0)}\rangle. \qquad (12.2.43)$$

As we can see, the presence of the remote point charge removes the degeneracy of the orbitals of the isolated atom, $|\psi_{2s}^{(0)}\rangle$ and $|\psi_{2p_z}^{(0)}\rangle$, while inducing their "mixing" into so-called "sp" hybrid orbitals, $|\psi_{2sp,\pm}^{(0)}\rangle = \frac{1}{\sqrt{2}}(|\psi_{2s}^{(0)}\rangle \pm |\psi_{2p_z}^{(0)}\rangle)$, associated with polarized probability densities, either toward or away from the point charge (see Fig. 12.2.2). Notice that the different energies attributed to each hybrid orbital are in accord with

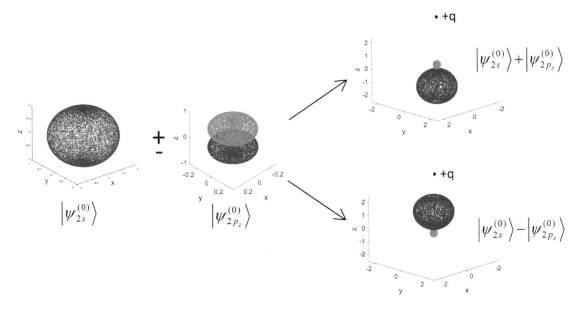

Figure 12.2.2   Angular distribution functions for degenerate $|\psi_{2s}^{(0)}\rangle$ and $|\psi_{2p_z}^{(0)}\rangle$ states of an isolated hydrogen-like atom (a) and "sp" hybridization orbitals (b) obtained by introducing a remote point charge $+q$. The angular distribution functions are as defined in Fig. 10.3.1, where the different shades in each plot add information corresponding to changes in the sign of the wave function, for which the probability density is presented.

the electrostatic interaction between the point charge and the polarized electronic charge distribution. For example, when the remote charge located at $z_q > 0$ is positive, $q > 0$, the lower energy state corresponds to $||\psi_{2s}^{(0)}\rangle - |\psi_{2p_z}^{(0)}\rangle|^2$. Since $|\psi_{2s}^{(0)}\rangle$ and $|\psi_{2p_z}^{(0)}\rangle$ interfere constructively (destructively) at large positive (negative) z, the (negatively charged) electronic probability density is polarized toward the positive point charge (see Ex. 12.2.11), which reduces the electrostatic energy. Similarly, the higher energy state corresponds to $||\psi_{2s}^{(0)}\rangle + |\psi_{2p_z}^{(0)}\rangle|^2$, where the interference between $|\psi_{2s}^{(0)}\rangle$ and $|\psi_{2p_z}^{(0)}\rangle$ polarizes the (negatively charged) electronic probability density away from the remote positive charge.

Notice that the hybridized "sp" function is asymmetric with respect to reflection through the (x,y) plane, which reflects the reduced symmetry of the Hamiltonian in the presence of the point charge. Importantly, while the energy difference between the hybridized orbitals, $\Delta E = \frac{Ke^2q}{R_q^2}\frac{6a_0}{Z}$, decays to zero as $1/R_q^2$ with increasing distance to the point charge, the shape of the corresponding orbitals is invariant to changes in R_q in this asymptotic regime and reflects the broken symmetry of the Hamiltonian. Indeed, any distribution of charges along the z-axis that breaks the symmetry of reflection through the (x,y) plane would result in a non-diagonal matrix $\mathbf{H}_1$ with the same hybridized orbitals as its eigenvectors. (In contrast, for example, two identical charges positioned at $+z_q$ and $-z_q$ that maintain the reflection symmetry would correspond to a diagonal matrix $\mathbf{H}_1$, and consequently $|\psi_{2s}^{(0)}\rangle$ and $|\psi_{2p_z}^{(0)}\rangle$ would remain degenerate, where the hybridized orbitals would have no particular meaning).

Exercise 12.2.11   *Use the explicit form of $|\psi_{2s}^{(0)}\rangle$ and $|\psi_{2p_z}^{(0)}\rangle$ in Eq. (12.2.35), and show that far from the nucleus ($r > 2a_0/Z$) the probability density associated with $|\psi_{2s}^{(0)}\rangle - |\psi_{2p_z}^{(0)}\rangle$ is larger above the (x,y) plane, namely $p(x,y,|z|) \geq \rho(x,y,-|z|)$, and the result is reversed for $|\psi_{2s}^{(0)}\rangle + |\psi_{2p_z}^{(0)}\rangle$.*

Different hybridizations are obtained when several point charges are distributed at different points in space, where the atomic orbitals are mixed differently according to the geometrical arrangement of the charges (more generally, according to the symmetry group of the perturbed system Hamiltonian). Consider, for example, two point charges, the first positioned at a fixed distance R_q from the origin along the (positive) z-axis, $\mathbf{R}_1 = (0,0,R_q)$, and the second positioned at a fixed distance αR_q along the positive x-axis, $\mathbf{R}_2 = (\alpha R_q, 0, 0)$. For simplicity, we shall consider in this case only the asymptotic limit, $\alpha R_q, R_q >> a_0/Z$, where the distances between the point charges and the atom are sufficiently large that the perturbation operator can be well approximated by its first-order Taylor expansion as a function of $\mathbf{r}$ near the origin (Eq. (12.2.34)). In the present case, the perturbation operator obtains the form

$$\hat{H}_1 = C(\mathbf{R}_1, \mathbf{R}_2) + \frac{-Ke^2q}{R_q}\left(1 + \frac{1}{\alpha}\right) + \frac{-Ke^2q}{R_q^2}z + \frac{-Ke^2q}{\alpha^2 R_q^2}x, \qquad (12.2.44)$$

where $C(\mathbf{R}_1, \mathbf{R}_2) \equiv \frac{Ke^2 Zq}{|\mathbf{R}_1|} + \frac{Ke^2 Zq}{|\mathbf{R}_2|} + \frac{Ke^2 q^2}{|\mathbf{R}_1 - \mathbf{R}_2|}$ is the electrostatic energy associated with the clamped nucleus and point charges, which is independent of the electron coordinates. As discussed in the previous example, the terms linear in $\mathbf{r}$ do not contribute to the first-order correction for the ground-state energy, which therefore reads $E_1^{(1)} = \langle \psi_{1s}^{(0)} | \hat{H}_1 | \psi_{1s}^{(0)} \rangle = C(\mathbf{R}_1, \mathbf{R}_2) + \frac{-Ke^2 q}{R_q}(1 + \frac{1}{\alpha})$. These terms do contribute, however, to the first-order corrections to the degenerate excited-state energy. Particularly, using the integrals, $\langle \psi_{2s}^{(0)} | \hat{z} | \psi_{2p_z}^{(0)} \rangle = \langle \psi_{2s}^{(0)} | \hat{x} | \psi_{2p_x}^{(0)} \rangle = -3a_0/Z$ (see Ex. 12.2.12), the matrix representation of $\hat{H}_1$ in the basis for the degenerate excited states, Eq. (12.2.35), reads

$$\mathbf{H}_1 = C(\mathbf{R}_1, \mathbf{R}_2) \begin{bmatrix} 1 & 0 & 0 & 0 \\ 0 & 1 & 0 & 0 \\ 0 & 0 & 1 & 0 \\ 0 & 0 & 0 & 1 \end{bmatrix} + \frac{-Ke^2 q}{R_q} \begin{bmatrix} 1 + \frac{1}{\alpha} & -\frac{3a_0}{ZR_q} & 0 & -\frac{3a_0}{\alpha^2 ZR_q} \\ -\frac{3a_0}{ZR_q} & 1 + \frac{1}{\alpha} & 0 & 0 \\ 0 & 0 & 1 + \frac{1}{\alpha} & 0 \\ -\frac{3a_0}{\alpha^2 ZR_q} & 0 & 0 & 1 + \frac{1}{\alpha} \end{bmatrix}.$$

$$(12.2.45)$$

Two of the corresponding eigenvalues of $\mathbf{H}_1$ are associated with the same correction to the energy, $C(\mathbf{R}_1, \mathbf{R}_2) + \frac{-Ke^2 q}{R_q}(1 + \frac{1}{\alpha})$, which corresponds to two degenerate eigenstates, where the other two are different, $C(\mathbf{R}_1, \mathbf{R}_2) + \frac{-Ke^2 q}{R_q}\left(1 + \frac{1}{\alpha} \pm \frac{3a_0}{ZR_q}\sqrt{1 + 1/\alpha^4}\right)$. Setting, for example, $\alpha = 1/3^{1/4}$ (see Ex. 12.2.12), the two degenerate eigenstates are $|\psi_{2p_y}^{(0)}\rangle$ and $\sqrt{\frac{3}{4}}|\psi_{2p_z}^{(0)}\rangle - \frac{1}{2}|\psi_{2p_x}^{(0)}\rangle$, where the other two are $\frac{1}{\sqrt{2}}|\psi_{2s}^{(0)}\rangle \pm \frac{1}{\sqrt{8}}|\psi_{2p_z}^{(0)}\rangle \pm \sqrt{\frac{3}{8}}|\psi_{2p_x}^{(0)}\rangle$.

Exercise 12.2.12   *(a) Use the explicit form of the hydrogen-like orbitals (Eq. (12.2.35)) to show that $\langle \psi_{2s}^{(0)} | \hat{z} | \psi_{2p_z}^{(0)} \rangle = \langle \psi_{2s}^{(0)} | \hat{x} | \psi_{2p_x}^{(0)} \rangle = \langle \psi_{2s}^{(0)} | \hat{y} | \psi_{2p_y}^{(0)} \rangle = -3a_0/Z$. (b) Diagonalize the matrix $\mathbf{H}_1$ for $\alpha = 1/3^{1/4}$ to show that the eigenvalues are $C(\mathbf{R}_1, \mathbf{R}_2) + \frac{-Ke^2 q}{R_q}(1 + 3^{1/4}); C(\mathbf{R}_1, \mathbf{R}_2) + \frac{-Ke^2 q}{R_q}\left(1 + 3^{1/4} \pm \frac{6a_0}{ZR_q}\right)$. (c) Obtain the corresponding first-order corrected atomic orbitals.*

The Stark Effect

Another example for electrostatically induced hybridization of atomic orbitals is the Stark effect [12.2]. Consider a hydrogen-like atom centered at the origin of coordinate space, subject to a static electric field along the z-axis. The Hamiltonian takes the form $\hat{H} = \hat{H}_0 + \hat{H}_1$, where $\hat{H}_0$ is the hydrogen-like atom Hamiltonian (Eq. (12.2.29)), and $\hat{H}_1$ accounts for the interaction of the atomic dipole with the electric field, $\hat{H}_1 = -\hat{\mu} \cdot \mathbf{E}$. Identifying the z-axis with the direction of the electric field, $\mathbf{E} = (0, 0, E_z)$, and using the atom's dipole operator, $\hat{\mu} = -|e|\hat{r}$, the atom-field interaction reads $\hat{H}_1 = E_z|e|\hat{z}$. For weak fields, perturbation theory can be applied for calculating the energy levels and stationary wave functions for the atom in the presence of the field. Before implementing the theory, we note that in general we need to be careful when considering a perturbation operator that diverges asymptotically. Indeed, for $E_z \neq 0$, the local operator, $E_z|e|\hat{z}$, diverges in the position representation as $z \to \pm\infty$, which implies that sufficiently far from the origin, it dominates over $\hat{H}_0$, and a different partitioning of the Hamiltonian needs to be invoked for using perturbation theory. Moreover, since $\hat{H}_1 = E_z |e| \hat{z}$ does not support bound states, its eigenstates are very different from the bound states of the atom. We must therefore restrict the linear dependence of the field to a finite coordinate space (e.g., the interior of a finite range capacitor) and to small field intensities, for which the expected corrections to the eigenvalues of the full Hamiltonian remain small in comparison to their level spacings (see Fig. 12.1.1).

Starting from **the ground state** of the atom, $|\psi_{1s}^{(0)}\rangle$, the first-order correction to the energy takes the following form (see Eq. (12.2.35) for the Cartesian coordinate representation):

$$E_1^{(1)} = \langle \psi_{1s}^{(0)} | \hat{H}_1 | \psi_{1s}^{(0)} \rangle = E_z |e| \int |\psi_{1s}(\mathbf{r})|^2 z\, d\mathbf{r} = 0. \qquad (12.2.46)$$

Since $|\psi_{1s}(\mathbf{r})|^2$ and z are, respectively, even and odd functions of z, the integral vanishes, and so does the first-order correction to the energy. Moreover, the second-order correction to the energy, as defined in Eq. (12.1.17), depends on the squares of matrix elements of the perturbation operator, and therefore we can conclude that the dependence of the ground state energy on the field intensity (E_z) will be at least quadratic, which is indeed the case (the quadratic Stark effect).

Turning to **the first excited state** of the atom, we encounter a 4-fold degeneracy. Again, it is convenient to choose the orthonormal set of real-valued functions, $\{\psi_{2s}, \psi_{2p_x}, \psi_{2p_y}, \psi_{2p_z}\}$ (see Eq. (12.2.35)), as a basis for the degenerate subspace, where the matrix representation of the perturbation operator reads

$$
\mathbf{H}_1 =
\begin{bmatrix}
\langle \psi_{2s}^{(0)} | \hat{H}_1 | \psi_{2s}^{(0)} \rangle & \langle \psi_{2s}^{(0)} | \hat{H}_1 | \psi_{2p_z}^{(0)} \rangle & \langle \psi_{2s}^{(0)} | \hat{H}_1 | \psi_{2p_y}^{(0)} \rangle & \langle \psi_{2s}^{(0)} | \hat{H}_1 | \psi_{2p_x}^{(0)} \rangle \\
\langle \psi_{2p_z}^{(0)} | \hat{H}_1 | \psi_{2s}^{(0)} \rangle & \langle \psi_{2p_z}^{(0)} | \hat{H}_1 | \psi_{2p_z}^{(0)} \rangle & \langle \psi_{2p_z}^{(0)} | \hat{H}_1 | \psi_{2p_y}^{(0)} \rangle & \langle \psi_{2p_z}^{(0)} | \hat{H}_1 | \psi_{2p_x}^{(0)} \rangle \\
\langle \psi_{2p_y}^{(0)} | \hat{H}_1 | \psi_{2s}^{(0)} \rangle & \langle \psi_{2p_y}^{(0)} | \hat{H}_1 | \psi_{2p_z}^{(0)} \rangle & \langle \psi_{2p_y}^{(0)} | \hat{H}_1 | \psi_{2p_y}^{(0)} \rangle & \langle \psi_{2p_y}^{(0)} | \hat{H}_1 | \psi_{2p_x}^{(0)} \rangle \\
\langle \psi_{2p_x}^{(0)} | \hat{H}_1 | \psi_{2s}^{(0)} \rangle & \langle \psi_{2p_x}^{(0)} | \hat{H}_1 | \psi_{2p_z}^{(0)} \rangle & \langle \psi_{2p_x}^{(0)} | \hat{H}_1 | \psi_{2p_y}^{(0)} \rangle & \langle \psi_{2p_x}^{(0)} | \hat{H}_1 | \psi_{2p_x}^{(0)} \rangle
\end{bmatrix}
$$

$$
=
\begin{bmatrix}
0 & \gamma & 0 & 0 \\
\gamma & 0 & 0 & 0 \\
0 & 0 & 0 & 0 \\
0 & 0 & 0 & 0
\end{bmatrix},
\qquad (12.2.47)
$$

where $\gamma = -3a_0/Z$ (see Ex. 12.2.12 for the relevant integrals). The first-order corrections to the excited state energy in the presence of the field are the eigenvalues of this matrix, $E_2^{(1)} = 0, 0, \gamma, -\gamma$, where the corresponding eigenvectors define corrected zero-order atomic orbitals. The zero eigenvalue remains doubly degenerate, and the corresponding orbitals are any linear combination of the orthonormal states, $|\psi_{2p_x}^{(0)}\rangle$ and $|\psi_{2p_y}^{(0)}\rangle$. The two other eigenvalues, $E_2^{(1)} = \pm\gamma$, correspond to hybrid "sp" orbitals, $\frac{1}{\sqrt{2}}(|\psi_{2s}^{(0)}\rangle \pm |\psi_{2p_z}^{(0)}\rangle)$, as illustrated in Fig. 12.2.3. Notice that the electron densities are pointing toward opposite directions, parallel and antiparallel to the field direction, in

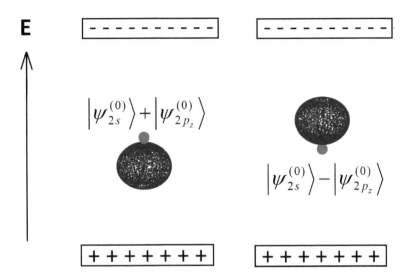

Figure 12.2.3 Angular distribution functions for "sp" hybridization orbitals obtained by introducing a hydrogen-like atom into a constant electric field along the z direction. The angular distribution functions are as defined in Fig. 10.3.1, where the different shades in each plot add information corresponding to changes in the sign of the wave function, for which the probability density is presented.

consistency with the corresponding changes in the orbital energies. As we can see, the corrections to the excited state energy in the presence of the field are linear in the field intensity (the linear stark effect).

Envelope Function Approximations in Quantum Wells

In Chapters 5 and 6 we discussed energy quantization in quantum wells. These structures were idealized in several aspects, one being ignorance of the detailed atomistic structure of the underlying material composing the structure. Indeed, the typical length of quantum wells ($\sim$ 10–1000 nm) is large in comparison to the typical unit cell of most lattices ($\sim$ 0.1–1 nm), and therefore, for a qualitative understanding of the electron energy quantization phenomenon (the quantum size effect), models neglecting the atomistic structure are sufficient. A quantitative treatment of the "multi-scale" Schrödinger equation, should account consistently for the boundary conditions derived from the size, shape, and environment of the quantum well (or the nanoparticle), as well as for the short-range potential energy alternations on the atomic scale. The inherent length-scale separation enables the implementation of approximate and yet accurate solutions to the full Schrödinger equation for the quantum structure. The most common approaches, such as the $k \cdot p$ perturbation theory, relate specifically to the basis (Bloch) functions associated with the infinitely periodic lattice structure. The interested reader is directed to more advanced discussions of this approach in other textbooks [12.3]. Here the atomistic details of the structure are regarded as a small perturbation. As we have already seen, perturbation theory provides accurate approximation, as long as the matrix elements of the perturbation operator are small in magnitude in comparison to the energy level spacings of the zero-order Hamiltonian. In a typical nanostructure, the quantum size effect means that the level spacings, derived from the confining potential energy function at the boundaries, increase with reducing structure size, to a point where the level spacing is sufficiently large to invoke a perturbative treatment of the underlying lattice structure.

Let our zero-order Hamiltonian for an idealized quantum well correspond to the "particlein-a-box" model studied in Section 5.3, with the zero-order eigenfunctions and eigenvalues:

$$\langle x|\psi_n\rangle = \sqrt{\frac{2}{L}}\sin\left(\frac{n\pi}{L}x\right) \quad;\quad E_n = \frac{\hbar^2\pi^2 n^2}{2mL^2} \quad;\quad n=1,2,3\ldots, \tag{12.2.48}$$

where m and L are the particle's mass and the box length, respectively.

The perturbation operator corresponds to an alternating potential energy with a lattice period (a unit cell), which we set to L/k, where k is an integer (of the order of 10–10^3, for typical nanostructure). For simplicity, we consider a harmonic modulation function,

$$\hat{H}_1 = \alpha\cos\left(k\hat{x}\frac{2\pi}{L}\right), \tag{12.2.49}$$

where α is the amplitude of the potential energy modulation.

We can readily see that the first-order correction to the energy vanishes for any $n\neq k$ (Ex. 12.2.13),

$$E^{(1)}_{n\neq k} = \langle\psi_n|\hat{H}_1|\psi_n\rangle = 0, \tag{12.2.50}$$

whereas the second-order correction reads

$$E^{(2)}_n = \sum_{\substack{n'=1\\n'\neq n}}^{\infty}\frac{\left|\left\langle\psi^{(0)}_{n'}\left|\hat{H}_1\right|\psi^{(0)}_n\right\rangle\right|^2}{E^{(0)}_n - E^{(0)}_{n'}} = \frac{-\alpha^2}{16k}\frac{2mL^2}{\hbar^2\pi^2}\left\{\begin{array}{ll}\frac{1}{k+n}+\frac{1}{k-n} & ;\quad n\neq k,2k\\ \frac{1}{k+n} & ;\quad n=k,2k\end{array}\right. . \tag{12.2.51}$$

Similarly, we can obtain the first-order correction to the full Hamiltonian eigenfunctions (See Eq. (12.1.14) and Ex. 12.2.13),

$$\left|\psi^{(1)}_n\right\rangle = \sum_{\substack{n'=1\\n'\neq n}}^{\infty}\frac{\left\langle\psi^{(0)}_{n'}\left|\hat{H}_1\right|\psi^{(0)}_n\right\rangle}{E^{(0)}_n - E^{(0)}_{n'}}\left|\psi^0_{n'}\right\rangle$$

$$\psi^{(1)}_n(x) = \frac{\alpha}{8k}\sqrt{\frac{2}{L}}\left(\frac{2mL^2}{\hbar^2\pi^2}\right)\left\{\begin{array}{ll}\frac{-1}{n+k}\sin\left(\frac{(n+2k)\pi x}{L}\right) & ;\ n=k,2k\\ \frac{-1}{n+k}\sin\left(\frac{(n+2k)\pi x}{L}\right)+\frac{1}{n-k}\sin\left(\frac{(n-2k)\pi x}{L}\right) & ;\ n\neq k,2k.\end{array}\right.$$
$$\tag{12.2.52}$$

Exercise 12.2.13 *Use the explicit form of the particle-in-a-box eigenstates (Eq. (12.2.48)) and the perturbation operator, Eq. (12.2.49), to show the following: (a) For any $n,n'>0$,*
$$\langle\psi_{n'}|\hat{H}_1|\psi_n\rangle = \frac{\alpha}{2}\left\{\begin{array}{ll}\delta_{n',n+2k}-\delta_{n',2k-n} & ;\ 2k>n\\ \delta_{n',n+2k}+\delta_{n',n-2k} & ;\ 2k\leq n\end{array}\right.$$ *(b) The first-order corrections to the energy vanish unless n=k (Eq. (12.2.50)). (c) The second-order correction to the energy*

is given in terms of Eq. (12.2.51). (d) The first-order correction to the function is given in terms of Eq. (12.2.52).

It is instructive to discuss the effect of the alternating lattice potential on the lowest eigenstates of the confining quantum well potential, $n \sim 1 - 10$, which should correspond to the energy levels accessible by charge carriers in a typical material. For typical quantum wells (or nanoparticles) the atomic lattice period is about two to three orders of magnitude smaller than the confining potential length, which means that we are interested in the regime where $n << k$. In this limit, the de Broglie wavelength associated with the quantum number n, namely $\lambda_n = 2L/n$ (see Eq. (1.1.3)), is much larger than the unit cell length, L/k. In Fig. 12.2.4 the exact eigenfunctions in the presence of the perturbation are compared to the zero-order particle-in-a-box functions. As we can see, two length scales are apparent. A slowly varying envelope corresponding to the zero-order function, and rapid oscillations at a wavelength set by the rapidly changing (lattice) potential. This form is captured already using the first-order correction to the wave function, based on Eq. (12.2.52), which yields in the limit $n << k$ (Ex. 12.2.14),

$$\psi_n^{(0)}(x) + \psi_n^{(1)}(x) \approx \sqrt{\frac{2}{L}} \sin\left(\frac{n\pi x}{L}\right)\left[1 - \frac{\alpha}{E_{2k}^{(0)}}\cos\left(\frac{2k\pi x}{L}\right)\right]. \qquad (12.2.53)$$

As we can see, a measure for the perturbation effect is the ratio between the oscillating potential energy amplitude, α, and the corresponding kinetic energy for a particle in the box with a de Broglie wavelength at the same oscillation period (a quantum

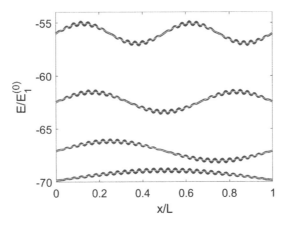

Figure 12.2.4 Stationary wave functions ($\psi_n(x)$, $n = 1, 2, 3, 4$, displaced from each other according to their respective energy levels) for a "particle-in-a-box" in the presence of an oscillating potential energy, $\alpha \cos(2k\pi x/L)$. The box contains 30 oscillation periods ($k = 30$), and the potential amplitude is, $\alpha = 0.2E_{2k}^{(0)}$. The thin line corresponds to the perturbation-free box, and the thick line represents the exact wave function in the presence of the perturbation (calculated numerically) and/or the first-order approximation by perturbation theory (Eq. (12.2.53)), which are indistinguishable in the given plot resolution. Notice that while the perturbation induces a significant reduction of the energy levels, the effect on the level spacings and on the wavefunctions is relatively minor.

number $2k$). For $\alpha \ll E_{2k}^{(0)}$ the explicit expression for the wave function indeed reveals a slowly varying envelope function ($\sqrt{\frac{2}{L}} \sin\left(\frac{n\pi x}{L}\right)$, in the present case) multiplied by a rapidly oscillating function at the lattice periodicity. The similarity between the solutions in the perturbed and unperturbed systems in this limit is also apparent from the corrections to the energy, which approaches a nearly constant value in this limit (see Eq. (12.2.51)), $E_n^{(2)} \xrightarrow[n \ll k]{} \frac{-\alpha^2}{2E_{2k}^{(0)}}$, with a negligible effect on the energy level spacings in the perturbed system (see Ex. 4.6.6).

Exercise 12.2.14 *Derive Eq. (12.2.53) from Eq. (12.2.52) in the limit $n << k$.*

In the opposite limit, $n >> k$, the de Broglie wavelength associated with the quantum number n, becomes short in comparison to the unit cell length, L/k. As the kinetic energy of the particle-in-a-box increases with n, the perturbation associated with the potential energy oscillations becomes negligible (for any finite α). This is indeed reflected in the vanishing perturbative corrections to the energy levels, Eq. (12.2.51), and wave functions, Eq. (12.2.52), $E_n^{(2)} \xrightarrow[n\to\infty]{} 0$, $|\psi_n^{(1)}\rangle \xrightarrow[n\to\infty]{} 0$.

A most interesting regime is associated with the intermediate quantum numbers, $n \approx k$, where the length scale separation breaks down (i.e., the half de Broglie wavelength is of the order of the unit cell length, $\lambda_n/2 \sim L/k$). The perturbative correction to the energy (Eq. (12.2.51)) is dominated in this regime by the term $E_n^{(2)} \approx \frac{-\alpha^2}{16k} \frac{2mL^2}{\hbar^2\pi^2} \frac{1}{k-n}$ and therefore changes its sign from negative to positive, as the quantum number changes from $n < k$ to $n > k$, respectively. This discontinuity in the corrections to the energy is a signature of the emergence of an energy gap formation in the exact spectrum of the full Hamiltonian. The gap is characteristic of the Schrödinger equation for an infinitely periodic lattice, as will be discussed in Chapter 14.

We now turn to a similar study of a two-dimensional potential energy well. The zero-order Hamiltonian corresponds to a particle in an infinite two-dimensional box (see Section 5.5),

$$\hat{H}_0 = \frac{1}{2m}(\hat{p}_x^2 + \hat{p}_y^2) + V(\hat{x}, \hat{y}), \qquad (12.2.53)$$

where $V(x,y) = \begin{cases} 0; & 0 \le x \le L_x, 0 \le y \le L_y \\ \infty; & elsewhere \end{cases}$. The corresponding zero-order eigenvalues and eigenfunctions are identified by two quantum numbers in this case, $n_x, n_y = 1, 2, \ldots$:

$$E_{n_x,n_y}^{(0)} = \frac{\hbar^2\pi^2}{2m}\left[\left(\frac{n_x}{L_x}\right)^2 + \left(\frac{n_y}{L_y}\right)^2\right] \quad ;$$

$$\psi_{n_x,n_y}^{(0)}(x,y) = \sqrt{\frac{2}{L_x}}\sqrt{\frac{2}{L_y}}\sin\left(\frac{n_x\pi x}{L_x}\right)\sin\left(\frac{n_y\pi y}{L_y}\right). \qquad (12.2.54)$$

The two-dimensional lattice corresponds to an alternating potential energy with a two-dimensional unit cell of dimensions $(L_x/k_x, L_y/k_y)$. In the simplest case, the potential energy function obtains a product form,

$$\hat{H}_1 = \alpha \cos\left(k_x \hat{x} \frac{2\pi}{L_x}\right) \cos\left(k_y \hat{y} \frac{2\pi}{L_y}\right). \tag{12.2.55}$$

Consequently, the matrix elements of the perturbation operator in the basis of $\hat{H}_0$ eigenstates become products of one-dimensional integrals,

$$\langle \psi_{n_x,n_y} | \hat{H}_1 | \psi_{n'_x,n'_y} \rangle$$
$$= \alpha \left(\frac{2}{L_x} \int_0^{L_x} \sin\left(\frac{n_x \pi x}{L_x}\right) \cos\left(\frac{k_x 2\pi x}{L_x}\right) \sin\left(\frac{n'_x \pi x}{L_x}\right) dx \right)$$
$$\left(\frac{2}{L_y} \int_0^{L_y} \sin\left(\frac{n_y \pi y}{L_y}\right) \cos\left(\frac{k_y 2\pi y}{L_y}\right) \sin\left(\frac{n'_y \pi y}{L_y}\right) dy \right) \tag{12.2.56}$$

where each one-dimensional integral obtains the form (see Ex. 12.2.13)

$$\frac{2}{L} \int_0^L dx \sin\left(\frac{n'\pi x}{L}\right) \cos\left(\frac{2k\pi x}{L}\right) \sin\left(\frac{n\pi x}{L}\right) = \begin{cases} 2k > n & ; (\delta_{n',n+2k} - \delta_{n',2k-n})/2 \\ 2k \le n & ; (\delta_{n',n+2k} + \delta_{n',n-2k})/2. \end{cases}$$

While calculating the corrections to the energy levels and wave functions we must notice that, unlike in the one-dimensional case, some eigenvalues of the zero-order Hamiltonian can be degenerate and should be treated accordingly. However, here we restrict the discussion to the low-lying energy levels, where both $n_x \ll k_x$ and $n_y \ll k_y$. In this regime, both diagonal and non-diagonal matrix elements of the perturbation vanish, $\langle \psi_{n_x,n_y} | \hat{H}_1 | \psi_{n'_x,n'_y} \rangle |_{n_x,n'_x \ll k_x; n_y,n'_y \ll k_y} = 0$, such that the first-order corrections to the energy vanish for any zero-order eigenstate,

$$E_{n_x,n_y}^{(1)} = 0. \tag{12.2.57}$$

The second-order corrections to the energy include contributions also from high quantum numbers, and therefore do not vanish:

$$E_{n_x,n_y}^{(1)} = \sum_{\substack{n'_x=1 \\ n'_x \ne n_x}}^{\infty} \sum_{\substack{n'_y=1 \\ n'_y \ne n_y}}^{\infty} \frac{\left| \langle \psi_{n_x,n_y} | \hat{H}_1 | \psi_{n'_x,n'_y} \rangle \right|^2}{E_{n_x,n_y}^{(0)} - E_{n'_x,n'_y}^{(0)}}. \tag{12.2.58}$$

Restricting the discussion again to the typical parameter regime, where $n_x \ll k_x$ and $n_y \ll k_y$, we obtain (Ex. 12.2.15)

$$E_{n_x,n_y}^{(2)} \approx -\frac{\alpha^2}{4} \frac{2m}{\hbar^2 \pi^2} \left(\frac{k_x^2}{L_x^2} + \frac{k_y^2}{L_y^2} \right)^{-1} = -\frac{\alpha^2}{E_{2k_x,2k_y}^{(0)}} \tag{12.2.59}$$

Similarly, the first-order corrections to the eigenfunctions read

$$|\psi_{n_x,n_y}^{(1)}\rangle = \sum_{\substack{n'_x=1 \\ n'_x \ne n_x}}^{\infty} \sum_{\substack{n'_y=1 \\ n'_y \ne n_y}}^{\infty} \frac{\langle \psi_{n'_x,n'_y} | \hat{H}_1 | \psi_{n_x,n_y} \rangle}{E_{n_x,n_y}^{(0)} - E_{n'_x,n'_y}^{(0)}} |\psi_{n'_x,n'_y}\rangle, \tag{12.2.60}$$

where for both $n_x \ll k_x$ and $n_y << k_y$ (see Ex. (12.2.15)) we obtain

$$\psi_{n_x,n_y}^{(0)}(x,y) + \psi_{n_x,n_y}^{(1)}(x,y) \approx \sqrt{\frac{2}{L_x}}\sqrt{\frac{2}{L_y}} \sin\left(\frac{n_x \pi x}{L_x}\right) \sin\left(\frac{n_y \pi y}{L_y}\right)$$

$$\left[1 - \frac{2\alpha}{E_{2k_x,2k_y}^{(0)}} \cos\left(\frac{2k_x \pi x}{L_x}\right) \cos\left(\frac{2k_y \pi y}{L_y}\right)\right]. \qquad (12.2.61)$$

As in the one-dimensional case, the two-dimensional wave functions reflect two length scales. Each wave is composed of a slowly varying envelope function, multiplied by a rapidly oscillating function reflecting the underlying unit cell structure.

Exercise 12.2.15   *Use the explicit form of the zero-order solutions (Eq.(12.2.54))and the perturbation matrix elements, Eq. (12.2.56), to derive the results of Eqs. (12.2.59, 12.2.61).*

12.3 The Variation Approach

As we saw in the previous section, a successful implementation of perturbation theory requires an identification of a suitable zero-order approximation to the full Hamiltonian, which is sufficiently close to it in some sense, and yet significantly easier to handle. When applicable, perturbation theory is the method of choice, providing accurate energy levels and wave functions, as well as physical insight. However, approximate solutions to the Schrödinger equation are needed also when perturbation theory as previously outlined becomes too cumbersome to be applicable. In this chapter we introduce the variation approach to the solution of the Schrödinger equation. The approach is robust and it provides numerous algorithms and numerical procedures for calculating energy levels and wave functions in complex, many-particle, nanoscale systems (atoms, molecules, etc).

The most important concept behind the different variation methods is that of trial wave functions. The latter are subject to constraints imposed by the nature of the physical system (variables, dimensions, boundary conditions, etc.), as well as to artificial constraints, which reduce the space of wave functions to a smaller subspace within the full Hilbert space. The trial functions are then optimized within the reduced subspace in order to obtain an approximation as close as possible to the exact full-space solution.

Consider, for example, a bounded particle in a one-dimensional coordinate space (x). All the proper functions having a specific analytic form can be the trial functions. Introducing a dependence of these functions on a parameter, σ, for example, $f(x,\sigma)$, the variation of σ can yield an optimal solution in the sense of being closest to an exact solution. A different example of a space of trial functions corresponds to linear combinations of a known set of functions, for example, $f(x,a_1,a_2) = a_1 \varphi_1(x) + a_2 \varphi_2(x)$. Here, the expansion coefficients, a_1 and a_2, can be varied for optimizing the solution. When considering multidimensional (or many-particle) coordinate systems, the trial space can be associated with additional constraints. For example, it can include all possible products of normalized one-coordinate functions, and so on. In what follows we shall discuss such examples in detail. First, we introduce the general guidelines for optimizing the trial function for a given system, characterized by the Hamiltonian, $\hat{H}$. For convenience, we shall use Dirac's notations, referring to the corresponding state vectors.

We wish to approximate the solutions of the time-independent Schrödinger equation, $\hat{H}|\psi\rangle = E|\psi\rangle$, by introducing a trial solution, denoted as a vector, $|\tilde{\psi}\rangle$. The energy associated with the trial vector is termed the variational energy, and it is defined as the expectation value of the Hamiltonian,

$$\tilde{\varepsilon} \equiv \frac{\langle \tilde{\psi}|\hat{H}|\tilde{\psi}\rangle}{\langle \tilde{\psi}|\tilde{\psi}\rangle}. \tag{12.3.1}$$

We now define an "error vector" associated with the trial function,

$$|\phi\rangle \equiv \hat{H}|\tilde{\psi}\rangle - \tilde{\varepsilon}|\tilde{\psi}\rangle. \tag{12.3.2}$$

If the trial function coincides with an exact Hamiltonian eigenstate, the error is shown to strictly vanish. Otherwise, the error vector can be used to assess the deviation of $|\tilde{\psi}\rangle$ from an exact solution, where an optimal $|\tilde{\psi}\rangle$ is associated with a minimal error. Invoking a Galerkin approach [12.4], an optimal state in the space of trial functions (the variation space) is reached when the corresponding error vector, $|\phi\rangle$ value becomes orthogonal to any infinitesimal change in the trial function, $|\delta\tilde{\psi}\rangle$. Namely, within the constraints associated with the variation space, the error cannot be further minimized by any infinitesimal change in $|\tilde{\psi}\rangle$,

$$\langle \delta\tilde{\psi}|\phi\rangle|_{\tilde{\psi}_{opt}} = 0, \quad \text{for any } \delta\tilde{\psi} \tag{12.3.3}$$

Using Eq. (12.3.2), *the condition for an optimal trial function reads*

$$\langle \delta\tilde{\psi}|\hat{H} - \tilde{\varepsilon}|\tilde{\psi}\rangle|_{\tilde{\psi}_{opt}} = 0, \quad \text{for any } \delta\tilde{\psi} \tag{12.3.4}$$

As a concrete example, let us consider a space of trial functions that depend on a continuous parameter, σ, where variation in $|\tilde{\psi}\rangle$ is associated with variation of this parameter. Particularly, an infinitesimal change is defined as

$$|\delta\tilde{\psi}\rangle \equiv \left|\frac{\partial}{\partial\sigma}\tilde{\psi}\right\rangle d\sigma . \tag{12.3.5}$$

In this case, the general optimization condition, Eq. (12.3.4), amounts to (see Ex. 12.3.1)

$$\frac{\partial\tilde{\varepsilon}}{\partial\sigma}\bigg|_{\sigma_{opt}} = 0. \tag{12.3.6}$$

Namely, *the optimal value of σ corresponds to a stationary point of the expectation value,* $\tilde{\varepsilon}(\sigma)$.

Exercise 12.3.1   *(a) Show that for $|\delta\tilde{\psi}\rangle \equiv \left|\frac{\partial}{\partial\sigma}\tilde{\psi}\right\rangle d\sigma$, the general condition for the optimal trial function (Eq. (12.3.4)) obtains the form $\left\langle \frac{\partial}{\partial\sigma}\tilde{\psi}\right|\hat{H} - \tilde{\varepsilon}|\tilde{\psi}\rangle = 0$. (b) Use* the Hermiticity of $\hat{H}$ to show that this condition leads to $\frac{\partial}{\partial \sigma}\langle \tilde{\psi}|\hat{H}|\tilde{\psi}\rangle = \tilde{\varepsilon}\frac{\partial}{\partial \sigma}\langle \tilde{\psi}|\tilde{\psi}\rangle$. (c) Use the result of (b) and the definition of the variational energy (Eq. (12.3.1)) to obtain Eq. (12.3.6).

The Variation Principle

A critical aspect of the variation approach is the fundamental relation between the optimal variational energy $\tilde{\varepsilon}$, associated with a given variation space, $|\tilde{\psi}\rangle$, and the exact energy levels of a given system. Recalling that the system Hamiltonian eigenstates constitute a complete orthogonal system of vectors that span the full Hilbert space, any trial function in that space can be expanded using the complete set,

$$\hat{H}|\varphi_n\rangle = E_n|\varphi_n\rangle \quad ; \quad |\tilde{\psi}\rangle = \sum_{n=0}^{\infty} a_n|\varphi_n\rangle, \tag{12.3.7}$$

where $n = 0$ denotes here the ground (minimal energy) state of the system. Consequently, the variational energy is a weighted average of the exact Hamiltonian eigenvalues (see Eq. (11.7.2)), and as such, it is larger or equal to the ground state energy,

$$\tilde{\varepsilon} = \frac{\langle \tilde{\psi}|\hat{H}|\tilde{\psi}\rangle}{\langle \tilde{\psi}|\tilde{\psi}\rangle} = \frac{\sum_{n=0}^{\infty}|a_n|^2 E_n}{\sum_{n=0}^{\infty}|a_n|^2} \geq \frac{\sum_{n=0}^{\infty}|a_n|^2 E_0}{\sum_{n=0}^{\infty}|a_n|^2} = E_0. \tag{12.3.8}$$

It follows that the variational energy $\tilde{\varepsilon}$ is an upper bound to the exact ground-state energy. Therefore, the optimal trial function with respect to the ground state of a system is the one that minimizes $\tilde{\varepsilon}$. Referring again to the concrete example defined in Eq. (12.3.5), the optimization condition derived from a Galerkin condition, Eq. (12.3.6), is indeed a necessary condition for a minimum of the variational energy as a function of the variation parameter, that is, a minimum of $\tilde{\varepsilon}(\sigma)$.

Given a variation space, the variation principle as expressed in Eq. (12.3.8) provides a guideline for searching the optimal approximation to the ground state of a given system. Indeed, any variation leading to a decrease in $\tilde{\varepsilon}$ improves on the trial function. Additionally, re-minimization of $\tilde{\varepsilon}$ within an extended variation space (by removing constraints) cannot increase the optimal value, and will typically reduce it, yielding an improved approximation. This is the main idea beyond powerful algorithms (as, e.g., the linear variation method, to be discussed in what follows), which can converge efficiently toward the exact ground-state solution.

The variation principle is not restricted to approximating the ground state of a system. If the ground state $|\psi_0\rangle$ is known exactly, it is instructive to use trial functions that are orthogonal to it, namely $|\tilde{\psi}\rangle = |\tilde{\varphi}\rangle - |\psi_0\rangle\langle\psi_0|\tilde{\varphi}\rangle$. In this case we can readily see (Ex. 12.3.2) that $\tilde{\varepsilon}$ is an upper bound to the first excited-state energy, $\tilde{\varepsilon} \geq E_1$, and similarly, a trial function $|\tilde{\psi}\rangle = |\tilde{\varphi}\rangle - \sum_{n=0}^{N}|\psi_n\rangle\langle\psi_n|\tilde{\varphi}\rangle$ (where the counting index $n = 1, 2, \ldots, N$ covers the set of all the eigenstates associated with an energy smaller than E_{N+1}) yields a variation energy that is an upper bound to the exact energy level, E_{N+1}.

Exercise 12.3.2 *A trial function is defined as $|\tilde{\psi}\rangle = |\tilde{\varphi}\rangle - \sum_{n=0}^{N} |\psi_n\rangle\langle\psi_n|\tilde{\varphi}\rangle$, where $|\psi_n\rangle$ are all the orthonormal eigenstates of the corresponding system Hamiltonian, $\hat{H}|\psi_n\rangle = E_n|\psi_n\rangle$, with energy smaller than E_{N+1}. (a) Show that $|\tilde{\psi}\rangle$ is orthogonal to $\{|\psi_n\rangle\}$, for $n = 0, 1, \ldots, N$. (b) Use the definition of the variational energy (Eq. (12.3.1)) and follow Eq. (12.3.8) to show that $\tilde{\varepsilon} \geq E_{N+1}$.*

Nonlinear Variation

To demonstrate a concrete implementation of the variation principle, we chose an example of a particle in an effective semi-infinite one-dimensional quantum well, as described in Fig. 12.3.1 and Eq. (12.3.9). The exact bound states for this model can be readily obtained by solving the Schrödinger equation, which translates to solving a homogeneous linear equation. (See the discussion of a similar problem in Sections 6.2, 6.3.) Here, however, we demonstrate a variation approach to the same problem.

$$V(x) = \begin{cases} x < 0 \, ; \, \infty \\ 0 \leq x \leq L \, ; \, 0 \\ x > L \, ; \, V_0 \end{cases} \tag{12.3.9}$$

Motivated by the known analytic solution for the ground state of a similar problem, that is, a particle in an infinite box, $\psi_1(x) = \sqrt{\frac{2}{L}}\sin\left(\frac{\pi x}{L}\right)$, and yet realizing that when the external potential, V_0, is finite, the wave function should penetrate the classically forbidden region extending beyond the box dimensions, $x > L$, we can attempt to use trial functions that are normalized particle-in-a-box functions, associated with a variable "box-length," $L \to L + \Delta$:

$$\tilde{\psi}(\Delta;x) = \begin{cases} x < 0 \, ; \, 0 \\ 0 \leq x \leq L+\Delta \, ; \, \sqrt{\frac{2}{L+\Delta}}\sin\left(\frac{\pi x}{L+\Delta}\right). \\ x > L+\Delta \, ; \, 0 \end{cases} \tag{12.3.10}$$

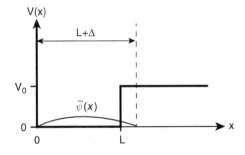

Figure 12.3.1 A semi-infinite quantum well potential (thick line) and a trial function $\tilde{\psi}(x)$ (thin line) chosen for approximating the ground-state energy. The trial function is taken to be the exact ground state of an infinite box potential in which the box length varies to account approximately for the wave function penetration in the region to the right of the semi-infinite quantum well.

The box extension parameter, $\Delta \geq 0$, is regarded here as the variation parameter. To find the optimal effective box length, we seek to minimize the variational energy by satisfying the stationarity condition (Eq. (12.3.6)),

$$\left.\frac{d\tilde{\varepsilon}(\Delta)}{d\Delta}\right|_{\Delta_{opt}} = 0. \tag{12.3.11}$$

The variational energy (Eq. (12.3.1)) is the expectation value of the system Hamiltonian, which reads in the present case

$$\tilde{\varepsilon}(\Delta) = \frac{2}{L+\Delta}\int_0^{L+\Delta} dx \sin\left(\frac{\pi x}{L+\Delta}\right)\left[\frac{-\hbar^2}{2m}\frac{\partial^2}{\partial x^2}+V(x)\right]\sin\left(\frac{\pi x}{L+\Delta}\right). \tag{12.3.12}$$

The explicit expression for $\tilde{\varepsilon}(\Delta)$ simplifies by assuming that the ground state energy is lying "deep" inside the potential energy well; namely, it is much smaller than V_0. Consequently, the required extension of the box length is expected to be small with respect to the box length itself, namely $\Delta << L+\Delta$. In this case (see Ex. (12.3.3)),

$$\tilde{\varepsilon}(\Delta) \cong \frac{\hbar^2 \pi^2}{2m(L+\Delta)^2}+V_0\frac{2\pi^2}{3}\frac{\Delta^3}{(L+\Delta)^3}. \tag{12.3.13}$$

According to the optimization condition, Eq. (12.3.11), the optimal box extension length reads (Ex. (12.3.4))

$$\Delta_{opt} = \frac{L}{2\alpha}(1+\sqrt{1+4\alpha}) \quad ; \quad \alpha \equiv V_0/(\hbar^2/2mL^2). \tag{12.3.14}$$

When the ground-state energy is much smaller than V_0, the parameter α is much larger than 1, and the optimal box extension can be approximated as $\Delta_{opt} \approx \frac{L}{\sqrt{\alpha}}$, where, as can be anticipated, Δ_{opt} becomes smaller for larger V_0 values. In particular, $\Delta_{opt} \to 0$ as $V_0 \to \infty$, and the corresponding variational energy approaches the exact value of the ground-state energy of an infinite potential well, $\tilde{\varepsilon}(\Delta) \xrightarrow[V_0\to\infty]{} \frac{\hbar^2\pi^2}{2mL^2}$.

Exercise 12.3.3 *(a) Show that the exact expression for the variational energy, as defined in Eq. (12.3.12), reads $\tilde{\varepsilon}(\Delta) = \frac{\hbar^2\pi^2}{2m(L+\Delta)^2}+\frac{V_0\Delta}{L+\Delta}+\frac{V_0}{2\pi}\sin\left[2\pi\left(1-\frac{\Delta}{L+\Delta}\right)\right]$. (b) Obtain the approximation, Eq. (12.3.13), by expanding the result for $\Delta << L+\Delta$.*

Exercise 12.3.4 *Obtain the optimal value of the box extension parameter, Δ_{opt} (Eq. (12.3.14)), according to the variation principle. (Recall that $\Delta \geq 0$.)*

Exercise 12.3.5 *Recall the definition of the penetration length, Eq. (6.2.4), for a particle in a finite potential well, $\gamma = \frac{\hbar}{\sqrt{2m(V_0-E)}}$, and show that when the ground-state energy is much smaller than V_0, the optimal value, Δ_{opt}, increases with the penetration length (as might be expected).*

Exercise 12.3.6 *Calculate numerically the exact ground-state energies for a particle in a semi-infinite potential energy well, corresponding to $V_0/E_1^{(\infty)} = 10,100,1000,10000$ (Eq. (12.3.9)), using the approach described in Section 6.3. Show that the respective*

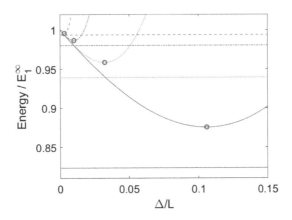

Figure 12.3.2 Variational calculations of the ground-state energies for a set of semi-infinite quantum well potentials (see Eq. (12.3.9)) with $V_0/E_1^{(\infty)} = 10, 100, 1000, 10000$ (corresponding to solid, dotted, dot-dashed, and dashed lines, respectively). The curved lines are the variational energies plotted as functions of the box extension parameter, Δ. For each quantum well, the optimal variational approximation (marked by a circle) is shown to be greater than the exact ground-state energy (marked by the horizontal lines; see Ex. 12.3.6). The optimal variational approximations follow the correct qualitative trend of increasing ground-state energy with increasing V_0, and approach the value $E_1^{(\infty)}$, which corresponds to an infinite box potential. Moreover, owing to the specific choice of the trial function, the relative error in the variational approximation decreases with increasing V_0, as the trial function approaches the exact ground-state wave function.

ground-state energies are 0.823310, 0.939163, 0.980165, 0.993664, in units of $E_1^{(\infty)} = \hbar^2\pi^2/(2mL^2)$.

For any finite value of V_0, the optimal variational energy is only an upper bound for the exact ground-state energy of the particle in the finite potential well, as illustrated in Fig. 12.3.2. It is interesting to notice that the exact ground-state energy for any finite V_0 must be smaller than the ground-state energy for an infinite potential well of the same length, namely smaller than $E_1^{(\infty)} = \frac{\hbar^2\pi^2}{2mL^2}$. This result holds, since the variation principle means that $E_1^{(V_0)} \leq \tilde{\varepsilon}(\Delta)$ for any Δ. Since $\tilde{\varepsilon}(0) = E_1^{(\infty)}$, we have $E_1^{(V_0)} \leq E_1^{(\infty)}$. The equality corresponds to $V_0 \to \infty$, and we can therefore conclude that for any finite V_0, we have $E_1^{(V_0)} < E_1^{(\infty)}$.

The Method of Linear Variation

It is most useful to extend the variation space by considering trial functions of several parameters, for example, $|\tilde{\psi}(c_1, c_2, \ldots, c_N)\rangle$. Infinitesimal changes in these parameters correspond to infinitesimal changes in $|\tilde{\psi}\rangle$ or $\langle\tilde{\psi}|$,

$$|\delta\tilde{\psi}\rangle = \sum_{n=1}^{N} \left|\frac{\partial}{\partial c_n}\tilde{\psi}\right\rangle dc_n \quad ; \quad \langle\delta\tilde{\psi}| = \sum_{n=1}^{N} dc_n \left\langle\frac{\partial}{\partial c_n}\tilde{\psi}\right|, \qquad (12.3.15)$$

which generalizes Eq. (12.3.5). The equations for the optimal parameter values are derived from the general optimization condition, $\langle\delta\tilde{\psi}|\hat{H} - \tilde{\varepsilon}|\tilde{\psi}\rangle|_{\tilde{\psi}_{opt}} = 0$, for any $|\delta\tilde{\psi}\rangle$

(see Eq. (12.3.4)), which reads in this case $\sum_{n=1}^{N} dc_n \left\langle \frac{\partial}{\partial c_n} \tilde{\psi} \middle| \hat{H} - \tilde{\varepsilon} \middle| \tilde{\psi} \right\rangle \bigg|_{\tilde{\psi}_{opt}} = 0$ for any dc_n.
This yields the following set of N equations,

$$\left\langle \frac{\partial}{\partial c_n} \tilde{\psi} \middle| \hat{H} - \tilde{\varepsilon} \middle| \tilde{\psi} \right\rangle \bigg|_{\mathbf{c}^{(opt)}} = 0 \quad ; \quad n = 1, 2, \ldots, N, \tag{12.3.16}$$

where the optimal trial function is associated with the optimal parameter set, $\mathbf{c}^{(opt)} = (c_1^{(opt)}, c_2^{(opt)}, \ldots, c_N^{(opt)})$. Extending the derivation in Ex. 12.3.1 to several variables, the optimization conditions (Eq. (12.3.16)) lead to a set of equations (see Ex. 12.3.7),

$$\frac{\partial \tilde{\varepsilon}}{\partial c_n} \bigg|_{\mathbf{c}^{(opt)}} = 0 \quad ; \quad n = 1, 2, \ldots, N. \tag{12.3.17}$$

Exercise 12.3.7 *Generalize the derivation of the optimization condition for a single variation parameter, Ex. 12.3.1, to the case of several variation parameters and show that Eq. (12.3.17) is obtained from Eq. (12.3.16).*

A most powerful approach to variation in a multiple-parameter space is the method of linear variation. The idea is to expand the trial functions as linear combination of a known set of functions within the relevant space of physical states and to regard the expansion coefficients as variation parameters. In this case the trial function takes the generic form,

$$|\tilde{\psi}\rangle = \sum_{n'=1}^{N} c_{n'} |\varphi_{n'}\rangle, \tag{12.3.18}$$

where the functions, $\{|\varphi_1\rangle, |\varphi_2\rangle, \ldots, |\varphi_N\rangle\}$, are termed the basis set for the variation space. Taking the partial derivatives with respect to the expansion coefficients, we obtain

$$\left| \frac{\partial}{\partial c_n} \tilde{\psi} \right\rangle = |\varphi_n\rangle \quad ; \quad n = 1, 2, \ldots, N. \tag{12.3.19}$$

The corresponding bra reads $\left\langle \frac{\partial}{\partial c_n} \tilde{\psi} \right| = \langle \varphi_n |$, where the set of equations for the optimal coefficients (Eq. (12.3.16)) take a particularly simple form,

$$\langle \varphi_n | \hat{H} - \tilde{\varepsilon} | \tilde{\psi} \rangle |_{\mathbf{c}^{(opt)}} = 0 \quad ; \quad n = 1, 2, \ldots, N. \tag{12.3.20}$$

Substitution of the trial function, Eq. (12.3.18), we obtain a system of N linear equations for the N optimal expansion coefficients,

$$\sum_{n'=1}^{N} \langle \varphi_n | \hat{H} - \tilde{\varepsilon} | \varphi_{n'} \rangle c_{n'}^{(opt)} = 0 \quad ; \quad n = 1, 2, \ldots, N. \tag{12.3.21}$$

By defining two matrices, $\mathbf{H}$ and $\mathbf{S}$, of dimension, $N \times N$, whose elements are

$$\begin{aligned} H_{n,n'} &= \langle \varphi_n | \hat{H} | \varphi_{n'} \rangle \\ S_{n,n'} &= \langle \varphi_n | \varphi_{n'} \rangle \end{aligned} \quad , \tag{12.3.22}$$

the system of linear equations can be identified as a generalized eigenvalue equation,

$$\sum_{n'=1}^{N} [H_{n,n'} - \tilde{\varepsilon} S_{n,n'}] c_{n'}^{(opt)} = 0 \quad ; \quad n = 1, 2, \ldots, N, \tag{12.3.23}$$

or, in matrix notations, it is often termed the secular equation,

$$[\mathbf{H} - \tilde{\varepsilon} \mathbf{S}] \mathbf{c}^{(opt)} = 0. \tag{12.3.24}$$

The matrix $\mathbf{H}$ is the representation of the Hamiltonian within (namely, its projection onto) the finite-dimensional variation space. The matrix $\mathbf{S}$ is often referred to as the overlap matrix, since its entries are the overlap integrals (Eq. (4.6.1)) between the different basis functions. Notice that the algebraic equation has generally multiple eigenvalues. According to the variation principle, Eq. (12.3.8), the lowest eigenvalue, $\tilde{\varepsilon}_0 = \min(\tilde{\varepsilon}_1, \tilde{\varepsilon}_2, \ldots, \tilde{\varepsilon}_N)$, is the best approximation (from above) to the exact ground-state energy of the system,

$$\tilde{\varepsilon}_0 \geq E_0. \tag{12.3.25}$$

The eigenvector corresponding to $\tilde{\varepsilon}_0$, namely $\mathbf{c}_0^{(opt)} = (c_{1,0}^{(opt)}, c_{2,0}^{(opt)}, \ldots, c_{N,0}^{(opt)})$, defines the optimal approximation to the ground-state wave function within the variation space,

$$|\tilde{\psi}_0\rangle = \sum_{n'=1}^{N} c_{n',0}^{(opt)} |\varphi_{n'}\rangle. \tag{12.3.26}$$

When the set of basis functions is orthonormal, $\mathbf{S}$ becomes an identity matrix, and consequently, the secular equation, Eq. (12.3.24), simplifies to

$$\mathbf{S} = \mathbf{I} \quad ; \quad \mathbf{H} \mathbf{c}^{(opt)} = \tilde{\varepsilon} \mathbf{c}^{(opt)} \tag{12.3.27}$$

In this case, the expansion coefficients can be interpreted as the projections of $|\tilde{\psi}\rangle$ on the basis functions, namely $c_n = \langle \varphi_n | \tilde{\psi} \rangle$. It is instructive to define a projection operator onto the variation space, $\hat{P}_N \equiv \sum_{n'=1}^{N} |\varphi_{n'}\rangle\langle\varphi_{n'}|$, where $\langle\varphi_{n'}|\varphi_{n''}\rangle = \delta_{n',n''}$. Using the identity, $\hat{P}_N^2 = \hat{P}_N$, we can rewrite Eq. (12.3.27) for the optimal trial function as $[\hat{P}_N \hat{H} \hat{P}_N] \hat{P}_N |\tilde{\psi}\rangle = \tilde{\varepsilon} \hat{P}_N |\tilde{\psi}\rangle$, which means that the optimal approximations for the trial function and the variational energy in a given linear variation space are obtained by solving the eigenvalue equation for the projection of the full Hamiltonian ($\hat{P}_N \hat{H} \hat{P}_N$) onto that space. Extending the variation space by increasing N toward the entire space, $\hat{P}_N \xrightarrow[N \to \infty]{} \hat{I}$ this eigenvalue equation coincides with the exact Schrödinger equation, $\hat{H}|\tilde{\psi}\rangle = \tilde{\varepsilon}|\tilde{\psi}\rangle$, which means that $|\tilde{\psi}\rangle \xrightarrow[N \to \infty]{} |\psi\rangle$, and $\tilde{\varepsilon} \xrightarrow[N \to \infty]{} E$, where E is the exact eigenvalue of $\hat{H}$. Indeed, optimizing the expansion coefficients within an increasingly larger variation space, the eigenvalues and eigenvectors of $\hat{P}_N \hat{H} \hat{P}_N$ converge to the exact ones,

$$\lim_{N \to \infty} |\tilde{\psi}\rangle\big|_{\mathbf{c}^{(opt)}} = \sum_{n'=1}^{\infty} c_{n'}^{(opt)} |\varphi_{n'}\rangle = |\psi\rangle \quad ; \quad \langle\varphi_{n'}|\varphi_{n''}\rangle = \delta_{n',n''}$$

$$\lim_{N \to \infty} \tilde{\varepsilon}(\mathbf{c}^{(opt)}) = \lim_{N \to \infty} \frac{\langle \tilde{\psi} | \hat{H} | \tilde{\psi} \rangle}{\langle \tilde{\psi} | \tilde{\psi} \rangle}\bigg|_{\mathbf{c}^{(opt)}} = E. \tag{12.3.28}$$

Importantly, Eq. (12.3.28) is not restricted to the ground state of the system, and applies to any of the eigenvectors of Eq. (12.3.27) in the $N \to \infty$ limit. Therefore, the set of eigenvectors of Eq. (12.3.27) provides a set of approximations to the energy levels $\{\tilde{\varepsilon}_m\}$ and corresponding wave functions, $|\tilde{\psi}_m\rangle = \sum\limits_{n'=1}^{N} c_{n',m}^{(opt)} |\varphi_{n'}\rangle$, which converge, respectively, to the exact energy levels and eigenfunctions of the Hamiltonian as $N \to \infty$.

The eigenvalues and expansion coefficients can be obtained by efficient numerical matrix diagonalization techniques [12.5], which make the linear variation a most powerful approach for obtaining approximate solutions to the Schrödinger equation. Notice that the matrix representation of the Hamiltonian operator for any N is Hermitian, that is, $\mathbf{H}^{*t} = \mathbf{H}$. Consequently, its eigenvectors are an orthonormal basis set for the corresponding N-dimensional vector space, which means that the set of approximated functions, $|\tilde{\psi}_m\rangle$, is therefore also orthonormal: $\langle \tilde{\psi}_m | \tilde{\psi}_{m'} \rangle = \delta_{m,m'}$. (See Ex. 12.1.6 for a similar consideration.)

To demonstrate a concrete implementation of the linear variation principle, we chose the analytically solvable model of a one-dimensional harmonic oscillator,

$$\hat{H} = \frac{\hbar\Omega}{2}[\hat{y}^2 + \hat{p}_y^2], \tag{12.3.29}$$

where $\hat{y}$ and $\hat{p}_y$ are the corresponding dimensionless position and momentum operators (see Eqs. (8.5.8–8.5.11)), and Ω is the classical oscillator frequency. The energy levels for this system are analytically known, namely $E_n = \hbar\Omega \left(n + \frac{1}{2}\right)$, for $n = 0, 1, 2 \ldots$. For demonstrating the linear variation method, we ignore our knowledge of the exact solutions and follow the procedure set by Eqs. (12.3.18, 12.3.22, 12.3.27, 12.3.28) to obtain optimal approximated solutions.

Let us start by choosing a basis set for expanding the eigenfunctions. In the present example we chose a finite set of particle-in-a-box wave functions (see Eq. (5.3.9)),

$$\langle y | \varphi_n \rangle = \begin{cases} 0 & ; \quad -L/2 > y \\ \sqrt{\frac{2}{L}} \sin\left[\frac{n\pi}{L}(y + L/2)\right] & ; \quad -L/2 \leq y \leq L/2 \quad ; \quad n = 1, 2, 3, \ldots, N \, . \\ 0 & ; \quad L/2 < y \end{cases}$$

$$\tag{12.3.30}$$

Notice that the functions are displaced such that they are centered at the minimum of the potential energy well at $y = 0$ and are confined to the region $-L/2 \leq y \leq L/2$, where L is the box length in dimensionless units. (Also notice that the boundary conditions on the basis functions, $\langle y | \varphi_n \rangle \xrightarrow[y \to \pm L/2]{} 0$ are only adequate for expanding functions that are confined to the same region, and therefore, accurate approximations to the eigenfunctions of the harmonic oscillator Hamiltonian, associated with increasingly large quantum numbers, can only be obtained by increasing L accordingly.) The next step is to construct the N-dimensional matrices $\mathbf{H}$ and $\mathbf{S}$, whose elements are given as

$$H_{n,n'}(L) = \frac{\hbar\Omega}{L} \int\limits_{-L/2}^{L/2} dy \sin\left[\frac{n\pi}{L}(y + L/2)\right] \left[y^2 - \frac{d^2}{dy^2}\right] \sin\left[\frac{n'\pi}{L}(y + L/2)\right]$$

$$S_{n,n'}(L) = \frac{2}{L} \int\limits_{-L/2}^{L/2} dy \sin\left[\frac{n\pi}{L}(y + L/2)\right] \sin\left[\frac{n'\pi}{L}(y + L/2)\right] = \delta_{n,n'}$$

$$n, n' \in 1, 2, 3, \ldots, N. \tag{12.3.31}$$

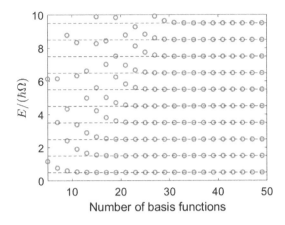
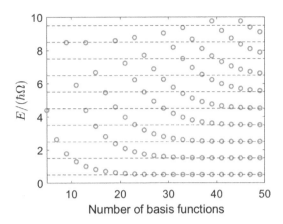

Figure 12.3.3 Linear variation calculations of the lowest energy levels of a harmonic oscillator, using particle-in-a-box basis functions, with box lengths $L = 20$ and $L = 40$ in the left and right plots, by matrix diagonalization are marked as circles, and the exact eigenvalues are marked as dashed horizontal lines.

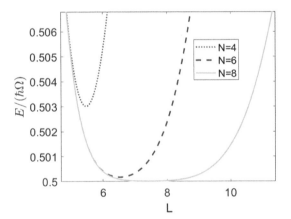

Figure 12.3.4 The variational ground-state energy of a harmonic oscillator, optimized using the linear variation method with $N = 4, 6, 8$ particle-in-a-box basis functions. By regarding the box length (L) as a nonlinear variation parameter, the variational energy can be minimized to obtain an optimal approximation to the exact ground-state energy ($E = 0.5\hbar\Omega$) within each variation space.

Since the basis functions are orthonormal, the variational approximations for the eigenvalues and eigenvectors are obtained by solving the corresponding secular equation, Eq. (12.3.27), namely finding the roots of the corresponding determinant,

$$|\mathbf{H} - \tilde{\varepsilon}\mathbf{I}| = 0. \tag{12.3.32}$$

In Fig. 12.3.3 the eigenvalues of $\mathbf{H}$, as obtained by numerical matrix diagonalization, are plotted as functions of the size of the variation space (N) for two choices of the box length, L. As N increases, the spectrum of the finite size matrices converges (from above) toward the exact spectrum of the Hamiltonian (Eq. (12.3.28)). Notice that the variational approximations for the energy levels for a given N depend on the box length, L. Indeed, for any (fixed) N, the trial function depends on L, which can be optimized as an additional nonlinear variation parameter. This is demonstrated in Fig. 12.3.4 for the oscillator ground state energy.

The applications of the linear variation approach are far-reaching. Apart from being a powerful computational tool, enabling us to map the solution of multidimensional differential eigenvalue equations onto matrix algebra problems, the representation of quantum states in terms of finite-dimensional vectors is intuitive and conceptually useful. Associating a basis set with physically meaningful states, where the system eigenstates are associated with linear combinations of these states, assigns physical meaning to the expansion coefficients and helps to obtain insight into the system under study. As we shall see in the coming chapters, fundamental concepts such as molecular orbitals are based on linear combinations of atomic orbitals, which are remarkably useful for describing charge delocalization between different atoms to form chemical bonds in molecules. Similar concepts are used for the description of delocalized band states in solid crystals as linear combinations of localized atomic orbitals. These concepts are useful for understanding macroscopic bulk properties of different solid materials on the basis of their microscopic unit cell structure.

The Mean-Field Approximation

In the examples discussed so far, the space of trial functions was characterized by a set of parameters, optimized according to the variation principle. Here we consider a somewhat broader definition of the variation space, especially tailored to systems of multidimensional coordinates (e.g., many-particle systems), in which the physical states are defined in tensor product spaces (see Section 11.6). The resulting "mean-field approximation" is a powerful approach to multidimensional quantum mechanical systems. In many-electron systems it provides the foundation for the concept of electronic orbitals, a most successful tool for interpreting the structure of many-electron atoms and of the periodic table of the elements (see Chapter 13). Electronic orbitals are most useful also in the description of molecular and crystalline materials (see Chapter 14), as well as in approximate descriptions of processes on the nanoscale, involving electron transfer, electron energy transfer, light–matter interactions, and more (see Chapters 18–20). In other many-particle systems, involving bosons, the mean-field approximation yields the Gross–Pitaevskii equation, often used to describe the phenomenon of Bose–Einstein condensation [12.6]. In all cases, when valid, the mean-field approximation enables us to discuss a many-body system, whose complexity increases exponentially with the number of particles, in terms of a set of coupled single-particle systems, which are much more tractable. Moreover, this approximation provides intuitive interpretation of the underlying physics from a single-particle perspective. For systems with effectively strong interparticle coupling, the mean-field approximation breaks down and cannot account for all aspects of the correlated many particles. Nevertheless, it is still useful as a starting point for corrections by perturbation theory or for generating basis sets for extended linear variation treatments.

For simplicity, we start our discussion with the case of a two-dimensional coordinate space, corresponding, for example, to two particles in a one-dimensional space or to a single-particle in a two-dimensional space. The generalization to the multidimensional/many-particle cases follows. Specifically, let us consider a Hamiltonian of the generic form,

$$\hat{H} \equiv \hat{H}_1 + \hat{H}_2 + \hat{V}_{1,2}, \tag{12.3.33}$$

where x_1 and x_2 are two Cartesian coordinates. $\hat{H}_1$ and $\hat{H}_2$ are operators in the respective single-coordinate subspaces, where their representations in the two-coordinate space are $\hat{H}_1 \otimes \hat{I}_2$ and $\hat{I}_1 \otimes \hat{H}_2$, respectively, and $\hat{V}_{1,2}$ is a coupling operator within the two spaces. In the absence of such a coupling (setting $\hat{V}_{1,2}$ to zero), the eigenstates of $\hat{H}$ are product states, namely $(\hat{H}_1 \otimes \hat{I} + \hat{I} \otimes \hat{H}_2)|\varphi\rangle \otimes |\chi\rangle = (\lambda_\varphi + \lambda_\chi)|\varphi\rangle \otimes |\chi\rangle$ (see Eq. (11.6.24)). *In general, however, when $\hat{V}_{1,2} \neq 0$, the exact eigenstates of the Hamiltonian cannot be expressed as single products, $|\psi\rangle \neq |\varphi\rangle \otimes |\chi\rangle$. If the two coordinates correspond to different particles, the particles are said to be entangled in this case, which means that their properties cannot be determined independently (the particles are correlated).* Instead, the general expression for the Hamiltonian eigenstates in the presence of a finite interaction ($\hat{V}_{1,2} \neq 0$) involves a sum of products (see Eqs. (11.6.8–11.6.12)),

$$[\hat{H}_1 + \hat{H}_1 + \hat{V}_{1,2}]|\psi\rangle = E|\psi\rangle \quad ; \quad |\psi\rangle = \sum_{n_1,n_2} \psi_{n_1,n_2}|\alpha_{n_1}\rangle \otimes |\beta_{n_2}\rangle, \tag{12.3.34}$$

where the sets $\{|\alpha_{n_1}\rangle\}$ and $\{|\beta_{n_2}\rangle\}$ span the single-coordinate spaces, $\hat{I}_1 = \sum_{n_1} |\alpha_{n_1}\rangle\langle\alpha_{n_1}|$, and $\hat{I}_2 = \sum_{n_2} |\beta_{n_2}\rangle\langle\beta_{n_2}|$.

In spite of the inherent correlation between the two coordinates within each Hamiltonian eigenstate for $\hat{V}_{1,2} \neq 0$, in many applications the coupling operator is sufficiently small in some sense, where it is useful to seek for approximations in the form of a single products to the Hamiltonian eigenstates. For this purpose, we may refer to the space of all possible products of normalized proper states as the variation space. Each trial state is therefore formulated as

$$|\tilde{\psi}\rangle \equiv |\tilde{\varphi}\rangle \otimes |\tilde{\chi}\rangle, \tag{12.3.35}$$

where

$$\langle\tilde{\varphi}|\tilde{\varphi}\rangle = 1 \quad ; \quad \langle\tilde{\chi}|\tilde{\chi}\rangle = 1. \tag{12.3.36}$$

Any infinitesimal variation in $|\tilde{\psi}\rangle$, subject to the constrained product form, amounts to either $|\delta\tilde{\chi}\rangle$ or $|\delta\tilde{\varphi}\rangle$, namely

$$|\delta\tilde{\psi}\rangle = |\delta\tilde{\varphi}\rangle \otimes |\tilde{\chi}\rangle + |\tilde{\varphi}\rangle \otimes |\delta\tilde{\chi}\rangle \quad ; \quad \langle\delta\tilde{\psi}| = \langle\delta\tilde{\varphi}| \otimes \langle\tilde{\chi}| + \langle\tilde{\varphi}| \otimes \langle\delta\tilde{\chi}|. \tag{12.3.37}$$

Invoking the general variation optimization condition, $\langle\delta\tilde{\psi}|[\hat{H} - \tilde{\varepsilon}]|\tilde{\psi}\rangle = 0$ for any $\langle\delta\tilde{\psi}|$ (Eq. (12.3.4)), and using Eq. (12.3.37), two coupled equations are obtained,

$$\langle\delta\tilde{\varphi}| \otimes \langle\tilde{\chi}|[\hat{H} - \tilde{\varepsilon}]|\tilde{\varphi}\rangle \otimes |\tilde{\chi}\rangle = 0 \quad ; \quad \langle\tilde{\varphi}| \otimes \langle\delta\tilde{\chi}|[\hat{H} - \tilde{\varepsilon}]|\tilde{\varphi}\rangle \otimes |\tilde{\chi}\rangle = 0, \tag{12.3.38}$$

where the optimal product states, $|\bar{\varphi}\rangle$ and $|\tilde{\chi}\rangle$, must satisfy these equations for any infinitesimal changes, $\langle\delta\bar{\varphi}|$ and $\langle\delta\tilde{\chi}|$. This means that (see Ex. 12.3.8, for an explicit derivation)

$$\langle\tilde{\chi}_{opt}|\hat{H}|\tilde{\chi}_{opt}\rangle|\bar{\varphi}_{opt}\rangle = \tilde{\varepsilon}|\bar{\varphi}_{opt}\rangle \quad ; \quad \langle\bar{\varphi}_{opt}|\hat{H}|\bar{\varphi}_{opt}\rangle|\tilde{\chi}_{opt}\rangle = \tilde{\varepsilon}|\tilde{\chi}_{opt}\rangle. \tag{12.3.39}$$

The optimal states, $|\bar{\varphi}_{opt}\rangle$ and $|\tilde{\chi}_{opt}\rangle$, turn out to be eigenvectors of effective Hamiltonian operators in their respective single-coordinate spaces, associated with the eigenvalue, $\tilde{\varepsilon}$. These effective Hamiltonians, $\langle\tilde{\chi}_{opt}|\hat{H}|\tilde{\chi}_{opt}\rangle \equiv \hat{H}_1^{(\tilde{\chi})}$ and $\langle\bar{\varphi}_{opt}|\hat{H}|\bar{\varphi}_{opt}\rangle \equiv \hat{H}_2^{(\bar{\varphi})}$, are obtained by partial averaging over vectors ($|\bar{\varphi}_{opt}\rangle$ or $|\tilde{\chi}_{opt}\rangle$) in the complementary vector space (see Ex. 12.3.8). Considering the generic form of the full Hamiltonian, Eq. (12.3.33), we obtain

$$\begin{aligned}\langle\tilde{\chi}_{opt}|\hat{H}|\tilde{\chi}_{opt}\rangle &= \hat{H}_1 + \langle\tilde{\chi}_{opt}|\hat{V}_{1,2}|\tilde{\chi}_{opt}\rangle + \langle\tilde{\chi}_{opt}|\hat{H}_2|\tilde{\chi}_{opt}\rangle\\ \langle\bar{\varphi}_{opt}|\hat{H}|\bar{\varphi}_{opt}\rangle &= \hat{H}_2 + \langle\bar{\varphi}_{opt}|\hat{V}_{1,2}|\bar{\varphi}_{opt}\rangle + \langle\bar{\varphi}_{opt}|\hat{H}_1|\bar{\varphi}_{opt}\rangle\end{aligned}. \tag{12.3.40}$$

The terms $\langle\tilde{\chi}_{opt}|\hat{H}_2|\tilde{\chi}_{opt}\rangle$ and $\langle\bar{\varphi}_{opt}|\hat{H}_1|\bar{\varphi}_{opt}\rangle$ are scalars corresponding to expectation values in the one-dimensional coordinate spaces. Therefore, they merely shift the eigenvalues in Eq. (12.3.39), which can be reformulated as

$$\begin{aligned}[\hat{H}_1 + \langle\tilde{\chi}_{opt}|\hat{V}_{1,2}|\tilde{\chi}_{opt}\rangle]|\bar{\varphi}_{opt}\rangle &= \tilde{\varepsilon}_{\bar{\varphi}_{opt}}|\bar{\varphi}_{opt}\rangle\\ [\hat{H}_2 + \langle\bar{\varphi}_{opt}|\hat{V}_{1,2}|\bar{\varphi}_{opt}\rangle]|\tilde{\chi}_{opt}\rangle &= \tilde{\varepsilon}_{\tilde{\chi}_{opt}}|\tilde{\chi}_{opt}\rangle\end{aligned}, \tag{12.3.41}$$

where

$$\begin{aligned}\tilde{\varepsilon}_{\bar{\varphi}_{opt}} &\equiv \tilde{\varepsilon} - \langle\tilde{\chi}_{opt}|\hat{H}_2|\tilde{\chi}_{opt}\rangle\\ \tilde{\varepsilon}_{\tilde{\chi}_{opt}} &\equiv \tilde{\varepsilon} - \langle\bar{\varphi}_{opt}|\hat{H}_1|\bar{\varphi}_{opt}\rangle\end{aligned}. \tag{12.3.42}$$

The coupled eigenvalue equations, Eq. (12.3.41), are termed the mean-field equations. The "mean-field" in this context refers to the way in which the coupling between the two coordinate subspaces is accounted for. Instead of the fully detailed $\hat{V}_{1,2}$, which appears in the full multidimensional Schrödinger equation, the mean-field equations associate each single coordinate space with an reduced effective equation, in which the coupling to the complementary space is represented as an effective field. These fields are derived by partially averaging over the complementary space, hence the operators $\langle\tilde{\chi}_{opt}|\hat{V}_{1,2}|\tilde{\chi}_{opt}\rangle$ and $\langle\bar{\varphi}_{opt}|\hat{V}_{1,2}|\bar{\varphi}_{opt}\rangle$ are termed the "mean-field operators." Notice that the two equations are coupled, in the sense that the field appearing in one of them depends on the solution of the other equation. In practice, such nonlinear coupled equations are often solved by guessing an initial form of $|\bar{\varphi}_{opt}\rangle$ and/or $|\tilde{\chi}_{opt}\rangle$ and improving iteratively, until a self-consistent solution for the two equations is reached. The converged mean-fields are therefore sometimes referred to as the self-consistent-field (SCF), and the solutions to Eq. (12.3.41) are referred to as the SCF solutions.

Exercise 12.3.8 *A two-dimensional coordinate system is associated with an identity operator, $\hat{I} = \hat{I}_1 \otimes \hat{I}_2 = \sum_{n_1}|\alpha_{n_1}\rangle\langle\alpha_{n_1}| \otimes \sum_{n_2}|\beta_{n_2}\rangle\langle\beta_{n_2}|$, where $\{|\alpha_{n_1}\rangle\}$ and $\{|\beta_{n_2}\rangle\}$ are complete orthonormal systems in the respective spaces. A general operator in the full space reads (see Eq. (11.6.14)) $\hat{A} = \sum_{n_1,n_1'}\sum_{n_2,n_2'}A_{n_1,n_1',n_2,n_2'}|\alpha_{n_1}\rangle\langle\alpha_{n_1'}| \otimes |\beta_{n_2}\rangle\langle\beta_{n_2'}|$. Given*

two vectors in the respective one-dimensional coordinated spaces, $|\tilde{\varphi}\rangle = \sum\limits_{n_1} \tilde{\varphi}_{n_1}|\alpha_{n_1}\rangle$ *and*
$|\tilde{\chi}\rangle = \sum\limits_{n_2} \tilde{\chi}_{n_2}|\beta_{n_2}\rangle$, *show that:*

(a) $\langle\tilde{\chi}|\hat{A}|\tilde{\varphi}\rangle \otimes |\tilde{\chi}\rangle = \hat{A}_1^{(\tilde{\chi})}|\tilde{\varphi}\rangle$, *where*

$$\hat{A}_1^{(\tilde{\chi})} = \sum_{n_1, n_1', n_2, n_2'} A_{n_1, n_1', n_2, n_2'} \langle\tilde{\chi}|\beta_{n_2}\rangle\langle\beta_{n_2'}|\tilde{\chi}\rangle |\alpha_{n_1}\rangle\langle\alpha_{n_1'}| = \langle\tilde{\chi}|\hat{A}|\tilde{\chi}\rangle.$$

(b) $\langle\tilde{\varphi}|\hat{A}|\tilde{\varphi}\rangle \otimes |\tilde{\chi}\rangle = \hat{A}_2^{(\tilde{\varphi})}|\tilde{\chi}\rangle$, *where*

$$\hat{A}_2^{(\tilde{\varphi})} = \sum_{n_1, n_1', n_2, n_2'} A_{n_1, n_1', n_2, n_2'} \langle\tilde{\varphi}|\alpha_{n_1}\rangle\langle\alpha_{n_1'}|\tilde{\varphi}\rangle |\beta_{n_2}\rangle\langle\beta_{n_2'}| = \langle\tilde{\varphi}|\hat{A}|\tilde{\varphi}\rangle.$$

(c) Use the results of (a) and (b) to show that $\langle\tilde{\chi}|[\hat{H}-\tilde{\varepsilon}]|\tilde{\varphi}\rangle \otimes |\tilde{\chi}\rangle = 0 \Rightarrow \hat{H}_1^{(\tilde{\chi})}|\tilde{\varphi}\rangle = \tilde{\varepsilon}|\tilde{\varphi}\rangle$
and $\langle\tilde{\varphi}|[\hat{H}-\tilde{\varepsilon}]|\tilde{\varphi}\rangle \otimes |\tilde{\chi}\rangle = 0 \Rightarrow \hat{H}_2^{(\tilde{\varphi})}|\tilde{\chi}\rangle = \tilde{\varepsilon}|\tilde{\chi}\rangle.$

The solution to the mean-field (or the SCF) equations provides the variationally optimal product state approximations to the eigenstates of the two-dimensional Hamiltonian,

$$|\psi\rangle \approx |\tilde{\psi}_{opt}\rangle \equiv |\tilde{\varphi}_{opt}\rangle \otimes |\tilde{\chi}_{opt}\rangle, \tag{12.3.43}$$

where the corresponding variational energy reads (see Ex. 12.3.9)

$$\tilde{\varepsilon} = \langle\tilde{\psi}|\hat{H}|\tilde{\psi}\rangle = \tilde{\varepsilon}_{\tilde{\varphi}_{opt}} + \tilde{\varepsilon}_{\tilde{\chi}_{opt}} - \langle\tilde{\psi}|\hat{V}_{1,2}|\tilde{\psi}\rangle. \tag{12.3.44}$$

Exercise 12.3.9 *Use Eq. (12.3.41) to derive the relation in Eq. (12.3.44).*

The generalization of the product state approximation to N-dimensional Hamiltonians is straightforward. Considering the Hamiltonian,

$$\hat{H} = \left[\sum_{n=1}^{N} \hat{H}_n\right] + \hat{V}_{1,2,\dots,N}, \tag{12.3.45}$$

the mean-field product trial state reads

$$|\tilde{\psi}\rangle \equiv |\tilde{\varphi}_{opt,1}\rangle \otimes |\tilde{\varphi}_{opt,2}\rangle \otimes \cdots \otimes |\tilde{\varphi}_{opt,N}\rangle, \tag{12.3.46}$$

where the set $\{|\tilde{\varphi}_{opt,n}\rangle\}$ is the solutions of the following coupled SCF equations,

$$[\hat{H}_n + \hat{H}_n^{(\tilde{\varphi}_{opt,1}, \tilde{\varphi}_{opt,2}, \dots \tilde{\varphi}_{opt,n-1}, \dots \tilde{\varphi}_{opt,n+1}, \dots, \tilde{\varphi}_{opt,N})}]|\tilde{\varphi}_{opt,n}\rangle = \tilde{\varepsilon}_{\tilde{\varphi}_{opt,n}}|\tilde{\varphi}_{opt,n}\rangle \quad ; \quad n = 1, 2, \dots, N, \tag{12.3.47}$$

where

$$\tilde{\varepsilon}_{\tilde{\varphi}_{opt,n}} = \tilde{\varepsilon} - \sum_{\substack{n'=1 \\ n'\neq n}}^{N} \langle\tilde{\varphi}_{opt,n'}|\hat{H}_{n'}|\tilde{\varphi}_{opt,n'}\rangle. \tag{12.3.48}$$

$$\hat{H}_n^{(\tilde{\varphi}_{opt,1}, \tilde{\varphi}_{opt,2}, \dots \tilde{\varphi}_{opt,n-1}, \tilde{\varphi}_{opt,n+1}, \dots, \tilde{\varphi}_{opt,N})} = \langle\tilde{\varphi}_{opt,1}| \otimes \dots \otimes \langle\tilde{\varphi}_{opt,n-1}| \otimes \langle\tilde{\varphi}_{opt,n+1}| \cdots \otimes$$
$$\langle\tilde{\varphi}_{opt,N}|\hat{V}_{1,2,\dots,N}|\tilde{\varphi}_{opt,1}\rangle \otimes \dots |\tilde{\varphi}_{opt,n-1}\rangle \otimes |\tilde{\varphi}_{opt,n+1}\rangle \cdots \otimes |\tilde{\varphi}_{opt,N}\rangle, \tag{12.3.49}$$

and the variational energy reads

$$\tilde{\mathcal{E}} = \sum_{n=1}^{N} \tilde{\mathcal{E}}_{\tilde{\varphi}_{opt,n}} - [N-1]\langle \tilde{\psi} | \hat{V}_{1,2,\ldots,N} | \tilde{\psi} \rangle. \tag{12.3.50}$$

Notice that the variational energy is different from the sum of eigenvalues as obtained from the different SCF equations. Indeed, the expectation value of the coupling operator (times $N-1$) must be subtracted from the sum of SCF eigenvalues to obtain the total state energy. This is attributed to overcounting of the interaction operator within the SCF equations, in which this operator appears N times. (This does not apply in the absence of interaction; see Ex. 12.3.10.)

Exercise 12.3.10   *Show that in the absence of coupling, namely when $\hat{V}_{1,2,\ldots,N} = 0$, the variational energy equals the sum of independent variational energies in the one-dimensional coordinate subspaces, namely $\tilde{\mathcal{E}} = \sum_{n=1}^{N} \tilde{\mathcal{E}}_{\tilde{\varphi}_{opt,n}} = \sum_{n=1}^{N} \langle \tilde{\varphi}_{opt,n} | \hat{H}_n | \tilde{\varphi}_{opt,n} \rangle.$*

According to the variation principle, the energy associated with the ground-state solutions of the coupled SCF equations is an upper bound for the exact ground-state energy of the system. Denoting the ground-state eigenvalue and eigenstate of each equation (Eq. (12.3.47)) as $\tilde{\mathcal{E}}_{opt,n}^{(0)}$ and $|\tilde{\varphi}_{opt,n}^{(0)}\rangle$, respectively, the optimal approximation for the system's ground state by the mean field approximation is $|\tilde{\psi}_0\rangle \equiv |\tilde{\varphi}_{opt,1}^{(0)}\rangle \otimes |\tilde{\varphi}_{opt,2}^{(0)}\rangle \otimes \cdots \otimes |\tilde{\varphi}_{opt,N}^{(0)}\rangle$, where the corresponding energy reads

$$\tilde{\mathcal{E}}_0 = \sum_{n=1}^{N} \tilde{\mathcal{E}}_{opt,n}^{(0)} - [N-1]\langle \tilde{\psi}_0 | \hat{V}_{1,2,\ldots,N} | \tilde{\psi}_0 \rangle \geq E_0. \tag{12.3.51}$$

As a concrete implementation of the mean-field approximation, let us consider a system of two electrons bound to a positive nucleus in an isolated He atom. Setting the nucleus at rest, the Hamiltonian for the two electrons obtains the form (Eq. (3.4.10))

$$\hat{H} = \frac{-\hbar^2}{2m_e}\Delta_{\mathbf{r}_1} + \frac{-\hbar^2}{2m_e}\Delta_{\mathbf{r}_2} + \frac{-KZe^2}{|\mathbf{r}_1|} + \frac{-KZe^2}{|\mathbf{r}_2|} + \frac{Ke^2}{|\mathbf{r}_1 - \mathbf{r}_2|}. \tag{12.3.52}$$

$\mathbf{r}_1$ and $\mathbf{r}_2$ are the two electron position vectors, K is Coulomb's constant, m_e and e are the electron mass and charge, and $Z = 2$ is the number of protons in the atomic nucleus. Defining $\hat{H}_n = \frac{-\hbar^2}{2m_e}\Delta_{\mathbf{r}_n} + \frac{-KZe^2}{|\mathbf{r}_n|}$ and $\hat{V}_{1,2} = \frac{Ke^2}{|\mathbf{r}_1 - \mathbf{r}_2|}$, the Hamiltonian, Eq. (12.3.52), is of the generic form, Eqs. (12.3.33, 12.3.45).

A mean-field approximation for the two-electron function constrains the variational space of trial functions to products of single-electron functions,

$$\tilde{\psi}(\mathbf{r}_1, \mathbf{r}_2) = \tilde{\varphi}(\mathbf{r}_1)\tilde{\chi}(\mathbf{r}_2). \tag{12.3.53}$$

In analogy to the single-electron (hydrogen-like) atom, each single-electron function within a many-electron Hamiltonian is termed "an atomic orbital." The approximation of the many-electron function as a product of orbitals is referred to as the Hartree product [12.7].

The optimal single-particle states are obtained by solving the self-consistent mean-field eigenvalue equations (Eq. (12.3.41)). Using the position representation, the variationally optimized single-particle wave functions are denoted as $\tilde{\varphi}_{opt}(\mathbf{r}_1) = \langle \phi_{\mathbf{r}_1} | \tilde{\varphi}_{opt} \rangle$

and $\tilde{\chi}_{opt}(\mathbf{r}_2) = \langle \phi_{\mathbf{r}_2} | \tilde{\chi}_{opt} \rangle$, and the mean-field equations take the explicit form (see Ex. 12.3.11)

$$
\left[\frac{-\hbar^2}{2m_e} \Delta_{\mathbf{r}_1} + \frac{-KZe^2}{|\mathbf{r}_1|} + \int d\mathbf{r}_2 |\tilde{\chi}_{opt}(\mathbf{r}_2)|^2 \frac{Ke^2}{|\mathbf{r}_1 - \mathbf{r}_2|} \right] \tilde{\varphi}_{opt}(\mathbf{r}_1) = \tilde{\varepsilon}_{\tilde{\varphi}_{opt}} \tilde{\varphi}_{opt}(\mathbf{r}_1)
$$
$$
\left[\frac{-\hbar^2}{2m_e} \Delta_{\mathbf{r}_2} + \frac{-KZe^2}{|\mathbf{r}_2|} + \int d\mathbf{r}_1 |\tilde{\varphi}_{opt}(\mathbf{r}_1)|^2 \frac{Ke^2}{|\mathbf{r}_1 - \mathbf{r}_2|} \right] \tilde{\chi}_{opt}(\mathbf{r}_2) = \tilde{\varepsilon}_{\tilde{\chi}_{opt}} \tilde{\chi}_{opt}(\mathbf{r}_2)
$$

(12.3.54)

Each orbital is therefore an eigenfunction of an effective single-particle Hamiltonian. The latter is a sum of a "hydrogen-like" term, which corresponds to the interaction between an electron and the bare atomic nucleus, and an additional potential energy term, which corresponds to the mean field interaction with the other electron in the atom. For example, the interaction experienced by electron 1, $V_{eff}(\mathbf{r}_1) \equiv \int d\mathbf{r}_2 |\tilde{\chi}_{opt}(\mathbf{r}_2)|^2 \frac{Ke^2}{|\mathbf{r}_1 - \mathbf{r}_2|}$, corresponds to the averaged interaction between that electron, positioned at $\mathbf{r}_1$, and electron 2, which is distributed over the entire space according to a probability density function, $|\tilde{\chi}_{opt}(\mathbf{r}_2)|^2$. In other words, the two-electron interaction energy, $\frac{Ke^2}{|\mathbf{r}_1 - \mathbf{r}_2|}$, is replaced by an effective single-electron electrostatic potential at $\mathbf{r}_1$, obtained by averaging over the charge distribution associated with the other electron. A similar description holds also for the second electron, where the two effective potentials experienced by the two electrons depend on the orbitals that define the spatial electronic charge distributions, and, in turn, the shape and energy of each orbital depends on the effective mean-field potentials that modify the single-electron Hamiltonians with respect to the hydrogen like-model. Consequently, the orbitals of He are different from the hydrogen-like orbitals. Qualitatively, the presence of an additional electron within the atom, which corresponds to a distribution of a negative charge, screens to some extent the positive charge of the nucleus. This effect results in higher single-electron energies and more diffused orbitals in comparison to the orbitals of a hydrogen-like atom with the same nuclear charge.

Exercise 12.3.11 (a) Use the identity operator in the full space, $\int d\mathbf{r}'_1 \int d\mathbf{r}'_2 |\phi_{\mathbf{r}'_1}\rangle\langle\phi_{\mathbf{r}'_1}| \otimes |\phi_{\mathbf{r}'_2}\rangle\langle\phi_{\mathbf{r}'_2}|$, and the definition of $\hat{V}_{1,2}$ according to Eq. (12.3.52) to show that

$$
\langle \phi_{\mathbf{r}_1} | \otimes \langle \tilde{\chi}_{opt} | \hat{H}_{1,2} | \tilde{\varphi}_{opt} \rangle \otimes | \tilde{\chi}_{opt} \rangle = \int d\mathbf{r}_2 |\tilde{\chi}_{opt}(\mathbf{r}_2)|^2 \frac{Ke^2}{|\mathbf{r}_1 - \mathbf{r}_2|} \tilde{\varphi}_{opt}(\mathbf{r}_1)
$$
$$
\langle \tilde{\varphi}_{opt} | \otimes \langle \phi_{\mathbf{r}_2} | \hat{H}_{1,2} | \tilde{\varphi}_{opt} \rangle \otimes | \tilde{\chi}_{opt} \rangle = \int d\mathbf{r}_1 |\tilde{\varphi}_{opt}(\mathbf{r}_1)|^2 \frac{Ke^2}{|\mathbf{r}_1 - \mathbf{r}_2|} \tilde{\chi}_{opt}(\mathbf{r}_2).
$$

(b) Use the definitions of $\hat{H}_1$ and $\hat{H}_2$ according to Eq. (12.3.52), and the coordinate representation of the kinetic energy operator ($\langle \phi_{\mathbf{r}} | \hat{T}_{\mathbf{r}} | \psi \rangle = \frac{-\hbar^2}{2m} \Delta \psi(\mathbf{r})$; see Ex. 11.5.4) to show that

$$
\langle \phi_{\mathbf{r}_1} | \hat{H}_1 | \tilde{\varphi}_{opt} \rangle = \left[\frac{-\hbar^2}{2m_e} \Delta_{\mathbf{r}_1} + \frac{-KZe^2}{|\mathbf{r}_1|} \right] \tilde{\varphi}_{opt}(\mathbf{r}_1)
$$
$$
\langle \phi_{\mathbf{r}_2} | \hat{H}_2 | \tilde{\chi}_{opt} \rangle = \left[\frac{-h^2}{2m_e} \Delta_{\mathbf{r}_2} + \frac{-KZe^2}{|\mathbf{r}_2|} \right] \tilde{\chi}_{opt}(\mathbf{r}_2).
$$

(c) Use the results of (a) and (b) to show that the generic mean-field equations translate to Eq. (12.3.54) in the case of the He atom Hamiltonian (Eq. (12.3.52)).

So far, our discussion has ignored the fact that the two electrons are identical particles. The full consequences of this fact are quite remarkable, as will be discussed in detail in the next chapter. Here, we only point out that the identity of the two electrons is reflected in the symmetry of the Hamiltonian for permutation between $\mathbf{r}_1$ and $\mathbf{r}_2$ (see Eq. (12.3.52)); namely, the Hamiltonian commutes with the particle permutation operator, and the two-particle probability density must be invariant to their permutation, namely $|\psi(\mathbf{r}_2,\mathbf{r}_1)|^2 = |\psi(\mathbf{r}_1,\mathbf{r}_2)|^2$. It follows that the exact Hamiltonian eigenstates are subject to a symmetry constraint, $\psi(\mathbf{r}_2,\mathbf{r}_1) \propto \psi(\mathbf{r}_1,\mathbf{r}_2)$. Imposing this requirement on the space of single product trial functions, we conclude that the two optimal orbitals must be identical, namely (see Ex. 12.3.12)

$$\tilde{\varphi}_{opt}(\mathbf{r}) \propto \tilde{\chi}_{opt}(\mathbf{r}). \tag{12.3.55}$$

Consequently, the two seemingly different coupled self-consistent equations are in fact identical in the present case and amount to a single nonlinear equation for the electronic orbital, denoted as $\tilde{\varphi}_{opt}(\mathbf{r})$:

$$\left[\frac{-\hbar^2}{2m_e}\Delta_\mathbf{r} + \frac{-KZe^2}{|\mathbf{r}|} + \int d\mathbf{r}' |\tilde{\varphi}_{opt}(\mathbf{r}')|^2 \frac{Ke^2}{|\mathbf{r}-\mathbf{r}'|} \right] \tilde{\varphi}_{opt}(\mathbf{r}) = \tilde{\varepsilon}_{\tilde{\varphi}_{opt}} \tilde{\varphi}_{opt}(\mathbf{r}). \tag{12.3.56}$$

The lowest-energy solution of this nonlinear eigenvalue equation corresponds to an orbital, $\tilde{\varphi}_{opt,0}(\mathbf{r})$, and its corresponding orbital energy, $\tilde{\varepsilon}_{\tilde{\varphi}_{opt},0}$. The Hartree (orbital) approximation for the ground state of the He atom therefore reads

$$\tilde{\psi}_0(\mathbf{r}_1,\mathbf{r}_2) = \tilde{\varphi}_{opt,0}(\mathbf{r}_1)\tilde{\varphi}_{opt,0}(\mathbf{r}_2), \tag{12.3.57}$$

where the corresponding variational energy is (see Eq. (12.3.44))

$$\tilde{\varepsilon} = 2\tilde{\varepsilon}_{\tilde{\varphi}_{opt},0} - Ke^2 \int d\mathbf{r} \int d\mathbf{r}' \frac{|\tilde{\varphi}_{opt,0}(\mathbf{r}')|^2 |\tilde{\varphi}_{opt,0}(\mathbf{r})|^2}{|\mathbf{r}-\mathbf{r}'|} \tag{12.3.58}$$

The variational energy as obtained by a numerical solution of Eq. (12.3.56) is -77.9 eV [12.8]. As can be expected in view of the screening effect previously discussed, this value is significantly higher than the energy of two noninteracting electrons in a hydrogen-like atom with $Z = 2$, which amounts to -108.8 eV (see Ex. 12.3.13). Moreover, the mean-field ground state energy is reasonably close to the exact energy, which can be measured by ionization experiments and equals -79 eV. While being reasonably successful in calculation of the ground state for He, the Hartree product fails to predict the properties of atoms with more than two electrons. Indeed, the Hartree product ignores a crucial effect on many-electron systems, which is derived from electron spin. The existence of spin and its consequences in many-electron systems will be addressed in Chapter 13.

Exercise 12.3.12 *Let us define a permutation operator, $\hat{P}_{1,2}f(\mathbf{r}_1,\mathbf{r}_2) = f(\mathbf{r}_2,\mathbf{r}_1)$. (a) Show that any eigenfunction of this operator must satisfy $\psi(\mathbf{r}_2,\mathbf{r}_1) = \alpha\psi(\mathbf{r}_1,\mathbf{r}_2)$, where α is a scalar. (b) Show that if an eigenfunction of $\hat{P}_{1,2}$ is a product, $\psi(\mathbf{r}_1,\mathbf{r}_2) = g(\mathbf{r}_1)h(\mathbf{r}_2)$, then $g(\mathbf{r}) \propto h(\mathbf{r})$.*

Exercise 12.3.13 *Calculate the exact ground state energy of the Hamiltonian,*

$$\hat{H} = \frac{-\hbar^2}{2m_e}\Delta_{\mathbf{r}_1} + \frac{-\hbar^2}{2m_e}\Delta_{\mathbf{r}_2} + \frac{-KZe^2}{|\mathbf{r}_1|} + \frac{-KZe^2}{|\mathbf{r}_2|},$$

which corresponds to the He atom, when the electron–electron interaction is completely ignored. Refer as needed to Ex. 4.3.4 and to the known solution for a hydrogen-like atom (see Chapter 10).

Bibliography

[12.1] S. S. Shaik and P. C. Hiberty, "A Chemist's Guide to Valence Bond Theory" (John Wiley & Sons, 2007).

[12.2] P. S. Epstein, "The Stark effect from the point of view of Schrödinger's quantum theory," Physical Review 28, 695 (1926).

[12.3] C. Kittel, "Quantum Theory of Solids" (John Wiley & Sons, New York, 1987).

[12.4] Y. Saad, "Iterative Methods for Sparse Linear Systems," 2nd ed. (SIAM, 2003).

[12.5] G. H. Golub and C.F. Van Loan, "Matrix Computations" (The John Hopkins University Press, 1989).

[12.6] A. I. Streltsov, O. E. Alon, and L. S. Cederbaum, "General variational many-body theory with complete self-consistency for trapped bosonic systems," Physical Review A 73, 063626 (2006).

[12.7] D. R. Hartree, "The Calculation of Atomic Structures" (John Wiley & Sons, 1957).

[12.8] I. N. Levine, "Quantum Chemistry" (Pearson, 2014).

13 Many-Electron Systems

13.1 The Electron Spin

So far, we have discussed nonrelativistic quantum mechanics, its postulates, and some of its applications in nano- and subnanoscale physics. As it turns out, this is not enough for understanding important phenomena, including the structure of many-electron systems, which is the topic of the present chapter. Indeed, the explanation of observations regarding the nature of atoms, such as periodical-like changes in the properties of atoms of different elements (the periodic table), requires additional accounting for the relativistic nature of physical reality. Remarkably, however, in the weak relativistic regime (namely, when a classical treatment would involve particle velocities significantly smaller than the speed of light) a fairly accurate description of reality is obtained by extending the nonrelativistic framework of quantum mechanics (representing the states of physical systems, the observables, and measurement outcomes as outlined in the previous chapters) with an additional postulate, which associates "spin" to the elementary particles. By the end of this chapter, we shall see that the existence of the electron spin is indeed crucial for understanding the structure of many-electron atoms, and particularly the periodic-like alternation of physical properties and chemical reactivity of atoms depending on their atomic number (the number of protons in the atomic nucleus). Here we start by becoming familiar with the existence of the electron spin. This property is most directly manifested in the presence of an external magnetic field.

We start by recalling that in classical mechanics, a particle of mass μ and charge q, rotating in a circle, induces a magnetic moment perpendicular to the rotation plane. This moment, $\boldsymbol{\mu}$, is proportional to the particle's angular momentum,

$$\boldsymbol{\mu} = \frac{q}{2\mu}\mathbf{L}. \tag{13.1.1}$$

In the presence of an external magnetic filed, $\mathbf{B}$, the particle's energy changes according to the projection of its magnetic moment on the external field,

$$H_B = -\boldsymbol{\mu} \cdot \mathbf{B} = \frac{-q}{2\mu}\mathbf{L} \cdot \mathbf{B}. \tag{13.1.2}$$

For an electron of charge $-e$ and a magnetic field directed along the z-axis, $\mathbf{B} = (0,0,B_z)$, the change of energy amounts to $H_B = \frac{e}{2m_e}L_z B_z$. Replacing the classical angular momentum with its corresponding quantum mechanical operator, the Hamiltonian for an electron in the magnetic field obtains an additional term,

$$\hat{H}_{B_z} = \frac{e}{2m_e} B_z \hat{L}_z. \tag{13.1.3}$$

Let us focus, for example, on an electron in a hydrogen-like atom. In the absence of a magnetic field, the energy eigenstates associated with a given angular momentum (and a given principal quantum number, n) are degenerate (i.e., $\psi_{n,l,m}$, with $m = -l, \ldots, 0, \ldots l$; see Chapters 9 and 10). The introduction of a magnetic field, Eq. (13.1.3), removes this degeneracy, where the energy levels are split according to their magnetic quantum number, m, $\hat{H}_{B_z}|\psi_{n,l,m}\rangle = \frac{eB_z}{2m_e} m\hbar |\psi_{n,l,m}\rangle = m\mu_B B_z|\psi_{n,l,m}\rangle$. The magnitude of the effect is set by $\mu_B \equiv \frac{\hbar e}{2m_e}$, named "the Bohr magneton" after Niels Bohr. *The splitting of the energy levels in the magnetic field is manifested in the well-known Zeeman effect, observed also in many-electron hydrogen-like atoms.* Indeed, multiple splitting of atomic spectral lines has been observed in experiments since the end of the nineteenth century [13.1]. However, it became clear that magnetic fields induce splitting of atomic energy levels beyond what could be attributed to the quantization of the electron's angular momentum. This "anomalous" Zeeman effect was the first indication that the magnetic moments of electrons are not simply determined by their orbital angular momentum. The most compelling evidence for the nature of the magnetic moment of electrons was first revealed in the Stem–Gerlach experiment from 1922, which demonstrated for the first time that, even when its orbital angular momentum is zero, the electron still has a magnetic moment. In that experiment [13.2], neutral silver atoms were scattered through an inhomogeneous magnetic field. Unexpectedly, the single atomic beam split into two parts while passing through the field, indicating two remarkable facts: (i) Although electrically neutral, the silver atoms experienced magnetic forces when passing through a magnetic field. (ii) The magnetic forces were of two different signs, as indicated by the splitting into two beams.

The first observation is attributed to the existence of a net magnetic moment in a neutral silver atom, which exerts a force on the atomic center of mass as it passes through an inhomogeneous magnetic field. A more detailed analysis of the structure of silver atoms later related the net magnetic moment to that of an (unpaired) electron in the atom. Notably, within the orbital approximation (to be discussed in Section 13.3), the unpaired electron in the ground state of a silver atom is associated with a 4s-type orbital, hence with a zero orbital angular momentum. Since a magnetic moment of a charged particle is expected to be proportional to its angular momentum (see Eq. (13.1.1)), the orbital angular momentum cannot account for the magnetic moment, where another source for angular momentum must be present, unrelated to the particle's orbital. This additional angular momentum of particles is referred to as an "intrinsic particle spin." In nonrelativistic quantum mechanics, where particle positions are associated with a three-dimensional coordinate space, the existence of spin is an additional postulate. (It is worthwhile to mention, however, that spin is inherent in Dirac's equation for an electron [13.3], which extends the Schrödinger equation to include relativistic corrections.)

A no less remarkable observation in the Stern–Gerlach experiment was the splitting of the particle's energy was into *two* distinctive states. Attributing the splitting to the quantum Zeeman effect, via Eq. (13.1.3), the number of split states in a magnetic field should reflect the number of different values of the magnetic quantum number, m. Following the arguments outlined in Section 9.2 for orbital angular momenta, this number must be $2l + 1$, where $l = 0, 1, \ldots$ is the angular quantum number (to comply with the boundary conditions for proper stationary solutions of the Schrödinger equation in the three-dimensional coordinate space). Consequently, for any orbital angular momentum, $2l + 1$ must be an odd number. An even number of m values points to half-integer angular quantum numbers, which, again, cannot be attributed to an orbital angular momentum. In particular, the splitting into two beams in the Stern–Gerlach experiment, namely, $2l + 1 = 2$, is consistent with an angular quantum number, $l = 1/2$, where $m = -l, \ldots, l$ amounts to $m = -1/2, 1/2$. (Remarkably, mathematical considerations based solely on the Hermiticity and commutation relations of the angular momentum operators, irrespective of any coordinate representation, suggest that normalizable angular momentum eigenvectors can indeed be associated with either even or odd angular quantum numbers [4.3].) Finally, a quantitative analysis reveals that the magnitude of the energy splitting attributed to the intrinsic electron spin is different from what can be expected for an orbital angular momentum: instead of the Bohr magneton, μ_B , the splitting was associated with $g\mu_B$, where $g = 2.0034$ (the electron's gyromagnetic factor) is characteristic to the electron spin.

It is therefore necessary to introduce ***another postulate to nonrelativistic quantum mechanics that assumes the existence of an intrinsic spin for each particle. This measurable property is associated with angular momentum operators, analogues to the orbital angular momentum operators***. To distinguish from the orbital angular momentum operator, $\hat{L} = (\hat{L}_x, \hat{L}_y, \hat{L}_z)$, the spin operator is denoted as

$$\hat{S} = (\hat{S}_x, \hat{S}_y, \hat{S}_z). \tag{13.1.4}$$

Notice that although the spin operates in a "spin-space" different form the coordinate space, it is a vector in the coordinate space, with spatial components satisfying angular momentum commutation relations (see Eqs. (3.3.3, 3.3.5)) namely

$$[\hat{S}_x, \hat{S}_y] = i\hbar\hat{S}_z$$
$$[\hat{S}_y, \hat{S}_z] = i\hbar\hat{S}_x$$
$$[\hat{S}_z, \hat{S}_x] = i\hbar\hat{S}_y$$

$$[\hat{S}^2, \hat{S}_x] = [\hat{S}^2, \hat{S}_y] = [\hat{S}^2, \hat{S}_z] = 0. \tag{13.1.5}$$

Defining spin ladder operators,

$$\hat{S}_+ \equiv \hat{S}_x + i\hat{S}_y \quad ; \quad \hat{S}_- \equiv \hat{S}_x - i\hat{S}_y \tag{13.1.6}$$

and using the commutation relations, Eq. (13.1.5), we can readily verify that the following equations hold (see Ex. 9.2.2):

$$[\hat{S}_+, \hat{S}_-] = 2\hbar\hat{S}_z \quad ; \quad [\hat{S}_z, \hat{S}_\pm] = \pm\hbar\hat{S}_\pm \quad ; \quad [\hat{S}^2, \hat{S}_\pm] = 0$$

$$\hat{S}^2 - \hat{S}_z^2 = \frac{1}{2}(\hat{S}_+\hat{S}_- + \hat{S}_-\hat{S}_+). \tag{13.1.7}$$

According to the standard quantum mechanical postulates (see Chapter 11), the spin-state of a particle is associated with an abstract vector, namely, a Dirac ket. Considering, for example, a single electron, the Stern–Gerlach experiment indicates that its spin quantum number (denoted traditionally as s instead of l) is $s = 1/2$, with two possible projections on a selected axis, corresponding to the magnetic quantum numbers (denoted m_s instead of m), $m_s = \pm 1/2$. It is conventional to denote these two spin states $|\sigma_{1/2}\rangle = |\alpha\rangle$ and $|\sigma_{-1/2}\rangle = |\beta\rangle$, where, by their definition, these states are proper simultaneous eigenstates of $\hat{S}^2$ and $\hat{S}_z$,

$$\begin{array}{ll} \hat{S}^2|\alpha\rangle = \hbar^2 s(s+1)|\alpha\rangle = \hbar^2\frac{3}{4}|\alpha\rangle \quad ; & \hat{S}^2|\beta\rangle = \hbar^2 s(s+1)|\beta\rangle = \hbar^2\frac{3}{4}|\beta\rangle \\ \hat{S}_z|\alpha\rangle = \hbar m_s|\alpha\rangle = \hbar\frac{1}{2}|\alpha\rangle & \hat{S}_z|\beta\rangle = \hbar m_s|\beta\rangle = -\hbar\frac{1}{2}|\beta\rangle \end{array} . \tag{13.1.8}$$

Since the Hermitian operator, $\hat{S}_z$, has only two eigenvalues, the corresponding eigenstates ($|\alpha\rangle$ and $|\beta\rangle$) are a complete orthonormal system, which spans the space of spin states of a single electron:

$$\langle\alpha|\alpha\rangle = \langle\beta|\beta\rangle = 1 \quad ; \quad \langle\alpha|\beta\rangle = 0. \tag{13.1.9}$$

Using the perfect analogy to normalized angular momentum eigenvectors (see Section 9.2), the spin states satisfy the following relations (Ex. 13.1.1):

$$\begin{array}{ll} \hat{S}_+|\beta\rangle = \hbar|\alpha\rangle \quad ; & \hat{S}_+|\alpha\rangle = 0 \\ \hat{S}_-|\alpha\rangle = \hbar|\beta\rangle \quad ; & \hat{S}_-|\beta\rangle = 0 \\ \hat{S}_x|\alpha\rangle = \frac{\hbar}{2}|\beta\rangle \quad ; & \hat{S}_x|\beta\rangle = \frac{\hbar}{2}|\alpha\rangle \\ \hat{S}_y|\alpha\rangle = i\frac{\hbar}{2}|\beta\rangle \quad ; & \hat{S}_y|\beta\rangle = -i\frac{\hbar}{2}|\alpha\rangle. \end{array} \tag{13.1.10}$$

Using $|\alpha\rangle$ and $|\beta\rangle$ as a basis for the single-electron spin space, any electronic spin state can be expressed as

$$|\sigma\rangle = c_\alpha|\alpha\rangle + c_\beta|\beta\rangle. \tag{13.1.11}$$

It is useful to represent the spin states and the spin operators as vectors and matrices in the two-dimensional vector space. Projecting Eq. (13.1.11) onto the basis vectors, we obtain $c_\alpha = \langle\alpha|\sigma\rangle, c_\beta = \langle\beta|\sigma\rangle$, where the vector representation reads

$$\begin{bmatrix} \langle\alpha|\sigma\rangle \\ \langle\beta|\sigma\rangle \end{bmatrix} = \begin{bmatrix} c_\alpha \\ c_\beta \end{bmatrix}. \tag{13.1.12}$$

Introducing the identity,

$$\hat{I} = |\alpha\rangle\langle\alpha| + |\beta\rangle\langle\beta|, \tag{13.1.13}$$

any operator in the single-electron spin space obtains the form

$$\hat{A} = A_{\alpha,\alpha}|\alpha\rangle\langle\alpha| + A_{\alpha,\beta}|\alpha\rangle\langle\beta| + A_{\beta,\alpha}|\beta\rangle\langle\alpha| + A_{\beta,\beta}|\beta\rangle\langle\beta|, \tag{13.1.14}$$

which corresponds to the matrix representation,

$$\begin{bmatrix} A_{\alpha,\alpha} & A_{\alpha,\beta} \\ A_{\beta,\alpha} & A_{\beta,\beta} \end{bmatrix} = \begin{bmatrix} \langle\alpha|\hat{A}|\alpha\rangle & \langle\alpha|\hat{A}|\beta\rangle \\ \langle\beta|\hat{A}|\alpha\rangle & \langle\beta|\hat{A}|\beta\rangle \end{bmatrix}. \tag{13.1.15}$$

Specifically, the matrix representations of the three components of the single-electron spin vector operator read (see Ex. 13.1.2)

$$\mathbf{S}_z = \frac{\hbar}{2}\begin{bmatrix} 1 & 0 \\ 0 & -1 \end{bmatrix} \quad ; \quad \mathbf{S}_x = \frac{\hbar}{2}\begin{bmatrix} 0 & 1 \\ 1 & 0 \end{bmatrix} \quad ; \quad \mathbf{S}_y = \frac{\hbar}{2}\begin{bmatrix} 0 & -i \\ i & 0 \end{bmatrix}. \tag{13.1.16}$$

These matrices are related to the Pauli matrices, named after Wolfgang Pauli (who also introduced the spin-related exclusion principle, to be discussed in Section 13.3):

$$\boldsymbol{\sigma}_z = \begin{bmatrix} 1 & 0 \\ 0 & -1 \end{bmatrix} \quad ; \quad \boldsymbol{\sigma}_x = \begin{bmatrix} 0 & 1 \\ 1 & 0 \end{bmatrix} \quad ; \quad \boldsymbol{\sigma}_y = \begin{bmatrix} 0 & -i \\ i & 0 \end{bmatrix}$$

$$\boldsymbol{\sigma}_+ = \begin{bmatrix} 0 & 1 \\ 0 & 0 \end{bmatrix} \quad ; \quad \boldsymbol{\sigma}_- = \begin{bmatrix} 0 & 0 \\ 1 & 0 \end{bmatrix}. \tag{13.1.17}$$

Since the representation of the identity is exact, the relations between the spin operators (Eqs. (13.1.5–13.1.7)) as well as their operations on spin states (Eqs. (13.1.8, 13.1.10)) are reproduced by the corresponding matrix–matrix and matrix–vector multiplications (see Ex. 13.1.3).

Exercise 13.1.1 *The spin states $|\alpha\rangle$ and $|\beta\rangle$ are common eigenstates of $\hat{S}^2$ and $\hat{S}_z$ (see Eq. (13.1.8)). Denoting the common eigenstates by the respective quantum numbers, $|s,m_s\rangle$, one can identify $|\alpha\rangle = \left|\frac{1}{2},\frac{1}{2}\right\rangle$ and $\left|\beta\right\rangle = \left|\frac{1}{2},\frac{-1}{2}\right\rangle$.*

(a) *Use the definition of the spin ladder operators ($\hat{S}_\pm$, Eq. (13.1.6)) and their commutation relations (Eq. (1.3.7)) to show that $\hat{S}_\pm|s,m_s\rangle \propto |s,m_s\pm 1\rangle$. Particularly, show that $\hat{S}_+|\beta\rangle \propto |\alpha\rangle$ and $\hat{S}_-|\alpha\rangle \propto |\beta\rangle$.*

(b) *Use the commutation relation (Eq. (13.1.7)) to show that $\hat{S}_+|\alpha\rangle = 0$ and $\hat{S}_-|\beta\rangle = 0$. (Show that $\langle\alpha|\hat{S}_-\hat{S}_+|\alpha\rangle = \langle\beta|\hat{S}_+\hat{S}_-|\beta\rangle = 0$.)*

(c) *Normalize the vectors $\hat{S}_+|\beta\rangle$ and $\hat{S}_-|\alpha\rangle$, and show that the normalized vectors satisfy the relations $\hat{S}_+|\beta\rangle = \hbar|\alpha\rangle$ and $\hat{S}_-|a\rangle = \hbar|\beta\rangle$.*

Exercise 13.1.2 *Use the properties of the spin eigenstates in Eqs. (13.1.8, 13.1.10) and derive the matrix representations of the spin operators in Eq. (13.1.16).*

Exercise 13.1.3 *(a) Verify that the commutation relations between the spin operators (Eqs. (13.1.5–13.1.7)) are satisfied by the spin matrices (Eqs. (13.1.16–13.1.17)). (b) Verify that the results in Eqs. (13.1.8, 13.1.10) are obtained by matrix–vector multiplications of the appropriate spin matrices (Eqs. (13.1.16–13.1.17)) on the basis vectors $\begin{bmatrix} 1 \\ 0 \end{bmatrix}$ and $\begin{bmatrix} 0 \\ 1 \end{bmatrix}$, corresponding to the states $|\alpha\rangle$ and $|\beta\rangle$, respectively.*

A full description of the physical state of an electron (within nonrelativistic quantum mechanics) corresponds to a vector in an extended space, which is a tensor product of the coordinate and spin spaces. This means that the vectors are spanned by a complete orthonormal set of product states (see Section 11.6),

$$|\Phi\rangle \equiv \sum_{\sigma \in \alpha, \beta} \sum_{n=1}^{\infty} c_{\sigma,n} |\varphi_n\rangle \otimes |\sigma\rangle. \tag{13.1.18}$$

Here $\{|\alpha\rangle, |\beta\rangle\}$ and $\{|\varphi_n\rangle\}$ are complete orthonormal systems of state vectors in the spin and coordinate spaces, respectively. The two spaces are in fact coupled by spin–orbit coupling operators, derived from relativistic corrections (Dirac's equation). In atoms, this coupling manifests in additional splitting of atomic energy levels, even in the absence of an external magnetic field, due to the magnetic interaction between the orbital electronic degrees of freedom and the intrinsic magnetic moments of electrons and the nuclei (the fine and hyperfine structures, respectively). When the relevant classical velocities are far below the speed of light, spin–orbit couplings induce small corrections to the energy level spacings associated with the electrostatic interaction terms, which enables us to treat them, when required, using perturbation theory. In "heavy" atoms, however, the internal electrons are associated with high effective velocities, and spin–orbit coupling becomes important. (The interested reader may want to address other textbooks for a comprehensive coverage of this issue [4.3].)

When spin–orbit coupling is completely neglected, the electron spin is revealed only in the presence of an external magnetic field. The electromagnetic field enters the single-electron Hamiltonian in terms of the scalar and vector potentials, $\Phi(\mathbf{r})$ and $\mathbf{A}(\mathbf{r})$, respectively [13.4], where the magnetic field is the curl of the vector potential, $\mathbf{B} = \nabla \times \mathbf{A}(\mathbf{r})$. Introducing the corresponding quantum mechanical operators, the "Pauli Hamiltonian" is obtained for an electron in the field:

$$\hat{H} = \frac{1}{2m_e}[\hat{\mathbf{p}} - e\mathbf{A}(\hat{\mathbf{r}})]^2 - e\Phi(\hat{\mathbf{r}}) + \frac{e}{m_e}\mathbf{B}(\hat{\mathbf{r}}) \cdot \hat{\mathbf{S}}. \tag{13.1.19}$$

The Zeeman effect associated with the interaction of the orbital angular momentum with the magnetic field (Eqs. 13.1.2, 13.1.3) is derived from the kinetic energy term, $\frac{1}{2m_e}[\hat{\mathbf{p}} - e\mathbf{A}(\hat{\mathbf{r}})]^2$, where the spin interaction with the magnetic field appears as a separate term, $\frac{e}{m_e}\mathbf{B}(\hat{\mathbf{r}}) \cdot \hat{\mathbf{S}}$. Notice that the latter interaction has a twice larger pre-factor in comparison to the interaction of the orbital angular momentum with the field (compare to Eq. (13.1.3).) According to the Pauli Hamiltonian, the Zeeman splitting associated with the spin is proportional to a scaled Bohr magneton, $g_e \mu_B$, where $g_e = 2$. This value for the electron's "gyromagnetic factor" is indeed close to the measured one, $g_e = 2.0034$.

By setting the static potential energy to $-e\Phi = \frac{-KZe^2}{r}$, the Pauli Hamiltonian (Eq. (13.1.10)) corresponds to an electron in a hydrogen-like atom and a magnetic field. Identifying the magnetic field direction with the z-axis, $\frac{e}{m_e}\mathbf{B}(\hat{\mathbf{r}}) \cdot \hat{\mathbf{S}} = \frac{e}{m_e}B_z(\hat{\mathbf{r}}) \cdot \hat{S}_z$, the eigenstates of the Pauli Hamiltonian are readily seen to be products of spin and orbital states, characterized by four quantum numbers,

$$|\Phi_{n,l,m,m_s}\rangle \equiv |\psi_{n,l,m}\rangle \otimes |\sigma_{m_s}\rangle, \tag{13.1.20}$$

where $|\sigma_{1/2}\rangle \equiv |\alpha\rangle$, and $|\sigma_{-1/2}\rangle \equiv |\beta\rangle$. In atomic physics and chemistry, the vector $|\Phi_{n,l,m,m_s}\rangle$ is often termed a "spin-orbital," associated with a well-defined information on both the orbital and the spin state of the electron. As we shall discuss, the concept of spin-orbitals is very useful also when approximating the state of many-electron systems, and many-electron atoms in particular. Notice that in the absence of a magnetic field, the two spin states for an electron are degenerate. In that case the Hamiltonian eigenstates can be any superposition of the two spin states, $|\Phi\rangle \equiv |\psi_{n,l,m}\rangle \otimes (c_\alpha|\alpha\rangle + c_\beta|\beta\rangle)$. A measurement process, such as a Stern–Gerlach experiment, can induce collapse of the superposition state, $|\Phi\rangle$, into either an $|\alpha\rangle$ or a $|\beta\rangle$ spin state. Indeed, considering a beam of particles entering an inhomogeneous magnetic field along the z-axis, the center of mass of each particle experiences a force depending on the projection of its magnetic moment on that axis, which is different for the $|\alpha\rangle$ and $|\beta\rangle$ spin states. The result is splitting of the particles into two distinguishable beams, according to their spin state.

13.2 Spin and Identical Particles

The existence of spin at the single-particle level has a profound effect on systems composed of many identical particles. In the following sections, we shall see how this manifests itself in the electronic structure of different materials, starting from the electronic structure of many-electron atoms, and its reflection in the periodic table of the elements. We start this section by discussing some general properties of systems of identical particles in quantum mechanics, and then we introduce another postulate of nonrelativistic quantum mechanics, which imposes additional limitations on proper wave functions (or, generally, the physical state) of systems of many identical particles according to their spins.

Let us consider first two particles, marked as "1" and "2." If the particles are physically identical, exchanging their identities (namely, replacing their marks) does not affect any measurable property of the system. As we shall see, in quantum mechanics this fact is associated with remarkable consequences. Let us associate the two-particle system with a wave function, $\psi(x_1, x_2)$. Since the particles are identical, the probability density for locating the first particle at the position x_1 and the second at position x_2 must be equal to the probability of locating the first particle at x_2 and the second at x_1, which means

$$|\psi(x_1, x_2)|^2 = |\psi(x_2, x_1)|^2. \tag{13.2.1}$$

An immediate consequence of Eq. (13.2.1) is that a physically proper two-particle function must fulfill the condition

$$\psi(x_1, x_2) = e^{i\varphi} \psi(x_2, x_1), \tag{13.2.2}$$

where φ is a real-valued parameter. Namely, *a permutation of the particle indexes, $1 \leftrightarrow 2$, changes the two-particle wave function by up to a multiplication by a complex*

number of a unit norm. Since physical states (e.g., solutions to the Schrödinger equation) are invariant to a multiplication by a constant, both $\psi(x_2,x_1)$ and $\psi(x_1,x_2)$ are equivalent descriptions of the same state of the two identical particles. Nevertheless, as discussed in the following sections, nature teaches us that the constant associated with a permutation of identical particles (φ in Eq. (13.2.1)) cannot be chosen arbitrarily.

We now turn to the formal definition of the two-particle permutation operator. In the abstract two-particle vector space, each physical sate is associated with a linear combination of tensor products (see Eq. (11.6.12)),

$$|\psi\rangle = \sum_{n_1,n_2} \psi_{n_1,n_2} |\varphi_{n_1}\rangle \otimes |\varphi_{n_2}\rangle, \tag{13.2.3}$$

where $|\varphi_{n_i}\rangle$ is a single-particle basis state in the subspace corresponding to the ith particle. Since the particles are identical, so are their respective single-particle subspaces, which means that identical basis sets can be invoked for the two subspaces,

$$\{|\varphi_{n_1}\rangle\} = \{|\varphi_{n_2}\rangle\}. \tag{13.2.4}$$

The permutation operator, $\hat{P}_{1,2}$, exchanges the basis states in any given product state,

$$\hat{P}_{1,2}|\varphi_{n_1}\rangle \otimes |\varphi_{n_2}\rangle \equiv |\varphi_{n_2}\rangle \otimes |\varphi_{n_1}\rangle. \tag{13.2.5}$$

where, as we can readily verify (Ex. 13.2.1), $\hat{P}_{1,2}$ is Hermitian, namely, for any $|\psi\rangle$ and $|\chi\rangle$,

$$\langle\chi|\hat{P}_{1,2}|\psi\rangle = \langle\psi|\hat{P}_{1,2}|\chi\rangle^*, \tag{13.2.6}$$

as well as unitary,

$$(\hat{P}_{1,2})^\dagger \hat{P}_{1,2} = (\hat{P}_{1,2})^2 = \hat{I}. \tag{13.2.7}$$

Generalizing the condition in Eq. (13.2.1) to any representation means that in a system of identical particles the probability of finding particle "1" in a state $|\varphi_{n_1}\rangle$ and particle "2" in a state $|\varphi_{n_2}\rangle$ must be equal to the probability of finding particle "1" in a state $|\varphi_{n_2}\rangle$ and particle "2" in a state $|\varphi_{n_1}\rangle$. Consequently, any physical state, $|\psi\rangle$, of a system of identical particles must satisfy the condition

$$|(\langle\varphi_{n_2}| \otimes \langle\varphi_{n_1}|)|\psi\rangle|^2 = |(\langle\varphi_{n_1}| \otimes \langle\varphi_{n_2}|)|\psi\rangle|^2, \tag{13.2.8}$$

from which follows

$$(\langle\varphi_{n_1}| \otimes \langle\varphi_{n_2}|)\hat{P}_{1,2}|\psi\rangle = e^{i\varphi}(\langle\varphi_{n_1}| \otimes \langle\varphi_{n_2}|)|\psi\rangle, \tag{13.2.9}$$

where φ is real-valued. Since the latter holds for any $|\varphi_{n_1}\rangle$ and $|\varphi_{n_2}\rangle$, it means that the state $|\psi\rangle$ must be an eigenstate of the permutation operator, $\hat{P}_{1,2}|\psi\rangle = e^{i\varphi}|\psi\rangle$. However, considering that $\hat{P}_{1,2}$ is Hermitian (Eq. (13.2.6)), its eigenvalues in the space of proper states must be real. Consequently,

$$\hat{P}_{1,2}|\psi\rangle = \pm|\psi\rangle. \tag{13.2.10}$$

Returning to the coordinate representation, Eq. (13.2.2) is refined as follows:

$$\psi(x_1,x_2) = \langle x_1,x_2|\psi\rangle = \pm\langle x_1,x_2|\hat{P}_{1,2}|\psi\rangle = \pm\langle x_2,x_1|\psi\rangle = \pm\psi(x_2,x_1), \tag{13.2.11}$$

which means that ***the wave function of a system of two identical particles must be either symmetric or antisymmetric for permutation of the particle indexes***.

The generalization to the case of many (N) particles is straightforward. Defining a basis of products of identical single-particle basis sets, the state vector of a system of identical particles obtains the form

$$|\psi\rangle = \sum_{n_1,n_2,\ldots,n_N} \psi_{n_1,n_2,\ldots,n_N} |\varphi_{n_1}\rangle \otimes |\varphi_{n_2}\rangle \otimes \cdots \otimes |\varphi_{n_N}\rangle. \qquad (13.2.12)$$

A pair permutation (transposition) $i \leftrightarrow j$ between the ith and jth particles is represented here by an operator, $\hat{P}_{i,j}$, defined by its operation on any product state,

$$\hat{P}_{i,j}|\varphi_{n_1}\rangle \cdots \otimes |\varphi_{n_i}\rangle \otimes \cdots \otimes |\varphi_{n_j}\rangle \otimes \cdots \otimes |\varphi_{n_N}\rangle = |\varphi_{n_1}\rangle \cdots \otimes |\varphi_{n_j}\rangle \otimes \cdots \otimes |\varphi_{n_i}\rangle \otimes \cdots |\varphi_{n_N}\rangle. \qquad (13.2.13)$$

The general condition for any pair permutation on a system of identical particles (Eq. (13.2.8)) obtains the form

$$|(\langle\varphi_{n_1}| \otimes \langle\varphi_{n_2}| \otimes \cdots \otimes \langle\varphi_{n_N}|)\hat{P}_{i,j}|\psi\rangle|^2 = |(\langle\varphi_{n_1}| \otimes \langle\varphi_{n_2}| \otimes \cdots \otimes \langle\varphi_{n_N}|)|\psi\rangle|^2, \qquad (13.2.14)$$

which means that ***the state of a system of identical particles must be an eigenvector of any pair permutation operator, with an eigenvalue, ± 1***,

$$\hat{P}_{i,j}|\psi\rangle = \pm|\psi\rangle. \qquad (13.2.15)$$

Moreover, as we shall discuss, nature tells us that ***the two different eigenvalues correspond to two different types of particles***. Depending on the particle's type, the state $|\psi\rangle$ is associated with only one eigenvalue of $\hat{P}_{i,j}$, which is either $+1$ or -1, and is the same for all (i, j) pairs.

Exercise 13.2.1 *Consider a system of three identical particles. The three-particle space is spanned by the complete set of products of single-particle states, $|\varphi_{n_1}\rangle \otimes |\varphi_{n_2}\rangle \otimes |\varphi_{n_3}\rangle$, associated with the quantum numbers n_1, n_2, n_3, respectively for the particle indexes 1, 2, and 3. The permutation between particles 1 and 2 is associated with the operator $\hat{P}_{1,2}$, defined as $\hat{P}_{1,2}|\varphi_{n_1}\rangle \otimes |\varphi_{n_2}\rangle \otimes |\varphi_{n_3}\rangle = |\varphi_{n_2}\rangle \otimes |\varphi_{n_1}\rangle \otimes |\varphi_{n_3}\rangle$. Show that:*

(a) The operator $\hat{P}_{1,2}$ is Hermitian, namely $\langle\chi|\hat{P}_{1,2}|\psi\rangle = \langle\psi|\hat{P}_{1,2}|\chi\rangle^*$, for any three particle states, $|\psi\rangle = \sum_{n_1,n_2,n_3} \psi_{n_1,n_2,n_3}|\varphi_{n_1}\rangle \otimes |\varphi_{n_2}\rangle \otimes |\varphi_{n_3}\rangle$, and $|\chi\rangle = \sum_{n_1,n_2,n_3} \chi_{n_1,n_2,n_3}|\varphi_{n_1}\rangle \otimes |\varphi_{n_2}\rangle \otimes |\varphi_{n_3}\rangle$.

(b) The operator $\hat{P}_{1,2}$ is unitary, namely $(\hat{P}_{1,2})^\dagger \hat{P}_{1,2} = \hat{I}$.

Exercise 13.2.2 *The following symmetry property of a many-particle observable, $\hat{O}$, to exchange of particle indexes, $\hat{O}_{1,\ldots i,\ldots,j,\ldots,N} = \hat{O}_{1,\ldots,j,\ldots,i,\ldots,N}$, means that the following identity holds for any two tensor product basis states:*

$$\langle\varphi_{n_1'}| \cdots \otimes \langle\varphi_{n_i'}| \otimes \cdots \otimes \langle\varphi_{n_j'}| \otimes \cdots \langle\varphi_{n_N'}|\hat{O}|\varphi_{n_1}\rangle \cdots \otimes |\varphi_{n_i}\rangle \otimes \cdots \otimes |\varphi_{n_j}\rangle \otimes \cdots |\varphi_{n_N}\rangle$$
$$= \langle\varphi_{n_1'}| \cdots \otimes \langle\varphi_{n_j'}| \otimes \cdots \otimes \langle\varphi_{n_i'}| \otimes \cdots \langle\varphi_{n_N'}|\hat{O}|\varphi_{n_1}\rangle \cdots \otimes |\varphi_{n_j}\rangle \otimes \cdots \otimes |\varphi_{n_i}\rangle \otimes \cdots |\varphi_{n_N}\rangle.$$

Use the definition of the permutation operator, Eq. (13.2.13), to show that such an observable commutes with any permutation operator, namely $[\hat{P}_{i,j}, \hat{O}_{1,\ldots i,\ldots,j,\ldots,N}] = 0$.

The discussion so far has referred to any state of a system of many identical particles. Here we focus on properties of the *Hamiltonian eigenstates* of such systems. Notice that when particles are physically identical, any system observable is invariant to their index permutations. Without loss of generality, we consider the system Hamiltonian; but the discussion based on Eq. (13.2.16) applies in fact to any observable in a system of identical particles. Denoting the Hamiltonian of an N-particle system as $\hat{H}_{1,2,...,N}$, the identity of the ith and jth particles is expressed in the symmetry property,

$$\hat{H}_{1,...i,...,j,...,N} = \hat{H}_{1,...,j,...,i,...,N}, \tag{13.2.16}$$

which means that the Hamiltonian commutes with the pair permutation operators (see Ex. 13.2.2),

$$[\hat{P}_{i,j}, \hat{H}_{1,...i,...,j,...,N}] = 0. \tag{13.2.17}$$

It immediately follows that if $|\psi\rangle$ is a Hamiltonian eigenstate,

$$\hat{H}_{1,...i,...,j,...,N}|\psi\rangle = E|\psi\rangle, \tag{13.2.18}$$

then $\hat{P}_{i,j}|\psi\rangle$ is also a Hamiltonian eigenstate, corresponding to the same eigenvalue,

$$\hat{H}_{1,...i,...,j,...,N}(\hat{P}_{i,j}|\psi\rangle) = E(\hat{P}_{i,j}|\psi\rangle), \tag{13.2.19}$$

where $|\psi\rangle$ and $\hat{P}_{i,j}|\psi\rangle$ are degenerate states. Since this argument can be repeated for all possible permutations (including general permutations, namely sequences of pair permutations), there is a set of $N!$ degenerate states associated with any eigenvalue, E, which corresponds to all the different arrangements of the N particle indexes. Therefore, any Hamiltonian eigenstate can be written as a linear combination,

$$\hat{H}_{1,2,...,N}\left(\sum_{n=1}^{N!} a_n\hat{R}_n|\psi\rangle\right) = E\left(\sum_{n=1}^{N!} a_n\hat{R}_n|\psi\rangle\right). \tag{13.2.20}$$

(Here, $\hat{R}_n = \prod_{i,j>i=1}^{N} (\hat{P}_{i,j})^{p_{i,j}^{(n)}}$ creates the nth unique order of the N indexes. It is an ordered sequence of operations over all pairs of indexes (i, j), where each $p_{i,j}^{(n)}$ obtains the value zero or one.) Importantly, while the general superposition, $|\Psi\rangle = \sum_{n=1}^{N!} a_n\hat{R}_n|\psi\rangle$, is an eigenstate of the identical particles Hamiltonian, it does not necessarily meet the condition Eq. (13.2.15) on the identical particles state. (This is sometimes referred to as the "exchange degeneracy" problem [4.3].) Indeed, any physical eigenstate $|\Psi\rangle$ of the many-particle Hamiltonian must also be an eigenstate of all the permutation operators corresponding to two identical particles, with the same eigenvalue, either $+1$ or -1. In order to properly fulfil these requirements on $|\Psi\rangle$, the expansion coefficients $\{a_n\}$ must be properly selected: we can readily see (Ex. 13.2.3) that setting all the coefficients identical to each other and normalizing, namely, $a_n^{(S)} = \frac{1}{\sqrt{N!}}$, results in a symmetric many-particle state, $|\Psi_S\rangle$, associated with the eigenvalue $+1$ ***for any pair permutation***. Similarly, an antisymmetric many-particle state, $|\Psi_A\rangle$, associated with the eigenvalue -1 ***for any pair permutation***, can be obtained by setting $a_n^{(A)} =$

$(-1)^{\sum\limits_{i=1}^{N}\sum\limits_{j>i}^{N}P_{i,j}^{(n)}}\frac{1}{\sqrt{N!}}$, which means that the expansion coefficients of all the different index arrangements are identical in magnitude, but may differ in sign. Particularly, when the index arrangement is associated with an odd number of pair permutations (with respect to a common reference order), the respective expansion coefficient obtains a minus sign. The two systematic choices of expansion coefficients can be cast in the form of two operators,

$$\hat{S} \equiv \frac{1}{\sqrt{N!}} \sum_{n=1}^{N!} \prod_{i,j>i=1}^{N} (\hat{P}_{i,j})^{p_{j,j}^{(n)}}$$

$$\hat{A} \equiv \frac{1}{\sqrt{N!}} \sum_{n=1}^{N!} (-1)^{\sum\limits_{i=1}^{N}\sum\limits_{j>i}^{N}p_{i,j}^{(n)}} \prod_{i,j>i=1}^{N} (\hat{P}_{i,j})^{p_{i,j}^{(n)}} \qquad (13.2.21)$$

$\hat{S}$ and $\hat{A}$ are termed, respectively, the symmetrizer and the antisymmetrizer. As we have detailed, the states obtained by applying these operators on a state $|\psi\rangle$, namely,

$$|\Psi_s\rangle = \hat{S}|\psi\rangle$$

$$|\Psi_A\rangle = \hat{A}|\psi\rangle, \qquad (13.2.22)$$

are symmetric or antisymmetric, respectively, under any pair permutation (see Exs. 13.2.3, 13.2.4, 13.2.5),

$$\hat{P}_{i,j}|\Psi_S\rangle = |\Psi_S\rangle$$

$$\hat{P}_{i,j}|\Psi_A\rangle = -|\Psi_A\rangle. \qquad (13.2.23)$$

Notice that while our discussion was motivated by the search for Hamiltonian eigenstates that additionally satisfy Eq. (13.2.15) (or Eq. (13.2.23)), the result is not restricted to Hamiltonian eigenstate; namely, ***the operators $\hat{S}$ or $\hat{A}$ map any state of a system of many identical particles onto a state that is symmetric or antisymmetric under permutations.*** Also notice that the states $|\Psi_S\rangle$ and $|\Psi_A\rangle$ are associated with different eigenvalues of the same Hermitian operator (see Ex. 13.2.1), which means that ***the subspaces of symmetric and antisymmetric states with respect to permutation of identical particles are orthogonal subspaces.*** Finally, our definition of the operators $\hat{S}$ or $\hat{A}$ assures that they preserve the norm of the state to which they are applied (unless the result is a trivial improper state). Alternative normalizations of these operators that differ by a constant multiplication (e.g., to fulfil the properties of projection operators) are also commonly used [4.3].

Exercise 13.2.3 *The operator $\hat{R}_n = \prod\limits_{i,j>i=1}^{N} (\hat{P}_{i,j})^{p_{i,j}^{(n)}}$ is a sequence of permutation operators, defined by a specific vector of scalars $(p_{1,2}^{(n)}, p_{1,3}^{(n)}, p_{2,3}^{(n)}, \ldots)$ with entries 0 or 1. Each $\hat{R}_n$ corresponds to one of the N! unique arrangements of the particle indexes. The symmetrizer and the antisymmetrizer operators are defined as sums over all possible arrangements with appropriate coefficients, as follows: $\hat{S}|\psi\rangle \equiv \frac{1}{\sqrt{N!}} \sum\limits_{n=1}^{N!} \hat{R}_n|\psi\rangle$, and, $\hat{A}|\psi\rangle \equiv \frac{1}{\sqrt{N!}} \sum\limits_{n=1}^{N!} (-1)^{\sum\limits_{i=1}^{N}\sum\limits_{j>i}^{N}p_{i,j}^{(n)}} \hat{R}_n|\psi\rangle$, respectively. Show that $\hat{P}_{i,j}\hat{S}|\psi\rangle \equiv \hat{S}|\psi\rangle$ and $\hat{P}_{i,j}\hat{A}|\psi\rangle \equiv -\hat{A}|\psi\rangle$ by showing that the following holds for any tensor product basis vector:*

$$\langle\varphi_{n_1}|\otimes\langle\varphi_{n_2}|\otimes\cdots\otimes\langle\varphi_{n_N}|\hat{P}_{i,j}\hat{S}|\psi\rangle=\langle\varphi_{n_1}|\otimes\langle\varphi_{n_2}|\otimes\cdots\otimes\langle\varphi_{n_N}|\hat{S}|\psi\rangle$$
$$\langle\varphi_{n_1}|\otimes\langle\varphi_{n_2}|\otimes\cdots\otimes\langle\varphi_{n_N}|\hat{P}_{i,j}\hat{A}|\psi\rangle=-\langle\varphi_{n_1}|\otimes\langle\varphi_{n_2}|\otimes\cdots\otimes\langle\varphi_{n_N}|\hat{A}|\psi\rangle.$$

Exercise 13.2.4 *The symmetrizer and antisymmetrizer operators applied to a generic two-particle state,* $|\psi\rangle=\sum\limits_{n_1,n_2}\psi_{n_1,n_2}|\varphi_{n_1}\rangle\otimes|\varphi_{n_2}\rangle$, *yield*

$$|\psi_S\rangle=\hat{S}|\psi\rangle=\frac{1}{\sqrt{2}}\sum_{n_1,n_2}\psi_{n_1,n_2}|\varphi_{n_1}\rangle\otimes|\varphi_{n_2}\rangle+\frac{1}{\sqrt{2}}\hat{P}_{1,2}\sum_{n_1,n_2}\psi_{n_1,n_2}|\varphi_{n_1}\rangle\otimes|\varphi_{n_2}\rangle$$
$$=\frac{1}{\sqrt{2}}\sum_{n_1,n_2}\psi_{n_1,n_2}(|\varphi_{n_1}\rangle\otimes|\varphi_{n_2}\rangle+|\varphi_{n_2}\rangle\otimes|\varphi_{n_1}\rangle)$$

and

$$|\psi_A\rangle=\hat{A}|\psi\rangle=\frac{1}{\sqrt{2}}\sum_{n_1,n_2}\psi_{n_1,n_2}|\varphi_{n_1}\rangle\otimes|\varphi_{n_2}\rangle-\frac{1}{\sqrt{2}}\hat{P}_{1,2}\sum_{n_1,n_2}\psi_{n_1,n_2}|\varphi_{n_1}\rangle\otimes|\varphi_{n_2}\rangle$$
$$=\frac{1}{\sqrt{2}}\sum_{n_1,n_2}\psi_{n_1,n_2}(|\varphi_{n_1}\rangle\otimes|\varphi_{n_2}\rangle-|\varphi_{n_2}\rangle\otimes|\varphi_{n_1}\rangle).$$

Show that $\hat{S}|\psi\rangle$ *and* $\hat{A}|\psi\rangle$ *are indeed eigenstates of the permutation operator,* $\hat{P}_{1,2}$ *(Eq. (13.2.23)). What are the corresponding eigenvalues?*

Exercise 13.2.5 *The symmetrizer and antisymmetrizer operators applied to a generic three-particle state,* $|\psi\rangle=\sum\limits_{n_1,n_2,n_3}\psi_{n_1,n_2,n_3}|\varphi_{n_1}\rangle\otimes|\varphi_{n_2}\rangle\otimes|\varphi_{n_3}\rangle$, *yield*

$$\hat{S}|\psi\rangle=\frac{1}{\sqrt{6}}\sum_{n_1,n_2,n_3}\psi_{n_1,n_2,n_3}(|\varphi_{n_1}\rangle\otimes|\varphi_{n_2}\rangle\otimes|\varphi_{n_3}\rangle+|\varphi_{n_2}\rangle\otimes|\varphi_{n_1}\rangle\otimes|\varphi_{n_3}\rangle+|\varphi_{n_1}\rangle\otimes|\varphi_{n_3}\rangle\otimes$$
$$|\varphi_{n_2}\rangle+|\varphi_{n_3}\rangle\otimes|\varphi_{n_2}\rangle\otimes|\varphi_{n_1}\rangle+|\varphi_{n_2}\rangle\otimes|\varphi_{n_3}\rangle\otimes|\varphi_{n_1}\rangle+|\varphi_{n_3}\rangle\otimes|\varphi_{n_1}\rangle\otimes|\varphi_{n_2}\rangle)$$
$$\hat{A}|\psi\rangle=\frac{1}{\sqrt{6}}\sum_{n_1,n_2,n_3}\psi_{n_1,n_2,n_3}(|\varphi_{n_1}\rangle\otimes|\varphi_{n_2}\rangle\otimes|\varphi_{n_3}\rangle-|\varphi_{n_2}\rangle\otimes|\varphi_{n_1}\rangle\otimes|\varphi_{n_3}\rangle-|\varphi_{n_1}\rangle\otimes|\varphi_{n_3}\rangle\otimes$$
$$|\varphi_{n_2}\rangle-|\varphi_{n_3}\rangle\otimes|\varphi_{n_2}\rangle\otimes|\varphi_{n_1}\rangle+|\varphi_{n_2}\rangle\otimes|\varphi_{n_3}\rangle\otimes|\varphi_{n_1}\rangle+|\varphi_{n_3}\rangle\otimes|\varphi_{n_1}\rangle\otimes|\varphi_{n_2}\rangle).$$

Show that $\hat{S}|\psi\rangle$ *and* $\hat{A}|\psi\rangle$ *are indeed eigenstates of the two particle permutation operators,* $\hat{P}_{1,2},\hat{P}_{2,3}$, *and* $\hat{P}_{1,3}$ *(Eq. (13.2.23)). What are the corresponding eigenvalues?*

The general considerations so far allow for either symmetric or antisymmetric proper states under permutation of identical particle pairs. Evidence from experiments show, however, that nature distinguishes between these two options, where particles are divided into two distinctive groups, bosons, and fermions. Particles whose many-particle states are symmetric with respect to pair permutation of identical particles are termed bosons, whereas particles whose many-particle states are antisymmetric with respect to pair permutation are termed fermions. Moreover, what seems to distinguish bosons from fermions relates to the nature of their intrinsic spin. At the level of elementary particles, bosons (such as photons, mesons, phonons, etc.) are associated with an integer spin quantum number, $s=0,1,2,\ldots$, whereas fermions (such as electrons, protons, neutrons, etc.) are associated with a half-integral spin quantum number, $s = 1/2, 3/2, \ldots$. Larger particles, composed out of elementary ones, can be classified into bosons and fermions according to their net intrinsic spin, defined by the addition of the spin vectors of their elementary particles. Consequently, for example, the hydrogen nucleus, ^1H, is a proton, hence a fermion associated with $s = 1/2$, while the deuterium nucleus, ^2H, is a boson owing to the addition of the proton and neutron spins ($s = 1$). Similarly, the nucleus ^3He is a fermion, while ^4He is a boson, and so on.

In summary, physically proper state vectors for systems composed of identical particles are subject to additional requirements, derived in part from general consideration with respect to their information content (Eqs. (13.2.1, 13.2.8, 13.2.14)) and in part from experimental observations on many-particle systems. These requirements were not included in our discussion of the postulates of quantum mechanics in Chapter 11. They can be summarized as an additional postulate:

A proper physical state (or a wave function) of a system composed of identical particles must be either symmetric (unchanged) or antisymmetric (flip sign) with respect to any permutation of two identical particle indexes. The symmetric and antisymmetric cases correspond to permutation of identical bosons (associated with integer spin quantum number) and fermions (associated with a half-integral spin quantum number), respectively.

13.3 The Electronic Structure of Many-Electron Atoms

A direct consequence of the existence of the electron spin is the electronic structure of many-electron atoms. Indeed, electrons are fermions, and the many-electron state of any atom must be antisymmetric to any pair permutation (transposition) of electron indexes. In Chapter 10 we became familiar with hydrogen-like atoms, composed of a nucleus and a single electron. The quantized energy levels for the electron and the spatial distribution of its charge around the nucleus were associated with single-electron wave functions, namely orbitals. In a nonrelativistic framework, the electron spin is independent of its spatial orbital (see Section 13.1), and it should be undetectable in the absence of an external magnetic field. Consequently, the (orthonormal) Hamiltonian eigenstates in a single-electron atom can be associated with a spin-orbital product state, $|\Phi_{n,l,m,m_s}\rangle \equiv |\psi_{n,l,m}\rangle \otimes |\sigma_{m_s}\rangle$, where n, l, m are the principal, angular, and magnetic quantum numbers, and $m_s = \pm 1/2$ is the spin quantum number (see Eq. (13.1.20)).

Turning to many-electron atoms, one would have liked to make use of the knowledge acquired by solving the Schrödinger equation for a single-electron atom by maintaining the concept of orbitals and associating each electron with its own spin-orbital. If this was indeed possible, an N-electron state would have been a single tensor product over the single-electron spaces, for example, $|\Psi\rangle = |\Phi_{n_1,l_1,m_1,m_{s_1}}\rangle \otimes |\Phi_{n_2,l_2,m_2,m_{s_2}}\rangle \otimes \cdots \otimes |\Phi_{n_N,l_N,m_N,m_{s_N}}\rangle$. However, there are two issues that exclude such single products from being a correct description of many-electron atoms. First, since electrons are fermions, the many-electron state must be antisymmetric to any permutation of two electron indexes. If all the electrons would be associated with identical spin-orbitals, the many-electron product state would be symmetric (invariant) under index permutation, whereas if two electrons would be associated with different spin-orbitals, the product state would be asymmetric, namely neither symmetric nor antisymmetric, with respect to their index permutation. Therefore, a single product state violates the antisymmetry requirement for fermions. The second issue is that electrons interact with each other. Therefore, even if the consequences of their being identical particles are ignored, their many-body Hamiltonian is non-separable (see Eq. (3.4.10) for the explicit form of an atomic Hamiltonian in the nucleus reference frame, which is not a sum over single-electron terms), which means that the exact Hamiltonian eigenstates are not single products.

In spite of these difficulties, spin-orbitals are very often used within theoretical approximations to many-electron systems, including atoms, molecules, and extended materials. The concept of orbitals is indeed useful to obtain a tractable description of such complex systems. The widespread usage of orbitals is based on the following solutions to the two issues just discussed: First, by applying the antisymmetrizer (see Section 13.2) to a single product state, a proper state can be generated, which is a superposition of products, fulfilling the requirement of antisymmetry with respect to pair permutations (a Slater determinant). Second, the variational principle (see Section 12.3) enables us to optimize the spatial orbitals used for approximating the many-electron state, while accounting for the electron–electron interactions. This is carried out within a self-consistent (mean-field) approximation (the Hartree–Fock equations), as discussed in what follows.

Slater's Determinant and Pauli's Exclusion Principle

An elegant way to maintain the concept of single-particle states in many-electron atoms, while maintaining the antisymmetry of the many-electron state with respect to pair permutations, was proposed by Slater [13.5].

The electrons are distributed among spin-orbitals, which account for both the spatial and spin state of each electron. Invoking the coordinate representation for the spatial orbitals, assigning the ith electron to the jth spin-orbital is denoted as

$$\langle i|\Phi_j\rangle \equiv \varphi_j(\mathbf{r}_i) \cdot \sigma_j(i). \tag{13.3.1}$$

Assuming an effective centro-symmetric field (an effective central potential) for each electron in the nucleus reference frame, each spatial orbital obtains the form $\varphi_j(\mathbf{r}) \propto R_{n,l}^{(Z,N)}(r)Y_{l,m}(\theta,\varphi)$, where the radial functions $\{R_{n,l}^{(Z,N)}(r)\}$ are different from those of the hydrogen-like atom and change from one atom to another (see what follows). $\sigma_j(i)$ is the spin state, which can be either $\sigma_{m_s=1/2}(i)$ or $\sigma_{m_s=-1/2}(i)$. The index j therefore stands for a set of four quantum numbers, $j \leftrightarrow (n,l,m,m_s)$.

The N-electron state is expressed in terms of the spin orbitals, as a determinant rather than as a single product. The columns and rows correspond, respectively, to the spin orbital and electron indexes:

$$\langle 1,2,\ldots,N \mid \Psi \rangle \equiv \frac{1}{\sqrt{N!}} \begin{vmatrix} \langle 1 \mid \Phi_1 \rangle & \langle 1 \mid \Phi_2 \rangle & \langle 1 \mid \Phi_3 \rangle & \cdots & \langle 1 \mid \Phi_N \rangle \\ \langle 2 \mid \Phi_1 \rangle & \langle 2 \mid \Phi_2 \rangle & \langle 2 \mid \Phi_3 \rangle & \cdots & \langle 2 \mid \Phi_N \rangle \\ \langle 3 \mid \Phi_1 \rangle & \langle 3 \mid \Phi_2 \rangle & \langle 3 \mid \Phi_3 \rangle & \cdots & \langle 3 \mid \Phi_N \rangle \\ \vdots & \vdots & \vdots & \ddots & \vdots \\ \langle N \mid \Phi_1 \rangle & \langle N \mid \Phi_2 \rangle & \langle N \mid \Phi_3 \rangle & \cdots & \langle N \mid \Phi_N \rangle \end{vmatrix}. \qquad (13.3.2)$$

Notice that the determinant is an explicit linear combination of all possible permutations of the N electrons in the different N spin-orbitals, with alternating signs. Clearly, *the determinant is anti-symmetric to a permutation of two electron indexes, since the latter amounts to exchanging two rows, which changes the sign of the determinant*. One can readily verify that the slater determinant is equivalent to the antisymmetrizer operation on a direct product of spin orbital states, namely

$$\langle 1,2,\ldots,N|\Psi \rangle = \langle 1,2,\ldots,N|\hat{A}(|\Phi_1\rangle \otimes |\Phi_2\rangle \otimes \cdots \otimes |\Phi_N\rangle), \qquad (13.3.3)$$

where, $\hat{A}$ is defined in Eq. (13.2.21).

A remarkable property of this many-electron state is that it becomes improper when two (or more) of the spin orbitals are identical. Indeed, setting $|\Phi_j\rangle = |\Phi_{j'}\rangle$ in Eq. (13.3.2), the determinant has two identical columns and therefore vanishes identically, meaning that the state is improper. This restriction is the grounds for *Pauli's exclusion principle (named after Wolfgang Pauli): In a many-electron atom, any spin orbital can occupy up to a single electron, or, any spatial orbital can occupy up to two electrons, associated with a different spin quantum number. Alternatively, two electrons cannot be associated with the same set of four quantum numbers.*

The description of many-electron systems in terms of determinants composed out of single-particle states extends beyond the realm of atoms. It is useful also in molecules and extended many-atom systems, to be discussed in the following chapters. Particularly, Pauli's exclusion principle sets the foundations for the Fermi–Dirac distribution of fermions (electrons, in particular) over the single-particle energy states, which determines macroscopic optical and electronic properties of different materials, to be discussed in Chapter 20.

The Hartree–Fock Approximation

In Chapter 12 we became familiar with the mean-field approach for the description of many-particle systems. The idea is to approximate the many particle states in terms of single products of single-particle states and to use the variation principle for optimizing the latter. In the context of an N-electron atom, a naive implementation of this approach would correspond to a product of orthonormal atomic spin orbitals as a variational trial state (a Hartree product, as discussed in Section 12.3 for the case of He),

$$|\tilde{\Psi}_H\rangle = |\tilde{\Phi}_1\rangle \otimes |\tilde{\Phi}_2\rangle \otimes \cdots \otimes |\tilde{\Phi}_N\rangle \qquad (13.3.4)$$

$$\langle \tilde{\Phi}_k|\tilde{\Phi}_{k'}\rangle = \delta_{k,k'}, \qquad (13.3.5)$$

where $|\Phi_k\rangle \equiv |\varphi_k\rangle \otimes |\sigma_k\rangle$ are specific spin orbitals, corresponding to a set of four quantum numbers, $k \leftrightarrow (n, l, m, m_s)$. However, such a Hartree product is incompatible with the requirement for antisymmetry with respect to any pair permutation. A proper trial state would be, instead, the corresponding antisymmetrized state, namely the determinant,

$$|\tilde{\Psi}\rangle = \hat{A}|\tilde{\Psi}_H\rangle, \tag{13.3.6}$$

where $\hat{A}$ is the antisymmetrizer (Eq. (13.2.21)).

The task is therefore to optimize the trial state, $|\tilde{\Psi}\rangle$. According to the variational principle, the optimal trial function is obtained when (Eq. (12.3.4))

$$\langle \delta\tilde{\Psi}|(\hat{H} - \varepsilon)|\tilde{\Psi}\rangle = 0 \tag{13.3.7}$$

for any infinitesimal change, $|\delta\tilde{\Psi}\rangle$. ε is the variational energy, $\varepsilon = \langle \tilde{\Psi}|\hat{H}|\tilde{\Psi}\rangle$, and $\hat{H}$ is the many-electron Hamiltonian, which has the generic form (see Eq. (3.4.10))

$$\hat{H} = \sum_{j=1}^{N} \hat{h}_j + \sum_{j'>j=1}^{N} \hat{w}_{j,j'}. \tag{13.3.8}$$

$\hat{h}_j$ is the single-electron (hydrogen-like) operator in the jth particle subspace (Within each subspace, the operation is identical for all the electrons, $\hat{h} = \frac{-\hbar^2}{2m_e}\Delta_{\mathbf{r}} + \frac{-KZe^2}{|\hat{\mathbf{r}}|}$), where $\hat{w}_{j,j'}$ is the electron–electron interaction operator, in the subspace of the jth and j'th electrons. (Within each subspace, the operation is identical for all the electron pairs, $\hat{w} = \frac{Ke^2}{|\hat{\mathbf{r}}-\hat{\mathbf{r}}'|}$).

The targets for optimization are the spin-orbitals $\{|\tilde{\Phi}_k\rangle\}$. We therefore restrict the changes in the trial state to changes in the orbitals. (In fact, we shall constrain the spins such that only the spatial parts are changed; see what follows),

$$\left|\delta\tilde{\Psi}\right\rangle = \hat{A}\left|\delta\tilde{\Psi}_H\right\rangle = \sum_{k=1}^{N} \hat{A}\left|\tilde{\Phi}_1\right\rangle \otimes \cdots \otimes \left|\delta\tilde{\Phi}_k\right\rangle \otimes \cdots \otimes \left|\tilde{\Phi}_N\right\rangle. \tag{13.3.9}$$

Consequently, the optimization condition, Eq. (13.3.7), reads $\langle \delta\tilde{\Psi}_H|\hat{A}(\hat{H}-\varepsilon)\hat{A}|\tilde{\Psi}_H\rangle = 0$. Using the facts that (see Ex. 13.3.1) the antisymmetrizer is Hermitian, it commutes with the system Hamiltonian (Eq. (13.3.8)),

$$[\hat{A}, \hat{H}] = 0, \tag{13.3.10}$$

and it satisfies the identity, $\hat{A}^2 = \sqrt{N!}\hat{A}$, the optimization condition (Eq. (13.3.7)) simplifies to

$$\langle \delta\tilde{\Psi}_H|(\hat{H} - \varepsilon)\hat{A}|\tilde{\Psi}_H\rangle = 0. \tag{13.3.11}$$

Using Eq. (13.3.9) for $|\delta\tilde{\Psi}\rangle$ and requiring the condition in Eq. (13.3.11) to hold independently for any $|\delta\tilde{\Phi}_k\rangle$, we obtain the following set of N coupled equations (for $k = 1, 2, \ldots, N$),

$$(\langle \tilde{\Phi}_1| \otimes \cdots \otimes \langle \delta\tilde{\Phi}_k| \otimes \cdots \otimes \langle \tilde{\Phi}_N|)(\hat{H} - \varepsilon)\hat{A}(|\tilde{\Phi}_1\rangle \otimes |\tilde{\Phi}_2\rangle \otimes \cdots \otimes |\tilde{\Phi}_N\rangle) = 0, \tag{13.3.12}$$

where, the orthogonality of the single-particle states (Eq. (13.3.5)) is constrained by requiring

$$\langle \delta\tilde{\Phi}_k | \tilde{\Phi}_{k'} \rangle |_{k \neq k'} = 0. \tag{13.3.13}$$

Substitution of the Hamiltonian, Eq. (13.3.8), in Eq. (13.3.12), utilizing its structure in the many-electron tensor product space, the optimal set of normalized single-particle states $\{|\tilde{\Phi}_k\rangle\}$ satisfies the following condition, for any infinitesimal change, $|\delta\tilde{\Phi}_k\rangle$,

$$\langle \delta\tilde{\Phi}_k | \hat{h}_k - \varepsilon_k | \tilde{\Phi}_k \rangle + \sum_{j=1}^{N} [\langle \delta\tilde{\Phi}_k| \otimes \langle \tilde{\Phi}_j | \hat{w}_{k,j} \left[|\tilde{\Phi}_k\rangle \otimes |\tilde{\Phi}_j\rangle - |\tilde{\Phi}_j\rangle \otimes |\tilde{\Phi}_k\rangle \right]] = 0, \tag{13.3.14}$$

where ε_k is a scalar (see Ex. 13.3.2).

Exercise 13.3.1 *The antisymmetrizer $\hat{A}$ is defined in Eq. (13.2.21). (a) Show that $\hat{A} = \hat{A}^\dagger$. (b) Given the symmetry of the many-electron Hamiltonian to permutations (Eqs. 13.2.16, 13.2.17), show that the antisymmetrizer commutes with the Hamiltonian, $[\hat{A}, \hat{H}_{1,\dots i,\dots,j,\dots,N}] = 0$. (c) Show that $\hat{A}^2 = \sqrt{N!}\hat{A}$.*

Exercise 13.3.2 *Given the antisymmetrizer, $\hat{A}$, the single-particle operators in the jth particle subspaces, $\{\hat{h}_j\}$, and the pair interactions $\{\hat{w}_{j,j'}\}$ in the subspace of the jth and j'th particles, prove the following identities:*

(a) $((\langle \tilde{\Phi}_1| \otimes \cdots \otimes \langle \delta\tilde{\Phi}_k| \otimes \cdots \otimes \langle \tilde{\Phi}_N|) \hat{A}(|\tilde{\Phi}_1\rangle \otimes |\tilde{\Phi}_2\rangle \otimes \cdots \otimes |\tilde{\Phi}_N\rangle)) = \frac{1}{\sqrt{N!}} \langle \delta\tilde{\Phi}_k | \tilde{\Phi}_k \rangle$

(b) $((\langle \tilde{\Phi}_1| \otimes \cdots \otimes \langle \delta\tilde{\Phi}_j| \otimes \cdots \otimes \langle \tilde{\Phi}_N|) \sum_{j'=1}^{N} \hat{h}_{j'} \hat{A}(|\tilde{\Phi}_1\rangle \otimes |\tilde{\Phi}_2\rangle \otimes \cdots \otimes |\tilde{\Phi}_N\rangle))$

$$= \frac{1}{\sqrt{N!}} \left[\sum_{\substack{j'=1 \\ (j' \neq j)}}^{N} \langle \tilde{\Phi}_{j'} | \hat{h}_{j'} | \tilde{\Phi}_{j'} \rangle \right] \langle \delta\tilde{\Phi}_j | \tilde{\Phi}_j \rangle + \frac{1}{\sqrt{N!}} \langle \delta\tilde{\Phi}_j | \hat{h}_j | \tilde{\Phi}_j \rangle$$

(c) $((\langle \tilde{\Phi}_1| \otimes \cdots \otimes \langle \delta\tilde{\Phi}_j| \otimes \cdots \otimes \langle \tilde{\Phi}_N|) \sum_{j' > j''=1}^{N} \hat{w}_{j',j''} \hat{A}(|\tilde{\Phi}_1\rangle \otimes |\tilde{\Phi}_2\rangle \otimes \cdots \otimes |\tilde{\Phi}_N\rangle))$

$$= \frac{1}{\sqrt{N!}} \langle \delta\tilde{\Phi}_j | \tilde{\Phi}_j \rangle \left[\sum_{\substack{j' > j''=1 \\ j',j'' \neq j}}^{N} \langle \tilde{\Phi}_{j'}| \otimes \langle \tilde{\Phi}_{j''} | \hat{w}_{j',j''} | \tilde{\Phi}_{j'} \rangle \otimes |\tilde{\Phi}_{j''}\rangle \right.$$

$$\left. - \langle \tilde{\Phi}_{j'}| \otimes \langle \tilde{\Phi}_{j''} | \hat{w}_{j',j''} | \tilde{\Phi}_{j''} \rangle \otimes |\tilde{\Phi}_{j'}\rangle \right] + \frac{1}{\sqrt{N!}} \sum_{j' \neq j=1}^{N} \left[\langle \delta\tilde{\Phi}_j| \otimes \langle \tilde{\Phi}_{j'} | \hat{w}_{j,j'} | \tilde{\Phi}_j \rangle \otimes |\tilde{\Phi}_{j'}\rangle \right.$$

$$\left. - \langle \delta\tilde{\Phi}_j| \otimes \langle \tilde{\Phi}_{j'} | \hat{w}_{j,j'} | \tilde{\Phi}_{j'} \rangle \otimes |\tilde{\Phi}_j\rangle \right].$$

Use the identities, (a), (b), and (c), and Eqs. (13.3.8, 13.3.12) to show that

$$\langle \delta\tilde{\Phi}_k | \hat{h}_k | \tilde{\Phi}_k \rangle + \sum_{j=1}^{N} [\langle \delta\tilde{\Phi}_k| \otimes \langle \tilde{\Phi}_j | \hat{w}_{k,j} [|\tilde{\Phi}_k\rangle \otimes |\tilde{\Phi}_j\rangle - |\tilde{\Phi}_j\rangle \otimes |\tilde{\Phi}_k\rangle]] = \varepsilon_k \langle \delta\tilde{\Phi}_k | \tilde{\Phi}_k \rangle, where,$$

$$\varepsilon_k = \varepsilon - \sum_{j' \neq k = 1}^{N} \langle \tilde{\Phi}_{j'} | \hat{h}_{j'} | \tilde{\Phi}_{j'} \rangle$$

$$- \sum_{\substack{j' > j'' = 1 \\ j', j'' \neq k}}^{N} [\langle \tilde{\Phi}_{j'} | \otimes \langle \tilde{\Phi}_{j''} | \hat{w}_{j,j''} | \tilde{\Phi}_{j'} \rangle \otimes | \tilde{\Phi}_{j''} \rangle - \langle \tilde{\Phi}_{j'} | \otimes \langle \tilde{\Phi}_{j''} | \hat{w}_{j',j''} | \tilde{\Phi}_{j''} \rangle \otimes | \tilde{\Phi}_{j'} \rangle].$$

Each pair interaction, $\hat{w}_{k,j}$, corresponds to an operator in a two-particle Hilbert space. To obtain reduced representations of these operators in a single-particle space, we make use of the following tensor product identities:

$$\langle \delta \tilde{\Phi}_k | \otimes \langle \tilde{\Phi}_j | = \langle \delta \tilde{\Phi}_k | [\hat{I} \otimes \langle \tilde{\Phi}_j |]$$
$$|\tilde{\Phi}_k \rangle \otimes |\tilde{\Phi}_j \rangle = [\hat{I} \otimes |\tilde{\Phi}_j \rangle] |\tilde{\Phi}_k \rangle$$
$$|\tilde{\Phi}_j \rangle \otimes |\tilde{\Phi}_k \rangle = [|\tilde{\Phi}_j \rangle \otimes \hat{I}] |\tilde{\Phi}_k \rangle.$$

Notice that each term in square brackets is a vector in one of the single-particle spaces and an operator in the other single-particle space. Therefore, objects of the form, $[\cdots] \hat{w}_{k,j} [\cdots]$, are operators in a single-particle space. Using these identities in Eq. (13.3.14), we obtain

$$\langle \delta \tilde{\Phi}_k | \{\hat{h}_k - \varepsilon_k + \sum_{j=1}^{N} [\hat{I} \otimes \langle \tilde{\Phi}_j |] \hat{w}_{k,j} [\hat{I} \otimes |\tilde{\Phi}_j \rangle] - [\hat{I} \otimes \langle \tilde{\Phi}_j |] \hat{w}_{k,j} [|\tilde{\Phi}_j \rangle \otimes \hat{I}]\} |\tilde{\Phi}_k \rangle = 0. \quad (13.3.15)$$

We now recall that the single-particle states are spin-orbitals, namely, products of the type

$$|\tilde{\Phi}_j \rangle = |\tilde{\varphi}_j \rangle \otimes |\sigma_{m_{s,j}} \rangle, \quad (13.3.16)$$

where the spin states $(|\sigma_{m_{s,j}} \rangle)$ are fixed to either $|\sigma_{1/2} \rangle$ or $|\sigma_{-1/2} \rangle$ for each j, such that the optimization targets are only the spatial orbitals, $\langle \mathbf{r} | \tilde{\varphi}_j \rangle$. Utilizing the fact that $\hat{h}_k$ and $\hat{w}_{k,j}$ operate only on the spatial orbitals, Eq. (13.3.14) leads to (see Ex. 13.3.3.)

$$\langle \delta \tilde{\varphi}_k | \left\{ \hat{h}_k - \varepsilon_k + \sum_{j=1}^{N} [\hat{I} \otimes \langle \tilde{\varphi}_j |] \hat{w}_{k,j} [\hat{I} \otimes |\tilde{\varphi}_j \rangle] - \delta_{m_{s,j}, m_{s,k}} [\hat{I} \otimes \langle \tilde{\varphi}_j |] \hat{w}_{k,j} [|\tilde{\varphi}_j \rangle \otimes \hat{I}] \right\} |\tilde{\varphi}_k \rangle = 0.$$
$$(13.3.17)$$

Since this condition must hold for any $\langle \delta \tilde{\varphi}_k |$, the optimal set, $\{|\tilde{\varphi}_k \rangle\}$, are solutions of the following nonlinear equations, known as the canonical Hartree–Fock equations [13.6], [13.7],

$$[\hat{h}_k + \hat{J}_k - \hat{K}_k] |\tilde{\varphi}_k \rangle = \varepsilon_k |\tilde{\varphi}_k \rangle. \quad (13.3.18)$$

The operator $\hat{h}_k + \hat{J}_k - \hat{K}_k$ is known as the Fock operator, where the operators $\hat{J}_k$ and $\hat{K}_k$ are the "mean-field" operators that capture the electron–electron interactions within the single-particle subspaces,

$$\hat{J}_k = \sum_{j=1}^{N} [\hat{I} \otimes \langle \tilde{\varphi}_j |] \hat{w}_{k,j} [\hat{I} \otimes |\tilde{\varphi}_j \rangle] \quad (13.3.19)$$

$$\hat{K}_k = \sum_{j=1}^{N} \delta_{m_{s,j},m_{s,k}} [\hat{I} \otimes \langle \tilde{\varphi}_j |] \hat{w}_{k,j} [|\tilde{\varphi}_j\rangle \otimes \hat{I}]. \tag{13.3.20}$$

Notice that $\hat{K}_k$ depends explicitly on the spin states $(m_{s,k})$ associated with the kth orbital, $|\tilde{\varphi}_k\rangle$. Therefore, the Fock operator is in principle different for orbitals associated with the two different spin states, $m_{s,k} = 1/2$ and $m_{s,k} = -1/2$. (Things are simplified, e.g., in cases where the number of electrons is even, and each spatial orbital is populated twice, with two different spin states. In this case, the Fock operator is identical for the two spin states of each spatial orbital.)

Introducing the identity operator in the single-particle space, using the coordinate representation, $\hat{I} = \int d\mathbf{r} |\phi_\mathbf{r}\rangle \langle \phi_\mathbf{r}|$ (see Eq. (11.6.6)), the Coulomb interaction in the two-particle spaces obtains a diagonal form (Ex. 13.3.4),

$$\hat{w}_{k,j} = \int d\mathbf{r} \int d\mathbf{r}' \frac{Ke^2}{|\mathbf{r}-\mathbf{r}'|} [|\phi_\mathbf{r}\rangle \langle \phi_\mathbf{r}|]_k \otimes [|\phi_{\mathbf{r}'}\rangle \langle \phi_{\mathbf{r}'}|]_j, \tag{13.3.21}$$

where the square brackets are added to specify products between different single-particle spaces. The operators $\hat{J}_k$ and $\hat{K}_k$ therefore obtain the explicit forms (Ex. 13.3.5)

$$\hat{J}_k = \int d\mathbf{r}' \frac{Ke^2}{|\hat{\mathbf{r}}-\mathbf{r}'|} \sum_{j=1}^{N} |\tilde{\varphi}_j(\mathbf{r}')|^2 \tag{13.3.22}$$

$$\hat{K}_k = \int d\mathbf{r} \int d\mathbf{r}' \sum_{j=1}^{N} \delta_{m_{s,j},m_{s,k}} \tilde{\varphi}_j^*(\mathbf{r}') \tilde{\varphi}_j(\mathbf{r}) \frac{Ke^2}{|\mathbf{r}-\mathbf{r}'|} |\phi_\mathbf{r}\rangle \langle \phi_{\mathbf{r}'}|. \tag{13.3.23}$$

Exercise 13.3.3 *Use the product form of each spin-orbital (Eq. (13.3.16)) to obtain Eq. (13.3.17) from Eq. (13.3.15).*

Exercise 13.3.4 *The operator $\hat{w}_{k,j}$ is a two-particle operator confined to the subspace of the kth and jth particles, which is diagonal in the two-particle coordinate representation, namely*

$$[\langle \phi_\mathbf{r}|]_k \otimes [\langle \phi_{\mathbf{r}'}|]_j \cdot \hat{w}_{k,j} \cdot [|\phi_{\mathbf{r}''}\rangle]_k \otimes [|\phi_{\mathbf{r}'''}\rangle]_j = \frac{Ke^2}{|\mathbf{r}-\mathbf{r}'|} \delta(\mathbf{r}-\mathbf{r}'') \delta(\mathbf{r}'-\mathbf{r}''').$$

Introducing identity operators in the corresponding single-particle subspaces, $[\hat{I}_\mathbf{r}]_k \otimes [\hat{I}_\mathbf{r}]_j \cdot \hat{w}_{k,j} \cdot [\hat{I}_\mathbf{r}]_k \otimes [\hat{I}_\mathbf{r}]_j$, derive Eq. (13.3.21) for $\hat{w}_{k,j}$.

Exercise 13.3.5 *The single-particle operators, $\hat{J}_k$ and $\hat{K}_k$, are defined in Eq. (13.3.19) and Eq. (13.3.20), respectively. Use Eq. (13.3.21) to derive the explicit coordinate representations of these operators, (Eq. (13.3.22) and Eq. (13.3.23)), respectively.*

The operator $\hat{J}_k$ is the Coulomb operator, encountered already within the Hartree product approximation (See Eq. (12.3.54)). As one can see, this operator corresponds to a mean-field obtained by averaging the repulsive interaction between the kth electron and the probability density function, $\rho(\mathbf{r}') = \sum_{j=1}^{N} |\tilde{\varphi}_j(\mathbf{r}')|^2$, derived from the spatial orbitals of all the electrons in the system. One can notice that the kth electron is included in the sum over j, where a so-called "self-interaction" of an electron with itself appears to be overcounted. Nevertheless, in Eq. (13.3.18) the contribution to $\hat{J}_k$ associated with $j = k$ is canceled out by an identical contribution, but at opposite sign, to the operator $\hat{K}_k$ (see Eq. (13.3.24)). Hence, the Hartree–Fock equations are free of the self-interaction.

The operator $\hat{K}_k$ is the electronic exchange operator. As we can see, it operates on single-particle states, but transfers states from all the jth spaces into the kth space. Moreover, as a single-particle operator, it is nonlocal, that is, non-diagonal in the coordinate representation, hence its operation on an orbital induces delocalization in the coordinate space (which is reminiscent of the effect of a kinetic energy operator). While there is no classical analog for the exchange operator, it can be associated with a position uncertainty induced by the fact that each electron is simultaneously "shared" by all the populated orbitals.

We now return to Eq. (13.3.18), invoking the coordinate representation. Recalling the explicit form of the single-electron operators, $\langle \phi_{\mathbf{r}} | \hat{h}_k | \tilde{\varphi}_k \rangle = \left[\frac{-\hbar^2}{2m_e} \Delta_{\mathbf{r}} + \frac{-KZe^2}{|\mathbf{r}|} \right] \tilde{\varphi}_k(\mathbf{r})$ (see Eqs. (11.6.42, 11.6.47)), we substitute Eqs. (13.3.22, 13.3.23) in Eq. (13.3.18) to obtain the Hartree–Fock equations in the coordinate representation,

$$
\begin{aligned}
&\left[\frac{-\hbar^2}{2m_e} \Delta_{\mathbf{r}} + \frac{-KZe^2}{|\mathbf{r}|} \right] \tilde{\varphi}_k(\mathbf{r}) + \sum_{j=1}^{N} \int d\mathbf{r}' \frac{Ke^2}{|\mathbf{r}-\mathbf{r}'|} |\tilde{\varphi}_j(\mathbf{r}')|^2 \tilde{\varphi}_k(\mathbf{r}) \\
&- \sum_{j=1}^{N} \delta_{m_{s,j}, m_{s,k}} \int d\mathbf{r}' \tilde{\varphi}_j^*(\mathbf{r}') \tilde{\varphi}_j(\mathbf{r}) \frac{Ke^2}{|\mathbf{r}-\mathbf{r}'|} \tilde{\varphi}_k(\mathbf{r}') = \varepsilon_k \tilde{\varphi}_k(\mathbf{r})
\end{aligned} \qquad (13.3.24)
$$

The optimal spatial orbitals, $\{\tilde{\varphi}_k(\mathbf{r})\}$, are shown to be the eigenfunctions of an integro-differential operator (the Fock operator in the coordinate representation), where each spin orbital is associated with a corresponding orbital energy, ε_k. Importantly, the equation is nonlinear, where, by its definition, the Fock operator requires knowledge of the set of N spin-orbitals populated by N electrons of the system. The solution of the Hartree–Fock equations is therefore carried out iteratively. An initial guess is invoked for the set of orbitals, and a new set is obtained by solving the integro-differential equations. This process is repeated until the orbitals, orbital energies, and mean-field operator are self-consistently reproduced.

In practice, there are several different implementations of the Hartree–Fock method for calculating the optimal orbitals. The interested reader is directed to an extensive complementary literature on this subject. Here we briefly mention that on top of the orthogonality condition, Eq. (13.3.5), additional constraints are imposed while solving Eq. (13.3.24). Spherical averaging of the mean-field operators results in a centrally symmetric Fock operator, which means that the angular parts of the atomic orbitals in the many-electron atom are eigenvalues of the electron's angular momentum operator, $\hat{L}^2$, as in any Schrödinger equation with a central potential (see Section 9.1). Consequently, *the angular dependence of the spatial atomic orbitals in a many-electron atom is the same as in a hydrogen-like atom*. Additionally, *the angular momentum quantum numbers are affecting also the radial dependence of the orbitals*, via the centrifugal single-electron potential (Eq. (10.2.8)). The index "j," which characterizes each spin-orbital, stands for a set of four quantum numbers, $j \leftrightarrow (n, l, m, m_s)$, where

$$|\tilde{\Phi}_j\rangle = |\tilde{\varphi}_{n,l,m}\rangle \otimes |\sigma_{m_s}\rangle \quad ; \quad \langle \mathbf{r}|\tilde{\varphi}_{n,l,m}\rangle = R_{n,l}^{(Z,N)}(r)Y_{l,m}(\theta,\varphi). \tag{13.3.25}$$

The radial part, $R_{n,l}^{(Z,N)}(r)$, depends on the identity of the atom, namely the number of protons in the nucleus, Z, and the number of electrons, N. Usually, only the radial part is variationally optimized. While optimizing the orbitals, we can impose that each spatial orbital is populated twice with two different spin states (for an even N, in a so-called, "closed shell" atom). In this case, the method is referred to as "Restricted Hartree–Fock" [13.7]. Alternatively, this restriction can be relaxed. For practical purposes, the trial radial orbitals are often expanded in a set of "primitive" basis functions. The orbital optimization amounts to optimizing the expansion coefficients, via the method of linear variation (see Section 12.3), namely by solving iteratively an algebraic generalized eigenvalue problem in which the Fock operator replaces the Hamiltonian.

The iterative procedure for solving the Hartree–Fock equations seems similar in nature to the SCF solution of the Hartree equation (Eq. (12.3.54)). However, the latter ignores the requirement on the many-electron wave function to be antisymmetric with respect to pair permutation and therefore, does not constrain the number of electrons populating each spin-orbital. In contrast, when the Hartree–Fock equations are solved, the N electrons are populated in N different (orthogonal) spin orbitals, such that the many-electron wave function (the slater determinant, Eq. (13.3.2)) remains proper and does not vanish identically. Notice that in principle, the Hermitian Fock operator has an infinite number of eigenstates. The selection of specific eigenstates of the Fock operator to be populated by the electrons in the many-electron ground state is guided by the criterion of minimizing the variational energy (Eq. (12.3.8)),

$$\varepsilon = \langle \tilde{\Psi}|\hat{H}|\tilde{\Psi}\rangle, \tag{13.3.26}$$

where $|\tilde{\Psi}\rangle$ is the normalized slater determinant associated with N selected spin-orbitals,

$$|\tilde{\Psi}\rangle = \hat{A}|\tilde{\Phi}_1\rangle \otimes |\tilde{\Phi}_2\rangle \otimes \cdots \otimes |\tilde{\Phi}_N\rangle. \tag{13.3.27}$$

Substituting the Hamiltonian, Eq. (13.3.8), one can readily express the variational energy (or, the "Hartree–Fock energy") in terms of the optimal orbitals (see Ex. 13.3.6),

$$\varepsilon = \sum_{k=1}^{N} \langle \tilde{\varphi}_k|\hat{h}_k|\tilde{\varphi}_k\rangle + \frac{1}{2}\sum_{k=1}^{N}\sum_{j=1}^{N} \langle \tilde{\varphi}_k| \otimes \langle \tilde{\varphi}_j|\hat{w}_{k,j}[|\tilde{\varphi}_k\rangle \otimes |\tilde{\varphi}_j\rangle - \delta_{m_{s,j},m_{s,k}}|\tilde{\varphi}_j\rangle \otimes |\tilde{\varphi}_k\rangle]. \tag{13.3.28}$$

Exercise 13.3.6 *Using the properties of the antisymmetrizer (Eq. (13.3.10), and $\hat{A}^2 = \sqrt{N!}\hat{A}$), derive Eq. (13.3.28).*

Importantly, the Hartree–Fock energy is different from a sum of the populated orbital energies. The latter are obtained by taking the expectation value of the Fock operator (Eq. (13.3.18)), with respect to each specific orbital,

$$\varepsilon_k = \langle \tilde{\varphi}_k|[\hat{h}_k + \hat{J}_k - \hat{K}_k]|\tilde{\varphi}_k\rangle$$

$$= \langle \tilde{\varphi}_k|\hat{h}_k|\tilde{\varphi}_k\rangle + \sum_{j=1}^{N} \langle \tilde{\varphi}_k| \otimes \langle \tilde{\varphi}_j|\hat{w}_{k,j}[|\tilde{\varphi}_k\rangle \otimes |\tilde{\varphi}_j\rangle - \delta_{m_{s,j},m_{s,k}}|\tilde{\varphi}_j\rangle \otimes |\tilde{\varphi}_k\rangle]', \tag{13.3.29}$$

where the sum of the occupied orbital energies reads

$$\sum_{k=1}^{N} \varepsilon_k = \sum_{k=1}^{N} \langle \tilde{\varphi}_k | \hat{h}_k | \tilde{\varphi}_k \rangle + \sum_{k=1}^{N} \sum_{j=1}^{N} \langle \tilde{\varphi}_k | \otimes \langle \tilde{\varphi}_j | \hat{w}_{k,j} [| \tilde{\varphi}_k \rangle \otimes | \tilde{\varphi}_j \rangle - \delta_{m_{s,j}, m_{s,k}} | \tilde{\varphi}_j \rangle \otimes | \tilde{\varphi}_k \rangle].$$

(13.3.30)

Subtracting Eq. (13.3.28) from Eq. (13.3.30), the difference between the variational energy and the sum over orbital energies reads

$$\sum_{k=1}^{N} \varepsilon_k - \varepsilon = \frac{1}{2} \sum_{k=1}^{N} \sum_{j=1}^{N} \langle \tilde{\varphi}_k | \otimes \langle \tilde{\varphi}_j | \hat{w}_{k,j} [| \tilde{\varphi}_k \rangle \otimes | \tilde{\varphi}_j \rangle - \delta_{m_{s,j}, m_{s,k}} | \tilde{\varphi}_j \rangle \otimes | \tilde{\varphi}_k \rangle].$$
(13.3.31)

Indeed, the kth orbital energy (Eq. (13.3.18)) includes the Coulomb and exchange pair interactions between this orbital and any other orbital. When summing up all the orbital energies, each specific pair interaction appears twice, since it appears in the calculation of two of the orbital energies. Therefore, the overcounted interaction energy needs to be subtracted from the sum of orbital energies, to relate correctly to the variational energy. Remarkably, however, when it comes to selecting the N orbitals (eigenstates of the Fock operator) for which the Hartree–Fock energy is minimal, it is usually sufficient to minimize the corresponding sum over orbital energies. This is known as *the "Aufbau principle": in the ground state of an atom, the N electrons usually occupy the N spin orbitals corresponding to the lowest possible sum of orbital energies*. Put differently, the orbitals are "filled" from the lowest orbital energy and upwards, subject to the Pauli's exclusion principle, namely, up to two electrons of different spin states per each spatial orbital. The success of the orbital energies in predicting the electronic population in the ground state relies on the fact that the energy differences between nearby orbital energies are typically large in magnitude in comparison to the overcounted interaction energy (Eq. (13.3.31)), such that the sum of orbital energies dominates the Hartree–Fock energy. Indeed, in cases where the orbital energies get close to each other (examples are discussed in what follows), the overcounted interaction energy may become dominant. Orbital populations then deviate from the Aufbau principle, and only a careful calculation of the ground state energy can determine the set of spin orbitals to be populated.

As it turns out, the mean-field approximation that leads to the Hartree–Fock approximation seems to provide a reliable picture with respect to the electronic structure of many-electron atoms at their ground state. Other mean-field approaches, such as the Thomas–Fermi (TF) model [13.8] and the density functional theory [13.9], provide similar qualitative trends, but much of the understanding of the electronic structure, and particularly the orbital approximation, is naturally related to the Hartree–Fock approximation. A rigorous quantitative description of the ground-state energy and the many-electron wave function requires, however, "post mean-field" treatments. The latter are based either on perturbation theory, where the difference between the full Hamiltonian and the Fock operator is regarded as a perturbation, or on the linear variation approach, where the exact wave function is approximated as a linear combination of different determinants, each corresponding to a different selection of N

spin-orbitals in which the N electrons are populated (configuration-interaction and coupled cluster approaches [13.10]).

The Periodic Table of the Elements

We are now ready for a comparative review of the structure of different many-electron atoms, within the realm of the orbital (Hartree–Fock) approximation, the Pauli exclusion principle, and the Aufbau principle as introduced in the preceding discussion. We recall that this idealized picture ignores relativistic corrections, which become important for the "heavier" atoms, associated with large atomic numbers ($> \sim 100$). Let us discuss several principles that apply to all the atoms within this framework.

I. *The many-electron wave function at the atomic ground state is approximated as a Slater-determinant*, where the N electrons are populated in N spin-orbitals that minimize the Hartree–Fock energy.

II. *Each spin-orbital is a product $\varphi(r, \theta, \varphi) \otimes |\sigma_{m_s}\rangle$, where $\varphi(r, \theta, \varphi)$ is the spatial orbital and $|\sigma_{m_s}\rangle$ is the electronic spin state*, associated with the quantum number, m_s, which obtains one of the two values, $m_s = -1/2, 1/2$.

III. The spatial orbitals obtain the form $\varphi(r, \theta, \varphi) = Y_{l,m}(\theta, \varphi) R_{n,l}^{(Z,N)}(r)$. The quantum numbers (l, m) are constrained, as in the case of the hydrogen-like atom, namely $m = -l, -l+1, \ldots, 0 \ldots, l-1, l$. *Given the quantum numbers, l and m, the orbital's angular dependence is "universal" for all the atoms.* (In case of degeneracy with respect to l, a set of degenerate functions $\{Y_{l,m}(\theta, \varphi)\}$ can be replaced by an orthonormal set of their linear combinations.)

IV. The radial wave functions depend on the angular quantum, l, and on a "main" quantum number, n, as in the hydrogen-like atom, where $n = l+1, l+2$, These radial functions are not "universal": *Different atoms, associated with different numbers of protons (Z) and/or electrons (N), would be associated with different radial functions, $R_{n,l}^{(Z,N)}(r)$, for the same n and l.*

V. *Each orbital corresponds to an orbital-energy, $\varepsilon_{n,l}^{(Z,N)}$*. While for a hydrogen-like atom the orbitals associated with different angular quantum numbers (l) are degenerate, *in a many-electron atom the orbital energies depend on both n and l*. Indeed, solutions to the radial Schrödinger equation, which correspond to different l values, are degenerate when the electron is subject to the bare atomic nucleus (see Eq. (10.2.8)). This degeneracy is removed when the hydrogen-like Hamiltonian is replaced by the Fock operator (Eq. (13.3.24)), in which the presence of many electrons is taken into account in terms of additional mean-field operators. This effect is schematically represented in Fig. 13.3.1, where the energy levels associated with a quantum number n in the hydrogen-like atom are shown to split into groups of energy levels (often termed energy shells) in a many-electron atom. The values of the orbital energies $\{\varepsilon_{n,l}^{(Z,N)}\}$ change from one atom to another, as illustrated schematically in Fig. 13.3.2. As the atomic number (Z) increases, the orbital energy, $\varepsilon_{n,l}^{(Z,N)}$, tends to decrease in response to the increasing attractive positive nucleus charge (which is not fully compensated for by the corresponding addition of electrons). Importantly, the effect is different for different orbitals, since the increase in the number of electrons changes the repulsive mean-field in a nontrivial way. Notice particularly that even the order of the atomic orbitals according to their energy may differ between different atoms. (For example, the $4s$ ($n = 4$, $l = 0$) and $3d$ ($n = 3$, $l = 2$) orbital energies cross each other, as Z is changed.) Nevertheless, *for a fixed quantum number, n (namely, within a given energy shell), the orbital energy increases with increasing angular quantum number, l.* This observed trend can be qualitatively explained by combining classical electrostatic arguments with our earlier acquaintance with the model of the hydrogen-like atom. First, we recall that as in classical electrostatics, the typical distance between the electron and the nucleus in a hydrogen-like atom increases with increasing orbital energy (see Fig. 10.3.3). Assuming this trend to hold also for many-electron atoms, we can conclude that electrons associated with internal energy shells (smaller n values) are closer to the nucleus (in a probabilistic, quantum mechanical, sense) than electrons associated with external energy shells. Consequently, electrons in orbitals associated with a given quantum number n experience the positive nucleus charge as well as a mean-field owing to the electrons in the inner shell orbitals. This repulsive mean-field effectively screens the positive nucleus charge, resulting in a less positive "effective core charge." Comparing now hydrogen-like orbitals associated with the same quantum number n, but with different quantum numbers l (Fig. 10.3.4), we notice that in the hydrogen-like atom, lower l values are associated with a larger probability density at distances in the vicinity of the nucleus. If this is the case also in a many-electron atom, it means that electrons associated with lower l values have increasing probability of "penetrating" beyond the "screen" of the inner shell electrons, into the vicinity of the nucleus. Consequently, electrons in orbitals of lower l values experience a more positive (attractive) effective core charge, and the corresponding orbital energy is accordingly lower. For example, in Fig. 13.3.2, one can see that in an atom like Ca, the energy of the $4s$ orbital ($l = 0$) drops below that of the $3d$ orbital ($l = 2$), although in terms of the main quantum number (n) $4s$ belongs to a higher energy shell. The relatively low energy of the $4s$ orbital can be attributed to the "penetration" of this orbital to the vicinity of the nucleus.

VI. *The ground state of each neutral atom is approximated by an "electronic configuration," which represents a single Slater determinant that minimizes the many-electron energy.* Examples of the electronic configuration for some elements and their correspondence to a Slater determinant are presented in Figs. 13.3.3 and 13.3.4. Usually, the orbitals are populated from lower to higher orbital energy (*the Aufbau principle*), where each spatial orbital can be populated by up to two electrons (*Pauli's principle*). In some exceptional cases (see, e.g., the example of scandium (Sc), with $Z = 21$), the requirement of minimizing the many-electron energy does not comply with the Aufbau principle, as the difference between nearby orbital energies becomes small with respect to the overcounted interaction energy (see Eq. (13.3.31)). When an orbital energy is degenerate, namely, when there is more

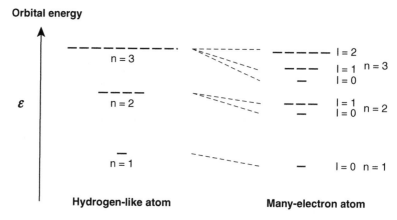

Figure 13.3.1 A schematic representation of the orbital energies and their degeneracies in a hydrogen-like atom and in many-electron atoms. Each dark horizontal line represents an orbital. In a hydrogen-like atom, all the orbitals associated with a given n are degenerate. The degeneracy is removed by electron–electron interactions in many-electron atoms, where the orbital energy depends on both quantum numbers, n and l.

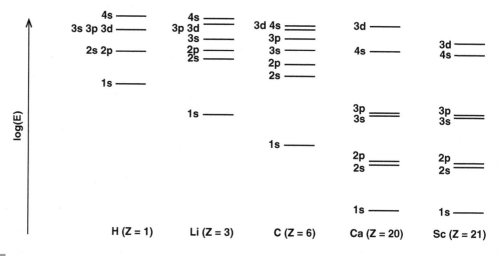

Figure 13.3.2 A schematic representation of the changes in the orbital energies for the lowest-energy orbitals: $1s, 2s, 2p, 3s, 3p, 3d, 4s$, and in their order, for a few selected atoms corresponding to the atomic numbers 1, 3, 6, 20, 21.

than one spatial orbital to be populated at the same orbital energy, the lowest energy state would correspond to population of one electron at each spatial orbital, at the same electronic spin state. An example for the case of carbon is illustrated in Fig. 13.3.5. This principle (**Hund's rule**, after Friedrich Hund) is attributed to the exchange interaction between electrons of the same spin, which lowers the many-particle energy, when the electrons are delocalized among the degenerate orbitals. This effect will be discussed in the next section.

Z	Symbol	Electronic Configuration
1	H	$1s^1$
2	He	$1s^2$
3	Li	$1s^2\,2s^1$
4	Be	$1s^2\,2s^2$
5	B	$1s^2\,2s^2\,2p^1$
6	C	$1s^2\,2s^2\,2p^2$
7	N	$1s^2\,2s^2\,2p^3$
8	O	$1s^2\,2s^2\,2p^4$
9	F	$1s^2\,2s^2\,2p^5$
10	Ne	$1s^2\,2s^2\,2p^6$
11	Na	$1s^2\,2s^2\,2p^6\,3s^1$
⋮	⋮	⋮
20	Ca	$1s^2\,2s^2\,2p^6\,3s^2\,3p^6\,4s^2$
21	Sc	$1s^2\,2s^2\,2p^6\,3s^2\,3p^6\,4d^1\,4s^2$

Figure 13.3.3 The electronic configuration of some selected neutral atoms of different elements.

When it comes to chemical reactivity of atoms, their tendency to form chemical bonds (see Chapter 14) or to exchange electrons with their surroundings, the populated atomic orbitals at the outer shell are the most relevant. According to the previously outlined principles of electronic population of the atomic orbitals, which apply to all atoms in their ground electronic state, each energy shell can accommodate a finite number of electrons (2 for n = 1, 8 for n = 2, 18 for n = 3, etc.). As the atomic number increases from one atom to another, fully occupied shells get "closed" and new shells "open". Consequently, different atoms are associated with similar electronic population in their outer shell orbitals. Remarkably, atoms of elements associated with similar outer shell electronic configuration show similarity in their properties. This is the basis for the observed "periodic" trends observed in the properties of different elements along the

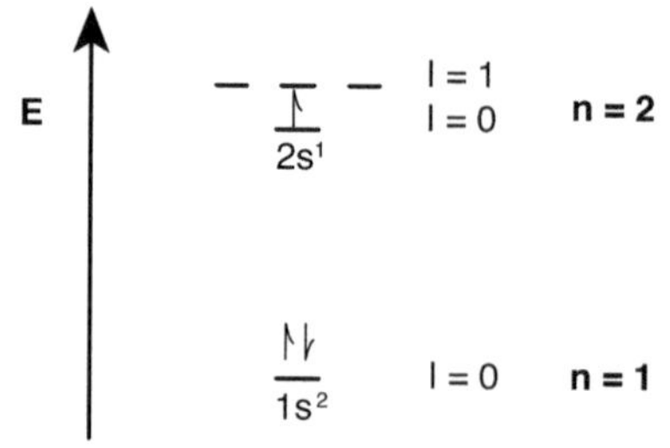

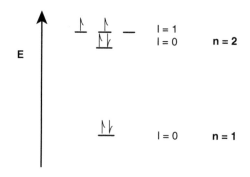

$$\psi(1,2,3) = \frac{1}{\sqrt{3!}} \begin{vmatrix} Y_{0,0}(\theta_1,\varphi_1)R_{1,0}^{(3,3)}(r_1)\alpha(1) & Y_{0,0}(\theta_1,\varphi_1)R_{1,0}^{(3,3)}(r_1)\beta(1) & Y_{0,0}(\theta_1,\varphi_1)R_{2,0}^{(3,3)}(r_1)\alpha(1) \\ Y_{0,0}(\theta_2,\varphi_2)R_{1,0}^{(3,3)}(r_2)\alpha(2) & Y_{0,0}(\theta_2,\varphi_2)R_{1,0}^{(3,3)}(r_2)\beta(2) & Y_{0,0}(\theta_2,\varphi_2)R_{2,0}^{(3,3)}(r_2)\alpha(2) \\ Y_{0,0}(\theta_3,\varphi_3)R_{1,0}^{(3,3)}(r_3)\alpha(3) & Y_{0,0}(\theta_3,\varphi_3)R_{1,0}^{(3,3)}(r_3)\beta(3) & Y_{0,0}(\theta_3,\varphi_3)R_{2,0}^{(3,3)}(r_3)\alpha(3) \end{vmatrix}$$

Figure 13.3.4 A detailed view of the electronic configuration of Li $(Z = N = 3)$, $1s^2 2s^1$, and the Slater determinant it corresponds to. The different spin states are denoted as $\langle i | \sigma_{1/2} \rangle = \alpha(i)$ and $\langle i | \sigma_{-1/2} \rangle = \beta(i)$, and are conventionally marked by half arrows up and down, respectively.

Figure 13.3.5 A detailed view of the electronic configuration of a carbon atom $(C, Z = N = 6)$, namely, $1s^2 2s^2 2p^2$. The higher occupied atomic orbitals are degenerate p-type orbitals, where the minimal energy determinant corresponds to two electrons with the same spin state, singly occupying two different orbitals. Notice that different determinants corresponding to different choices of occupying two out of the three p orbitals are all associated with the same variational (Hartree–Fock) energy.

periodic table of the elements. Consider, for example, the group of elements known as the alkali metals. They are all highly reactive metals, associated with relatively low first ionization potential and relatively small atomic radii. Inspecting their electronic configuration reveals that all neutral alkali atoms have a single electron at their external shell s-type orbital, for example, Li: $[1s^2]2s^1$, Na: $[1s^2 2s^2 2p^6]3s^1$, K: $[1s^2 2s^2 2p^6 3s^2 3p^6]4s^1$, and so on. Similarly, the group of alkaline earth metals (Be, Mg, Ca, Sr, etc.) are all associated with an outer shell electronic configuration of ns^2. The halogens group (F, Cl, Br, I, etc.) are nonmetals, associated with the outer shell configuration, ns^{2}np^5 The

1 H																	2 He
3 Li	4 Be											5 B	6 C	7 N	8 O	9 F	10 Ne
11 Na	12 Mg											13 Ai	14 Si	15 P	16 S	17 Ci	18 Ar
19 K	20 Ca	21 Sc	22 Ti	23 V	24 Cr	25 Mn	26 Fe	27 Co	28 Ni	29 Cu	30 Zn	31 Ga	32 Ge	33 As	34 Se	35 Br	36 Kr
37 Rb	38 Sr	39 Y	40 Zr	41 Nb	42 Mo	43 Tc	44 Ru	45 Rh	46 Pd	47 Ag	48 Cd	49 In	50 Sn	51 Sb	52 Te	53 I	54 Xe
55 Cs	56 Ba	-- 71 Lu	72 Hf	73 Ta	74 W	75 Re	76 Os	77 Ir	78 Pt	79 Au	80 Hg	81 Ti	82 Pb	83 Bi	84 Po	85 At	86 Rn
87 Fr	88 Ra	-- 103 Lr	104 Rf	105 Db	106 Sg	107 Bh	108 Hs	109 Mt	110 Ds	111 Rg	112 Cn	113 Uut	114 Uuq	115 Uup	116 Uuh	117 Uus	118 Uuo

57 La	58 Ce	59 Pr	60 Nd	61 Pm	62 Sm	63 Eu	64 Gd	65 Tb	66 Dy	67 Ho	68 Er	69 Tm	70 Yb
89 Ac	90 Th	91 Pa	92 U	93 Np	94 Pu	95 Am	96 Cm	97 Bk	98 Cf	99 Es	100 Fm	101 Md	102 No

Figure 13.3.6   The periodic table of the elements

noble gas atoms (He, Ne, Ar, Kr, etc.) are all associated with the outer shell configuration, $ns^2 np^6$, and so on. Each such group (or "family") of elements corresponds to a column in the periodic table of the elements. The members within each family have the same outer shell electronic configuration in their neutral state, but they differ in the number of filled energy shells. Consequently, their atomic numbers (and the numbers of electrons) differ by electron numbers corresponding to occupation of an entire shell. These numbers can take the values, $2n^2 = 2, 8, 18, 32, \ldots$, which indeed correspond to the observed "period lengths" in the periodic table of the elements (Fig. 13.3.6). It is remarkable that these numbers are related to the constraints set on the angular and magnetic quantum numbers to obtain proper solutions to the Schrödinger equation for a single-electron atom.

13.4 Two-Electron Spin-Orbital Functions – Singlet and Triplet States

When discussing the spin of a single electron, it was emphasized that in nonrelativistic quantum mechanics, the space of physical states is a tensor product of the spin and coordinate spaces. This idea was adopted within the concept of spin-orbitals for many-electron system, in the sense that each spin-orbital is a product of spatial and spin states. Nevertheless, while a Hartree product of spin orbitals can always be factorized into a single product of a many-electron spatial function (wave function) and a many-electron spin function, the antisymmetrizer operation on a Hartree product generates a superposition of such products, in which the many-electron spin and orbital states may become entangled; namely, the many-electron state cannot be factorized into a single product of spin and spatial states. Such determinants do comply with being antisymmetric under permutation of electrons between spin-orbitals, but they include "artificial correlation" (at least within nonrelativistic quantum mechanics) between the spin and spatial coordinates. Even an approximate nonrelativistic description of the physical states of many-electron systems must therefore extend beyond the realm of single determinants. In this section we focus on systems of two electrons, introducing the corresponding product states between two-electron "spin functions" (known as the singlet and triplet states) and spatial two-electron functions, while emphasizing their relation to Slater determinants. Generalization to systems of N electrons will be discussed only qualitatively. A quantitative analysis of many-electron spin functions can be found elsewhere [13.11].

Let us consider a system of two electrons (a "helium-like" atom). The single-electron states are spin orbitals, $|\Phi_j\rangle = |\varphi_j\rangle \otimes |\sigma_j\rangle$, where in the coordinate representation (Eq. (13.3.1)), each spin orbital has the form

$$\langle i|\Phi_j\rangle \equiv \varphi_j(\mathbf{r}_i) \cdot \sigma_j(i). \tag{13.4.1}$$

We now focus on a minimal variational space, in which the number of spatial orbitals is limited to two. These spatial-orbitals are denoted as $\varphi_1(\mathbf{r}_i)$ and $\varphi_2(\mathbf{r}_i)$, where the spin state per electron is either $\alpha(i)$ or $\beta(i)$, such that there are four different spin-orbitals in total:

$$\varphi_1(\mathbf{r}_i)\alpha(i)$$
$$\varphi_2(\mathbf{r}_i)\alpha(i)$$
$$\varphi_1(\mathbf{r}_i)\beta(i)$$
$$\varphi_2(\mathbf{r}_i)\beta(i).$$

Since only two out of the four spin-orbitals are populated in a two-electron system, six ($\binom{4}{2}$) different two-electron determinants can be defined (see also Fig. 13.4.1):

$$\begin{aligned}
\Psi_{1\alpha,1\beta} &= \tfrac{1}{\sqrt{2}}\varphi_1(\mathbf{r}_1)\varphi_1(\mathbf{r}_2)[\alpha(1)\beta(2)-\beta(1)\alpha(2)] \\
\Psi_{2\alpha,2\beta} &= \tfrac{1}{\sqrt{2}}\varphi_2(\mathbf{r}_1)\varphi_2(\mathbf{r}_2)[\alpha(1)\beta(2)-\beta(1)\alpha(2)] \\
\Psi_{1\alpha,2\alpha} &= \tfrac{1}{\sqrt{2}}[\varphi_1(\mathbf{r}_1)\varphi_2(\mathbf{r}_2)-\varphi_2(\mathbf{r}_1)\varphi_1(\mathbf{r}_2)]\alpha(1)\alpha(2) \\
\Psi_{1\beta,2\beta} &= \tfrac{1}{\sqrt{2}}[\varphi_1(\mathbf{r}_1)\varphi_2(\mathbf{r}_2)-\varphi_2(\mathbf{r}_1)\varphi_1(\mathbf{r}_2)]\beta(1)\beta(2) \\
\Psi_{1\alpha,2\beta} &= \tfrac{1}{\sqrt{2}}[\varphi_1(\mathbf{r}_1)\varphi_2(\mathbf{r}_2)\alpha(1)\beta(2)-\varphi_2(\mathbf{r}_1)\varphi_1(\mathbf{r}_2)\beta(1)\alpha(2)] \\
\Psi_{1\beta,2\alpha} &= \tfrac{1}{\sqrt{2}}[\varphi_1(\mathbf{r}_1)\varphi_2(\mathbf{r}_2)\beta(1)\alpha(2)-\varphi_2(\mathbf{r}_1)\varphi_1(\mathbf{r}_2)\alpha(1)\beta(2)].
\end{aligned} \tag{13.4.2}$$

We recall that the Hamiltonian for the two-electron atom reads (Eq. (13.3.8))

$$\hat{H} = \hat{h}_1 + \hat{h}_2 + \hat{w}_{1,2}. \tag{13.4.3}$$

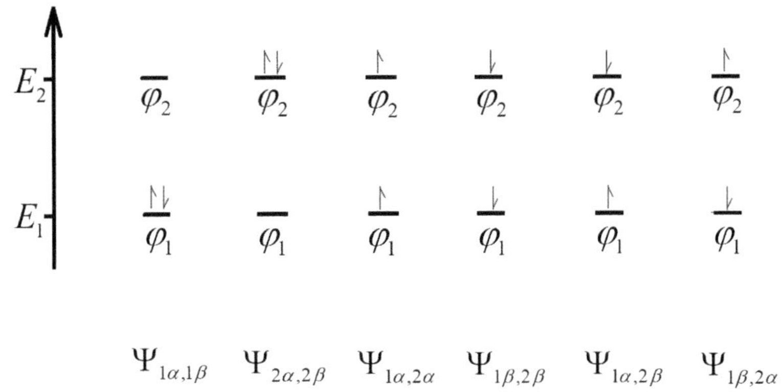

Figure 13.4.1 Six possible Slater determinants for two electrons in a space of two spatial orbitals (φ_1 and φ_2). The electronic spin states are marked by half arrows up and down, for α and β, respectively.

In an atom, $\hat{h}_1$ and $\hat{h}_2$ are the single-electron (hydrogen-like) Hamiltonians, which read in the coordinate representation: $\hat{h}_j \equiv \hat{h}_{\mathbf{r}_j} = \frac{-\hbar^2}{2m_e}\Delta_{\mathbf{r}_j} + \frac{-KZe^2}{|\mathbf{r}_j|}$, where $\hat{w}_{1,2} = \frac{Ke^2}{|\mathbf{r}_1 - \mathbf{r}_2|}$ is the electron–electron interaction. The variational energy associated with each determinant is given as $\varepsilon = \langle \Psi | \hat{H} | \Psi \rangle$, which can be readily expressed as (see Ex. 13.4.1)

$$\varepsilon_{1\alpha,1\beta} = 2E_1 + J_{1,1}$$
$$\varepsilon_{2\alpha,2\beta} = 2E_2 + J_{2,2}$$
$$\varepsilon_{1\alpha,2\alpha} = \varepsilon_{1\beta,2\beta} = E_1 + E_2 + J_{1,2} - K_{1,2} \qquad (13.4.4)$$
$$\varepsilon_{1\alpha,2\beta} = \varepsilon_{1\beta,2\alpha} = E_1 + E_2 + J_{1,2}.$$

Here E_1 and E_2 are integrals corresponding to the single-particle Hamiltonians,

$$E_1 = \int d\mathbf{r}\varphi_1^*(\mathbf{r})\hat{h}_{\mathbf{r}}\varphi_1(\mathbf{r})$$
$$E_2 = \int d\mathbf{r}\varphi_2^*(\mathbf{r})\hat{h}_{\mathbf{r}}\varphi_2(\mathbf{r}) \qquad (13.4.5)$$

$\{J_{i,j}\}$ correspond to the Coulomb integrals,

$$J_{i,j} = Ke^2 \int d\mathbf{r} \int d\mathbf{r}' \frac{|\varphi_i(\mathbf{r})|^2 |\varphi_j(\mathbf{r}')|^2}{|\mathbf{r} - \mathbf{r}'|}, \qquad (13.4.6)$$

and $K_{1,2}$ corresponds to the exchange integrals,

$$K_{1,2} = Ke^2 \int d\mathbf{r} \int d\mathbf{r}' \frac{\varphi_1^*(\mathbf{r})\varphi_2^*(\mathbf{r}')\varphi_2(\mathbf{r})\varphi_1(\mathbf{r}')}{|\mathbf{r} - \mathbf{r}'|}. \qquad (13.4.7)$$

Exercise 13.4.1 *For each determinant in Eq. (13.4.2), (a) Calculate the appropriate energy $\varepsilon = \langle \Psi | \hat{H} | \Psi \rangle$ in Eq. (13.4.4) by expressing it in terms of the integrals given in Eqs. (13.4.5–13.4.7). (b) Calculate the orbital energy for each of the relevant spin orbitals, $\varphi_1(\mathbf{r}_i)\alpha(i)$, $\varphi_2(\mathbf{r}_i)\alpha(i)$ $\varphi_1(\mathbf{r}_i)\beta(i)$, or $\varphi_2(\mathbf{r}_i)\beta(i)$, using Eq. (13.3.29). (c) Verify that the relation between the total energy and the sum over orbital energies (Eq. (13.3.31)) holds.*

In the first two determinants, $\Psi_{1\alpha,1\beta}$ and $\Psi_{2\alpha,2\beta}$ (Eq. (13.4.2)), the two electrons occupy the same spatial orbital with different spin states. Each two-electron determinant is naturally factorized in these cases into a product of two-electron spatial and spin functions, where the spatial function is symmetric and the spin function is antisymmetric to permutation of the electron indexes. One can readily verify that the two-electron spin state, $[\alpha(1)\beta(2) - \beta(1)\alpha(2)]$, is an eigenstate of the two-electron spin operators, $\hat{S}_z = \hat{S}_{z_1} + \hat{S}_{z_2}$ and $\hat{S}^2 = (\hat{S}_{x_1} + \hat{S}_{x_2})^2 + (\hat{S}_{y_1} + \hat{S}_{y_2})^2 + (\hat{S}_{z_1} + \hat{S}_{z_2})^2$, associated with the quantum numbers $m_s = 0$ and $s = 0$, respectively (see Ex. 13.4.2). Such a spin state is termed a singlet state, owing to the single possible projection of the total spin on the z-axis.

When the energy difference between the single-particle energies is large with respect to all the Coulomb and exchange integrals, namely $|E_2 - E_1| >> J_{1,1}, J_{2,2}, J_{1,2}, K_{1,2}$, it is easy to see (Eq. (13.4.4)) that double occupation of the orbital with the lowest single-particle energy, that is, the determinant, $\Psi_{1\alpha,1\beta}$, is the lowest-energy state of the two-particle system in the two-orbital space (see Fig. 13.4.2). Indeed, considering a two-electron atom (He, or He-like atom), the ground state configuration corresponds to $1s^2$, where the orbital of lowest single-particle energy is identified as a $1s$-type function. Notice that the spatial state of the two electrons in the variational ground state of the He atom (a product of two identical orbitals) coincides with the Hartree solution, Eq. (12.3.57), which indeed gives a reasonable approximation of the atom's ground-state energy (see Section 12.3). The spin state of the two electrons, which is missing in the Hartree approximation, has no effect on the ground-state energy. Nevertheless, the antisymmetry of the singlet spin state with respect to permutation of the two electrons is essential. The double occupation of the two electrons in the same (lowest-energy) orbital means that the spatial two-electron state is symmetric with respect to permutation, and the entire state is antisymmetric only due to the spin state. In a broader context, the case of He-like atoms is indicative with respect to the ground state of any atom according to the Aufbau principle: whenever the energy gap to the next orbital energy is large with respect to all the Coulomb and exchange integrals, the electrons will first be paired with different spins in the orbital with the lowest available orbital energy.

Given that the ground state of the system corresponds to $\Psi_{1\alpha,1\beta}$, the next two determinants, $\Psi_{1\alpha,2a}$ and $\Psi_{1\beta,2\beta}$ (see Eq. (13.4.2)), correspond to the first excited electronic state (Fig. 13.4.2). As before, the total energy corresponds to the sum of single-particle energies and the Coulomb interaction term between the two orbitals, but the identical electronic spins in these configurations allow for an additional contribution to the energy, in terms of an exchange interaction term (see Eq. (13.4.4)). Notice that exchange interaction lowers the state energy, owing to its nonlocal nature (see the discussion of the exchange operator, $\hat{K}_k$, in Section 13.3). Like in $\Psi_{1\alpha,1\beta}$ and $\Psi_{2\alpha,2\beta}$, the determinants $\Psi_{1\alpha,2a}$ and $\Psi_{1\beta,2\beta}$ are, in fact, products of two-electron spatial and spin functions, but in this case, the spin function is symmetric and the spatial function is antisymmetric to permutation of the electrons' indexes. The two-electron spin states, $[\alpha(1)\alpha(2)]$ or $[\beta(1)\beta(2)]$, are eigenstates of the two-electron spin operators, $\hat{S}_z$ and

$\hat{S}^2$, corresponding to the total spin quantum numbers, $s = 1$ and $m_s = 1$ or $m_s = -1$, respectively (see Ex. 13.4.2). These spin states are termed triplet states, since the total spin associated with $s = 1$ has three possible projections on the z-axis, $m_s = 0, \pm 1$ (the two-electron spin state associated with $m_s = 0$ is discussed in what follows).

The next two determinants, $\Psi_{1\alpha,2\beta}$ and $\Psi_{1\beta,2\alpha}$, are associated with yet higher energies (see Eq. (13.4.4) and Fig. 13.4.2). Since each spatial orbital is paired with a different electronic spin state in each spin-orbital, the exchange interaction vanishes, and the two-electron energy is consequently higher. Importantly, unlike in the case of the other determinants, $\Psi_{1\alpha,2\beta}$ and $\Psi_{1\beta,2\alpha}$ are not single products of two-electron spatial and spin functions, namely, the spin and spatial states are entangled. Moreover, these two determinants are eigenstates of $\hat{S}_z$, corresponding to $m_s = 0$, but they are not eigenstates of the total spin operator, $\hat{S}^2$, which implies that the two-electron spin is not well-defined. Indeed, while both $\Psi_{1\alpha,2\beta}$ and $\Psi_{1\beta,2\alpha}$ satisfy the requirement for antisymmetry with respect to electron pair permutation, each of them is an incomplete description of the physical state. Given that the two electrons have different spins, there is no way to distinguish which spin is correlated with each one of the orbitals, as suggested by the construction of $\Psi_{1\alpha,2\beta}$ or $\Psi_{1\beta,2\alpha}$. A complete description would give an equal probability to the two possible configurations, namely, it would be a superposition of the two determinants with equal probability weights,

$$\Psi_\pm = \frac{1}{\sqrt{2}}[\Psi_{1\alpha,2\beta} \pm \Psi_{1\beta,2\alpha}]. \tag{13.4.8}$$

(We skip here a detailed explanation for the restriction of the coefficients to ± 1 rather than to any phase. It follows from the Hermiticity of the spin permutation operator, and arguments similar to ones used in the derivation of Eqs. (13.2.5–13.2.11), but for the spin permutation only.)

A proper description of two-electron states for electrons of different spins extends, therefore, beyond the space of single determinants. In this sense, the states Ψ_+ and Ψ_- are "post-Hartree Fock," that is, configuration-interaction approximations to the exact two electron states. We can readily see that the states Ψ_+ and Ψ_- are products of two-electron spatial and spin functions (Ex. 13.4.3),

$$\Psi_\pm = \frac{1}{\sqrt{4}}[\varphi_1(\mathbf{r}_1)\varphi_2(\mathbf{r}_2) \mp \varphi_2(\mathbf{r}_1)\varphi_1(\mathbf{r}_2)][\alpha(1)\beta(2) \pm \beta(1)\alpha(2)], \tag{13.4.9}$$

where in Ψ_+ the spin function is symmetric and the spatial function is antisymmetric with respect to permutation of the electron indexes, and vice versa in Ψ_-. Also here, the two-electron spin states, $[\alpha(1)\beta(2) + \beta(1)\alpha(2)]$ and $[\alpha(1)\beta(2) - \beta(1)\alpha(2)]$, are eigenstates of the two-electron spin operators, $\hat{S}_z$ and $\hat{S}^2$, corresponding to the total spin quantum numbers, $m_s = 0$, $s = 1$, and $m_s = 0$, $s = 0$, respectively (see Ex.13.4.2). Ψ_+ and Ψ_- therefore correspond to triplet and singlet spin states, respectively. The total energy, $\varepsilon = \langle \Psi | \hat{H} | \Psi \rangle$, associated with these two-electron states reads (see Ex. 13.4.4 and Fig. 13.4.2)

$$\begin{aligned}\mathcal{E}_+ &= E_1 + E_2 + J_{1,2} - K_{1,2}\\ \mathcal{E}_- &= E_1 + E_2 + J_{1,2} + K_{1,2}\end{aligned}. \tag{13.4.10}$$

Ψ_+ has the same energy as the other two triplet states, $\Psi_{1\alpha,2\alpha}$ and $\Psi_{1\beta,2\beta}$, which corresponds to the same antisymmetric spatial two-electron function. Ψ_- is associated with a higher energy, which corresponds to the symmetric spatial function, $\frac{1}{\sqrt{2}}[\varphi_1(\mathbf{r}_1)\varphi_2(\mathbf{r}_2) + \varphi_2(\mathbf{r}_1)\varphi_1(\mathbf{r}_2)]$. A qualitative argument attributes the increase in energy to increasing electrostatic repulsion between the electrons, which diverges when the two electrons are positioned at the same location in coordinate space. The probability density for this to occur depends on the spatial wave function, where it is finite in Ψ_- (the singlet state), $|\varphi_1(\mathbf{r}_1)\varphi_2(\mathbf{r}_2) + \varphi_2(\mathbf{r}_1)\varphi_1(\mathbf{r}_2)|^2 \xrightarrow[\mathbf{r}_2 \to \mathbf{r}_1]{} 4|\varphi_2(\mathbf{r}_1)\varphi_1(\mathbf{r}_1)|^2$, but vanishes in the triplet state, $|\varphi_1(\mathbf{r}_1)\varphi_2(\mathbf{r}_2) - \varphi_2(\mathbf{r}_1)\varphi_1(\mathbf{r}_2)|^2 \xrightarrow[\mathbf{r}_2 \to \mathbf{r}_1]{} 0$.

In summary, the four different spin-functions for two electrons are the singlet state,

$$\frac{1}{\sqrt{2}}[\alpha(1)\beta(2) - \beta(1)\alpha(2)],$$

and the triplet states,

$$[\alpha(1)\alpha(2)]$$

$$\frac{1}{\sqrt{2}}[\alpha(1)\beta(2) + \beta(1)\alpha(2)]$$

$$[\beta(1)\beta(2)].$$

The singlet and triplet spin functions are common eigenfunctions of the total two-electron spin operators, $\hat{S}^2$ and $\hat{S}_z$, associated with the quantum numbers, $s = m_s = 0$ for the singlet, and $s = 1, m_s = 1, 0, -1$ for the triplet (Ex. 13.4.2):

$$\hat{S}^2[\alpha(1)\beta(2) - \beta(1)\alpha(2)] = 0\hbar^2[\alpha(1)\beta(2) - \beta(1)\alpha(2)] \tag{13.4.11}$$

$$\hat{S}^2 \left\{ \begin{array}{c} \alpha(1)\alpha(2) \\ \alpha(1)\beta(2) + \beta(1)\alpha(2) \\ \beta(1)\beta(2) \end{array} \right\} = 2\hbar^2 \left\{ \begin{array}{c} \alpha(1)\alpha(2) \\ \alpha(1)\beta(2) + \beta(1)\alpha(2) \\ \beta(1)\beta(2) \end{array} \right\} \tag{13.4.12}$$

$$\hat{S}_z[\alpha(1)\beta(2) - \beta(1)\alpha(2)] = 0\hbar[\alpha(1)\beta(2) - \beta(1)\alpha(2)] \tag{13.4.13}$$

$$\begin{aligned}\hat{S}_z\alpha(1)\alpha(2) &= \hbar\alpha(1)\alpha(2)\\ \hat{S}_z[\alpha(1)\beta(2) + \beta(1)\alpha(2)] &= 0\\ \hat{S}_z\beta(1)\beta(2) &= -\hbar\beta(1)\beta(2).\end{aligned} \tag{13.4.14}$$

When two electrons occupy a single spatial orbital, the spin state is a singlet, and the two-electron state corresponds to $\Psi_{1\alpha,1\beta}$ (or $\Psi_{2\alpha,2\beta}$), see Eq. (13.4.2) and Fig. 13.4.1. The antisymmetry to permutation is attributed to the spin function in this case.

When two electrons occupy two different spatial orbitals, there are four proper two-electron states. The three lower-energy states correspond to triplet states, associated with an antisymmetric spatial function and a symmetric spin function with respect to permutation of the electron indexes,

$$\left\{ \begin{array}{c} \Psi_{1\alpha,2\alpha} \\ \Psi_{+} \\ \Psi_{1\beta,2\beta} \end{array} \right\} = \frac{1}{\sqrt{2}} [\varphi_1(\mathbf{r}_1)\varphi_2(\mathbf{r}_2) - \varphi_2(\mathbf{r}_1)\varphi_1(\mathbf{r}_2)] \left\{ \begin{array}{c} \alpha(1)\alpha(2) \\ \frac{1}{\sqrt{2}}[\alpha(1)\beta(2) + \beta(1)\alpha(2)] \\ \beta(1)\beta(2) \end{array} \right\}.$$

(13.4.15)

The higher-energy state corresponds to a singlet state, associated with an antisymmetric spin function and a symmetric spatial function, with respect to permutation of the electron indexes:

$$\Psi_{-} = \frac{1}{\sqrt{4}}[\varphi_1(\mathbf{r}_1)\varphi_2(\mathbf{r}_2) + \varphi_2(\mathbf{r}_1)\varphi_1(\mathbf{r}_2)][\alpha(1)\beta(2) - \beta(1)\alpha(2)].$$

(13.4.16)

Exercise 13.4.2 *Show that the following four spin states of two electrons,*

$$\alpha(1)\alpha(2)$$
$$\beta(1)\beta(2)$$
$$\alpha(1)\beta(2) + \beta(1)\alpha(2)$$
$$\alpha(1)\beta(2) - \beta(1)\alpha(2);$$

are eigenfunctions of the two-electron spin operators:

$$\hat{S}_z = \hat{S}_{z_1} + \hat{S}_{z_2}$$
$$\hat{S}^2 = (\hat{S}_{x_1} + \hat{S}_{x_2})^2 + (\hat{S}_{y_1} + \hat{S}_{y_2})^2 + (\hat{S}_{z_1} + \hat{S}_{z_2})^2.$$

What are the respective eigenvalues and spin quantum numbers for each of the spin states?

Exercise 13.4.3 *Substitute the relevant determinants, defined in Eq. (13.4.2), in Eq. (13.4.8) to derive Eq. (13.4.9).*

Exercise 13.4.4 *Express the energy $\varepsilon = \langle\Psi|\hat{H}|\Psi\rangle$ of Ψ_{+} and Ψ_{-}, as defined in Eq. (13.4.9), in terms of the integrals defined in Eq. (13.4.5–13.4.7) (obtain Eq. (13.4.10)).*

When the two single-particle energies associated with the two orbitals $E_1 = \langle\varphi_1|\hat{h}_1|\varphi_1\rangle$ and $E_2 = \langle\varphi_2|\hat{h}_2|\varphi_2\rangle$ are well separated, namely $E_2 - E_1 >> J_{1,1}, J_{2,2}, J_{1,2}, K_{1,2}$, the situation corresponds to a He-like atom, where φ_1 and φ_2 can be identified with the 1s and 2s type orbitals. A summary of the different two-electron states for a He-like atom is presented in Fig. 13.4.2. The singlet and triplet spin stats are marked by (S) and (T) respectively and are sorted according to their energy, for $E_2 - E_1 >> J_{1,1}, J_{2,2}, J_{1,2}, K_{1,2}$. The ground state is a singlet spin state, the first excited state is a triplet (threefold degenerate in the absence of a magnetic field), where the next two excitations are again singlet states. Let us emphasize that while we considered explicitly only a He-like atom, similar considerations apply to the ground and first excited states of many-electron "closed shell" atoms (or molecules, in fact), where the ground state corresponds to an even number of paired electrons, and the first

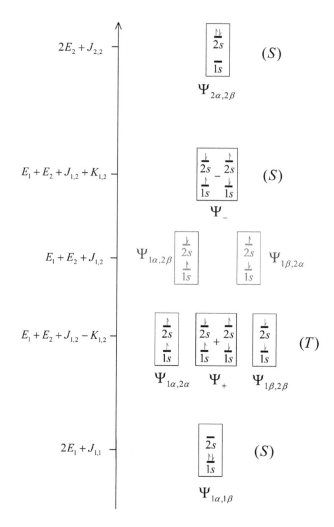

$2E_2 + J_{2,2}$ —

(S)

$\Psi_{2\alpha,2\beta}$

$E_1 + E_2 + J_{1,2} + K_{1,2}$ —

(S)

Ψ_-

$E_1 + E_2 + J_{1,2}$ — $\Psi_{1\alpha,2\beta}$ ... $\Psi_{1\beta,2\alpha}$

$E_1 + E_2 + J_{1,2} - K_{1,2}$ —

(T)

$\Psi_{1\alpha,2\alpha}$ Ψ_+ $\Psi_{1\beta,2\beta}$

$2E_1 + J_{1,1}$ —

(S)

$\Psi_{1\alpha,1\beta}$

Figure 13.4.2 A schematic representation of the ground and first excited states of a "helium-like" atom within the manifold of the two lowest-energy orbitals. The proper states that correspond to uncorrelated products of spin and orbital functions are marked in black. (Slater determinants that artificially correlate spin and orbitals are marked in gray.) The different states are sorted according to the two-electron energy.

excitations involve single- or two-electron transitions to the next empty orbital (according to its single-particle energy). In particular, the ground state of such systems is a singlet, where the first excitations are associated with a triplet and a singlet state, with the triplet being lower in energy due to the stabilizing effect of the exchange interaction.

We now turn to the case where two electrons are populated in two degenerate spatial orbitals. In this case, $E_1 = E_2 \equiv E$, and consequently, $|E_2 - E_1| << J_{1,1}, J_{2,2}, J_{1,2}, K_{1,2}$. For simplicity we shall additionally assume that the different Coulomb integrals are identical in the degenerate manifold, namely $J_{1,1} = J_{2,2} = J_{1,2} \equiv J$. In this case (see

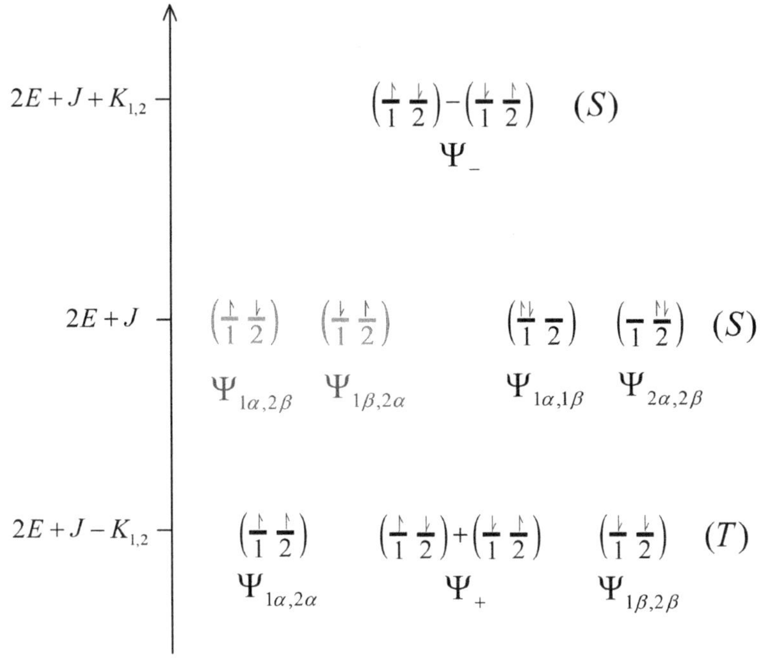

Figure 13.4.3 A schematic representation of the single determinant as well as proper uncorrelated products of spin and orbital functions, for two degenerate orbitals. The different states are sorted according to the two-electron energy, demonstrating Hund's rule, where the triplet state corresponds to the lowest two-electron energy. (Slater determinants that artificially correlate spin and orbitals are marked in gray.)

Fig. 13.4.3) the lowest two-electron energy is associated with the triplet spin states, where the two electrons are associated with the same spin, and therefore with a stabilizing exchange interaction. Also in this case, a similar consideration applies to the many-electron atoms and leads to **Hund's rule: in a many-electron atom where several electrons occupy several degenerate orbitals at a given single-particle energy, the ground state corresponds to a maximal number of unpaired electrons occupying the degenerate orbitals, with the same electronic spin.** Considering, for example, the configuration of carbon in its ground state (see Fig. (13.3.5)), two electrons occupy two of the three degenerate carbon orbitals associated with $n = 2, l = 1$. In the space of these two spatial orbitals, the ground state corresponds to the triplet state (as in Fig. 13.4.3).

Bibliography

[13.1] P. Zeeman, "The Effect of Magnetisation on the Nature of Light Emitted by a Substance", Nature 55, 347 (1897).

[13.2] W. Gerlach and O. Stem, "Der experimentelle Nachweis der Richtungsquantelung im Magnetfeld," Zeitschrift für Physik 9, 349 (1922).

[13.3] B. Thaller, "The Dirac Equation" (Springer Science Berlin Heidelberg, 1992).

[13.4] J. D. Jackson, "Classical Electrodynamics," 3rd ed. (Wiley, 1999).

[13.5] J. C. Slater, "The theory of complex spectra," Phys. Rev. 34, 1293 (1929).

[13.6] V. Fock, "Naherungsmethode zur Losung des quantenmechanischen Mehrkorperproblems," Zeitschrift für Physik 61, 126 (1930).

[13.7] A. Szabo and N. S. Ostlund, "Modern Quantum Chemistry: Introduction to Advanced Electronic Structure Theory" (Dover, 1996).

[13.8] R. Latter, "Atomic energy levels for the Thomas-Fermi and Thomas-Fermi-Dirac potential," Physical Review 99, 510 (1955).

[13.9] R. G. Parr and W. Yang, "Density Functional Theory of Atoms and Molecules" (Oxford University Press, 1989).

[13.10] R. J. Bartlett and M. Musiał, "Coupled-cluster theory in quantum chemistry," Reviews of Modern Physics 79, 291 (2007).

[13.11] R. Pauncz, "Spin Eigenfunctions: Construction and Use" (Springer, 1979).

14 Many-Atom Systems

14.1 The "Chemical Space" and Many-Atom Systems

The most relevant systems in nanoscale science extend beyond the realm of isolated atoms. The interaction between atoms and the possibility to form chemical bonds give rise to a variety of chemical compounds, ranging from metals to nonmetals, from molecular materials to crystals and to periodic lattices. The so-called "chemical space" of possible combinations of small organic molecules was estimated to exceed an astronomic number of 10^{60} [14.1]. Remarkably, the versatility of the different "allowed" forms of stable compounds can be fairly understood within the postulates and concepts of quantum mechanics. Moreover, macroscopic properties of different materials, such as chemical and mechanical stability, electric conductivity, observed color, and more, are directly related to the interaction between the atoms on the nanoscale, which is well described by quantum mechanics.

In this chapter we invoke a "bottom-up" approach for extending the treatment of many-electron atoms into many-atom systems. Based on their electronic structure, detailed quantitative information can be obtained about the properties of different materials, using elaborate atomistic-level electronic structure calculations. Our focus here will be on a qualitative description of quantum mechanical concepts and principles. First, we shall introduce concepts that explain the formation of a "chemical bond" between two atoms. These concepts will then be invoked for understanding the principles underlying the structure and the properties of extended many-atomic systems (solid lattices). In Chapter 13 we saw that the properties of many-electron systems have remarkable consequences when several electrons are bound to a nucleus within a single atom. In particular, Pauli's exclusion "forces" electrons to populate single-particle states of increasing single-particle energy, such that high-energy electrons are partially "shielded" from the bare nucleus. As it turns out, when atoms are brought to proximity, the external electrons of one atom are attracted to the atomic cores of other atoms, which tends to "spontaneously" delocalize the electron over several atoms. In other words, the stationary solutions of the Schrödinger equation become delocalized (this phenomenon is reminiscent of the spatial delocalization observed in quantum wells in Sections 6.4 and 6.5). Just how much these electrons tend to delocalize, or be shared among different atoms, differs from one element to another, and from one chemical compound to another. However, the same principles that determine the electronic configuration of a single atom are applicable also to groups of atoms coupled to each other, as happens in molecules or in extended atomic lattices. In particular, the orbital approximation can be extended to account for single-particle functions (orbitals) that are delocalized among different atoms. These orbitals are filled by paired electrons of different spins, typically from low to high orbital energy, up to some highest occupied orbitals. Just as the properties of a single atom are determined to a large extent by the electrons' occupancy at the external energy shell, the properties of more-complex materials, such as molecules or lattices, can be correlated with the properties of the highest occupied (or lowest unoccupied) orbitals (sometimes referred to as the frontier orbitals).

As a concrete example, consider the property of electrical conductivity, which turns out to be dramatically different between metals and ionic lattices. *Why does a metallic lattice (e.g., Na) conduct electric current, whereas an ionic compound of the same metal (e.g., NaCl) insulates?* Macroscopic conductivity depends on the mobility of charge carriers, namely their ability to accelerate (gain kinetic energy) in response to an electric potential bias. As discussed in the following sections, this ability depends on the electronic configuration within the many-atom material. Materials that conduct are those in which the energy gap between the highest occupied orbitals and the lowest unoccupied orbitals is small (in comparison to the available thermal energy) or nonexisting (as the in the case of a partially filled energy band). Therefore, electrons in these materials can accelerate in response to any potential bias. In contrast, materials in which the energy gap to an unoccupied orbital exceeds by far the thermal energy will be insulating and will start conducting only at some finite bias or at increasing temperatures.

14.2 The Born–Oppenheimer Approximation

We start from the definition of a many-atom system as a system composed of n nuclei and N electrons. The nuclei are denoted by $\alpha = 1, 2, \ldots, n$, where each nucleus has an electric charge, $Z_\alpha |e|$ (Z_a is the atomic number of the element, namely the number of protons in the nucleus), and a mass, m_α (the sum of proton and neutron masses). Denoting the position vector of the ith electron and the αth nucleus in a three-dimensional Cartesian coordinate reference system as $\mathbf{r}_i = (x_i, y_i, z_i)$ and $\mathbf{R}_\alpha = (x_\alpha, y_\alpha, z_\alpha)$, respectively, the quantum Hamiltonian of a general many-atom system in the position representation is given by Eq. (3.4.11) (see also Fig. 3.4.2).

The complexity of solving the Schrödinger equation with this many-body Hamiltonian can be reduced by noticing that the mass of a typical nucleus is three to five orders of magnitude larger than the mass of an electron. Using classical mechanics terms, since the forces between the particles are electrostatic and do not depend on the particle masses, the lighter electrons are expected to move much faster than the heavier nuclei, in response to their mutual interaction. Consequently, a timescale separation argument is expected to hold in this case. Namely, on the relevant timescale for electronic dynamics, the forces on the electrons owing to the presence of the nuclei can be approximated as emerging from static nuclei at some fixed positions. On the relevant timescale for nuclear dynamics, the forces on the nuclei can be approximated as emerging from a static field, attributed to the averaged position of the electrons in the system. In quantum mechanical terms, "fast" and "slow" dynamics correspond, respectively, to "large" and "small" energy gaps between energy eigenvalues of the underlying Hamiltonian (see Chapter 15). Analogy with classical mechanics suggests that the spectrum of the Hamiltonian should involve densely packed eigenstates that differ mostly by the state of the nuclei, whereas eigenstates that differ significantly in the state of the electrons will be farther apart in energy. While not always correct (see the discussion at the end of this section), these time and energy scale separation arguments suggest that when solving the full Schrödinger equation, the electrons and nuclei could be treated differently. This is the essence of the adiabatic treatment of the electronic structure of many-atom systems, attributed to Born and Oppenheimer [14.2].

Let us start by collecting all the electronic and nuclear coordinates into two respective "super vectors," defined as

$$\mathbf{r} \equiv (\mathbf{r}_1, \mathbf{r}_2, \ldots, \mathbf{r}_N)$$
$$\mathbf{R} \equiv (\mathbf{R}_1, \mathbf{R}_2, \ldots, \mathbf{R}_n) . \tag{14.2.1}$$

We would have liked to be able to solve the full stationary Schrödinger equation for the system. Invoking the position representation this equation reads

$$\hat{H}_{\mathbf{R},\mathbf{r}} \Psi(\mathbf{R}, \mathbf{r}) = E \Psi(\mathbf{R}, \mathbf{r}). \tag{14.2.2}$$

The full (nonrelativistic) Hamiltonian for a many-atom system (see Eq. (3.4.11)) can be written in a compact form,

$$\hat{H}_{\mathbf{R},\mathbf{r}} = \hat{T}_{\mathbf{R}} + \hat{T}_{\mathbf{r}} + \hat{V}_{\mathbf{R}} + \hat{V}_{\mathbf{R},\mathbf{r}} + \hat{V}_{\mathbf{r}}, \tag{14.2.3}$$

where

$$\hat{T}_{\mathbf{R}} = \sum_{\alpha=1}^{n} \frac{-\hbar^2}{2m_\alpha} \nabla_{\mathbf{R}_\alpha} \cdot \nabla_{\mathbf{R}_\alpha}$$

$$\hat{T}_{\mathbf{r}} = \sum_{i=1}^{N} \frac{-\hbar^2}{2m_e} \Delta_{\mathbf{r}_i}$$

$$\hat{V}_{\mathbf{R}} = \sum_{\beta>\alpha}^{n} \sum_{\alpha=1}^{n} \frac{Ke^2 Z_\alpha Z_\beta}{|\mathbf{R}_\alpha - \mathbf{R}_\beta|} \tag{14.2.4}$$

$$\hat{V}_{\mathbf{r}} = \sum_{j>i}^{N} \sum_{i=1}^{N} \frac{Ke^2}{|\mathbf{r}_i - \mathbf{r}_j|}$$

$$\hat{V}_{\mathbf{R},\mathbf{r}} = -\sum_{\alpha=1}^{n} \sum_{i=1}^{N} \frac{KZ_\alpha e^2}{|\mathbf{R}_\alpha - \mathbf{r}_i|} .$$

$\hat{T}_{\mathbf{R}}$ and $\hat{T}_{\mathbf{r}}$, are the kinetic energy operators, corresponding, respectively, to the nuclei and the electrons. $\hat{V}_{\mathbf{R}}$ and $\hat{V}_{\mathbf{r}}$ are the electrostatic repulsion terms corresponding, respectively, to all nucleus pairs and all electron pairs, and $\hat{V}_{\mathbf{R},\mathbf{r}}$ is the potential energy of electrostatic attraction between all electrons and nuclei.

The first step in the Born–Oppenheimer (BO) treatment of this equation is to define the "electronic Hamiltonian,"

$$\hat{H}_{elec}(\mathbf{R}) = \hat{H}_{\mathbf{R},\mathbf{r}} - \hat{T}_{\mathbf{R}}, \tag{14.2.5}$$

in which the kinetic energy of the nuclei is subtracted from the full Hamiltonian. The electronic Hamiltonian can be regarded as a differential operator in the space of the electron coordinates, which depends on the nuclei positions, $\mathbf{R}$, only parametrically. (Notice that the term $\hat{V}_{\mathbf{R}}$, which corresponds to the interaction between the nuclei, is conventionally included within the electronic Hamiltonian, although in the electronic space, it corresponds to an identity operator multiplied by a function of $\mathbf{R}$.) The electronic Hamiltonian can be rightly viewed as a straightforward generalization of an atomic Hamiltonian, where there are many electrons, and instead of a single nucleus, the system contains several (or many) nuclei, distributed in coordinate space at some fixed geometry, defined by the parameter $\mathbf{R}$.

The electronic Hamiltonian is a Hermitian operator in the electronic coordinate space, for any given positions of the nuclei, $\mathbf{R}$. Therefore, the eigenstates of this operator span the space of the electronic physical states (see Section 11.4). Considering the $\mathbf{R}$-dependent eigenvalue equation (conventionally termed, "the electronic Schrödinger equation"),

$$\hat{H}_{elec}(\mathbf{R})\phi_l(\mathbf{R},\mathbf{r}) = \varepsilon_l(\mathbf{R})\phi_l(\mathbf{R},\mathbf{r}), \tag{14.2.6}$$

there exists a set $\{\phi_l(\mathbf{R},\mathbf{r})\}$ of orthonormal functions with respect to integration over the multidimensional electronic space (termed "the electronic functions"), where for any $\mathbf{R}$,

$$\int \phi_l^*(\mathbf{R},\mathbf{r})\phi_{l'}(\mathbf{R},\mathbf{r})d\mathbf{r} = \delta_{l,l'}. \tag{14.2.7}$$

Any proper function of $\mathbf{r}$ can be therefore expanded as a linear combination of $\{\phi_l(\mathbf{R},\mathbf{r})\}$. Since this holds for any $\mathbf{R}$, it follows that any proper function of both $\mathbf{R}$ and $\mathbf{r}$ can be expanded using the sets $\{\phi_l(\mathbf{R},\mathbf{r})\}$ with appropriate expansion coefficients. In particular, this holds for the exact eigenstates of the full Hamiltonian, $\hat{H}_{\mathbf{R},\mathbf{r}}$, namely, for the exact solutions to the Schrödinger equation, Eq.(14.2.2), which can therefore be expanded as (the "BO expansion")

$$\Psi(\mathbf{R},\mathbf{r}) = \sum_l \chi_l(\mathbf{R})\phi_l(\mathbf{R},\mathbf{r}). \tag{14.2.8}$$

The linear expansion coefficients are a vector of functions $(\chi_1(\mathbf{R}), \chi_2(\mathbf{R}), \ldots)$ in the nuclear coordinate space, conventionally termed "the nuclear functions." The exact solution is formally expressed as an infinite (truncated, in practice) sum of products of "electronic functions" and "nuclear functions."

An exact solution to the full Schrödinger equation can therefore, in principle, be obtained in two steps. In the first step, complete orthogonal sets of electronic functions, $\{\phi_l(\mathbf{R},\mathbf{r})\}$, should be calculated for each and every (relevant) nuclei positions vector,

R. In the subsequent step, the expansion coefficients, $\{\chi_l(\mathbf{R})\}$, should be obtained for each solution, $\Psi(\mathbf{R}, \mathbf{r})$. An equation for the proper expansion coefficients is obtained by substitution of the expansion, Eq. (14.2.2), in the Schrödinger equation, Eq. (14.2.2), with the full Hamiltonian, $\hat{H}_{\mathbf{R},\mathbf{r}} = \hat{H}_{elec}(\mathbf{R}) + \hat{T}_{\mathbf{R}}$:

$$\left[\hat{H}_{elec}(\mathbf{R}) + \hat{T}_{\mathbf{R}}\right] \sum_l \chi_l(\mathbf{R})\phi_l(\mathbf{R}, \mathbf{r}) = E \sum_l \chi_l(\mathbf{R})\phi_l(\mathbf{R}, \mathbf{r}). \tag{14.2.9}$$

Projecting the equation on every specific electronic wave function, $\phi_m(\mathbf{R}, \mathbf{r})$, by integrating over the electronic coordinate space, using Eqs. (14.2.6, 14.2.7), we obtain the following coupled linear equations for the components of the vector, $\{\chi_m(\mathbf{R})\}$,

$$\sum_l \left[\hat{H}_{\mathbf{R}}\right]_{m,l} \chi_l(\mathbf{R}) = E\chi_m(\mathbf{R}). \tag{14.2.10}$$

$[\hat{H}_{\mathbf{R}}]_{m,l}$ is an operator in the nuclear coordinate space, obtained by integrating over the electronic coordinates,

$$\left[\hat{H}_{\mathbf{R}}\right]_{m,l} \equiv \int d\mathbf{r}\, \phi_m^*(\mathbf{R}, \mathbf{r}) \left[\hat{H}_{elec}(\mathbf{R}) + \hat{T}_{\mathbf{R}}\right] \phi_l(\mathbf{R}, \mathbf{r}) = \varepsilon_m(\mathbf{R})\delta_{m,l} + \left[\hat{T}_{\mathbf{R}}\right]_{m,l}, \tag{14.2.11}$$

where we used the fact that the set $\{\phi_l(\mathbf{R}, \mathbf{r})\}$ is an orthonormal set of eigenstates of the electronic Hamiltonian (Eqs. (14.2.6, 14.2.7), see Ex. 14.2.1).

Exercise 14.2.1 *Obtain Eqs. (4.2.10, 4.2.11) from Eq. (4.2.9) by multiplying from the left by $\phi_m^*(\mathbf{R}, \mathbf{r})$ and integrating over the electronic coordinates. Use the fact that the electronic functions, $\{\phi_l(\mathbf{R}, \mathbf{r})\}$, are the orthonormal (Eq. (14.2.7)) eigenfunctions of the electronic Hamiltonian (Eq. (14.2.6)).*

A closer look into the matrix representation of the nuclear kinetic energy operator in the basis of the electronic wave functions, denoted as $[\hat{T}_{\mathbf{R}}]_{m,l}$, is essential for defining the BO approximation. First, for simplicity of notation, we define mass-scaled coordinates for each nucleus, $\tilde{\mathbf{R}}_\alpha = \sqrt{\frac{2m_\alpha}{\hbar^2}} \mathbf{R}_\alpha$, and an extended gradient vector in the entire nuclear coordinate space,

$$\nabla_{\tilde{\mathbf{R}}} \equiv \left(\frac{\partial}{\partial \tilde{x}_1}, \frac{\partial}{\partial \tilde{y}_1}, \frac{\partial}{\partial \tilde{z}_1}, \ldots, \frac{\partial}{\partial \tilde{x}_\alpha}, \frac{\partial}{\partial \tilde{y}_\alpha}, \frac{\partial}{\partial \tilde{z}_\alpha}, \ldots, \frac{\partial}{\partial \tilde{x}_n}, \frac{\partial}{\partial \tilde{y}_n}, \frac{\partial}{\partial \tilde{z}_n} \right)^t,$$

such that the nuclear kinetic energy (see Eq. (14.2.4)) obtains the form

$$\hat{T}_{\mathbf{R}} = \sum_{\alpha=1}^{n} \frac{-\hbar^2}{2m_\alpha} \nabla_{\mathbf{R}_\alpha} \cdot \nabla_{\mathbf{R}_\alpha} \equiv -\nabla_{\tilde{\mathbf{R}}} \cdot \nabla_{\tilde{\mathbf{R}}}. \tag{14.2.12}$$

The matrix elements of the kinetic energy in the basis of electronic functions therefore reads (see Ex. 14.2.2)

$$[\hat{T}_{\mathbf{R}}]_{m,l} = -\int d\mathbf{r}\phi_m^*(\mathbf{R},\mathbf{r})\nabla_{\tilde{\mathbf{R}}} \cdot \nabla_{\tilde{\mathbf{R}}}\phi_l(\mathbf{R},\mathbf{r})$$

$$= \delta_{m,l}\hat{T}_{\mathbf{R}} - 2\left(\int d\mathbf{r}\phi_m^*(\mathbf{R},\mathbf{r})\nabla_{\tilde{\mathbf{R}}}\phi_l(\mathbf{R},\mathbf{r})\right)\cdot\nabla_{\tilde{\mathbf{R}}} - \left(\int d\mathbf{r}\phi_m^*(\mathbf{R},\mathbf{r})\nabla_{\tilde{\mathbf{R}}}\cdot\nabla_{\tilde{\mathbf{R}}}\phi_l(\mathbf{R},\mathbf{r})\right)$$

$$(14.2.13)$$

$$\equiv \delta_{m,l}\hat{T}_{\mathbf{R}} - 2\langle\phi_m(\mathbf{R})|\nabla_{\tilde{\mathbf{R}}}\phi(\mathbf{R})\rangle\cdot\nabla_{\tilde{\mathbf{R}}} - \langle\phi_m(\mathbf{R})|\nabla_{\tilde{\mathbf{R}}}\cdot\nabla_{\tilde{\mathbf{R}}}\phi_l(\mathbf{R})\rangle,$$

where in the last step we introduced Dirac's notations for the states in the electronic coordinate space, which depend parametrically on $\mathbf{R}$. Using Eq. (14.2.13) for $[\hat{T}_{\mathbf{R}}]_{m,l}$, we rewrite Eqs. (14.2.10, 14.2.11) as

$$\sum_l \left[\hat{H}_{\mathbf{R}}^{(0)} + \hat{H}_{\mathbf{R}}^{(1)}\right]_{m,l}\chi_l(\mathbf{R}) = E\chi_m(\mathbf{R}), \qquad (14.2.14)$$

where

$$\left[\hat{H}_{\mathbf{R}}^{(0)}\right]_{m,l} = \delta_{m,l}\left[\varepsilon_m(\mathbf{R}) + \hat{T}_{\mathbf{R}}\right], \qquad (14.2.15)$$

and

$$\left[\hat{H}_{\mathbf{R}}^{(1)}\right]_{m,l} = -2\langle\phi_m(\mathbf{R})|\nabla_{\tilde{\mathbf{R}}}\phi_l(\mathbf{R})\rangle\cdot\nabla_{\tilde{\mathbf{R}}} - \langle\phi_m(\mathbf{R})|\nabla_{\tilde{\mathbf{R}}}\cdot\nabla_{\tilde{\mathbf{R}}}\phi_l(\mathbf{R})\rangle. \qquad (14.2.16)$$

The "zero-order" Hamiltonian, $\hat{H}_{\mathbf{R}}^{(0)}$, is purely diagonal in the electronic state representation, where $\hat{H}_{\mathbf{R}}^{(1)}$ is non-diagonal (see Fig. 14.2.1).

Exercise 14.2.2   (a) *Use the definition of the gradient operator, $\nabla_{\mathbf{R}_\alpha} \equiv \left(\frac{\partial}{\partial x_\alpha}, \frac{\partial}{\partial y_\alpha}, \frac{\partial}{\partial z_\alpha}\right)^t$, to show that $\nabla_{\mathbf{R}_\alpha}\cdot\nabla_{\mathbf{R}_\alpha}[\phi_l(\mathbf{R},\mathbf{r})\chi_l(\mathbf{R})] = 2[\nabla_{\mathbf{R}_\alpha}\phi_l(\mathbf{R},\mathbf{r})]\cdot\nabla_{\mathbf{R}_\alpha}\chi_l(\mathbf{R}) + \phi_l(\mathbf{R},\mathbf{r})\nabla_{\mathbf{R}_\alpha}\cdot\nabla_{\mathbf{R}_\alpha}\chi_l(\mathbf{R}) + [\nabla_{\mathbf{R}_\alpha}\cdot\nabla_{\mathbf{R}_\alpha}\phi_l(\mathbf{R},\mathbf{r})]\chi_l(\mathbf{R}).$*

(b) *Use the definition of the nuclear kinetic energy operator in Eq. (14.2.4), the result of (a), and the orthonormality of the electronic functions (Eq. (14.2.7)) to show that*

$$[\hat{T}_{\mathbf{R}}]_{m,l}\chi_l(\mathbf{R}) = \sum_{\alpha=1}^n \frac{-\hbar^2}{m_\alpha}\left[\int d\mathbf{r}\phi_m^*(\mathbf{R},\mathbf{r})\nabla_{\mathbf{R}_\alpha}\phi_l(\mathbf{R},\mathbf{r})\right]\cdot\nabla_{\mathbf{R}_\alpha}\chi_l(\mathbf{R})$$

$$+ \sum_{\alpha=1}^n \frac{-\hbar^2}{2m_\alpha}\left[\int d\mathbf{r}\phi_m^*(\mathbf{R},\mathbf{r})\nabla_{\mathbf{R}_\alpha}\cdot\nabla_{\mathbf{R}_\alpha}\phi_l(\mathbf{R},\mathbf{r})\right]\chi_l(\mathbf{R})$$

$$+ \delta_{m,l}\sum_{\alpha=1}^n \frac{-\hbar^2}{2m_\alpha}\nabla_{\mathbf{R}_\alpha}\cdot\nabla_{\mathbf{R}_\alpha}\chi_l(\mathbf{R}).$$

(c) *Use the result of (b) and the definition of $\nabla_{\tilde{\mathbf{R}}}$ to obtain Eq. (14.2.13).*

While exact, Eqs. (14.2.14–14.2.16) are not particularly useful. Firstly, their implementation requires preknowledge of all the electronic functions $\{\phi_m(\mathbf{R},\mathbf{r})\}$ (an infinite set of multidimensional functions, in principle), and secondly, due to the non-diagonal terms, the nuclear functions associated with different electronic functions are all coupled to each other, such that a solution to the Schrödinger equation (Eq. (14.2.9)), means a simultaneous calculation of the entire set, $\{\chi_m(\mathbf{R})\}$.

$$
\hat{\mathbf{H}}_{\mathbf{R}} =
\begin{bmatrix}
\left[\hat{H}_{\mathbf{R}}^{(0)}\right]_{1,1} & 0 & 0 & \cdots \\
0 & \left[\hat{H}_{\mathbf{R}}^{(0)}\right]_{2,2} & 0 & \cdots \\
0 & 0 & \left[\hat{H}_{\mathbf{R}}^{(0)}\right]_{3,3} & \cdots \\
\vdots & \vdots & \vdots & \ddots
\end{bmatrix}
+
\begin{bmatrix}
\left[\hat{H}_{\mathbf{R}}^{(1)}\right]_{1,1} & \left[\hat{H}_{\mathbf{R}}^{(1)}\right]_{1,2} & \left[\hat{H}_{\mathbf{R}}^{(1)}\right]_{1,3} & \cdots \\
\left[\hat{H}_{\mathbf{R}}^{(1)}\right]_{2,1} & \left[\hat{H}_{\mathbf{R}}^{(1)}\right]_{2,2} & \left[\hat{H}_{\mathbf{R}}^{(1)}\right]_{2,3} & \cdots \\
\left[\hat{H}_{\mathbf{R}}^{(1)}\right]_{3,1} & \left[\hat{H}_{\mathbf{R}}^{(1)}\right]_{3,2} & \left[\hat{H}_{\mathbf{R}}^{(1)}\right]_{3,3} & \cdots \\
\vdots & \vdots & \vdots & \ddots
\end{bmatrix}
$$

Figure 14.2.1 The matrix representation of the full Hamiltonian in the basis of the electronic functions (see Eq. 14.2.11). The "partial" integration (over the electronic space) associates each matrix element, $\left[\hat{H}_{\mathbf{R}}\right]_{m,l} = \left[\hat{H}_{\mathbf{R}}^{(0)}\right]_{m,l} + \left[\hat{H}_{\mathbf{R}}^{(1)}\right]_{m,l}$, with an operator in the nuclear coordinate space.

A remarkable simplification is obtained by invoking the BO approximation (or the "adiabatic approximation"), assuming that $\hat{H}_{\mathbf{R}}^{(1)}$ can be neglected next to $\hat{H}_{\mathbf{R}}^{(0)}$. Namely, matrix elements involving first and second derivatives of the electronic functions, $\{\phi_m(\mathbf{R}, \mathbf{r})\}$, with respect to the nuclear coordinates $\{\mathbf{R}\}$ are assumed to be null:

$$
\int d\mathbf{r}\, \phi_m^*(\mathbf{R}, \mathbf{r})\, \hat{T}_{\mathbf{R}} \phi_l(\mathbf{R}, \mathbf{r}) \chi_l(\mathbf{R}) \approx \delta_{m,l} \hat{T}_{\mathbf{R}} \chi_l(\mathbf{R}). \tag{14.2.17}
$$

In fact, it is sufficient to assume that the vector of nonadiabatic couplings, $\langle \phi_m(\mathbf{R}) | \nabla_{\tilde{\mathbf{R}}} \phi_l(\mathbf{R}) \rangle$, vanishes for any m and l (see Ex. 14.2.3). Setting $\left[\hat{H}_{\mathbf{R}}^{(1)}\right]_{m,l} = 0$, Eq. (14.2.14) is replaced by

$$
\sum_l \left[\hat{H}_{\mathbf{R}}^{(0)}\right]_{m,l} \chi_l(\mathbf{R}) = E \chi_m(\mathbf{R}). \tag{14.2.18}
$$

Since $\hat{H}_{\mathbf{R}}^{(0)}$ is diagonal (see Eq. (14.2.15) and Fig. 14.2.1), the equations for different electronic states become decoupled, where a "nuclear equation" is obtained for each m, independently of all the other electronic states:

$$
\left[\hat{T}_{\mathbf{R}} + \varepsilon_m(\mathbf{R})\right] \chi_m(\mathbf{R}) = E \chi_m(\mathbf{R}). \tag{14.2.19}
$$

The result is cast in a form of an effective stationary Schrödinger equation for $\chi_m(\mathbf{R})$, where *the role of the effective potential energy in the coordinate space of the nuclei is played by the corresponding eigenvalue of the electronic Hamiltonian (Eq. (14.2.6))*, $\varepsilon_m(\mathbf{R})$. Each such equation has multiple solutions, which we denote here by an additional index, "k." Therefore, each eigenstate of $\hat{H}_{\mathbf{R}}^{(0)}$ (as defined by Eq. (14.2.18)) is associated with two indexes: $E_{m,k}$, where m accounts for the electronic function, $\phi_m(\mathbf{R}, \mathbf{r})$, and the combination (m, k) specifies a nuclear function, $\chi_m^{(m,k)}(\mathbf{R})$. The latter is an eigenstate of Eq. (14.2.19), which corresponds to the energy $E_{m,k}$,

$$
\left[\hat{T}_{\mathbf{R}} + \varepsilon_m(\mathbf{R}) - E_{m,k}\right] \chi_m^{(m,k)}(\mathbf{R}) = 0. \tag{14.2.20}
$$

The block-diagonal form of $\hat{H}_{\mathbf{R}}^{(0)}$ (Eq. (14.2.18)), means that its eigenvector, which corresponds to the eigenvalue, $E_{m,k}$, can be chosen as

$$
\chi_l^{(m,k)}(\mathbf{R}) = \delta_{m,l} \chi_m^{(m,k)}(\mathbf{R}). \tag{14.2.21}
$$

(This form is obligatory, in fact, if the spectrum of $\hat{H}_{\mathbf{R}}^{(0)}$ is nondegenerate.) Consequently, the exact BO expansion, Eq. (14.2.8), of the solution to the full Schrödinger equation (Eq. (14.2.2)) is approximated by a single product,

$$\Psi(\mathbf{R},\mathbf{r}) \cong \Psi_{m,k}(\mathbf{R},\mathbf{r}) = \phi_m(\mathbf{R},\mathbf{r})\chi_m^{(m,k)}(\mathbf{R}), \qquad (14.2.22)$$

which corresponds to the approximated eigenvalue,

$$E \cong E_{m,k}. \qquad (14.2.23)$$

Eq. (14.2.22) is conventionally referred to as the "Born–Oppenheimer Ansatz" for the solution of the Schrödinger equation for the many-atom system.

In summary, the BO approximation associates each energy level of the many-atom system, $E_{m,k}$, with a product state, $\Psi_{m,k}(\mathbf{R},\mathbf{r}) = \phi_m(\mathbf{R},\mathbf{r})\chi_m^{(m,k)}(\mathbf{R})$. The index m accounts for an electronic function, $\phi_m(\mathbf{R},\mathbf{r})$, which is one of the eigenfunctions of the electronic Hamiltonian (Eq. (14.2.6)). The combination of indexes (m,k) accounts for a nuclear function, $\chi_m^{(m,k)}(\mathbf{R})$, which is one of the eigenfunctions of the nuclear Schrödinger equation, Eq. (14.2.20), in which the electronic energy, $\varepsilon_m(\mathbf{R})$ plays the role of a potential energy in the nuclear coordinate space. A consecutive solution of the electronic and nuclear equations thus provides approximations for the stationary states of the Shroedinger equation for the entire system.

The emerging physical picture is that the states of the nuclei are dictated by the states of the electrons, where each electronic state defines a different potential energy "landscape" for the nuclei. In general, the number of nuclear degrees of freedom can be very large ($3n$ for an n-atom system), and the potential energy landscape is a function of a multidimensional vector $\mathbf{R}$. Each electronic state index, m, therefore defines a multidimensional "adiabatic Potential Energy Surface" (PES), $\varepsilon_m(\mathbf{R})$. Derivatives of the PES with respect to the nuclear coordinates define the forces on the nuclei, where the minima in the PES correspond to stable nuclear geometries and the maxima correspond to unstable geometries (transition states between stable nuclear configurations).

An illustrative example is given for the case of two atoms ($n = 2$) in Fig. 14.2.2. Notice that the electronic Hamiltonian (Eq. (14.2.5)) is invariant in this case to translation of the diatomic center of mass (three degrees of freedom) and to rotations in the center-of-mass frame (two degrees of freedom). Therefore, out of six nuclear degrees of freedom, only one, corresponding to the interatomic distance R, affects the electronic energy. Each adiabatic PES is therefore a *potential energy curve* in this case, $\varepsilon_m(R)$. The lower energy curve in Fig. 14.2.2 corresponds to a typical electronic ground state of a diatomic molecule, $\varepsilon_0(R)$, as discussed in Chapters 8 and 9 (see Fig. 8.4.1), in the context of molecular vibrations. There, we assumed the existence of a potential energy function, $V(R)$, which obtains an asymptotic constant value at large interatomic distances, increases in the opposite limit, where the interatomic distance gets shorter, and in between, obtains a minimum at some finite interatomic distance. Within the BO approximation, this potential energy curve is identified with the dependence of the electronic ground state energy on the interatomic distance, $V(R) = \varepsilon_0(R)$. In the next section we shall discuss this curve quantitatively. However, we have already acquired tools to understand it qualitatively. At large interatomic distances the electrons will

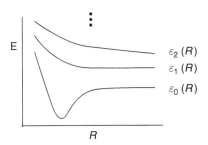

Figure 14.2.2 A schematic representation of electronic energies for a diatomic system. Each electronic energy is an adiabatic potential energy curve for the nuclei, as a function of the interatomic distance. When the potential energy curve has a minimum at the electronic ground state, the two atoms can form a stable molecule.

be exposed primarily to one of the nuclei or the other, such that their energy will depend only weakly on the internuclear distance. In contrast, as the internuclear distance approaches zero, the electronic energy must increase, not only because of the internuclear repulsion (recall that the latter is conventionally counted as a part of the electronic energy), but primarily since attempting to "fuse" the electrons from two atoms into one "united atom" necessitates the occupation of orbitals associated with higher single-particle energies, according to the Aufbau principle, as a consequence of Pauli's exclusion principle (see Chapter 13). The existence of a minimum of $\varepsilon_0(R)$ at some finite inter-atomic distance is not "universal." In fact, it depends on the identity of the two atoms, which determine whether the diatomic molecule is a stable entity or not. Only when such a minimum exists can a stable "chemically bonded" diatomic molecule be identified. The factors that govern the shape of the potential energy curve at the electronic ground state or different diatomic molecules, and therefore determine their stability with respect to decomposition into isolated atoms, will be analyzed in the next section. Importantly, even when the electronic ground state energy has a minimum at some nuclear configuration, R_0, corresponding to a stable molecule, higher potential energy curves, corresponding to excited electronic states, may not have a minimum, as illustrated in Fig. 14.2.2. When the electrons are excited to such a repulsive potential energy curve, the internuclear force, $-d\varepsilon_m(R)/dR$, is always in the direction of increasing R, resulting in bond cleavage, or molecular dissociation. Such electronic excitations are associated, for example, with collisions with other particles or with an interaction with electromagnetic radiation (see Section 18.3). It is emphasized that the principles described here are not limited to diatomic molecules. They often apply to chemical bonds in many-atom systems as well. Examples are photochemical reactions, induced by UV radiation [14.3] or by electronscattering (e.g., damages to molecular DNA following collisions with low-energy electrons [14.4]).

Breakdown of the Born–Oppenheimer Approximation

Within the BO approximation, the electronic state of the system determines the potential energy and therefore the forces experienced by the nuclei. The effect of the nuclei on the electrons is restricted to the parametric dependence of the electronic energy on the nuclear configuration in space. Therefore, changes in the nuclear geometry cannot induce transitions between different electronic states ("nonadiabatic" transitions), and the electronic quantum numbers are "adiabatically" conserved. An exact treatment should account also for the effect of the nuclear kinetic energy (the nuclear position uncertainty in the stationary states, or, otherwise, the nuclear "motion") on the electronic state, which should correspond to superpositions of the different eigenstates of the electronic equation, $\{\phi_m(\mathbf{R},\mathbf{r})\}$. Formally, the coupling between these states is an immediate consequence of the non-diagonal operator, $\hat{H}_{\mathbf{R}}^{(1)}$, in Eq. (14.2.14), which was neglected in the BO approximation. The neglected terms (see Eq. (14.2.16)) depend (directly, or indirectly; see Ex. 14.2.3) on the magnitude of the nonadiabatic coupling vectors, which corresponds to the derivatives of the electronic wave functions with respect to the different nuclear coordinates,

$$\mathbf{D}_{m,l}(\mathbf{R}) \equiv \langle \phi_m(\mathbf{R})|\nabla_{\tilde{\mathbf{R}}}\phi_l(\mathbf{R})\rangle. \tag{14.2.24}$$

Exercise 14.2.3   *The nonadiabatic terms in the nuclear Hamiltonian are given by* $\left[\hat{H}_{\mathbf{R}}^{(1)}\right]_{m,l}$ *in Eq. (14.2.16). Follow (a–d) to show that a sufficient condition for the vanishing of* $\left[\hat{H}_{\mathbf{R}}^{(1)}\right]_{m,l}$ *is the vanishing of the nonadiabatic coupling vector,* $\mathbf{D}_{i,j}(\mathbf{R}) = \langle \phi_i(\mathbf{R})|\nabla_{\tilde{\mathbf{R}}}\phi_j(\mathbf{R})\rangle$, *for any i and j.*

(a) Show that $\langle \phi_m(\mathbf{R})|\nabla_{\tilde{\mathbf{R}}} \cdot \nabla_{\tilde{\mathbf{R}}}\phi_l(\mathbf{R})\rangle = [\nabla_{\tilde{\mathbf{R}}} \cdot \mathbf{D}_{m,l}(\mathbf{R})] - \langle \nabla_{\tilde{\mathbf{R}}}\phi_m(\mathbf{R})| \cdot |\nabla_{\tilde{\mathbf{R}}}\phi_l(\mathbf{R})\rangle.$
(b) For any $\mathbf{R}$*, introduce a complete orthonormal system of electronic functions and show that*

$$\langle \nabla_{\tilde{\mathbf{R}}}\phi_m(\mathbf{R})| \cdot |\nabla_{\tilde{\mathbf{R}}}\phi_l(\mathbf{R})\rangle = \sum_k \langle \nabla_{\tilde{\mathbf{R}}}\phi_m(\mathbf{R})|\phi_k(\mathbf{R})\rangle \cdot \mathbf{D}_{k,l}(\mathbf{R}).$$

(c) Use Eq. (14.2.7) to prove that $\langle \nabla_{\tilde{\mathbf{R}}}\phi_m(\mathbf{R})|\phi_k(\mathbf{R})\rangle = -\mathbf{D}_{m,k}(\mathbf{R}).$
(d) Show that $\left[\hat{H}_{\mathbf{R}}^{(1)}\right]_{m,l} = -2\mathbf{D}_{m,l}(\mathbf{R}) \cdot \nabla_{\tilde{\mathbf{R}}} - [\nabla_{\tilde{\mathbf{R}}} \cdot \mathbf{D}_{m,l}(\mathbf{R})] - \sum_k \mathbf{D}_{m,k}(\mathbf{R}) \cdot \mathbf{D}_{k,l}(\mathbf{R}).$

Recalling the definition of $\nabla_{\tilde{\mathbf{R}}}$ (see Eq. (14.2.12)), the derivative couplings should decrease as the nuclear masses increase, namely $\mathbf{D}_{m,l}(\mathbf{R}) \propto \{1/\sqrt{m_\alpha}\}$. One might have expected that the large mass ratio between the nuclei and the electrons would be sufficient for justifying the neglection of the derivative coupling [14.2], but this is not the case. One can see (Ex. 14.2.4) that the off-diagonal derivative couplings can be expressed as

$$\mathbf{D}_{m,l}(\mathbf{R}) = \frac{\langle \phi_m(\mathbf{R})|(\nabla_{\tilde{\mathbf{R}}}\hat{H}_{elec}(\mathbf{R}))|\phi_l(\mathbf{R})\rangle}{\varepsilon_l(\mathbf{R}) - \varepsilon_m(\mathbf{R})} \quad ; \quad l \neq m. \tag{14.2.25}$$

The numerator contains the matrix element of the vector operator, $\nabla_{\tilde{\mathbf{R}}}\hat{H}_{elec}(\mathbf{R})$, which is proportional to the forces on the nuclear degrees of freedom due to the electron–nucleus and nucleus–nucleus interactions. These matrix elements are finite (except for pathological cases). The denominator is the difference between the two adiabatic PES, corresponding to the m and l electronic states. Therefore, the derivative coupling,

$\mathbf{D}_{m,l}(\mathbf{R})$, diverges whenever the two PES cross each other; namely, when $\varepsilon_l(\mathbf{R}) \rightarrow \varepsilon_m(\mathbf{R})$, and consequently the BO approximation is no longer valid.

Exercise 14.2.4 *Derive Eq. (14.2.25) by calculating* $\nabla_{\tilde{\mathbf{R}}} \langle \phi_m(\mathbf{R}) | \hat{H}_{elec}(\mathbf{R}) | \phi_l(\mathbf{R}) \rangle$, *using the properties of the electronic functions, Eqs. (14.2.6, 14.2.7).*

The Non-Crossing Rule and Conical Intersections

It is important to zoom into the properties of the electronic Hamiltonian as two electronic energies (or two adiabatic PES) come close together. As the energy gap between two PES becomes much smaller than the gaps between them to any other PES, one can attempt to approximate the exact electronic Hamiltonian in a reduced electronic Hilbert space, spanned by two orthonormal basis states (e.g., $\phi_l(\mathbf{R}_0, \mathbf{r})$ and $\phi_m(\mathbf{R}_0, \mathbf{r})$ at some $\mathbf{R}_0$ for which the respective PES are close). When this approximation holds, the electronic Hamiltonian is mapped on a two-level system model (see Section 12.2), with the matrix elements, $H_{l,m}(\mathbf{R}) = \langle \phi_l(\mathbf{R}_0) | \hat{H}_{elec}(\mathbf{R}) | \phi_m(\mathbf{R}_0) \rangle$. The energy gap between the electronic energies, namely the eigenvalues of this matrix, reads (see Eqs. (12.2.3–12.2.6)) $|\varepsilon_l(\mathbf{R}) - \varepsilon_m(\mathbf{R})| = \sqrt{(H_{l,l}(\mathbf{R}) - H_{m,m}(\mathbf{R}))^2 + 4|H_{l,m}(\mathbf{R})|^2}$. A strict crossing, namely $|\varepsilon_l(\mathbf{R}) - \varepsilon_m(\mathbf{R})| = 0$, means that two independent conditions should hold at the same point, $\mathbf{R}$: $H_{l,l}(\mathbf{R}) - H_{m,m}(\mathbf{R}) = 0$ and $|H_{l,m}(\mathbf{R})| = 0$. Two cases should be distinguished. In the first case, $H_{l,m}(\mathbf{R})$ vanishes identically, irrespective of $\mathbf{R}$. This happens when $\phi_l(\mathbf{R}, \mathbf{r})$ and $\phi_m(\mathbf{R}, \mathbf{r})$ are nondegenerate eigenfunctions of a symmetry operator that commutes with $\hat{H}_{elec}(\mathbf{R})$. In this case surface crossing requires fulfilment of a single equation, $H_{l,l}(\mathbf{R}) - H_{m,m}(\mathbf{R}) = 0$, which can be met by a proper choice of a single parameter. Consequently, given an f-dimensional PES, only $f-1$ nuclear degrees of freedom can be chosen freely while meeting the condition for crossing another PES; namely, the crossing is limited to a surface of dimension $f-1$. In the second case, $\phi_l(\mathbf{R}, \mathbf{r})$ and $\phi_m(\mathbf{R}, \mathbf{r})$ are degenerate with respect to all the symmetry operators. Therefore, $H_{l,m}(\mathbf{R})$ is not guaranteed to vanish, and the fulfilment of the two independent conditions, $H_{l,l}(\mathbf{R}) - H_{m,m}(\mathbf{R}) = 0$ and $|H_{l,m}(\mathbf{R})| = 0$, requires a proper choice of two independent parameters. Consequently, only $f-2$ nuclear positions can be chosen freely to meet the crossing condition; namely, the crossing of two f-dimensional PES, is limited to a surface of dimension $f-2$ in this case. Considering the example of a diatom ($n = 2$), illustrated in Fig. 14.2.2, the electronic Hamiltonian is invariant to translation of the center of mass, and to rotation about the line connecting the two centers of the atoms, and therefore the PES depends on a single nuclear coordinate, namely the interatomic distance. Since $f = 1$ in this case, different potential energy curves (one-dimensional surfaces) corresponding to different rotational quantum numbers (with respect to rotation around the molecular axis) can cross only at a single point (a zero-dimensional space corresponding to $f-1 = 0$). Remarkably, when the two curves correspond to the same rotational quantum number, two independent parameters are needed to assure crossing; and for $f = 1$, one has $f-2 < 0$, which means that crossing cannot occur. This result is known as the "non-crossing rule" due to von Neumann and Wigner [14.5]. Extending the same argument for polyatomic systems, two PES associated with the same symmetry eigenstates can only cross at a single point when $f = 2$ (or, $f - 2 = 0$, which corresponds to "a conical intersection" [14.6]), or along a line, when $f = 3$ (or, $f - 2 = 1$), and so forth. Overall, intersections between two (or more) different PES become ubiquitous as the number of atoms in the system increases. Notice that although the crossing dimensions are always smaller than the dimensions of the full nuclear coordinate space (f), and the probability for finding the nuclei strictly at crossing points is therefore of zero measure in the nuclear coordinate space, the nonadiabatic coupling vector ($\mathbf{D}_{m,l}(\mathbf{R})$, Eq. (14.2.25)) can obtain arbitrarily large values also at the vicinity of the crossing points. *Therefore, the BO approximation is valid only when the relevant electronic energies (or the nuclear PES) are sufficiently far in energy from each other, at the relevant nuclear positions, namely at nuclear positions associated with a substantial probability density.* This situation is not always met in practice.

From the point of view of perturbation theory, the BO approximation should hold strictly only when the following condition is met (see Eq. (12.1.29)):

$$| \int d\mathbf{R} \chi_{m,k}^*(\mathbf{R}) \left[\hat{H}_{\mathbf{R}}^{(1)} \right]_{m,l} \chi_{l,k'}(\mathbf{R}) | << |E_{m,k} - E_{l,k'}|, \qquad (14.2.26)$$

where $E_{m,k} = \int d\mathbf{R} \chi_{m,k}^*(\mathbf{R}) \left[\hat{H}_{\mathbf{R}}^{(0)} \right]_{m,m} \chi_{m,k}(\mathbf{R})$. Namely, the matrix elements of $\hat{H}_{\mathbf{R}}^{(1)}$ need to be negligible in comparison to the difference between the energy levels associated with $\hat{H}_{\mathbf{R}}^{(0)}$ (see the related discussion in Section 12.1). While the condition, Eq. (14.2.26), is not so transparent, it is still clear that the right-hand side increases as the energy difference between the nuclear PES, $\varepsilon_l(\mathbf{R})$ and $\varepsilon_m(\mathbf{R})$, increases, which is in agreement with the conclusion derived from Eq. (14.2.25). As we shall see in Chapter 15, "larger" energy gaps between stationary solutions of the Schrödinger equation are translated to "faster" dynamics in the system (higher underlying frequencies). We may therefore argue that the validity of the BO approximation relies on the electrons being much faster than the nuclei. A quantitative analysis of this argument is found, for example, in the Landau–Zenner model for nonadiabatic transitions [14.7].

14.3 Covalent Bonds: the Hydrogen Molecular Ion, H_2^+

Perhaps the most interesting phenomenon associated with many-atom systems is the spontaneous tendency of atoms to form chemical bonds, namely, stable geometrical arrangements in which the distances between neighboring atoms are of the order of atomic radii. Chemical bonds range from molecules to solid crystals and metals. We start by considering the simplest example for chemical bonding, which is the formation of the hydrogen molecular ion, composed of two atomic hydrogen nuclei (two protons) and a single electron. Being a single-electron system, the discussion of many-electron aspects can be avoided, and the concept of orbital (single-electron wave functions) becomes exact. As we shall see, however, the concepts derived for this single-electron system are most useful when discussing the general case of many-electron molecules, which we postpone to the next section. This is similar to the role played by atomic orbitals of a hydrogen-like atom, in analyzing many-electron atoms within the orbital approximation.

Following the Born–Oppenheimer framework (see Section 14.2), we first freeze the position of the nuclei in space and attempt to solve the stationary Schrödinger equation for the electronic Hamiltonian. Setting the number of nuclei to $n = 2$, with $Z_\alpha = Z_\beta = 1$, and the number of electrons to $N = 1$, the electronic Hamiltonian (Eqs. (14.2.3–14.2.5)) reads in this case

$$\hat{H}_{elec}(\mathbf{R}_\alpha, \mathbf{R}_\beta) = \frac{-\hbar^2}{2m_e}\Delta_\mathbf{r} + \frac{Ke^2}{|\mathbf{R}_\alpha - \mathbf{R}_\beta|} - \frac{Ke^2}{|\mathbf{R}_\alpha - \mathbf{r}|} - \frac{Ke^2}{|\mathbf{R}_\beta - \mathbf{r}|}. \tag{14.3.1}$$

Denoting the fixed distance between the two nuclei as $|\mathbf{R}_\alpha - \mathbf{R}_\beta| \equiv R$, it is most convenient to set the origin of the frame of reference at the center of mass and to position the two nuclei along the z-axis, where $\mathbf{R}_\alpha = (0,0,-R/2) = \frac{-R}{2}\mathbf{k}$, and $\mathbf{R}_\beta = (0,0,R/2) = \frac{R}{2}\mathbf{k}$. Within these definitions one can readily see that the electronic Hamiltonian depends only on a single nuclear parameter, which is the internuclear distance, R (see Fig. 14.3.1):

$$\hat{H}_{elec}(R) = \frac{-\hbar^2}{2m_e}\Delta_\mathbf{r} + \frac{Ke^2}{R} - \frac{Ke^2}{|\mathbf{r} + \mathbf{k}R/2|} - \frac{Ke^2}{|\mathbf{r} - \mathbf{k}R/2|}. \tag{14.3.2}$$

The cylindrical symmetry of the diatomic molecule is reflected in the independence of its electronic Hamiltonian on the angle φ, which means that

$$[\hat{H}_{elec}(R), \hat{L}_z] = 0, \tag{14.3.3}$$

where $\hat{L}_z = -i\hbar\frac{\partial}{\partial\varphi}$ is the corresponding angular momentum operator. Additionally, the identity of the two nuclei means that the Hamiltonian commutes with the inversion operator,

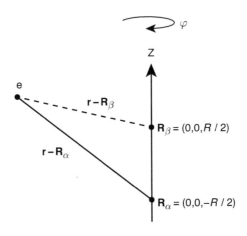

Figure 14.3.1   The coordinate system for the electronic Hamiltonian in the hydrogen molecular ion.

$$[\hat{H}_{elec}(R), \hat{P}_{x,y,z}] = 0, \tag{14.3.4}$$

where $\hat{P}_{x,y,z} f(x,y,z) = f(-x,-y,-z)$.

We are interested in the ground state of the electronic Hamiltonian. Rather than following analytic solutions to the electronic Schrödinger equation for H_2^+, as a specific case of a "two-center problem" [14.8], we follow a generic approach, readily extendable to more-complex molecules. The approach is based on the method of linear variation (see Section 12.3), where the wave functions are approximated as linear combinations of known basis functions, and the coefficients are variationally optimized. A natural choice is to expand the single-electron wave functions of the many-atom system in a basis of atomic orbitals of the single (isolated) atoms. This approach is therefore termed Molecular Orbitals as Linear Combinations of Atomic Orbitals (or MO-LCAO).

Focusing on H_2^+, a minimal basis for expanding the electronic ground state of the molecule at a finite internuclear distance, R, would correspond to the atomic orbitals of the independent atoms at their ground state, namely two 1s atomic orbitals ($\psi_{1,0,0}$), centered at the positions of the two nuclei:

$$\begin{aligned} \varphi_\alpha(R, \mathbf{r}) &\equiv \psi_{1,0,0}(\mathbf{r} + \mathbf{k}R/2) \\ \varphi_\beta(R, \mathbf{r}) &\equiv \psi_{1,0,0}(\mathbf{r} - \mathbf{k}R/2) \end{aligned} \tag{14.3.5}$$

The ansatz for the variational trial functions (Eq. (12.3.18)) therefore reads

$$\tilde{\phi}(R, \mathbf{r}) = c_\alpha(R)\varphi_\alpha(R, \mathbf{r}) + c_\beta(R)\varphi_\beta(R, \mathbf{r}). \tag{14.3.6}$$

The optimal coefficients and the corresponding energy levels are obtained according to the linear variation method by solving the corresponding secular equation (Eq. (12.3.24)),

$$[\mathbf{H}(R) - \tilde{\varepsilon}(R)\mathbf{S}(R)] = 0$$

$$\begin{bmatrix} H_{\alpha,\alpha}(R) - \tilde{\varepsilon}(R)S_{\alpha,\alpha}(R) & H_{\alpha,\beta}(R) - \tilde{\varepsilon}(R)S_{\alpha,\beta}(R) \\ H_{\beta,\alpha}(R) - \tilde{\varepsilon}(R)S_{\beta,\alpha}(R) & H_{\beta,\beta}(R) - \tilde{\varepsilon}(R)S_{\beta,\beta}(R) \end{bmatrix} \begin{bmatrix} c_\alpha(R) \\ c_\beta(R) \end{bmatrix} = \begin{bmatrix} 0 \\ 0 \end{bmatrix}. \tag{14.3.7}$$

The matrices $\mathbf{H}(R)$ and $\mathbf{S}(R)$ are, respectively, the Hamiltonian and overlap matrices (see Eq. (12.3.22)). Due to the symmetry of the electronic Hamiltonian (Eq. (14.3.4) in particular) and its Hermiticity, the following relations hold:

$$\begin{aligned} H_{\alpha,\alpha}(R) &= \langle \varphi_\alpha(R)|\hat{H}_{elec}(R)|\varphi_\alpha(R)\rangle = H_{\beta,\beta}(R) \\ H_{\beta,\alpha}(R) &= \langle \varphi_\beta(R)|\hat{H}_{elec}(R)|\varphi_\alpha(R)\rangle = H_{\alpha,\beta}(R) \end{aligned} \tag{14.3.8}$$

The diagonal matrix elements of $\mathbf{S}(R)$ are the normalization integrals of the atomic orbitals, where

$$\begin{aligned} S_{\beta,\beta}(R) &= S_{\alpha,\alpha}(R) = 1 \\ S_{\beta,\alpha}(R) &= \langle \varphi_\beta(R)|\varphi_\alpha(R)\rangle = S_{\alpha,\beta}(R) \equiv S(R) \end{aligned} \tag{14.3.9}$$

and we used the fact that the atomic orbitals are real. The two variational energy levels are obtained by solving the secular equation, Eq. (14.3.7),

$$\tilde{\varepsilon}_\pm(R) = \frac{H_{\alpha,\alpha}(R) \pm H_{\beta,\alpha}(R)}{1 \pm S(R)}, \tag{14.3.10}$$

where the corresponding eigenvectors obey the conditions (Ex. 14.3.1),

$$\frac{c_\beta^\pm(R)}{c_\alpha^\pm(R)} = \pm 1. \tag{14.3.11}$$

The normalized variational approximations to the electronic wave functions that correspond to $\tilde{\varepsilon}_\pm(R)$ are therefore (Ex. 14.3.2)

$$\tilde{\phi}_\pm(R,\mathbf{r}) = \frac{1}{\sqrt{2(1 \pm S(R))}} [\varphi_\alpha(R,\mathbf{r}) \pm \varphi_\beta(R,\mathbf{r})]. \tag{14.3.12}$$

Exercise 14.3.1 *Solve the secular equation, Eq. (14.3.7): (a) Express the eigenvalues as functions of $H_{\alpha,\alpha}(R), H_{\beta,\alpha}(R), S(R)$ (use Eqs. (14.3.8, 14.3.9)) by calculating the roots of the determinant,*

$$\begin{vmatrix} H_{\alpha,\alpha}(R) - \tilde{\varepsilon}(R)S_{\alpha,\alpha}(R) & H_{\alpha,\beta}(R) - \tilde{\varepsilon}(R)S_{\alpha,\beta}(R) \\ H_{\beta,\alpha}(R) - \tilde{\varepsilon}(R)S_{\beta,\alpha}(R) & H_{\beta,\beta}(R) - \tilde{\varepsilon}(R)S_{\beta,\beta}(R) \end{vmatrix} = 0.$$

(b) For each eigenvalue obtain the ratio between the elements $c_\beta(R)$ and $c_\alpha(R)$ of the respective eigenvector (Eq. (14.3.11)).

Exercise 14.3.2 *The variational approximations for the two lowest eigenfunctions of the electronic Hamiltonian are expressed as linear combinations of atomic orbitals, according to Eq. (14.3.6). Using the result, Eq. (14.3.11), for the corresponding relations between the expansion coefficients, show that the normalized wave functions, $\tilde{\phi}_\pm(R,\mathbf{r})$, are given by Eq. (14.3.12).*

Clearly, the (variational) solutions to the electronic Schrödinger equation depend on the interatomic distance. To analyze this dependence, we turn to explicit evaluation of the matrix elements, $H_{\alpha,\alpha}(R), H_{\beta,\alpha}(R)$ and $S(R)$. Using Eq. (14.3.5) for the atomic orbitals and Eq. (14.3.2) for the electronic Hamiltonian, we obtain (see Ex. 14.3.3)

$$H_{\alpha,\alpha}(R) = \frac{Ke^2}{R} + E_{1s} + C(R) \quad ; \quad H_{\beta,\alpha}(R) = \left(\frac{Ke^2}{R} + E_{1s}\right)S(R) + A(R), \tag{14.3.13}$$

where

$$E_{1s} = \int d\mathbf{r}\,\psi_{1,0,0}(\mathbf{r}) \left[\frac{-\hbar^2}{2m_e}\Delta_\mathbf{r} - \frac{Ke^2}{|\mathbf{r}|}\right] \psi_{1,0,0}(\mathbf{r})$$

$$C(R) = -Ke^2 \int d\mathbf{r} \frac{|\psi_{1,0,0}(\mathbf{r})|^2}{|\mathbf{r} - kR|}$$

$$A(R) = -Ke^2 \int d\mathbf{r} \frac{\psi_{1,0,0}(\mathbf{r} - kR)\psi_{1,0,0}(\mathbf{r})}{|\mathbf{r} - kR|} \tag{14.3.14}$$

$$S(R) = \int d\mathbf{r}\,\psi_{1,0,0}(\mathbf{r} - kR)\psi_{1,0,0}(\mathbf{r})$$

E_{1s} is simply the ground state energy of a hydrogen atom, $E_{1s} = -R_H$ (see Section 10.2). The terms $C(R)$ and $A(R)$ are known as the single-electron "Coulomb integral" and "exchange integral," respectively (to be distinguished from the two-electron Coulomb and exchange integrals, Eqs. (13.4.6, 13.4.7)), and $S(R)$ is simply the overlap integral between two atomic $1s$ orbitals centered at the two nuclei. The explicit dependence of these terms on R can be readily obtained by integrating over the electronic coordinates. We already encountered the Coulomb integral when discussing the effect of a remote point of charge $q|e|$ on the stationary states of a hydrogen atom, using perturbation theory (see Section 12.2). To carry out these integrals, it is convenient to follow the same practice, namely, to change variables to elliptical coordinates, which yields (see Ex. 14.3.4)

$$S(R) = e^{-R/a_0}\left[\frac{R^2}{3a_0^2} + \frac{R}{a_0} + 1\right],\tag{14.3.15}$$

$$C(R) = \frac{-Ke^2}{R}\left[1 - \left(\frac{R}{a_0} + 1\right)e^{\frac{-2R}{a_0}}\right],\tag{14.3.16}$$

$$A(R) = \frac{-Ke^2}{a_0}\left[e^{-R/a_0}\left(\frac{R}{a_0} + 1\right)\right],\tag{14.3.17}$$

where a_0 is Bohr's radius.

Exercise 14.3.3 *Obtain the expressions in Eqs. (14.3.13, 14.3.14) for the integrals, $H_{\alpha,\alpha}(R)$ and $H_{\beta,\alpha}(R)$, as defined in Eq. (14.3.8). Use Eq. (14.3.5) and Fig. 10.2.2 for the explicit form of the atomic orbitals, and Eq. (14.3.2) for the electronic Hamiltonian. Change integration variables to comply with the definitions of the integrals in Eq. (14.3.14).*

Exercise 14.3.4 *To derive the results, Eqs. (14.3.15–14.3.17), for the integrals defined in Eq. (14.3.14), first follow steps (a) and (b) of Ex. 12.2.8. Then:*

(a) *Set $f(x,y,z) = -Ke^2\frac{|\psi_{1,0,0}(\mathbf{r})|^2}{|\mathbf{r}-k\mathbf{R}|}$ and obtain the result for $C(R)$ in Eq. (14.3.16). (This amounts to setting the charge to $q = 1$ in the result of Ex. 12.2.8.)*

(b) *Set $f(x,y,z) = -Ke^2\frac{\psi_{1,0,0}(\mathbf{r}-k\mathbf{R})\psi_{1,0,0}(\mathbf{r})}{|\mathbf{r}-k\mathbf{R}|}$ and obtain the result for $A(R)$ in Eq. (14.3.17).*

(c) *Set $f(x,y,z) = \psi_{1,0,0}(\mathbf{r}-k\mathbf{R})\psi_{1,0,0}(\mathbf{r})$ and obtain the result for $S(R)$ in Eq. (14.3.15).*

(d) *Show that $H_{\alpha,\alpha}(R) = -R_H + 2R_H\left(1 + \frac{a_0}{R}\right)e^{-2R/a_0}$ and $H_{\alpha,\beta}(R) = R_H\left(\frac{2a_0}{R} - 1 - \frac{7R}{3a_0}\right.$*
 $\left.- \frac{R^2}{3a_0^2}\right)e^{-R/a_0}$.

(e) *Show that as the interatomic distance exceeds the size of a single atom, that is, for $R > a_0$, the coupling matrix element is negative, namely $H_{\alpha,\beta}(R) = -|H_{\alpha,\beta}(R)|$.*

It is important to notice that the Coulomb integral, $C(R)$, has a clear classical meaning. It describes the attraction of an electron density corresponding to the ground state of a hydrogen atom to another positive charge associated with a second nucleus (a proton in this case). Indeed, as the interatomic distance increases far beyond the size of the atom, namely $R \gg a_0$, and the second nucleus becomes external to the electronic density (which decays as $\propto e^{-2R/a_0}$), the Coulomb integral reads $C(R) \rightarrow \frac{-Ke^2}{R}$, which is the classical attraction energy between a proton and an electron. In contrast, the exchange integral, $A(R)$, has no classical analogue, since it appears as if the electron is delocalized between the two atomic orbitals, while interacting explicitly with only one of the two nuclei. This integral decays to zero as $R \gg a_0$, similarly to the overlap integral $(S(R))$ between two atomic orbitals, each centered at another nucleus (see Eqs. (14.3.15, 14.3.17)).

The calculated variational electronic energies, $\tilde{\varepsilon}_+(R)$ and $\tilde{\varepsilon}_-(R)$, according to Eqs. (14.3.10, 14.3.13, 14.3.15–14.3.17), are plotted in Fig. 14.3.2 as functions of the internuclear distance. Remarkably, while approximate, the qualitative results of this treatment are correct. The electronic ground state energy, $\tilde{\varepsilon}_+(R)$, has a point of minimum at a finite internuclear distance, R_0, suggesting the existence of a stable chemical bond, at a corresponding "bond-length," R_0. Within the BO approximation, a minimum along the electronic energy curve corresponds to a "potential energy well," which confines the nuclear motion and supports bound vibrational states (see Chapter 8). For interatomic distances smaller than R_0 the electronic energy increases and diverges asymptotically, $\tilde{\varepsilon}_+(R) \xrightarrow[R \to 0]{} Ke^2/R$ (Ex. 14.3.5), owing to the electrostatic repulsion between the two positively charged nuclei (protons). For interatomic distances larger than R_0, the electronic energy increases, but up to a constant asymptotic value, $\tilde{\varepsilon}_+(R) \xrightarrow[R \to \infty]{} -R_H$, which is Rydberg's constant (Ex. 14.3.5). This limit corresponds to a "dissociated" H_2^+ molecule, $H_2^+ \to H + H^+$, composed of a neutral hydrogen atom in its ground state, and a nucleus (proton) at an infinite distance. The "dissociation energy" is the difference between the asymptotic energy $(-R_H)$ and the ground state energy, $E_{+,0}$, of the bound molecular ion (see Fig. 14.3.2). Within the BO approximation, $E_{+,0}$ corresponds to the zero-point energy of the nuclei (see Sections 8.3 and 8.4) at the nuclear potential energy well, set by $\tilde{\varepsilon}_+(R)$. The corresponding ground state wave function (Eq. (14.2.22)) reads $\tilde{\Psi}_{+,0}(R,\mathbf{r}) = \tilde{\phi}_+(R,\mathbf{r})\tilde{\chi}_+^{(0)}(R)$, where $\tilde{\phi}_+(R,\mathbf{r})$ is the electronic function corresponding to $\tilde{\varepsilon}_+(R)$, and $\tilde{\chi}_+^{(0)}(R)$ is the nuclear wave function of minimal energy.

Exercise 14.3.5   *(a) Use Eqs. (14.3.13, 14.3.15–14.3.17) to show that as the interatomic distance goes to zero, the electronic energies reflect the classical electrostatic repulsion between the nuclei, $\tilde{\varepsilon}_\pm(R) \xrightarrow[R \to 0]{} \frac{Ke^2}{R}$. (Show that $H_{\alpha,\alpha}(R) = R_H \left(-3 + \frac{2a_0}{R} + o(R^2)\right)$,
$H_{\beta,\alpha}(R) = R_H \left(-3 + \frac{2a_0}{R} - \frac{R}{3a_0} + o(R^2)\right)$, and $S(R) = 1 - \frac{R^2}{6a_0^2} + o(R^3)$). (b) Show that as the interatomic distance becomes infinite, the electronic energies converge to the ground state energy of an isolated hydrogen atom, $\tilde{\varepsilon}_\pm(R) \xrightarrow[R \to \infty]{} -R_H$.*

Exercise 14.3.6   *(a) Use Eqs. (14.3.10, 14.3.13) to show that the difference between the two electronic energies reads $\tilde{\varepsilon}_+(R) - \tilde{\varepsilon}_-(R) = 2\frac{A(R) - S(R)C(R)}{1 - S(R)^2}$. (b) Use Eqs. (14.3.15–14.3.17) to show that as the internuclear distance becomes large in comparison to atomic sizes, $R \gg a_0$, the energy splitting between the electronic states, $\tilde{\varepsilon}_+(R) - \tilde{\varepsilon}_-(R)$, decays exponentially, as $\propto \frac{R}{a_0}e^{-R/a_0}$, and becomes proportional to the exchange integral, $A(R)$.*

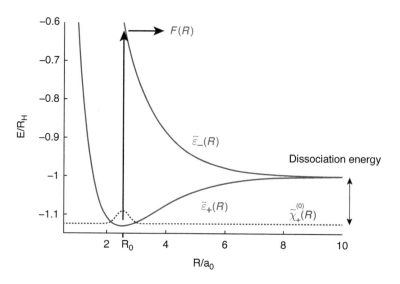

Figure 14.3.2 The two lowest electronic energies, $\tilde{\varepsilon}_+(R)$ and $\tilde{\varepsilon}_-(R)$, obtained by the linear variation approach within the BO approximation for the H_2^+ molecular ion, are presented as solid curves. The electronic ground state has a minimum at $R_0 = 2.49a_0$, where $\tilde{\varepsilon}_+(R_0) = -1.13R_H$, suggesting the existence of a stable molecular ion. The corresponding vibrational ground state wave function, $\tilde{\chi}_+^{(0)}(R)$, and the corresponding nuclear zero-point energy, $E_{+,0}$, are presented by the dotted line. The bond dissociation energy in the electronic ground state is represented by the double arrow. The electronic excitation energy at R_0 is represented by the vertical arrow, and the repulsive force between the nuclei on the excited electronic states is represented schematically by the horizontal arrow.

In the excited electronic state, the electronic energy, $\tilde{\varepsilon}_-(R)$ does not have a minimum. Consequently, the nuclear potential energy curve is repulsive for any R. This means that the force on the nuclei, $F(R) = -d\tilde{\varepsilon}_-(R)/dR$, is always in the direction of increasing R. An electronic excitation of H_2^+ into its first excited state would therefore correspond to a dissociation of the chemical bond.

The energy gap between the two electronic energies closes as $R \gg a_0$, where the two states approach their asymptotic value, $-R_H$. As already discussed, this limit corresponds to a dissociated molecule, where the electron seems to "pick" one of the nuclei to form a hydrogen atom, and the interaction of this neutral atom with a remote point charge decays asymptotically to zero. This description is qualitatively correct, but it is somewhat naïve, since, as discussed in what follows, at any finite R the symmetry of the system (e.g., Eq. (14.3.4)) implies that the electronic stationary states are associated with equal probability density for the electron to be located near the two identical nuclei. The situation is reminiscent of a particle in a symmetric double well potential, discussed in Section 6.4, where the two nuclei define two potential energy wells for the electron. The "tunneling splitting" between the two lowest energy levels in this double well system indeed corresponds to these delocalized states. A quantitative calculation of $\tilde{\varepsilon}_+(R) - \tilde{\varepsilon}_-(R)$ shows that for $R \gg a_0$; the energy splitting is proportional to the single-electron exchange integral, $A(R)$ (see Ex. 14.3.6)), which is attributed to the nonclassical delocalization of the electronic wave function between

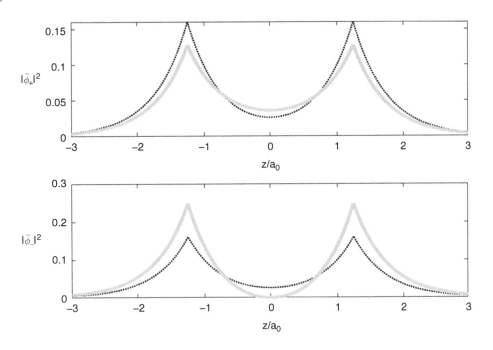

Figure 14.3.3 Solid lines: Probability densities for the bonding ($|\tilde{\phi}_+(R_0, \mathbf{r})|^2$) and antibonding ($|\tilde{\phi}_-(R_0, \mathbf{r})|^2$) molecular orbitals for H_2^+, plotted as functions of the electronic coordinate $\mathbf{r} = (0, 0, z)$, for the internuclear distance, $R_0 = 2.49a_0$. Dotted lines: Probability densities for a $1s$ atomic orbital, evenly split ("classically delocalized") between the two nuclei.

the two nuclei. Importantly, this exchange (delocalization) effect is the main source for the relative stabilization (lower energy) of the molecule in comparison to its dissociated components. This principle generalizes to more-complex many-atom systems as well.

We now turn to discussing the correlation between the properties of the different electronic energy levels, $\tilde{\varepsilon}_+(R)$ and $\tilde{\varepsilon}_-(R)$, and the corresponding molecular orbitals, $\tilde{\phi}_\pm(R, \mathbf{r})$, as given in Eq. (14.3.12) and presented in Fig. 14.3.3. The two probability densities, $|\tilde{\phi}_\pm(R, \mathbf{r})|^2$, are even functions of z; namely, they are evenly delocalized between the two nuclei (see Fig. 14.3.3). Nevertheless, they differ from each other in a way that reflects the different molecular orbitals (see Eq. (14.3.12)). In $\tilde{\phi}_+(R, \mathbf{r})$ the two atomic orbitals appear with the same sign, whereas they have opposite signs in $\tilde{\phi}_-(R, \mathbf{r})$. Put differently, in accordance with the inversion symmetry of the Hamiltonian (Eq. (14.3.4)), both molecular orbitals are eigenfunctions of the inversion symmetry operator, but they correspond to different eigenvalues; namely, $\tilde{\phi}_\pm(R, \mathbf{r})$ correspond to the eigenvalues ± 1 (even and odd functions of z).

The origin of the energy splitting between the two states, and the fact that $\tilde{\phi}_+(R, \mathbf{r})$ is associated with the lower energy can be qualitatively explained by considering the difference between the orbitals and its influence on the electrostatic energy of the system. What destabilizes the "chemical bond" is the repulsion between the two positively charged nuclei. The presence of a negatively charged electron may compensate for that by its mutual attraction to the two nuclei, as long as it is positioned between the nuclei (a screening effect). In H_2^+, optimal stabilization is obtained when the electron is right in the middle between the two positive nuclei. Inspecting the probability densities, $|\tilde{\phi}_+(R,\mathbf{r})|^2$ and $|\tilde{\phi}_-(R,\mathbf{r})|^2$, in the region between the nuclei, namely, in the plane perpendicular to the bond axis ($z = 0$), we can readily see that they are different. While in $|\tilde{\phi}_+(R,\mathbf{r})|^2$ the two atomic orbitals interfere constructively, building up finite probability density in that plane between the nuclei, the destructive interference between the atomic orbitals in $|\tilde{\phi}_-(R,\mathbf{r})|^2$ yields zero probability density in that plane ($z = 0$ is a "nodal plane" in this case). The effect is emphasized in Fig. 14.3.3 by comparing $|\tilde{\phi}_+(R,\mathbf{r})|^2$ and $|\tilde{\phi}_-(R,\mathbf{r})|^2$ to the probability density of an atomic orbital, evenly split between the two nuclei (a "classically delocalized" orbital, $\frac{1}{2}|\varphi_\alpha(R,\mathbf{r})|^2 + \frac{1}{2}|\varphi_\beta(R,\mathbf{r})|^2$). As we can see, the probability density near the $z = 0$ plane is enhanced in $|\tilde{\phi}_+(R,\mathbf{r})|^2$ and suppressed in $|\tilde{\phi}_-(R,\mathbf{r})|^2$ in comparison to the "classically delocalized" atomic orbital. Since the electrostatic energy decreases in correlation to the increase of the probability density in this plane, and vice versa, the energy associated with $\tilde{\phi}_+(R,\mathbf{r})$ decreases, and the energy associated with $\tilde{\phi}_-(R,\mathbf{r})$ increases in comparison to an atomic orbital energy (see Fig. 14.3.2). As we have highlighted, this energy difference is attributed primarily to the quantum mechanical single-electron exchange integral.

The decrease/increase in molecular orbital energies in which the two atomic orbitals appear in the same/opposite sign is universal and applies in general. Consequently, *molecular orbitals that do not have any nodal plane (do not change their sign) perpendicular to the line connecting the two nuclei are termed "bonding orbitals," while molecular orbitals that have such a nodal plane are termed "antibonding orbitals."*

14.4 Linear Combinations of Atomic Orbitals (LCAO) in Molecules

As we saw in the previous section, insight was gained on the nature of the chemical bond in the single-electron molecule, H_2^+, by invoking the BO approximation and the linear variation approach. The idea of MO-LCAO can be readily extended to different molecules as well as to extended many-atom systems, including atomic lattices. In the discussion of H_2^+, the variational calculation was limited to a "minimal basis" composed of the two 1s orbitals associated with the two hydrogen nuclei. However, even for this simple molecule the variational space must be extended (see Section 12.3, and Eq. (12.3.28) in particular) to approach the exact ground state energy (as a function of the internuclear distance) and to obtain higher electronically excited states. When it comes to many-electron molecules, extending the linear variation space becomes necessary. Recalling that electrons are fermions, Pauli's exclusion principle (see Section 13.3) limits the occupation of each spin orbital to a single electron. Consequently, the number of (linearly independent) spin-orbitals needed for approximating the ground state of a many-electron molecule must be larger than (or at least equal to) the number of electrons, and the space of linear variation must increase accordingly.

A natural way to increase the number of calculated molecular orbitals within a linear variation approximation is to increase the basis by adding more atomic orbitals of the atoms composing the molecule. Denoting the different atoms by their indexes, $\alpha = 1, 2, \ldots, n$, a set of N_α atomic orbitals associated with the αth atom is denoted as $\{\varphi_{\alpha,1}(\mathbf{r} - \mathbf{R}_\alpha), \varphi_{\alpha,2}(\mathbf{r} - \mathbf{R}_\alpha), \ldots, \varphi_{a,N_\alpha}(\mathbf{r} - \mathbf{R}_\alpha)\}$, where $\varphi_{a,n_\alpha}(\mathbf{r} - \mathbf{R}_\alpha)$ is an "atomic orbital" centered around the nucleus position, $\mathbf{R}_\alpha$. The atomic orbitals are associated with the known atomic angular function (of types s, p, d, etc.; see Chapter 13) and a fixed radial function (e.g., an exponent or a Gaussian, multiplied by a polynomial [13.6]). The linear variation approximations for the *molecular orbitals*, $\tilde{\phi}_1^{MO}(\mathbf{r}), \tilde{\phi}_2^{MO}(\mathbf{r}), \tilde{\phi}_3^{MO}(\mathbf{r}), \ldots$, therefore read

$$\tilde{\phi}_k^{MO}(\mathbf{r}) = \sum_{\alpha=1}^{n} \sum_{n_\alpha=1}^{N_\alpha} c_{\alpha,n_\alpha}^{(k)} \varphi_{a,n_\alpha}(\mathbf{r} - \mathbf{R}_\alpha), \qquad (14.4.1)$$

where the variational parameters to be optimized are the expansion coefficients, $\{c_{\alpha,n_\alpha}^{(k)}\}$. The ground state of the electronic Hamiltonian corresponds to a Slater determinant, namely an antisymmetrized product of molecular orbitals (Eqs. (13.3.2, 13.3.3)), where the optimal orbitals are self-consistent solutions to the nonlinear Hartree–Fock equations (Eq. (13.3.24)),

$$\begin{aligned}
&\left[\frac{-\hbar^2}{2m_e}\Delta_{\mathbf{r}} + \sum_{\alpha=1}^{n} \frac{-KZe^2}{|\mathbf{r} - \mathbf{R}_\alpha|}\right] \tilde{\phi}_k^{MO}(\mathbf{r}) + \sum_{j=1}^{N} \int d\mathbf{r}' \frac{Ke^2}{|\mathbf{r} - \mathbf{r}'|} |\tilde{\phi}_j^{MO}(\mathbf{r}')|^2 \tilde{\phi}_k^{MO}(\mathbf{r}) \\
&- \sum_{j=1}^{N} \delta_{m_{s,j}, m_{s,k}} \int d\mathbf{r}' \left(\tilde{\phi}_j^{MO}(\mathbf{r}')\right)^* \tilde{\phi}_j^{MO}(\mathbf{r}) \frac{Ke^2}{|\mathbf{r} - \mathbf{r}'|} \tilde{\phi}_k^{MO}(\mathbf{r}') = \varepsilon_k \tilde{\phi}_k^{MO}(\mathbf{r})
\end{aligned} \qquad (14.4.2)$$

The iterative solution of these equations corresponds to calculations of the eigenstates of the Fock operator (readjusted in each iteration) instead of the Hamiltonian. This is done by representing the Fock operator as a matrix in the variational space of the selected atomic orbital. Since atomic orbitals corresponding to different atoms are generally non-orthogonal (unless for specific symmetry considerations; see the following discussion), their overlap must be accounted for, such that the secular equation obtained by the linear variation principle is a generalized eigenvalue problem (Eq. (12.3.24)) for the molecular orbitals' coefficients $\{c_{\alpha,n_\alpha}^{(k)}\}$ and the corresponding orbital energies, $\{\varepsilon_k\}$ (the Roothaan equations [13.6]).

The number of molecular orbitals obtained as solutions to the Hartree–Fock equations is limited by the number of basis functions. This number typically exceeds the number of spatial orbitals ($N/2$ or $(N+1)/2$) needed to construct a single Slater determinant for the N electrons in the molecule. The choice of the orbitals to be occupied in the electronic ground state is guided by minimizing the variational (Hartree–Fock) many-electron energy (see Eq. (13.3.28)). As in the case of atoms, for a rough estimate it is sufficient to minimize the sum over molecular orbital energies. This means that in the ground state the orbitals are "filled" from the lowest orbital energy and upward, subject to the Pauli's exclusion principle, namely, up to two electrons of different spin states per spatial orbital (the Aufbau principle). As in the case of atoms, however, there are exceptions to this general rule when the energy differences between nearby molecular orbital energies are small in comparison to the overcounted interaction energy (see Eq. (13.3.31), and the related discussion). In the case of degeneracy of molecular orbitals, Hund's rule applies, namely the ground state corresponds to a maximal number of unpaired electrons occupying the degenerate orbitals (see the discussion in Section 13.4).

LCAO for Atom Pairs

Calculations based on the BO and the Hartree–Fock approximations as just outlined often provide a qualitatively correct account for the electronic energy of molecules in their ground state, at near-equilibrium geometries. Focusing first on atom pairs, the orbital energy levels and their ground state occupation are presented schematically in Fig. 14.4.1 for some homonuclear pairs. Graphical representations of some of the molecular orbitals are presented in Fig. 14.4.2. The orbitals are classified into "bonding" and "antibonding" (marked by a star), where an antibonding orbital corresponds to the existence of a nodal plane between the nuclei, perpendicular to the bond axis. Other classifications correspond to the symmetry properties of the molecular orbitals. Identifying the z-axis as the line connecting the two atomic centers, the electronic Hamiltonian, and hence the Fock operator, commute with the angular momentum operator, $\hat{L}_z$, for any diatomic. When the atoms are identical, this also holds for the inversion operator, $\hat{P}_{x,y,z}$ (as in the case of H_2^+; see Eqs. (14.3.3, 14.3.4)). Consequently, the molecular orbitals (eigenfunctions of the Fock operator) can be identified with

$$
\begin{array}{ccc}
\sigma_u^*(2p)\ \underline{} & \sigma_u^*(2p)\ \underline{} & \sigma_u^*(2p)\ \underline{} \\
\pi_g^*(2p)\ \underline{}\ \ \underline{} & \pi_g^*(2p)\ \underline{}\ \ \underline{} & \pi_g^*(2p)\underline{\uparrow\downarrow}\ \ \underline{\uparrow\downarrow} \\[4pt]
\sigma_g(2p)\ \underline{} & \sigma_g(2p)\underline{\uparrow\downarrow} & \pi_u(2p)\underline{\uparrow\downarrow}\ \ \underline{\uparrow\downarrow} \\
\pi_u(2p)\underline{\uparrow}\ \ \underline{\uparrow} & \pi_u(2p)\underline{\uparrow\downarrow}\ \ \underline{\uparrow\downarrow} & \sigma_g(2p)\underline{\uparrow\downarrow} \\[4pt]
\sigma_u^*(2s)\underline{\uparrow\downarrow} & \sigma_u^*(2s)\underline{\uparrow\downarrow} & \sigma_u^*(2s)\underline{\uparrow\downarrow} \\
\sigma_g(2s)\underline{\uparrow\downarrow} & \sigma_g(2s)\underline{\uparrow\downarrow} & \sigma_g(2s)\underline{\uparrow\downarrow} \\[4pt]
\sigma_u^*(1s)\underline{\uparrow\downarrow} & \sigma_u^*(1s)\underline{\uparrow\downarrow} & \sigma_u^*(1s)\underline{\uparrow\downarrow} \\
\sigma_g(1s)\underline{\uparrow\downarrow} & \sigma_g(1s)\underline{\uparrow\downarrow} & \sigma_g(1s)\underline{\uparrow\downarrow} \\[6pt]
B_2 & N_2 & F_2 \\[4pt]
(BO=1) & (BO=3) & (BO=1)
\end{array}
$$

Figure 14.4.1 The order of the molecular orbitals and their occupation by electrons at the ground state of three homonuclear atom pairs, B_2, N_2 and F_2. The different orbitals are marked according to their rotational quantum number (σ, π) and their symmetry with respect to inversion (g, u), and according to the atomic orbital types dominating their LCAO expansions (namely s or p type). A star corresponds to an antibonding orbital (see Fig. 14.4.2). The bond order (BO), which corresponds to the excess of electrons in bonding orbitals, is marked for each atom pair.

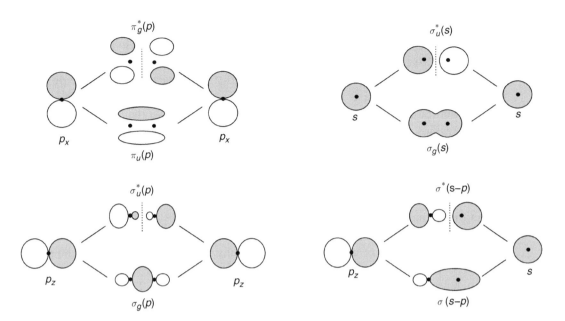

Figure 14.4.2 Schematic illustrations of molecular orbitals (middle) and the atomic orbitals (sides) dominating their LCAO expansions. The plots represent the electron probability densities, where the different colors refer to changes in the sign of the underlying functions. The vertical dotted lines mark nodal planes between the nuclei (small black dots), which correspond to antibonding orbitals. The different molecular orbitals are marked according to their rotational quantum number (σ, π) and their symmetry with respect to inversion $(g, u,$ when relevant), and according to the atomic orbital types dominating their LCAO expansions.

eigenfunctions of the operators $\hat{L}_z$ and $\hat{P}_{x,y,z}$, and can be classified according to the respective eigenvalues of these operators. The Greek letters $\sigma, \pi, \delta, \ldots$ correspond to the angular momentum quantum numbers, $|m| = 0, 1, 2, \ldots$ respectively, whereas g and u stand for the eigenvalues of the inversion operator, $+1$ and -1, which correspond, respectively, to its even ("gerade") and odd ("ungerade") eigenfunctions.

As we can see in Fig. 14.4.1, the same orbital types (e.g., σ_g^*, or π_u, and so forth) are common to the different homonuclear atom pairs, which reflect their common symmetry properties. The spatial distribution of the electronic probability density in each of the molecular orbitals as well as the orbital energies depend, as expected, on the specific atoms involved (see, e.g., Fig. 5.7 in Ref. [14.9] for the relative energies of the different molecular orbitals in different homonuclear atom pairs). The general trend is that the orbital energies decrease with increasing atomic number, which correlates with the decrease in the corresponding atomic orbital energies (see, e.g., Fig. 13.3.2). The order of the orbital energies (and therefore their corresponding electronic population) may change, however, between different atomic pairs. For example, the π_u orbitals in B_2 are lower in energy then the σ_g orbital (which means that the total spin state of the molecule in its ground state is a triplet, via Hund's rule), whereas in F_2 their order is reversed.

The set of occupied molecular orbitals in each atomic pair (and at each interatomic distance), is determined by minimizing the variational (Hartree–Fock) energy. The prospects for the formation of a "chemical bond" between the atoms can be extracted from the dependence of the electronic ground-state energy on the nuclear positions. When a minimum exists at some finite interatomic distance, a stable chemical bond between the atoms can be formed around that distance (see the discussion of H_2^+ in the previous section). A rough indicator for the existence of a chemical bond can be gained by a simple inspection of the occupation numbers of electrons in bonding versus antibonding orbitals. Referring to the discussion of H_2^+ (Fig. 14.3.3), bonding orbitals are associated with lower energy in comparison to the "classically delocalized" atomic orbitals, and therefore, electronic population in these orbitals contributes to stabilization of the chemical bond, while antibonding orbitals are associated with higher energies, and their population destabilizes the chemical bond. The "bond order" is defined as the sum of electron pairs populating bonding orbitals minus the sum of pairs populating antibonding orbitals. Whenever the bond order is larger than zero, the rough prediction is for the existence of a stable chemical bond. For example, according to this prediction, the diatomic pairs, N_2, O_2, F_2, would be stable molecules (associated with bond orders, 3, 2,1, respectively), whereas the noble gas dimer Ne_2 has a zero bond order and would not exist as a stable molecule at any interatomic distance. This is indeed the case.

Further insight into the nature of the chemical bond in different molecules can be gained by inspecting the contribution of the different atomic orbitals to each molecular orbital. Given a molecular orbital, $\tilde{\phi}_k^{MO}(\mathbf{r})$, this information is encoded in the

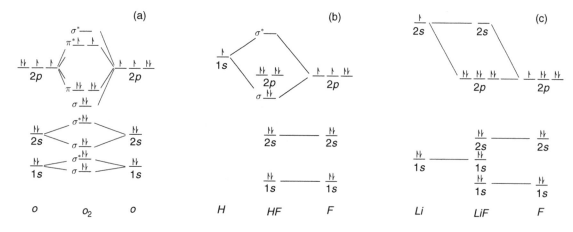

Figure 14.4.3 Orbital energy levels are presented in the middle of each plot (a, b, and c), between the corresponding atomic orbital energy levels. The orbital populations in the electronic ground state are presented by the half arrows. (a), (b), and (c) represent typical cases of "purely covalent," "polar," and "ionic" chemical bonds, according to the degree of localization of the orbitals in which the bonding electrons are populated. The schemes are not in scale, to emphasize qualitative differences.

expansion coefficients vector, $\left\{ c_{\alpha,n\alpha}^{(k)} \right\}$ (see Eq. (14.4.1)). Analysis of differentdiatomic molecules reveals that each molecular orbital is primarily dominated by only a few atomic orbitals, typically two, one from each atom, which correspond to the same quantum number for rotation around the molecular (z) axis ($|m|$), and the same (or nearly the same) atomic orbital energy. Before analyzing the origin for these two conditions, let us discuss a few examples, illustrated in Fig. 14.4.3. The molecular and atomic orbitals are presented according to their (relative) orbital energies, where each molecular orbital is correlated to the atomic orbitals, which dominate its expansion (Eq. (14.4.1)).

In a homonuclear molecule, such as O_2 (Fig. 14.4.3 (a)), each molecular orbital is shown to be dominated by a pair of atomic orbitals (one from each atom) of the same orbital energy and the same rotational quantum number, $|m|$. Notice that both s-type and p_z-type atomic orbitals correspond to $|m| = 0$, and therefore the corresponding molecular orbitals are of σ-type. The two degenerate pairs of p_x and p_y atomic orbitals both correspond to $|m| = 1$, which correspond to the two degenerate pairs of π-type molecular orbitals (see Fig. 14.4.2 for graphical illustrations of these orbitals).

In a heteronuclear molecule, such as HF (Fig. 14.4.3 (b)), the two sets of atomic orbitals corresponding to H and F do not match in energy. Nevertheless, the atomic orbital energies of $1s$ (H) and $2p(F)$ are sufficiently close (in the sense that will be elaborated in what follows). Out of the three degenerate $2p$ atomic orbitals of F, only the $2p_z$ orbital is associated with the same rotational quantum number ($|m| = 0$) as the $1s$ of H, such that two σ-type orbitals (bonding and antibonding) are formed, where only the bonding orbital turns out to be occupied at the molecular electronic ground state. Each of the other "molecular orbitals" of HF that are occupied turns out to be dominated by a single atomic orbital, whose association as molecular orbitals is essentially only formal. (These orbitals are often termed "nonbonding" molecular orbitals.) According to this MO analysis, what stabilizes the chemical bond in the molecule HF is the single electron pair populated in the single bonding molecular orbital, which corresponds to a bond of order 1.

A qualitatively different picture emerges for the pair LiF (Fig. 14.4.3 (c)). In this case, the two sets of atomic orbitals do not even nearly match in any of their relevant orbital energies. It turns out that all the populated "molecular orbitals" in the electronic ground state of LiF are dominated by single atomic orbitals, associated with either Li or F (formally, "nonbonding" molecular orbitals). The population of these orbitals at the ground electronic state should correspond to the minimal variational (Hartree–Fock) energy (roughly, to the Aufbau principle). This leads to a remarkable result when the two atoms are close to each other. The electronic ground state corresponds to a "transfer" of an electron from a high-lying $2s$ orbital of Li into an atomic orbital $2p$ of F, which is unpopulated in an isolated F atom. What stabilizes the "chemical bond" in LiF is therefore the Coulomb attraction between the positively charged Li^+ (which lost an electron) and the negatively charged F^- (which gained an electron). In contrast to the chemical bonds in O_2 or HF, which are often referred to as molecular, or covalent bonds, the chemical bond in LiF is termed "ionic."

Ionic bonds are commonly formed between an atom with a relatively high energy of its highest occupied atomic orbital (a low "ionization potential"), which is typical for metal atoms, and an atom with a relatively low energy of its lowest unoccupied atomic orbital (a high "electron affinity"), which is typical for nonmetal atoms. Ionic materials are indeed formed when electrons are transferred from metal to nonmetal atoms, resulting in bond formation between oppositely charged ions. The attraction between oppositely charged atoms stabilizes the formation of extended lattices (beyond atomic pairs).

In pure nonmetal materials, ionic bonds cannot be formed. When all the atomic ionization potentials are relatively high, the required energy for charge transfer from one atom to another can be provided only at interatomic distances, which are small in comparison to the typical size of each atom, and therefore, are typically inaccessible. In this absence of "full charging," bonds between nonmetal atoms are covalent, where the bond stabilization is attributed to charge delocalization between the nuclei (as discussed in Section 14.3, for H_2^+).

In pure metals, the stabilization energy attributed to electronic delocalization among the different atoms increases to the extent that it typically exceeds the gain in energy attributed to charging the atoms. Charge delocalization among many atoms further stabilizes the formation of extended atomic lattices, in which atoms are bounded by "metallic bonds" based on electron delocalization over the many-atom system.

The difference between ionic, molecular, and metallic bonds leads to dramatic effects on the properties of the respective materials. One example is electrical conductance, which will be addressed in Section 14.5. Specifically, ionic materials are typically insulators, and metals are conductors. Semiconductors are chemical compounds of atoms on the borderline between metals and nonmetals (such as silicon).

Effective LCAO: Symmetry and Energy Compatibility

The examples presented in Fig. 14.4.3 are representative for diatomic pairs in general, in the sense that each molecular orbital is primarily dominated by two atomic orbitals, one from each atom (in the case of bonding and antibonding molecular orbitals), or by a single atomic orbital (in the case of nonbonding orbitals). This may seem surprising at first, since each molecular orbital is formally expanded in a linear variation space, which includes numerous atomic orbitals from the two atoms (Eq. (14.4.1)). *Why then are most of the atomic orbitals "filtered out" in each molecular orbital?* The reason is that for the atomic orbitals to be effectively coupled within a molecular orbital, they must fit in their symmetry and their orbital energy, as explained in what follows.

The symmetry requirement is straightforward. Considering first the single-electron Hamiltonian of any diatomic, its cylindrical symmetry means that it is invariant to rotation around the bond (z) axis (see Fig. 14.3.1)), and therefore it commutes with the electron's rotation operator, $\hat{L}_z$ (Eq. (14.3.3)). Since by construction each atomic orbital can be chosen as an eigenfunction of $\hat{L}_z$ (notice that while the two atom centers are displaced from each other along the z-axis, the angular coordinate φ is identical in the two atoms; see Fig. 14.3.1), the Hamiltonian matrix elements between atomic orbitals associated with different eigenvalues of $\hat{L}_z$ must vanish identically (see Ex. 14.4.1). Since this rule applies to all the atomic orbitals, the matrix representation of the Hamiltonian in the linear variational space spanned by such atomic orbitals is block-diagonal, where each block corresponds to a specific eigenvalue of $\hat{L}_z$, associated with $m = 0, +/-1, +/-2, \ldots.$ Consequently, each molecular orbital (defined by an eigenvector of this block-diagonal matrix, $c^{(k)}_{\alpha,n_\alpha}$) is associated with a specific eigenvalue of $\hat{L}_z$, where atomic orbitals associated with different m values "don't mix."(Notice that when the atomic orbitals are selected as linear combinations of eigenfunction associated with m and -m, each block corresponds to |m|.) The same argument holds in the many-electron system, where the method of linear variation is applied to the nonlinear Fock operator. The matrix representation of this operator in the basis of atomic orbitals is also a block diagonal, as long as the cylindrical symmetry is imposed on the mean-field operators at each step of the self-consistent solution of the Hartree–Fock equations.

Similar considerations apply beyond diatomic molecules, namely in complex polyatomic systems. When the electronic Hamiltonian commutes with a set of symmetry operators, the atomic orbitals can be chosen to comply with specific eigenvalues of these symmetry operators. Such "symmetry-adopted" atomic orbitals lead to block-diagonal representations of the polyatomic molecular Hamiltonian (or Fock operator) in the variational basis, which divides the variational calculation of molecular orbitals into subspaces associated with specific eigenvalues of the different symmetry operators. (In group theory terms, each such subgroup of orbitals corresponds to an irreducible representation of the symmetry group defined by the symmetry operators [14.9].) This has important practical implementation, as it helps to reduce the computational effort associated with the variational solution by diagonalizing the different blocks independently.

Exercise 14.4.1   *Let the operator $\hat{H}$ commute with the Hermitian operator $\hat{A}$, and let $|\varphi_1\rangle$ and $|\varphi_2\rangle$ be two eigenvectors of $\hat{A}$, associated with two different eigenvalues, $\hat{A}|\varphi_1\rangle = \alpha_1|\varphi_1\rangle$, $\hat{A}|\varphi_2\rangle = \alpha_2|\varphi_2\rangle$, $\alpha_1 \neq \alpha_2$. (a) Show that the vectors $\hat{H}|\varphi_1\rangle$ and $\hat{H}|\varphi_2\rangle$ are eigenvectors of $\hat{A}$, corresponding to the eigenvalue, α_1 and α_2, respectively. (b) Use the Hermiticity of $\hat{A}$ to prove that $H_{1,2} = \langle \varphi_2|\hat{H}|\varphi_1\rangle = 0$.*

A second requirement for the creation of bonding and antibonding molecular orbitals as linear combinations of atomic orbitals is that the atomic orbitals must be sufficiently close in their orbital energies. *Just how close the atomic orbitals need to be* is not easy to define in the most general case. Nevertheless, it is possible to justify the requirement for energy matching and even to quantify it in terms of the molecular parameters in most cases.

For simplicity we start from a "tight-binding" model for a single electron in a system of two positively charged atomic cores at a fixed interatomic distance. The Hamiltonian obtains the generic form,

$$\hat{H} = \hat{T}_{\mathbf{r}} + \hat{V}_\alpha + \hat{V}_\beta, \qquad (14.4.3)$$

where $\hat{V}_\alpha \equiv V(\mathbf{r} - \mathbf{R}_\alpha)$ and $\hat{V}_\beta \equiv V(\mathbf{r} - \mathbf{R}_\beta)$ are potential energy operators (potential energy wells), corresponding to the attraction of the electron to two atomic cores, denoted by the indexes α and β, and $\hat{T}_\mathbf{r}$ is the kinetic energy operator. We wish to obtain molecular orbitals for this system, within a linear variation space of two specific atomic orbitals,

$$|\phi_k\rangle = c_\alpha^{(k)}|\varphi_a\rangle + c_\beta^{(k)}|\varphi_\beta\rangle, \tag{14.4.4}$$

where $|\varphi_\alpha\rangle$ and $|\varphi_\beta\rangle$, are approximations to bound states of the "isolated" atoms,

$$[\hat{T} + \hat{V}_\alpha]|\varphi_\alpha\rangle \approx \varepsilon_\alpha|\varphi_\alpha\rangle \\ [\hat{T} + \hat{V}_\beta]|\varphi_\beta\rangle \approx \varepsilon_\beta|\varphi_\beta\rangle \tag{14.4.5}$$

We now restrict the discussion to cases in which the spatial probability densities associated with $|\varphi_\alpha\rangle$ and $|\varphi_\beta\rangle$ are localized in the respective potential energy wells, such that the distance between the wells exceeds significantly the widths of these probability density distributions (see Fig. 14.4.4). In this case, one can invoke an approximation that states that the amplitude of $|\varphi_\alpha\rangle$ vanishes in the region where $\hat{V}_\beta \neq 0$, and similarly, the amplitude of $|\varphi_\beta\rangle$ vanishes in the region where $\hat{V}_\alpha \neq 0$. Additionally, the spatial overlap between $|\varphi_\alpha\rangle$ and $|\varphi_\beta\rangle$ is approximated as null. While this set of assumptions is never strictly met or any given Hamiltonian, it leads to insightful and useful discrete models for binding between the atoms, often referred to as "tight-binding" models. Formally, the following set of conditions is assumed to hold simultaneously:

$$\langle\varphi_\alpha|\hat{V}_\beta|\varphi_\alpha\rangle = \langle\varphi_\beta|\hat{V}_\alpha|\varphi_\beta\rangle = \langle\varphi_\beta|\hat{V}_\alpha|\varphi_\alpha\rangle = \langle\varphi_\beta|\hat{V}_\beta|\varphi_\alpha\rangle = 0, \tag{14.4.6}$$

$$\langle\varphi_\beta|\varphi_\alpha\rangle = 0. \tag{14.4.7}$$

According to the method of linear variation, the optimal expansion coefficients for the variational trial function, Eq. (14.4.4), are obtained by solving the secular equation (see Eq. (12.3.24)),

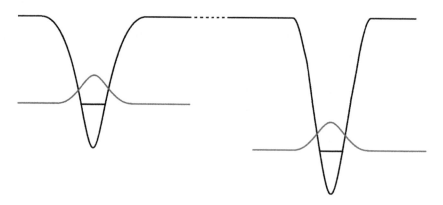

A schematic illustration of probability density distributions confined in two remote potential energy wells. The tight binding model assumptions become relevant when the distance between the two potential well minima is much larger than the standard deviations of the probability density distributions in each potential well.

$$\begin{bmatrix} H_{\alpha,\alpha} - \varepsilon_k & H_{\alpha,\beta} \\ H_{\alpha,\beta}^* & H_{\beta,\beta} - \varepsilon_k \end{bmatrix} \begin{bmatrix} c_\alpha^{(k)} \\ c_\beta^{(k)} \end{bmatrix} = \begin{bmatrix} 0 \\ 0 \end{bmatrix}, \tag{14.4.8}$$

where $H_{\alpha,\alpha} = \langle \varphi_\alpha | \hat{H} | \varphi_\alpha \rangle$, $H_{\beta,\beta} = \langle \varphi_\beta | \hat{H} | \varphi_\beta \rangle$, and $H_{\alpha,\beta} = \langle \varphi_\alpha | \hat{H} | \varphi_\beta \rangle = \langle \varphi_\beta | \hat{H} | \varphi_\alpha \rangle^*$. Solving this equation amounts to finding the eigenvalues and eigenvectors of atwo-level system Hamiltonian (see Eqs. (12.2.3–12.2.8)). The two eigenvalues,corresponding to the molecular orbital energies, therefore read

$$\varepsilon_1 = \frac{H_{\alpha,\alpha} + H_{\beta,\beta}}{2} + \frac{H_{\alpha,\alpha} - H_{\beta,\beta}}{2} \sqrt{1 + \frac{4|H_{\alpha,\beta}|^2}{(H_{\alpha,\alpha} - H_{\beta,\beta})^2}}$$

$$\varepsilon_2 = \frac{H_{\alpha,\alpha} + H_{\beta,\beta}}{2} - \frac{H_{\alpha,\alpha} - H_{\beta,\beta}}{2} \sqrt{1 + \frac{4|H_{\alpha,\beta}|^2}{(H_{\alpha,\alpha} - H_{\beta,\beta})^2}} \tag{14.4.9}$$

The respective eigenvectors, which correspond to the molecular orbitals, are

$$|\phi_1\rangle = \sqrt{\frac{\xi+1}{2\xi}} |\varphi_\alpha\rangle - \sqrt{\frac{\xi-1}{2\xi}} |\varphi_\beta\rangle \quad ; \quad |\phi_2\rangle = \sqrt{\frac{\xi-1}{2\xi}} |\varphi_\alpha\rangle + \sqrt{\frac{\xi+1}{2\xi}} |\varphi_\beta\rangle, \tag{14.4.10}$$

where $\xi \equiv \sqrt{1 + 4|H_{\alpha,\beta}|^2/(H_{\alpha,\alpha} - H_{\beta,\beta})^2}$, and where, without loss of generality, these solutions correspond to the typical case (see, e.g., the example of H_2^+ (Ex. 14.3.4e)) in which the coupling matrix element is real and negative, namely $H_{\alpha,\beta} = -|H_{\alpha,\beta}|$.

The eigenvalues and eigenvectors depend critically on the ratio between the interatomic coupling matrix element and the difference between the two "local" matrix elements at the two atoms, $2|H_{\alpha,\beta}|/|H_{\alpha,\alpha} - H_{\beta,\beta}|$. Recalling that $|\varphi_\alpha\rangle$ and $|\varphi_\beta\rangle$ are approximated eigenstates of the "isolated" atoms (see Eq. (14.4.5)), $|H_{\alpha,\alpha} - H_{\beta,\beta}|$ can be regarded as the energy difference between the two atomic orbital energies, and the critical parameter reads

$$2|H_{\alpha,\beta}|/|H_{\alpha,\alpha} - H_{\beta,\beta}| \approx 2|H_{\alpha,\beta}|/|\varepsilon_\alpha - \varepsilon_\beta|. \tag{14.4.11}$$

It is instructive to examine the two limiting cases:

Case I: The atomic orbitals match in energy, $\varepsilon_\alpha \approx \varepsilon_\beta$, where $|\varepsilon_\alpha - \varepsilon_\beta| \ll 2|H_{\alpha,\beta}|$.

In this case (see Fig. 14.4.5 (a)) we can set $4|H_{\alpha,\beta}|^2/(H_{\alpha,\alpha} - H_{\beta,\beta})^2 \gg 1$ in Eqs. (14.4.9, 14.4.10), such that the approximated molecular orbital energies (Eq. (14.4.9)) read (see Ex. 14.4.2)

$$\varepsilon_1 \approx \frac{H_{\alpha,\alpha} + H_{\beta,\beta}}{2} + |H_{\alpha,\beta}| \quad ; \quad \varepsilon_2 \approx \frac{H_{\alpha,\alpha} + H_{\beta,\beta}}{2} - |H_{\alpha,\beta}|, \tag{14.4.12}$$

where the corresponding eigenvectors are

$$|\phi_1\rangle \cong \sqrt{\frac{1}{2}} |\varphi_\alpha\rangle - \sqrt{\frac{1}{2}} |\varphi_\beta\rangle \quad ; \quad |\phi_2\rangle \cong \sqrt{\frac{1}{2}} |\varphi_\alpha\rangle + \sqrt{\frac{1}{2}} |\varphi_\beta\rangle. \tag{14.4.13}$$

As we can see, each molecular orbital is shown to be delocalized over the space of the two atoms, and therefore the molecular orbitals are significantly different from

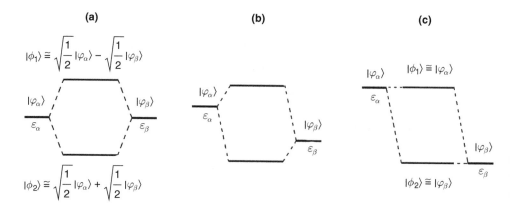

Figure 14.4.5 A schematic illustration of the effect of energy matching between atomic orbitals, on the LCAO in an atom pair. (a) Perfect energy matching resulting in perfect orbital delocalization between the two atoms. (b) The intermediate case, where the mismatch in energy is of the order of the coupling energy. (c) The limit where the energy mismatch exceeds by far the coupling energy, resulting in localized pair orbitals.

the localized atomic orbitals. Considering the sign of the atomic orbital coefficients within each molecular orbital, the linear combinations, $|\phi_2\rangle$ and $|\phi_1\rangle$, correspond to bonding and antibonding molecular orbitals, respectively. This difference is expressed also in the molecular orbital energies, ε_2 and ε_1, which are substantially different from the atomic orbital energies, where $\varepsilon_2 << \varepsilon_\alpha \approx \varepsilon_\beta$ and $\varepsilon_1 >> \varepsilon_\alpha \approx \varepsilon_\beta$, as illustrated in Fig. 14.4.5 (a). Notice the similarity of this result to the case of H_2^+ that was studied in Section 14.3, where $\varepsilon_{1,2}$ and $|\phi_{1,2}\rangle$ are reminiscent of $\tilde{\varepsilon}_{-,+}$ and $|\tilde{\phi}_{-,+}\rangle$, respectively.

Exercise 14.4.2   *Use the condition for "energy matching,"* $2|H_{\alpha,\beta}| >> |H_{\alpha,\alpha} - H_{\beta,\beta}|$, *in Eqs. (14.4.9, 14.4.10) to obtain the approximations for the orbital energies and coefficients in Eqs. (14.4.12, 14.4.13).*

Case II: The atomic orbitals mismatch in energy, $\varepsilon_\alpha \neq \varepsilon_\beta$, *where* $|\varepsilon_\alpha - \varepsilon_\beta| >> 2|H_{\alpha,\beta}|$.

Setting $4|H_{\alpha,\beta}|^2/(H_{\alpha,\alpha} - H_{\beta,\beta})^2 << 1$ in Eqs. (14.4.9, 14.4.10) leads in this case to the following approximations for the molecular orbital energies (Ex. 14.4.3),

$$\varepsilon_1 \approx H_{\alpha,\alpha} + \frac{|H_{\alpha,\beta}|^2}{H_{\alpha,\alpha} - H_{\beta,\beta}} \quad ; \quad \varepsilon_2 \approx H_{\beta,\beta} - \frac{|H_{\alpha,\beta}|^2}{H_{\alpha,\alpha} - H_{\beta,\beta}}, \tag{14.4.14}$$

and to the corresponding molecular orbitals (Eq. (14.4.10)),

$$|\phi_1\rangle \cong |\varphi_\alpha\rangle - \frac{|H_{\alpha,\beta}|}{H_{\alpha,\alpha} - H_{\beta,\beta}}|\varphi_\beta\rangle \cong |\varphi_\alpha\rangle \quad ; \quad |\phi_2\rangle \cong \frac{|H_{\alpha,\beta}|^2}{H_{\alpha,\alpha} - H_{\beta,\beta}}|\varphi_\alpha\rangle + |\varphi_\beta\rangle \cong |\varphi_\beta\rangle.$$
$$\tag{14.4.15}$$

In this case each "molecular orbital" remains nearly localized at one of the two atoms and essentially identifies with the corresponding atomic orbital (see Fig. 14.4.5(c)). This localization is expressed also in the molecular orbital energies, which are nearly equal to the atomic orbital energies. The orbitals $|\phi_2\rangle$ and $|\phi_1\rangle$ correspond in this case to nonbonding orbitals, which do not induce electronic localization and therefore hardly affect the stability of the chemical bond between the atoms.

Exercise 14.4.3 *Use the condition for "energy mismatching," $2|H_{\alpha,\beta}| << |H_{\alpha,\alpha} - H_{\beta,\beta}|$, in Eqs. (14.4.9, 14.4.10) to obtain the approximations for the orbital energies and coefficients in Eqs. (14.4.14, 14.4.15).*

Within the tight-binding model, the condition for effective mixing of two atomic orbitals to give delocalized (bonding, or antibonding) molecular orbitals is indeed unambiguous. It can be cast into a single requirement, namely, the difference between the atomic orbital energies, $|\varepsilon_\alpha - \varepsilon_\beta|$, must be small in comparison to the coupling energy, $|H_{\alpha,\beta}|$. In realistic many-atom systems the tight-binding model assumptions (Eqs. (14.4.6, 14.4.7)) do not hold strictly, and additional care must also be taken to many-electron interactions. Nevertheless, by properly redefining the parameters, $|H_{\alpha,\beta}|$, and $|\varepsilon_\alpha - \varepsilon_\beta|$, the energy-matching criterion is useful also beyond the realm of elementary tight-binding models.

Relaxing the assumptions outlined in Eqs. (14.4.6, 14.4.7), still within the single-electron framework, leads to a generalized eigenvalue equation instead of Eq. (14.4.8). The equation for the optimal variational expansion coefficients reads in this case

$$\begin{bmatrix} H_{\alpha,\alpha} - \varepsilon_k & H_{\alpha,\beta} - \varepsilon_k s \\ H_{\alpha,\beta} - \varepsilon_k s & H_{\beta,\beta} - \varepsilon_k \end{bmatrix} \begin{bmatrix} c_\alpha^{(k)} \\ c_\beta^{(k)} \end{bmatrix} = \begin{bmatrix} 0 \\ 0 \end{bmatrix}. \tag{14.4.16}$$

Here, $s = \langle \varphi_\beta | \varphi_\alpha \rangle$ is the overlap between the two atomic orbitals, and

$$\begin{aligned} H_{\alpha,\alpha} &= \varepsilon_\alpha + \langle \varphi_\alpha | \hat{V}_\beta | \varphi_\alpha \rangle \\ H_{\beta,\beta} &= \varepsilon_\beta + \langle \varphi_\beta | \hat{V}_\alpha | \varphi_\beta \rangle \\ H_{\alpha,\beta} &= \varepsilon_\alpha s + \langle \varphi_\alpha | \hat{V}_\beta | \varphi_\beta \rangle. \end{aligned} \tag{14.4.17}$$

To extract the proper conditions for obtaining effective linear combinations that correspond to bonding and antibonding orbitals (as in case I), we enforce the two expansion coefficients to be equal in magnitude (which corresponds to "perfect delocalization"),

$$c_\alpha^{(k)} = \pm c_\beta^{(k)}. \tag{14.4.18}$$

Along with Eq. (14.4.16), this condition leads to the following equations for the molecular orbital energies (Ex. 14.4.4),

$$\varepsilon_\pm = \frac{H_{\alpha,\alpha} \pm H_{\alpha,\beta}}{1 \pm s} = \frac{H_{\beta,\beta} \pm H_{\alpha,\beta}}{1 \pm s}, \tag{14.4.19}$$

from which one immediately obtains the condition $H_{\alpha,\alpha} - H_{\beta,\beta} = 0$, namely

$$\varepsilon_\alpha - \varepsilon_\beta + \langle \varphi_\alpha | \hat{V}_\beta | \varphi_\alpha \rangle - \langle \varphi_\beta | \hat{V}_\alpha | \varphi_\beta \rangle = 0. \tag{14.4.20}$$

This condition generalizes the condition for matching atomic orbital energies, as discussed for the tight-binding model ($|\varepsilon_\alpha - \varepsilon_\beta| << 2|H_{\alpha,\beta}|$), since in order to have $H_{\alpha,\alpha} - H_{\beta,\beta} \approx 0$, in addition to $\varepsilon_\alpha - \varepsilon_\beta \approx 0$, the term $\langle \varphi_\alpha | \hat{V}_\beta | \varphi_\alpha \rangle - \langle \varphi_\beta | \hat{V}_\alpha | \varphi_\beta \rangle$, which is completely ignored in the tight-binding approximation, must also be sufficiently small. Another apparent difference from the tight-binding model is that the orbital energies in Eq. (14.4.19) are not equally displaced in energy with respect to $H_{\alpha,\alpha}(H_{\beta,\beta})$. Referring again to the typical case where $H_{\alpha,\beta} = -|H_{\alpha,\beta}|$ (e.g., Ex. 14.3.4e), the stabilization (energy decrease) of the bonding molecular orbital, ε_+, is shown to be less in absolute magnitude than the destabilization (energy increase) of the anti-bonding molecular orbital, ε_- (see Ex. 14.4.4). In spite of these refinements beyond the tight-binding assumptions, we can see that as long as the overlap between the atomic orbitals remains substantially smaller than 1, and consistently, the integrals $\langle \varphi_\alpha | \hat{V}_\beta | \varphi_\alpha \rangle, \langle \varphi_\beta | \hat{V}_\alpha | \varphi_\beta \rangle, \langle \varphi_\alpha | \hat{V}_\beta | \varphi_\beta \rangle$ remain small in comparison to $\langle \varphi_\alpha | \hat{H} | \varphi_\beta \rangle$, the condition for delocalized molecular orbitals requires proximity in the atomic orbital energies.

Exercise 14.4.4   *(a) Show that solutions to the secular equation, Eq. (14.4.16), which are perfectly delocalized (defined in Eq. (14.4.18)), correspond to the orbital energies, $\varepsilon_\pm$, as given in Eq. (14.4.19). (b) Show that the orbital energies in Eq. (14.4.19) are not equally displaced in energy with respect to $H_{\alpha,\alpha}(H_{\beta,\beta})$, namely $|\varepsilon_+ - H_{\alpha,\alpha}| < |\varepsilon_- - H_{\alpha,\alpha}|$.*

Similar considerations apply also when many-electron interactions are accounted for, within the Hartree–Fock approximation. The atomic and molecular orbitals are eigenstates of the corresponding Fock operator, which includes the mean-field terms in addition to the single-electron terms in the Hamiltonian. The secular equation for the optimal variation coefficients consequently involves the matrix elements of the nonlinear molecular Fock operator. In this case, the condition for effective linear combination of atomic orbitals into delocalized molecular orbitals accounts also for matrix elements of the mean-field operators; hence, it is more cumbersome. Nevertheless, the requirements for proximity in energy of the atomic orbitals applies similarly.

The Simple Hückel Model for Conjugated Polyene Molecules

The preceding tight-binding analysis suggests that given a proper symmetry, degenerate atomic orbitals from different atoms should form delocalized molecular orbitals even in the presence of the smallest interatomic coupling. This observation is indeed useful for qualitative understanding of the electronic structure of certain types of polyatomic molecules. A remarkable example is the class of organic molecules, termed conjugated polyenes (see Fig. 14.4.6). These molecules contain chains of carbon atoms bonded to each other (and to hydrogen atoms) by σ-type bonds in a molecular plane. Each carbon atom has an additional electron that, in the absence of coupling to other atoms, would be occupied in a p-type atomic orbital, polarized in the perpendicular direction to the molecular plane. When the carbon atomic nuclei are clumped (within the BO approximation) at equal interatomic distances between neighboring atoms, each of the (degenerate) p-type atomic orbitals is coupled to its neighbors. Consequently, instead of (π-type) molecular orbitals, shared by distinctive pairs of atoms, the molecular orbitals tend to delocalize over the entire atomic chain. These orbitals are often referred to as "conjugated π orbitals," and the molecule is termed a conjugated polyene.

A simple single-electron model Hamiltonian that accounts for the molecular orbitals in such conjugated polyenes was proposed by Hückel [14.10]. This model can be derived from an effective single-electron Schrödinger equation (or the Hartree–Fock equations) by invoking a set of assumptions. Each molecular orbital is approximated as a linear combination of the p-type atomic orbitals, $\{|\varphi_n\rangle\}$, associated with the different carbon atoms in the system. For a conjugated system of N carbon atoms, $n = 1, 2, \ldots, N$, this reads

$$|\phi_k^{MO}\rangle = \sum_{n=1}^{N} c_n^{(k)} |\varphi_n\rangle. \tag{14.4.21}$$

The optimal expansion coefficients and the corresponding molecular orbital energies are obtained as the eigenvectors and eigenvalues of a model Hamiltonian, subject to tight-binding assumptions. Namely, the overlap between the atomic orbitals is assumed to be null,

$$S_{n,m} = \langle \varphi_n | \varphi_m \rangle = \delta_{n,m}, \tag{14.4.22}$$

the diagonal ("on-site") matrix elements of the single-particle Hamiltonian are assumed to depend locally on the carbon atom, independently of its position along the chain, and obtain the same value for all carbon atoms,

$$H_{n,n} = \varepsilon_\alpha, \tag{14.4.23}$$

and the electronic coupling between different sites is restricted to the nearest neighboring atoms and denoted β, where $\beta = -|\beta|$, in line with the common interatomic electronic interaction (see, e.g., Ex. 14.3.4e). Again, the nearest-neighbor couplings between any two carbon atoms are assumed to be uniform,

$$H_{n,m} = \begin{cases} \beta \, ; \, n \text{ and } m \text{ are nearest neighbors} \\ 0 \, ; \, otherwise \end{cases}. \tag{14.4.24}$$

In Fig. 14.4.7, the secular equations for the molecular orbitals energy levels and expansion coefficients are presented for two conjugated molecules of six carbon atoms, hexatriene and benzene. The different carbon atoms are indexed according to their position in each molecule. The apparent difference between these two molecules is that hexatriene is linear, whereas benzene is cyclic. This difference is reflected also in the electronic properties of these molecules, where, for example, in the electronic ground states, benzene absorbs electromagnetic radiation at shorter wavelengths in comparison to hexatriene, suggesting a difference in the electronic structure of the two molecules. This difference and its relation to the molecular geometry can indeed be revealed already at the level of Hückel's tight-binding model, as discussed in what follows.

(a)

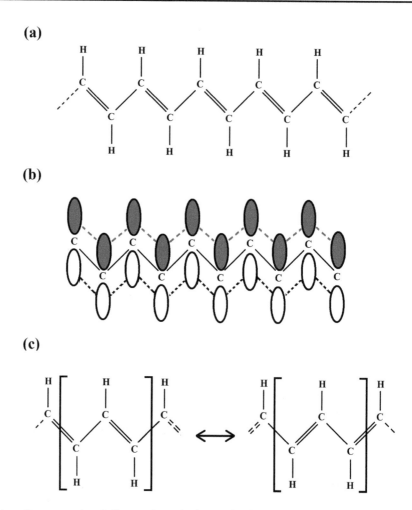

(b)

(c)

Figure 14.4.6 (a) A schematic representation of a linear conjugated polyene molecule with "π-type bonds" between carbon atoms. (b) The underlying "p-type" atomic orbitals with electronic density polarized perpendicularly to the plane defined by the atom nuclei. (c) Illustration of two equivalent arrangements of "π bonds" between neighbouring carbon atoms in a uniform infinite chain.

The molecular geometry is reflected in different "boundary conditions" at the terminal atoms in each chain. In the linear case, each terminal atom (corresponding to index 1 or 6) has only one neighbor ("reflecting" boundary conditions), whereas in the cyclic case, these atoms are not different from any other atom along the chain (periodic boundary conditions), where each of them has two neighbors. Within the Hückel model, the periodic boundary conditions are included in the Hamiltonian matrix in terms of two additional nonzero entries, $H_{1,6} = H_{6,1} = \beta$ (see Fig. 14.4.7). The consequences with respect to the orbitals and the orbital energy levels are readily revealed from the solutions to the secular equations (see Ex. 14.4.5):

$$\begin{bmatrix} \varepsilon_\alpha - \varepsilon_k & \beta & 0 & 0 & 0 & 0 \\ \beta & \varepsilon_\alpha - \varepsilon_k & \beta & 0 & 0 & 0 \\ 0 & \beta & \varepsilon_\alpha - \varepsilon_k & \beta & 0 & 0 \\ 0 & 0 & \beta & \varepsilon_\alpha - \varepsilon_k & \beta & 0 \\ 0 & 0 & 0 & \beta & \varepsilon_\alpha - \varepsilon_k & \beta \\ 0 & 0 & 0 & 0 & \beta & \varepsilon_\alpha - \varepsilon_k \end{bmatrix} \begin{bmatrix} c_1^{(k)} \\ c_2^{(k)} \\ c_3^{(k)} \\ c_4^{(k)} \\ c_5^{(k)} \\ c_6^{(k)} \end{bmatrix} = \begin{bmatrix} 0 \\ 0 \\ 0 \\ 0 \\ 0 \\ 0 \end{bmatrix}$$

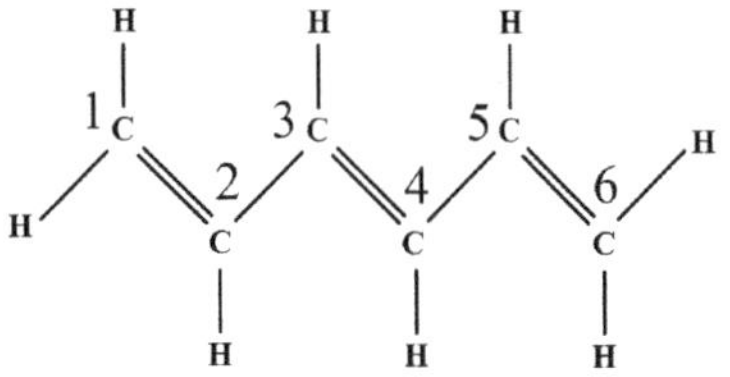

$$\begin{bmatrix} \varepsilon_\alpha - \varepsilon_k & \beta & 0 & 0 & 0 & \beta \\ \beta & \varepsilon_\alpha - \varepsilon_k & \beta & 0 & 0 & 0 \\ 0 & \beta & \varepsilon_\alpha - \varepsilon_k & \beta & 0 & 0 \\ 0 & 0 & \beta & \varepsilon_\alpha - \varepsilon_k & \beta & 0 \\ 0 & 0 & 0 & \beta & \varepsilon_\alpha - \varepsilon_k & \beta \\ \beta & 0 & 0 & 0 & \beta & \varepsilon_\alpha - \varepsilon_k \end{bmatrix} \begin{bmatrix} c_1^{(k)} \\ c_2^{(k)} \\ c_3^{(k)} \\ c_4^{(k)} \\ c_5^{(k)} \\ c_6^{(k)} \end{bmatrix} = \begin{bmatrix} 0 \\ 0 \\ 0 \\ 0 \\ 0 \\ 0 \end{bmatrix}$$

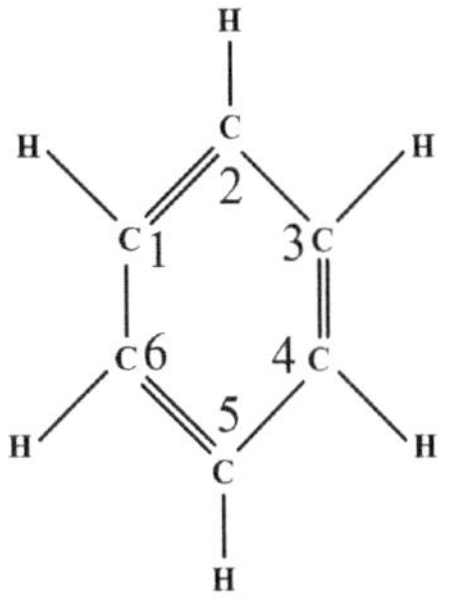

Figure 14.4.7 Two six-membered polyene molecules and the corresponding secular equations for the atomic orbital coefficients within the π-system. Notice the difference in connectivity: atoms 1 and 6 are neighbors in benzene, but not in hexatriene, which corresponds to additional entries in the respective Hamiltonian matrix.

For a uniform linear chain of N sites, the orbital energy levels obtain the form

$$\varepsilon^{(k)} = \varepsilon_\alpha + 2\beta \cos\left(\frac{\pi k}{(N+1)}\right) \quad ; \quad k = 1,2,\ldots,N, \tag{14.4.25}$$

and the corresponding normalized expansion coefficients are

$$c_n^{(k)} = \sqrt{\frac{2}{N+1}} \sin\left(\frac{n\pi k}{(N+1)}\right). \tag{14.4.26}$$

For a cyclic uniform chain of N sites ($N > 2$), the orbital energy levels obtain the form

$$\varepsilon^{(k)} = \varepsilon_\alpha + 2\beta \cos\left(\frac{2\pi k}{N}\right) \quad ; \quad k = 1,2,\ldots,N, \tag{14.4.27}$$

and the corresponding normalized expansion coefficients are

$$c_n^{(k)} = \frac{1}{\sqrt{N}} e^{-i\frac{2\pi kn}{N}}, \tag{14.4.28}$$

where $c_n^{(k)} = c_{n+N}^{(k)}$.

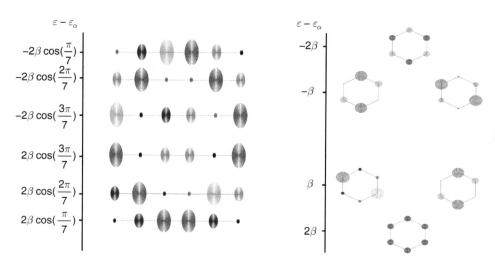

Figure 14.4.8   Illustration of delocalized molecular orbitals for the π electrons in hexatriene and benzene according to the simple Hückel model, at the corresponding orbital energies, expressed in terms of the atomic (on-site) orbital energies ε_α and the interatomic, nearest-neighbor coupling matrix element, β. The area of the circle around each atom center is proportional to the squared coefficient of the respective atomic orbital in the expansion of each molecular orbital. The different shades correspond to sign alternations in each molecular orbital.

Exercise 14.4.5

(a) The secular equation for the Hückel model for a uniform linear chain reads (see Fig. 14.4.7)

$$\begin{cases} (\varepsilon_\alpha - \varepsilon_k)c_1^{(k)} + \beta c_2^{(k)} = 0 & ; \quad n = 1 \\ \beta c_{n-1}^{(k)} + (\varepsilon_\alpha - \varepsilon_k)c_n^{(k)} + \beta c_{n+1}^{(k)} = 0 & ; \quad 1 < n < N. \\ \beta c_{N-1}^{(k)} + (\varepsilon_\alpha - \varepsilon_k)c_N^{(k)} = 0 & ; \quad n = N \end{cases}$$

Show that the orbital energies (Eq. (14.4.25)) and coefficients (Eq. (14.4.26)) satisfy these secular equations. Show that $\sum_{n=1}^{N} (c_n^{(k)})^2 = 1$.

(b) The secular equation for the Hückel model for a uniform cyclic chain reads (see Fig. 14.4.7)

$$\begin{cases} (\varepsilon_\alpha - \varepsilon_k)c_1^{(k)} + \beta c_2^{(k)} + \beta c_N^{(k)} = 0 & ; \quad n = 1 \\ \beta c_{n-1}^{(k)} + (\varepsilon_\alpha - \varepsilon_k)c_n^{(k)} + \beta c_{n+1}^{(k)} = 0 & ; \quad 1 < n < N. \\ \beta c_1^{(k)} + \beta c_{N-1}^{(k)} + (\varepsilon_\alpha - \varepsilon_k)c_N^{(k)} = 0 & ; \quad n = N \end{cases}$$

Show that the orbital energies (Eq. (14.4.27)) and coefficients (Eq. (14.4.28)) satisfy these secular equations. Show that $\sum_{n=1}^{N} (c_n^{(k)})^2 = 1$.

The results of the Hückel model for the conjugated π orbitals in hexatriene and benzene are illustrated in Fig. 14.4.8. The orbital energies are calculated according to

Eqs. (14.4.25, 14.4.27) by setting $N = 6$. As we can see, the differences in the molecular geometry are reflected in the orbital energies. Particularly, degenerate π orbitals appear only in the cyclic system (benzene), which is reminiscent of the degeneracy encountered for a particle-on-a-ring energy, discussed in Section 5.4. Another apparent difference between the two systems is the energy gap between the Highest Occupied Molecular Orbital (HOMO) and the Lowest Unoccupied Molecular Orbital (LUMO). The HOMO and LUMO can be identified in each case by populating the molecular orbitals of the conjugate π system according to the Pauli exclusion and the Aufbau principles (see Chapter 13), recalling that each carbon atom contributes a single electron to the system. The Pauli exclusion principle means that at the electronic ground state only half of the orbitals will be populated by electrons. Inspecting the results of the Hückel model, the gap between the HOMO and LUMO energies is indeed larger for the cyclic molecule. Since this gap defines the minimal electronic excitation energy, the model prediction is that the excitation energy in benzene should be larger than the excitation energy in hexatriene. This is in qualitative agreement with the measured absorption spectra of the two molecules (in the UV regime of the electromagnetic radiation), where benzene absorbs at shorter wavelengths.

The results of the Hückel model (Eqs. (14.4.25, 14.4.27)) reveal another important aspect of conjugated π systems, which characterizes also extended solid-state systems (to be discussed in the next section): ***The spacing between neighboring orbital energies decreases as the number of sites in the system increases. This is another manifestation of the quantum size effect***, introduced in Chapter 5, where the "size" of the system is correlated here with the number of coupled atomic sites. For a linear chain, this principle reads (see Ex. 14.4.6)

$$\varepsilon^{(k+1)} - \varepsilon^{(k)} = 2\beta \left\{ \cos \left[\frac{(k+1)\pi}{(N+1)} \right] - \cos \left[\frac{k\pi}{(N+1)} \right] \right\} \xrightarrow[N\to\infty]{} 0, \tag{14.4.29}$$

whereas for a cyclic chain we have

$$\varepsilon^{(k+1)} - \varepsilon^{(k)} = 2\beta \left\{ \cos \left[\frac{2(k+1)\pi}{N} \right] - \cos \left[\frac{2k\pi}{N} \right] \right\} \xrightarrow[N\to\infty]{} 0. \tag{14.4.30}$$

Exercise 14.4.6 *Prove the asymptotic results in Eqs. (14.4.29, 14.4.30).*

14.5 Extended Systems and Energy Band Formation

The perception of chemical bonding within the BO and the orbital approximations is extendable also to solid crystals, composed of different materials (insulators, conductors, or semiconductors) with different types of chemical bonds (ionic, molecular, or metallic). The number of atoms in such systems can be astronomically large, and yet the principles outlined so far are useful for analyzing the physical properties (e.g., electric conductivity, interaction with radiation) as well as chemical properties (stability, reactivity) of such extended systems. An important characteristic of crystalline materials is their periodic structure at their ground (minimal energy) state. The invariance of the electron-pair interaction to translation implies that the periodicity is attributed to the underlying single-electron potential energy. Consequently, the entire crystal can be mapped by discrete translations of a unit-cell Hamiltonian, containing a finite number of atoms (or order 1–100, typically). This simplifies considerably the theoretical description within the orbital approximation by restricting each orbital to satisfy the same periodic boundary conditions. Notice, however, that at the crystal boundaries (surfaces), the translational symmetry is broken, and the finite dimensions of the crystal must be accounted for, giving rise to quantum size effects on top of the effect of the underlying crystal unit cell. This boundary effect is particularly important for nanocrystals (see the discussion of envelope functions in Section 12.2). In this section we focus on extended, infinitely periodic crystals, representing bulk materials. For simplicity, our introduction to the subject will be restricted to one spatial dimension, where the generalization to a three-dimensional lattice will be mentioned only briefly.

Let us envision an infinitely periodic linear chain of atoms as appear in an ordered crystal (see Fig. 14.5.1). Each atom is associated with a single positive nucleus (clumped at its minimal energy position, within a BO framework) and several (many) electrons. If the electron–electron interaction is neglected, each electron in the system can be regarded separately as a particle in a system of identical multiple potential wells. As we shall see, the single-particle states (orbitals) in such a system are delocalized over the entire lattice (see also Section 6.5). The same picture holds if electron–electron interactions are accounted or within a mean-field (orbital) approximation, where the Fock operator (Eqs. (13.3.18–13.3.20)) obtains the lattice periodicity. In what follows we first analyze the exact solutions to the Schrödinger equation for a single particle in a continuous, infinitely periodic, one-dimensional potential, introducing Bloch's theorem, and then we apply a variational LCAO approach within the tight-binding approximation to the same problem.

The Bloch Theorem

The Schrödinger equation for a particle of mass m in a periodic one-dimensional potential, with a lattice period, a, along the x coordinate, $-\infty < x < \infty$, reads

$$\left[\frac{-\hbar^2}{2m} \frac{\partial^2}{\partial x^2} + V(x) - E \right] \psi(x) = 0 \quad ; \quad V(x) = V(x+a), \tag{14.5.1}$$

where the periodic potential energy can be expanded as a Fourier series,

$$V(x) = \sum_{n=-\infty}^{\infty} V_n e^{\frac{i2\pi n}{a} x}. \tag{14.5.2}$$

It is useful in this case to expand the proper solutions to the Schrödinger equation as (Fourier) integrals over an orthonormal set of one-dimensional plane waves (Eq. (11.5.22)),

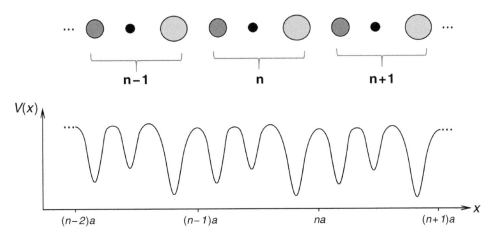

Figure 14.5.1   Illustration of an infinitely periodic atomic lattice (three periods are presented), and a corresponding effective single-particle potential with the lattice period (a).

$$\psi(x) = \frac{1}{\sqrt{2\pi}} \int_{-\infty}^{\infty} e^{ikx}\overline{\psi}(k)dk \quad ; \quad \overline{\psi}(k) = \frac{1}{\sqrt{2\pi}} \int_{-\infty}^{\infty} e^{-ikx}\psi(x)dx, \tag{14.5.3}$$

where $k = p_x/\hbar$ is the wave vector that corresponds to the momentum, p_x. Substituting the expansion $\psi(x)$ in the Schrödinger equation (Eq. (14.5.1)) and projecting on a plane wave, e^{ikx}, we obtain (see Ex. 14.5.1)

$$\sum_{n=-\infty}^{\infty} \left[V_n + \delta_{n,0} \left(\frac{\hbar^2 k^2}{2m} - E \right) \right] \overline{\Psi} \left(k - \frac{2\pi}{a}n \right) = 0. \tag{14.5.4}$$

The differential equation in the coordinate representation therefore translates into an algebraic equation in the wave vector representation, where, according to Eq. (14.5.4), the value of $\overline{\psi}(k)$ at each point in k-space depends on its value at a discrete (infinite) set of points, $k - \frac{2\pi}{a}n$, where n is an integer $(-\infty < n < \infty)$. To obtain these values, the Schrödinger equation can be projected on the corresponding full set of plane waves, $\{e^{i(k-\frac{2\pi}{a}n)x}\}$, which yields a closed (infinite) set of coupled equations for $\overline{\psi}(k)$ at the set of points (see Ex. 14.5.1):

$$\sum_{n'=-\infty}^{\infty} \left[V_{n'-n} + \delta_{n,n'} \left(\frac{\hbar^2}{2m} \left(k - \frac{2\pi}{a}n \right)^2 - E \right) \right] \overline{\Psi} \left(k - \frac{2\pi}{a}n' \right) = 0 \quad ; \quad -\infty < n < \infty. \tag{14.5.5}$$

Recalling the integral definition of $\overline{\psi}(k)$ (Eq. (14.5.3)) and changing the integration variable, we can readily see that the functions, $\psi(x)$ and $e^{-ika}\psi(x+a)$, correspond precisely to the same set of values, $\{\overline{\psi}\left(k - \frac{2\pi n}{a}\right), -\infty < n < \infty\}$, where

$$\overline{\Psi} \left(k - \frac{2\pi n}{a} \right) = \frac{1}{\sqrt{2\pi}} \int_{-\infty}^{\infty} e^{i\frac{2\pi}{a}nx} e^{-ikx} \psi(x)dx = \frac{1}{\sqrt{2\pi}} \int_{-\infty}^{\infty} e^{i\frac{2\pi}{a}nx} e^{-ikx} e^{-ika} \psi(x+a)dx, \tag{14.5.6}$$

which means that the corresponding solutions to the Schrödinger equation are indistinguishable; namely, each solution to the Schrödinger equation with a periodic potential energy (Eqs. (14.5.1–14.5.2) satisfies an identity,

$$\psi(x+a) = e^{ika}\psi(x). \tag{14.5.7}$$

This result is referred to as the Bloch theorem [14.11]. It means that the single-particle wave functions (the orbitals) are delocalized over the entire coordinate space and are "periodic up to a phase." The phase, e^{ika}, is defined by a "Bloch wave vector," k, and it accumulates from one unit cell to another along the lattice. Each solution to the Schrödinger equation is therefore denoted by its corresponding wave vector, namely $\psi_k(x)$.

Formulating equivalently the Bloch theorem, the function $u_k(x) \equiv e^{-ikx}\psi_k(x)$ is periodic in the coordinate space with the lattice period, a:

$$\psi_k(x) = e^{ikx}u_k(x) \quad ; \quad u_k(x) = u_k(x+a). \tag{14.5.8}$$

Expanding the periodic function in a Fourier series,

$$u_k(x) = \sum_{n=-\infty}^{\infty} u_n^{(k)} e^{i\frac{2\pi}{a}nx}, \tag{14.5.9}$$

the expansion coefficients $\{u^{(k)}\}$ are identified as (see Ex. 14.5.1)

$$u_n^{(k)} = \frac{1}{\sqrt{2\pi}}\overline{\Psi}\left(k + \frac{2\pi n}{a}\right), \tag{14.5.10}$$

which can be obtained explicitly by solving the Bloch equations, Eq. (14.5.5), where, in practice, a finite truncation of the infinite summation is often sufficiently accurate. Notice that the same derivation applies to a three-dimensional periodic lattice, where the wave functions $\psi_k(x)$ and $u_k(x)$ become three-dimensional, and k is replaced by the three-dimensional wave vector (e.g., $\mathbf{k} = (k_x, k_y, k_z)$) defined according to the parameters of the three-dimensional unit cell.

An important consequence of Eqs. (14.5.8–14.5.10) is that solutions to the Schrödinger equation associated with the Bloch wave vectors, k and $k + \frac{2\pi}{a}n$, are identical (see Ex. 14.5.1):

$$\psi_k(x) = \psi_{k+2\pi n/a}(x). \tag{14.5.11}$$

This means that all the physically distinguishable values of the Bloch wave vector, namely the set of k values that correspond to linearly independent solutions of the Schrödinger equation, are confined within a finite interval (termed the first Brillouin zone [12.3]):

$$-\frac{\pi}{a} < k \le \frac{\pi}{a}. \tag{14.5.12}$$

For each value of k, explicit solutions to the Schrödinger equation depend on the periodic functions $u_k(x)$, namely on the corresponding sets of Fourier coefficients, $\{u_n^{(k)}\}$. The latter can be calculated using the relation, Eq. (14.5.10). Rearranging Eq. (14.5.5), the sets $\{u_n^{(k)}\}$ comprise the eigenvectors of an algebraic eigenvalue equation,

$$\sum_{n'=-\infty}^{\infty} H_{n,n'} u_{n'}^{(k)} = E u_n^{(k)} \quad ; \quad -\infty < n < \infty, \qquad (14.5.13)$$

where

$$H_{n,n'} = \left[V_{n-n'} + \delta_{n,n'} \frac{\hbar^2}{2m} \left(k + \frac{2\pi n}{a} \right)^2 \right]. \qquad (14.5.14)$$

Assigning an additional quantum number to the different solutions of this eigenvalue equation, $l = 0, 1, \ldots$ (a nonnegative integer, for simplicity and without loss of generality), each eigenvalue and a corresponding eigenfunction of the one-dimensional Schrödinger equation for the particle in an infinitely periodic potential is associated with two quantum numbers (clearly, more quantum numbers are needed at higher dimensions):

$$\psi_{k,l}(x) = e^{ikx} u_{k,l}(x) \leftrightarrow E_{k,l} \quad ; \quad l = 0, 1, \ldots ; \quad -\frac{\pi}{a} < k \le \frac{\pi}{a}, \qquad (14.5.15)$$

where $u_{k,l}(x) = \sum_{n=-\infty}^{\infty} u_n^{(k,l)} e^{i\frac{2\pi}{a}nx}$. *The spectrum of the periodic Hamiltonian can thus be divided into energy bands. Each band is characterized by a specific value of the quantum number l and a continuum of states associated with the quantum number k.* When the lattice potential is associated with "well-separated" and "deep" potential energy wells, each of the low energy bands can be correlated with a specific bound state of a single isolated potential well, mapped into a continuous band, as the potential is periodically continued. An example for the onset of such energy bands in a system of multiple potential energy wells was given in Section 6.5; see Fig. 6.5.1. Notice that this simple picture breaks down when the distance between neighboring wells becomes comparable to the "width" of the particle's probability density distribution within an isolated well. In that case the energy width of the relevant bands broadens to the extent that different bands may overlap. Notice also that there are no "bound states" within an infinitely periodic lattice potential, in the sense that the wave function does not vanish (except for at isolated points) in the entire coordinate space.

Exercise 14.5.1

(a) Derive Eq. (14.5.4) by substituting the plane wave expansion of $\psi(x)$ in the Schrödinger equation with a periodic potential, Eqs. (14.5.1, 14.5.2), projecting the result on the plane wave, e^{ikx}, and using the representation of the delta function, $\delta(k) = \frac{1}{2\pi} \int_{-\infty}^{\infty} e^{ikx} dx$.

(b) Derive Eq. (14.5.5) by substituting the plane wave expansion of $\psi(x)$ in the Schrödinger equation with a periodic potential, Eqs. (14.5.1, 14.5.2), projecting the result on the plane wave, $e^{i(k-\frac{2\pi}{a}n)x}$, and using the representation of the delta function, $\delta(k) = \frac{1}{2\pi} \int_{-\infty}^{\infty} e^{ikx} dx$.

(c) Use the definition, $u_k(x) \equiv e^{-ikx} \psi_k(x)$, and Eq. (14.5.7) to derive Eq. (14.5.8).

(d) Obtain Eq. (14.5.10) by substituting Eqs. (14.5.8, 14.5.9) in Eq. (14.5.6).

(e) Use the general form of the solutions to the Schrödinger equation for a periodic potential (Eq. (14.5.7)) to prove Eq. (14.5.11).

LCAO for Extended Systems

We now turn to an approximate LCAO description of a periodic one-dimensional lattice composed of a repeating unit cell of several (different, in general) atoms (see Fig. 14.5.1). The periodicity assures that the same atoms appear in neighboring cells, which means that their atomic orbitals perfectly match in their energy as well as in their symmetry properties. Therefore, effective linear combinations of the atomic orbitals are expected between different cells, and since all the cells are coupled, the orbitals in this structure are expected to extend over the entire lattice, in accordance with the Bloch theorem. Models based on extended "lattice orbitals," approximated as linear combinations of atomic orbitals, are often adequate for qualitative description of the different electronic properties of different types of materials, as we demonstrate in the following discussion.

Assuming that the effective potential energy well attributed to each atomic core confines the electronic density associated with the atomic orbitals, such that its standard deviation is much smaller than the interatomic distance between neighboring atoms, a tight-binding model Hamiltonian can be invoked for the chain (see Fig. 14.4.4 and Eqs. (14.4.3–14.4.7)). Importantly, the model assumes that the interatomic interactions are "short-ranged" in the sense that the atoms within each unit cell interact only with the two nearest-neighboring cells (otherwise, the "cell" size must be extended until the relevant interactions are confined within the nearest-neighbors range). For simplicity, and without loss of generality, our examples will invoke a stronger assumption, namely that interatomic interactions are restricted to nearest neighbors both between and within the cells (as in the Hückel model for finite chains, discussed in Section 14.4 (Eqs. (14.4.21–14.4.24)). Additionally, we shall consider a minimal basis of a single atomic orbital from each atom along the chain. This is justified when the selected atomic orbitals are well separated in energy from the other atomic orbitals within each atom (in comparison to the interatomic interaction matrix elements).

Denoting each unit cell by an integer, $-\infty < n < \infty$, and the atoms within the unit cell by another integer, $m = 1, 2, \ldots, M$ (see Fig. 14.5.1), each atom, and hence each atomic orbital, $|\varphi_{m,n}\rangle$, is identified by these two indexes. The lattice orbitals are introduced as linear combinations of the atomic orbitals,

$$|\phi_l\rangle = \sum_{n=-\infty}^{\infty} \sum_{m=1}^{M} c_{m,n}^{(l)} |\varphi_{m,n}\rangle, \qquad (14.5.16)$$

where the expansion coefficients and the corresponding lattice orbital energies, ε_l, are obtained according to the linear variation method as the eigenvectors and eigenvalues of the (infinite-dimensional) secular equation (Eq. 12.3.24):

$$\mathbf{H}\mathbf{c}^{(l)} = \varepsilon_l \mathbf{S}\mathbf{c}^{(l)}. \qquad (14.5.17)$$

The matrix representation of the system Hamiltonian and the overlap matrix are subject to the tight-binding model assumptions (Eqs. (14.4.5–14.4.7)), which take the explicit form,

$$S_{(n',m'),(n,m)} \equiv \langle \varphi_{n',m'} | \varphi_{n,m} \rangle = \delta_{n,n'} \delta_{m,m'}, \tag{14.5.18}$$

$$H_{(n',m'),(n,m)} \equiv \begin{cases} \alpha_m & ; \quad n = n', m = m' \\ \beta_{m',m} & ; \quad (n',m') \text{ and } (n,m) \text{ correspond to nearest neighbors} \\ 0 & ; \qquad otherwise. \end{cases}$$
$$\tag{14.5.19}$$

Notice that the lattice periodicity means that the entries of $\mathbf{H}$ depend only on the internal indexes within each cell, and not on the cell index,

$$\mathbf{H} = \begin{bmatrix}
\ddots & \beta_{M-1,M} & & & & & & & \\
\beta_{M-1,M} & \alpha_M & \beta_{M,1} & & & & & & \\
& \beta_{M,1} & \alpha_1 & \beta_{1,2} & & & & & \\
& & \beta_{1,2} & \alpha_2 & \beta_{2,3} & & & & \\
& & & \beta_{2,3} & \ddots & & & & \\
& & & & & & \beta_{M-1,M} & & \\
& & & & & \beta_{M-1,M} & \alpha_M & \beta_{M,1} & \\
& & & & & & \beta_{M,1} & \alpha_1 & \beta_{1,2} \\
& & & & & & & \beta_{1,2} & \ddots
\end{bmatrix}. \tag{14.5.20}$$

The Hamiltonian matrix can therefore be conveniently represented as a block matrix,

$$\mathbf{H} = \begin{bmatrix}
\ddots & \beta & & & \\
\beta^t & \alpha & \beta & & \\
& \beta^t & \alpha & \beta & \\
& & \beta^t & \alpha & \beta \\
& & & \beta^t & \ddots
\end{bmatrix}, \tag{14.5.21}$$

where the matrices α and β are defined in the space of a single unit cell,

$$\alpha = \begin{bmatrix}
\alpha_1 & \beta_{1,2} & & & \\
\beta_{1,2} & \alpha_2 & \beta_{2,3} & & \\
& \beta_{2,3} & \ddots & & \\
& & & \alpha_{M-1} & \beta_{M-1,M} \\
& & & \beta_{M-1,M} & \alpha_M
\end{bmatrix} \quad ; \quad \beta = \begin{bmatrix}
0 & 0 & 0 & 0 & 0 \\
0 & 0 & 0 & 0 & 0 \\
0 & 0 & \ddots & & \\
0 & 0 & & 0 & 0 \\
\beta_{M,1} & 0 & & 0 & 0
\end{bmatrix}. \tag{14.5.22}$$

Information attributed to the chemical composition of the material is encoded here in the on-site atomic orbital energies, $\alpha_1, \alpha_2, \ldots, \alpha_M$, and in the intersite couplings, $\beta_{1,2}, \beta_{2,3}, \ldots, \beta_{M-1,M}, \beta_{M,1}$.

One can readily verify (Ex. 14.5.2) that the eigenvector elements that satisfy the secular equation (Eq. (14.5.17)) are of a product form,

$$c_{m,n}^{(l)} = u_m^{(l,\tilde{k})} e^{ikn} \quad ; \quad u_m^{(l,\tilde{k})} = u_{m+M}^{(l,\tilde{k})}, \tag{14.5.23}$$

where the vector $(u_1^{(l,\tilde{k})}, u_2^{(l,\tilde{k})}, \ldots, u_M^{(l,\tilde{k})}) \equiv \mathbf{u}^{(l,\tilde{k})}$ is the solution of a reduced eigenvalue equation within the space of a single unit cell,

$$\left[\boldsymbol{\alpha} + \boldsymbol{\beta}' e^{-i\tilde{k}} + \boldsymbol{\beta} e^{i\tilde{k}} \right] \mathbf{u}^{(l,\tilde{k})} = \varepsilon_l(\tilde{k}) \mathbf{u}^{(l,\tilde{k})} \quad ; \quad l = 1, 2, \ldots, M. \tag{14.5.24}$$

The result is nothing but a discrete version of Bloch's theorem, Eq. (14.5.8). The eigen-vectors to the secular equation, Eq. (14.5.17), are products of a phase factor, e^{ikn}, which depends only on the "external" unit cell index, n, and a vector $\mathbf{u}^{(l,\tilde{k})}$, which is independent of the cell index; namely, its elements are replicated periodically along the entire lattice with the unit cell periodicity (M). Notice that the dimensionless parameter $\tilde{k}$ is continuous. However, the solutions to the secular equation are invariant to changes $\tilde{k} \to \tilde{k} + 2\pi n$, which means that the independent solutions are confined to a finite range, $-\pi < \tilde{k} < \pi$. Moreover, dividing by the unit cell length, $k = \tilde{k}/a$ and associating each cell with a position along the real coordinate axis, $x_n = na$, we can identify e^{ikn} with the continuous Bloch wave, e^{ikx}:

$$e^{ikn} = e^{ikna} = e^{ikx_n} \quad ; \quad -\pi/a < k \le \pi/a. \tag{14.5.25}$$

Each eigenvalue and eigenvector of the infinite chain Hamiltonian are therefore defined by two indexes, $\tilde{k}$ and l. The spectrum of the lattice Hamiltonian is composed of continuous energy bands, where each of the M eigenvalues of the effective unit cell Hamiltonian, that is, $\varepsilon_1(\tilde{k}), \varepsilon_1(\tilde{k}), \ldots, \varepsilon_M(\tilde{k})$ (see Eq. (14.5.24)), is a continuous function of the Bloch wave vector $\tilde{k}$. The corresponding eigenvectors are Bloch waves, extending through the entire lattice, multiplied by a periodic vector, with the unit cell period, which depends on both $\tilde{k}$ and l.

Exercise 14.5.2

(a) *Defining the "unit cell" coefficient vector,* $\mathbf{c}_n^{(l)} = (c_{1,n}^{(l)}, c_{2,n}^{(l)}, \ldots, c_{M,n}^{(l)})$ *and using Eq. (14.5.18) for the matrix* $\mathbf{S}$ *and Eq. (14.5.21) for the Hamiltonian,* $\mathbf{H}$, *show that the secular equation (Eq. (14.5.17)) reads* $\boldsymbol{\beta}' \mathbf{c}_{n-1}^{(l)} + [\boldsymbol{\alpha} - \varepsilon_l \mathbf{I}] \mathbf{c}_n^l + \boldsymbol{\beta} \mathbf{c}_{n+1}^{(l)} = 0$.

(b) *Use the ansatz* $\mathbf{c}_n^{(l)} = \mathbf{u}^{(l,\tilde{k})} e^{ikn}$, *where* $\mathbf{u}^{(l,\tilde{k})} \equiv (u_1^{(l,\tilde{k})}, u_2^{(l,\tilde{k})}, \ldots, u_M^{(l,\tilde{k})})$, *to show that the unit cell vectors,* $\{\mathbf{u}^{(l,\tilde{k})}\}$, *are the eigenvectors of a finite-dimensional Hermitian matrix,* $[\boldsymbol{\alpha} + \boldsymbol{\beta}' e^{-i\tilde{k}} + \boldsymbol{\beta} e^{i\tilde{k}}] \mathbf{u}^{(l,\tilde{k})} = \varepsilon_l(\tilde{k}) \mathbf{u}^{(l,\tilde{k})}$, *for* $l = 1, 2, \ldots, M$.

Conductors and Insulators

Let us consider a specific example, in which there are only two atoms in the unit cell of a periodic lattice $(M = 2)$. We shall denote the corresponding two atomic orbital energies of the neighboring atoms as $\alpha_1, \alpha_2 = \Delta, -\Delta$, and the coupling matrix element

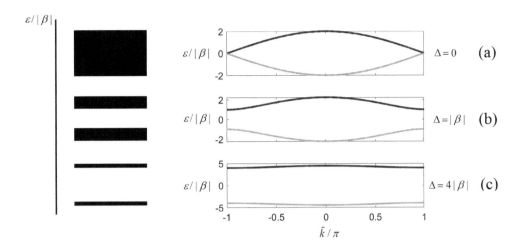

Figure 14.5.2 Band energies for different periodic lattices with two atoms per unit cell ($M = 2$). (a) The two atomic orbitals are identical ($\Delta = 0$), yielding a single band with a bandwidth, $4|\beta|$. (b) The difference in the atomic orbital energies is of the order of the interatomic coupling matrix element, yielding two bands, separated by 2Δ. (c) The difference in atomic energies is much larger than the interatomic coupling matrix element, resulting in much narrower bands, separated by 2Δ.

between them as β. The corresponding α and β matrices composing the infinite lattice Hamiltonian (Eqs. (14.5.21, 14.5.22)) are

$$\alpha = \begin{bmatrix} \Delta & \beta \\ \beta & -\Delta \end{bmatrix} \quad ; \quad \beta = \begin{bmatrix} 0 & 0 \\ \beta & 0 \end{bmatrix}, \tag{14.5.26}$$

and accordingly, the reduced (single-cell) eigenvalue equation for the band energies and eigenfunctions (Eq. (14.5.24)) obtains the form

$$\begin{bmatrix} \Delta - \varepsilon_l(\tilde{k}) & \beta\left(1 + e^{i\tilde{k}}\right) \\ \beta\left(1 + e^{-i\tilde{k}}\right) & -\Delta - \varepsilon_l(\tilde{k}) \end{bmatrix} \begin{bmatrix} u_1^{(l,\tilde{k})} \\ u_2^{(l,\tilde{k})} \end{bmatrix} = \begin{bmatrix} 0 \\ 0 \end{bmatrix}. \tag{14.5.27}$$

The two eigenvalues for each wave vector $\tilde{k}$ therefore read (Ex. 14.5.3)

$$\varepsilon_{\pm}(\tilde{k}) = \pm\sqrt{\Delta^2 + 4\beta^2 \cos^2(\tilde{k}/2)} \quad ; \quad -\pi < \tilde{k} \le \pi. \tag{14.5.28}$$

The spectrum of the periodic lattice with $M = 2$ is shown to be grouped into two continuous bands. Since the function $\cos^2(\tilde{k}/2)$ is bounded between 0 and 1, each band has a finite energy range,

$$\begin{aligned} |\Delta| &< \varepsilon_+(\tilde{k}) < \sqrt{\Delta^2 + 4\beta^2} \\ -\sqrt{\Delta^2 + 4\beta^2} &< \varepsilon_-(\tilde{k}) < -|\Delta| \end{aligned} \quad ; \quad -\pi < \tilde{k} \le \pi. \tag{14.5.29}$$

As we can see, the bands are separated by an energy gap between $-|\Delta|$ and $|\Delta|$ (see Fig. 14.5.2).

Exercise 14.5.3 *Calculate the eigenvalues, $\varepsilon_l(\tilde{k})$, of the secular equation, Eq. (14.5.27), and obtain the results in Eq. (14.5.28).*

The population of the lattice orbitals by electrons at the ground state depends on the number of electrons populating the corresponding atomic orbitals. As an example, let us consider a case in which the relevant atomic orbital is "half field" in the ground state of each single atom (this would be the case, e.g., for the highest occupied (valence) orbitals of alkali metal atoms); namely, each atomic orbital contributes a single electron to the system of lattice orbitals. Since the number of independent eigenvectors of the secular equation (lattice orbitals) equals the number of atomic orbitals, and since the Aufbau principle means that the orbitals are doubly occupied from the lowest energy and upward, only half of the lattice orbitals would be populated at the electronic ground state. In the present model, this implies that only the lowest of the two energy bands would be populated with electrons.

Now let us consider the possibility of electric conductance in the different cases. Net flow of charges through the lattice necessitates their acceleration, namely an increase in their kinetic energy. In the example of a fully occupied lower energy band, this would require overcoming the energy "band gap" to the higher unoccupied energy levels. In cases where this gap is much larger than the thermal energy or the applied bias voltage, electric conductance would be blocked, and the material would behave as an insulator. If, on the other hand, the energy gap is relatively small (or vanishes), the material would behave as a conductor. The presence of an "energy gap" in the spectrum near the highest occupied lattice orbital therefore has dramatic consequences with respect to the electric conductance of the material. In what follows we discuss the relation between the formation of an energy gap and the chemical composition of the underlying material (e.g., ionic vs. metallic), using the theoretical model of a one-dimensional atomic lattice at equal interatomic distances, with half-filled orbitals. Two extreme cases are analyzed in detail, corresponding to a conductor and an insulator.

Case I: A uniform unit cell (a conductor model)

In the first case we consider a hypothetical chain of clamped identical atoms, in which all the interatomic distances and nearest-neighbor coupling matrix elements are the same along the entire chain. The identity of the atoms is imposed by setting to zero the difference in their atomic orbital energies, namely, $\Delta = 0$ in Eq. (14.5.26). As one can readily see from Eqs. (14.5.28, 14.5.29), the two bands merge in this case into a single continuous band (see Fig. 14.5.2(a)). Defining $\tilde{k}/2 = \tilde{\tilde{k}}$, the entire spectrum depends on a single quantum number, where Eq. (14.5.28) compactly reduces to

$$\varepsilon(\tilde{\tilde{k}}) = 2\beta \cos(\tilde{\tilde{k}}) \quad ; \quad -\pi < \tilde{\tilde{k}} \le \pi, \tag{14.5.30}$$

where the bandwidth equals $4|\beta|$. Each $\tilde{\tilde{k}}$ defines an eigenvector of the 2×2 unit-cell Hamiltonian (see Eq. (14.5.27)), where, in this case, the coefficients of the two atomic orbitals within the unit cell relate to each other as (see Ex. 14.5.4)

$$u_2^{(\tilde{\tilde{k}})}/u_1^{(\tilde{\tilde{k}})} = e^{-i\tilde{\tilde{k}}}. \tag{14.5.31}$$

The identity of the two atoms within the unit cell, therefore, has two consequences. First, all the lattice orbitals are evenly distributed (perfectly delocalized) among the two atoms in each unit cell, in the sense that

$$|u_1^{(\tilde{k})}|^2 = |u_2^{(\tilde{k})}|^2. \tag{14.5.32}$$

Second, there is no energy gap between the highest occupied lattice orbital and the lowest unoccupied lattice orbital, which implies that this model system corresponds to a conductor.

Notice that, as expected, this model resembles that of a uniform chain of N identical sites, subject to periodic boundary conditions, in the limit $N \to \infty$. Setting the on-site energy to zero, the eigenvalues read (Eq. (14.4.27)) $\varepsilon_l = 2\beta \cos\left(\frac{2\pi l}{N}\right)$, with $l = 1, 2, \ldots, N$. As $N \to \infty$, the argument $2\pi l/N$ covers continuously a single period of the cosine function, which coincides with Eq. (14.5.30). The corresponding eigenvectors of a uniform chain under periodic boundary conditions (Eq. (14.4.28)) also satisfy a similar relation to Eq. (14.5.31), $c_{m+1}^{(l)}/c_m^{(l)} = e^{-i\frac{2\pi l}{N}}$.

Case II: A nonuniform unit cell (an insulator model)

In the second limiting case we consider the two atoms within each unit cell to be "very different" from each other, in the sense that the difference in their on-site atomic orbital energies exceeds by far the interatomic coupling matrix element. Choosing $\Delta > 0$, we assume

$$\Delta \gg |\beta|. \tag{14.5.33}$$

The two bands (Eq. (14.5.28)) are separated in this case by an energy gap (see Figs. 14.5.2 (b), 14.5.2 (c)), and we obtain the form

$$\varepsilon_{\pm}(\tilde{k}) \approx \pm\Delta\left(1 + \frac{2\beta^2}{\Delta^2}\cos^2(\tilde{k}/2)\right) \quad ; \quad -\pi < \tilde{k} \leq \pi, \tag{14.5.34}$$

where each bandwidth is $\sim 2\beta^2/\Delta$.

Substitution of the band energies in the secular equation for the lattice orbitals (Eq. (14.5.27)) reveals a remarkable difference between the orbitals in the different bands. Particularly, the ratio between the coefficients associated with the two atomic orbitals within the unit cell differs in the two bands, where (see Ex. 14.5.4))

$$\frac{u_2^{(+,\tilde{k})}}{u_1^{(+,\tilde{k})}} = \frac{\beta}{\Delta}e^{-i\tilde{k}/2}\cos(\tilde{k}/2) \quad ; \quad \frac{u_1^{(-,\tilde{k})}}{u_2^{(-,\tilde{k})}} = -\frac{\beta}{\Delta}e^{i\tilde{k}/2}\cos(\tilde{k}/2). \tag{14.5.35}$$

Since $|\Delta| \gg |\beta|$, this means that

$$|u_1^{(+,\tilde{k})}|^2 \gg |u_2^{(+,\tilde{k})}|^2 \quad ; \quad |u_1^{(-,\tilde{k})}|^2 \ll |u_2^{(-,\tilde{k})}|^2 \tag{14.5.36}$$

In the upper energy band, $\varepsilon_+(\tilde{k})$, the orbitals are primarily localized on the first site in each cell, corresponding to the atomic orbital energy, Δ, whereas in the lower energy band, $\varepsilon_-(\tilde{k})$, the orbitals are primarily localized on the second site in each cell, corresponding to the atomic orbital energy, $-\Delta$. The result is not surprising, considering that effective linear combinations of atomic orbitals require matching in their energies.

Regarding each unit cell as an atom pair, a difference in the two atomic orbital energies that exceeds the coupling matrix element between them means that the "molecular
orbitals" within the pair would remain localized on the respective atoms (as in an
ionic material; see Fig. 14.4.5(c), and the related discussion). Notice, however, that
in the lattice, identical atoms belonging to neighboring unit cells are still coupled indirectly through the other atom in the unit cell. The effective coupling matrix element
between the identical atoms is relatively small, and in this case equals $\beta_{eff} \approx \beta^2/(2\Delta)$
(see Ex. 14.5.5). Yet, since the respective orbitals are degenerate in energy, this coupling results in delocalized lattice orbitals along the entire lattice. Each kind of atom
is therefore associated with a relatively narrow band, whose width is $\sim 4\beta_{eff} = 2\beta^2/\Delta$,
where the two bands are separated by an energy gap, 2Δ. From a chemical point of
view, the situation depicted in this model $(\Delta \gg |\beta|)$ is typical to cases in which the
valence orbitals of the different atoms within the unit cell are far in energy from each
other, which is commonly the case in ionic crystals (see, e.g., Fig. 14.4.3c). In the half-
filling model previously discussed, the lower of the two bands is fully occupied at the
electronic ground state, and when the energy gap to the next (conductance) band is
much larger than the thermal energy and/or an applied bias voltage, these materials
are insulators.

Exercise 14.5.4 *(a) Use Eq. (14.5.30) for the band energies (corresponding to $\Delta = 0$)
in the secular equation for the unit cell coefficients (Eq. (14.5.27)) to obtain the relation between the coefficients as given in Eq. (14.5.31) (recall that $\tilde{k}/2 = \tilde{\tilde{k}}$). (b) Use
Eq. (14.5.34) for the two band energies (corresponding to $\Delta \gg |\beta|$) in the secular
equation for the unit cell coefficients (Eq. (14.5.27)) to obtain the relations between
the coefficients, as given in Eq. (14.5.35).*

Exercise 14.5.5 *Consider a tight-binding model Hamiltonian for two degenerate atomic
orbitals coupled indirectly through a third, nondegenerate atomic orbital,*

$$H = \begin{bmatrix} \Delta & \beta & 0 \\ \beta & -\Delta & \beta \\ 0 & \beta & \Delta \end{bmatrix}.$$

*In the case $\Delta \gg |\beta|$ the corrections to the energy of the two degenerate states can be
calculated using perturbation theory, where $H = H_0 + V$:*

$$H_0 = \begin{bmatrix} \Delta & 0 & 0 \\ 0 & -\Delta & 0 \\ 0 & 0 & \Delta \end{bmatrix} \quad ; \quad V = \begin{bmatrix} 0 & \beta & 0 \\ \beta & 0 & \beta \\ 0 & \beta & 0 \end{bmatrix}.$$

*Denoting the local atomic orbitals as $\{|\varphi_n\rangle, n = 1,2,3\}$, we chose a symmetric and an
antisymmetric linear combination as two degenerate zero-order vectors, $|\psi_s^{(0)}\rangle = (|\varphi_1\rangle +
|\varphi_3\rangle)/\sqrt{2}$, $|\psi_a^{(0)}\rangle = (|\varphi_1\rangle - |\varphi_3\rangle)/\sqrt{2}$, and a third, localized eigenvector, $|\psi_2^{(0)}\rangle = |\varphi_2\rangle$.*

(a) Show that the first-order corrections to the two degenerate state energies vanish.
(b) Show that the second-order correction to the antisymmetric state energy vanishes.

(c) *Show that the second-order correction to the symmetric state energy reads* $E_s^{(2)} = \frac{\beta^2}{\Delta}$.

(d) *Show that the resulting energy splitting between* $|\psi_s^{(0)}\rangle$ *and* $|\psi_a^{(0)}\rangle$, *induced by the coupling to* $|\psi_2^{(0)}\rangle$, *is equivalent to the splitting induced by a direct coupling matrix element,* $\beta_{eff} = \frac{\beta^2}{2\Delta}$, *within an effective two-state Hamiltonian,*

$$H_{eff} = \begin{bmatrix} 0 & \beta_{eff} \\ \beta_{eff} & 0 \end{bmatrix}.$$

Bibliography

[14.1] P. Kirkpatrick and C. Ellis, "Chemical space," Nature 432, 823 (2004).

[14.2] M. Born and R. Oppenheimer, "Zur Quantentheorie der Molekeln," Annals der Physik 389, 457 (1927).

[14.3] N. J. Turro, "Modern Molecular Photochemistry" (University Science Books, 1991).

[14.4] J. Simons, "How do low-energy (0.1–2 eV) electrons cause DNA-strand breaks?," Accounts of Chemical Research 39, 772 (2006).

[14.5] J. von Neumann and E. P. Wigner, "Über das Verhalten von Eigenwerten bei adiabatischen Prozessen," Physikalische Zeitschrift 30, 467 (1929).

[14.6] G. A. Worth and L. S. Cederbaum, "Beyond Born-Oppenheimer: Molecular dynamics through a conical intersection," Annual Review of Physical Chemistry 55, 127 (2004).

[14.7] C. Zenner, "Non-adiabatic crossing of energy levels," Proceedings of the Royal Society of London A 137, 696 (1932).

[14.8] C. A. Coulson and P. D. Robinson, "Wave functions for the hydrogen atom in spheroidal coordinates I: The derivation and properties of the functions," Proceedings of the Physical Society 71, 815 (1958).

[14.9] G. L. Miessler, P. J. Fischer and D. A. Tarr, "Inorganic Chemistry" (Pearson, 2014).

[14.10] E. Hückel, "Quantentheoretische Beiträge zum Benzolproblem," Zeitschrift für Physik 70, 204 (1931).

[14.11] F. Bloch, "Über die Quantenmechanik der Elektronen in Kristallgittern," Zeitschrift für Physik 52, 555 (1929).

15   Quantum Dynamics

15.1 Time-Independent Hamiltonians

So far, our attention has almost entirely been dedicated to solutions of the time-independent Schrödinger equation, $\hat{H}|\psi\rangle = E|\psi\rangle$. There are two good reasons for that. First, the information needed to describe the properties of a system at equilibrium is fully contained in its stationary states, namely the eigenstates of its Hamiltonian. In previous chapters we often focused on the state of minimal energy, referred to as the ground state; however, as we shall discuss in Chapter 16, this statement holds also for any equilibrium state. A second motivation for focusing on the eigenstates of the system Hamiltonian relates to systems out of equilibrium, when dynamical processes are of interest. This is the case when inherent processes within the system need to be analyzed, or when the transient response of an equilibrated system to external perturbations is under study. In these cases, the system is prepared (namely, driven to) a nonstationary state, and dynamics is induced. The time-dependent Schrödinger equation (see Eq. (4.1.1) and postulate 3, Eq. (11.1.12)) determines that infinitesimal changes to the state of the system in time are attributed directly to the system Hamiltonian, which is identified as the "generator of motion,"

$$i\hbar\frac{\partial}{\partial t}|\psi(t)\rangle = \hat{H}|\psi(t)\rangle. \tag{15.1.1}$$

Restricting ourselves first to a time-independent Hamiltonian, its eigenstates are a complete orthonormal system for expanding the physical state vector of the system, at any point in time:

$$|\psi(t)\rangle = \sum_n a_n(t)|\varphi_n\rangle \quad ; \quad \{\hat{H}|\varphi_n\rangle = \varepsilon_n|\varphi_n\rangle\}. \tag{15.1.2}$$

(Without loss of generality, we consider here a discrete set of nondegenerate eigenstates.) The information with respect to the dynamics in the system is therefore encoded in the Hamiltonian eigenstates and in the expansion coefficients. These coefficients are obtained by substitution of the expansion in the Schrödinger equation, Eq. (15.1.1). Projecting on the bra states, $\{\langle\varphi_m|\}$, yields for any m

$$i\hbar\frac{\partial}{\partial t}a_m(t) = \sum_n a_n(t)\langle\varphi_m|\hat{H}|\varphi_n\rangle = \varepsilon_m a_m(t). \tag{15.1.3}$$

The coefficients are the solutions of this differential equation, which read

$$a_m(t) = a_m(0)e^{\frac{-i\varepsilon_m t}{\hbar}}, \tag{15.1.4}$$

and the state of the system at any time (Eq. (15.1.2)) reads

$$|\psi(t)\rangle = \sum_n a_n(0)e^{\frac{-i\varepsilon_n t}{\hbar}}|\varphi_n\rangle. \tag{15.1.5}$$

Any solution to the time-dependent Schrödinger equation with a time-independent Hamiltonian is therefore a linear combination of the stationary solutions to this equation (see Ex. 15.1.1 and Sections 4.3–4.6 for an introduction to stationary solutions):

$$|\psi(t)\rangle = \sum_n a_n(0)|\psi_n(t)\rangle \quad ; \quad |\psi_n(t)\rangle = e^{\frac{-i\varepsilon_n t}{\hbar}}|\varphi_n\rangle. \tag{15.1.6}$$

Projecting the expansion at $t = 0$ on the bra states, $\{\langle\varphi_m|\}$, yields for any m

$$a_m(0) = \langle\varphi_m|\psi(0)\rangle. \tag{15.1.7}$$

As we can see, the expansion coefficients are fully defined by the initial state of the system (its "preparation"). *Any dynamical process in a system with a time-independent Hamiltonian therefore corresponds to a superposition of time-dependent stationary waves, often termed a "coherent wave packet" (see also Section 4.2). Time evolution of a state of a system can hence be regarded as a time-dependent interference between its Hamiltonian eigenstates.*

Exercise 15.1.1   *A stationary solution to the time-dependent Schrödinger equation is defined as $|\psi_n(t)\rangle = e^{\frac{-i\varepsilon_n t}{\hbar}}|\varphi_n\rangle$, where $|\varphi_n\rangle$, is an eigenstate of the system Hamiltonian, $\hat{H}|\varphi_n\rangle = \varepsilon_n|\varphi_n\rangle$ (see Section 4.3). Show that any linear combination of stationary solutions is also a solution to the time-dependent Schrödinger equation with the same time-independent Hamiltonian, $i\hbar\frac{\partial}{\partial t}|\psi(t)\rangle = \hat{H}|\psi(t)\rangle$.*

We now turn to evaluation of the time-dependence of the results of measurements performed on a system undergoing time evolution. We first recall that the state $|\psi(t)\rangle$ corresponds to a statistical ensemble of different realizations of the system, and that *the statistical distribution of any measurable quantity at time t can be expressed in terms of an expectation value of a suitable Hermitian operator,*

$$\langle\psi(t)|\hat{O}|\psi(t)\rangle \equiv O(t). \tag{15.1.8}$$

For example, the probability of measuring a specific (nondegenerate) eigenvalue χ of an operator $\hat{A}$ is obtained by defining a projection operator into the corresponding eigenstate, $\hat{O}_\chi \equiv |\chi\rangle\langle\chi|$, where the time-dependence of the probability measurement reads (see Section 11.1, postulate 4)

$$P_\chi(t) = |\langle\chi|\psi(t)\rangle|^2 = \langle\psi(t)|\chi\rangle\langle\chi|\psi(t)\rangle \equiv \langle\psi(t)|\hat{O}_\chi|\psi(t)\rangle = O_\chi(t). \tag{15.1.9}$$

Similarly, the probability of measuring any eigenvalue χ of an operator $\hat{A}$ within a specified part of its spectrum (e.g., the set $\{\chi\}$) is obtained by a projector into the subspace of specified corresponding eigenstates, $\hat{O}_{\{\chi\}} \equiv \sum_{\chi\in\{\chi\}} |\chi\rangle\langle\chi|$, where

$$P_{\{\chi\}}(t) = \sum_{\chi \in \{\chi\}} |\langle \chi | \psi(t) \rangle|^2 = \sum_{\chi \in \{\chi\}} \langle \psi(t) | \chi \rangle \langle \chi | \psi(t) \rangle \equiv \langle \psi(t) | \hat{O}_{\{\chi\}} | \psi(t) \rangle = O_{\{\chi\}}(t).$$

(15.1.10)

Eq. (15.1.8) clearly holds also for a measurement of the statistical average of any dynamical variable represented by an operator $\hat{A}$ (see Eq. (11.7.1)),

$$\langle A(t) \rangle = \langle \psi(t) | \hat{A} | \psi(t) \rangle \equiv A(t),$$

(15.1.11)

where we recall (Eq. (11.7.2)) that any (Hermitian) $\hat{A}$ can be formally expressed as a weighted sum of projection operators into its eigenstates, $\hat{A} = \sum_{\chi} \chi | \chi \rangle \langle \chi | = \sum_{\chi} \chi \hat{O}_{\chi}$.

Since any Hermitian operator ($\hat{O}$, Eq. (15.1.8)), can be represented in terms of the Hamiltonian eigenstates (Section 11.2) as defined in Eq. (15.1.2),

$$\hat{O} = \sum_{n,m} O_{m,n} | \varphi_m \rangle \langle \varphi_n |,$$

(15.1.12)

the time-dependence of the statistical distribution of any measurable quantity on the system can be expressed in terms of the stationary states of the system, namely, as a superposition of oscillatory waves, whose frequencies are defined by the energy differences between the Hamiltonian eigenvalues,

$$\omega_{n,m} \equiv \frac{\varepsilon_n - \varepsilon_m}{\hbar}.$$

(15.1.13)

This can be readily verified by explicit calculation of the expectation value of $\hat{O}$, using the expansion of $|\psi(t)\rangle$ in the Hamiltonian eigenstates, Eq. (15.1.5) (see Ex. 15.1.2),

$$O(t) = \sum_{n,m} \gamma_{n,m} e^{-i\omega_{n,m}t}.$$

(15.1.14)

The set of coefficients $\{\gamma_{n,m}\}$ depends on the initial preparation of the system, namely, on the set of initial expansion coefficients $\{a_n(0)\}$ (see Eq. (15.1.7)) and on the selected observable to be measured (via the matrix elements of the corresponding operator, $\{o_{m,n}\}$),

$$\gamma_{n,m} = a_m^*(0) a_n(0) o_{m,n}.$$

(15.1.15)

Notice that in the case of a stationary state, $a_n(0) = \delta_{n,m}$ and consequently $\gamma_{n,m} \propto \delta_{n,m}$, where any observable is time-independent (see Eq. (15.1.14) and Sections 4.3, 4.4). Also notice that in the general (nonstationary) case, *system observables corresponding to operators that commute with the system Hamiltonian are constants of motion.* This can be readily seen from Eqs. (15.1.1, 15.1.8),

$$\frac{d}{dt} \langle \psi(t) | \hat{O} | \psi(t) \rangle = \langle \psi(t) | \frac{i}{\hbar} [\hat{H}, \hat{O}] | \psi(t) \rangle.$$

(15.1.16)

We can therefore summarize that dynamical processes in a system, characterized by a time-independent Hamiltonian, are fully described in terms of the Hamiltonian eigenstates, and particularly, on their projections on the initial state. Nonstationary solutions to the time-dependent Schrödinger equations are merely specific superpositions of stationary ones. Notice, however, that although formally correct, the representation of the system dynamics in terms of the Hamiltonian eigenstates in not always practical or useful. In many cases of interest, the entire spectrum of the Hamiltonian is too difficult to compute and moreover, not necessarily needed, since the initial state often "filters" only a relevant subspace of Hamiltonian eigenstates (see Eq. (15.1.7)). It is therefore useful to relate directly to the time-dependent Schrödinger equation without going through the solution of the time-independent equation.

Exercise 15.1.2 *Given a complete orthonormal system of the Hamiltonian eigenstates, $\hat{H}|\varphi_n\rangle = \varepsilon_n|\varphi_n\rangle$, any operator in the system's Hilbert space can be represented according to Eq. (15.1.12), and any state of the system, $|\psi(t)\rangle$, can be expanded as in Eq. (15.1.6). (a) Show that the time-dependence of any observable obtains the form of Eq. (15.1.14) with $\gamma_{n,m}$ as defined in Eq. (15.1.15). (b) Show that $\gamma_{n,m} = \gamma_{m,n}^*$.*

The Two-Level System (a "Qubit")

As the simplest example, let us consider here the dynamics of a quantum two-level system, often referred to as a single quantum bit (a qubit). In Chapter 12 we encountered the TLS Hamiltonian, its eigenvalues, and its eigenvectors. We recall that in this case the Hilbert space of physical states is spanned by two orthonormal states, $|\chi_1\rangle$ and $|\chi_2\rangle$ (Eq. (12.2.1)), and the matrix representation of the (Hermitian) Hamiltonian operator in this basis obtains the generic form (Eq. (12.2.3))

$$\begin{bmatrix} \langle\chi_1|\hat{H}|\chi_1\rangle & \langle\chi_1|\hat{H}|\chi_2\rangle \\ \langle\chi_2|\hat{H}|\chi_1\rangle & \langle\chi_2|\hat{H}|\chi_2\rangle \end{bmatrix} = \begin{bmatrix} \varepsilon_1 & \gamma \\ \gamma^* & \varepsilon_2 \end{bmatrix}. \tag{15.1.17}$$

Defining $\bar{\varepsilon} \equiv (\varepsilon_1 + \varepsilon_2)/2$, $\Delta \equiv (\varepsilon_1 - \varepsilon_2)/2$, and $\alpha \equiv \sqrt{1 + |\gamma|^2/\Delta^2}$, the eigenvalues and eigenvectors of the TLS Hamiltonian, $\hat{H}|\varphi_\pm\rangle = E_\pm|\varphi_\pm\rangle$, read (See Ex. 12.2.1)

$$E_\pm = \bar{\varepsilon} \pm \sqrt{\Delta^2 + |\gamma|^2} \quad ; \quad |\varphi_\pm\rangle = a_1^{(\pm)}|\chi_1\rangle + a_2^{(\pm)}|\chi_2\rangle, \tag{15.1.18}$$

where

$$a_1^{(\pm)} = \sqrt{\frac{\alpha \pm 1}{2\alpha}} \quad ; \quad a_2^{(\pm)} = \pm\frac{|\gamma|}{\gamma}\sqrt{\frac{\alpha \mp 1}{2\alpha}}. \tag{15.1.19}$$

Without loss of generality, let us identify an "initial" state with one of the two orthonormal basis states,

$$|\psi(0)\rangle = |\chi_1\rangle. \tag{15.1.20}$$

Following the general derivation (Eqs. 15.1.1–15.1.5), the state of the system at any other time can be represented as a superposition of the stationary solutions to the time-dependent Schrödinger equation (Eqs. (15.1.6, 15.1.7)),

$$|\psi(t)\rangle = \langle\varphi_+|\chi_1\rangle e^{\frac{-iE_+t}{\hbar}}|\varphi_+\rangle + \langle\varphi_-|\chi_1\rangle e^{\frac{-iE_-t}{\hbar}}|\varphi_-\rangle. \tag{15.1.21}$$

Alternatively, the time-dependent state can be expanded in terms of the two basis states,

$$|\psi(t)\rangle = c_1(t)|\chi_1\rangle + c_2(t)|\chi_2\rangle, \tag{15.1.22}$$

where (Ex. 15.1.3)

$$c_1(t) = |\langle \varphi_+|\chi_1\rangle|^2 e^{\frac{-iE_+t}{\hbar}} + |\langle \varphi_-|\chi_1\rangle|^2 e^{\frac{-iE_-t}{\hbar}}$$

$$c_2(t) = \langle \varphi_+|\chi_1\rangle\langle \chi_2|\varphi_+\rangle e^{\frac{-iE_+t}{\hbar}} + \langle \varphi_-|\chi_1\rangle\langle \chi_2|\varphi_-\rangle e^{\frac{-iE_-t}{\hbar}}. \qquad (15.1.23)$$

Using the explicit expressions, Eqs. (15.1.18, 15.1.19), and the orthonormality of the basis states, we obtain

$$\langle \varphi_\pm|\chi_1\rangle = \sqrt{\frac{\alpha \pm 1}{2\alpha}} \quad ; \quad \langle \varphi_\pm|\chi_2\rangle = \pm\frac{\gamma}{|\gamma|}\sqrt{\frac{\alpha \mp 1}{2\alpha}} \qquad (15.1.24)$$

and therefore (Ex. 15.1.3)

$$c_1(t) = e^{\frac{-iE_+t}{\hbar}}\left(\frac{\alpha+1}{2\alpha} + \frac{\alpha-1}{2\alpha}e^{\frac{i(E_+-E_-)t}{\hbar}}\right)$$

$$c_2(t) = \frac{|\gamma|}{\gamma}e^{\frac{-iE_+t}{\hbar}}\frac{\sqrt{\alpha^2-1}}{2\alpha}\left(1 - e^{\frac{i(E_+-E_-)t}{\hbar}}\right). \qquad (15.1.25)$$

Exercise 15.1.3 *(a) Given the expansion of the state of a TLS, $|\psi(t)\rangle$ in terms of its stationary states (Eq. (15.1.21)) and the expansion of the stationary states in terms of the basis states (Eq. 15.1.18), derive Eqs. (15.1.22, 15.1.23). (b) Use the explicit expressions for the projections of the TLS stationary states on the basis states in terms of the TLS Hamiltonian parameters (Eqs. (15.1.18, 15.1.19)) to derive Eq. (15.1.25).*

Any operator in the Hilbert space of the TLS obtains the form

$$\hat{O} = o_{1,1}|\chi_1\rangle\langle\chi_1| + o_{1,2}|\chi_1\rangle\langle\chi_2| + o_{2,1}|\chi_2\rangle\langle\chi_1| + o_{2,2}|\chi_2\rangle\langle\chi_2| \qquad (15.1.26)$$

The corresponding observable, $\langle \psi(t)|\hat{O}|\psi(t)\rangle$, can therefore be expressed in terms of the time-dependent expansion coefficients of the state of the system (Eq. (15.1.22)),

$$O(t) = o_{1,1}|c_1(t)|^2 + o_{2,2}|c_2(t)|^2 + 2\mathrm{Re}[o_{1,2}c_1^*(t)c_2(t)]. \qquad (15.1.27)$$

Using Eq. (15.1.25) in Eq. (15.1.27), we can see that any TLS observable is either constant in time or oscillates at a single frequency (Ex. 15.1.4), corresponding to the difference between the two Hamiltonian eigenvalues,

$$\omega = \frac{E_+ - E_-}{\hbar} = \frac{\sqrt{(\varepsilon_1 - \varepsilon_2)^2 + 4|\gamma|^2}}{\hbar}. \qquad (15.1.28)$$

Notice that this is a specific case of the general expansion of any observable, Eq. (15.1.14).

Exercise 15.1.4 *Given the time-dependent expansion coefficients for the TLS state, Eqs. (15.1.22, 15.1.25), and the general expansion of a TLS observable, Eq. (15.1.27), show that the TLS observables are either time-independent, or oscillating at a single frequency, $\omega = \frac{E_+ - E_-}{\hbar}$.*

An important observable is the "survival probability" of an initial state, which is the probability of finding a system at its initial state as a function of time,

$$P(t) = |\langle \psi(0)|\psi(t)\rangle|^2 = \langle \psi(0)|\psi(t)\rangle\langle\psi(t)|\psi(0)\rangle. \tag{15.1.29}$$

Recalling that the initial state can be identified with one of the basis states, for example, $|\psi(0)\rangle = |\chi_1\rangle$, the survival probability can be expressed explicitly in terms of the TLS Hamiltonian parameters (Ex. 15.1.5):

$$P(t) = |c_1(t)|^2 = 1 - \frac{4|\gamma|^2}{(\varepsilon_1 - \varepsilon_2)^2 + 4|\gamma|^2}\sin^2\left(\frac{t\sqrt{(\varepsilon_1 - \varepsilon_2)^2 + 4|\gamma|^2}}{2\hbar}\right). \tag{15.1.30}$$

$P(t)$ is uniquely determined by two parameters, $|\gamma|$ and $|\varepsilon_1 - \varepsilon_2|$. In applications of the TLS model, $|\gamma|$ is often referred to as the interaction strength between the two eigenstates of the "unperturbed" diagonal TLS Hamiltonian, $|\chi_1\rangle$ and $|\chi_2\rangle$, whose energy difference is $|\varepsilon_1 - \varepsilon_2|$. The survival probability of the state $|\chi_1\rangle$ is shown to oscillate between unity and $1 - \frac{4|\gamma|^2}{(\varepsilon_1-\varepsilon_2)^2+4|\gamma|^2}$ at a frequency $\omega = \sqrt{(\varepsilon_1 - \varepsilon_2)^2 + 4|\gamma|^2}/\hbar$ (Eq. (15.1.28)). The dependence of the oscillation frequency and amplitude on $|\gamma|$ and $|\varepsilon_1 - \varepsilon_2|$ is demonstrated in Fig. 15.1.1.

It is instructive to analyze in detail two limiting cases. First, when $|\gamma| >> \varepsilon_1 - \varepsilon_2|$, the two eigenstates of the TLS Hamiltonian are nearly equally delocalized over the two basis states, namely $|\varphi_\pm\rangle \approx \frac{1}{\sqrt{2}}|\chi_1\rangle \pm \frac{|\gamma|}{\gamma}\frac{1}{\sqrt{2}}|\chi_2\rangle$ (see Ex. 15.1.6). The survival probability in this limit (see Eq. (15.1.30)) is approximated as

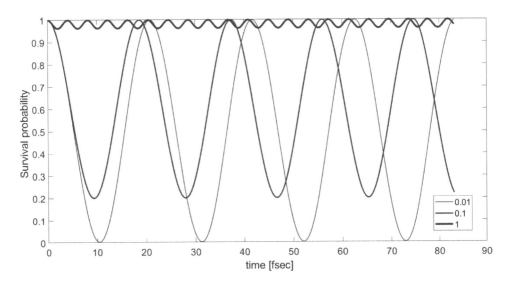

Figure 15.1.1 The survival probability as a unction of time for a TLS (a single qubit). The interaction strength parameter is fixed at $\gamma = 0.1$ eV, where the level spacing changes, $|\varepsilon_1 - \varepsilon_2| = 0.01, 0.1, 1$ eV, as indicated on the plot. For $|\varepsilon_1 - \varepsilon_2| << \gamma$ the probability oscillates between zero and one, whereas for $|\varepsilon_1 - \varepsilon_2| \gg \gamma$, the state is nearly stationary, and the survival probability remains close to unity at all times.

$$P_{|\gamma|>>|\varepsilon_1-\varepsilon_2|}(t) \approx 1 - \sin^2(t|\gamma|/\hbar); \tag{15.1.31}$$

namely, it oscillates between zero and one. Indeed, the initial state is "far from stationary" in the sense that it has similar projections on two nondegenerate stationary states, resulting in "beating" of the state $|\psi(t)\rangle$ between the two states $|\chi_1\rangle$ and $|\chi_2\rangle$. This is reminiscent, for example, of the beating of the probability density in a symmetric double quantum well, as discussed in Chapter 6 (see Fig. 6.4.3).

In the opposite limit, $|\gamma| << |\varepsilon_1 - \varepsilon_2|$, the TLS Hamiltonian is "nearly diagonal," where the two eigenstates are well approximated as $|\varphi_+\rangle \approx |\chi_1\rangle$, and $|\varphi_-\rangle \approx |\chi_2\rangle$ (see Ex. 15.1.7). In this case, the approximation for the survival probability (see Eq. (15.1.30)) reads

$$P_{|\gamma|<<|\varepsilon_1-\varepsilon_2|}(t) \approx 1 - \frac{4|\gamma|^2}{(\varepsilon_1-\varepsilon_2)^2} \sin^2\left(\frac{t(\varepsilon_1-\varepsilon_2)}{2\hbar}\right). \tag{15.1.32}$$

Also in this case, the probability is oscillatory in time, but the oscillation amplitude is much smaller than unity ($P(t) \approx 1$ at all times), which means "localization" of $|\psi(t)\rangle$ at its initial state $|\chi_1\rangle$ (see Fig. 15.1.1). Indeed, in this limit the initial state nearly coincides with one of the Hamiltonian eigenstates ($|\chi_1\rangle \approx |\varphi_+\rangle$), and therefore the system is in a nearly stationary state.

Exercise 15.1.5   *Use Eq. (15.1.25) for $c_1(t)$, and the definition $\alpha \equiv \sqrt{1+|\gamma|^2/\Delta^2}$ to derive Eq. (15.1.30).*

Exercise 15.1.6   *Use Eqs. (15.1.18, 15.1.19) for the stationary states of the TLS to show that in the strong interaction limit, $|\gamma| >> |\varepsilon_1 - \varepsilon_2|$, the states are approximated as $|\varphi_\pm\rangle \approx \frac{1}{\sqrt{2}}|\chi_1\rangle \pm \frac{|\gamma|}{\gamma}\frac{1}{\sqrt{2}}|\chi_2\rangle$ (recall the definition, $\alpha \equiv \sqrt{1+4|\gamma|^2/(\varepsilon_1-\varepsilon_2)^2}$).*

Exercise 15.1.7   *Use Eqs. (15.1.18, 15.1.19) for the stationary states of the TLS and show that in the weak interaction limit, $|\gamma| << \varepsilon_1 - \varepsilon_2|$, the states are approximated as $|\varphi_+\rangle \approx |\chi_1\rangle$ and $|\varphi_-\rangle \approx |\chi_2\rangle$ (recall the definition, $\alpha \equiv \sqrt{1+4|\gamma|^2/(\varepsilon_1-\varepsilon_2)^2}$).*

15.2 Unitary Evolution

An additional motivation for solving the time-dependent equation directly is that we often encounter Hamiltonians that are explicitly time-dependent. Since the Hamiltonian eigenvalues and eigenstates are also time-dependent in this case, stationary solutions to the full Schrödinger equation cannot be defined in general. Importantly, in the Schrödinger representation of quantum mechanics, the full Hamiltonian of an energy-conserving system is strictly time-independent, and any time-dependence is attributed only to the physical states. Nevertheless, the Schrödinger equation formally holds also for Hamiltonians that depend explicitly on time, $\hat{H} \mapsto \hat{H}(t)$:

$$i\hbar \frac{\partial}{\partial t}|\psi(t)\rangle = \hat{H}(t)|\psi(t)\rangle. \tag{15.2.1}$$

Time-dependent Hamiltonians can be derived exactly, by applying unitary time-dependent transformations to the Schrödinger equation (see the discussion of the interaction and the Heisenberg picture representations in Section 15.3), or as effective field operators within solvers of the time-dependent schroedinger equation [15.1] [15.2] [15.3], or in approximate mean-field and mixed quantum–classical [15.4] approaches. In the latter, the Hamiltonian corresponds to a reduced system, where "external" degrees of freedom, not accounted for explicitly within the system, appear as time-dependent fields, namely, time-dependent parameters in the system Hamiltonian.

Solutions to the time-dependent Schrödinger equation are often formulated in terms of a "time-evolution operator." Without loss of generality, if the "initial" time is t', the solution at any time t is related to $|\psi(t')\rangle$ by a time-evolution operator (or the "propagator"), $\hat{U}(t,t')$, and vice versa:

$$|\psi(t)\rangle = \hat{U}(t,t')|\psi(t')\rangle, \tag{15.2.2a}$$

$$|\psi(t')\rangle = \hat{U}(t',t)|\psi(t)\rangle = \hat{U}^{-1}(t,t')|\psi(t)\rangle. \tag{15.2.2b}$$

The Schrödinger equation relates $\hat{U}(t,t')$ to the system Hamiltonian. Substitution of $|\psi(t)\rangle$ (Eq. (15.2.2a) in the Schrödinger equation, Eq. (15.2.1), yields

$$i\hbar \frac{\partial}{\partial t}\hat{U}(t,t')|\psi(t')\rangle = \hat{H}(t)\hat{U}(t,t')|\psi(t')\rangle. \tag{15.2.3}$$

Since this identity holds for any initial state, $|\psi(t')\rangle$, in the Hilbert space of the system, an equation of motion is obtained for the time-evolution operator itself:

$$\frac{\partial}{\partial t}\hat{U}(t,t') = \frac{-i}{\hbar}\hat{H}(t)\hat{U}(t,t'). \tag{15.2.4}$$

A fundamental property of the time-evolution operator is its unitarity,

$$\hat{U}^{\dagger}(t,t') = \hat{U}^{-1}(t,t'). \tag{15.2.5}$$

This is trivially satisfied when $t = t'$, since according to Eq. (15.2.2)

$$\hat{U}(t',t') = \hat{I}. \tag{15.2.6}$$

For $t \neq t'$, the unitarity of the time-evolution operator is attributed to the Hermiticity of the system Hamiltonian. Taking the Hermitian conjugate of Eq. (15.2.4) (see also Ex. 11.2.3), we have

$$\frac{\partial}{\partial t}\hat{U}^{\dagger}(t,t') = \frac{i}{\hbar}\hat{U}^{\dagger}(t,t')\hat{H}(t). \tag{15.2.7}$$

Using Eqs. (15.2.4, 15.2.7), we can readily conclude that $\hat{U}^{\dagger}(t,t')\hat{U}(t,t')$ is a constant of motion (see Ex. 15.2.1),

$$\frac{\partial}{\partial t}\hat{U}^{\dagger}(t,t')\hat{U}(t,t') = 0. \tag{15.2.8}$$

Given the initial condition, Eq. (15.2.6), the unitarity is hence proved for any t:

$$\hat{U}^{\dagger}(t,t')\hat{U}(t,t') = \hat{I} \Leftrightarrow \hat{U}^{\dagger}(t,t') = \hat{U}^{-1}(t,t'). \tag{15.2.9}$$

An immediate reflection of the unitarity of the time-evolution operator is the conservation of the initial norm for any solution to the time-dependent Schrödinger equation,

$$\langle \psi(t)|\psi(t)\rangle = \langle \psi(t')|\hat{U}^\dagger(t,t')\hat{U}(t,t')|\psi(t')\rangle = \langle \psi(t')|\psi(t')\rangle. \qquad (15.2.10)$$

Exercise 15.2.1   *Use Eqs. (15.2.4, 15.2.7) for the time-derivative of the time-evolution operator to show that $\frac{\partial}{\partial t}\hat{U}^\dagger(t,t')\hat{U}(t,t') = 0$.*

The explicit expression for the time-evolution operator depends on the system Hamiltonian. For a time-independent Hamiltonian, one can readily see that Eqs. (15.2.4, 15.2.6) for the time-evolution operator are consistent with an exponential function of $\hat{H}$ (see Section 3.4 for the definition of functions of operators):

$$\hat{U}(t,t') = e^{\frac{-i}{\hbar}\hat{H}(t-t')}. \qquad (15.2.11)$$

When the Hamiltonian depends explicitly on time, $\hat{H} \mapsto \hat{H}(t)$, the time-evolution operator, $\hat{U}(t,t')$, can be expressed by integrating the Schrödinger equation (Eq. (15.2.4)) from t' to any other time:

$$\hat{U}(t,t') = \hat{U}(t',t') + \int_{t'}^{t} d\tau \frac{-i}{\hbar}\hat{H}(\tau)\hat{U}(\tau,t'). \qquad (15.2.12)$$

Iterative substitutions of $\hat{U}(\tau,t')$ under the time integral, using the initial condition, $\hat{U}(t',t') = \hat{I}$ (Eq. (15.2.6)), yields the infinite Dyson series expansion for the time-evolution operator [15.5]:

$$\hat{U}(t,t') = \hat{I} + \frac{-i}{\hbar}\int_{t'}^{t} d\tau'\hat{H}(\tau') + \left(\frac{-i}{\hbar}\right)^2 \int_{t'}^{t} d\tau'\hat{H}(\tau')\int_{t'}^{\tau'} d\tau''\hat{H}(\tau'')$$

$$+ \left(\frac{-i}{\hbar}\right)^3 \int_{t'}^{t} d\tau'\hat{H}(\tau')\int_{t'}^{\tau'} d\tau''\hat{H}(\tau'')\int_{t'}^{\tau''} d\tau'''\hat{H}(\tau''') + \dots. \qquad (15.2.13)$$

Notice that in each term the product of the Hamiltonians at different times is ordered; for example, for forward propagation, $t > t'$, we have $t > \tau' > \tau'' > \tau''' \dots > t'$. This is important since, in general, the time-dependent Hamiltonian evaluated at one time does not commute with the Hamiltonian evaluated at another time.

In specific cases, where the Hamiltonians at different times $\hat{H}(t)$ and $\hat{H}(t')$ commute (for any t and t'), the integrals can be reordered and recollected as a Taylor expansion of an exponential function. In this case, the time-evolution operator obtains the form

$$[\hat{H}(\tau),\hat{H}(\tau')] = 0 \Rightarrow \hat{U}(t,t') = e^{\frac{-i}{\hbar}\int_{t'}^{t}\hat{H}(\tau)d\tau}. \qquad (15.2.14)$$

Other formulations of the time-evolution operator as an exponential operator exist also in the general case of time-dependent Hamiltonians by extending the Hilbert space of the system artificially to include an additional "time-coordinate." In this case the time-evolution operator is an exponential operator of a time-dependent Hamiltonian in the extended space, $\exp\left(-\frac{it}{\hbar}\left[\hat{H}(t') - i\hbar\frac{\partial}{\partial t'}\right]\right)$, and the solutions to the Schrödinger equation in the physical space are obtained by a proper projection of an initial state at $t = 0$ onto the extended space, and back projection of the final states (at $t \neq 0$) onto the physical space [15.6].

15.3 Time-Dependent Unitary Transformations

The representation of the state of the system depends on the selected basis set, where a change of representation is equivalent to a unitary operation on the state (see Eq. (11.2.15)). It is often convenient to choose a unitary transformation that changes in time, namely, to invoke a time-dependent reference frame for representing the physical state of a time-evolving system. Denoting an abstract state of a system as $|\psi(t)\rangle$, the corresponding unitarily transformed state then reads

$$|\tilde{\psi}(t)\rangle = \hat{S}(t)|\psi(t)\rangle \quad ; \quad \hat{S}(t)\hat{S}^{\dagger}(t) = \hat{I} \tag{15.3.1}$$

Using the Schrödinger equation, Eq. (15.2.1), for $|\psi(t)\rangle$, we can readily obtain a corresponding equation of motion (a transformed Schrödinger equation) for $|\tilde{\psi}(t)\rangle$:

$$i\hbar\frac{\partial}{\partial t}|\tilde{\psi}(t)\rangle = \hat{\tilde{H}}(t)|\tilde{\psi}(t)\rangle \quad ; \quad \hat{\tilde{H}}(t) = \hat{S}(t)\hat{H}(t)\hat{S}^{\dagger}(t) - i\hbar\hat{S}(t)\left(\frac{\partial}{\partial t}\hat{S}^{\dagger}(t)\right), \tag{15.3.2}$$

where $\hat{\tilde{H}}(t)$ is the transformed Hamiltonian (see Ex. 15.3.1).

Exercise 15.3.1 *The state of the system, $|\psi(t)\rangle$, is associated with a solution to the time-dependent Schrödinger equation, $i\hbar\frac{\partial}{\partial t}|\psi(t)\rangle = \hat{H}(t)|\psi(t)\rangle$. A transformed state, $|\tilde{\psi}(t)\rangle$, is related to $|\psi(t)\rangle$ via a unitary transformation, as defined in Eq. (15.3.1). Express the time-derivative of $|\tilde{\psi}(t)\rangle$ in terms of the operation of a transformed Hamiltonian on $|\tilde{\psi}(t)\rangle$, as defined in Eq. (15.3.2).*

In calculations of measurable quantities, the same unitary transformation must apply consistently to the operator representing the measurement (see Eq. (11.2.13)). For generality we consider here system operators that can depend inherently on time, namely $\hat{O} \mapsto \hat{O}(t)$. The result of any time-dependent measurement on the system can therefore be formulated as (see Eqs. (15.1.8–15.1.11)) $O(t) = \langle\tilde{\psi}(t)|\hat{O}(t)|\tilde{\psi}(t)\rangle$. This expression is invariant to the introduction of unitary operators,

$$O(t) = \langle\psi(t)|\hat{O}(t)|\psi(t)\rangle = \langle\psi(t)|\hat{S}^{\dagger}(t)\hat{S}(t)\hat{O}(t)\hat{S}^{\dagger}(t)\hat{S}(t)|\psi(t)\rangle = \langle\tilde{\psi}(t)|\hat{\tilde{O}}(t)|\tilde{\psi}(t)\rangle, \tag{15.3.3}$$

where the transformed operator that corresponds to the unitarily transformed state, $|\tilde{\psi}(t)\rangle$, can therefore be identified as

$$\hat{\tilde{O}}(t) = \hat{S}(t)\hat{O}(t)\hat{S}^{\dagger}(t). \tag{15.3.4}$$

The Interaction Picture Representation

A commonly practiced unitary transformation leads to the *"interaction picture representation"* of the system dynamics. Consider a Hamiltonian that is a sum of a time-independent term ($\hat{H}_0$) and an "interaction" term, which, generally, can depend inherently on time:

$$\hat{H}(t) = \hat{H}_0 + \hat{V}(t). \tag{15.3.5}$$

It is convenient in this case to choose an inverse propagator that corresponds to the time-independent $\hat{H}_0$ as a unitary transformation, namely

$$\hat{S}(t) \equiv \hat{U}_0^\dagger(t,0) = e^{\frac{i}{\hbar}\hat{H}_0 t}. \tag{15.3.6}$$

The transformed state, denoted as $|\psi_I(t)\rangle$, is referred to as the interaction picture representation of the state of the system,

$$|\psi_I(t)\rangle \equiv e^{\frac{i}{\hbar}\hat{H}_0 t}|\psi(t)\rangle. \tag{15.3.7}$$

The corresponding equation of motion for $|\psi_I(t)\rangle$ obtains the form (see Ex. 15.3.2)

$$i\hbar\frac{\partial}{\partial t}|\psi_I(t)\rangle = \hat{V}_I(t)|\psi_I(t)\rangle \quad ; \quad \hat{V}_I(t) = e^{\frac{i}{\hbar}\hat{H}_0 t}\hat{V}(t)e^{\frac{-i}{\hbar}\hat{H}_0 t}, \tag{15.3.8}$$

where $\hat{V}_I(t)$ is the interaction picture representation of the interaction operator.

The transformation of the Schrödinger equation into the interaction picture is equivalent to replacing the original Hamiltonian by an effective time-dependent Hamiltonian, $\hat{H}^{(I)}(t) \equiv \hat{V}_I(t) = e^{\frac{i}{\hbar}\hat{H}_0 t}\hat{V}(t)e^{\frac{-i}{\hbar}\hat{H}_0 t}$, which holds regardless of any time-dependence of the original interaction term, $\hat{V}(t)$. Consequently, the effective time-evolution operator in the interaction picture, to be denoted as $\hat{U}^{(I)}(t,0)$, is not a simple exponential operator, but rather a Dyson series expansion in powers of $\hat{H}^{(I)}(t)$ (see Eq. (15.2.13)):

$$|\psi_I(t)\rangle \equiv \hat{U}^{(I)}(t,0)|\psi_I(0)\rangle$$

$$\frac{\partial}{\partial t}\hat{U}^{(I)}(t,0) = \frac{1}{i\hbar}\hat{H}^{(I)}(t)\hat{U}^{(I)}(t,0) \quad ; \quad \hat{H}^{(I)}(t) \equiv e^{\frac{i}{\hbar}\hat{H}_0 t}\hat{V}(t)e^{\frac{-i}{\hbar}\hat{H}_0 t}$$

$$\hat{U}^{(I)}(t,0) = \hat{I} + \frac{-i}{\hbar}\int_0^t dt'\hat{H}^{(I)}(t') + \left(\frac{-i}{\hbar}\right)^2\int_0^t dt'\hat{H}^{(I)}(t')\int_0^{t'} dt''\hat{H}^{(I)}(t'') + \dots. \tag{15.3.9}$$

As we shall discuss, the interaction representation is particularly useful when the interaction operator can be regarded as a small perturbation, and consequently the Dyson expansion can be truncated at a low order.

Given any system observable, $\hat{O}(t)$, and the definition of the transformed state (Eqs. (15.3.6, 15.3.7)), the time evolution of the corresponding observable in the interaction picture is given by Eq. (15.3.3):

$$O(t) = \langle\psi_I(t)|\hat{O}_I(t)|\psi_I(t)\rangle. \tag{15.3.10}$$

The corresponding interaction picture representation of the operator reads

$$\hat{O}_I(t) \equiv e^{\frac{i}{\hbar}\hat{H}_0 t}\hat{O}(t)e^{\frac{-i}{\hbar}\hat{H}_0 t}, \tag{15.3.11}$$

and the equation of motion for $\hat{O}_I(t)$ is therefore

$$\frac{\partial}{\partial t}\hat{O}_I(t) = \frac{i}{\hbar}[\hat{H}_0, \hat{O}_I(t)] + \left[\frac{\partial}{\partial t}\hat{O}(t)\right]_I. \qquad (15.3.12)$$

Notice that the transformed operator $\hat{O}_I(t)$ depends on time even when the system operator is inherently time-independent, $\hat{O}(t) \mapsto \hat{O}$.

Exercise 15.3.2 *The state of the system, $|\psi(t)\rangle$, is associated with a solution to the time-dependent Schrödinger equation, $i\hbar\frac{\partial}{\partial t}|\psi(t)\rangle = \hat{H}(t)|\psi(t)\rangle$, where $\hat{H}(t) = \hat{H}_0 + \hat{V}(t)$. A transformed state, $|\psi_I(t)\rangle$, is related to $|\psi(t)\rangle$ via a unitary transformation, as defined in Eq. (15.3.7). Express the time derivative of $|\psi_I(t)\rangle$ in terms of the operation of the transformed interaction operator, as defined in Eq. (15.3.8).*

The Heisenberg Picture

Another commonly practiced time-dependent unitary transformation of the state leads to the "***Heisenberg picture representation***" of the system dynamics. Let $\hat{H}(t)$ be the system Hamiltonian, and let $\hat{U}(t,0)$ be the corresponding time-evolution operator (propagator) associated with $\hat{H}(t)$. One can then choose the inverse of the full system's propagator as a time-dependent unitary transformation, namely

$$\hat{S}(t) \equiv \hat{U}^\dagger(t,0). \qquad (15.3.13)$$

The transformed state obtained in this case is referred to as the Heisenberg picture representation of the state, which reads

$$|\psi_H\rangle \equiv \hat{U}^\dagger(t,0)|\psi(t)\rangle = \hat{U}^\dagger(t,0)\hat{U}(t,0)|\psi(0)\rangle = |\psi(0)\rangle. \qquad (15.3.14)$$

As we can see, in the Heisenberg picture the state vectors are time-independent, which can be readily verified using Eq. (15.2.8) (see Ex. 15.2.1).

Concerning system observables, Eq. (15.3.3) together with the definitions in Eqs. (15.3.13, 15.3.14)) means that the time-dependence of any observable is given as

$$O(t) = \langle\psi(t)|\hat{O}(t)|\psi(t)\rangle = \langle\psi(0)|\hat{O}_H(t)|\psi(0)\rangle, \qquad (15.3.15)$$

where the Heisenberg picture representation of any system operator, $\hat{O}_H(t)$, reads

$$\hat{O}_H(t) \equiv \hat{U}^\dagger(t,0)\hat{O}(t)\hat{U}(t,0), \qquad (15.3.16)$$

and specifically for a time-independent Hamiltonian, we obtain $\hat{O}_H(t) \equiv e^{\frac{i}{\hbar}\hat{H}t}\hat{O}(t)e^{\frac{-i}{\hbar}\hat{H}t}$. The corresponding equation of motion for $\hat{O}_H(t)$ is readily obtained from the Schrödinger equation for $\hat{U}(t,0)$ (see Ex. 15.3.3):

$$\frac{\partial}{\partial t}\hat{O}_H(t) = \frac{i}{\hbar}[\hat{H}_H(t), \hat{O}_H(t)] + \left[\frac{\partial}{\partial t}\hat{O}(t)\right]_H. \qquad (15.3.17)$$

Notice that (as in the interaction picture) the transformed operator $\hat{O}_H(t)$ depends on time even when the system operator is inherently time-independent, $\hat{O}(t) \mapsto \hat{O}$. In the latter case, the Heisenberg equation simplifies:

$$\hat{O}(t) \mapsto \hat{O} \quad \Rightarrow \quad \frac{\partial}{\partial t}\hat{O}_H(t) = \frac{i}{\hbar}[\hat{H}_H(t), \hat{O}_H(t)]. \tag{15.3.18}$$

Namely, the generator of motion for an operator in the Heisenberg picture is its commutator with the system Hamiltonian. Any observable represented by an operator that commutes with the Hamiltonian is therefore a constant of motion, namely time-independent. When the system Hamiltonian itself does not depend explicitly on time, we can readily see that its Heisenberg picture representation is equal to the Hamiltonian itself, namely

$$\hat{H}(t) \mapsto \hat{H} \quad \Rightarrow \quad \hat{H}_H(t) = \hat{H}. \tag{15.3.19}$$

Exercise 15.3.3   *Use Eqs. (15.2.4, 15.2.7) to derive Eq. (15.3.17) from Eq. (15.3.16).*

15.4 Quantum-Classical Correspondence

The Heisenberg picture associates the dynamics of a system with time-dependent observables rather than with time-dependent state vectors. This concept is similar to the picture invoked in classical mechanics, and therefore it provides a convenient framework for analyzing the differences as well as the correspondence between quantum and classical mechanics. Consider, for example, a point particle of mass m, in a one-dimensional coordinate system, associated with the generic (time-independent) Hamiltonian

$$\hat{H} = \frac{\hat{p}^2}{2m} + V(\hat{x}) \quad ; \quad V(\hat{x}) = \sum_{n=0}^{\infty} a_n \hat{x}^n, \tag{15.4.1}$$

where $\hat{p}$ and $\hat{x}$ are the canonical momentum and position operators (Eqs. (3.2.1, 3.2.2)). Transforming to the Heisenberg picture representation (Eq. (15.3.16)), the corresponding operators are time-dependent,

$$\hat{p}_H(t) \equiv e^{\frac{i}{\hbar}\hat{H}t} \hat{p} e^{\frac{-i}{\hbar}\hat{H}t} \quad ; \quad \hat{x}_H(t) \equiv e^{\frac{i}{\hbar}\hat{H}t} \hat{x} e^{\frac{-i}{\hbar}\hat{H}t} \tag{15.4.2}$$

and the corresponding equations of motion for $\hat{p}_H(t)$ and $\hat{x}_H(t)$, according to Eq. (15.3.18), read (see Ex. 15.4.1)

$$\frac{\partial}{\partial t}\hat{x}_H(t) = \frac{i}{\hbar}[\hat{H}, \hat{x}_H(t)] = \frac{i}{\hbar}[\hat{H}, \hat{x}]_H(t) = \frac{1}{m}\hat{p}_H(t)$$

$$\frac{\partial}{\partial t}\hat{p}_H(t) = \frac{i}{\hbar}[\hat{H}, \hat{p}_H(t)] = \frac{i}{\hbar}[\hat{H}, \hat{p}]_H(t) = -V'(\hat{x}_H(t)), \tag{15.4.3}$$

where $V'(\alpha) = \frac{d}{d\alpha}V(\alpha)$.

Exercise 15.4.1 *(a) Given the commutation relation,* $\left[\hat{x}, \frac{\partial}{\partial x}\right] = -1$*, prove that,* $\left[\hat{x}^n, \frac{\partial}{\partial x}\right] = -n\hat{x}^{n-1}$*. (b) Given the definitions of the canonical position and momentum operators, Eqs. (3.2.1, 3.2.2), and using the solution to (a), derive the results in Eq. (15.4.3).*

As we can see, the equations of motion for the Heisenberg operators are identical to Hamilton's equations of classical mechanics [3.1] for the corresponding classical observables, namely, $\dot{x} = p/m$, $\dot{p} = -V'(x)$. However, while in classical mechanics the solutions $x(t)$ and $p(t)$ correspond to directly measurable quantities, in quantum mechanics the latter can be obtained only by a statistical average over an ensemble of systems represented by the quantum state (namely expectation values; see Chapter 11). To derive equations for the quantum mechanical measurable quantities, we must first introduce the (time-independent) state of the system in the Heisenberg picture, $|\psi_H\rangle$, where the Heisenberg equations of motion, Eq. (15.4.3), lead to

$$\frac{\partial}{\partial t}\langle\psi_H|\hat{x}_H(t)|\psi_H\rangle = \frac{1}{m}\langle\psi_H|\hat{p}_H(t)|\psi_H\rangle \tag{15.4.4}$$

$$\frac{\partial}{\partial t}\langle\psi_H|\hat{p}_H(t)|\psi_H\rangle = -\langle\psi_H|V'[\hat{x}_H(t)]|\psi_H\rangle. \tag{15.4.5}$$

This result is the Heisenberg picture representation of the Ehrenfest theorem (see Ex. (11.7.5)). Taking the derivative of the potential energy with respect to $\hat{x}_H(t)$,

$$[V'[\hat{x}_H(t)] = \sum_{n=1}^{\infty} a_n n\hat{x}_H^{n-1}(t), \tag{15.4.6}$$

and recalling the identification of measurable quantities with expectation values (Eq. (15.3.3)), and particularly

$$p(t) \equiv \langle\psi_H|\hat{p}_H(t)|\psi_H\rangle \quad ; \quad x^n(t) \equiv \langle\psi_H|\hat{x}_H^n(t)|\psi_H\rangle, \tag{15.4.7}$$

we obtain the following equations of motion for the quantum mechanical expectation values:

$$\frac{\partial}{\partial t}x(t) = \frac{1}{m}p(t)$$

$$\frac{\partial}{\partial t}p(t) = -\sum_{n=1}^{\infty} a_n nx^{n-1}(t). \tag{15.4.8}$$

Notice that, in general, $x^n(t) = \langle\psi_H|\hat{x}_H^n(t)|\psi_H\rangle \neq [\langle\psi_H|\hat{x}_H(t)|\psi_H\rangle]^n = [x(t)]^n$. Therefore, the term $\sum_{n=1}^{\infty} a_n nx^{n-1}(t)$ cannot be replaced in general by $\frac{d}{dx(t)}V(x(t))$. This is the essence of the difference between the quantum and classical equations of motion. Indeed, in classical mechanics the observable $x^n(t)$ identifies with $[x(t)]^n$, such that the second equation in Eq. (15.4.8) can be written as $\frac{\partial}{\partial t}p(t) = -\frac{d}{dx(t)}V(x(t))$. Consequently, in classical mechanics, given that $x(0)$ and $p(0)$ are known at an initial time ($t = 0$), the measurables $x(t)$ and $p(t)$ are uniquely defined at any time. In contrast, in quantum mechanics the equations for the expectation values $x(t)$ and $p(t)$ depend also on the higher moments, $\{\hat{x}^n(t) = \langle\psi_H|\hat{x}_H^n(t)|\psi_H\rangle\}$. In the general case, the additional equations of motion for these higher moments result in an infinite hierarchy of coupled equations, whose formal solution requires an infinite set of corresponding initial conditions. (This is reminiscent of the information contained in the continuous wave function, $\psi_H(x)$.)

The time evolution of the position and momentum observables is therefore different in quantum and classical mechanics. Notice, however, that the classical and quantum expressions may coincide. This is the case for quadratic potential energy functions in the particle's position,

$$V(\hat{x}) = a_0 + a_1\hat{x} + a_2\hat{x}^2 \quad ; \quad V'(\hat{x}) = a_1 + 2a_2\hat{x}, \tag{15.4.9}$$

where the Heisenberg equations of motion (Eqs. (15.4.3)) and the corresponding Hamilton's equations become linear in the position and momentum variables. The quantum equations for the observables (Eqs. (15.4.7, 15.4.8)) read in this case

$$\frac{\partial}{\partial t}x(t) = \frac{1}{m}p(t)$$

$$\frac{\partial}{\partial t}p(t) = -a_1 - 2a_2 x(t), \tag{15.4.10}$$

which are identical to the classical Hamilton equations. Two important cases fall into this category. The first is the motion of a free particle, which corresponds to $a_2 = 0$, and the second is the harmonic oscillator. Owing to the universality of the harmonic approximation in quantum systems (also in the multidimensional case [see Section 8.2]), the quantum–classical correspondence in the harmonic oscillator model is the basis for semiclassical approaches for approximate treatments of quantum dynamics in multidimensional systems. In what follows we consider the dynamics of a free particle and of the harmonic oscillator in some detail.

Gaussian Wave Packets

The quantum–classical correspondence is interesting, since it provides a rigorous framework for associating wave function dynamics in quantum mechanics with the motion of particles in classical mechanics. Returning to a point particle of mass m in a one-dimensional coordinate space, an intuitive (as well as optimal, as will be discussed in what follows) mapping of the particle's properties onto a wave function is in terms of a normalized Gaussian wave packet. In Chapter 4 we discussed qualitatively the dynamics of a Gaussian wave packet, whereas here we add the quantitative analysis. In the position representation, the initial wave function obtains the form (see Eq. (4.2.1))

$$\psi(x,0) = \langle x|\psi_0\rangle = \left(\frac{1}{2\pi\sigma^2}\right)^{1/4} e^{\frac{-(x-x_0)^2}{4\sigma^2}} e^{ip_0 x/\hbar}, \tag{15.4.11}$$

where we can readily verify that the parameters x_0 and σ correspond to the expectation value of the particle's position, $\langle\psi_0|\hat{x}|\psi_0\rangle = x_0$, and to the position uncertainty,

$\sqrt{\langle\psi_0|\hat{x}^2|\psi_0\rangle - \langle\psi_0|\hat{x}|\psi_0\rangle^2} = \sigma$ (see Ex. 2.3.1). Using the transformation to the momentum representation,

$$\langle p|\psi_0\rangle = \frac{1}{\sqrt{2\pi\hbar}}\int dx\langle x|\psi_0\rangle e^{-ipx/\hbar} \quad;\quad \langle x|\psi_0\rangle = \frac{1}{\sqrt{2\pi\hbar}}\int dp\langle p|\psi_0\rangle e^{ipx/\hbar}, \quad (15.4.12)$$

the initial Gaussian wave packet obtains the form

$$\overline{\Psi}(p,0) = \langle p|\psi_0\rangle = \left(\frac{2\sigma^2}{\pi\hbar^2}\right)^{1/4} e^{\frac{-(p-p_0)^2}{(\hbar/\sigma)^2}} e^{-i(p-p_0)x_0/\hbar}, \quad (15.4.13)$$

where one can readily verify that the parameter p_0 corresponds to the expectation value of the particle's momentum, $\langle\psi_0|\hat{p}|\psi_0\rangle = p_0$, and the momentum uncertainty reads $\sqrt{\langle\psi_0|\hat{p}^2|\psi_0\rangle - \langle\psi_0|\hat{p}|\psi_0\rangle^2} = \frac{\hbar}{2\sigma}$. Consequently, the Gaussian wave packet $|\psi_0\rangle$ is a state of minimum uncertainty,

$$\left(\sqrt{\langle\psi_0|\hat{p}^2|\psi_0\rangle - \langle\psi_0|\hat{p}|\psi_0\rangle^2}\right)\left(\sqrt{\langle\psi_0|\hat{x}^2|\psi_0\rangle - \langle\psi_0|\hat{x}|\psi_0\rangle^2}\right) = \frac{\hbar}{2}, \quad (15.4.14)$$

(see Eq. (11.7.4), Ex. 11.7.10). Thereby, this state provides an optimal mapping of a classical particle onto a quantum state. We are interested in the dynamics of the quantum system from an initial state $|\psi_0\rangle$. The time evolution depends, of course, on the underlying system's Hamiltonian, where we focus on the cases of a free particle and a harmonic oscillator.

Case I: The "free particle"

The free-particle Hamiltonian reads

$$\hat{H} = \frac{\hat{p}^2}{2m} + a_0. \quad (15.4.15)$$

The corresponding Heisenberg equations, Eqs. (15.4.3), therefore obtain the form

$$\frac{\partial}{\partial t}\hat{x}_H(t) = \frac{1}{m}\hat{p}_H(t) \quad;\quad \frac{\partial}{\partial t}\hat{p}_H(t) = 0. \quad (15.4.16)$$

Integrating these equations from $t = 0$, for the initial conditions, $\hat{p}_H(0) = \hat{p}$ and $\hat{x}_H(0) = \hat{x}$, we obtain the Heisenberg operators,

$$\hat{p}_H(t) = \hat{p} \quad;\quad \hat{x}_H(t) = \hat{x} + \frac{\hat{p}}{m}t. \quad (15.4.17)$$

For the Gaussian wave packet, $|\psi_0\rangle$ (Eqs. (15.4.11–15.4.13)), the corresponding evolution of the expectation values, $x(t) = \langle\psi_0|\hat{x}_H(t)|\psi_0\rangle$ and $p(t) = \langle\psi_0|\hat{p}_H(t)|\psi_0\rangle$, is readily obtained:

$$p(t) = p_0 \quad;\quad x(t) = x_0 + \frac{p_0}{m}t. \quad (15.4.18)$$

Indeed, ***the quantum mechanical result for the position and momentum expectation values coincides in this case with the classical trajectory of a free particle*** (see also Fig. 4.2.1).

It is instructive to also follow the standard deviations of the position and momentum distributions for the Gaussian wave packet. For this purpose, we consider the time evolution of the second moments. Using Eq. (15.4.17), we obtain

$$\hat{x}_H^2(t) = \left(\hat{x} + \frac{\hat{p}}{m}t\right)^2 = \hat{x}^2 + \frac{\hat{p}^2}{m^2}t^2 + \frac{\hat{x}\hat{p} + \hat{p}\hat{x}}{m}t$$

$$\hat{p}_H^2(t) = \hat{p}^2 \tag{15.4.19}$$

Taking the expectation values with respect to $|\psi_0\rangle$, we obtain the corresponding quantum observables (Ex. 15.4.2),

$$x^2(t) = \sigma^2 + x_0^2 + \frac{t^2}{m^2}\frac{\hbar^2}{4\sigma^2} + \frac{t^2}{m^2}p_0^2 + \frac{2t}{m}x_0 p_0$$

$$p^2(t) = p_0^2 + \frac{h^2}{4\sigma^2}, \tag{15.4.20}$$

which amount to the following standard deviations in the momentum and position distributions (Ex. 15.4.3):

$$\sqrt{x^2(t) - [x(t)]^2} = \sigma\sqrt{1 + \frac{t^2}{m^2}\frac{\hbar^2}{4\sigma^4}}$$

$$\sqrt{p^2(t) - [p(t)]^2} = \frac{h}{2\sigma}. \tag{15.4.21}$$

While the momentum uncertainly is fixed in time, which corresponds to the absence of a net force on the free particle, the position uncertainty is shown to grow in time. This is expected on account of the initial distribution of momenta, leading to dispersion and hence broadening of the distribution in the position representation. The dispersion rate is shown to decrease with decreasing width of the momentum distribution and with increasing particle mass. The broadening of the position distribution for a fixed momentum distribution means that the uncertainty product grows in time:

$$\sqrt{x^2(t) - [x(t)]^2}\sqrt{p^2(t) - [p(t)]^2} = \frac{\hbar}{2}\sqrt{1 + \frac{t^2}{m^2}\frac{\hbar^2}{4\sigma^4}} \geq \frac{\hbar}{2}. \tag{15.4.22}$$

The minimal uncertainty state is therefore only transient for the free particle (see also Fig. 4.2.1).

Exercise 15.4.2 *In Eq. (15.4.19) the time-dependent Heisenberg operators, $\hat{x}_H(t)$ and $\hat{p}_H(t)$, are expressed in terms of the operators, $\hat{x}^2$, $\hat{p}^2$, $\hat{x}\hat{p}$, and $\hat{p}\hat{x}$. (a) Express the expectation values of $\hat{x}^2$, $\hat{p}^2$, $\hat{x}\hat{p}$, and $\hat{p}\hat{x}$ in terms of the parameters x_0, p_0, σ of the Gaussian wave packet (Eq. (15.4.11)). (b) Obtain the expressions for the quantum mechanical expectation values, $x^2(t) = \langle\psi_0|\hat{x}_H^2(t)|\psi_0\rangle$ and $p^2(t) = \langle\psi_0|\hat{p}_H^2(t)|\psi_0\rangle$, in Eq. (15.4.20).*

Exercise 15.4.3 *Use Eqs. (15.4.18, 15.4.20) to derive Eq. (15.4.21).*

For completeness, we turn to the time evolution of the initial Gaussian wave packet, Eq. (15.4.11), in the Schrödinger picture for a free particle. First, we notice that the time-evolution operator (Eq. (15.2.11)) is most conveniently applied in this case, in the momentum representation,

$$\overline{\Psi}(p,t) = \langle p | e^{-it\frac{\hat{p}^2}{2m\hbar}} | \psi_0 \rangle = \left(\frac{2\sigma^2}{\pi\hbar^2} \right)^{1/4} e^{\frac{-(p-p_0)^2}{(\hbar/\sigma)^2}} e^{-i(p-p_0)x_0/\hbar} e^{\frac{-ip^2t}{2m\hbar}}. \tag{15.4.23}$$

The time-dependence of the state in the position representation is obtained by expanding in the momentum eigenstates basis, using the relation, Eq. (15.4.12),

$$\psi(x,t) = \frac{1}{\sqrt{2\pi\hbar}} \int_{-\infty}^{\infty} dp \overline{\Psi}(p,t) e^{ipx/\hbar}$$

$$= \frac{1}{\sqrt{2\pi\hbar}} \left(\frac{2\sigma^2}{\pi\hbar^2} \right)^{1/4} \int_{-\infty}^{\infty} dp e^{\frac{-(p-p_0)^2}{(\hbar/\sigma)^2}} e^{-i(p-p_0)x_0/\hbar} e^{ipx/\hbar} e^{\frac{-ip^2t}{2m\hbar}}, \tag{15.4.24}$$

Notice that this expression can be regarded as a superposition of the set of stationary states for a free particle $\left\{ \frac{1}{\sqrt{2\pi\hbar}} e^{ipx/\hbar} e^{\frac{-ip^2t}{2m\hbar}} \right\}$, with $\langle p | \psi_0 \rangle$ as the expansion coefficients (see Eq. (15.1.5)). Evaluating the integral in Eq. (15.4.24), the time-dependent probability density is shown to remain a Gaussian function (see Ex. 15.4.4),

$$|\psi(x,t)|^2 = \sqrt{\frac{1}{2\pi \left[\sigma^2 + \frac{\hbar^2 t^2}{4\sigma^2 m^2} \right]}} e^{\frac{-\left(x - x_0 - \frac{p_0 t}{m} \right)^2}{2\left[\sigma^2 + \frac{\hbar^2 t^2}{4\sigma^2 m^2} \right]}}. \tag{15.4.25}$$

The center of the position space distribution (corresponding to the position expectation value) is indeed shown to follow the classical trajectory (Eq. (15.4.18)), and the distribution width (corresponding to the position uncertainty) indeed increases in time according to Eq. (15.4.21).

Exercise 15.4.4 *Use the momentum space representation of the time-dependent Gaussian wave packet, Eq. (15.4.24), change the integration variable, $p' = p - p_0$, and use the identity, $\int_{-\infty}^{\infty} dp' e^{-zp'^2} e^{ip'x} = \sqrt{\frac{\pi}{z}} e^{\frac{-x^2}{4z}}$ for a complex-valued z to obtain an explicit expression for the time-evolution of the Gaussian wave packet for a free particle,*

$$\langle x | \psi(t) \rangle = \left(e^{\frac{-ip_0^2 t}{2m\hbar}} e^{ip_0 x/\hbar} \right) \sqrt{\frac{\hbar}{2\pi}} \left(\frac{2\sigma^2}{\pi\hbar^2} \right)^{1/4} \sqrt{\frac{\pi}{\left[\sigma^2 + \frac{i\hbar t}{2m} \right]}} e^{\frac{-\left(x - x_0 - \frac{p_0 t}{m} \right)^2}{4\left[\sigma^2 + \frac{i\hbar t}{2m} \right]}}, \text{ and the correspond-}$$

ing probability density, Eq. (15.4.25).

Case II: The Harmonic oscillator

The Hamiltonian of a one-dimensional harmonic oscillator reads (see Chapter 8)

$$\hat{H} = \frac{\hat{p}^2}{2m} + \frac{m\omega^2 \hat{x}^2}{2}, \tag{15.4.26}$$

and the corresponding Heisenberg equations, Eq. (15.4.3), for this Hamiltonian obtain the form

$$\frac{\partial}{\partial t}\hat{x}_H(t) = \frac{1}{m}\hat{p}_H(t) \quad ; \quad \frac{\partial}{\partial t}\hat{p}_H = -m\omega^2\hat{x}_H. \tag{15.4.27}$$

The explicit expressions for the position and momentum operators in the Heisenberg picture are readily obtained by integrating these equations from $t = 0$ (see Ex. 15.4.5), with the initial conditions, $\hat{p}_H(0) = \hat{p}$ and $\hat{x}_H(0) = \hat{x}$:

$$\hat{x}_H(t) = \hat{x}\cos(\omega t) + \frac{\hat{p}}{m\omega}\sin(\omega t) \quad ; \quad \hat{p}_H(t) = -m\omega\hat{x}\sin(\omega t) + \hat{p}\cos(\omega t). \tag{15.4.28}$$

Using the relations, $x_0 = \langle\psi_0|\hat{x}|\psi_0\rangle$ and $p_0 = \langle\psi_0|\hat{p}|\psi_0\rangle$, for the Gaussian wave packet ($|\psi_0\rangle$, as defined in Eqs. (15.4.11–15.4.13)), the time evolution of the expectation values, $x(t) = \langle\psi_0|\hat{x}_H(t)|\psi_0\rangle$ and $p(t) = \langle\psi_0|\hat{p}_H(t)|\psi_0\rangle$ reads

$$x(t) = x_0\cos(\omega t) + \frac{p_0}{m\omega}\sin(\omega t) \quad ; \quad p(t) = -m\omega x_0\sin(\omega t) + p_0\cos(\omega t). \tag{15.4.29}$$

As we can see, *the quantum mechanical result for the position and momentum expectation values coincides with the classical trajectory of a particle in a harmonic potential energy well (Eqs. (15.4.9, 15.4.10); see also Fig. 4.2.3).*

To follow the standard deviations of the position and momentum distributions for the Gaussian wave packet in the harmonic oscillator model, we consider the time evolution of the corresponding second moments. Using Eq. (15.4.28), we immediately obtain

$$\hat{x}_H^2(t) = \hat{x}^2\cos^2(\omega t) + \frac{\hat{p}^2}{m^2\omega^2}\sin^2(\omega t) + \frac{\hat{x}\hat{p} + \hat{p}\hat{x}}{m\omega}\cos(\omega t)\sin(\omega t)$$
$$\hat{p}_H^2(t) = m^2\omega^2\hat{x}^2\sin^2(\omega t) + \hat{p}^2\cos^2(\omega t) - m\omega(\hat{x}\hat{p} + \hat{p}\hat{x})\cos(\omega t)\sin(\omega t). \tag{15.4.30}$$

Taking the expectation values with respect to the wave packet, $|\psi_0\rangle$, we obtain (Ex. 15.4.6)

$$x^2(t) = (x_0^2 + \sigma^2)\cos^2(\omega t) + \frac{1}{m^2\omega^2}\left(p_0^2 + \frac{\hbar^2}{4\sigma^2}\right)\sin^2(\omega t) + \frac{2x_0p_0}{m\omega}\cos(\omega t)\sin(\omega t)$$

$$p^2(t) = m^2\omega^2(x_0^2 + \sigma^2)\sin^2(\omega t) + \left(p_0^2 + \frac{\hbar^2}{4\sigma^2}\right)\cos^2(\omega t) - 2m\omega x_0p_0\sin(\omega t)\cos(\omega t),$$

$$\tag{15.4.31}$$

and finally, using Eqs. (15.4.29, 15.4.31), the standard deviations are given as (Ex. 15.4.6)

$$\sqrt{x^2(t) - [x(t)]^2} = \sqrt{\sigma^2\cos^2(\omega t) + \frac{\hbar^2}{4\sigma^2 m^2\omega^2}\sin^2(\omega t)}$$

$$\sqrt{p^2(t) - [p(t)]^2} = \sqrt{m^2\omega^2\sigma^2\sin^2(\omega t) + \left(\frac{\hbar^2}{4\sigma^2}\right)\cos^2(\omega t)}. \tag{15.4.32}$$

In general, the momentum and position uncertainties are shown to oscillate in time at the oscillator frequency.

Exercise 15.4.5   *Obtain the time-dependent Heisenberg operators for the harmonic oscillator (Eq. (15.4.28)) by solving Eq. (15.4.27) for the initial conditions, $\hat{x}_H(0) = \hat{x}$, $\hat{p}_H(0) = \hat{p}$.*

Exercise 15.4.6   *Use the results of Ex. 15.4.2(a) for the expectation values of $\hat{x}^2$, $\hat{p}^2$, $\hat{x}\hat{p}$, and $\hat{p}\hat{x}$ as functions of the parameters x_0, p_0, σ of the Gaussian wave packet to obtain the expressions for the quantum mechanical expectation values, $x^2(t) = \langle \psi_0 | \hat{x}_H^2(t) | \psi_0 \rangle$ and $p^2(t) = \langle \psi_0 | \hat{p}_H^2(t) | \psi_0 \rangle$, in Eq. (15.4.31). (b) Obtain the expressions for the standard deviations of the momentum and position distributions in Eq. (15.4.32).*

The Coherent State

A particularly interesting case of a Gaussian wave packet dynamics under the Harmonic oscillator Hamiltonian corresponds to the parameter $\sigma = \sqrt{\frac{\hbar}{2m\omega}}$. In this case, the standard deviations are constant in time (see Eq. (15.4.32)), such that the minimal uncertainty product is maintained at any time:

$$\sigma = \sqrt{\frac{\hbar}{2m\omega}} \Rightarrow \sqrt{x^2(t) - [x(t)]^2}\sqrt{p^2(t) - [p(t)]^2} = \frac{\hbar}{2}. \tag{15.4.33}$$

The solution of the Schrödinger equation for a harmonic oscillator corresponding to a minimal uncertainty Gaussian wave packet is referred to as a coherent state. Since the position and momentum expectation values follow a classical trajectory, while the standard deviations in the position and momentum do not change in time, the correspondence between quantum and classical mechanics is maximized in this case. Indeed, coherent states are the basis for semiclassical approximations in atomic and molecular physics as well as in quantum field theory [15.7]. Recalling the definition of the annihilation operator for the harmonic oscillator (Eq. (8.5.10)),

$$\hat{a} \equiv \sqrt{\frac{m\omega}{2\hbar}}\hat{x} + i\sqrt{\frac{1}{2m\omega\hbar}}\hat{p}, \tag{15.4.34}$$

we can readily see that the initial Gaussian wave packet, Eq. (15.4.11), is an eigenfunction of the creation operator (Ex. 15.4.7),

$$\hat{a}|\psi_0\rangle = \sqrt{\frac{m\omega}{2\hbar}}\left(x_0 + \frac{ip_0}{m\omega}\right)|\psi_0\rangle. \tag{15.4.35}$$

The real and imaginary parts of the eigenvalue correspond, respectively, to the (dimensionless) initial position and momentum expectation values of the Gaussian wave packet. A coherent state can therefore be uniquely defined by a complex eigenvalue of $\hat{a}$, denoted as α:

$$|\psi_0\rangle \equiv |\alpha\rangle \quad ; \quad \alpha \equiv \sqrt{\frac{1}{2}}\left(\sqrt{\frac{m\omega}{\hbar}}x_0 + i\frac{p_0}{\sqrt{\hbar m\omega}}\right). \tag{15.4.36}$$

Again, for completeness, we turn to evaluation of the time evolution of the coherent state (defined in terms of the Gaussian wave packet, Eq. (15.4.11)) in the Schrödinger picture. This is most conveniently done by expanding the state in the basis of the harmonic oscillator Hamiltonian eigenstates, defined by the equation $\hbar\omega\left(\hat{a}^\dagger\hat{a}+\frac{1}{2}\right)|\varphi_n\rangle = \hbar\omega\left(n+\frac{1}{2}\right)|\varphi_n\rangle, n=0,1,2,\ldots$ (see Eqs. (8.5.12, 8.5.13)). This expansion reads (Ex. 15.4.8)

$$|\alpha\rangle = e^{i\Phi}e^{\frac{-|\alpha^2|}{2}}\sum_{n=0}^\infty \frac{\alpha^n}{\sqrt{n!}}|\varphi_n\rangle, \tag{15.4.37}$$

where Φ is an arbitrary phase. The time-dependence of the state is obtained by applying the time-evolution operator,

$$|\psi(t)\rangle = e^{\frac{-it}{\hbar}\hbar\omega(\hat{a}^\dagger\hat{a}+1/2)}|\alpha\rangle = e^{i\Phi}e^{\frac{-|\alpha^2|}{2}}\sum_{n=0}^\infty \frac{\alpha^n}{\sqrt{n!}}e^{\frac{-it}{\hbar}\hbar\omega(n+1/2)}|\varphi_n\rangle. \tag{15.4.38}$$

Identifying, $\alpha(t)\equiv\alpha e^{-i\omega t}$, and using the relation, Eq. (15.4.37), the result can be cast as (Ex. 15.4.9)

$$|\psi(t)\rangle = e^{\frac{-i\omega t}{2}}e^{i\Phi(t)}|\alpha(t)\rangle = e^{\frac{-i\omega t}{2}}e^{i\Phi(t)}|\alpha e^{-i\omega t}\rangle. \tag{15.4.39}$$

Consequently, up to a global phase, the time evolution of the coherent state amounts to a time-periodic change in the parameter, $\alpha\mapsto\alpha(t)$. Using Eq. (15.4.36), the real and imaginary parts of α are related to position and momentum expectation values of the Gaussian wave packet, namely

$$\begin{aligned}x(t) &= \sqrt{\frac{2\hbar}{m\omega}}\text{Re}[\alpha(t)]\\ p(t) &= \sqrt{2m\omega\hbar}\text{Im}[\alpha(t)]\end{aligned}. \tag{15.4.40}$$

These values are shown to coincide with the classical trajectory, as obtained in the Heisenberg picture (Eq. (15.4.29); see Ex. 15.4.10). The value of the phase, $\Phi(t)$, is uniquely determined by the Schrödinger equation (Ex. 15.4.11), where the corresponding position representation of the coherent state at any time reads

$$\begin{aligned}\psi(x,t) &= \left(\frac{m\omega}{\pi\hbar}\right)^{1/4}e^{\frac{-m\omega(x-x(t))^2}{2\hbar}}e^{ip(t)x/\hbar}e^{i\Phi(t)}\\ \Phi(t) &= \frac{m\omega}{4\hbar}\left[x_0^2-\left(\frac{p_0}{m\omega}\right)^2\right]\sin(2\omega t)-\frac{x_0p_0}{2\hbar}\cos(2\omega t)-\frac{\omega t}{2}\end{aligned}. \tag{15.4.41}$$

Exercise 15.4.7 *Use the position representation of the annihilation operator, $\hat{a}=\sqrt{\frac{m\omega}{2\hbar}}x+\sqrt{\frac{\hbar}{2m\omega}}\frac{\partial}{\partial x}$, and of the coherent state, $\psi(x,0)=\left(\frac{m\omega}{\pi\hbar}\right)^{1/4}e^{\frac{-m\omega(x-x_0)^2}{2\hbar}}e^{ip_0x/\hbar}$, to show that the coherent state is an eigenstate of the annihilation operator, Eq. (15.4.35).*

Exercise 15.4.8 *Derive the expansion of the coherent state in terms of the harmonic oscillator eigenstates, Eq. (15.4.37). Apply the annihilation operator to the formal expansion, $\hat{a}|\alpha\rangle = \sum_{n=0}^\infty\gamma_n\hat{a}|\varphi_n\rangle$, recalling that $\hat{a}|\varphi_n\rangle=\sqrt{n}|\varphi_{n-1}\rangle$ (Eq. (8.5.3)). Then show that $\gamma_{m+1}\sqrt{m+1}=\alpha\gamma_m$. Use the normalization condition $\langle\alpha|\alpha\rangle=1$ to show that $|\langle\varphi_0|\alpha\rangle|^2=e^{-|\alpha|^2}$, and prove the identity, Eq. (15.4.37).*

Exercise 15.4.9 *Using the identity $e^{-i\omega nt}=(e^{-i\omega t})^n$, rewrite Eq. (15.4.38) as Eq. (15.4.39).*

Exercise 15.4.10   *Use Eqs. (15.4.36, 15.4.40) to derive the quantum mechanical position and momentum expectation values for the coherent state. Show that these results identify with a classical trajectory of a corresponding particle.*

Exercise 15.4.11   *Show that* $\psi(x,t)$*, as defined in Eq. (15.4.41), is a solution of the time-dependent Schrödinger equation for the harmonic oscillator,* $i\hbar\frac{\partial}{\partial t}\psi(x,t) = \hbar\omega\left[\hat{a}^{\dagger}\hat{a} + \frac{1}{2}\right]\psi(x,t)$*, where* $\hat{a} = \sqrt{\frac{m\omega}{2\hbar}}x + \sqrt{\frac{\hbar}{2m\omega}}\frac{\partial}{\partial x}$*.*

15.5 Transition Probabilities and Transition Rates

The values of measurable quantities in any experiment depend on the initial preparation of the system. In this sense the initial preparation at time t' is somewhat equivalent to the final measurement at time t. Experiments and their interpretations are often formulated in terms of **transition probabilities**, which consider both the initial and the final states of the system. In this section we shall account for transition probabilities associated with a single solution of the time-dependent Schrödinger equation (a coherent ensemble of systems), and in Chapter 17 we shall extend the discussion to mixed states (to be introduced in Chapter 16). Without loss of generality, we shall associate the preparation time with $t' = 0$.

Let the state of a system at the initial time $(t' = 0)$ be

$$|\psi(0)\rangle = |\chi_i\rangle. \tag{15.5.1}$$

According to the postulates of quantum mechanics, the probability amplitude for the system to be in a state $|\chi_f\rangle$ at any time t is the projection of the state of the system on $|\chi_f\rangle$ ("the Green's function," $g_{f,i}(t,0) \equiv \langle\chi_f|\hat{U}(t,0)|\chi_i\rangle$). The corresponding transition probability is the absolute square of this projection,

$$P_{i\rightarrow f}(t) = |\langle\chi_f|\hat{U}(t,0)|\chi_i\rangle|^2. \tag{15.5.2}$$

Here, $\hat{U}(t,0)$ is the unitary time-evolution operator of the system (see Section 15.2).

It is often convenient to reformulate the transition probability in terms of the correlator between two projection operators, referring to the initial and final states. For this purpose, let us start by defining the trace of an operator. Let $\{|\varphi_n\rangle\}$ be a complete orthonormal system of vectors in the system Hilbert space, where $\hat{I} = \sum_n |\varphi_n\rangle\langle\varphi_n|$ is the identity operator. (For simplicity we refer here to a discrete set, but the same applies also in the continuous case (see Section 11.3).) The trace of an operator $\hat{O}$ in that space is defined as (see Ex. 15.5.1)

$$tr\{\hat{O}\} \equiv \sum_n \langle\varphi_n|\hat{O}|\varphi_n\rangle. \tag{15.5.3}$$

Using this definition, the transition probability (Eq. 15.5.2) can be formulated as (Ex. 15.5.2)

$$P_{i\rightarrow f}(t) = tr\{\hat{U}^{\dagger}(t,0)|\chi_f\rangle\langle\chi_f|\hat{U}(t,0)|\chi_i\rangle\langle\chi_i|\}. \tag{15.5.4}$$

Defining projection operators into the initial and final states,

$$\hat{P}^{(i)} \equiv |\chi_i\rangle\langle\chi_i| \quad ; \quad \hat{P}^{(f)} \equiv |\chi_f\rangle\langle\chi_f|, \tag{15.5.5}$$

and recalling the definition (Eq. (15.3.16)) of the time-dependent operators in the Heisenberg picture, $\hat{A}_H(t) = \hat{U}^\dagger(t,0)\hat{A}\hat{U}(t,0)$, the transition probability corresponds to the trace of a correlator between $\hat{P}_H^{(i)}(0)$ and $\hat{P}_H^{(f)}(t)$:

$$P_{i\to f}(t) = tr\{\hat{P}_H^{(f)}(t)\hat{P}_H^{(i)}(0)\}. \tag{15.5.6}$$

Exercise 15.5.1 *Using the definition of the trace of an operator, Eq. (15.5.3), (a) prove the following identities: $tr\{\hat{A}\hat{B}\} = tr\{\hat{B}\hat{A}\}$, $tr\{\hat{A}^\dagger\} = tr\{\hat{A}\}^*$, $tr\{[\hat{A},\hat{B}]\} = 0$. (b) Show that the trace of an operator is invariant to a similarity transformation of the operator, $tr\{\hat{S}^{-1}\hat{A}\hat{S}\} = tr\{\hat{A}\}$. (c) Show that the trace of an operator is invariant to a unitary transformation of the operator, $tr\{\hat{U}^\dagger\hat{A}\hat{U}\} = tr\{\hat{A}\}$. (d) Show that the trace of an operator is independent of the basis in which the operator is represented, $tr\{\hat{A}\} \equiv \sum_n\langle\varphi_n|\hat{A}|\varphi_n\rangle = \sum_m\langle\chi_m|\hat{A}|\chi_m\rangle$, where $\{|\varphi_n\rangle\}$ and $\{|\chi_m\rangle\}$ are complete orthonormal systems in the relevant Hilbert space, $\hat{I} = \sum_n|\varphi_n\rangle\langle\varphi_n| = \sum_m|\chi_m\rangle\langle\chi_m|$. (e) Show that the trace of a tensor product of operators in a multidimensional space, $\hat{A}_1 \otimes \hat{A}_2 \otimes \cdots \otimes \hat{A}_N$, is a product of traces over the subspaces, $tr\{\hat{A}_1 \otimes \hat{A}_2 \otimes \cdots \otimes \hat{A}_N\} = tr\{\hat{A}_1\} \cdot tr\{\hat{A}_2\}\cdots tr\{\hat{A}_N\}$ (recall that the multidimensional space is spanned by a complete set of tensor product states (Eq. (11.6.12)).*

Exercise 15.5.2 *Use Eqs. (15.5.2, 15.5.3) and the identity operator $\hat{I} = \sum_n|\varphi_n\rangle\langle\varphi_n|$ to derive Eq. (15.5.4).*

In many cases of interest (see, e.g., Chapter 18) the change of the transition probability in time is the quantity that is being measured directly. The corresponding theoretical quantity of interest is the transition rate, namely the time-derivative of the transition probability,

$$k_{i\to f}(t) \equiv \frac{d}{dt}P_{i\to f}(t). \tag{15.5.7}$$

Using the explicit expressions for the transition probability (Eqs. (15.5.4, 15.5.6)), the transition rate reads (Ex. 15.5.3)

$$\frac{d}{dt}P_{i\to f}(t) = 2\,\mathrm{Re}\,tr\{\hat{U}^\dagger(t,0)|\chi_f\rangle\langle\chi_f|\frac{d}{dt}\hat{U}(t,0)|\chi_i\rangle\langle\chi_i|\}, \tag{15.5.8}$$

or, in the Heisenberg picture representation,

$$\frac{d}{dt}P_{i\to f}(t) = tr\left\{\frac{d}{dt}\hat{P}_H^{(f)}(t)\hat{P}_H^{(i)}(0)\right\}. \tag{15.5.9}$$

Exercise 15.5.3 *Use Eqs. (15.5.4, 15.5.6) to derive Eqs. (15.5.8, 15.5.9).*

The formally exact and general expressions for the transition probability and the transition rate rely on the exact time-evolution operator for the system. In practice, the time-evolution operator associated with the full-system Hamiltonian is usually too complicated to evaluate. Moreover, in many cases of interest its evaluation can be avoided. This is the case where the full Hamiltonian can be decomposed into a zero-order Hamiltonian ($\hat{H}_0$) whose propagator is simple to evaluate, and an interaction (time-dependent, in principle) that is "small" in some specified sense, namely $\hat{H} = \hat{H}_0 + \hat{V}(t)$. In this case it is useful to evaluate the transition probabilities and transition rates between eigenstates of the zero-order Hamiltonian using the approximated (but practical) time-dependent perturbation theory, as will be discussed in the next section.

Here we still focus on the exact expressions for transition probabilities and transition rates for Hamiltonians of the generic form, $\hat{H} = \hat{H}_0 + \hat{V}(t)$. It is most convenient in this case to transform the states and operators into the interaction picture (see Section 15.3), where the time-dependent states are transformed according to Eq. (15.3.7), and the corresponding propagator, $\hat{U}^{(I)}(t,0)$, is defined in Eq. (15.3.9). Using the relation between $\hat{U}(t,0)$ and $\hat{U}^{(I)}(t,0)$, the exact transition probability from the state $|\chi_i\rangle$ at $t = 0$ to the state $|\chi_f\rangle$ at time t reads (see Eq. (15.5.2) and Ex. 15.5.4)

$$P_{i\to f}(t) = |\langle \chi_f | e^{\frac{-it}{\hbar}\hat{H}_0} \hat{U}^{(I)}(t,0) | \chi_i \rangle|^2. \tag{15.5.10}$$

It is often the case that the final state of interest is an eigenstate of $\hat{H}_0$ (see Chapter 17). In this case the expression for the transition probability simplifies to (see Ex. (15.5.4))

$$P_{i\to f}(t) = |\langle \chi_f | \hat{U}^{(I)}(t,0) | \chi_i \rangle|^2 \quad ; \quad \hat{H}_0 | \chi_f \rangle = \varepsilon_f | \chi_f \rangle. \tag{15.5.11}$$

Alternatively, defining the projection operators, $\hat{P}_i = |\chi_i\rangle\langle\chi_i|$ and $\hat{P}_f = |\chi_f\rangle\langle\chi_f|$, and using the trace definition, the transition probability reads (see Ex. 15.5.4)

$$P_{i\to f}(t) = tr\{\hat{U}^{\dagger(I)}(t,0)\hat{P}_f\hat{U}^{(I)}(t,0)\hat{P}_i\} \quad ; \quad \hat{H}_0 | \chi_f \rangle = \varepsilon_f | \chi_f \rangle. \tag{15.5.12}$$

Exercise 15.5.4 *Let the time-dependent state of a system be $|\psi(t)\rangle \equiv \hat{U}(t,0)|\chi_i\rangle$. (a) Use the transformation of the state to the interaction picture, $|\psi_I(t)\rangle \equiv e^{\frac{i}{\hbar}\hat{H}_0 t}|\psi(t)\rangle$ (Eq. (15.3.7)), and the definition of the interaction picture propagator, $|\psi_I(t)\rangle \equiv \hat{U}^{(I)}(t,0)|\chi_i\rangle$, (Eq. (15.3.9)), to show that $|\langle \chi_f | \hat{U}(t,0) | \chi_i \rangle|^2 = |\langle \chi_f | e^{\frac{-it}{\hbar}\hat{H}_0}\hat{U}^{(I)}(t,0) | \chi_i \rangle|^2$. (b) Given that $|\chi_f\rangle$ is an eigenvector of $\hat{H}_0$, show that $|\langle \chi_f | \hat{U}(t,0) | \chi_i \rangle|^2 = |\langle \chi_f | \hat{U}^{(I)}(t,0) | \chi_i \rangle|^2$. (c) Given the definitions $\hat{P}_i = |\chi_i\rangle\langle\chi_i|$ and $\hat{P}_f = |\chi_f\rangle\langle\chi_f|$, and using the trace definition (Eq. (15.5.3)), derive Eq. (15.5.12) from Eq. (15.5.11).*

Taking the time-derivative of the transition probability, Eq. (15.5.11), using the expression for the time-derivative of $\hat{U}^{(I)}(t,0)$ (Eq. (15.3.9)), the exact transition rate from $|\chi_i\rangle$ at $t = 0$ to an $\hat{H}_0$ eigenstate, $|\chi_f\rangle$ at time t reads (Ex. 15.5.5)

$$k_{i\to f}(t) = \frac{1}{\hbar}2\,\mathrm{Im}\langle \chi_i | \hat{U}^{\dagger(I)}(t,0) | \chi_f \rangle \langle \chi_f | \hat{V}_I(t) \hat{U}^{(I)}(t,0) | \chi_i \rangle, \tag{15.5.13}$$

where $\hat{V}_I(t) = e^{\frac{it}{\hbar}\hat{H}_0}\hat{V}(t)e^{\frac{-it}{\hbar}\hat{H}_0}$. This result can also be expressed as

$$k_{i\to f}(t) = \frac{1}{\hbar}2\,\mathrm{Im}\,tr\{\hat{P}_i\hat{U}^{\dagger(I)}(t,0)\hat{P}_f\hat{V}_I(t)\hat{U}^{(I)}(t,0)\}. \tag{15.5.14}$$

Exercise 15.5.5   *(a) Take the time-derivative of the transition probability, Eq. (15.5.11),*
$P_{i \to f}(t) = \langle \chi_i | \hat{U}^{\dagger(I)}(t,0) | \chi_f \rangle \langle \chi_f | \hat{U}^{(I)}(t,0) | \chi_i \rangle$, *using Eq. (15.3.9) for the time derivative of $\hat{U}^{(I)}(t,0)$, to show that*

$$\frac{\partial}{\partial t} P_{i \to f}(t) = \frac{-1}{i\hbar} \langle \chi_i | \hat{U}^{\dagger(I)}(t,0) e^{\frac{it}{\hbar}\hat{H}_0} \hat{V}(t) e^{\frac{-it}{\hbar}\hat{H}_0} | \chi_f \rangle \langle \chi_f | \hat{U}^{(I)}(t,0) | \chi_i \rangle$$

$$+ \frac{1}{i\hbar} \langle \chi_i | \hat{U}^{\dagger(I)}(t,0) | \chi_f \rangle \langle \chi_f | e^{\frac{it}{\hbar}\hat{H}_0} \hat{V}(t) e^{\frac{-it}{\hbar}\hat{H}_0} \hat{U}^{(I)}(t,0) | \chi_i \rangle.$$

(b) Recall that $\langle \chi_i | \hat{A} | \chi_i \rangle = \langle \chi_i | \hat{A}^{\dagger} | \chi_i \rangle^*$ and show that

$$\frac{\partial}{\partial t} P_{i \to f}(t) = \frac{2}{\hbar} \text{Im} \, \langle \chi_i | \hat{U}^{\dagger(I)}(t,0) | \chi_f \rangle \langle \chi_f | e^{\frac{it}{\hbar}\hat{H}_0} \hat{V}(t) e^{\frac{-it}{\hbar}\hat{H}_0} \hat{U}^{(I)}(t,0) | \chi_i \rangle.$$

15.6 Time-Dependent Perturbation Theory

In most experiments the preparation and the measurement of the state of a system are
carried out in an asymptotic regime (in space and/or in time), where the full-system
Hamiltonian coincides with some zero-order Hamiltonian, denoted $\hat{H}_0$. *Of prime interest is therefore to evaluate transition probabilities between different eigenstates of $\hat{H}_0$,
owing to the presence of an interaction term in the full Hamiltonian*, which obtains the
general form

$$\hat{H}(\lambda,t) = \hat{H}_0 + \lambda \hat{V}(t). \tag{15.6.1}$$

The parameter $0 \leq \lambda \leq 1$ is introduced as a continuous scalar pre-factor where $\lambda = 0$
and $\lambda = 1$ correspond, respectively, to the zero-order and full Hamiltonians. Notice
that in some cases the time-dependent interaction corresponds, for example, to the
effect of an external (semiclassical) time-dependent field on the system. In other cases,
the time-dependence can be introduced formally by "turning on" an otherwise time-
independent interaction at some time, $\hat{V}(t) = h(t-t')\hat{V}(t')$, where $h(\tau)$ is the Heaviside
step function. While formally relevant for any $\hat{H}_0$ and $\hat{V}(t)$, the interest in the transition
probability between eigenstates of $\hat{H}_0$ is primarily in cases where the interaction that
drives the transition is a small perturbation to $\hat{H}_0$. It is most convenient in this case
to invoke the interaction picture, where the full time-evolution operator is given as
the Dyson series (Eq. (15.3.9)). First, the interaction picture involves the "interaction-
free" propagator, $\hat{U}_0(t,t') = \exp(-i(t-t')\hat{H}_0/\hbar)$, which yields only trivial phase factors
(see what follows) when applied to eigenstates of $\hat{H}_0$. Second, when the interaction is
sufficiently small and/or for sufficiently short times, the Dyson series can be truncated
at low orders and still provide accurate results.

We are interested in evaluating the transition probability between two different
(orthogonal) eigenstates of the zero-order Hamiltonian, namely

$$\hat{H}_0 | \chi_i \rangle = \varepsilon_i | \chi_i \rangle \quad ; \quad \hat{H}_0 | \chi_f \rangle = \varepsilon_f | \chi_f \rangle \tag{15.6.2}$$

$$\langle \chi_f | \chi_i \rangle = 0. \tag{15.6.3}$$

Without loss of generality, we shall set the initial (system preparation) time to $t = 0$. The transition amplitude from the state $|\chi_i\rangle$ at time $t = 0$ to the state $|\chi_f\rangle$ at time t can be readily formulated in terms of the interaction picture propagator (see Eq. (15.5.11)):

$$P_{i \to f}(t) = |g_{f,i}(\lambda, t)|^2 \quad ; \quad g_{f,i}(\lambda, t) \equiv \langle \chi_f | \hat{U}^{(I)}(t, 0) | \chi_i \rangle. \tag{15.6.4}$$

Using the Dyson expansion (Eq. (15.3.9)) for the time-evolution operator in the interaction picture, we obtain

$$g_{f,i}(\lambda, t) = \langle \chi_f | \chi_i \rangle + \frac{-i}{\hbar} \int_0^t dt' \langle \chi_f | \hat{H}^{(I)}(t') | \chi_i \rangle$$

$$+ \left(\frac{-i}{\hbar} \right)^2 \int_0^t dt' \langle \chi_f | \hat{H}^{(I)}(t') \int_0^{t'} dt'' \hat{H}^{(I)}(t'') | \chi_i \rangle + \dots, \tag{15.6.5}$$

where

$$\hat{H}^{(I)}(\tau) \equiv \lambda \hat{V}_I(\tau) = \lambda e^{\frac{i}{\hbar} \hat{H}_0 \tau} \hat{V}(\tau) e^{\frac{-i}{\hbar} \hat{H}_0 \tau}. \tag{15.6.6}$$

Regarding the transition amplitude as a function of λ, Eqs. (15.6.5, 15.6.6) can be identified as its series expansion in powers of λ,

$$g_{f,i}(\lambda, t) = \sum_{n=0}^{\infty} \lambda^n g_{f,i}^{(n)}(t), \tag{15.6.7}$$

where for $n > 0$,

$$g_{f,i}^{(1)}(t) = \frac{-i}{\hbar} \int_0^t dt_0 \langle \chi_f | \hat{V}_I(t_0) | \chi_i \rangle$$

$$g_{f,i}^{(2)}(t) = \left(\frac{-i}{\hbar} \right)^2 \int_0^t dt_1 \langle \chi_f | \hat{V}_I(t_1) \int_0^{t_1} dt_0 \hat{V}_I(t_0) | \chi_i \rangle$$

$$\vdots$$

$$g_{f,i}^{(n)}(t) = \left(\frac{-i}{\hbar} \right)^n \int_0^t dt_{n-1} \langle \chi_f | \hat{V}_I(t_{n-1}) \int_0^{t_{n-1}} dt_{n-2} \hat{V}_I(t_{n-2}) \cdots \int_0^{t_1} dt_0 \hat{V}_I(t_0) | \chi_i \rangle. \tag{15.6.8}$$

Approximating the expansion of the transition amplitude up to the nth-order is referred to as the nth-order time-dependent perturbation theory. The first-order approximation for the transition probability therefore reads

$$P_{i \to f}^{(1)}(t) \cong \left| \langle \chi_f | \chi_i \rangle + \lambda g_{f,i}^{(1)}(t) \right|^2. \tag{15.6.9}$$

We shall restrict the discussion to this first order, where the transition amplitude is linear in the interaction strength parameter; hence, observables depend quadratically on the interaction strength. This is sufficient for treating a wealth of phenomena attributed to linear response of the system to time-dependent perturbations (see Chapters 17–20). For the description of nonlinear response, attributed to higher orders in the Dyson expansion, the reader is directed to a complementary literature [15.8], [15.9]. **The explicit expression for the first-order transition probability between two orthogonal eigenstates of the zero-order Hamiltonian, $\hat{H}_0$ (Eqs. (15.6.2, 15.6.3)), due to the interaction term, $\lambda \hat{V}(t)$, obtains the form** (see Ex. 15.6.1, 15.6.2 and 15.6.3)

$$P_{i \to f}^{(1)}(t) = \frac{\lambda^2}{\hbar^2} \left| \int_0^t dt' e^{\frac{i}{\hbar}(\varepsilon_f - \varepsilon_i)t'} \langle \chi_f | \hat{V}(t') | \chi_i \rangle \right|^2 . \tag{15.6.10}$$

Exercise 15.6.1 *Use the definition of the first-order propagator in the interaction representation, $g_{f,i}^{(1)}(t) = \frac{-i}{\hbar} \int_0^t dt_0 \langle \chi_f | \hat{V}_I(t_0) | \chi_i \rangle$ (Eqs. (15.6.6, 15.6.8)), and Eq. (15.6.2) for the states $|\chi_i\rangle$ and $|\chi_f\rangle$, to show that $g_{f,i}^{(1)}(t) = \frac{-i}{\hbar} \int_0^t d\tau \langle \chi_f | e^{\frac{i}{\hbar}\varepsilon_f \tau} \hat{V}(\tau) e^{\frac{-i}{\hbar}\varepsilon_i \tau} | \chi_i \rangle.$*

Exercise 15.6.2 *Use the orthogonality of the initial and final states (Eq. (15.6.3)) and the result of Ex. 15.6.1 to derive the first-order transition probability, $P_{i \to f}^{(1)}(t)$ (Eq. 15.6.10), from Eq. (15.6.9).*

Exercise 15.6.3 *Follow the steps given here as an alternative derivation of the expression for the first-order transition probability, Eq. (15.6.10): Any solution to the time-dependent Schrödinger equation can be expanded at any time in the basis of the eigenstates of the zero-order Hamiltonian, $\{\hat{H}_0 | \varphi_n \rangle = \varepsilon_n | \varphi_n \rangle\}$, namely $|\psi(t)\rangle = \sum_n b_n(t) |\varphi_n\rangle \equiv \sum_n a_n(t) e^{\frac{-i\varepsilon_n t}{\hbar}} |\varphi_n\rangle$, where the projection of the system state on any eigenstate, $|\varphi_m\rangle$, is given as $P_m(t) = |\langle \varphi_m | \psi(t)\rangle|^2 = |b_m(t)|^2 = |a_m(t)|^2$. Show that substitution of this expansion in the time-dependent Schrödinger equation, $i\hbar \frac{\partial}{\partial t} |\psi(t)\rangle = \hat{H} |\psi(t)\rangle$, yields coupled equations for the expansion coefficients, $\{a_n(t)\}$: $i\hbar \sum_n \dot{a}_n(t) e^{\frac{-i\varepsilon_n t}{\hbar}} |\varphi_n\rangle = \lambda \sum_n \hat{V}(t) a_n(t) e^{\frac{-i\varepsilon_n t}{\hbar}} |\varphi_n\rangle$. Project this equation on the bra state $\langle \varphi_m |$, multiply by $e^{\frac{i\varepsilon_m t}{\hbar}}$, and integrate over time from 0 to t to obtain $a_m(t) = a_m(0) + \frac{\lambda}{i\hbar} \sum_n \int_0^t dt' \langle \varphi_m | \hat{V}(t') | \varphi_n \rangle e^{\frac{-i(\varepsilon_n - \varepsilon_m)t'}{\hbar}} a_n(t')$. Since this result means that $a_n(t') = a_n(0) + o(\lambda)$, the expression for the probability amplitude, $a_m(t)$, to first order in λ, reads $a_m(t) \cong a_m(0) + \frac{\lambda}{i\hbar} \sum_n \int_0^t dt' \langle \varphi_m | \hat{V}(t') | \varphi_n \rangle e^{\frac{-i(\varepsilon_n - \varepsilon_m)t'}{\hbar}} a_n(0)$. Choose the initial state as the ith eigenstate of $\hat{H}_0$, namely $a_m(0) = \delta_{m,i}$. Substitute this condition in the approximated expression for the probability amplitude and show that the probability to find the system in any other eigenstate ($m \neq i$) at time t reads $P_{i \to m}(t) = |a_m(t)|^2 \cong \frac{\lambda^2}{\hbar^2} \left| \int_0^t dt' \langle \varphi_m | \hat{V}(t') | \varphi_i \rangle e^{\frac{-i(\varepsilon_i - \varepsilon_m)t'}{\hbar}} \right|^2$, which reproduces the result, Eq. (15.6.10).*

It is important to assess the validity of the first-order approximation. First, we notice that this approximation coincides with the exact transition probability in the short time limit. This can be readily verified by expanding the integrands in the infinite Dyson series to their lowest order in powers of t (see Ex. 15.6.4), which yields

$$\lim_{t\to0}P^{(1)}_{i\to f}(t) = \lim_{t\to0}P_{i\to f}(t) = \frac{\lambda^2}{\hbar^2}|\langle\chi_f|\hat{V}(0)|\chi_i\rangle|^2 t^2. \qquad (15.6.11)$$

However, the maximal time in which the first-order approximation is valid depends on the Hamiltonian parameters. An upper bound for the validity can be obtained by recalling that the exact transition probability is bounded at all times, $0 \le P_{i\to f}(\lambda,t) \le 1$ (which follows immediately from the unitarity of the time evolution operator). Imposing this condition on the first-order approximation, at any time, $\left|P^{(1)}_{i\to f}(t)\right| < 1$, we obtain a necessary condition for the validity of Eq. (15.6.10):

$$\lambda < \frac{\hbar}{\left|\int_0^t dt' e^{\frac{i}{\hbar}(\varepsilon_f-\varepsilon_i)t'}\langle\chi_f|\hat{V}(t')|\chi_i\rangle\right|}. \qquad (15.6.12)$$

This general limitation translates to more transparent conditions on the Hamiltonian parameters (the interaction strength, in particular) when the interaction is constant in time (for $t \ge 0$), as discussed in what follows.

Exercise 15.6.4 *Prove that in the short time limit $(t \to 0)$, the exact transition probability increases quadratically in time: Start from the Dyson expansion (15.6.5, 15.6.6) of the probability amplitude for $|\chi_i\rangle \ne |\chi_f\rangle$ (where, $V_{m,n}(\tau) \equiv \langle\chi_m|\hat{V}(\tau)|\chi_n\rangle$),*

$$g_{f,i}(\lambda,t) = \frac{-i\lambda}{\hbar} \int_0^t dt' e^{\frac{i(\varepsilon_f-\varepsilon_i)t'}{\hbar}} V_{f,i}(t')$$

$$+ \sum_j \left(\frac{-i\lambda}{\hbar}\right)^2 \int_0^t dt' \int_0^{t'} dt'' e^{\frac{i(\varepsilon_f-\varepsilon_j)t'}{\hbar}} V_{f,j}(t') e^{\frac{i(\varepsilon_j-\varepsilon_i)t''}{\hbar}} V_{j,i}(t'') + \dots.$$

Expand the exponential functions and the interaction to their lowest order in time, for example, $e^{\frac{i(\varepsilon_f-\varepsilon_i)t'}{\hbar}} \approx \left[1 + \frac{i(\varepsilon_f-\varepsilon_i)t'}{\hbar}\right]$ and $V_{f,i}(t') \approx \left[V_{f,i}(0) + t'V'_{f,i}(0)\right]$, and show that $\lim_{t\to0}[g_{f,i}(\lambda,t)] = \frac{-i\lambda}{\hbar}V_{f,i}(0)t + o(t^2)$ and therefore $\lim_{t\to0}P_{i\to f}(t) = \frac{\lambda^2}{\hbar^2}|\langle\chi_f|\hat{V}(0)|\chi_i\rangle|^2 t^2$.

A Constant Perturbation

Let us focus now on cases in which the interaction is time-independent, except for its "turning on" at $t = 0$. The Hamiltonian is formally defined as in Eq. (15.6.1), where

$$\hat{V}(t) = \begin{cases} 0 \; ; \; t < 0 \\ \lambda\hat{V} \; ; \; t \ge 0 \end{cases}. \qquad (15.6.13)$$

Since the interaction is constant during the system propagation time $(t \ge 0)$, the general expression, Eq. (15.6.10), simplifies in this case (Ex. 15.6.5):

$$P^{(1)}_{i\to f}(t) = \frac{4\lambda^2|\langle\chi_f|\hat{V}|\chi_i\rangle|^2}{|\varepsilon_f-\varepsilon_i|^2} \sin^2\left(\frac{(\varepsilon_f-\varepsilon_i)t}{2\hbar}\right), \qquad (15.6.14)$$

where the transition probability is shown to oscillate in time at a frequency,

$$\omega_{f,i} \equiv \frac{\varepsilon_f - \varepsilon_i}{\hbar}. \tag{15.6.15}$$

The oscillation amplitude prefactor is shown to depend on the ratio between the coupling matrix element, $|\langle \chi_f | \hat{V} | \chi_i \rangle|$, and the energy gap between the two corresponding $\hat{H}_0$ eigenstates, $|\varepsilon_f - \varepsilon_i|$.

In the "strong interaction" limit, $2\lambda |\langle \chi_f | \hat{V} | \chi_i \rangle| \gg |\varepsilon_f - \varepsilon_i|$, the prefactor is much larger than unity. Consequently, the requirement of probability conservation, $|P_{i \to f}^{(1)}(t)| \le 1$, necessitates that $\sin^2 \left(\frac{(\varepsilon_f - \varepsilon_i)t}{2\hbar} \right) \ll 1$, which means that $\sin \left(\frac{(\varepsilon_f - \varepsilon_i)t}{2\hbar} \right) \approx \frac{(\varepsilon_f - \varepsilon_i)t}{2\hbar}$, and therefore

$$2\lambda |\langle \chi_f | \hat{V} | \chi_i \rangle| \gg |\varepsilon_f - \varepsilon_i| \Rightarrow t \le \frac{\hbar}{\lambda |\langle \chi_f | \hat{V} | \chi_i \rangle|}. \tag{15.6.16}$$

Indeed, in this parameter regime the validity of the first-order approximation is limited to the short time limit (Eq. (15.6.11)), in which the probability increases quadratically on time, irrespective of $|\varepsilon_f - \varepsilon_i|$.

In contrast, the condition for probability conservation is less strict in the "weak interaction limit," $2\lambda |\langle \chi_f | \hat{V} | \chi_i \rangle| \ll |\varepsilon_f - \varepsilon_i|$, where the oscillation amplitude of $P_{i \to f}^{(1)}(t)$ is much smaller than unity. This result is reminiscent of the condition for validity of the low-order time-independent perturbation theory (Eq. (12.1.29)), where the low-order expansion in powers of the interaction operator holds as long as the coupling matrix element between the eigenstates of $\hat{H}_0$ is small in comparison to the difference between the corresponding energy levels. Indeed, the first-order time-dependent perturbation theory can generally be valid for times much longer than the trivial short time limit (Eq. (15.6.11)). In the simplest case of a two-level system (to be discussed shortly), the validity of the approximation extends to a number of oscillation periods, which is as large as $|\varepsilon_1 - \varepsilon_2|^2 / |2\lambda \langle \chi_2 | \hat{V} | \chi_1 \rangle|^2$.

Exercise 15.6.5 *Obtain Eq. (15.6.14) for the transition probability amplitude from the general expression, Eq. (15.6.10), in the case where the interaction operator is constant through the propagation time, $\hat{V}(t)|_{t \ge 0} \mapsto \hat{V}$.*

It is instructive to test the results of the time-dependent perturbation theory by comparing them to exact results for the TLS model, introduced in Eq. (15.1.17). The zero-order Hamiltonian and the interaction correspond, respectively, to the diagonal and non-diagonal parts of the Hamiltonian, where the matrix representations of these operators in the basis of the two orthonormal states, $|\chi_1\rangle$ and $|\chi_2\rangle$, read

$$H_0 = \begin{bmatrix} \varepsilon_1 & 0 \\ 0 & \varepsilon_2 \end{bmatrix} \quad ; \quad V = \begin{bmatrix} 0 & \gamma \\ \gamma^* & 0 \end{bmatrix}. \tag{15.6.17}$$

Let us identify the initial and final states with the two basis states ($\hat{H}_0$-eigenstates),

$$|\chi_i\rangle = |\chi_1\rangle \quad ; \quad |\chi_f\rangle = |\chi_2\rangle, \tag{15.6.18}$$

where the interaction term is "switched on" at $t \geq 0$, with $\lambda = 1$. Since the two states span the Hilbert space of the TLS, the transition probability, $P_{1 \to 2}(t)$, is complementary to the survival probability, $P(t) \equiv P_{1 \to 1}(t)$, that was discussed in Section 15.1, namely $P_{1 \to 2}(t) + P_{1 \to 1}(t) = 1$. Using Eq. (15.1.30), the exact transition probability (for $\lambda = 1$) is given as

$$P_{1 \to 2}(t) = \frac{4|\gamma|^2}{(\varepsilon_1 - \varepsilon_2)^2 + 4|\gamma|^2} \sin^2 \left(\frac{t\sqrt{(\varepsilon_1 - \varepsilon_2)^2 + 4|\gamma|^2}}{2\hbar} \right). \tag{15.6.19}$$

Implementing the first-order perturbation theory expression, Eq. (15.6.14), for the case of the TLS, the approximated transition probability reads

$$P_{1 \to 2}^{(1)}(t) = \frac{4|\gamma|^2}{|\varepsilon_2 - \varepsilon_1|^2} \sin^2 \left(\frac{(\varepsilon_2 - \varepsilon_1)t}{2\hbar} \right). \tag{15.6.20}$$

First, we notice that (as in the general case, Eq. (15.6.11)) the first-order approximation coincides with the exact transition probability in the short time limit,

$$\lim_{t \to 0} P_{1 \to 2}^{(1)}(t) = \lim_{t \to 0} P_{1 \to 2}(t) = \frac{|\gamma|^2 t^2}{\hbar^2}. \tag{15.6.21}$$

This holds, importantly, regardless of the energy difference between the two eigenstates of $\hat{H}_0$ (Ex. 15.6.6). In the strong interaction limit, $|\gamma| \gg |\varepsilon_1 - \varepsilon_2|$, the exact transition probability oscillates at a frequency $\sim 2|\gamma|/\hbar$ (Eq. (15.1.31)), where the short time limit (Eq. (15.6.21)) is valid for times much shorter than the corresponding time period, $t \ll h/(2|\gamma|)$. In the weak interaction limit, $|\gamma| \ll |\varepsilon_1 - \varepsilon_2|$, the exact transition probability oscillates at a frequency $\sim |\varepsilon_2 - \varepsilon_1|/\hbar$ (Eq. (15.1.32)), where in this case the short time limit (Eq. (15.6.21)) is strictly valid only for $t \ll h/|\varepsilon_2 - \varepsilon_1| \ll h/|\gamma|$.

Second, we notice that in the weak interaction limit, $|\varepsilon_1 - \varepsilon_2| \gg 2|\gamma|$, the first-order approximation for the transition probability coincides with the exact result also beyond the short time limit. Indeed, Eq. (15.6.20) is readily obtained from the exact result, Eq. (15.6.19), when $|\varepsilon_1 - \varepsilon_2| \gg 2|\gamma|$ (see also Eq. (15.1.32)). Notice that the transition probability is much smaller than unity in this case, at any time, which is in line with the necessary condition for the validity of the first-order approximation (Eq. (15.6.16)).

It is important to notice, however, that even in the weak interaction limit, the accuracy of the first-order approximation is limited to finite times. The difference between the oscillation frequency in the exact and the approximate expressions ($\omega_{exact} = \sqrt{(\varepsilon_1 - \varepsilon_2)^2 + 4|\gamma|^2}/\hbar$ vs. $\omega = |\varepsilon_1 - \varepsilon_2|/\hbar$, respectively, see Eqs. (15.6.19, 15.6.20)) means that an error accumulates in the approximation. The error becomes critical when the numbers of completed oscillation periods in the exact and the approximated solutions differ by the order of a single period. This critical time, $t_c = 2\pi/(\omega_{exact} - \omega)$ (see Ex. 15.6.7), sets an upper bound on the times in which the first-order approximation is valid. In the weak interaction limit, this leads to the following validity condition:

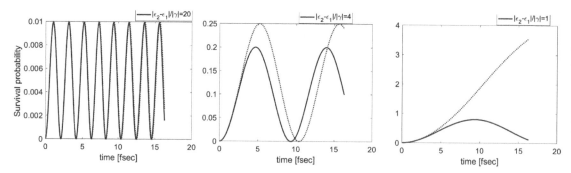

Figure 15.6.1 Comparison between the exact result (Eq. (15.6.19), solid) and the first-order approximation (Eq. (15.6.20), dotted) for the transition probability, in the different parameter regimes of the TLS model. The interaction strength parameter is fixed at $\gamma = 0.1$ eV, where the level spacing changes, $|\varepsilon_1 - \varepsilon_2| = 20\gamma, 4\gamma, \gamma$ (left, middle, and right plots, respectively). The accuracy of the perturbative expression is shown to be limited to the parameter regime, $|\varepsilon_1 - \varepsilon_2| >> \gamma$.

$$|\varepsilon_1 - \varepsilon_2| \gg 2|\gamma| \quad ; \quad t << \frac{h|\varepsilon_1 - \varepsilon_2|}{2|\gamma|^2} = \left(\frac{2\pi}{\omega}\right)\frac{|\varepsilon_1 - \varepsilon_2|^2}{2|\gamma|^2}. \tag{15.6.22}$$

Notice that in this limit, the number of oscillation periods $(2\pi/\omega)$) for which the first-order approximation is valid can still be very large, $n \sim |\varepsilon_1 - \varepsilon_2|^2/(2|\gamma|^2)$, which demonstrates the relevance of the first-order perturbation treatment to relatively "long time" dynamics in this case (see Fig. 15.6.1).

Exercise 15.6.6   *Show that for times much shorter than the oscillation period, both the exact and the approximate expressions for the TLS transition probability (Eq. (15.6.19), and Eq. (15.6.20), respectively) converge to quadratic time-dependence of Eq. (15.6.21).*

Exercise 15.6.7   *The approximate and exact expressions for the TLS transition probability are given by Eqs. (15.6.20) and (15.6.19), respectively. (a) Show that $P_{1\to2}^{(1)}(t) = \frac{1}{2}(\alpha^2 - 1)[1 - \cos(\omega t)]$ and $P_{1\to2}(t) = \frac{\alpha^2-1}{2\alpha^2}[1 - \cos(\omega\alpha t)]$, where $\omega = \frac{|\varepsilon_1-\varepsilon_2|}{\hbar}$ and $\alpha \equiv \sqrt{1+4|\gamma|^2/(\varepsilon_1 - \varepsilon_2)^2}$. (b) The oscillation frequencies of the approximate and the exact expressions are ω and $\omega\alpha$, respectively. At a certain time, t_c, the approximate solution completes n oscillation periods, whereas the exact solution completes $n+1$ periods. Show that $t_c = \frac{2\pi}{\omega\alpha-\omega}$.*

Monochromatic Driving

An important class of time-dependent perturbations is associated with the interaction of electromagnetic radiation with matter. When the radiation field can be regarded semiclassically, its effect on the system corresponds to a time-dependent interaction term in the system Hamiltonian, and when the field intensity is sufficiently low, this interaction can be treated in the linear response regime, namely, using first-order perturbation theory. More specifically, a coherent laser field is often represented by a single wavelength (monochromatic radiation), where in the long wavelength limit, the system–field interaction amounts a time-periodic driving of the system's electric dipole. Motivated by this class of applications (to be discussed in more detail in Chapter 18), let us consider here a generic Hamiltonian,

$$\hat{H}(t) = \hat{H}_0 + \lambda \hat{\mu} \sin(\Omega t). \tag{15.6.23}$$

The field-free Hamiltonian is $\hat{H}_0$, where Ω and λ are, respectively, the frequency and intensity of the driving field, and $\hat{\mu}$ is the system dipole operator that interacts with the field.

As before, we are interested in the transition probability between two eigenstates of the zero-order Hamiltonian, namely from $|\chi_i\rangle$ (corresponding to the initial state of the system at $t = 0$) to a final state, $|\chi_f\rangle$, at time t, where $\langle \chi_f | \chi_i \rangle = 0$. The energy difference between the initial and the final states defines a frequency, $\omega_{f,i} \equiv \frac{\varepsilon_f - \varepsilon_i}{\hbar}$. Implementing the general formula of first-order perturbation theory, Eq. (15.6.10), we obtain in this case (Ex. 15.6.8)

$$P_{i \to f}^{(1)}(t) = \frac{\lambda^2 |\langle \chi_f | \hat{\mu} | \chi_i \rangle|^2}{4\hbar^2} \left| \frac{e^{i(\omega_{f,i}+\Omega)t} - 1}{\omega_{f,i} + \Omega} - \frac{e^{i(\omega_{f,i}-\Omega)t} - 1}{\omega_{f,i} - \Omega} \right|^2. \tag{15.6.24}$$

The denominators in the right-hand side already suggest that the transition probability obtains maximal values when the driving field frequency (times $\hbar$) matches the energy difference between the initial and final $\hat{H}_0$ eigenstates, $\Omega \approx \pm \frac{\varepsilon_f - \varepsilon_i}{\hbar}$. This is known as the "resonance condition" for energy exchange between the system and the field,

$$\varepsilon_f \cong \varepsilon_i \pm \hbar \Omega, \tag{15.6.25}$$

where the plus and minus signs correspond, respectively, to energy absorption and emission by the system. Near the resonance, it is useful to define the detuning parameter, $\Delta \equiv \Omega - |\omega_{f,i}|$. In the limit of "small detuning", namely $|\Delta| << \Omega$, the expression for the transition probability simplifies, since only one of the terms corresponding to either emission $\left(\frac{e^{i(\omega_{f,i}+\Omega)t}-1}{\omega_{f,i}+\Omega} \right)$ or absorption $\left(\frac{e^{i(\omega_{f,i}-\Omega)t}-1}{\omega_{f,i}-\Omega} \right)$ is dominant, and the relative contribution of the other term can be neglected. This approximation is often referred to as the rotating wave approximation, where the time-dependent function, $\sin(\Omega t) = (e^{i\Omega t} - e^{-i\Omega t})/(2i)$, is replaced by a single rotating wave, namely either $e^{i\Omega t}/(2i)$ or $e^{-i\Omega t}/(2i)$, neglecting the other (counterrotating) wave. Within the rotating wave approximation, the transition probability simplifies to (Ex. 15.6.9)

$$P_{i \to f}^{(1)}(t) \cong \frac{\lambda^2 |\langle \chi_f | \hat{\mu} | \chi_i \rangle|^2 t^2}{4\hbar^2} \left[\frac{\sin[\Delta t/2]}{\Delta t/2} \right]^2 \tag{15.6.26}$$

$$\Delta = \begin{cases} \Omega - \omega_{f,i} & ; \quad \varepsilon_i < \varepsilon_f \\ \Omega + \omega_{f,i} & ; \quad \varepsilon_i > \varepsilon_f \end{cases}. \tag{15.6.27}$$

Strictly on resonance, namely for $\Delta = 0$, the transition probability is shown to increase quadratically in time, $P_{i \to f}^{(1)}(t) \cong \frac{\lambda^2 |\langle \chi_f | \hat{\mu} | \chi_i \rangle|^2 t^2}{4\hbar^2}$, which means that the transition rate, $k^{(1)}_{i \to f}(t) = \frac{\partial}{\partial t} P_{i \to f}(t)^{(1)}(t)$, increases linearly in time. (We recall, however, that probability conservation imposes an upper bound, $P^{(1)}_{i \to f}(t) \leq 1$, which limits the validity of the approximation to $t << \frac{2\hbar}{|\lambda||\langle \chi_f|\hat{\mu}|\chi_i \rangle|}$.)

Off-resonance, where $0 < |\Delta| << \Omega$, the quadratic increase of the transition probability at short times, $t << 1/|\Delta|$, is modulated by the squared sinc function. The transition rate reads (Ex. 15.6.10)

$$k^{(1)}_{i \to f}(t) = \frac{\lambda^2 |\langle \chi_f|\hat{\mu}|\chi_i \rangle|^2}{2\hbar^2} \frac{\sin[\Delta t]}{\Delta}. \tag{15.6.28}$$

For $t >> 1/|\Delta|$, we can approximate the result by considering the limit, $t \to \infty$, where $\frac{\sin[\Delta t]}{\Delta} \xrightarrow{t \to \infty} \pi \delta(\Delta)$, such that the rate expression converges to

$$k^{(1)}_{i \to f}(t) \xrightarrow{t >> 1/|\Delta|} \frac{\pi \lambda^2}{2\hbar^2} |\langle \chi_f|\hat{\mu}|\chi_i \rangle|^2 \delta(\Delta) \quad ; \quad |\Delta| \neq 0. \tag{15.6.29}$$

Recalling the definition of Dirac's delta, this result means that in the long time limit the transition rate vanishes for off-resonant frequencies; namely, field-induced transitions between $\hat{H}_0$ eigenstates are restricted by the resonance conditions, $\varepsilon_f = \varepsilon_i \pm \hbar\Omega$.

Exercise 15.6.8 *Show that for an interaction term, $\hat{V} = \hat{\mu}\sin(\Omega t)$, the transition probability to first order in λ (Eq. (15.6.10)) reads as Eq. (15.6.24).*

Exercise 15.6.9 *Derive Eq. (15.6.26) from Eq. (15.6.24) for $|\Delta| << \Omega$.*

Exercise 15.6.10 *Derive Eq. (15.6.28) for the transition rate by taking the time-derivative of the transition probability as given in Eq. (15.6.26).*

Bibliography

[15.1] H.-D. Meyer, U. Manthe, and L. S. Cederbaum, "The multi-configurational time-dependent Hartree approach," Chemical Physics Letters 165, 73 (1990).

[15.2] H. Wang and M. Thoss, "Multilayer formulation of the multiconfiguration time-dependent Hartree theory," The Journal of Chemical Physics 119, 1289 (2003).

[15.3] O. E. Alon, A. I. Streltsov, and L. S. Cederbaum, "Multiconfigurational time-dependent Hartree method for bosons: Many-body dynamics of bosonic systems," Physical Review A 77, 033613 (2008).

[15.4] R. Car and M. Parrinello, "Unified approach for molecular dynamics and density-functional theory," Physical Review Letters 55, 2471 (1985).

[15.5] J. J. Sakurai and J. Napolitano, "Modern Quantum Mechanics," 2nd ed. (Addison Wesley, 2011).

[15.6] U. Peskin and Nimrod Moiseyev, "The solution of the time-dependent Schrödinger equation by the (t, t') method: Theory, computational algorithm and applications," The Journal of Chemical Physics 99, 4590 (1993).

[15.7] J. R. Klauder and B.-S. Skagerstam, "Coherent states: applications in physics and mathematical physics" (World Scientific, 1985).

[15.8] S. Mukamel, "Principles of nonlinear optical spectroscopy" (Oxford University Press, 1999).

[15.9] D. J. Tannor, "Introduction to quantum mechanics: A time-dependent perspective" (University Science Books, 2007).

16 Incoherent States

16.1 Mixed Ensembles

According to the postulates of quantum mechanics (see Chapter 11), the outcome of a single measurement on a given system is inherently nondeterministic, even when the physical state of the system is well defined. The state corresponds to a statistical ensemble, and each measurement samples a possible realization of that state. So far, we discussed physical states associated with a single wave function (a single solution to the Schrödinger equation, or a single vector in the Hilbert space of the system). The ensemble that corresponds to such a state is termed "coherent" or "pure." Given the state vector, $|\psi(t)\rangle$, the statistical distribution of any measurable quantity can be expressed in terms of an expectation value of a corresponding operator (see Eq. (15.1.8)):

$$O_{\{|\psi\rangle\}}(t) = \langle\psi(t)|\hat{O}|\psi(t)\rangle. \qquad (16.1.1)$$

In many cases, however, the experimental apparatus (at the preparation and/or the measurement stage) cannot resolve pure physical states. Instead, the states that are measured correspond to a "mixed" or an "incoherent" statistical ensemble, where different parts of the ensemble are associated with different vectors in the system Hilbert space, for example, $|\psi_1(t)\rangle$, $|\psi_2(t)\rangle$, $|\psi_3(t)\rangle$,..., with well-defined relative statistical weights, $w_1, w_2, w_3,...$, where

$$\sum_i w_i = 1 \quad ; \quad w_i \geq 0 \quad ; \quad \langle\psi_i(t)|\psi_i(t)\rangle = 1. \qquad (16.1.2)$$

The statistical distribution of any measurable quantity depends on the relative weights of the different coherent states within the ensemble, in addition to the inherent statistical distribution attributed to each coherent state. The expectation value for a mixed ensemble is therefore a weighted average [15.5],

$$O_{\{|\psi_1\rangle,|\psi_2\rangle,|\psi_3\rangle,...\}}(t) = \sum_i w_i\langle\psi_i(t)|\hat{O}|\psi_i(t)\rangle. \qquad (16.1.3)$$

It is important to notice that the mixed ensemble (which refers to a "mixed state") cannot be associated with a single state vector in the Hilbert space of the system. A naïve attempt to associate a mixed state with a linear combination of the pure states composing the ensemble, namely $|\Psi(t)\rangle \equiv \sum_i a_i|\psi_i(t)\rangle$, would generally not yield the correct expectation values for system observables, as defined by Eq. (16.1.3). Indeed, the expectation values for the state $|\Psi(t)\rangle$, $O_{\{\sum_n a_n|\psi_n\rangle\}}(t) = \langle\Psi(t)|\hat{O}|\Psi(t)\rangle$, cannot coincide with $O_{\{|\psi_1\rangle,|\psi_2\rangle,|\psi_3\rangle,...\}}(t)$ for any system operator $\hat{O}$. It is sufficient to notice that $O_{\{\sum_n a_n|\psi_n\rangle\}}(t) = \sum_i |a_i|^2 \langle\psi_i(t)|\hat{O}|\psi_i(t)\rangle + \sum_{i>j} 2\,\mathrm{Re}[a_j^* a_i \langle\psi_j(t)|\hat{O}|\psi_i(t)\rangle]$, where the off-diagonal contributions (associated with $j \neq i$) are always absent in $O_{\{|\psi_1\rangle,|\psi_2\rangle,|\psi_3\rangle,...\}}(t)$. Indeed, the off-diagonals are "coherences" (or "phase relations") between the different coherent states, introduced artificially in this case by construction of the linear combination state, $|\Psi(t)\rangle$ (an elaborate definition of coherences is discussed in what follows).

As a concrete example for the inherent difference between a mixed and a pure state, let us consider a system of noninteracting spin-half particles in a mixed ensemble, in which half of the particles are in a spin state $|\alpha\rangle$, and the other half in a spin $|\beta\rangle$. Recalling that $\hat{S}_z|\alpha\rangle = \frac{\hbar}{2}|\alpha\rangle$ and $\hat{S}_z|\beta\rangle = \frac{-\hbar}{2}|\beta\rangle$ (Eq. (13.1.10)), we can readily see that the spin polarization in the z-direction (the expectation value of $\hat{S}_z$) vanishes for this ensemble. Implementing Eq. (16.1.3) with $\hat{O} = \hat{S}_z$, $|\psi_1(t)\rangle = |\alpha\rangle$, $|\psi_2(t)\rangle = |\beta\rangle$, $w_1 = w_2 = 0.5$, we indeed obtain $S_z = 0.5\langle\alpha|\hat{S}_z|\alpha\rangle + 0.5\langle\beta|\hat{S}_z|\beta\rangle = 0.5\frac{\hbar}{2} + 0.5\frac{-\hbar}{2} = 0$. Interestingly, the same result is obtained for an ensemble in which all the particles are associated with a pure superposition state, $|\Psi\rangle = \frac{1}{\sqrt{2}}|\alpha\rangle + \frac{1}{\sqrt{2}}|\beta\rangle$, where $S_z = \langle\Psi|\hat{S}_z|\Psi\rangle = 0$. Indeed, in both the mixed $\{0.5|\alpha\rangle, 0.5|\beta\rangle\}$ and the pure $\{\frac{1}{\sqrt{2}}|\alpha\rangle + \frac{1}{\sqrt{2}}|\beta\rangle\}$ states, the same probability (50%) is obtained for measuring the particle spin in the positive or negative direction along the z-axis. Nevertheless, these states are in fact different and can be distinguished by a different experiment. Let us consider, for example, a measurement of the spin polarization along the x-direction, namely the expectation value of $\hat{S}_x$. Using $\hat{S}_x|\alpha\rangle = \frac{\hbar}{2}|\beta\rangle$, and $\hat{S}_x|\beta\rangle = \frac{\hbar}{2}|\alpha\rangle$ (Eq. (13.1.10)), we obtain different results for the mixed and the pure states, namely $S_x = 0$ and $S_x = \frac{\hbar}{2}$, respectively (see Ex. 16.1.1). Indeed, in the pure state, each particle is associated with a specific coherent superposition of the different spin states $|\alpha\rangle$ and $|\beta\rangle$, which happens to be an eigenstate of $\hat{S}_x$, whereas in the mixed state, each particle is either at a state $|\alpha\rangle$ or $|\beta\rangle$, and therefore, none of the particles is in an eigenstate of $\hat{S}_x$. (In fact, each particle is in some coherent superposition of the two $\hat{S}_x$ eigenstates.) We can therefore conclude that a superposition state such as $\frac{1}{\sqrt{2}}|\alpha\rangle + \frac{1}{\sqrt{2}}|\beta\rangle$ is indeed different from the mixed ensemble $\{0.5|\alpha\rangle, 0.5|\beta\rangle\}$, although for some measurements the pure and mixed states would give the same result. In the pure (superposition) state, there are coherence relations between the different spin states that are nonexisting within the mixed state. Moreover, these coherences depend on the superposition coefficients, where different coefficients affect the values of different observables (see Ex. 16.1.1).

Exercise 16.1.1 *(a) A mixed state corresponds to an ensemble of noninteracting spin-half particles, of which half are in a spin state $|\alpha\rangle$, and the other half in a spin state $|\beta\rangle$, where $w_\alpha = w_\beta = 0.5$. Use Eqs. (13.1.9, 13.1.10) to compute the expectation value of the three different components of the single particle spin vector in this state, $S_i = w_\alpha\langle\alpha|\hat{S}_i|\alpha\rangle + w_\beta\langle\beta|\hat{S}_i|\beta\rangle (i \in x, y, z)$, and show that the spin vector orientation in this state is random, $S_x = S_y = S_z = 0$. (b) Repeat the calculation of the spin vector components for a pure state in which all the particles are associated with a superposition state,*

$|\Psi\rangle = \frac{1}{\sqrt{2}}|\alpha\rangle + \frac{1}{\sqrt{2}}|\beta\rangle$. *Show that in this case* $S_y = S_z = 0$, $S_x = \hbar/2$, *namely, the spin is polarized along the x-axis. (c) Repeat the calculation of the spin vector components for a pure state in which all the particles are associated with another superposition state,* $|\Psi\rangle = \frac{1}{\sqrt{2}}|\alpha\rangle + i\frac{1}{\sqrt{2}}|\beta\rangle$. *Show that in this case* $S_x = S_z = 0$, $S_y = \hbar/2$, *namely, the spin is polarized along the y-axis.*

16.2 The Density Operator

In the previous section we introduced mixed states of a physical system, which cannot be represented in terms of a single vector in the system's Hilbert space. It is desirable to find a general state representation that captures the statistical distributions of any measured property, and yet has the same formal structure for both mixed and pure states. For this purpose, it is convenient to express the expectation value of any operator in terms of a trace. Using the trace definition (Eq. (15.5.3) and Ex. 16.2.1), Eq. (16.1.1) for the expectation value of a pure state can be rewritten as

$$O(t) = \langle \psi(t)|\hat{O}|\psi(t)\rangle = tr\{\hat{O}|\psi(t)\rangle\langle\psi(t)|\}, \tag{16.2.1}$$

where, for the case of a mixed state, Eq. (16.1.3) leads to

$$O(t) = \sum_i w_i \langle \psi_i(t)|\hat{O}|\psi_i(t)\rangle = tr\{\hat{O}\sum_i w_i|\psi_i(t)\rangle\langle\psi_i(t)|\}. \tag{16.2.2}$$

In both cases the expectation value of $\hat{O}$ is a trace of a product of $\hat{O}$ and another operator that contains the information with respect to the state of the system. Defining the system's density operator,

$$\hat{\rho}(t) \equiv \sum_i w_i|\psi_i(t)\rangle\langle\psi_i(t)|, \tag{16.2.3}$$

a general expression for the expectation value, applicable for all states, is revealed:

$$O(t) = tr\{\hat{\rho}(t)\hat{O}\}, \tag{16.2.4}$$

where a pure state is a specific case in which the statistical weights vanish except for one, for example, $w_i = \delta_{i,i_0}$, where $\hat{\rho}(t) = |\psi_{i_0}(t)\rangle\langle\psi_{i_0}(t)|$.

Some important properties of the density operator are immediately apparent. First, since the relative weights of the different states in the ensemble sum to unity, Eq. (16.1.2), the trace of $\hat{\rho}(t)$ is unity:

$$tr\{\hat{\rho}(t)\} = 1. \tag{16.2.5}$$

Second, as an operator in the system Hilbert space, $\hat{\rho}(t)$ is Hermitian; namely, for any state vectors, $|\varphi\rangle$ and $|\chi\rangle$, we have

$$\langle\chi|\hat{\rho}(t)|\varphi\rangle = \langle\varphi|\hat{\rho}(t)|\chi\rangle^*. \tag{16.2.6}$$

The eigenvalues of the density operator are therefore real. Moreover, according to the definition of the density operator, its eigenvalues are nonnegative (Ex. 16.2.2)):

$$\hat{\rho}(t)|\chi_n(t)\rangle = \lambda_n(t)|\chi_n(t)\rangle \quad ; \quad 0 \le \lambda_n(t) \le 1. \tag{16.2.7}$$

Exercise 16.2.1 *Introduce the identity operator, $\hat{I} = \sum_n |\varphi_n\rangle\langle\varphi_n|$, into the general definition of a measurable quantity (Eq. (16.1.3)), and use the definition of the trace of an operator, Eq. (15.5.3), to show that $\sum_i w_i\langle\psi_i(t)|\hat{O}|\psi_i(t)\rangle = tr\{\sum_i w_i|\psi_i(t)\rangle\langle\psi_i(t)|\hat{O}\} = tr\{\hat{\rho}(t)\hat{O}\}$.*

Exercise 16.2.2 *Use the generic structure of the density operator, Eq. (16.2.3), to show that its eigenvalues are nonnegative; namely, if $\hat{\rho}(t)|\phi\rangle = \eta|\phi\rangle(where\langle\phi|\phi\rangle = 1)$, then $\eta \ge 0$.*

The matrix representation of the density operator in any set of vectors spanning the system's Hilbert space is referred to as the density matrix, for example, using, $\hat{I} = \sum_n |\varphi_n\rangle\langle\varphi_n|$, we have

$$\hat{\rho} = \sum_{n,m} |\varphi_n\rangle\langle\varphi_n|\hat{\rho}|\varphi_m\rangle\langle\varphi_m| = \sum_{n,m} \rho_{n,m}|\varphi_n\rangle\langle\varphi_m|, \tag{16.2.8}$$

where $\rho_{n,m}$ are the matrix elements of the density matrix in the selected basis,

$$\rho_{n,m} = [\boldsymbol{\rho}]_{n,m}. \tag{16.2.9}$$

The diagonal elements, $\{\rho_{n,n}\}$, are referred to as the "relative populations" of the basis states, namely, $\rho_{n,n}$ is the probability for measuring the system in the state, $|\varphi_n\rangle$. Using the definition, $\rho_{n,n} = \langle\varphi_n|\hat{\rho}|\varphi_n\rangle = \sum_i w_i\langle\varphi_n|\psi_i(t)\rangle\langle\psi_i(t)|\varphi_n\rangle = \sum_i w_i|\langle\varphi_n|\psi_i(t)\rangle|^2$, we can readily see that the relative populations are nonnegative, $0 \le \rho_{n,n} \le 1$, where, according to Eq. (16.2.5),

$$\sum_n \rho_{n,n} = 1. \tag{16.2.10}$$

The off-diagonal elements of the density matrix, $\rho_{n,m}$, where $m \ne n$, are referred to as the "coherences" between the basis states, $\rho_{n,m} = \langle\varphi_n|\hat{\rho}|\varphi_m\rangle = \sum_i w_i\langle\varphi_n|\psi_i(t)\rangle\langle\psi_i(t)|\varphi_m\rangle = \sum_i w_i\langle\varphi_n|\psi_i(t)\rangle\langle\varphi_m|\psi_i(t)\rangle^*$. These entries are shown to depend on the projections of the states composing the ensemble on the basis states, $\{\langle\varphi_n|\psi_i\rangle\}$, including the phases of these complex-valued projections. As we can readily see, *the populations and coherences of the density matrix are basis-dependent and in general vary upon basis set transformations (see Ex. 16.2.3)*. Particularly, since the density operator is Hermitian, it can be diagonalized, and therefore any density operator can be represented in a basis in which the coherences between the basis states are null.

Exercise 16.2.3 *In Ex. 16.1.1 we discussed the difference between two ensembles of spin-half particles. The first corresponded to a random spin orientation, where half of the particles are found in a spin state $|\alpha\rangle$, and the other half in a spin state $|\beta\rangle$. The other ensemble* *corresponded to spin polarization along the x direction, where all the particles are in a superposition state, $|\Psi\rangle = \frac{1}{\sqrt{2}}|\alpha\rangle + \frac{1}{\sqrt{2}}|\beta\rangle$. The density operators corresponding to the two ensembles are $\hat{\rho}_R = 0.5|\alpha\rangle\langle\alpha| + 0.5|\beta\rangle\langle\beta|$ and $\hat{\rho}_P = |\Psi\rangle\langle\Psi|$. (a) Show that the density matrices corresponding to these two ensembles in the basis of the two spin states are*

$$\rho_R = \begin{bmatrix} 0.5 & 0 \\ 0 & 0.5 \end{bmatrix} \text{ and } \rho_P = \begin{bmatrix} 0.5 & 0.5 \\ 0.5 & 0.5 \end{bmatrix},$$ *respectively. (b) Change the basis from*

$|\alpha\rangle$ *and* $|\beta\rangle$ *($\hat{S}_z$-eigenstates) into the two eigenstates of* $\hat{S}_x$, *namely,* $\frac{1}{\sqrt{2}}|\alpha\rangle + \frac{1}{\sqrt{2}}|\beta\rangle$ *and* $\frac{1}{\sqrt{2}}|\alpha\rangle - \frac{1}{\sqrt{2}}|\beta\rangle$. *Show that the matrix representation of the random ensemble is invariant to the transformation, whereas the matrix representation of the pure state becomes diagonal in this basis,* $\rho_P = \begin{bmatrix} 1 & 0 \\ 0 & 0 \end{bmatrix}$.

16.3 Liouville's Space

By its definition, a density operator does not correspond to any specific system observable but rather to a state of the system. Yet, this state representation is an operator, not a vector, in the standard Hilbert space of the system. It is instructive, however, to associate density operators with elements in a vector space, named Liouville's space. We can readily verify that system operators are indeed vectors in a space over the field of complex numbers, which is closed with respect to addition of operators and to their multiplications by scalar complex numbers, and where the set of properties defining a vector space are fulfilled. Defining an inner product between two operators $\hat{A}$ and $\hat{B}$ in this space,

$$(\hat{A}, \hat{B}) \equiv tr\{\hat{A}^\dagger \hat{B}\}, \tag{16.3.1}$$

the norm of a vector is also defined,

$$\|\hat{A}\| = tr\{\hat{A}^\dagger \hat{A}\}, \tag{16.3.2}$$

and the vector space (the Liouville space) can be identified as a Hilbert space. The mapping of an operator in the standard system Hilbert space onto a vector in Liouville's space is straightforward. Given any complete orthonormal system of vectors in the standard Hilbert space, for example, $\{|\varphi_n\rangle\}$, the set of all linearly independent outer products between these vectors, $\{|\varphi_n\rangle\langle\varphi_m|\}$, is a complete orthonormal system of operators that spans the Liouville space (see Exs. 16.3.1 and 16.3.2).

The "matrix representation" of any operator in the standard Hilbert space of the system (e.g., Eq. (16.2.8)) is also its "vector representation" in the Liouville space; namely, any system operator can be expanded as a linear combination of the set $\{|\varphi_n\rangle\langle\varphi_m|\}$:

$$\hat{A} = \sum_{m,n} A_{n,m}\hat{\varphi}_{n,m} \quad ; \quad \hat{\varphi}_{n,m} \equiv |\varphi_n\rangle\langle\varphi_m|. \tag{16.3.3}$$

Under the inner product, Eq. (16.3.1), the set $\{\hat{\varphi}_{n,m}\}$ is orthonormal, and $A_{n,m}$ is the projection of $\hat{A}$ on the basis vector $\hat{\varphi}_{n,m}$:

$$tr\{\hat{\varphi}_{n,m}^\dagger \hat{\varphi}_{n',m'}\} = \delta_{n,n'}\delta_{m,m'} \quad ; \quad A_{n,m} = tr\{\hat{\varphi}_{n,m}^\dagger, \hat{A}\}. \tag{16.3.4}$$

Exercise 16.3.1 *Let the set of vectors $\{|\varphi_n\rangle\}$ be a complete orthonormal system spanning a Hilbert pace, where $\langle\varphi_m|\varphi_n\rangle = \delta_{m,n}$. The outer products between these vectors, $\{\hat{\varphi}_{m,n} = |\varphi_m\rangle\langle\varphi_n|\}$, are operators in that space. (a) Associating each operator, $\hat{\varphi}_{m,n}$, with a vector in a (Liouville) vector space with an inner product, $(\hat{A},\hat{B}) = tr\{\hat{A}^\dagger\hat{B}\}$, show that $\{\hat{\varphi}_{m,n}\}$ is an orthonormal set, namely $(\hat{\varphi}_{m,n}, \hat{\varphi}_{m',n'}) = \delta_{m,m'}\delta_{n,n'}$. (b) Show that any operator in the original Hilbert space can be expanded as a linear combination of the set $\{\hat{\varphi}_{m,n}\}$, $\hat{A} = \sum_{m,n} A_{m,n}\hat{\varphi}_{m,n}$, where $A_{m,n}$ is the inner product of $\hat{A}$ with $\hat{\varphi}_{m,n}$, $A_{m,n} = (\hat{\varphi}_{m,n},\hat{A}) = tr\{\hat{\varphi}_{m,n}^\dagger\hat{A}\}$.*

Exercise 16.3.2 *Consider the space of two-dimensional matrices, $\begin{bmatrix} x & y \\ z & w \end{bmatrix}$, where x, y, z and w are complex-valued numbers. (a) Show that the set of Pauli matrices (Eq. (13.1.17)) and the identity matrix,*

$$\boldsymbol{\sigma}_z = \begin{bmatrix} 1 & 0 \\ 0 & -1 \end{bmatrix}; \ \boldsymbol{\sigma}_x = \begin{bmatrix} 0 & 1 \\ 1 & 0 \end{bmatrix}; \ \boldsymbol{\sigma}_y = \begin{bmatrix} 0 & -i \\ i & 0 \end{bmatrix}; \ \mathbf{I} = \begin{bmatrix} 1 & 0 \\ 0 & 1 \end{bmatrix},$$

compose an orthogonal set of vectors in this space, under the inner product, $(\mathbf{A},\mathbf{B}) = tr\{\mathbf{A}^\dagger\mathbf{B}\}$. (b) Show that the corresponding normalized basis vectors under this inner product read

$$\boldsymbol{\sigma}_z = \frac{1}{\sqrt{2}}\begin{bmatrix} 1 & 0 \\ 0 & -1 \end{bmatrix}; \ \boldsymbol{\sigma}_x = \frac{1}{\sqrt{2}}\begin{bmatrix} 0 & 1 \\ 1 & 0 \end{bmatrix}; \ \boldsymbol{\sigma}_y = \frac{1}{\sqrt{2}}\begin{bmatrix} 0 & -i \\ i & 0 \end{bmatrix}; \ \mathbf{I} = \frac{1}{\sqrt{2}}\begin{bmatrix} 1 & 0 \\ 0 & 1 \end{bmatrix}.$$

(c) Show that any two-by-two matrix $\begin{bmatrix} x & y \\ z & w \end{bmatrix}$ can be written as a linear combination of these matrices, with expansion coefficients given by its inner product with the basis vectors in (b).

Operations on a Hilbert space operator can be represented in terms of Liouville's space "super operators." Consider the ordered sequence of operations $\hat{B}$ and $\hat{A}$. Using a complete orthonormal set $\{|\varphi_n\rangle\}$ in the standard Hilbert space, the matrix elements of $\hat{B}\hat{A}$ read

$$\langle\varphi_{n'}|\hat{B}\hat{A}|\varphi_{m'}\rangle = \sum_n \langle\varphi_{n'}|\hat{B}|\varphi_n\rangle\langle\varphi_n|\hat{A}|\varphi_{m'}\rangle = \sum_{m,n}\langle\varphi_{n'}|\hat{B}|\varphi_n\rangle\delta_{m',m}\langle\varphi_n|\hat{A}|\varphi_m\rangle. \tag{16.3.5}$$

Regarding the result as a linear map from the Liouville space vector whose elements are $A_{n,m}$ to another vector whose elements are $[\hat{B}\hat{A}]_{n',m'}$, we can readily identify the matrix representation of the corresponding Liouville space "super-operator," $B^{(L)}_{(n',m'),(n,m)}$:

$$[\hat{B}\hat{A}]_{n',m'} = \sum_{m,n} B^{(L)}_{(n',m'),(n,m)} A_{n,m} \quad ; \quad B^{(L)}_{(n',m'),(n,m)} = B_{n',n}\delta_{m',m}. \tag{16.3.6}$$

Similarly, the alternative sequence of Hilbert space operators $\hat{A}\hat{B}$ can also be formulated as a Liouville space super operator on $\hat{A}$:

$$\langle \varphi_{n'}|\hat{A}\hat{B}|\varphi_{m'}\rangle = \sum_{m}\langle \varphi_{n'}|\hat{A}|\varphi_{m}\rangle\langle \varphi_{m}|\hat{B}|\varphi_{m'}\rangle = \sum_{m,n}\langle \varphi_{m}|\hat{B}|\varphi_{m'}\rangle\delta_{n',n}\langle \varphi_{n}|\hat{A}|\varphi_{m}\rangle \quad (16.3.7)$$

$$[\hat{A}\hat{B}]_{n',m'} = \sum_{m,n}B^{(R)}_{(n',m'),(n,m)}A_{n,m} \quad ; \quad B^{(R)}_{(n',m'),(n,m)} = \delta_{n',n}B_{m,m'}. \quad (16.3.8)$$

Notice that each vector element (corresponding to the projection on basis operator, $|\varphi_{n}\rangle\langle\varphi_{m}|$; see Ex. (16.3.1)) in Liouville's space is characterized by two indexes, and each super-matrix element is characterized by four indexes. The dimensions of Liouville's space vectors and super-matrices are therefore the squares of the dimensions of the corresponding Hilbert space objects. Particularly, the matrix representations of the Liouville super operators in Eqs. (16.3.6, 16.3.8) can be formally expressed as tensor products, $\mathbf{B}^{(L)} = \mathbf{B} \otimes \mathbf{I}$ and $\mathbf{B}^{(R)} = \mathbf{I} \otimes \mathbf{B}^{t}$ (see Eq. (11.6.18)).

16.4 The Liouville–von Neumann Equation

We now turn to the time evolution of density operators in Liouville's space. Each of the pure states defining a mixed ensemble (see Eq. (16.2.3)) is a physical state-vector in the system's Hilbert space, whose time evolution is given by the time-dependent Schrödinger equation. Using $i\hbar\frac{\partial}{\partial t}|\psi_{i}(t)\rangle = \hat{H}(t)|\psi_{i}(t)\rangle$, we readily obtain an equation of motion for the density operator, in terms of the underlying system Hamiltonian (Ex. 16.4.1),

$$\frac{\partial}{\partial t}\hat{\rho}(t) = \frac{-i}{\hbar}[\hat{H}(t),\hat{\rho}(t)]. \quad (16.4.1)$$

The result is known as the Liouville–von Neumann equation for the density operator. Notice that in spite of some structural resemblance (up to the sign of the commutator), there is no relation between this equation and the Heisenberg equation of motion. Indeed, the Heisenberg picture representation of the density operator is $\hat{\rho}_{H}(t) = \hat{\rho}_{H}(0)$, where $\frac{\partial}{\partial t}\hat{\rho}_{H}(t) = 0$ (see Ex. 16.4.2).

Exercise 16.4.1 *Use the Schrödinger equation for the coherent states, $i\hbar\frac{\partial}{\partial t}|\psi_{i}(t)\rangle = \hat{H}(t)|\psi_{i}(t)\rangle$, and the Hermitian conjugated equation to show that the time derivative of the density operator (Eq. (16.2.3)) is given by Eq. (16.4.1).*

Exercise 16.4.2 *(a) Show that by its definition, the density operator (Eq. (16.2.3)) can be formulated using the time-evolution operator (Eq. (15.2.2)), $\hat{\rho}(t) = \hat{U}(t,0)\hat{\rho}(0)\hat{U}^{\dagger}(t,0)$. (b) Show that the Heisenberg picture representation (Eq. (15.3.16)) of the density operator, $\hat{\rho}_{H}(t)$, is time-independent. (c) Use Eq. (15.3.17) to show that the time derivative of $\hat{\rho}_{H}(t)$ vanishes at all times.*

Notice the analogy of Liouville's equation for the density operator $\hat{\rho}(t)$, Eq. (16.4.1), to the time-dependent Schrödinger equation for a pure state, $i\hbar\frac{\partial}{\partial t}|\psi(t)\rangle = \hat{H}(t)|\psi(t)\rangle$.

In both cases the time-derivative of the state vector is identified with an operation of a suitable operator (generator of motion) in the relevant vector space. In the case of the Schrödinger equation, the space is the standard system's Hilbert space, and the generator of motion in that space is the Hamiltonian. In the Liouville equation for the density operator, the vector space is Liouville's space, where the generator of motion is the commutator with the Hamiltonian. The latter is often denoted as the Liouvillian "super-operator," $\hat{L}(t)$:

$$\hat{L}(t)\hat{\rho}(t) \equiv [\hat{H}(t),\hat{\rho}(t)], \qquad (16.4.2)$$

where the Liouville–von Neumann equation reads

$$i\hbar\frac{\partial}{\partial t}\hat{\rho}(t) = \hat{L}(t)\hat{\rho}(t). \qquad (16.4.3)$$

Exercise 16.4.3   *Given a finite basis for the system's Hilbert space, where operators are represented as $N \times N$ matrices, use Eqs. (16.3.6, 16.3.8) to show that the Liouville super-operator can be represented as a matrix of dimensions $N^2 \times N^2$, $\mathbf{L} = \mathbf{H} \otimes \mathbf{I} - \mathbf{I} \otimes \mathbf{H}^t$, where $\mathbf{H}$ and $\mathbf{I}$ are, respectively, the matrix representation of the Hamiltonian and the identity (use the matrix tensor product definition, Eqs. (11.6.17, 11.6.18), to show that $[\hat{H}\hat{\rho} - \hat{\rho}\hat{H}]_{n',m'} = \sum\limits_{m,n} (\mathbf{H} \otimes \mathbf{I} - \mathbf{I} \otimes \mathbf{H}^t)_{(n',m'),(n,m)}\rho_{n,m}).$*

16.5 Equilibrium States

We now focus on the equilibrium state of a system. When the external constraints enable a system to reach a stationary state, with no net flow of particles or energy inside the system or at any of its boundaries, the system is said to reach equilibrium. Unless specified differently (see Chapter 20 for the discussion of steady states in nonequilibrium), we shall assume that the boundary conditions are uniform such that stationary net flow through the boundaries is excluded, and we shall identify the equilibrium state with a stationary state, in which the probability distribution of any measurable property is time-independent. Since any measurable property is expressed in terms of the system's density operator, $\hat{\rho}(t)$, Eq. (16.2.4), the requirement of time-independence for any measurable property, $\hat{O}$, reads

$$\frac{\partial}{\partial t}tr\{\hat{\rho}(t)\hat{O}\} = tr\left\{\left[\frac{\partial}{\partial t}\hat{\rho}(t)\right]\hat{O}\right\} = 0. \qquad (16.5.1)$$

The equilibrium state is therefore associated with a time-independent system density operator, $\hat{\rho}^{(eq)}$, namely

$$\frac{\partial}{\partial t}\hat{\rho}^{(eq)} = 0. \qquad (16.5.2)$$

According to the Liouville–von Neumann equation (Eq. (16.4.1)), the time derivative of a density operator is proportional to its commutator with the system Hamiltonian, which means

$$[\hat{H}, \hat{\rho}^{(eq)}] = 0. \tag{16.5.3}$$

Notice that to comply with the definition of the equilibrium density operator (Eq. (16.5.2)), the commutator must vanish at all times. This means that *a strict equilibrium state is limited to time independent Hamiltonians*. Indeed, the equilibrium density operator for a mixed state can be regarded as a generalization of the stationary pure states in systems with time-independent Hamiltonians.

An important consequence of Eq. (16.5.3) is that the matrix representation of $\hat{\rho}^{(eq)}$ in the basis of the Hamiltonian eigenstates can be diagonal. In fact, off-diagonal matrix elements of $\hat{\rho}^{(eq)}$, attributed to coherences between nondegenerate Hamiltonian eigenstates, must vanish (see Ex. 16.5.1). Vanishing is not mandatory for coherences between degenerate eigenstates of the system Hamiltonian. However, since $\hat{\rho}^{(eq)}$ and $\hat{H}$ commute, they share a common set of eigenstates, in which their matrix representations are both diagonal (Section 11.7). Denoting this common set as $\{|\varphi_1\rangle, |\varphi_2\rangle, |\varphi_3\rangle, \ldots\}$, the equilibrium density operator obtains the form

$$\hat{\rho}^{(eq)} = \sum_n w_n |\varphi_n\rangle\langle\varphi_n|, \tag{16.5.4}$$

where (Ex. 16.5.1)

$$\hat{H}|\varphi_n\rangle = \varepsilon_n|\varphi_n\rangle \quad ; \quad \hat{\rho}^{(eq)}|\varphi_n\rangle = w_n|\varphi_n\rangle. \tag{16.5.5}$$

For simplicity, we refer to systems with a discrete energy spectrum, where the generalization to the case of continuous spectrum involves replacing summations by proper integrals (see Chapter 11), but otherwise it is conceptually similar.

The corresponding matrix representation of the density operator in the general case of a mixed equilibrium state therefore reads

$$\rho^{(eq)} = \begin{bmatrix} w_1 & & & \\ & w_2 & 0 & \\ & 0 & \ddots & \\ & & & w_N \end{bmatrix}. \tag{16.5.6}$$

Since $tr\{\rho^{(eq)}\} = 1$ (see Eq. (16.2.5)), the information with respect to the underlying equilibrium state of the system is fully encoded in the normalized distribution of the diagonal matrix elements, which reflects the relative weights of the system's stationary states. It is instructive to consider two limiting cases.

Exercise 16.5.1 *Given a time-independent system Hamiltonian, any pure state can be expanded in terms of the stationary solutions (Eq. (15.1.5)), $|\psi_i(t)\rangle = \sum_n a_n^{(i)} e^{\frac{-it}{\hbar}\varepsilon_n}|\varphi_n\rangle$, where $|\varphi_n\rangle$ are the Hamiltonian eigenstates, $\hat{H}|\varphi_n\rangle = \varepsilon_n|\varphi_n\rangle$. (a) Use this expansion and the definition of the system's density operator, Eq. (16.2.3), to show that the matrix representation of the density operator in the basis of the Hamiltonian eigenstates reads $\langle\varphi_{m'}|\hat{\rho}(t)|\varphi_m\rangle = \sum_i w_i a_{m'}^{(i)} \left(a_m^{(i)}\right)^* e^{\frac{-it}{\hbar}(\varepsilon_{m'}-\varepsilon_m)}$. (b) Show that the equilibrium requirement,*

$\frac{\partial}{\partial t} \langle \varphi_{m'} | \hat{\rho}(t) | \varphi_m \rangle = 0$, *means that off-diagonal matrix elements between nondegenerate Hamiltonian eigenstates* $(\varepsilon_m \neq \varepsilon_{m'})$ *must vanish identically. (c) Show that any density matrix that is diagonal in the basis of stationary states,* $\langle \varphi_{m'} | \hat{\rho}(t) | \varphi_m \rangle \propto \delta_{m',m}$, *must be time-independent. (d) Given the diagonal representation of the equilibrium density operator in the basis of Hamiltonian eigenstates* $\{ | \varphi_n \rangle \}$ *(Eq. (16.5.4)), show that* $\hat{\rho}^{(eq)} | \varphi_n \rangle = w_n | \varphi_n \rangle$.

When the equilibrium state corresponds to a delta distribution of weights, namely $w_n = \delta_{n,m}$, the density operator takes the form $\hat{\rho}^{(eq)} = | \varphi_m \rangle \langle \varphi_m |$, or

$$\rho^{(eq)} = \begin{bmatrix} \ddots & & & \\ & \ddots & & \\ & & 1 & \\ & & & \ddots \end{bmatrix}. \qquad (16.5.7)$$

This is the case of ***a stationary pure state, where the system at equilibrium is associated with a single Hamiltonian eigenstate***, namely a single stationary state.

In contrast, when the equilibrium state corresponds to a uniform distribution of weights, namely $w_n = 1/N$, the density operator takes the form $\hat{\rho}^{(eq)} = \frac{1}{N} \sum_{m=1}^{N} | \varphi_m \rangle \langle \varphi_m |$, or

$$\rho^{(eq)} = \frac{1}{N} \begin{bmatrix} 1 & & & \\ & 1 & & \\ & & 1 & \\ & & & \ddots \end{bmatrix}. \qquad (16.5.8)$$

In this case, the equilibrium density operator is proportional to the identity operator, where all the Hamiltonian eigenstates are equally populated. Remarkably, since the identity operator is invariant to unitary transformations, including basis set transformations (see Eq. (11.2.13)), *a system in this state is associated with equal populations of the eigenstates of any operator. Consequently, the statistical distribution of any measurable quantity is a random distribution, and the state is referred to as the random state, or the zero-information state. The random state is unique in that the coherences (off-diagonal elements) of the density matrix vanish regardless of the selected basis* (see Ex. 16.1.1 and Ex. 16.2.3).

A convenient measure for the "purity" (or the "information content") of the state of the system is the norm of the corresponding density operator. Recalling the definition, Eq. (16.3.2), and considering the Hermiticity of the density operator (Eq. (16.2.6)), the norm reads

$$\| \hat{\rho} \| = tr\{ \hat{\rho}^2 \}. \qquad (16.5.9)$$

As we can see, $0 \leq tr\{ \hat{\rho}^2 \} \leq 1$, where the upper limit, $tr\{ \hat{\rho}^2 \} \to 1$, corresponds to maximum purity (a delta distribution of weights, Eq. (16.5.7)), and the lower limit, $tr\{ \hat{\rho}^2 \} \to 0$, corresponds to a random state, Eq. (16.5.8), in the limit of an infinite-dimensional Hilbert space $(N \to \infty)$.

It is instructive to associate the loss of purity in the state of a system with a monotonically increasing entropy function. von Neumann introduced the following entropy measure which accounts for the "randomness" of mixed quantum ensembles:

$$S \equiv -tr\{\hat{\rho}\ln(\hat{\rho})\}. \tag{16.5.10}$$

For an equilibrium state (Eqs. (16.5.4, 16.5.5)), the von Neumann entropy therefore reads

$$S^{(eq)} = -\sum_n w_n \ln(w_n). \tag{16.5.11}$$

Indeed, for a pure state the entropy vanishes, $S^{(eq)} = 0$ (Ex. 16.5.2), whereas for a random ensemble, $S^{(eq)}$ obtains a maximal value (subject to the normalization condition), $S^{(eq)} = -\sum_n \frac{1}{N}\ln(\frac{1}{N}) = \ln(N)$; see Ex. 16.5.3.

Exercise 16.5.2 *Show that the von Neumann equilibrium entropy, Eq. (16.5.11) vanishes for a pure state, namely when the weights are either $w_n \to 0$ or $w_n \to 1$.*

Exercise 16.5.3 *The statistical weights at equilibrium are obtained by maximizing the von Neumann entropy, subject to constraints. For a random ensemble the only constraint is the normalization, $\sum_{n=1}^{N} w_n = 1$, which can be imposed in terms of a Lagrange multiplier λ. The function to be maximized in this case is $F(w_1, w_2, \ldots) = S^{(eq)}(w_1, w_2, \ldots) - \lambda \left[\sum_{n=1}^{N} w_n - 1\right]$. Apply the necessary condition for a maximum, $\{\frac{\partial}{\partial w_n} F(w_1, w_2, \ldots) = 0\}$, and show that the maximum is obtained when the weight is uniform for all the stationary states, namely, $w_n = e^{-(1+\lambda)}$. Determine the value of λ by the normalization constraint to show that the equilibrium weights in this case read $w_n = \frac{1}{N}$.*

According to the standard postulates of statistical mechanics, an equilibrium state of a system is a state of maximum entropy, within a given set of constraints. As practiced in quantum mechanics, a system is characterized by its Hamiltonian; but the state of the system (its equilibrium state, in particular) depends on the amount of energy and the number of particles available to the system. (Hamiltonians that correspond to system with varying numbers of particles will be discussed in Chapter 20.) The ability of the system to exchange particles and/or energy with its surroundings (the rest of the universe, or the "reservoir"), can be constrained in different ways. The precise form of the equilibrium density matrix, namely the distribution of the weights between the different stationary states $(w_1, w_2, \ldots, w_N)$, depends therefore on the imposed constraints. We can readily verify (Ex. 16.5.3) that the random state distribution (Eq. (16.5.8)) is obtained by maximizing the von Neumann entropy, $S^{(eq)}$, where the only constraint is the trivial normalization of the weights. Additional constraints lead to different results for the weights distribution. Two standard cases corresponding to the canonical and the grand canonical statistical ensembles are discussed in detail in what follows.

The Canonical Ensemble

A typical constraint on a system that exchanges energy with its surroundings at equilibrium refers to the amount of energy contained in the system. This constraint characterizes systems in contact with a thermal reservoir (a thermal bath) at a fixed temperature (in the thermodynamic limit, where the system–bath interaction strength is assumed to be weak). The constraint means that the energy, averaged over the statistical ensemble representing the state of the system, obtains a constant value, $tr\{\hat{H}\hat{\rho}^{(eq)}\} = U$. Expressing the density matrix in terms of the statistical weights of the stationary states of the system, we obtain

$$tr\{\hat{H}\hat{\rho}^{(eq)}\} = \sum_{n=1}^{N} w_n \varepsilon_n = U. \tag{16.5.12}$$

The equilibrium weights are obtained by maximizing the von Neumann entropy, subject to this constraint, as well as to the normalization constraint, $\sum_{n=1}^{N} w_n = 1$. Introducing corresponding Lagrange multipliers [15.5], β and λ, the functional to be maximized in this case reads

$$F(w_1, w_2, \ldots) = S^{(eq)}(w_1, w_2, \ldots) - \lambda \left[\left(\sum_{n=1}^{N} w_n\right) - 1\right] - \beta \left[\left(\sum_{n=1}^{N} w_n \varepsilon_n\right) - U\right]. \tag{16.5.13}$$

Using Eq. (16.5.11) and imposing the necessary conditions for a maximum, $\frac{\partial}{\partial w_n} F(w_1, w_2, \ldots) = 0$, for $n = 1, 2, \ldots, N$, we obtain

$$-\ln(w_n) - 1 - \lambda - \beta \varepsilon_n = 0, \tag{16.5.14}$$

which means $w_n = e^{-(1+\lambda+\beta\varepsilon_n)}$. The value of λ is set by the normalization constraint, which yields (Ex. 16.5.4)

$$w_n = \frac{e^{-\beta\varepsilon_n}}{\sum_{n'=1}^{N} e^{-\beta\varepsilon_{n'}}}. \tag{16.5.15}$$

The value of the other Lagrange multiplier, β, can be determined by writing the explicit expression for the ensemble-averaged energy,

$$U = \sum_{n=1}^{N} \frac{\varepsilon_n e^{-\beta\varepsilon_n}}{\sum_{n'=1}^{N} e^{-\beta\varepsilon_{n'}}} = -\frac{\partial}{\partial\beta} \ln\left(\sum_{n=1}^{N} e^{-\beta\varepsilon_n}\right). \tag{16.5.16}$$

This expression coincides with the familiar results for a canonical thermodynamic ensemble, $U = -k_B \frac{\partial \ln(Z)}{\partial(1/T)}$, where k_B is Boltzmann's constant, Z is the canonical partition function, and T is the absolute temperature. Using this relation to identify the expressions for the partition function and the temperature, the equilibrium weights for a canonical (thermal) ensemble are given as

$$w_n = \frac{1}{Z} e^{\frac{-\varepsilon_n}{k_B T}} \quad ; \quad Z = \sum_{n=1}^{N} e^{\frac{-\varepsilon_n}{k_B T}} \tag{16.5.17}$$

Recalling the definition of the equilibrium density matrix in the representation of the Hamiltonian eigenstates, Eqs. (16.5.4, 16.5.5), the canonical density operator itself reads (see Ex. 16.5.5)

$$\hat{\rho}^{(eq)} = \frac{e^{-\hat{H}/(k_B T)}}{tr\{e^{-\hat{H}/(k_B T)}\}}. \tag{16.5.18}$$

Notice that in the infinite temperature limit, the equilibrium state of the canonical ensemble approaches that of a random ensemble.

Exercise 16.5.4   *For a canonical ensemble, the weights of the stationary states are constrained, such that $\sum_{n=1}^{N} w_n = 1$ and $\sum_{n=1}^{N} w_n \varepsilon_n = U$. (a) Derive Eq. (16.5.14) by maximizing the von Neumann entropy subject to these constraints. Determine the value of λ by the normalization constraint to show that the equilibrium weights are given in this case by Eq. (16.5.15).*

Exercise 16.5.5   *Derive Eq. (16.5.18) by substitution of the result, Eq. (16.5.17), for the canonical ensemble in the general expression for the density matrix, Eq. (16.5.4), and by using Eq. (16.5.5).*

The Grand Canonical Ensemble

When the system can exchange particles (in addition to energy) with its surroundings, not only the energy but also the particle content in the system are constrained by the surroundings. This is the typical situation for systems in contact with particle reservoirs at fixed temperature and chemical potentials. For simplicity we shall restrict our discussion to systems with a single type of particle. The additional constraint means in this case that the number of particles per state of the system, averaged over the statistical ensemble, obtains a constant value, namely $tr\{\hat{N}\hat{\rho}^{(eq)}\} = N_0$. The particle number operator, $\hat{N}$, as well as Hamiltonians of systems with an indefinite number of particles will be introduced and discussed in Chapter 20 (in relation to "second quantization" for fermions). For the present discussion, all we need to know is that such operators exist and that any stationary state of the system is associated with a well-defined number of particles; namely, the Hamiltonian eigenstates $\{|\varphi_n\rangle\}$ are also eigenstates of the system particle number operator (see also Eq. 16.5.5):

$$\hat{N}|\varphi_n\rangle = N_n|\varphi_n\rangle. \tag{16.5.19}$$

Expressing the density matrix in this basis, the additional constraint on the number of particles reads

$$tr\{\hat{N}\hat{\rho}^{(eq)}\} = \sum_{n=1}^{N} w_n N_n = N_0. \tag{16.5.20}$$

To obtain the equilibrium weights $\{w_n\}$ in this equilibrium state, the von Neumann entropy is maximized subject to the constraints, Eqs. (16.5.12, 16.5.20), and the normalization constraint, $\sum_{n=1}^{N} w_n = 1$. Introducing the corresponding Lagrange multipliers, β, η, and λ, the functional to be maximized in this case reads

$$F(w_1, w_2, \ldots) = S^{(eq)}(w_1, w_2, \ldots) - \lambda \left[\left(\sum_{n=1}^{N} w_n \right) - 1 \right]$$
$$- \beta \left[\left(\sum_{n=1}^{N} w_n \varepsilon_n \right) - U \right] - \eta \left[\left(\sum_{n=1}^{N} w_n N_n \right) - N_0 \right]. \quad (16.5.21)$$

Imposing the necessary conditions for a maximum, $\frac{\partial}{\partial w_n} F(w_1, w_2, \ldots) = 0$, we obtain in this case $w_n = e^{-(1+\lambda+\beta\varepsilon_n+\eta N_n)}$, where the value of λ is set by the normalization condition (Ex. 16.5.6)

$$w_n = \frac{e^{-\beta\varepsilon_n - \eta N_n}}{\sum_{n'} e^{-\beta\varepsilon_{n'} - \eta N_{n'}}}. \quad (16.5.22)$$

Using Eq. (16.5.22) and defining $Z = \sum_{n'} e^{-\beta\varepsilon_{n'} - \eta N_{n'}}$, the ensemble-averaged energy (Eq. (16.5.12)) and particle number (Eq. (16.5.20)) can be rewritten as

$$U = -\frac{\partial}{\partial \beta} \ln(Z) \quad ; \quad N_0 = -\frac{\partial}{\partial \eta} \ln(Z). \quad (16.5.23)$$

Comparing the expressions to standard thermodynamic definitions of these quantities in a grand canonical ensemble, $U = -k_B \frac{\partial \ln(Z)}{\partial (1/T)}$, and $N_0 = k_B T \frac{\partial \ln(Z)}{\partial \mu}$, where Z is the partition function, T is the absolute temperature, and μ is the chemical potential, the Lagrange multipliers can be identified with thermodynamics quantities $\beta = 1/(k_B T)$ and $\eta = -\mu\beta$. Consequently, the statistical weights (Eq. (16.5.22)) for an equilibrium state corresponding to a grand canonical ensemble obtain the form

$$w_n = \frac{1}{Z} e^{\frac{-1}{k_B T}(\varepsilon_n - \mu N_n)} \quad ; \quad Z = \sum_{n=1}^{N} e^{\frac{-1}{k_B T}(\varepsilon_n - \mu N_n)}, \quad (16.5.24)$$

where the corresponding equilibrium density operator reads (see Ex. 16.5.7)

$$\hat{\rho}^{(eq)} = \frac{e^{\frac{-1}{k_B T}(\hat{H} - \mu\hat{N})}}{tr\left\{ e^{\frac{-1}{k_B T}(\hat{H} - \mu\hat{N})} \right\}} \quad (16.5.25)$$

Exercise 16.5.6 *For a grand canonical ensemble, the weights of the stationary states are constrained, such that $\sum_{n=1}^{N} w_n = 1$, $\sum_{n=1}^{N} w_n \varepsilon_n = U$, and $\sum_{n=1}^{N} w_n N_n = N_0$. Derive Eq. (16.5.22) by maximizing the von Neumann entropy subject to these constraints.*

Exercise 16.5.7 *Derive Eq. (16.5.25) by substitution of the result, Eq. (16.5.24), for the grand canonical ensemble in the general expression for the density matrix, Eq. (16.5.4), and by using Eqs. (16.5.5, 16.5.19).*

Quantum Rate Processes

In many applications to "real-life" problems, the rates of specific elementary processes are of interest. It may be the rate of charge hopping between impurities in a disordered material or between molecules in solution, the rate of electronic energy (exciton) transfer between chromophores in biological photosynthetic systems, or the rate of light absorption/emission by a nanocrystal; rates are of the essence. And yet, the emergence of a seemingly unidirectional process associated with a single rate constant (a single characteristic timescale) is a nontrivial consequence from the rigorous perspective of quantum mechanics. Indeed, such a description involves several approximations that are valid only in the limit of weak interaction and extensive averaging, which eventually lead to Markovian-like quantum dynamics. In this chapter we draw the connection between exact quantum dynamics and rate constants. Implementations of the general formulation (e.g., of Fermi's golden rule) to specific processes will be discussed in Chapter 18, and the connection to Markovian dynamics in open quantum systems will be discussed in Chapters 19 and 20.

17.1 Transition Rates between Pure States

We start by recalling that as far as the evolution of a single nonstationary solution to the Schrödinger equation is concerned, transition probabilities are inherently time-dependent. This is trivially true for systems in which the Hamiltonian depends explicitly on time, but even when the underlying Hamiltonian is time-independent (see Section 15.1), the transition probability from an initial state, $|\chi_i\rangle$, to another specific final state, $|\chi_f\rangle$, namely $P_{i \to f}(t) = |\langle \chi_f | \psi(t) \rangle|^2 = |\langle \chi_f | \hat{U}(t,0) | \chi_i \rangle|^2$, is a superposition of oscillatory waves in time (see Eqs. (15.1.9, 15.1.14)). Consequently, any (nonzero) transition rate between $|\chi_i\rangle$, and $|\chi_f\rangle$, defined as the time-derivative of the corresponding transition probability, $k_{i \to f}(t) = \frac{\partial P_{i \to f}(t)}{\partial t}$, also is time-dependent.

We now focus on the most relevant context in which transition rates are of interest, that is, when the initial and final states are associated with some zero-order Hamiltonian, $\hat{H}_0$. In a typical scenario, the experimental characterization of the initial and final states (namely, the preparation and the measurement) is defined in the framework of $\hat{H}_0$, where the experiment selects either a pure eigenstate, or a coherent superposition of eigenstates, or a mixed state of $\hat{H}_0$ eigenstates. The system Hamiltonian includes an additional interaction term, $\hat{H}(t) = \hat{H}_0 + \hat{V}(t)$, which induces transitions of populations and/or coherences between $\hat{H}_0$ eigenstates whose rates are of interest (several examples will be encountered in the following chapters). The interaction may be local in space (as in a scattering experiment) or in time (as in a pulsed laser excitation), and it is often sufficiently weak such that the characterization of the initial and final states in terms of $\hat{H}_0$ remains relevant. Moreover, the weak interaction (in a sense that will be elaborated below) merits the use of time-dependent perturbation theory for assessing the transition rates.

The exact rate of transition from an initial coherent state, $|\chi_i\rangle$, to a final eigenstate of the zero-order Hamiltonian, $\hat{H}_0|\chi_f\rangle = \varepsilon_f|\chi_f\rangle$, reads (see Eq. (15.5.13)) $k_{i\to f}(t) = \frac{2}{\hbar} \operatorname{Im}\langle\chi_i|\hat{U}^{\dagger(I)}(t,0)|\chi_f\rangle\langle\chi_f|\hat{V}_I(t)\hat{U}^{(I)}(t,0)|\chi_i\rangle$. Here $\hat{U}^{(I)}(t,0)$ is the interaction picture propagator (Eq. (15.3.9)), and $\hat{V}_I(t) = e^{\frac{it}{\hbar}\hat{H}_0}\hat{V}(t)e^{\frac{-it}{\hbar}\hat{H}_0}$. A first-order approximation for the rate is obtained by expanding the propagator up to first order in the Dyson series, $\hat{U}^{(I)}(t,0) \cong \hat{I} + \frac{-i}{\hbar}\int_0^t dt'V_I(t')$. Restricting to final states that are orthogonal to the initial state and keeping the nonvanishing terms to lowest order in the interaction (see Ex. 17.1.1), we obtain

$$k_{i\to f}^{(1)}(t) = \frac{2}{\hbar^2} \operatorname{Re}\int_0^t dt'\langle\chi_i|V_I(t')|\chi_f\rangle\langle\chi_f|\hat{V}_I(t)|\chi_i\rangle \quad ; \quad \langle\chi_f|\chi_i\rangle = 0. \qquad (17.1.1)$$

In the case where both the initial and the final states are $\hat{H}_0$-eigenstates, namely, $\hat{H}_0|\chi_i\rangle = \varepsilon_i|\chi_i\rangle$ and $\hat{H}_0|\chi_f\rangle = \varepsilon_f|\chi_f\rangle$, the rate reads (Ex. 17.1.1)

$$k_{i\to f}^{(1)}(t) = \frac{2}{\hbar^2}\operatorname{Re}\int_0^t d\tau e^{\frac{i}{\hbar}(\varepsilon_f-\varepsilon_i)\tau}\langle\chi_i|\hat{V}(t-\tau)|\chi_f\rangle\langle\chi_f|\hat{V}(t)|\chi_i\rangle. \qquad (17.1.2)$$

Notice that this result for the first-order transition rate can be derived directly as well, by taking the time-derivative of the first-order perturbation expression for the transition probability (Eq. 15.6.10); see Ex. 17.1.2.

As we can see, the first-order transition rate, Eq. (17.1.2), is time-dependent for two reasons. First, the interaction may depend on time explicitly, which appears as $\langle\chi_f|\hat{V}(t)|\chi_i\rangle$, and second, the time integrand is a function of τ, resulting in a time-dependent integral. The latter is inherent for transitions between any two orthogonal $\hat{H}_0$-eigenstates and applies also when the interaction is constant for $\tau \geq 0$, namely, for $\hat{V}(\tau) \mapsto \hat{V}$, where

$$k_{i\to f}^{(1)}(t) = \frac{2|\langle\chi_f|\hat{V}|\chi_i\rangle|^2}{\hbar^2} \operatorname{Re}\int_0^t e^{\frac{i}{\hbar}(\varepsilon_f-\varepsilon_i)\tau}d\tau = \frac{2|\langle\chi_f|\hat{V}|\chi_i\rangle|^2}{\hbar(\varepsilon_f-\varepsilon_i)}\sin\left(\frac{(\varepsilon_f-\varepsilon_i)t}{\hbar}\right). \qquad (17.1.3)$$

The state-to-state transition rate is shown to be time periodic in this case (Ex. 17.1.3).

Exercise 17.1.1 *(a) Use the first-order Dyson expansion, $\hat{U}^{(1)}(t,0) \cong \hat{I} + \frac{-i}{\hbar}\int_0^t dt'V_I(t')$, in the exact expression for the transition rate (Eq. (15.5.13)) to obtain the first-order* *approximation for the rate, Eq. (17.1.1). (b) Show that when the initial and final states are eigenstates of $\hat{H}_0$ the first-order approximation for the rate is given by Eq. (17.1.2).*

Exercise 17.1.2   *The first-order approximation for the transition probability is given by Eq. (15.6.10). Show that the rate expression, Eq. (17.1.2), is indeed the time derivative of the transition probability.*

Exercise 17.1.3   *Derive the expression for the time-dependent transition rate, Eq. (17.1.3), from Eq. (17.1.2) for a time-independent interaction operator, $\hat{V}(\tau) \mapsto \hat{V}$.*

17.2 The Emergence of a Rate Constant and Fermi's Golden Rule

In spite of the intrinsic time-dependence of transition rates between pure states, in applications of quantum theory we often encounter the notion of time-independent "rate constants" for transitions between different states of a system. Typical examples are absorption or emission of radiation, charge and/or energy transfer between molecules or chromophores, chemical reactions, and so forth. *How do rate constants come about in quantum mechanics?* To answer this question, we first need a proper definition of rate constants in the context in which they are used, namely within kinetic equations.

Let us consider the simplest kinetic equation for a transition between two complementary states of a given system, the initial state "i" and the final state "f":

$$\begin{aligned}
\dot{P}_i(t) &= -k_{i,f}P_i(t) + k_{f,i}P_f(t), \\
\dot{P}_f(t) &= -k_{f,i}P_f(t) + k_{i,f}P_i(t)
\end{aligned} \tag{17.2.1}$$

where $P_i(t)$ and $P_f(t)$ are probabilities for populating the state "i" and the state "f" respectively, subject to probability conservation, $P_i(t) + P_f(t) = 1$. A practical way to experimentally measure the rate constant $k_{i,f}$ in Eq. (17.2.1) is to prepare the system in the initial state "i," where $P_i(0) = 1$ and $P_f(0) = 0$, and to follow the rate of change in the populations at short times where the populations have not changed much. Approximating $P_i(t) \approx 1$ and $P_f(0) \approx 0$, Eq. (17.2.1) yields the following practical definition for the rate constant:

$$k_{i,f} \equiv \dot{P}_f(t)|_{t \ll 1/k_{i,f}}. \tag{17.2.2}$$

Indeed, within the realm of the kinetic equations, Eq. (17.2.1), the time-derivative of the transition probability is effectively constant for times much shorter than $1/k_{i,f}$, and therefore it can be identified with the rate constant.

The emergence of a constant transition rate at short times is a characteristic of the kinetic equations, which seems to be in apparent contrast with the quantum mechanical state-to-state transition rate, as given in Eq. (17.1.3). Indeed, in the short time limit the latter is shown to be linear in time, $k_{i \to f}^{(1)}(t) \xrightarrow[t \to 0]{} \frac{2|\langle \chi_f|\hat{V}|\chi_i\rangle|^2}{\hbar^2}t$ (which is also true

The change from an oscillatory transition probability in a two-level system (left) into a "unidirectional" decay in the case of an ensemble of final states (right).

for the exact result (Eq. (15.6.11)). At longer times, the solution of the kinetic equations, Eq. (17.2.1), that is, $P_i(t) = P_i(0)e^{-(k_{i,f}+k_{f,i})t} + \frac{k_{f,i}}{k_{i,f}+k_{f,i}}(1 - e^{-(k_{i,f}+k_{f,i})t})$, predicts an exponential decay of the state populations, again, in contrast with the apparent time-periodic oscillations in Eq. (17.1.3). In fact, it is only in the infinite time limit, that $k_{i \to f}^{(1)}(t)$ approaches a constant, namely, $\lim_{t \to \infty} k_{i \to f}^{(1)}(t) = \frac{2\pi}{\hbar}|\langle \chi_f|\hat{V}|\chi_i\rangle|^2 \delta(\varepsilon_f - \varepsilon_i)$. However, the validity of the first-order approximation at long times ($t >> h/|\varepsilon_f - \varepsilon_i|$) is limited to, $|\langle \chi_f|\hat{V}|\chi_i\rangle| << |\varepsilon_f - \varepsilon_i|$ (see Eq. (15.6.16)), for which $\delta(\varepsilon_f - \varepsilon_i) = 0$. The conclusion is that within its regime of validity, Eq. (17.1.3) associates a constant state-to-state rate with a vanishing rate.

Indeed, *transitions between two pure orthogonal states as expressed in Eqs. (17.1.2, 17.1.3) cannot be associated with a rate constant*. This, however, does not exclude the emergence of rate constants in other cases. In fact, transitions between two pure states can be rarely addressed in actual experiments, since the preparation of the "initial state" and the measurement of the "final state" can rarely address specific eigenstates of an underlying Hamiltonian. *In the more common situation, transitions occur between mixed ensembles of quantum states*. For example, consider an electron transfer event between two impurities in a solid or between two molecules in a solution. The measurement concerns the initial and final electronic states, but the change in electronic state involves a manifold of different states which depend, for example, on the positions of atoms in the environment (phonons in the solid, or the molecular dipoles in a solvent). These additional degrees of freedom affect the transfer rate, but they are not measured directly, since only the electronic state is concerned. The effect of specific states in the environment is therefore averaged by the measurement in an incoherent fashion, which merits a mixed state description (see Chapter 16).

To see how a rate constant may come about when a transition into a mixed ensemble of states is concerned, let us analyze the rate of transition from a single eigenstate ($|\chi_i\rangle$) of $\hat{H}_0$ into a finite discrete set of eigenstates of $\hat{H}_0$, ($\{|\chi_f\rangle\}$), orthogonal to $|\chi_i\rangle$. We seek the rate of transition to the entire final ensemble, regardless of the identity of specific final states (see Fig. 17.2.1). According to the postulates of quantum mechanics, the transition probability is additive in this case, $P_{i \to \{f\}}(t) = \sum_{f \in \{f\}} P_{i \to f}(t) = \sum_{f \in \{f\}} |\langle \chi_f|\psi(t)\rangle|^2$ (see Section 11.1, postulate 4), and consequently the transition rate to the ensemble is the sum of the individual state-to-state transition rates. In this section we shall focus on interaction operators, which remain constant in time after the initial time, namely, $\hat{V}(\tau) \mapsto \hat{V}$ for $\tau \geq 0$. Extension of the conclusions to the general case of a time-dependent interaction will be addressed in Section 17.3.

In the absence of an exact solution, we invoke the general result of first-order perturbation theory, $k_{i \to f}^{(1)}(t)$, remembering to keep track of the conditions for its validity. Using Eq. (17.1.3), the rate of transition to the entire final ensemble obtains the form

$$k_{i \to \{f\}}^{(1)}(t) = \sum_{f \in \{f\}} k_{i \to f}^{(1)}(t) = \frac{2}{\hbar^2} \int_0^t \sum_{f \in \{f\}} |\langle \chi_f | \hat{V} | \chi_i \rangle|^2 \cos(\omega_{f,i}\tau) d\tau, \qquad (17.2.3)$$

where $\omega_{f,i} \equiv \frac{\varepsilon_f - \varepsilon_i}{\hbar}$. We recall that when the coupling to a specific final state is large in comparison to the corresponding energy difference, namely $|\langle \chi_f | \hat{V} | \chi_i \rangle| >> |\hbar \omega_{f,i}|$, the approximation, $k_{i \to f}^{(1)}(t)$, is valid only in the short time limit, $t \leq \hbar / |\langle \chi_f | \hat{V} | \chi_i \rangle|$ (see Eq. (15.6.16)), where the transition probability is quadratic in time (see Eq. (15.6.11)), and the transition rate is linear in time, $k_{i \to f}^{(1)}(t) \cong \frac{2}{\hbar^2} |\langle \chi_f | \hat{V}(0) | \chi_i \rangle|^2 t$. The validity of Eq. (17.2.3) is therefore limited in time by the largest of the coupling matrix elements:

$$t \leq t_{\max} = \frac{\hbar}{\max(|\langle \chi_f | \hat{V} | \chi_i \rangle|)}. \qquad (17.2.4)$$

The transition rate, $k_{i \to \{f\}}^{(1)}(t)$, is expressed as a time integral (Eq. (17.2.3)) over a sum of cosine functions, weighted by the squared coupling matrix elements, $|\langle \chi_f | \hat{V} | \chi_i \rangle|^2$. Within its validity time-window ($t \leq t_{\max}$), the contributions of "strongly interacting" states ($|\langle \chi_f | \hat{V} | \chi_i \rangle| >> |\hbar \omega_{f,i}|$) to the integrand are nearly constants (which would amount to a rate that is linear in time). However, the contributions of "weakly interacting" states ($|\langle \chi_f | \hat{V} | \chi_i \rangle| << |\hbar \omega_{f,i}|$) are oscillatory, where we recall that $k_{i \to f}^{(1)}(t)$ can be valid for a large number of oscillation periods, as long as $|\langle \chi_f | \hat{V} | \chi_i \rangle| << |\hbar \omega_{f,i}|$ (see Section 15.6).

At short times, $\tau \to 0$, all the contributions to the summation in Eq. (17.2.3) interfere constructively. At longer times, however, dephasing of the different oscillations leads to cancellation of terms with opposite signs and to a decay of the integrand toward zero, as demonstrated in Fig. 17.2.2 for a specific model. The characteristic time in which the integrand decays to zero is set roughly by the highest oscillation frequency and corresponds roughly to half of its time period,

$$t_d \approx \frac{\pi}{\max(|\omega_{f,i}|)}. \qquad (17.2.5)$$

For times much longer than the decay time, $t_d << t$, the integrand nearly vanishes, and the time integral converges approximately to a constant value as a function of time. Hence, a "rate constant" emerges in this "long time" limit:

$$k_{i \to \{f\}}^{(1)}(t)|_{t >> t_d} \equiv k_{i \to \{f\}}. \qquad (17.2.6)$$

We can attempt to make a step forward at this point, claiming that since at long times the result of the integral, Eq. (17.2.3), does not depend on time, we can replace it by its infinite time limit to obtain

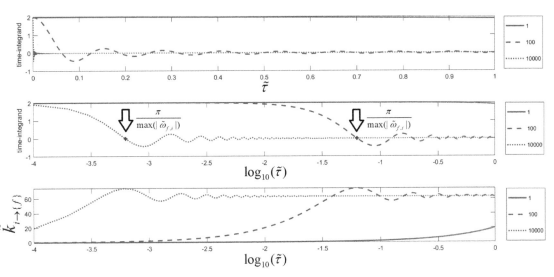

Figure 17.2.2 Numerical illustration of Eq. (17.2.3) for a concrete model. The initial state energy is set to zero, $\varepsilon_i = 0$, and the final ensemble includes N states in the energy range, $-\varepsilon_0 < \varepsilon_f < \varepsilon_0$, with a constant coupling matrix element, $|\langle \chi_f|\hat{V}|\chi_i\rangle| \equiv V$, and a constant nearest level spacing, $\Delta = 0.1\,V$. Using dimensionless parameters, $\tilde{\Delta} = \Delta/V$, $\tilde{\varepsilon}_0 = \varepsilon_0/V$, $\tilde{\omega}_{f,i} = \hbar\omega_{f,i}/V$, $\tilde{\tau} = \tau V/\hbar$, and $\tilde{k}(\tilde{\tau}) = \hbar k(t)/V$, Eq. (17.2.3) for this model obtains the form $\tilde{k}^{(1)}_{i\to\{f\}}(\tilde{t}) = 2\int_0^{\tilde{t}} \sum_{f=1}^{N} \cos(\tilde{\omega}_{f,i}\tilde{\tau})d\tilde{\tau}$. The upper graph depicts the normalized time integrand,

$\frac{2}{N}\sum_{f=1}^{N}\cos(\tilde{\omega}_{f,i}\tilde{\tau})$ as a function of $\tilde{\tau}$ in the range, $0 \leq \tilde{\tau} \leq 1$ (where perturbation theory may be valid). The integrand is shown to decay within a time period that shortens, as the bandwidth $(2\tilde{\varepsilon}_0)$ increases. (as N increases from 1 to 100 and to 10000 at a constant $\tilde{\Delta}$.) This is emphasized in the middle graph using a logarithmic timescale, where the characteristic decay times, $\pi/\max(\tilde{\omega}_{f,i}) = \pi/\tilde{\varepsilon}_0$, are marked by arrows. In the lower graph, the corresponding time integrals, $\tilde{k}^{(1)}_{i\to\{f\}}(\tilde{t})$, are plotted, demonstrating that a rate constant is obtained only as the bandwidth increases, $\tilde{\varepsilon}_0 \gg 1$. Notice that the rate constant converges with increasing bandwidth, $\tilde{k}^{(1)}_{i\to\{f\}}(\tilde{t}) \xrightarrow{N\to\infty} 2\pi/\tilde{\Delta}$, and hence, $k^{(1)}_{i\to\{f\}} \xrightarrow{N\to\infty} 2\pi V^2/(\hbar\Delta)$. This coincides with the result for a continuous spectrum (see Ex. 17.2.4).

$$k_{i\to\{f\}} \cong \frac{2}{\hbar^2}\int_0^{\infty} d\tau \sum_{f\in\{f\}} |\langle\chi_f|\hat{V}|\chi_i\rangle|^2 \cos(\omega_{f,i}\tau)$$

$$= \frac{1}{\hbar^2}\,\mathrm{Re}\int_{-\infty}^{\infty} d\tau \sum_{f\in\{f\}} |\langle\chi_f|\hat{V}|\chi_i\rangle|^2 e^{-i\omega_{f,i}\tau},$$

$$= \frac{2\pi}{\hbar}\sum_{f\in\{f\}} |\langle\chi_f|\hat{V}|\chi_i\rangle|^2 \delta(\varepsilon_i - \varepsilon_f) \tag{17.2.7}$$

where in the last step we used the representation of Dirac's delta, $2\pi\hbar\delta(\varepsilon_i - \varepsilon_f) = \int_{-\infty}^{\infty} \exp\left(\frac{-i(\varepsilon_f - \varepsilon_i)\tau}{\hbar}\right)d\tau$. This formal expression for the rate is referred to as "Fermi's golden rule." Dirac's delta attributes the transition to one of the final states, whose energy strictly matches the initial state energy. Recalling, however, that the infinite time limit is merely an idealization, where in fact there are restrictions on the times in which Eq. (17.2.3) is valid, as discussed in some detail in what follows, the Dirac delta should be thought of as an idealized limit of a broader distribution over the different final states. Nevertheless, Fermi's golden rule emphasizes that while at short times the initial state population is transferred to the entire ensemble of final states, on a longer timescale, where a constant rate emerges, the transition is restricted by energy conservation to final states whose energy matches the initial state energy. (This will be elaborated in the context of the Markovian approximation, to be discussed in Chapter 19; see, e.g., Ex. 19.2.8.) In what follows we discuss the limitations on the rate calculation by Eq. (17.2.3) and express these limitations in terms of restrictions on the parameters of the ensemble of final states, for which Fermi's golden rule is applicable.

First, we recall that by its definition, a rate constant is equal to the rate of probability transfer only in a short time limit, in which the initial probability hardly changes ($t << 1/k_{i \to \{f\}}$, Eq. (17.2.2)). To comply with the result that $k_{i \to \{f\}}^{(1)}(t)$ reaches a constant value only in the long time limit, $t_d << t$, a timescale separation is needed: Fast relaxation of the oscillations within $\sim t_d$ owing to multiple final states, followed by slow kinetics. The calculation of the "rate constant" by Eq. (17.2.3) is then relevant only within a given time window, $t_d << t << 1/k_{i \to \{f\}}$. For this time window to exist, we must have $t_d << 1/k_{i \to \{f\}}$, namely

$$t_d = \frac{\pi}{\max(|\omega_{f,i}|)} << \frac{1}{k_{i \to \{f\}}}. \tag{17.2.8}$$

This means that the rate of population transfer should be much smaller than the maximal transition frequency into the manifold of final states, $k_{i \to \{f\}} << \max|\omega_{f,i}|/\pi$. This condition on the final ensemble is often referred to as a *"wide-band" condition. It means that the maximal energy gap from the initial to a final state*, $\max(\varepsilon_f - \varepsilon_i) = \max(\hbar\omega_{f,i})$, *must be significantly larger than* $\hbar/k_{i \to \{f\}}$. As we shall discuss in Chapter 19, this wide-band condition also marks the emergence of Markovian dynamics within quantum mechanics.

Another time limit that restricts the validity of Fermi's golden rule is related to its derivation within first-order perturbation theory. Indeed, Eq. (17.2.4) sets an upper bound on the validity of Eq. (17.2.3). To converge to a rate constant within this validity window, we must have $t_d << t_{max}$, which means

$$t_d = \frac{\pi}{\max(|\omega_{f,i}|)} << \frac{\hbar}{\max(|\langle \chi_f | \hat{V} | \chi_i \rangle|)}. \tag{17.2.9}$$

This additional limitation restricts the coupling matrix elements between the initial state and each of the states within the final ensemble. The condition is often referred to as *the weak coupling limit, which means that the coupling matrix elements must be small in comparison to the energy bandwidth of the manifold of final states.*

Finally, there is another limit that may restrict the applicability of Eq. (17.2.3) to long times. The summation over cosine functions may revive its initial value (in full, or in part) owing to rephasing after the time period of the slowest oscillation, namely $t_r \approx \frac{2\pi}{\min(|\omega_{f,i}|)}$. The precise conditions for the emergence of revivals depend on the specifics of the frequency spectrum and will not be addressed here. Nevertheless, the integral over the sum of cosines can generally be approximated as a constant only up to some time, t_r. Notice that this is not a limitation if $\min(|\omega_{f,i}|) \to 0$ and $t_r \to \infty$. **This condition is readily fulfilled if the band of final state energies is continuous, and if the initial state energy is included within this band.** Nevertheless, for a discrete ensemble of final states, a finite t_r may restrict the applicability of Eq. (17.2.3). (Notice that this restriction is irrelevant when $t_r >> \min(1/k_{i \to \{f\}}, t_{max})$, namely, when $\min(|\omega_{f,i}|) << \max(\{|\langle \chi_f | \hat{V} | \chi_i \rangle| / \hbar\}, k_{i \to \{f\}})$, which means that ε_i coincides with $\{\varepsilon_f\}$ within a finite tolerance.)

We now consider explicitly a dense ensemble of final states, associated with a continuous spectrum of $\hat{H}_0$. Indeed, as discussed in the next chapter, "real-life" applications of Fermi's golden rule typically involve ensembles with "macroscopic" numbers of bound degrees of freedom, and/or open systems associated with continuous spectra. Formally, the discrete summation over the coupling matrix elements to the final states is replaced by an integral over a (positive) coupling function, $\lambda_{i,\{f\}}^2(\varepsilon_f) \equiv |\langle \chi_f(\varepsilon_f) | \hat{V} | \chi_i \rangle|^2$, which is a continuous function of the energy, multiplied by the energy-dependent density of final states, $\rho_{\{f\}}(\varepsilon) \equiv \frac{dn_{\{f\}}(\varepsilon)}{d\varepsilon}$ (see Eq. (5.5.8)). Defining a spectral density function,

$$J_{i,\{f\}}(\varepsilon) \equiv 2\pi \lambda_{i,\{f\}}^2(\varepsilon) \rho_{\{f\}}(\varepsilon), \qquad (17.2.10)$$

Eq. (17.2.3) for the transition rate obtains the form

$$k_{i \to \{f\}}^{(1)}(t) = \frac{1}{\pi \hbar^2} \int_0^t d\tau \int d\varepsilon J_{i,\{f\}}(\varepsilon) \cos((\varepsilon - \varepsilon_i)\tau / \hbar). \qquad (17.2.11)$$

The properties of the coupling between the initial state and the final ensemble are therefore fully encoded in the "spectral density" function. In particular, the time integrand in the rate expression, $\mathrm{Re}\, e^{-i\varepsilon_i \tau / \hbar} \int d\varepsilon J_{i,\{f\}}(\varepsilon) e^{i\varepsilon \tau / \hbar}$, can be identified with the Fourier transform of the spectral density. Within the limitations of perturbation theory (namely when the interaction is sufficiently weak, Eq. (17.2.9)), and when the spectral density is a sufficiently smooth and broad function of the energy (a wide band of final states, Eq. (17.2.8)), the time integrand decays sufficiently fast on the kinetics timescale and the upper limit of the time integral can be taken to infinity, leading to (Ex. 17.2.1):

$$k_{i \to \{f\}} \cong \lim_{t \to \infty} k_{i \to \{f\}}^{(1)}(t) = \frac{1}{\hbar} J_{i,\{f\}}(\varepsilon_i) = \frac{2\pi}{\hbar} \lambda_{i,\{f\}}^2(\varepsilon_i) \rho_{\{f\}}(\varepsilon_i). \qquad (17.2.12)$$

This is a central result, as it expresses the rate constant, when it exists, in terms of the properties of the final ensemble to which the initial state is coupled. In particular, **the rate increases linearly with the density of states and with the squared coupling matrix elements to the final states that match the initial state energy.** The latter can often be estimated to a good precision or obtained directly from experiments. In Chapter 18 we shall review some applications of Fermi's golden rule, demonstrating its usefulness and widespread usage in the context of nanoscale phenomena.

Notice that the result for a discrete ensemble, Eq. (17.2.7), is readily regained from Eq. (17.2.12) in the limit where the density of final states reveals the underlying discrete level structure. Setting the density of states to $\rho_{\{f\}}(\varepsilon) = \sum_f \delta(\varepsilon - \varepsilon_f)$, the spectral density reads $J_{i,\{f\}}(\varepsilon) \equiv \sum_f 2\pi\lambda_{i,\{f\}}^2(\varepsilon_f)\delta(\varepsilon - \varepsilon_f)$, which yields for the overall rate (Ex. 17.2.2)

$$k_{i \to \{f\}} = \frac{2\pi}{\hbar} \sum_f \lambda_{i,\{f\}}^2(\varepsilon_f)\delta(\varepsilon_i - \varepsilon_f). \tag{17.2.13}$$

Exercise 17.2.1 *One of the representations of Dirac's delta reads* $\delta(\varepsilon_i - \varepsilon_f) = \frac{1}{2\pi\hbar} \int\limits_{-\infty}^{\infty} d\tau \exp\left(\frac{-i(\varepsilon_f - \varepsilon_i)\tau}{\hbar}\right) d\tau$. *Use this representation to calculate the infinite time limit of the first-order rate, Eq. (17.2.11), as given in Eq. (17.2.12) (notice that the integrand in Eq. (17.2.11) is an even function of time).*

Exercise 17.2.2 *The spectral density for a discretely resolved density of states reads* $J_{i,\{f\}}(\varepsilon) \equiv \sum_f 2\pi\lambda_{i,\{f\}}^2(\varepsilon_f)\delta(\varepsilon - \varepsilon_f)$. *(a) Show that Eq. (17.2.13) is obtained directly from Eq. (17.2.12) in this case. (b) Derive Eq. (17.2.13) by substituting the given spectral density in Eq. (17.2.11) and taking the infinite time limit of the integral.*

Now let us consider an idealized model in which the spectral density is continuous and uniform over a finite energy band:

$$J_{i,\{f\}}(\varepsilon) = \begin{cases} J_0 & ; \quad |\varepsilon| \le \varepsilon_0. \\ 0 & ; \quad |\varepsilon| > \varepsilon_0 \end{cases} \tag{17.2.14}$$

Assuming the conditions for validity of the golden rule formula hold, Eq. (17.2.12) immediately yields in this case

$$k_{i \to \{f\}} = \begin{cases} J_0/\hbar & ; \quad \varepsilon_i \le \varepsilon_0, \\ 0 & ; \quad \varepsilon_i > \varepsilon_0 \end{cases} \tag{17.2.15}$$

which means that the decay rate is constant if the initial state energy overlaps with the energy band of the final states, and vanishes otherwise.

It is instructive (and straightforward, in this case) to return to the formulation of the rate as a time integral (Eq. (17.2.11)) and to inspect the validity of the infinite time limit leading to Eq. (17.2.12). Substitution of Eq. (17.2.14) in Eq. (17.2.11) yields

$$k_{i \to \{f\}} = \frac{J_0}{\pi\hbar^2} \int\limits_0^t d\tau \int\limits_{-\varepsilon_0}^{\varepsilon_0} d\varepsilon \cos((\varepsilon - \varepsilon_i)\tau/\hbar). \tag{17.2.16}$$

Performing first the energy integral, we obtain (Ex. 17.2.3)

$$k_{i \to \{f\}} = \frac{2J_0\varepsilon_0}{\pi\hbar^2} \int\limits_0^t d\tau \cos(\varepsilon_i\tau/\hbar)\frac{\sin(\varepsilon_0\tau/\hbar)}{(\varepsilon_0\tau/\hbar)}. \tag{17.2.17}$$

As we can see, the time integrand is a decaying function (sinc $(\varepsilon_0\tau/\hbar) \xrightarrow[\varepsilon_0\tau/\hbar \to \infty]{} 0$), multiplied by an oscillatory function of time, whose frequency is set by the initial state energy, ε_i. If the initial state energy is well outside the band of final states, $|\varepsilon_i| >> |\varepsilon_0|$, the oscillations are rapid on the decay timescale, which will lead to a vanishingly small rate constant. If, however, the initial state energy is at the center of the band, $|\varepsilon_i| << |\varepsilon_0|$, the oscillations are suppressed, and the rate is maximized. Let us analyze this regime by setting $\varepsilon_i = 0$, in which case the time integrand is just the sinc function. The first zero of sinc $(\varepsilon_0\tau/\hbar)$ is reached at time $\tau = \pi\hbar/\varepsilon_0$, which coincides with half a time period of the largest frequency in this model, namely $\max(\omega_{f,i}) = \varepsilon_0/\hbar$. This decay time is in line with our qualitative discussion of the decay of a discrete sum of cosine functions (see Eqs. (17.2.3, 17.2.5)), and with the numerical results demonstrated in Fig. 17.2.2. As the bandwidth of final state energy (namely, ε_0) increases, the decay time shortens. If the bandwidth becomes infinite ("the wide band limit"), the decaying function approaches $\delta(\tau)$, and therefore the integral converges to $\frac{J_0}{\hbar}$, namely to the golden rule result, Eqs. (17.2.12, 17.2.14), already at finite times (see Ex. 17.2.4). Notice that when bandwidth of the final ensemble is "infinite," the other conditions for the validity of the golden rule expression (Eqs. (17.2.8, 17.2.9)) are guaranteed to hold, namely $\max(|\omega_{f,i}|) >> \max(\{|\langle\chi_f|\hat{V}|\chi_i\rangle|/\hbar\}, k_{i\rightarrow\{f\}})$.

Exercise 17.2.3   *The energy integral in Eq. (17.2.11) is related to the Fourier transform of the spectral density, $\int d\varepsilon J_{i,\{f\}}(\varepsilon)\cos((\varepsilon - \varepsilon_i)\tau/\hbar) = \mathrm{Re}\, e^{-i\varepsilon_i\tau/\hbar}\int d\varepsilon J_{i,\{f\}}(\varepsilon)e^{i\varepsilon\tau/\hbar}$. Show that for the "square window" spectral density function, given in Eq. (17.2.14), the energy integral reads $2J_0\varepsilon_0\mathrm{Re}\, e^{\frac{-i\varepsilon_i\tau}{\hbar}}\frac{\sin(\varepsilon_0\tau/\hbar)}{\varepsilon_0\tau/\hbar}$.*

Exercise 17.2.4   *(a) One of the representations of Dirac's delta reads $\delta(x) = \lim_{\alpha\rightarrow\infty}\frac{\sin(\alpha x)}{\pi x}$. Use it to show that the infinite bandwidth limit of the transition rate in Eq. (17.2.17) is $k_{i\rightarrow\{f\}} = \frac{J_0}{\hbar}$. (b) In Fig. 17.2.2 numerical results are presented for a discrete model, in which the initial state energy is set to zero, $\varepsilon_i = 0$, and the final ensemble includes N states in the energy range, $-\varepsilon_0 < \varepsilon_f < \varepsilon_0$, with a constant coupling matrix element, $|\langle\chi_f|\hat{V}|\chi_i\rangle| \equiv V$ and a constant nearest level spacing, Δ. As N increases (at a constant level spacing), the rates calculated by Eq. (17.2.3) are shown to converge to $k_{i\rightarrow\{f\}}^{(1)}\xrightarrow[N\rightarrow\infty]{}2\pi V^2/(\hbar\Delta)$. Obtain this result analytically by replacing the discrete summation over final states by an integral with a constant density of states, $\rho = 1/\Delta$. Show that the discrete model coincides with the result of the continuous model (a) by identifying the spectral density, $J_0 = 2\pi V^2/\Delta$.*

We can therefore conclude that a rate constant for transition from an initial to a final state does emerge in quantum mechanics, as long as the final state is not a finely resolved pure state, but rather an ensemble of states that is dense and wide in energy (see Fig. 17.2.1). Moreover, the rate constant does not appear instantaneously, but only after a transient time of relaxation of the quantum oscillations between the initial state and individual final states at different energies. When the bandwidth of the final state energies is sufficiently large, this fast relaxation precedes population transfer between the initial and the final states, at a constant rate. The preceding derivation of the rate expression was based on perturbation theory, which restricted its applicability to the limit of weak coupling between the initial and final states, and to short times on the population transfer timescale, where transfer rate can be identified with the "rate constant." In Chapter 19 we shall follow an alternative derivation leading to the emergence of rate constants over the entire kinetics timescale, including the infinite time limit of relaxation to equilibrium. That derivation is based on a Markovian approximation, which is closely related to the "weak coupling" and "wide band" limits encountered in this chapter.

17.3 Thermal Rate Constants

In Section 17.2 we analyzed the rate of transition from a single pure state, typically an eigenstate of some meaningful zero-order Hamiltonian, into a group of other (final) eigenstates of the same Hamiltonian. Here we emphasize that in many cases of interest, the experimental preparation of an initial state does not select a single eigenstate, but rather a group of eigenstates. In some cases, the experiment prepares a coherent superposition of states of a relevant zero-order Hamiltonian, resulting in dynamics of both populations and coherences (see Chapter 16). In the most common case, however, the initial state corresponds to an incoherent (mixed) ensemble, in which the relative weights of different eigenstates are associated with a "quasi-equilibrium" state characterized by macroscopic parameters such as temperature, chemical potential, and so forth.

In this section we focus on the formulation of "thermal rates," where the initial ensemble corresponds to a quasi-equilibrium state, associated with a zero-order Hamiltonian, $\hat{H}_0$, and a temperature, T (see Eq. (16.5.18)). Particularly, we are interested in the rate of transition between two subsets of (orthonormal) $\hat{H}_0$-eigenstates, $\{i\} \equiv \{|\chi_i\rangle\}$ and $\{f\} \equiv \{|\chi_f\rangle\}$, where $\hat{H}_0|\chi_i\rangle = \varepsilon_i|\chi_i\rangle$ and, $\hat{H}_0|\chi_f\rangle = \varepsilon_f|\chi_f\rangle$. Without loss of generality, we shall assume now that every transition from $\{i\}$ is only to $\{f\}$ and vice versa, either since these sets complete each other to a unity, or since the interaction operator, $\hat{V}$, couples $\{i\}$ only to the states in $\{f\}$. The two sets are associated with the corresponding orthogonal projection operators,

$$\hat{P}_{\{i\}} = \sum_{i \in \{i\}} |\chi_i\rangle\langle\chi_i| \quad ; \quad \hat{P}_{\{f\}} = \sum_{f \in \{f\}} |\chi_f\rangle\langle\chi_f|, \tag{17.3.1}$$

where

$$\hat{P}_{\{i\}}^2 = \hat{P}_{\{i\}} \quad ; \quad \hat{P}_{\{i\}}\hat{P}_{\{f\}} = 0 \quad ; \quad \hat{P}_{\{f\}}^2 = \hat{P}_{\{f\}} \tag{17.3.2}$$

and

$$[\hat{P}_{\{i\}}, \hat{H}_0] = [\hat{P}_{\{f\}}, \hat{H}_0] = 0. \tag{17.3.3}$$

Denoting the probability of populating the two ensembles at time t as $P_{\{i\}}(t)$ and $P_{\{f\}}(t)$, we chose an initial state in which only the ensemble $\{i\}$ is populated, namely

$$P_{\{i\}}(0) = 1 \quad ; \quad P_{\{f\}}(0) = 0. \tag{17.3.4}$$

The definition of a "thermal" rate constant depends on the assumption that quasi-equilibrium is maintained within the initial ensemble. This is justified when the relaxation to equilibrium within the ensemble induced by the environment is fast on the kinetics timescale, and when the ensemble is sufficiently large, such that the statistical weight of nonequilibrium populations (fluctuations) is negligibly small. Without loss of generality (see Chapter 20), we restrict ourselves here to a canonical initial ensemble. The relative populations of the different $\hat{H}_0$-eigenstates corresponds to a density operator, $\hat{\rho}_0 = e^{-\hat{H}_0/(k_B T)}/Z_0$, where $Z_0 = tr\{e^{-\hat{H}_0/(k_B T)}\}$ (see Eq. (16.5.18)). The relative population of each specific state within the initial ensemble therefore reads

$$\frac{P_i(0)}{P_{\{i\}}(0)} = \frac{e^{-\varepsilon_i/(k_B T)}}{Z_{\{i\}}} \quad ; \quad Z_{\{i\}} \equiv \sum_{i \in \{i\}} e^{-\varepsilon_i/(k_B T)}, \tag{17.3.5}$$

where $Z_{\{i\}}$ is the partition function corresponding to the sub-ensemble of initial states. For the selected initial condition (Eq. (17.3.4)), the initial state populations are given as

$$P_i(0) = \frac{e^{-\varepsilon_i/(k_B T)}}{Z_{\{i\}}}. \tag{17.3.6}$$

In the short time limit, in which the initial populations do not change much, the transition rate from $\{i\}$ to $\{f\}$ is expressed in terms of a weighted average over state-specific rates:

$$k_{\{i\} \to \{f\}}(t) \cong \sum_{i \in \{i\}} P_i(0) \sum_{f \in \{f\}} \frac{\partial}{\partial t} P_{i \to f}(t) = \sum_{i \in \{i\}} P_i(0) \sum_{f \in \{f\}} k_{i \to f}(t). \tag{17.3.7}$$

These rates can be formulated exactly (Ex. 17.3.1) in terms of the time-evolution operator with the full Hamiltonian, $\hat{U}(t,0)$:

$$k_{i \to f}(t) = \frac{d}{dt} tr\{|\chi_f\rangle\langle\chi_f|\hat{U}(t,0)|\chi_i\rangle\langle\chi_i|\hat{U}^\dagger(t,0)\}. \tag{17.3.8}$$

Substituting Eqs. (17.3.6, 17.3.8) in Eq. (17.3.7), we obtain formally exact expressions for the initial probability transfer rate between the ensembles (Ex. 17.3.2),

$$\begin{aligned} k_{\{i\} \to \{f\}}(t) &= \frac{d}{dt} tr\{\hat{P}_{\{f\}}\hat{U}(t,0)\hat{\rho}_{\{i\}}(0)\hat{U}^\dagger(t,0)\} \\ &= \frac{d}{dt} tr\{\hat{P}_{\{f\}}\hat{\rho}_{\{i\}}(t)\}, \\ &= tr\left\{\frac{i}{\hbar}[\hat{H}, \hat{P}_H^{\{f\}}(t)]\hat{\rho}_{\{i\}}(0)\right\} \end{aligned} \tag{17.3.9}$$

where $\hat{\rho}_{\{i\}}(0)$ is the thermal density operator, projected onto the initial state,

$$\hat{\rho}_{\{i\}}(0) \equiv \frac{e^{-\hat{H}_0/(k_B T)}\hat{P}_{\{i\}}}{tr\{e^{-\hat{H}_0/(k_B T)}P_{\{i\}}\}}, \tag{17.3.10}$$

and $\hat{P}_H^{\{f\}}(t)$ is the Heisenberg picture representation of $\hat{P}_{\{f\}}$ (Eq. (15.3.16)).

Exercise 17.3.1   *Use Eq. (15.5.8) for the exact state-to-state transition rate, $k_{i \to f}(t)$, and derive Eq. (17.3.8).*

Exercise 17.3.2 *(a) Substitute Eqs. (17.3.6, 17.3.8) in Eq. (17.3.7), and use the definition of the projection operators to the initial and final ensembles, $\hat{P}_{\{i\}}$ and $\hat{P}_{\{f\}}$ (use Eqs. (17.3.1–17.3.3), and recall that $\sum_{i \in \{i\}} \langle \chi_i | \hat{A} | \chi_i \rangle = tr\{\hat{A}\hat{P}_{\{i\}}\}$) to derive the result for the transition rate between the two ensembles, $k_{\{i\} \to \{f\}}(t) = \frac{d}{dt} tr\{\hat{P}_{\{f\}} \hat{U}(t,0) \hat{\rho}_{\{i\}}(0) \hat{U}^\dagger(t,0)\}$. (b) Use this result and the definition of the time-dependent density operator (Ex. 16.4.2) to show that $k_{\{i\} \to \{f\}}(t) = \frac{d}{dt} tr\{\hat{P}_{\{f\}} \hat{\rho}_{\{i\}}(t)\}$. (c) Use the result of (a) and the definition of the Heisenberg picture representation of $\hat{P}_{\{f\}}$ (Eq. (15.3.16)) to show that $k_{\{i\} \to \{f\}}(t) = tr\{\frac{i}{\hbar}[\hat{H}, \hat{P}_H^{\{f\}}(t)]\hat{\rho}_{\{i\}}(0)\}$.*

Exact expressions for thermal rates are often used for the evaluation of chemical reaction rates or transport rates in condensed phases [17.1] [17.2]. Nevertheless, in many cases the description of the initial state in terms of an equilibrium density operator, attributed to a zero-order Hamiltonian, is experimentally relevant only when the interaction (that induces population transfer between $\hat{H}_0$-eigenstates) is sufficiently weak. Since this is a common case indeed, approximated expressions for thermal rates, based on perturbation theory, are also valid and commonly implemented.

Here we generalize the perturbative result obtained for the transition rate from a pure state to the case of an initial thermal ensemble. Starting from Eq. (17.3.7) for the "short-time" transition rate between the two ensembles, we replace the exact state-to-state transition rate by its first-order approximation,

$$k_{\{i\} \to \{f\}}(t) \cong \sum_{i \in \{i\}} P_i(0) k_{i \to \{f\}}^{(1)}(t). \tag{17.3.11}$$

Using Eq. (17.2.3) for $k_{i \to \{f\}}^{(1)}(t)$, and Eq. (17.3.6) for the thermal weights ($\{P_i(0)\}$), we obtain the first-order approximation for the thermal rate (see Ex. 17.3.3):

$$k_{\{i\} \to \{f\}}^{(1)}(t) \cong \frac{2}{\hbar^2} \, \text{Re} \int_0^t tr \left\{ \hat{\rho}_{\{i\}}(0) \hat{P}_{\{i\}} \hat{V} \hat{P}_{\{f\}} e^{\frac{i\hat{H}_0 \tau}{\hbar}} \hat{V} e^{\frac{-i\hat{H}_0 \tau}{\hbar}} \right\} d\tau. \tag{17.3.12}$$

Defining the projected interaction operator, $\hat{V}_{\{i\},\{f\}} \equiv \hat{P}_{\{f\}} \hat{V} \hat{P}_{\{i\}}$, the result can be compactly expressed as an integral over the interaction correlation function (Ex. 17.3.4),

$$k_{\{i\} \to \{f\}}^{(1)}(t) = \frac{2}{\hbar^2} \, \text{Re} \int_0^t tr \left\{ \hat{\rho}_{\{i\}}(0) \left[\hat{V}_{\{i\},\{f\}}^\dagger(0) \right]_I \left[\hat{V}_{\{i\},\{f\}}(\tau) \right]_I \right\} d\tau, \tag{17.3.13}$$

where $\left[\hat{V}_{\{i\},\{f\}}(\tau) \right]_I = e^{\frac{i\hat{H}_0 \tau}{\hbar}} \hat{V}_{\{i\},\{f\}} e^{\frac{-i\hat{H}_0 \tau}{\hbar}}$.

Exercise 17.3.3 *Use Eq. (17.2.3) for $k_{i \to \{f\}}^{(1)}(t)$, and Eq. (17.3.6) for the thermal weights ($\{P_i(0)\}$), to derive Eq. (17.3.12).*

Exercise 17.3.4 *Use the definition $\hat{V}_{\{i\},\{f\}} \equiv \hat{P}_{\{f\}} \hat{V} \hat{P}_{\{i\}}$ and the properties of the projection operators, $\hat{P}_{\{i\}}$ and $\hat{P}_{\{f\}}$ (Eqs. (17.3.1–17.3.3)), to derive Eq. (17.3.13) from Eq. (17.3.12).*

The convergence of the time integral depends on the cumulative convergence of the state-to-state rates, $\{k_{i\to\{f\}}^{(1)}(t)\}$, associated with the populated states in the initial ensemble. When the conditions leading to Fermi's golden rule expression (see Section 17.2) apply to each of these rates, the upper time limit of the integral in Eq. (17.3.12) can be taken to infinity. The thermal rate constant can be readily obtained by calculating the time integral,

$$k_{\{i\}\to\{f\}}(T) = \sum_{i\in\{i\}} \frac{e^{-\varepsilon_i/(k_BT)}}{Z_{\{i\}}} k_{i\to\{f\}} = \frac{2\pi}{\hbar} \sum_{i\in\{i\},f\in\{f\}} \frac{e^{-\varepsilon_i/(k_BT)}}{Z_{\{i\}}} |\langle\chi_f|\hat{V}|\chi_i\rangle|^2 \delta(\varepsilon_f - \varepsilon_i).$$

$$(17.3.14)$$

The result is equivalent to substitution of the infinite time limit of $k_{i\to\{f\}}^{(1)}(t)$ (see Eq. (17.2.7)) in Eq. (17.3.11), and using Eq. (17.3.6) for the thermal weights. *The rate constant to transfer from any state in the initial ensemble $\{|\chi_i\rangle\}$ to any state in the final ensemble is therefore given by the state-to-state rates, summed over the final states, and averaged over the thermal distribution within the initial ensemble.*

Golden rule expressions for thermal rates are at the center of numerous theories of quantum transport. A few important examples include the Marcus formula for charge transfer in molecular systems, the Förster formula for electronic energy transfer between chromophores, the theory of linear absorption and emission spectroscopies, and Redfield's theory for relaxation in open quantum systems. These will be addressed in some detail in the following chapters.

Notice that a thermal rate constant may also apply even when the interaction operator depends explicitly on time. To see this, we return to the most general definition of the first-order approximation for a state-to-state transition rate, Eq. (17.1.1). Using this expression and Eq. (17.3.6) for the thermal weights in Eq. (17.3.11), Eq. (17.3.13) generalizes to (Ex. 17.3.5)

$$k_{\{i\}\to\{f\}}^{(1)} = \lim_{t\to\infty} \frac{2}{\hbar^2} \text{Re} \int_0^t tr\left\{\hat{\rho}_{\{i\}}(0)\left[\hat{V}_{\{i\},\{f\}}^\dagger(t')\right]_I, \left[\hat{V}_{\{i\},\{f\}}(t)\right]_I\right\} dt', \quad (17.3.15)$$

where $[\hat{V}_{\{i\},\{f\}}(t)]_I = e^{\frac{i\hat{H}_0t}{\hbar}} \hat{V}_{\{i\},\{f\}}(t) e^{\frac{-i\hat{H}_0t}{\hbar}} = e^{\frac{i\hat{H}_0t}{\hbar}} \hat{P}_{\{f\}}\hat{V}(t)\hat{P}_{\{i\}} e^{\frac{-i\hat{H}_0t}{\hbar}}$. Whether the infinite time limit exists in this case depends on the system; but in many cases, the explicit time-dependence of the interaction does not contradict a convergence of the time integral (although Eq. (17.3.13) does not apply). In the next chapter we shall deal explicitly with time-dependent interactions when discussing the interaction of a molecule with electromagnetic radiation.

Exercise 17.3.5 *Use Eq. (17.1.1) for $k_{i\to f}^{(1)}(t)$, and Eq. (17.3.6) for the thermal weights ($\{P_i(0)\}$), to derive Eq. (17.3.15) from Eq. (17.3.11).*

We now turn to consider a typical case in which both the initial and the final ensembles of $\hat{H}_0$-eigenstates are associated with continuous energy spectra. Formally, the discrete summation over the initial and final states in Eq. (17.3.14) are is replaced by integrals over a (positive) continuous function of the initial and final energies, $\lambda_{\{i\},\{f\}}^2(\varepsilon_f, \varepsilon_i) \equiv |\langle\chi_f(\varepsilon_f)|\hat{V}|\chi_i(\varepsilon_i)\rangle|^2$, multiplied by the densities of the initial and final states, $\rho_{\{i\}}(\varepsilon) \equiv \frac{dn_{\{i\}}(\varepsilon)}{d\varepsilon}$ and $\rho_{\{f\}}(\varepsilon) \equiv \frac{dn_{\{f\}}(\varepsilon)}{d\varepsilon}$. Eq. (17.3.14) for the thermal rate therefore obtains the form

$$k_{\{i\}\to\{f\}}(T) = \frac{2\pi}{\hbar} \int d\varepsilon_i \int d\varepsilon_f \frac{e^{-\varepsilon_i/(k_B T)}}{Z_{\{i\}}} \lambda^2_{\{i\},\{f\}}(\varepsilon_f, \varepsilon_i) \rho_{\{i\}}(\varepsilon_i) \rho_{\{f\}}(\varepsilon_f) \delta(\varepsilon_f - \varepsilon_i).$$

(17.3.16)

Using the definition of the ensemble spectral density (Eq. (17.2.10)) and noticing that, by definition, $\lambda^2_{i,\{f\}}(\varepsilon_f) = \lambda^2_{\{i\},\{f\}}(\varepsilon_f, \varepsilon_i)$, the last equation can be rewritten as

$$k_{\{i\}\to\{f\}}(T) = \frac{1}{\hbar} \int d\varepsilon_i \frac{e^{-\varepsilon_i/(k_B T)}}{Z_{\{i\}}} J_{i,\{f\}}(\varepsilon_i) \rho_{\{i\}}(\varepsilon_i).$$

(17.3.17)

The thermal rate is shown to be a Boltzmann-weighted spectral "overlap integral" between the initial ensemble density of states and the spectral density of the final ensemble.

Finally, we consider the relation between the forward and backward rates for transitions between two thermal ensembles. The same arguments applied in deriving $k_{\{i\}\to\{f\}}(T)$ are applicable also for evaluating the rate of a transition from the ensemble $\{f\}$, initially populated at thermal quasi-equilibrium, into the ensemble $\{i\}$. The result readily follows:

$$k_{\{f\}\to\{i\}}(T) = \frac{2\pi}{\hbar} \int d\varepsilon_f \int d\varepsilon_i \frac{e^{-\varepsilon_f/(k_B T)}}{Z_{\{f\}}} \lambda^2_{\{f\},\{i\}}(\varepsilon_i, \varepsilon_f) \rho_{\{f\}}(\varepsilon_f) \rho_{\{i\}}(\varepsilon_i) \delta(\varepsilon_i - \varepsilon_f).$$

(17.3.18)

Using the symmetry of the coupling function (derived from the Hermiticity of the interaction),

$$\lambda^2_{\{f\},\{i\}}(\varepsilon_i, \varepsilon_f) = |\langle \chi_i(\varepsilon_i)|\hat{V}|\chi_f(\varepsilon_f)\rangle|^2 = |\langle \chi_f(\varepsilon_f)|\hat{V}|\chi_i(\varepsilon_i)\rangle|^2 = \lambda^2_{\{i\},\{f\}}(\varepsilon_f, \varepsilon_i), \quad (17.3.19)$$

we readily obtain (Ex. 17.3.6)

$$\frac{k_{\{i\}\to\{f\}}(T)}{k_{\{f\}\to\{i\}}(T)} = \frac{Z_{\{f\}}}{Z_{\{i\}}}.$$

(17.3.20)

Notice that at thermal equilibrium in the entire system associated with the zero-order Hamiltonian, the state populations within both $\{i\}$ and $\{f\}$ must be Boltzmann-distributed (see Eq. (16.5.17)). This means that the ratio between the two ensemble populations equals the ratio of their reduced partition functions (see Eq. (17.3.5)), namely $P^{(eq)}_{\{f\}}/P^{(eq)}_{\{i\}} = Z_{\{f\}}/Z_{\{i\}}$. Using Eq. (17.3.20), this means that

$$P^{(eq)}_{\{i\}} k_{\{i\}\to\{f\}}(T) = P^{(eq)}_{\{f\}} k_{\{f\}\to\{i\}}(T).$$

(17.3.21)

The result can be readily identified as the infinite time limit of the kinetic equations (Eq. (17.2.1); see Ex. 17.3.7), in which the rate constants correspond to $k_{\{i\}\to\{f\}}(T)$ and $k_{\{f\}\to\{i\}}(T)$. It means that at equilibrium the forward $(P^{(eq)}_{\{i\}} k_{\{i\}\to\{f\}}(T))$ and the backward $(P^{(eq)}_{\{f\}} k_{\{f\}\to\{i\}}(T))$ rates of population transfer between the two ensembles must be equal (the "detailed balance" condition). We may conclude that Fermi's golden rule rates describe the kinetic evolution of the ensemble all the way from its initial state (at which the rates are calculated) toward equilibrium (the infinite time limit). This will become apparent in Chapter 19, where the Markovian approximation and quantum master equations are introduced.

Exercise 17.3.6 *Replacing the role of the initial and final states, the rate of transition from the thermal ensemble $\{f\}$ to the ensemble $\{i\}$ is given by Eq. (17.3.18). Use the symmetry of the coupling function (Eq. (17.3.19)) to derive Eq. (17.3.20).*

Exercise 17.3.7 *Let us associate the relative populations of two ensembles, $P_{\{i\}}(t)$ and $P_{\{f\}}(t)$, with generic probability-conserving kinetic equations (see Eq. (17.2.1)),*

$$\dot{P}_{\{i\}}(t) = -k_{\{i\}\to\{f\}}P_{\{i\}}(t) + k_{\{f\}\to\{i\}}P_{\{f\}}(t)$$

$$\dot{P}_{\{f\}}(t) = -k_{\{f\}\to\{i\}}P_{\{f\}}(t) + k_{\{i\}\to\{f\}}P_{\{i\}}(t),$$

where $P_{\{i\}}(t) + P_{\{f\}}(t) = 1$. Show that $P_{\{i\}}(t) = P_{\{i\}}(0)e^{-(k_{\{i\}\to\{f\}}+k_{\{f\}\to\{i\}})t} + \frac{k_{\{f\}\to\{i\}}}{k_{\{i\}\to\{f\}}+k_{\{f\}\to\{i\}}}$ $(1 - e^{-(k_{\{i\}\to\{f\}}+k_{\{f\}\to\{i\}})t})$, where $\lim_{t\to\infty} \frac{P_{\{i\}}(t)}{P_{\{f\}}(t)} = \frac{k_{\{f\}\to\{i\}}}{k_{\{i\}\to\{f\}}}$.

Bibliography

[17.1] W. H. Miller, S. D. Schwartz, and J. W. Tromp, "Quantum mechanical rate constants for bimolecular reactions," The Journal of Chemical Physics 79, 4889 (1983).

[17.2] I. R. Craig, M. Thoss, and H. Wang, "Proton transfer reactions in model condensed-phase environments: Accurate quantum dynamics using the multilayer multiconfiguration time-dependent Hartree approach," The Journal of Chemical Physics 127, 144503 (2007).

18 Thermal Rates in a Bosonic Environment

18.1 The Spin-Boson Model

In this chapter we discuss a few implementations of Fermi's golden rule for elementary kinetic processes in molecular and nanoscale systems. We shall focus on processes involving charge and energy transfer that are essential to applications in electro-optics, photovoltaics, and nanoelectronics. Particularly, we shall discuss charge transfer between a donor and an acceptor (so-called redox centers), absorption and emission of electromagnetic radiation by electronic dipoles, and electronic energy transfer (exciton transfer). In all these cases the process is characterized by a measurable change in the electronic state populations within the underlying system (a molecule, a quantum dot, a defect in a lattice, and so forth). However, this change is coupled to mechanical degrees of freedom, attributed to the atomic nuclei. These nuclear degrees of freedom affect the electronic transition and sometimes provide the driving force for its occurrence. Therefore, they are an essential part of the system and must be accounted for in the relevant system Hamiltonian.

Being much lighter particles, the electrons are typically much faster than the atomic nuclei. This is also reflected in the differences in energy between electronic states (when the nuclei are clumped), which are typically much larger than differences in energy between nuclear (vibrational) states (when the electronic state is fixed), as discussed in Section 14.2. In the examples addressed here, two electronic states (e.g., "donor" and "acceptor" or "ground" and "excited") are assumed to be well separated in energy from other electronic states for all the relevant nuclear configurations. This implies that the electronic degree of freedom can be effectively associated with a (nuclear coordinate–dependent) two-level system Hamiltonian. Formally, since the space of 2×2 matrices is spanned by Pauli's spin matrices (see Eq. (13.1.17)), the electronic degree of freedom is mapped on a spin-half system.

The energy at each electronic state depends on the configuration of the relevant nuclei. Near the minimal energy configuration, the harmonic approximation can be invoked (see Chapter 8), such that the corresponding potential energy surface is approximated as a sum of harmonic oscillators (Eq. (8.2.4)). Anharmonic corrections are essential for the detailed nuclear dynamics. Nevertheless, for the purpose of characterizing the effect of the multidimensional nuclear system on electronic transitions, the harmonic approximation provides a reasonable starting point, where analytical results can be derived (see what follows). Within the harmonic model, nuclear excitations and de-excitations are associated with creation and annihilation of bosonic particles (see Dirac's ladder operators, Eqs. (8.5.3, 8.5.4)), where the corresponding operators satisfy the bosonic commutation relations (Eq. (8.5.7)).

The Hilbert space of the system composed of the electronic states and the nuclear coordinates is therefore a tensor product of the respective subspaces. Using the spin operators, $\hat{\sigma}_x$, $\hat{\sigma}_z$, and $\hat{I}$ (Eq. (13.1.17)), and the dimensionless position and momentum operators, $\{\hat{Q}_j\}$ and $\{\hat{P}_j\}$, corresponding to harmonic oscillators at frequencies $\{\omega_j\}$, the generic system is mapped on a "spin-boson" model Hamiltonian [18.1],

$$\hat{H} = \hat{I} \otimes \sum_{j=1}^{N_\omega} \frac{\hbar\omega_j}{2}(\hat{P}_j^2 + \hat{Q}_j^2) + \sigma_x \otimes \gamma \hat{I}_\mathbf{Q} + \sigma_z \otimes \Delta_E \hat{I}_\mathbf{Q} + \sigma_z \otimes \sum_{j=1}^{N_\omega} \hbar\omega_j \Delta_j \hat{Q}_j. \quad (18.1.1)$$

Using Pauli's matrix representations of the spin operators (Eq. (13.1.17)), this Hamiltonian can be expressed as a 2×2 matrix of nuclear space operators,

$$\hat{H} = \begin{bmatrix} \hat{H}_{1,\mathbf{Q}} & \gamma \hat{I}_\mathbf{Q} \\ \gamma \hat{I}_\mathbf{Q} & \hat{H}_{2,\mathbf{Q}} \end{bmatrix}. \quad (18.1.2)$$

The diagonal terms are nuclear space operators associated with each one of the electronic states. Each electronic state defines a multidimensional potential energy surface (PES), which is a sum over harmonic potential wells, where the equilibrium configuration of the jth oscillator is either $Q_j = -\Delta_j$ or $Q_j = +\Delta_j$, depending on whether the electronic state is "1" or "2" (see Ex. 18.1.1):

$$\hat{H}_{1,\mathbf{Q}} = \Delta_E + \sum_{j=1}^{N_\omega} \frac{\hbar\omega_j}{2}(\hat{P}_j^2 + \hat{Q}_j^2) + \hbar\omega_j \Delta_j \hat{Q}_j$$

$$\hat{H}_{2,\mathbf{Q}} = -\Delta_E + \sum_{j=1}^{N_\omega} \frac{\hbar\omega_j}{2}(\hat{P}_j^2 + \hat{Q}_j^2) - \hbar\omega_j \Delta_j \hat{Q}_j. \quad (18.1.3)$$

At each electronic state, the minimal energy is obtained when all the oscillator coordinates are set to their equilibrium configurations. Without loss of generality, those minimal energies are set to Δ_E and $-\Delta_E$ for the electronic states "1" and "2," respectively. The off-diagonal terms in Eq. (18.1.2) are the interstate electronic coupling operators. Within this model, the couplings are characterized by a single parameter, γ, and are assumed to be independent of the nuclear coordinates (proportional to an identity operator in the nuclear space).

In the following sections, the spin-boson model Hamiltonian is implemented for analyzing different scenarios. In each case, the physical meanings of the different model parameters and their relation to a realistic description of the underlying system are outlined.

Exercise 18.1.1 *Use the matrix representations of the spin operators (Eq.(13.1.17)) to identify the explicit form of the nuclear space Hamiltonians (Eq. (18.1.3)), as the diagonal elements of the spin-boson model Hamiltonian (Eq. (18.1.1)).*

18.2 Charge Transfer in a Polarizable Medium

Charge transfer is one of the most central elementary processes taking place on the nanoscale. It may be between two local defects in a solid, an interface and an adsorbate, two molecules in a solvent, or two chromophores embedded inside a protein. Charge transfer plays an essential role in the microscopic description of electric conductivity, chemical reactivity, and biological activities such as vision, respiration, and photosynthesis.

Without loss of generality, we shall focus on a generic electron transfer event between two electron binding sites, a "donor" and an "acceptor,"

$$D^- + A \rightarrow D + A^-. \tag{18.2.1}$$

Each "site" is typically an interacting many-electron system in its own right, such that the charge transfer is in fact between two many-electron states, where an "extra electron" is localized either at the donor or at the acceptor site. If these two electronic states are well separated in energy from all the other electronic states in the donor–acceptor system, one may attempt to describe the electron transfer process in terms of a two-level system. However, as we saw in Chapter 15, transition probability between two states in a two-level system is either zero or oscillatory in time. The emergence of an electron transfer "rate constant" is attributed in this case to the coupling of the electronic change of state to other degrees of freedom, associated with the surroundings.

As a concrete example, let us consider an electron transfer process occurring within a polarizable medium, where the environment degrees of freedom are identified with surrounding electric dipoles, as illustrated schematically in Fig. 18.2.1. The electric dipoles at the vicinity of the donor and/or acceptor sites respond to changes in the location of the extra charge by changing their favorable orientation (the orientation of minimal potential energy). Importantly, our interest is only in the location of the extra charge within the donor–acceptor system, and not in the specific orientations of each and every dipole in the environment. Nevertheless, as we shall see, the polarizable

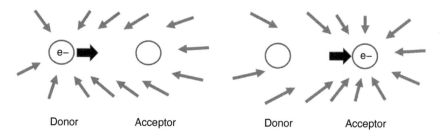

Donor   Acceptor   Donor   Acceptor

Figure 18.2.1   Electron transfer from a "donor" to an "acceptor" within a polarizable medium. The minimal energy orientations of the surrounding electric dipoles (marked as arrows) change in response to the change in the position of the electron charge.

medium plays a critical underlying role in determining the rate at which the electronic state (namely the location of the extra charge) changes.

Nonadiabatic Charge Transfer

Prior to formulating a microscopic model that accounts explicitly for both the electronic and the medium degrees of freedom, let us obtain some qualitative insight into the generic system as visualized in Fig. 18.2.1. In any "frozen" configuration of the surrounding dipoles, the extra electron is subject to two centers of attractive forces and hence can be qualitatively considered as a particle in some effective double well potential (see Section 6.4). The characteristics of this potential depend on the nature of the donor and acceptor, and also on the surrounding dipole orientations. This is illustrated in Fig. 18.2.2 for a single rotating dipole, positioned at the vicinity of the two positively charged centers. The dipole orientation is associated with a coordinate, q, where without loss of generality, the symmetric double well corresponds to the dipole vector being perpendicular to the line connecting the two centers of attraction ($q = 0$). Regarding the extra charge as a quantum particle in a double well potential, this symmetric configuration implies that the two lowest energy states are delocalized over the two potential energy wells (see Section 6.4). We shall denote the energy splitting between these corresponding energies as $V_{D,A}$. When the dipole rotates toward the right potential well ("the acceptor," see top panel), reflection symmetry is broken, and the two electronic states become localized in space: a state of lower energy in the right well and a state of higher energy in the left one. This trend continues as q increases until some optimal dipole orientation ($q = q_A$) is reached, in which the electronic energy at the right (acceptor) potential well reaches a minimum. From there on, namely, for $q > q_A$, the lower state energy also increases with increasing q. The same consideration applies to rotation of the dipole toward the left well ("the donor"; see bottom panel). The qualitative dependence of the two lowest energy levels on the dipole orientation is illustrated in Fig. 18.2.3.

Let us consider two limiting cases. When the energy splitting is large in comparison to the dipole-induced stabilization energy ($V_{D,A} > E_a$ as in Fig. 18.2.3 (a)), the electronic ground state is well separated in energy from the excited state for all the relevant dipole orientations. This implies that changes in the dipole orientation cannot provide sufficient energy for electronic excitation from the ground to the excited state; hence charge transfer between the potential wells is restricted to the ground electronic state, namely, it is "adiabatic" (limited to a single PES in the framework of the BO approximation). In the other limit, the tunneling splitting is small ($V_{D,A} \ll E_a$ as in Fig. 18.2.3 (b)), such that changes in the dipole orientation can transfer sufficient energy to electronic excitation. Hence charge transfer is nonadiabatic in this case and involves both the ground and the excited electronic states.

Our focus in what follows is on nonadiabatic charge transfer. This limit is consistent with a typical situation in which the donor and acceptor centers are spatially remote from each other, and the tunneling matrix element, $V_{D,A}$, falls exponentially with the donor–acceptor distance (see the discussion of the long-range single-electron exchange

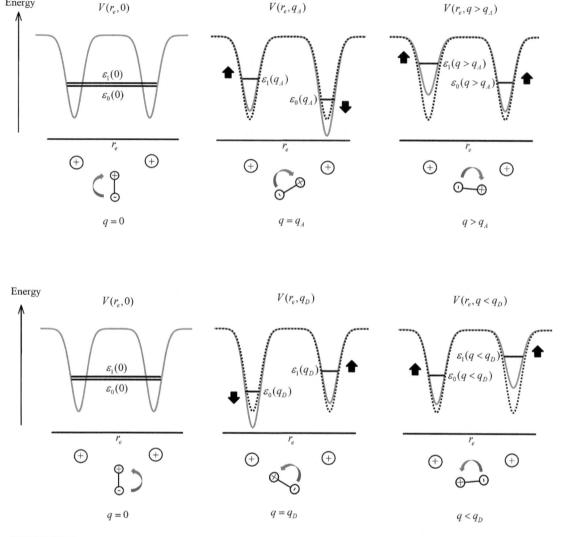

Figure 18.2.2 Changes to the effective "double well potential" for an electron in the presence of two fixed centers of positive charges, and a rotating electric dipole. In each plot the potential energy is plotted (solid line) as a function of the electronic coordinate, r_e, for a different dipole orientation. The dotted lines refer to a reference orientation in which the dipole is perpendicular to the line connecting the two fixed charge centers ($q = 0$). The top and bottom panels correspond, respectively, to rotation of the dipole to the right ($q > 0$) and to the left ($q < 0$).

integral in Section 14.3). It is instructive in these cases to analyze the charge transfer process in terms of localized "donor" and "acceptor" electronic eigenstates associated with the eigenstates of the two separated potential energy wells (often referred to as "diabatic states"). For most dipole orientations, these two localized states approximate well the ground and excited states of the double well potential (see, e.g., the asymmetric double well potentials in Fig. 18.2.2). However, near the "crossing point" (the symmetric configuration ($q = 0$) in Figs. 18.2.2 and 18.2.3), where the energydifference

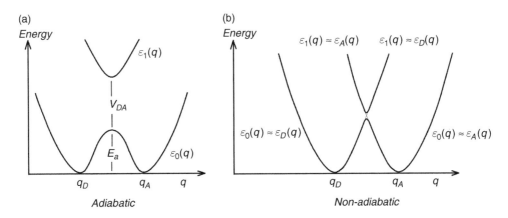

Figure 18.2.3 The two lowest energy levels of the effective electronic double well potential (see Fig. 18.2.2) as functions of the orientation of a nearby dipole (denoted by a coordinate q). The (a)/(b) plot corresponds to the adiabatic/nonadiabatic limit, at which the electronic coupling between the two potential wells is much larger/smaller than changes in the potential energy attributed to changes in the surrounding medium.

between the eigenstates of the separated donor and acceptor potential energy wells becomes smaller than the tunneling splitting $V_{D,A}$ associated with the corresponding double well potential, the exact electronic eigenstates are delocalized over the two wells and can be approximated in terms of effective linear combinations of the two localized states.

The Model Hamiltonian

A convenient framework for analyzing the charge transfer in the nonadiabatic limit is to use two localized electronic donor and acceptor states as a basis in which the full Hamiltonian can be represented. Denoting the electronic coordinates (of the many-electron donor–acceptor system) by $\mathbf{r} = \mathbf{r}_1, \mathbf{r}_2, \ldots$, and the coordinates of the environment (assuming the existence of multiple dipoles in the microscopic environment) by $\mathbf{q} \equiv q_1, q_2, \ldots$, the two basis states are the localized electronic states in the donor and acceptor potential wells, $\psi_D(\mathbf{r}, \mathbf{q})$ and $\psi_A(\mathbf{r}, \mathbf{q})$. The full system Hamiltonian can be decomposed as $\hat{H} = \hat{H}_e + \hat{T}_\mathbf{q}$, where $\hat{H}_e$ is the electronic Hamiltonian (see Section 14.2), which depends parametrically on the coordinates of the polarizable medium, $\mathbf{q}$, and $\hat{T}_\mathbf{q}$ is the respective kinetic energy. The matrix representation of the electronic Hamiltonian in the basis of localized electronic states is additionally assumed to be of a tight binding form (see Section 14.4), which means that the electronic Hamiltonian couples between $\psi_D(\mathbf{r}, \mathbf{q})$ and $\psi_A(\mathbf{r}, \mathbf{q})$, while their overlap integral vanishes. Again, this assumption is consistent with the nonadiabatic limit $V_{D,A} \ll E_a$ (see Fig. 18.2.3), where the two donor and acceptor centers are spatially remote from each other. The matrix elements of $\hat{H}_e$ are therefore denoted as

$$\int d\mathbf{r}\,\psi_D^*(\mathbf{r}, \mathbf{q})\hat{H}_e\,\psi_D(\mathbf{r}, \mathbf{q}) = \varepsilon_D(\mathbf{q}) \quad ; \quad \int d\mathbf{r}\,\psi_A^*(\mathbf{r}, \mathbf{q})\hat{H}_e\,\psi_A(\mathbf{r}, \mathbf{q}) = \varepsilon_A(\mathbf{q}) \qquad (18.2.2)$$

$$\int d\mathbf{r} \psi_D^*(\mathbf{r},\mathbf{q}) \hat{H}_e \psi_A(\mathbf{r},\mathbf{q}) = \int d\mathbf{r} \psi_A^*(\mathbf{r},\mathbf{q}) \hat{H}_e \psi_D(\mathbf{r},\mathbf{q}) = V_{DA}(\mathbf{q}), \qquad (18.2.3)$$

where

$$\int d\mathbf{r} \psi_D^*(\mathbf{r},\mathbf{q}) \psi_A(\mathbf{r},\mathbf{q}) = 0. \qquad (18.2.4)$$

An additional simplifying assumption is that the dependence of the localized electronic basis states, $\psi_D(\mathbf{r},\mathbf{q})$ and $\psi_A(\mathbf{r},\mathbf{q})$, on the dipole orientations can be disregarded, which is referred to as "the Condon approximation" [18.2]. Notice that the exact electronic eigenfunctions of the separated donor and acceptor Hamiltonians do depend on the dipole orientation. However, for sufficiently "deep" and "narrow" wells (that is, when the range of the potential well is much smaller than the inter-well separation), local perturbations to the potential wells may induce significant changes to the local energies (changes in the diagonal matrix elements of the electronic Hamiltonian, $\varepsilon_D(\mathbf{q})$ and $\varepsilon_A(\mathbf{q})$), but only negligible changes to the inter-well coupling, which depends on the remote "tails" of the electronic eigenfunctions. Consequently, the dependence of the inter-well coupling matrix element, $V_{D,A}(\mathbf{q})$, on $\mathbf{q}$ may be neglected:

$$V_{D,A}(\mathbf{q}) \approx V_{D,A}. \qquad (18.2.5)$$

Moreover, neglecting the dependence of the basis states on $\mathbf{q}$ and using Eq. (18.2.4), the representation of the dipoles' kinetic energy simplifies to

$$\int d\mathbf{r} \psi_D^*(\mathbf{r},\mathbf{q}) \hat{T}_{\mathbf{q}} \psi_D(\mathbf{r},\mathbf{q}) \approx \hat{T}_{\mathbf{q}} \quad ; \quad \int d\mathbf{r} \psi_A^*(\mathbf{r},\mathbf{q}) \hat{T}_{\mathbf{q}} \psi_A(\mathbf{r},\mathbf{q}) \approx \hat{T}_{\mathbf{q}} \qquad (18.2.6)$$

and

$$\int d\mathbf{r} \psi_D^*(\mathbf{r},\mathbf{q})) \hat{T}_{\mathbf{q}} \psi_A(\mathbf{r},\mathbf{q}) = \int d\mathbf{r} \psi_A^*(\mathbf{r},\mathbf{q}) \hat{T}_{\mathbf{q}} \psi_D(\mathbf{r},\mathbf{q}) \approx 0. \qquad (18.2.7)$$

It is convenient to introduce Dirac's notations for the $\mathbf{q}$-independent localized electronic basis states,

$$\psi_D(\mathbf{r}) \equiv \langle \mathbf{r}|D \rangle \quad ; \quad \psi_A(\mathbf{r}) \equiv \langle \mathbf{r}|A \rangle. \qquad (18.2.8)$$

Using Eqs. (18.2.2–18.2.8)), the matrix elements of the model Hamiltonian in the electronic basis read

$$\hat{H}_{D,\mathbf{q}} = \langle D|[\hat{T}_{\mathbf{q}} + \hat{H}_e(\mathbf{q})]|D \rangle = \hat{T}_{\mathbf{q}} + \varepsilon_D(\mathbf{q})$$
$$\hat{H}_{A,\mathbf{q}} = \langle A|[\hat{T}_{\mathbf{q}} + \hat{H}_e(\mathbf{q})]|A \rangle = \hat{T}_{\mathbf{q}} + \varepsilon_A(\mathbf{q})$$
$$\hat{H}_{D,A} = \hat{H}_{A,D} = \langle A|[\hat{T}_{\mathbf{q}} + \hat{H}_e(\mathbf{q})]|D \rangle = V_{DA}\hat{I}_{\mathbf{q}}. \qquad (18.2.9)$$

Notice that each electronic matrix element is an operator in the multidimensional dipole coordinates space. In matrix form, the model Hamiltonian reads

$$\hat{H} = \begin{bmatrix} \hat{H}_{D,\mathbf{q}} & V_{D,A}\hat{I}_{\mathbf{q}} \\ V_{D,A}\hat{I}_{\mathbf{q}} & \hat{H}_{A,\mathbf{q}} \end{bmatrix}. \qquad (18.2.10)$$

We now turn to formulating a concrete microscopic model for the polarizable medium. Considering a collection of independent dipole modes, $\mathbf{q} = q_1, q_2, q_3, \ldots$, the preferred orientation of each dipole changes as the electronic state changes (see Fig. 18.2.1). For small-amplitude motion around the preferred orientations at each of the two localized electronic states, the corresponding potential energy as a function of **q** can be described within a harmonic approximation (Eq. (8.2.4)). Attributing effective mass and frequency to each dipole mode coordinate, the multidimensional donor and acceptor Hamiltonians (Eq. (18.2.9)) obtain an explicit form,

$$\hat{H}_{D,q} = \varepsilon_D^0 + \sum_j \frac{\hat{p}_j^2}{2m_j} + \frac{1}{2}m_j\omega_j^2(\hat{q}_j - q_{j,D})^2$$

$$\hat{H}_{A,q} = \varepsilon_A^0 + \sum_j \frac{\hat{p}_j^2}{2m_j} + \frac{1}{2}m_j\omega_j^2(\hat{q}_j - q_{j,A})^2, \qquad (18.2.11)$$

where $\{q_{j,D}\}$ and $\{q_{j,A}\}$ are the dipoles' orientations corresponding to the potential energy minima, ε_D^0 and ε_A^0, at the donor and acceptor states, respectively. It is convenient to set the origin for each dipole coordinate to the average of the donor and acceptor minimal energy configurations, namely $(q_{j,D}+q_{j,A})/2$, and to transform into dimensionless position and momentum variables,

$$\hat{Q}_j \equiv \sqrt{\frac{m_j\omega_j}{\hbar}}\left(\hat{q}_j - \frac{q_{j,D}+q_{j,A}}{2}\right) \quad ; \quad \hat{P}_j \equiv \sqrt{\frac{1}{m_j\hbar\omega_j}}\hat{p}_j. \qquad (18.2.12)$$

The energy gap between the donor and acceptor potential energy minima is often referred to as the "driving force" for charge transfer, which is marked as

$$\varepsilon_D^0 - \varepsilon_A^0 \equiv 2\Delta_E. \qquad (18.2.13)$$

Using Eqs. (18.2.12, 18.2.13), the donor and acceptor Hamiltonians in Eq. (18.2.11) can be replaced by

$$\hat{H}_{D,Q} = \sum_j \frac{\hbar\omega_j}{2}\hat{P}_j^2 + E_D(\hat{Q}) \quad ; \quad E_D(\hat{Q}) = \Delta_E + \sum_j \frac{\hbar\omega_j}{2}\hat{Q}_j^2 + \hbar\omega_j\Delta_j\hat{Q}_j$$

$$\hat{H}_{A,Q} = \sum_j \frac{\hbar\omega_j}{2}\hat{P}_j^2 + E_A(\hat{Q}) \quad ; \quad E_A(\hat{Q}) = -\Delta_E + \sum_j \frac{\hbar\omega_j}{2}\hat{Q}_j^2 - \hbar\omega_j\Delta_j\hat{Q}_j, \qquad (18.2.14)$$

where $\hat{H}_{D,Q}$ and $\hat{H}_{A,Q}$ are identical to $\hat{H}_{D,q}$ and $\hat{H}_{A,q}$ up to an addition of a common constant (see Ex. (18.2.1)). The distance between the optimal dipole orientations at the donor and acceptor electronic states is a dimensionless measure for the coupling strength of each mode to the electronic degrees of freedom:

$$\Delta_j \equiv \sqrt{\frac{m_j\omega_j}{\hbar}}\left(\frac{q_{j,A}-q_{j,D}}{2}\right) = \frac{Q_{j,A}-Q_{j,D}}{2}. \qquad (18.2.15)$$

As one can readily see, the model for an "extra charge" in a donor–acceptor system coupled to a polarizable medium (Eqs. (18.2.10, 18.2.14)) is mapped on a spin-boson model Hamiltonian (Eqs. (18.1.1, 18.1.2)). Notice that this model is introduced under a set of approximations that overlook some of the complexities present in any realistic system. Specifically, the spin-boson model relies on the projection of the many-electron system onto only two electronic basis states, on the Condon approximation, and on the harmonic approximation, where the same set of modes at the same frequencies is assumed for the two electronic states. These simplifying assumptions clearly do not allow for quantitative predictions. Nevertheless, as demonstrated in what follows, the simple microscopic model reproduces the most important characteristics of charge transfer kinetics, and particularly the Marcus rate expression [18.2].

Exercise 18.2.1   *(a) Use Eqs. (18.2.12, 18.2.13, 18.2.15) to rewrite Eq. (18.2.11) as*

$$\hat{H}_{D,\mathbf{q}} = \Delta_E + \frac{\varepsilon_D^0 + \varepsilon_A^0}{2} + \sum_j \frac{\hbar\omega_j}{2}\Delta_j^2 + \sum_j \left(\frac{\hbar\omega_j}{2}\hat{P}_j^2 + \frac{\hbar\omega_j}{2}\hat{Q}_j^2 + \hbar\omega_j\Delta_j\hat{Q}_j \right)$$

$$\hat{H}_{A,\mathbf{q}} = -\Delta_E + \frac{\varepsilon_D^0 + \varepsilon_A^0}{2} + \sum_j \frac{\hbar\omega_j}{2}\Delta_j^2 + \sum_j \left(\frac{\hbar\omega_j}{2}\hat{P}_j^2 + \frac{\hbar\omega_j}{2}\hat{Q}_j^2 - \hbar\omega_j\Delta_j\hat{Q}_j \right).$$

(b) Show that Eq. (18.2.14) is obtained by setting the zero of energy to $\frac{\varepsilon_D^0 + \varepsilon_A^0}{2} + \sum_j \frac{\hbar\omega_j}{2}\Delta_j^2$.

The spin-boson model Hamiltonian for the donor–acceptor system (Eq. 18.2.10) can be conveniently decomposed into a zero-order Hamiltonian and an interaction,

$$\hat{H} = \hat{H}_0 + \hat{V}$$

$$\hat{H}_0 = \hat{H}_{D,\mathbf{Q}}|D\rangle\langle D| + \hat{H}_{A,\mathbf{Q}}|A\rangle\langle A| \quad ; \quad \hat{V} = V_{D,A}(|D\rangle\langle A| + |A\rangle\langle D|). \tag{18.2.16}$$

Each eigenstate of the zero-order Hamiltonian is a product state in the space of the electronic and the medium degrees of freedom (Ex. 18.2.2),

$$\hat{H}_0|D\rangle \otimes |\chi_{D,\mathbf{n}}\rangle = \varepsilon_{D,\mathbf{n}}|D\rangle \otimes |\chi_{D,\mathbf{n}}\rangle$$

$$\hat{H}_0|A\rangle \otimes |\chi_{A,\mathbf{m}}\rangle = \varepsilon_{A,\mathbf{m}}|A\rangle \otimes |\chi_{A,\mathbf{m}}\rangle, \tag{18.2.17}$$

where, $|D\rangle$ or $|A\rangle$ denote the electronic state, and $|\chi_{D,\mathbf{n}}\rangle$ and $|\chi_{A,\mathbf{m}}\rangle$ are eigenstates of the donor and acceptor **Q**-space Hamiltonians,

$$\hat{H}_{D,\mathbf{Q}}|\chi_{D,\mathbf{n}}\rangle = \varepsilon_{D,\mathbf{n}}|\chi_{D,\mathbf{n}}\rangle \quad ; \quad \hat{H}_{A,\mathbf{Q}}|\chi_{A,\mathbf{m}}\rangle = \varepsilon_{A,\mathbf{m}}|\chi_{A,\mathbf{m}}\rangle. \tag{18.2.18}$$

The vectors of quantum numbers, $\mathbf{n} = (n_1, n_2, n_3, \ldots)$ and $\mathbf{m} = (m_1, m_2, m_3, \ldots)$, correspond to multidimensional products of single harmonic oscillator eigenfunctions,

$$\langle\mathbf{Q}|\chi_{D,\mathbf{n}}\rangle = \varphi_{n_1}(Q_1 + \Delta_1) \cdot \varphi_{n_2}(Q_2 + \Delta_2) \cdot \varphi_{n_3}(Q_3 + \Delta_3)\ldots$$

$$\langle\mathbf{Q}|\chi_{A,\mathbf{m}}\rangle = \varphi_{m_1}(Q_1 - \Delta_1) \cdot \varphi_{m_2}(Q_2 - \Delta_2) \cdot \varphi_{m_3}(Q_3 - \Delta_3)\ldots, \tag{18.2.19}$$

where $\varphi_n(Q)$ is an eigenfunction of the one-dimensional harmonic oscillator Hamiltonian (Eqs. (8.3.7, 8.3.8)),

$$\frac{\hbar\omega}{2}(\hat{P}^2 + \hat{Q}^2)\varphi_n(Q) = \hbar\omega\left(n + \frac{1}{2}\right)\varphi_n(Q) \quad ; \quad n = 0, 1, 2\ldots. \tag{18.2.20}$$

The eigenvalues of the zero-order Hamiltonian therefore read (Ex. 18.2.2)

$$\varepsilon_{D,\mathbf{n}} = \Delta_E - \sum_j \frac{\hbar\omega_j}{2}\Delta_j^2 + \sum_j \hbar\omega_j\left(n_j + \frac{1}{2}\right)$$

$$\varepsilon_{A,\mathbf{m}} = -\Delta_E - \sum_j \frac{\hbar\omega_j}{2}\Delta_j^2 + \sum_j \hbar\omega_j\left(m_j + \frac{1}{2}\right). \tag{18.2.21}$$

Exercise 18.2.2 *For the zero-order Hamiltonian, $\hat{H}_0$, as defined in Eqs. (18.2.14, 18.2.16), show that the eigenvectors and eigenvalues are given by Eqs. (18.2.17–18.2.21).*

Thermal Charge Transfer Rates

We are interested in the thermal rate of charge transfer between the "donor" and the "acceptor" states in a system characterized by the full Hamiltonian, Eq. (18.2.16). Without loss of generality, we shall associate the initial state with the donor, and the final state with the acceptor, where the same considerations apply to the reverse process. More specifically, the initial and final states are identified with $\hat{H}_0$-eigenstates projected onto the donor and acceptor subspaces, respectively. The corresponding projection operators read (see Eq. (17.3.1))

$$\hat{P}_D = |D\rangle\langle D| \quad ; \quad \hat{P}_A = |A\rangle\langle A|, \tag{18.2.22}$$

where $\hat{P}_D^2 = \hat{P}_D$, $\hat{P}_D\hat{P}_A = 0$, $\hat{P}_A^2 = \hat{P}_A$, and $[\hat{P}_D, \hat{H}_0] = [\hat{P}_A, \hat{H}_0] = 0$ (see Eqs. (17.3.2, 17.3.3)). The definition of a thermal rate assumes an initial "quasi-equilibrium" donor state in which the relative weights of the different $\hat{H}_0$-eigenstates attributed to the donor correspond to a canonical statistical ensemble. Consequently, the occupation probability of any initial donor eigenstate is given by a Boltzmann distribution (Eq. (17.3.6)):

$$P_{D,\mathbf{n}}(0) = \frac{e^{-\varepsilon_{D,\mathbf{n}}/(k_B T)}}{Z_D} \quad ; \quad Z_D = \sum_{\mathbf{n}} e^{-\varepsilon_{D,\mathbf{n}}/(k_B T)}. \tag{18.2.23}$$

Alternatively, the initial density operator obtains the form (see Eq. (17.3.10))

$$\hat{\rho}_D(0) = \frac{e^{-\hat{H}_0/(k_B T)}\hat{P}_D}{tr\{e^{-\hat{H}_0,(k_B T)}\hat{P}_D\}}. \tag{18.2.24}$$

In the absence of interaction between the donor and acceptor manifolds of $\hat{H}_0$-eigenstates, this initial state would be stationary. The interaction couples, in principle, any donor state to the entire manifold of acceptor states, resulting in charge transfer from the donor to the acceptor. Provided that the state-to-state coupling is sufficiently small (see what follows), the overall transfer rate from the donor to the acceptor can be calculated by first-order perturbation theory. Using the general expression for the rate in terms of the interaction correlation function (Eqs. (17.3.12, 17.3.13)), we obtain in this case

$$k_{D\to A}^{(1)}(t) \cong \frac{2}{\hbar^2}\,\text{Re}\int_0^t tr\{\hat{\rho}_D(0)\hat{P}_D\hat{V}\hat{P}_A e^{\frac{i\hat{H}_0\tau}{\hbar}}\hat{V}e^{\frac{-i\hat{H}_0\tau}{\hbar}}\}d\tau. \tag{18.2.25}$$

Introducing explicitly the zero-order Hamiltonian and the interaction (Eq. (18.2.16)), and taking the (partial) trace over the electronic degrees of freedom, a compact expression is obtained in terms of the medium correlation function (Ex. 18.2.3):

$$k_{D\to A}^{(1)}(t) = 2\text{Re}\int_0^t c_{D,\mathbf{Q}}(\tau)d\tau \quad ; \quad c_{D,\mathbf{Q}}(\tau) = \frac{|V_{D,A}|^2}{\hbar^2}\,tr_{\mathbf{Q}}\left\{\frac{e^{-\hat{H}_{D,\mathbf{Q}}/(k_B T)}}{Z_D}e^{\frac{-i\hat{H}_{D,\mathbf{Q}}\tau}{\hbar}}e^{\frac{i\hat{H}_{A,\mathbf{Q}}\tau}{\hbar}}\right\}.$$

$$\tag{18.2.26}$$

Evaluating the trace using a complete orthonormal system of multidimensional harmonic oscillator eigenfunctions, we obtain (Ex. 18.2.4)

$$k_{D \to A}^{(1)}(t) = \frac{2}{\hbar^2} \sum_{\mathbf{n,m}} \frac{e^{-\varepsilon_{D,\mathbf{n}}/(k_B T)}}{Z_D} \mathrm{Re} \int_0^t e^{\frac{-i(\varepsilon_{D,\mathbf{n}} - \varepsilon_{A,\mathbf{m}})\tau}{\hbar}} |V_{D,A} \langle \chi_{D,\mathbf{n}} | \chi_{A,\mathbf{m}} \rangle|^2 d\tau. \qquad (18.2.27)$$

As discussed for the general case (Eqs. (17.2.3, 17.2.4)), this perturbative expression is valid only for times much shorter than the resonant state-to-state transition time,

$$t \ll \hbar / \max(|V_{D,A} \langle \chi_{D,\mathbf{n}} | \chi_{A,\mathbf{m}} \rangle|). \qquad (18.2.28)$$

Here $\langle \chi_{D,\mathbf{n}} | \chi_{A,\mathbf{m}} \rangle$ are multidimensional overlap integrals over the medium degrees of freedom (the "Franck–Condon (FC) factors"). Importantly, this condition is satisfied in typical cases of charge transfer within a polar medium over nanoscale distances, owing to the exponential decay of the electronic tunneling matrix element, $V_{D,A}$, with the donor–acceptor distance, and the fact that the products of FC factors are smaller than (or equal to) unity.

Exercise 18.2.3 *(a) Use the explicit expressions for the zero-order Hamiltonian (Eq. (18.2.16)) to show that $f(\hat{H}_0)|D\rangle = f(\hat{H}_{D,\mathbf{Q}})|D\rangle$ and $f(H_0)|A\rangle = f(H_{A,\mathbf{Q}})|A\rangle$, where f is an analytic function of its argument. (b) Use the results of (a), the interaction operator (Eq. (18.2.16)), and the definitions of the initial and final ensembles (Eqs. (18.2.22, 18.2.24)) to derive Eq. (18.2.26) from Eq. (18.2.25). Notice that the trace over the full electronic and nuclear space can be expressed as $tr\{\hat{O}\} = tr_{\mathbf{Q}}\{\langle D|\hat{O}|D\rangle + \langle A|\hat{O}|A\rangle\}$.*

Exercise 18.2.4 *Derive Eq. (18.2.27) from Eq. (18.2.26) by evaluating the trace over the nuclear space using a complete set of eigenstates of the multidimensional Hamiltonian, $\hat{H}_{D,\mathbf{Q}}$, and an identity operator, expressed in terms of $\hat{H}_{A,\mathbf{Q}}$-eigenstates (Eq. (18.2.18)).*

An explicit expression for the rate in terms of the microscopic model Hamiltonian can be derived directly from Eq. (18.2.26) by using the separability of the donor and acceptor Hamiltonians in the harmonic mode coordinates (see Eq. (18.2.14)). First, we make use of the fact that the multi-dimensional trace in the medium correlation function can be conveniently replaced by a product of traces over single-mode spaces (see Ex. 18.2.5), namely

$$k_{D \to A}^{(1)}(t) = \frac{2|V_{D,A}|^2}{\hbar^2} \mathrm{Re} \int_0^t e^{\frac{-i\tau}{\hbar} 2\Delta_E} \prod_j C_{D,j}(\tau) d\tau \quad ;$$

$$C_{D,j}(\tau) \equiv tr_{Q_j} \left\{ \frac{e^{\frac{-1}{k_B T} \hat{h}_{D,j}}}{Z_{D,j}} e^{\frac{-i\tau}{\hbar} \hat{h}_{D,j}} e^{\frac{i\tau}{\hbar} \hat{h}_{A,j}} \right\}. \qquad (18.2.29)$$

Here we defined the single-mode Hamiltonians at the donor and acceptor spaces,

$$\hat{h}_{D,j} = \frac{\hbar \omega_j}{2}(\hat{P}_j^2 + (\hat{Q}_j + \Delta_j)^2) \quad ; \quad \hat{h}_{A,j} = \frac{\hbar \omega_j}{2}(\hat{P}_j^2 + (\hat{Q}_j - \Delta_j)^2), \qquad (18.2.30)$$

where $Z_{D,j} = tr_{Q_j}\{e^{-\hat{h}_{D,j}/(k_B T)}\}$ is the partition function for the jth mode in the initial donor space. Using some algebra, the single-mode traces can be evaluated analytically (Ex. 18.2.6),

$$c_{D,j}(\tau) = e^{2i\Delta_j^2 \sin(\omega_j \tau)} e^{-2\Delta_j^2(1-\cos(\omega_j \tau))(1+2n(\omega_j))}, \qquad (18.2.31)$$

where $n(\omega_j)$ is the thermal occupation number of the jth mode (Ex. 18.2.7),

$$n(\omega_j) = \frac{1}{e^{\hbar\omega_j/(k_B T)} - 1}. \qquad (18.2.32)$$

Substitution in Eq. (18.2.29) yields the following explicit expression for the transfer rate,

$$k_{D\to A}^{(1)}(t) = \frac{2|V_{D,A}|^2}{\hbar^2} \, \mathrm{Re} \int_0^t e^{\frac{-i\tau}{\hbar} 2\Delta_E} e^{2i\sum_j \Delta_j^2 \sin(\omega_j\tau)} e^{-2\sum_j \Delta_j^2(1-\cos(\omega_j\tau))(1+2n(\omega_j))} d\tau. \qquad (18.2.33)$$

Exercise 18.2.5 (a) The donor partition function is defined as $Z_D = \sum_{\mathbf{n}} e^{-\varepsilon_{D,\mathbf{n}}/(k_B T)}$.

Use the definition of $\hat{H}_{D,\mathbf{Q}}$ (Eq. (18.2.14)) to show that $Z_D = e^{-\Delta_E/(k_B T)} e^{\sum_j \frac{\hbar\omega_j}{2}\Delta_j^2/(k_B T)}$

$Z_{D,1} Z_{D,2} Z_{D,3} \ldots$, where the jth mode partition function reads $Z_{D,j} \equiv \sum_{n_j} e^{-\frac{\hbar\omega_j}{k_B T}(n_j+\frac{1}{2})} =$

$\dfrac{e^{-\frac{\hbar\omega_j}{2k_B T}}}{1-e^{-\frac{\hbar\omega_j}{k_B T}}}$. (b) Show that $e^{-\frac{i\hat{H}_{D,\mathbf{Q}}\tau}{\hbar}} = e^{\frac{-i\tau}{\hbar}\Delta_E} e^{\frac{i\tau}{\hbar}\sum_j \frac{\hbar\omega_j}{2}\Delta_j^2} \prod_j e^{\frac{-i\tau}{\hbar}\frac{\hbar\omega_j}{2}(\hat{P}_j^2+(\hat{Q}_j+\Delta_j)^2)}$; $e^{\frac{i\hat{H}_{A,\mathbf{Q}}\tau}{\hbar}} =$

$e^{\frac{-i\tau}{\hbar}\Delta_E} e^{\frac{i\tau}{\hbar}\sum_j \frac{\hbar\omega_j}{2}\Delta_j^2} \prod_j e^{\frac{i\tau}{\hbar}\frac{\hbar\omega_j}{2}(\hat{P}_j^2+(\hat{Q}_j-\Delta_j)^2)}$. (c) Use the results of (a) and (b), and the definitions in Eq. (18.2.30) to derive Eq. (18.2.29) from Eq. (18.2.26). Recall that the trace of a tensor product of operators in a multidimensional space, $\hat{A}_1 \otimes \hat{A}_2 \otimes \cdots \otimes \hat{A}_N$, is a product, $tr\{\hat{A}_1 \otimes \hat{A}_2 \otimes \cdots \otimes \hat{A}_N\} = tr\{\hat{A}_1\} \cdot tr\{\hat{A}_2\} \cdots tr\{\hat{A}_N\}$ (Ex. 15.5.1).

Exercise 18.2.6 To prove the identity in Eq. (18.2.31), you can follow these steps: (a) Let $f(x)$ be an analytic function of x. Prove the identity: $e^{-\lambda \frac{\partial}{\partial x}} f(x) e^{\lambda \frac{\partial}{\partial x}} = f(x-\lambda)$. (b) Use the result of (a) to show that the single-mode Hamiltonians at the donor and acceptor states, as defined by Eq. (18.2.30), are transformations of a reference Hamiltonian, $\hat{h}_j = \frac{\hbar\omega_j}{2}(\hat{P}_j^2 + \hat{Q}_j^2)$, that is, $\hat{h}_{D,j} = e^{i\Delta_j \hat{P}_j} \hat{h}_j e^{-i\Delta_j \hat{P}_j}$, and $\hat{h}_{A,j} = e^{-i\Delta_j \hat{P}_j} \hat{h}_j e^{i\Delta_j \hat{P}_j}$. (c) Let $f(x)$ be an analytic function of x. Show that $f(\hat{h}_{D,j}) = e^{i\Delta_j \hat{P}_j} f(\hat{h}_j) e^{-i\Delta_j \hat{P}_j}$ and $f(\hat{h}_{A,j}) = e^{-i\Delta_j \hat{P}_j} f(\hat{h}_j) e^{i\Delta_j \hat{P}_j}$. (d) Expressing the dimensionless momentum operator in terms of Dirac's ladder operator, $\hat{P}_j = \frac{-i}{\sqrt{2}}(\hat{b}_j - \hat{b}_j^\dagger)$, and defining, $\hat{b}_j(\tau) = e^{\frac{i\tau}{\hbar}\hat{h}_j} \hat{b}_j e^{\frac{-i\tau}{\hbar}\hat{h}_j}$, show that $\hat{b}_j(\tau) = e^{-i\tau\omega_j} \hat{b}_j$. (e) Use the results of (c) and (d) to show that $c_{D,j}(\tau) =$

$tr_{Q_j}\left\{ \dfrac{e^{\frac{-1}{k_B T}\hat{h}_{D,j}}}{Z_{D,j}} e^{\frac{-i\tau}{\hbar}\hat{h}_{D,j}} e^{\frac{i\tau}{\hbar}\hat{h}_{A,j}} \right\} = tr_{Q_j}\left\{ \dfrac{e^{\frac{-1}{k_B T}\hat{h}_{D,j}}}{Z_{D,j}} e^{-\sqrt{2}\Delta_j[\hat{b}_j-\hat{b}_j^\dagger]} e^{-\sqrt{2}\Delta_j[e^{-i\tau\omega_j}\hat{b}_j - \hat{b}_j^\dagger e^{i\tau\omega_j}]} \right\}$. (f)

The Baker–Campbell–Hausdorff formula for two operators, $\hat{A}$ and $\hat{B}$, that commute with their commutator ($[\hat{A},[\hat{A},\hat{B}]] = [\hat{B},[\hat{A},\hat{B}]] = 0$) reads $e^{\hat{A}+\hat{B}} = e^{\hat{A}} e^{\hat{B}} e^{-[\hat{A},\hat{B}]/2}$. Use it and the commutator, $[\hat{b},\hat{b}^\dagger] = 1$, to show that $c_{D,j}(\tau) = e^{-2\Delta_j^2(1-e^{i\tau\omega_j})} tr_{Q_j}\left\{ \dfrac{e^{\frac{-1}{k_B T}\hat{h}_j}}{Z_{D,j}} e^{\sqrt{2}\Delta_j \hat{b}_j^\dagger(1-e^{i\tau\omega_j})} \right.$

$\left. \times e^{-\sqrt{2}\Delta_j \hat{b}_j(1-e^{-i\tau\omega_j})} \right\}$ (g) To evaluate the trace over the single-mode space, use a complete set of $\hat{h}_j$-eigenstates, $\hat{h}_j|\varphi_m\rangle = \hbar\omega_j(m+1/2)|\varphi_m\rangle$. Recalling that $\hat{b}_j|\varphi_m\rangle = \sqrt{m}|\varphi_{m-1}\rangle$, show that $c_{D,j}(\tau) = \dfrac{e^{-2\Delta_j^2(1-e^{i\tau\omega_j})}}{Z_{D,j}} \sum\limits_{m=0} e^{-\frac{\hbar\omega_j}{k_BT}(m+1/2)} \sum\limits_{n=0}^{m} [-2\Delta_j^2|(1-e^{i\tau\omega_j})|^2]^n \dfrac{m!}{(n!)^2(m-n)!}.$ (h)

The Laguerre polynomials of order m are defined as $L_m(x) = \sum\limits_{k=0}^{m} \binom{m}{k} \dfrac{(-1)^k}{k!}x^k$, and their generating function reads $\sum\limits_{m=0}^{\infty} t^m L_m(x) = \frac{1}{1-t}e^{-tx/(1-t)}$. Use it to show that $c_{D,j}(\tau) = e^{2i\Delta_j^2\sin(\omega_j\tau)}e^{-2\Delta_j^2(1-\cos(\omega_j\tau))(1+2n(\omega_j))}$, where $n(\omega_j) = \frac{1}{e^{\hbar\omega_j/(k_BT)}-1}$.

Exercise 18.2.7 *The averaged occupation number of a harmonic mode at a frequency, ω_j, and a temperature, T, is defined as $n(\omega_j) = \frac{1}{Z}\sum\limits_{n=0}^{\infty} n e^{-\hbar\omega_jn/(k_BT)}$, where $Z = \sum\limits_{n'=0}^{\infty} e^{-\hbar\omega_jn'/(k_BT)}$. Show that $n(\omega_j) = \dfrac{e^{-\hbar\omega_j/(k_BT)}}{1-e^{-\hbar\omega_j/(k_BT)}}$.*

The "Golden Rule" Rate

Let us return to the expression for the transition rate as a time-dependent integral, Eq. (18.2.27), in which the time integrand is a sum over multiple oscillatory terms at frequencies corresponding to the energy differences between $\hat{H}_0$-eigenstates in the donor and acceptor manifolds. As discussed in the general case (Eq. (17.2.3)), de-phasing of the oscillating terms may lead to their mutual cancelations after some characteristic time, and to a decay of the integrand to zero. When the decay time is within the validity window of perturbation theory (Eq. (17.2.9)) and much shorter than the "charge transfer time" (Eq. (17.2.8)), and when each initial state is embedded in a dense manifold of final states (where revivals of the initial state are suppressed), the rate converges to a constant value after the fast decay of the integrand. These conditions are typically met when the (nonvanishing) state-to-state coupling matrix elements are much smaller than the bandwidth of the corresponding energy differences, $\{\varepsilon_{D,n} - \varepsilon_{A,m}\}$, and when the frequency spectrum of the multidimensional surroundings medium includes the limit $\omega_j \to 0$, such that the energy spectrum of the initial and final states can be approximated as continuous. Given a rapid decay of the integrand, the upper time limit in Eq. (18.2.27) can be replaced by infinity, leading to Fermi's golden rule expression for the charge transfer "rate constant" (Ex. 18.2.8),

$$k_{D\to A} \cong \frac{2\pi}{\hbar} \sum_{n,m} \frac{e^{-\varepsilon_{D,n}/(k_BT)}}{Z_D}|V_{D,A}|^2|\langle\chi_{D,n}|\chi_{A,m}\rangle|^2\delta(\varepsilon_{D,n} - \varepsilon_{A,m}). \quad (18.2.34)$$

Within these conditions, the charge transfer rate is a weighted average over the thermal distribution of initial (donor) states and a sum over all final (acceptor) states. The dominant contributions to the rate are due to donor and acceptor states with matching energy (as imposed by Dirac's delta) and with nonvanishing FC factors. Notice that the amplitude of harmonic oscillator eigenfunctions at a given energy usually increases near the corresponding classical turning points (see Section 8.3). Consequently, the overlap integrals between the eigenfunctions of the donor and acceptor Hamiltonians are maximized when the classical turning points of the two potential energy surfaces coincide, namely, when the surfaces cross each other, $E_D(\mathbf{Q}) \approx E_A(\mathbf{Q})$ (see Fig. 18.2.4). The crossing surface is multidimensional and covers a wide energy range. Nevertheless, the thermal weights limit the number of initial states whose energy reaches the crossing surface. Hence, the charge transfer is a thermally activated process. The "activation energy" can be identified by further analysis, outlined in what follows.

Exercise 18.2.8   *Derive Eq. (18.2.34) by taking the upper limit of the time integral in Eq. (18.2.27) to infinity and using a suitable definition of Dirac's delta. Use the fact that the real part of the integrand is an even function of time.*

The Semiclassical Limit and Marcus Formula

In a typical scenario often met in charge transfer processes on the nanoscale, the polarizable medium is characterized by low-frequency modes. When their corresponding vibration quanta are much smaller than the thermal energy, a full account of the energy quantization of the surrounding modes is no longer critical, and semiclassical approximations can be justified. Here we shall refer to the high-temperature and low-frequency limits by deriving approximation for the general quantum mechanical charge transfer

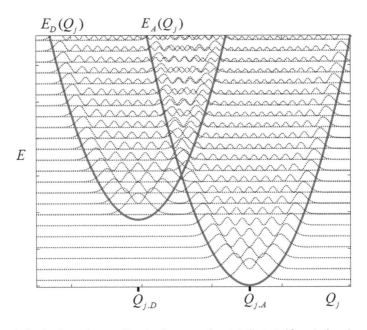

Figure 18.2.4   The role of Franck–Condon factors in controlling the charge transfer rate is illustrated for a single active environment mode. The two parabolic potential energy curves (solid lines) correspond to the donor and acceptor electronic states ($E_D(Q_j)$ and $E_A(Q_j)$; see Fig. 18.2.3). The nuclear probability densities (dotted lines), corresponding to the vibrational eigenfunctions, plotted at the respective energy levels, illustrate that the FC overlap between different states at the same energy is maximized at the vicinity of the crossing point between the two curves.

rate, Eq. (18.2.33). As we shall see, the result coincides with the celebrated semiclassical expression derived by Marcus [18.3].

First, let us consider the high-temperature limit, which implies that the thermal energy is much larger than the vibrational quantum for any mode of the polarizable medium, namely,

$$k_B T \gg \hbar \omega_j \quad \text{for any } j. \tag{18.2.35}$$

Within this limit, the thermal occupation number of the jth mode can be approximated as (see Eq. (18.2.32))

$$n(\omega_j) \approx \frac{k_B T}{\hbar \omega_j} \gg 1. \tag{18.2.36}$$

Second, and in line with the high-temperature limit, we restrict ourselves to low frequencies of the surrounding modes in the sense that their corresponding time periods are much larger than the decay time of the integrand in Eq. (18.2.33),

$$t_d \ll \frac{2\pi}{\omega_j} \quad \text{for any } j. \tag{18.2.37}$$

Let us recall that the decay of the time integrand requires the fulfilment of several conditions (weak state-to-state coupling to a dense and wide band of final states), in which Fermi's golden rule (Eq. (18.2.34)) is valid. When this holds, the trigonometric functions in Eq. (18.2.33) can be approximated by their lowest-order expansion in $\omega_j \tau$,

$$\sin(\omega_j \tau) \approx \omega_j \tau \quad ; \quad \cos(\omega_j \tau) \approx 1 - \omega_j^2 \tau^2 / 2. \tag{18.2.38}$$

Substituting Eqs. (18.2.36, 18.2.38) in Eq. (18.2.33) yields (Ex. 18.2.9)

$$k_{D \to A}^{(1)}(t) = \frac{2|V_{D,A}|^2}{\hbar^2} \, \text{Re} \int_0^t e^{\frac{-i\tau}{\hbar}(2\Delta_E - E_\lambda)} e^{\frac{-E_\lambda k_B T \tau^2}{\hbar^2}} d\tau. \tag{18.2.39}$$

Here the medium "reorganization energy" was introduced,

$$E_\lambda \equiv \sum_j 2\Delta_j^2 \omega_j \hbar = \sum_j \frac{\hbar \omega_j}{2}(2\Delta_j)^2, \tag{18.2.40}$$

as a global measure for the coupling between the electronic and the medium degrees of freedom (a measure for the polarizability of the surrounding medium). As we can see, the reorganization energy corresponds to the potential energy gained by changing the dipole orientations from their minimal energy orientation on a given (donor or acceptor) potential energy surface, to the orientation that would correspond to the minimal energy on the other (acceptor or donor) surface, namely (see Ex. 18.2.10 and Fig. 18.2.5)

$$E_\lambda = E_D(\mathbf{Q} = \Delta) - E_D(\mathbf{Q} = -\Delta) = E_A(\mathbf{Q} = -\Delta) - E_A(\mathbf{Q} = \Delta). \tag{18.2.41}$$

The time integrand in Eq. (18.2.39) is shown to be an oscillatory function, $e^{\frac{-i\tau}{\hbar}(2\Delta_E - E_\lambda)}$, multiplied by a Gaussian envelope, $e^{\frac{-E_\lambda k_B T \tau^2}{\hbar^2}}$, where the decay time can be identified as its standard deviation,

$$t_d \equiv \frac{\hbar}{\sqrt{2E_\lambda k_B T}}. \tag{18.2.42}$$

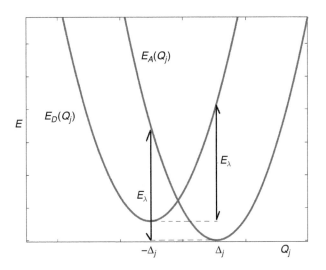

Figure 18.2.5 The reorganization energy, $E_\lambda \equiv 2\Delta_j^2 \omega_j \hbar$, attributed to a single environment mode. The two potential energy curves (solid lines) correspond to the donor and acceptor electronic states $(E_D(Q_j)$ and $E_A(Q_j))$; see Fig. 18.2.3).

Given that the time integral converges for $t \gg t_d$, the upper time-limit can be taken to infinity, which yields a compact expression for the thermal charge transfer rate constant (Ex. 18.2.11),

$$k_{D \to A} = |V_{D,A}|^2 \sqrt{\frac{\pi}{\hbar^2 k_B T E_\lambda}} e^{-\frac{(E_\lambda - 2\Delta_E)^2}{4 k_B T E_\lambda}}. \tag{18.2.43}$$

The regime of validity of the semiclassical approximation to Fermi's golden rule rate can be readily identified using Eqs. (18.2.37, 18.2.42), which yield

$$\frac{(\hbar \omega_j)^2}{k_B T E_\lambda} \ll 8\pi^2 \quad \text{for any } j. \tag{18.2.44}$$

This means that the approximation holds when the vibration quanta of the surrounding medium are small in comparison to both the thermal and the reorganization energies. Additionally, using Eq. (18.2.42) and the validity condition for first-order perturbation theory (Eq. (18.2.28)), similar restrictions are obtained with respect to the electronic coupling matrix element (Ex. 18.2.12),

$$\frac{|V_{D,A}|^2}{2 k_B T E_\lambda} \ll 1. \tag{18.2.45}$$

Exercise 18.2.9 *Use Eqs. (18.2.36, 18.2.38) in Eq. (18.2.33) to express the rate, Eq. (18.2.39), in terms of the reorganization energy, defined in Eq. (18.2.40).*

Exercise 18.2.10 *Show that Eq. (18.2.41) reproduces the reorganization energy as defined in Eq. (18.2.40), where $E_D(\hat{Q})$ and $E_A(\hat{Q})$ are the donor and acceptor potential energy surfaces, defined in Eq. (18.2.14).*

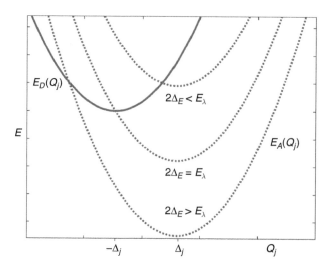

Figure 18.2.6 Illustration of the non-monotonic effect of the driving forces on the charge transfer rate. The solid line corresponds to a representative harmonic potential at the donor electronic state, where the three dotted lines represent different possibilities for the corresponding potential at the acceptor state, associated with different "driving forces," $2\Delta_E$. When $2\Delta_E < E_\lambda$, the energy at the crossing between the donor (solid) and the acceptor (dotted) potential energy surfaces is greater than the minimum energy at the donor state, and the charge transfer is thermally activated. When $2\Delta_E = E_\lambda$, the crossing identifies with the minimum at the donor potential energy, and the process is activation-less. When $2\Delta_E > E_\lambda$, the crossing is again higher in energy, and the charge transfer is again thermally activated (the inverted region).

Exercise 18.2.11 *Change the time limit in Eq. (18.2.39) to infinity (notice that the real part of the integrand is an even function of time) to obtain the result in Eq. (18.2.43). Use the identity,* $\int\limits_{-\infty}^{\infty} dk\, e^{-zk^2}\, e^{ikx} = \sqrt{\dfrac{\pi}{z}}\, e^{\frac{-x^2}{4z}}.$

Exercise 18.2.12 *(a) Show that the maximal state-to-state coupling matrix element between the donor and acceptor eigenstates is* $|V_{DA}|$, *and use Eqs. (18.2.28, 18.2.42) to obtain the validity condition, Eq. (18.2.45). (b) Show that this condition also assures that* $t_d \ll \dfrac{1}{k_{D\to A}}$ *(the wide band limit, Eq. (17.2.8)).*

Eq. (18.2.43) can be readily identified as the Marcus formula for the charge transfer rate [18.3], which depends on the electronic tunneling matrix element, $V_{D,A}$, the temperature, T, the reorganization energy, E_λ, and the driving force, $2\Delta_E$. The exponential factor is characteristic of a thermally activated process, where $k \propto e^{\frac{-E_a}{k_B T}}$. The corresponding activation energy can be readily identified as $E_a = \dfrac{(E_\lambda - 2\Delta_E)^2}{4E_\lambda}$. The maximal rate is obtained when the activation energy vanishes, that is, $E_\lambda = 2\Delta_E$ (the activation-less regime). This means that the rate changes non-monotonically as a function of the driving force (for a given reorganization energy), and as a function of the reorganization energy (for a given driving force), as illustrated in Fig. 18.2.6.

Notice that while the reorganization energy has a dramatic effect on the charge transfer kinetics, it does not affect the ratio between the forward and backward transfer rates

(Ex. 18.2.13), which is solely determined by the temperature and the "driving force" for charge transfer $(2\Delta_E)$,

$$\frac{k_{A \to D}}{k_{D \to A}} = e^{\frac{-2\Delta_E}{k_B T}} . \tag{18.2.46}$$

Indeed, since the direct sum of the donor and acceptor spaces is the entire Hilbert space, the ratio of the transition rates between them is uniquely related to their relative populations at equilibrium (the detailed balance condition, Eq. (17.3.21)), $\frac{k_{A \to D}}{k_{D \to A}} = \frac{P_D^{(eq)}}{P_A^{(eq)}}$. At equilibrium this ratio merely reflects the ratio between the reduced donor and acceptor partition functions (Eq. (17.3.20)), which is independent of the reorganization energy.

Exercise 18.2.13   *According to Eq. (18.2.43), the thermal rates, $k_{D \to A}$ and $k_{A \to D}$, differ only by the sign of the driving force. Use this to derive their ratio, Eq. (18.2.46).*

18.3 Radiation Absorption and Emission

The interaction of electromagnetic radiation with matter is fundamentally important from both basic and applicative science points of view. This interaction enables us to measure transitions between energy eigenstates of nanoscale materials (including atoms, molecules, quantum dots, and bulk), unraveling both the energy quantization and the underlying particle arrangement in the material (e.g., spatial and temporal probability densities and interference between probability amplitudes). Absorption and emission of electromagnetic radiation is an important mechanism of energy transfer to and from molecular and nanoscale systems. A remarkable example is photosynthesis, in which the energy absorbed from the sun light is channeled into a nanoscale reaction center and finally transformed into "energy-rich" chemical bonds. Applications engineered for energy storage and conversion (e.g. photovoltaic cells) or electro-optical devices are based on similar ideas in which electromagnetic radiation is converted into excitations (and thus triggers charge currents, chemical reactions, reorganization processes), or vice versa.

In this section we consider the elementary process in which an electronic transition within a given material (a molecule, as a specific example) is coupled to external electromagnetic radiation. Restricting ourselves to the weak coupling (linear response) regime, we shall use perturbation theory to describe the rate of electronic excitation (radiation absorption) or de-excitation (radiation emission), induced by coupling of the system to the radiation field. We shall reveal the dependence of these rates on the field parameters (intensity and frequency) as well as on the system parameters. In Section 15.6 we studied the effect of a monochromatic driving on state-to-state transitions within a generic system. Here this treatment is generalized to account or the case in which an electronic transition within a chromophore (typically, an atom, molecule, or a nanoparticle) is coupled to internal molecular vibrations as well as to

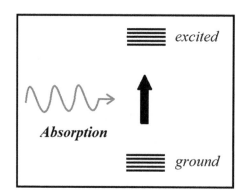 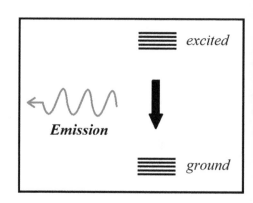

Figure 18.3.1 A schematic illustration of radiation-induced transitions. The groups of states correspond to the electronic ground and excited states of the system, where specific states within each group correspond to different nuclear states (in the system or in its surroundings) at a given electronic state.

a multidimensional environment at a finite temperature. In the presence of this coupling, measurements of radiation absorption and emission involve transitions between incoherent ensembles of states (see Fig. 18.3.1). As discussed in Chapter 17, this scenario is associated with the emergence of "rate constants" for these processes on their relevant timescale. As we shall see, these rates are smooth functions of the field frequency, unlike the delta peaks, which characterize transitions between pure states (see Eq. (15.6.28)).

The Model Hamiltonian

The zero-order Hamiltonian corresponds to the field-free system. Typically, the electronic transition involves the many-electron ground state and one of the excited electronic states. When these states are well separated in energy from all the other electronic states for the relevant nuclear configurations, one may attempt to describe the many-electron degrees of freedom in terms of a two-level model system. Moreover, when the two states are well separated from each other, the adiabatic (Born–Oppenheimer) approximation is justified (see Section 14.2). The system Hamiltonian can therefore be decomposed as

$$\hat{H}_0 = \hat{H}_e(\mathbf{q}) + \hat{T}_{\mathbf{q}}, \tag{18.3.1}$$

where $\hat{H}_e(\mathbf{q})$ is the electronic Hamiltonian (see Section 14.2), which depends parametrically on the nuclear coordinates, and $\hat{T}_{\mathbf{q}}$ is the respective kinetic energy. Denoting the electronic coordinates by $\mathbf{r} = \mathbf{r}_1, \mathbf{r}_2, \ldots$ and the nuclear coordinates by $\mathbf{q} \equiv q_1, q_2, \ldots$, the ground and excited states, denoted as $\psi_{gr}(\mathbf{r}, \mathbf{q})$ and $\psi_{ex}(\mathbf{r}, \mathbf{q})$, are eigenstates of the electronic Hamiltonian,

$$\hat{H}_e(\mathbf{q}) \psi_{gr}(\mathbf{r}, \mathbf{q}) = \varepsilon_{gr}(\mathbf{q}) \psi_{gr}(\mathbf{r}, \mathbf{q})$$
$$\hat{H}_e(\mathbf{q}) \psi_{ex}(\mathbf{r}, \mathbf{q}) = \varepsilon_{ex}(\mathbf{q}) \psi_{ex}(\mathbf{r}, \mathbf{q}). \tag{18.3.2}$$

These states are an orthonormal set, which spans the electronic space for any nuclear configuration:

$$\int d\mathbf{r}|\psi_{gr}(\mathbf{r},\mathbf{q})|^2 = \int d\mathbf{r}|\psi_{ex}(\mathbf{r},\mathbf{q})|^2 = 1 \quad ; \quad \int d\mathbf{r}\psi_{gr}^*(\mathbf{r},\mathbf{q})\psi_{ex}(\mathbf{r},\mathbf{q}) = 0. \quad (18.3.3)$$

Within the BO approximation, the $\mathbf{q}$-dependence of the electronic functions (namely, the derivative coupling) is assumed negligible, resulting in a diagonal representation of the nuclear kinetic energy operator (see Eq. (14.2.17)):

$$\int d\mathbf{r}\psi_{gr}^*(\mathbf{r},\mathbf{q})\hat{T}_{\mathbf{q}}\psi_{gr}(\mathbf{r},\mathbf{q}) \approx \hat{T}_{\mathbf{q}} \quad ; \quad \int d\mathbf{r}\psi_{ex}^*(\mathbf{r},\mathbf{q})\hat{T}_{\mathbf{q}}\psi_{ex}(\mathbf{r},\mathbf{q}) \approx \hat{T}_{\mathbf{q}}, \quad (18.3.4)$$

where

$$\int d\mathbf{r}\psi_{gr}^*(\mathbf{r},\mathbf{q})\hat{T}_{\mathbf{q}}\psi_{ex}(\mathbf{r},\mathbf{q}) = \int d\mathbf{r}\psi_{ex}^*(\mathbf{r},\mathbf{q})\hat{T}_{\mathbf{q}}\psi_{gr}(\mathbf{r},\mathbf{q}) \approx 0. \quad (18.3.5)$$

Invoking Dirac's notations for the $\mathbf{q}$-independent electronic states,

$$\psi_{gr}(\mathbf{r}) \equiv \langle \mathbf{r}|gr\rangle \quad ; \quad \psi_{ex}(\mathbf{r}) \equiv \langle \mathbf{r}|ex\rangle, \quad (18.3.6)$$

the matrix representation of the field-free Hamiltonian obtains the form

$$\hat{H}_0 = \hat{H}_{gr,\mathbf{q}}|gr\rangle\langle gr| + \hat{H}_{ex,\mathbf{q}}|ex\rangle\langle ex|, \quad (18.3.7)$$

where

$$\langle gr|[\hat{T}_{\mathbf{q}}+\hat{H}_e(\mathbf{q})]|gr\rangle = \hat{T}_{\mathbf{q}} + \varepsilon_{gr}(\mathbf{q}) = \hat{H}_{gr,\mathbf{q}}$$
$$\langle ex|[\hat{T}_{\mathbf{q}}+\hat{H}_e(\mathbf{q})]|ex\rangle = \hat{T}_{\mathbf{q}} + \varepsilon_{ex}(\mathbf{q}) = \hat{H}_{ex,\mathbf{q}}$$
$$\langle ex|[\hat{T}_{\mathbf{q}}+\hat{H}_e(\mathbf{q})]|gr\rangle = \langle gr|[\hat{T}_{\mathbf{q}}+\hat{H}_e(\mathbf{q})]|ex\rangle = 0. \quad (18.3.8)$$

For small-amplitude motion around the minimal energy configuration, the multidimensional nuclear potential energy surfaces, $\varepsilon_{gr}(\mathbf{q})$ and $\varepsilon_{ex}(\mathbf{q})$, can be approximated as harmonic (Eq. (8.2.4)). Transforming to a set of independent coordinates (normal modes; see Section 8.2), of masses, $\{m_j\}$, and frequencies, $\{\omega_j\}$, an explicit model for the multidimensional ground and excited state Hamiltonians (Eq. (18.3.7)) reads

$$\hat{H}_{gr,\mathbf{q}} = \varepsilon_{gr}^0 + \sum_j \frac{\hat{p}_j^2}{2m_j} + \frac{1}{2}m_j\omega_j^2(\hat{q}_j - q_{j,gr})^2$$

$$\hat{H}_{ex,\mathbf{q}} = \varepsilon_{ex}^0 + \sum_j \frac{\hat{p}_j^2}{2m_j} + \frac{1}{2}m_j\omega_j^2(\hat{q}_j - q_{j,ex})^2. \quad (18.3.9)$$

Here, $\{q_{j,gr}\}$ and $\{q_{j,ex}\}$ are the nuclear configurations corresponding to the potential energy minima values, ε_{gr}^0 and ε_{ex}^0, at the ground and the excited electronic states, respectively. (Notice that this simplified model assumes the same mode frequencies for the two potential energy surfaces.) It is convenient to set the origin for each coordinate to the average of the two minimal energy configurations, namely $(q_{j,gr} + q_{j,ex})/2$, and to transform to dimensionless position and momentum variables,

$$\hat{Q}_j \equiv \sqrt{\frac{m_j\omega_j}{\hbar}}\left(\hat{q}_j - \frac{q_{j,gr}+q_{j,ex}}{2}\right) \quad ; \quad \hat{P}_j \equiv \sqrt{\frac{1}{m_j\hbar\omega_j}}\hat{p}_j. \quad (18.3.10)$$

The gap between the minima of the ground and excited potential energy surfaces is often referred to as the "adiabatic excitation energy." Here we denote it as

$$\varepsilon_{gr}^0 - \varepsilon_{ex}^0 \equiv 2\Delta_E. \tag{18.3.11}$$

(Notice that by definition, $\varepsilon_{gr}^0 < \varepsilon_{ex}^0$, and therefore Δ_E is chosen to be negative.) The distance between the minimal energy configurations at the donor and acceptor electronic states is a dimensionless measure for the coupling strength of each mode to the electronic degrees of freedom,

$$\Delta_j \equiv \sqrt{\frac{m_j \omega_j}{\hbar}} \left(\frac{q_{j,ex} - q_{j,gr}}{2} \right) = \frac{Q_{j,ex} - Q_{j,gr}}{2}. \tag{18.3.12}$$

Using Eqs. (18.3.9–18.3.12), the ground and excited state Hamiltonians obtain the form (see Ex. 18.2.1 for an analogous derivation)

$$\hat{H}_{ex,\mathbf{Q}} = \sum_j \frac{\hbar \omega_j}{2} \hat{P}_j^2 + E_{ex}(\hat{\mathbf{Q}}) \quad ; \quad E_{ex}(\hat{\mathbf{Q}}) = -\Delta_E + \sum_j \frac{\hbar \omega_j}{2} \hat{Q}_j^2 - \hbar \omega_j \Delta_j \hat{Q}_j$$

$$\hat{H}_{gr,\mathbf{Q}} = \sum_j \frac{\hbar \omega_j}{2} \hat{P}_j^2 + E_{gr}(\hat{\mathbf{Q}}) \quad ; \quad E_{gr}(\hat{\mathbf{Q}}) = \Delta_E + \sum_j \frac{\hbar \omega_j}{2} \hat{Q}_j^2 + \hbar \omega_j \Delta_j \hat{Q}_j. \tag{18.3.13}$$

Consequently, the field-free molecular Hamiltonian obtains the form of a generic spin-boson model (see Eqs. (18.1.1–18.1.3)),

$$\hat{H}_0 = \hat{H}_{gr,\mathbf{Q}} |gr\rangle \langle gr| + \hat{H}_{ex,\mathbf{Q}} |ex\rangle \langle ex|. \tag{18.3.14}$$

We now turn to modeling the interaction of the system with the radiation field. A rigorous derivation based on quantum electrodynamics is beyond our scope here, but in the high photon-density (semiclassical) limit and within the long wavelength approximation (where the radiation wavelength is much larger than the spatial dimensions of the nanoscale emitter/absorber system) the presence of the radiation amounts to a time-dependent electric field, interacting with the system's dipole. Without loss of generality, we shall consider a monochromatic driving field, characterized by a frequency, Ω,

$$\mathbf{E}(t) = \mathbf{E}_0 \sin(\Omega t), \tag{18.3.15}$$

where the interaction term reads

$$\hat{V}(t) = -\mathbf{E}(t) \cdot \hat{\mathbf{d}}. \tag{18.3.16}$$

The dipole, $\hat{\mathbf{d}}$, is a vector operator, defined as a sum over all the charged particles in the system,

$$\hat{\mathbf{d}} = \sum_n \hat{\mathbf{R}}_n Z_n |e| - \sum_i \hat{\mathbf{r}}_i |e| \equiv \hat{\mathbf{d}}_\mathbf{q} + \hat{\mathbf{d}}_\mathbf{r}, \tag{18.3.17}$$

where $\mathbf{R}_n$ and Z_n are the position vector and the proton number of the nth nucleus. Defining the scalar operator,

$$\hat{\mu} = -\mathbf{E}_0 \cdot \hat{\mathbf{d}} \equiv \hat{\mu}_\mathbf{q} + \hat{\mu}_\mathbf{r}, \tag{18.3.18}$$

the matrix representation of the interaction operator in the basis of the ground and excited electronic states reads

$$\hat{V}(t) = \sin(\Omega t)[\langle gr|\hat{\mu}|gr\rangle |gr\rangle\langle gr| + \langle gr|\hat{\mu}|ex\rangle |gr\rangle\langle ex| + \langle ex|\hat{\mu}|gr\rangle |ex\rangle\langle gr|$$
$$+ \langle ex|\hat{\mu}|ex\rangle |ex\rangle\langle ex|]. \qquad (18.3.19)$$

Recalling that the dependence of the electronic wave functions on the nuclear coordinates is neglected within the framework of the Born–Oppenheimer approximation, the matrix elements of the electronic dipole between the electronic wave functions are also assumed to be independent of the nuclear coordinates (the Condon approximation [18.4]),

$$\langle ex|\hat{\mu}_{\mathbf{r}}|ex\rangle = \mu_{ex} \; ; \; \langle gr|\hat{\mu}_{\mathbf{r}}|gr\rangle = \mu_{gr}$$
$$\langle gr|\hat{\mu}_{\mathbf{r}}|ex\rangle = \langle ex|\hat{\mu}_{\mathbf{r}}|gr\rangle^* = \mu_{gr,ex}. \qquad (18.3.20)$$

Making use of Eq. (18.3.3), the matrix elements of the nuclear dipole operator read

$$\langle ex|\hat{\mu}_{\mathbf{q}}|ex\rangle = \langle gr|\hat{\mu}_{\mathbf{q}}|gr\rangle = \hat{\mu}_{\mathbf{q}}$$
$$\langle gr|\hat{\mu}_{\mathbf{q}}|ex\rangle = \langle ex|\hat{\mu}_{\mathbf{q}}|gr\rangle = 0. \qquad (18.3.21)$$

Notice that the transitions between the electronic ground and the excited states are attributed only to the off-diagonal elements (see what follows), referred to as the "transition dipoles." Indeed, when these matrix elements vanish, for example, due to specific symmetry properties of the electronic states, the system is transparent to the radiation.

The full Hamiltonian is the sum of the field-free system Hamiltonian (Eq. (18.3.14)) and the interaction term (Eq. (18.3.19)),

$$\hat{H} = \hat{H}_0 + \hat{V}(t). \qquad (18.3.22)$$

Notice that the zero-order Hamiltonian is analogous to $\hat{H}_0$ discussed in detail in Section 18.2 for the charge transfer scenario (Eqs. (18.2.17–18.2.21)). Using this analogy and replacing the donor and acceptor states, $|D\rangle$ and $|A\rangle$, by the ground and excited states, $|gr\rangle$ and $|ex\rangle$, each eigenstate of the zero-order Hamiltonian is a product of an electronic state and a nuclear state,

$$\hat{H}_0|gr\rangle \otimes |\chi_{gr,\mathbf{n}}\rangle = \varepsilon_{gr,\mathbf{n}}|gr\rangle \otimes |\chi_{gr,\mathbf{n}}\rangle$$
$$\hat{H}_0|ex\rangle \otimes |\chi_{ex,\mathbf{m}}\rangle = \varepsilon_{ex,\mathbf{m}}|ex\rangle \otimes |\chi_{ex,\mathbf{m}}\rangle. \qquad (18.3.23)$$

The nuclear states, $|\chi_{gr,\mathbf{n}}\rangle$ and $|\chi_{ex,\mathbf{m}}\rangle$, are eigenstates of the nuclear Hamiltonians, $\hat{H}_{gr,\mathbf{Q}}$ and $\hat{H}_{ex,\mathbf{Q}}$, respectively. These states are products of displaced harmonic oscillator eigenfunctions,

$$\langle \mathbf{Q}|\chi_{gr,\mathbf{n}}\rangle = \varphi_{n_1}(Q_1 + \Delta_1) \cdot \varphi_{n_2}(Q_2 + \Delta_2) \cdot \varphi_{n_3}(Q_3 + \Delta_3) \cdots$$
$$\langle \mathbf{Q}|\chi_{ex,\mathbf{m}}\rangle = \varphi_{m_1}(Q_1 - \Delta_1) \cdot \varphi_{m_2}(Q_2 - \Delta_2) \cdot \varphi_{m_3}(Q_3 - \Delta_3) \cdots , \qquad (18.3.24)$$

where the corresponding eigenvalues are

$$
\varepsilon_{gr,\mathbf{n}} = \Delta_E - \sum_j \frac{\hbar\omega_j}{2}\Delta_j^2 + \sum_j \hbar\omega_j\left(n_j + \frac{1}{2}\right)
$$

$$
\varepsilon_{ex,\mathbf{m}} = -\Delta_E - \sum_j \frac{\hbar\omega_j}{2}\Delta_j^2 + \sum_j \hbar\omega_j\left(m_j + \frac{1}{2}\right).
\tag{18.3.25}
$$

Thermal Absorption and Emission Rates

We are interested in the rate of field-induced transitions between the ground and the excited electronic states. Considering that the state-to-state interaction matrix elements, $\langle\chi_{gr,\mathbf{n}}|\langle gr|\hat{V}(t)|\chi_{ex,\mathbf{m}}\rangle|ex\rangle = \langle\chi_{gr,\mathbf{n}}|\chi_{ex,\mathbf{m}}\rangle\mu_{gr,ex}\sin(\Omega t)$, are proportional to the field intensity, $|\mathbf{E}_0|$ (see Eq. (18.3.18)) and to Franck–Condon factors (which are smaller than or equal to unity), first-order perturbation theory can become valid for sufficiently weak field intensities (see Eq. (15.6.12)). Using the general expression for thermal rates induced by explicitly time-dependent interactions (Eq. (17.3.15)), we obtain in this case (Ex. 18.3.1)

$$
k^{(1)}_{gr\to ex}(t) \cong \frac{2}{\hbar^2}\,\mathrm{Re}\int_0^t tr\{\hat{\rho}_{gr}(0)\hat{P}_{gr}e^{\frac{i\hat{H}_0 t'}{\hbar}}\hat{V}(t')e^{\frac{-i\hat{H}_0 t'}{\hbar}}\hat{P}_{ex}e^{\frac{i\hat{H}_0 t}{\hbar}}\hat{V}(t)e^{\frac{-i\hat{H}_0 t}{\hbar}}\}dt'
$$

$$
k^{(1)}_{ex\to gr}(t) \cong \frac{2}{\hbar^2}\,\mathrm{Re}\int_0^t tr\{\hat{\rho}_{ex}(0)\hat{P}_{ex}e^{\frac{i\hat{H}_0 t'}{\hbar}}\hat{V}(t')e^{\frac{-i\hat{H}_0 t'}{\hbar}}\hat{P}_{gr}e^{\frac{i\hat{H}_0 t}{\hbar}}\hat{V}(t)e^{\frac{-i\hat{H}_0 t}{\hbar}}\}dt',
\tag{18.3.26}
$$

where the initial and final states are associated with the electronic projection operators (see Eq. (17.3.1)),

$$
\hat{P}_{gr} = |gr\rangle\langle gr| \quad ; \quad \hat{P}_{ex} = |ex\rangle\langle ex|,
\tag{18.3.27}
$$

and, $\hat{P}_{gr}^2 = \hat{P}_{gr}$, $\hat{P}_{gr}\hat{P}_{ex} = 0$, $\hat{P}_{ex}^2 = \hat{P}_{ex}$, $[\hat{P}_{gr}, \hat{H}_0] = [\hat{P}_{ex}, \hat{H}_0] = 0$ (see Eqs. (17.3.2, 17.3.3)). An initial "quasi-equilibrium" state is assumed, in which the relative weights of the different $\hat{H}_0$-eigenstates correspond to a canonical statistical ensemble (see Section 17.3). The respective initial density operators for absorption and emission processes therefore read

$$
\hat{\rho}_{gr}(0) = \frac{1}{Z_{gr}}e^{-\hat{H}_0/(k_BT)}\hat{P}_{gr} \quad ; \quad \hat{\rho}_{ex}(0) = \frac{1}{Z_{ex}}e^{-\hat{H}_0/(k_BT)}\hat{P}_{ex},
\tag{18.3.28}
$$

where $Z_{gr} = tr\{e^{-\hat{H}_0/(k_BT)}\hat{P}_{gr}\}$ and $Z_{ex} = tr\{e^{-\hat{H}_0/(k_BT)}\hat{P}_{ex}\}$.

Exercise 18.3.1 *According to first-order perturbation theory, the rate of population transfer between an initial thermal ensemble and a final ensemble due to an explicitly time-dependent interaction is given by Eq. (17.3.15). Replace the generic projection operators as defined in Eqs. (17.3.1–17.3.4) by the relevant projection operators into the ground and excited electronic states in a chromophore (Eq. (18.3.27)) to obtain the absorption and emission rates in Eq. (18.3.26).*

Introducing explicitly the interaction (Eqs. (18.3.15–18.3.21)), the absorption and emission rates (Eq. (18.3.26)) obtain the more explicit form (Ex. 18.3.2),

$$k^{(1)}_{gr \to ex}(t) \cong 2 \, \mathrm{Re}\left[(1 - e^{-2i\Omega t})\int_0^t e^{i\Omega\tau}c_{gr}(\tau)d\tau + (1 - e^{2i\Omega t})\int_0^t e^{-i\Omega\tau}c_{gr}(\tau)d\tau\right]$$

$$k^{(1)}_{ex \to gr}(t) \cong 2 \, \mathrm{Re}\left[(1 - e^{-2i\Omega t})\int_0^t e^{i\Omega\tau}c_{ex}(\tau)d\tau + (1 - e^{2i\Omega t})\int_0^t e^{-i\Omega\tau}c_{ex}(\tau)d\tau\right],$$

$$(18.3.29)$$

where $c_{gr}(\tau)$ and $c_{ex}(\tau)$ are dipole–dipole correlation functions,

$$C_{gr}(\tau) \equiv \frac{1}{4\hbar^2}tr\{\hat{\rho}_{gr}(0)\hat{\mu}^{\dagger}_{ex,gr}(0)\hat{\mu}_{ex,gr}(\tau)\}$$

$$C_{ex}(\tau) \equiv \frac{1}{4\hbar^2}tr\{\hat{\rho}_{ex}(0)\hat{\mu}_{ex,gr}(0)\hat{\mu}^{\dagger}_{ex,gr}(\tau)\}$$

$$(18.3.30)$$

$$\hat{\mu}_{ex,gr}(\tau) = e^{\frac{i\hat{H}_0\tau}{\hbar}}\hat{P}_{ex}\hat{\mu}\hat{P}_{gr}e^{\frac{-i\hat{H}_0\tau}{\hbar}}. \qquad (18.3.31)$$

Considering the explicit form of the zero-order Hamiltonian (Eq. (18.3.14)), and taking the trace over the electronic degrees of freedom, these correlation functions are expressed in terms of the nuclear Hamiltonians (Ex. 18.3.3),

$$c_{gr}(\tau) = \frac{|\mu_{gr,ex}|^2}{4\hbar^2}tr_Q\left\{\frac{e^{-\hat{H}_{gr,Q}/(k_BT)}}{Z_{gr}}e^{\frac{i\hat{H}_{ex,Q}\tau}{\hbar}}e^{\frac{-i\hat{H}_{gr,Q}\tau}{\hbar}}\right\}$$

$$c_{ex}(\tau) = \frac{|\mu_{gr,ex}|^2}{4\hbar^2}tr_Q\left\{\frac{e^{-\hat{H}_{ex,Q}/(k_BT)}}{Z_{ex}}e^{\frac{i\hat{H}_{gr,Q}\tau}{\hbar}}e^{\frac{-i\hat{H}_{ex,Q}\tau}{\hbar}}\right\}. \qquad (18.3.32)$$

The trace over the nuclear degrees of freedom can be evaluated explicitly by introducing a complete orthonormal system of harmonic oscillator eigenfunctions. This leads to an expansion of the correlation functions (Eq. (18.3.32)) in terms of Franck–Condon overlap integrals (Ex. 18.3.4),

$$C_{gr}(\tau) = \frac{|\mu_{gr,ex}|^2}{4\hbar^2}\sum_{n,m}\frac{e^{-\varepsilon_{gr,n}/(k_BT)}}{Z_{gr}}|\langle\chi_{gr,n}|\chi_{ex,m}\rangle|^2 e^{\frac{-i(\varepsilon_{gr,n}-\varepsilon_{ex,m})\tau}{\hbar}}$$

$$C_{ex}(\tau) = \frac{|\mu_{gr,ex}|^2}{4\hbar^2}\sum_{n,m}\frac{e^{-\varepsilon_{ex,m}/(k_BT)}}{Z_{ex}}|\langle\chi_{gr,n}|\chi_{ex,m}\rangle|^2 e^{\frac{i(\varepsilon_{gr,n}-\varepsilon_{ex,m})\tau}{\hbar}} \qquad (18.3.33)$$

Exercise 18.3.2 *The absorption and emission rates of a field-driven chromophore are given in Eq. (18.3.26). Use the explicit form of the interaction, Eqs. (18.3.15–18.3.18), the projection operators to the ground and excited states, Eq. (18.3.27), and the decomposition of the field amplitude into rotating waves, $\sin(\Omega t) = (e^{i\Omega t} - e^{-i\Omega t})/(2i)$, to express the rates in terms of the dipole correlation functions (Eqs. (18.3.29–18.3.31)).*

Exercise 18.3.3 *(a) Given the explicit form of the zero-order Hamiltonian, Eq. (18.3.14), show that*

$$\langle ex|e^{\pm \frac{i\hat{H}_0\tau}{\hbar}}|ex\rangle = e^{\pm\frac{i\hat{H}_{ex,\mathbf{Q}}\tau}{\hbar}} \quad ; \quad \langle gr|e^{\pm\frac{i\hat{H}_0\tau}{\hbar}}|gr\rangle = e^{\pm\frac{i\hat{H}_{gr,\mathbf{Q}}\tau}{\hbar}}$$

$$\langle gr|e^{-\hat{H}_0/(k_BT)}|gr\rangle = e^{-\hat{H}_{gr,\mathbf{Q}}/(k_BT)} \quad ; \quad \langle ex|e^{-\hat{H}_0/(k_BT)}|ex\rangle = e^{-\hat{H}_{ex,\mathbf{Q}}/(k_BT)}$$

(b) Use the definition of the projection operators to the ground and excited electronic states (Eq. (18.3.27)), the result (a), and Eqs. (18.3.18, 18.3.20, 18.3.21) for the interaction matrix elements, to derive Eq. (18.3.32) from Eqs. (18.3.28, 18.3.30). Recall that the trace over the full electronic and nuclear space can be expressed as $tr\{\hat{O}\} = tr_\mathbf{Q}\{\langle gr|\hat{O}|gr\rangle + \langle ex|\hat{O}|ex\rangle\}$.

Exercise 18.3.4 *Derive Eq. (18.3.33) from Eq. (18.3.32) by evaluating the trace over the nuclear space using a complete set of eigenstates of the multidimensional Hamiltonian, $\hat{H}_{gr,\mathbf{Q}}$, and an identity operator, expressed in terms of $\hat{H}_{ex,\mathbf{Q}}$-eigenstates (as defined in Eqs. (18.3.23, 18.3.24)).*

As apparent from Eq. (18.3.33) for the dipole correlation functions, the absorption and emission rates (Eq. (18.3.29)) are sums over time integrals of the form $\int_0^t e^{\pm i(\varepsilon_{gr,\mathbf{n}}-\varepsilon_{ex,\mathbf{m}}\pm\hbar\Omega)\tau/\hbar}d\tau$. The dominant contributions to these sums are attributed to the smallest exponents, corresponding to resonances between the field frequency and the state-to-state transition frequency, $(\varepsilon_{ex,\mathbf{m}} - \varepsilon_{gr,\mathbf{n}})/\hbar \approx \Omega$. Taking this into consideration, each transition rate is dominated by only one of the rotating waves, $e^{\pm i\Omega\tau}$. Keeping only the dominant one is often referred to as the "Rotating Wave Approximation" (RWA) for the absorption and emission rates,

$$k^{(1)}_{gr\to ex}(t) \cong 2\,\mathrm{Re}\left[(1 - e^{2i\Omega t})\int_0^t e^{-i\Omega\tau}c_{gr}(\tau)d\tau\right]$$

$$k^{(1)}_{ex\to gr}(t) \cong 2\,\mathrm{Re}\left[(1 - e^{-2i\Omega t})\int_0^t e^{i\Omega\tau}c_{ex}(\tau)d\tau\right]. \qquad (18.3.34)$$

Notice that the time-dependence of the transition rates arises from the time integral as well as the pre-factor, $1 - e^{\pm 2i\Omega t}$. The latter oscillates around unity at twice the driving field frequency, which is typically much higher than the rate of change of the dominant contributions to the time integral, and in the weak coupling limit, much higher than the transition rate itself. Therefore, the contribution of the rapid oscillations averages to zero on the relevant timescales, and the RWA expressions for the measurable absorption and emission rates read (Ex. 18.3.5)

$$k^{(1)}_{gr\to ex}(t) \cong 2\,\mathrm{Re}\int_0^t e^{-i\Omega\tau}c_{gr}(\tau)d\tau$$

$$k^{(1)}_{ex\to gr}(t) \cong 2\,\mathrm{Re}\int_0^t e^{i\Omega\tau}c_{ex}(\tau)d\tau \qquad (18.3.35)$$

Exercise 18.3.5 *Within the rotating wave approximation, the interaction term in the Hamiltonian (Eq. (18.3.19)) can be replaced by* $\hat{V}(t) = \frac{e^{i\Omega t}}{2}\mu_{gr,ex}|gr\rangle\langle ex| + \frac{e^{-i\Omega t}}{2}\mu_{ex,gr}|ex\rangle\langle gr|$. *Use this interaction term in the rate expressions (Eq. (18.3.26)) to derive Eq. (18.3.35) directly.*

The "Golden Rule" Rates

The rate expressions (Eqs. (18.3.35, 18.3.32)) are analogous to those in Eq. (18.2.26) for charge transfer within the spin-boson model. Apart from the pre-factor in which the electronic tunneling matrix element is replaced by the transition dipole interaction, the harmonic correlation functions have precisely the same structure of sums over oscillatory functions (Eqs. (18.2.27, 18.3.33)). The remarkable difference, however, is that in the case of charge transfer the oscillation frequencies correspond directly to the energy differences between $\hat{H}_0$-eigenstates, whereas in the case of absorption/emission of radiation, these differences are displaced by the field frequency, $\pm\Omega$. As discussed in the general case (Eqs. (17.2.3, 17.2.8, 17.2.9)), as well as in the context of a multidimensional harmonic model (Eq. (18.2.34)), in the limits of weak state-to-state coupling between an initial thermal state and a dense and wide manifold of final states, the integrands in Eq. (18.3.35) decay within a time that is much shorter than the characteristic time of population transitions between the electronic states. Therefore, the upper time-limit in the expressions for the absorption and emission rates can be replaced by infinity, and these rates become constants, expressed in terms of Fermi's golden rule (Ex. 18.3.6),

$$k_{gr\to ex}(\Omega) = \frac{2\pi}{\hbar}\left|\frac{\mu_{gr,ex}}{2}\right|^2\sum_{\mathbf{n,m}}\frac{e^{-\varepsilon_{gr,\mathbf{n}}/(k_BT)}}{Z_{gr}}|\langle\chi_{gr,\mathbf{n}}|\chi_{ex,\mathbf{m}}\rangle|^2\delta(\varepsilon_{gr,\mathbf{n}} - \varepsilon_{ex,\mathbf{m}} + \hbar\Omega)$$

$$k_{ex\to gr}(\Omega) = \frac{2\pi}{\hbar}\left|\frac{\mu_{gr,ex}}{2}\right|^2\sum_{\mathbf{n,m}}\frac{e^{-\varepsilon_{ex,\mathbf{m}}/(k_BT)}}{Z_{ex}}|\langle\chi_{gr,\mathbf{n}}|\chi_{ex,\mathbf{m}}\rangle|^2\delta(\varepsilon_{gr,\mathbf{n}} - \varepsilon_{ex,\mathbf{m}} + \hbar\Omega) \tag{18.3.36}$$

Exercise 18.3.6 *Derive Eq. (18.3.36) by taking the upper limit of the time integral in Eqs. (18.3.33, 18.3.35) to infinity and by using a suitable definition of Dirac's delta. Use the fact that the real part of the integrand is an even function of time.*

Notice that the transition rates can be reformulated alternatively, in terms of the dipole–dipole correlation functions (Eq. (18.3.30)). Taking the time integral in Eq. (18.3.35) to infinity (in accordance with the validity of Fermi's golden rule), the rates are shown to be Fourier transforms of the dipole–dipole correlation functions (Ex. 18.3.7),

$$k_{gr\to ex}(\Omega) = \int_{-\infty}^{\infty} e^{-i\Omega\tau}c_{gr}(\tau)d\tau = \frac{1}{4\hbar^2}\int_{-\infty}^{\infty} e^{-i\Omega\tau}tr\{\hat{\rho}_{gr}(0)\hat{\mu}_{ex,gr}^{\dagger}(0)\hat{\mu}_{ex,gr}(\tau)\}d\tau$$

$$k_{ex\to gr}(\Omega) = \int_{-\infty}^{\infty} e^{i\Omega\tau}c_{ex}(\tau)d\tau = \frac{1}{4\hbar^2}\int_{-\infty}^{\infty} e^{i\Omega\tau}tr\{\hat{\rho}_{ex}(0)\hat{\mu}_{ex,gr}(0)\hat{\mu}_{ex,gr}^{\dagger}(\tau)\}d\tau, \tag{18.3.37}$$

which reproduces the expression for the spectral line shapes of linear response theory (see [18.7]).

Exercise 18.3.7   *The dipole correlation functions are defined in Eq. (18.3.30). (a) Show that $c_{gr}^*(\tau) = c_{gr}(-\tau)$ and $c_{ex}^*(\tau) = c_{ex}(-\tau)$. (b) Use the result (a) to show that $\mathrm{Re}[e^{-i\Omega\tau} c_{gr}(\tau)]$ and $\mathrm{Re}[e^{i\Omega\tau} c_{ex}(\tau)]$ are even functions of time, and that $\mathrm{Im}[e^{-i\Omega\tau} c_{gr}(\tau)]$ and $\mathrm{Im}[e^{i\Omega\tau} c_{ex}(\tau)]$ are odd functions of time. Use this result to derive Eq. (18.3.37) from Eq. (18.3.35), in the limit $t \to \infty$.*

According to Fermi's golden rule (Eq. (18.3.36)) the thermal absorption and emission rates are weighted averages over the initial thermal distributions of the initial nuclear states and a sum over all the final nuclear states. In analogy to the result for charge transfer (Eq. (18.2.34)), the energy delta selects state-to-state transitions in which the initial and final energies match each other. However, unlike in the case of charge transfer, where the matching condition is strict ($\varepsilon_{A,\mathbf{m}} = \varepsilon_{D,\mathbf{n}}$), the matching condition for radiation emission and absorption depends on the radiation frequency, $\varepsilon_{ex,\mathbf{m}} = \varepsilon_{gr,\mathbf{n}} + \hbar\Omega$. Of all transitions satisfying this condition, the dominating ones are those associated with maximal FC factors. While in the case of charge transfer the FC factors are maximal near crossings between the donor and acceptor potential energy surfaces (see Fig. 18.2.4), in the present case, the FC factors at a given field frequency are maximal near crossings between the ground and excited potential energy surfaces, displaced by $\hbar\Omega$, namely, $E_{ex}(\mathbf{Q}) \approx E_{gr}(\mathbf{Q}) + \hbar\Omega$, as illustrated in Fig. 18.3.2.

We now turn to deriving explicit expressions for the absorption and emission rates in terms of the microscopic model Hamiltonian parameters. For this purpose, we refer again to the close analogy between charge transfer (Eq. (18.2.26)) and field-driven transitions (Eqs. (18.3.35, 18.3.32)) within the spin-boson model. Using this analogy (Eqs. (18.2.29–18.2.33)), the nuclear correlation functions, $c_{gr}(\tau)$ and $c_{ex}(\tau)$, can be expressed explicitly in terms of the oscillator frequencies $\{\omega_j\}$ and coupling strengths, $\{\Delta_j\}$. The first-order approximations for the time-dependent thermal absorption and emission rates therefore read

$$k_{gr \to ex}^{(1)}(t) \cong \frac{|\mu_{gr,ex}|^2}{2\hbar^2} \, \mathrm{Re} \int_0^t e^{\frac{-i\tau}{\hbar}(2\Delta_E + \hbar\Omega)} e^{2i\sum_j \Delta_j^2 \sin(\omega_j\tau)} e^{-2\sum_j \Delta_j^2 (1-\cos(\omega_j\tau))(1+2n(\omega_j))} \, d\tau$$

$$k_{ex \to gr}^{(1)}(t) \cong \frac{|\mu_{gr,ex}|^2}{2\hbar^2} \, \mathrm{Re} \int_0^t e^{\frac{i\tau}{\hbar}(2\Delta_E + \hbar\Omega)} e^{2i\sum_j \Delta_j^2 \sin(\omega_j\tau)} e^{-2\sum_j \Delta_j^2 (1-\cos(\omega_j\tau))(1+2n(\omega_j))} \, d\tau.$$

$$(18.3.38)$$

The Semiclassical Limit

It is often the case that transitions between the ground and excited states involve coupling to low-frequency nuclear modes. This characterizes, for example, the coupling of an intramolecular electronic (or vibronic; see what follows) transition to surrounding dipoles in a polar solvent, or the coupling of a transition within an atomic or molecular impurity to phonon modes in a lattice. The influence of low-frequency modes on the field-induced transition rates can be readily accounted for by referring again to the close analogy to electron transfer processes, as analyzed in Section 18.2. Assuming first that all the nuclear modes addressed in Eq. (18.3.38) are associated with low frequencies (in comparison to the thermal energy (Eq. (18.2.35)), as well as to the relaxation rate of the relevant time integrand (Eq. (18.2.37)), the absorption and emission rates are approximated as (compare to Eq. (18.2.39))

$$k_{gr \to ex}^{(1)}(t) \cong \frac{|\mu_{gr,ex}|^2}{2\hbar^2} \ \mathrm{Re} \int_0^t e^{\frac{-i\tau}{\hbar}(2\Delta_E + \hbar\Omega - E_\lambda)} e^{\frac{-E_\lambda k_B T \tau^2}{\hbar^2}} d\tau$$

$$k_{ex \to gr}^{(1)}(t) \cong \frac{|\mu_{gr,ex}|^2}{2\hbar^2} \ \mathrm{Re} \int_0^t e^{\frac{-i\tau}{\hbar}(-2\Delta_E - \hbar\Omega - E_\lambda)} e^{\frac{-E_\lambda k_B T \tau^2}{\hbar^2}} d\tau. \qquad (18.3.39)$$

Here E_λ is the nuclear reorganization energy, as defined in Eq. (18.2.40). When the reorganization energy is sufficiently large with respect to the vibrational quanta $(\frac{(\hbar\omega_j)^2}{k_B T E_\lambda} \ll 8\pi^2$; see Eq. (18.2.44)), and with respect to the driven transition dipole $(\frac{|\mu_{gr,ex}|^2}{8k_B T E_\lambda} \ll 1$; see Eq. (18.2.45)), the time integral can be taken to infinity to obtain the Fermi's golden rule rates in the low-frequencies limit,

$$k_{gr \to ex}(\Omega) = \frac{|\mu_{gr,ex}|^2}{4} \sqrt{\frac{\pi}{\hbar^2 k_B T E_\lambda}} e^{\frac{-(E_\lambda - 2\Delta_E - \hbar\Omega)^2}{4k_B T E_\lambda}}$$

$$k_{ex \to gr}(\Omega) = \frac{|\mu_{gr,ex}|^2}{4} \sqrt{\frac{\pi}{\hbar^2 k_B T E_\lambda}} e^{\frac{-(E_\lambda + 2\Delta_E + \hbar\Omega)^2}{4k_B T E_\lambda}}. \qquad (18.3.40)$$

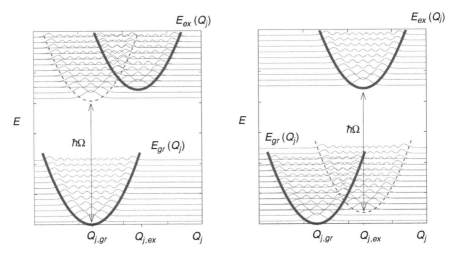

Figure 18.3.2 The role of Franck–Condon factors in controlling the absorption and emission rates is illustrated for a single active nuclear mode. In each plot, the two parabolic potential energy curves (thick solid lines) correspond to the ground and excited electronic states ($E_{gr}(Q_j)$ and $E_{ex}(Q_j)$; see Fig. 18.3.1). The nuclear probability densities (thin solid lines) associated with the vibrational eigenfunctions at each electronic state are plotted at the respective energy levels. For a given field frequency, Ω, the FC overlap between different states at the same energy is maximized near crossing points between two curves: in the case of absorption (let plot), the crossing between $E_{ex}(Q_j)$ and $E_{gr}(Q_j) + \hbar\Omega$, and in the case of emission (right plot), the crossing between $E_{gr}(Q_j)$ and $E_{ex}(Q_j) - \hbar\Omega$.

The absorption and emission rates as functions of the driving field frequency are proportional to the spectral lines, derived in linear response theory [18.7]. The semiclassical result (Eq. (18.3.40)) amounts to a Gaussian broadening of the transition lines, due to their coupling to the low-frequency vibrations. The line width (the standard deviation, $\sqrt{2k_BTE_\lambda}$) is shown to increase with increasing temperature and coupling strength (nuclear reorganization energy). The absorption and emission rates are peaked at different frequencies, $\Omega = (-2\Delta_E + E_\lambda)/\hbar$ and $\Omega = (-2\Delta_E - E_\lambda)/\hbar$, respectively, which deviate from resonance with the "adiabatic transition energy," $\Omega = -2\Delta_E/\hbar$ (see Eq. (18.3.11) and recall that $\Delta_E < 0$). Indeed, the resonant frequencies are higher (blue-shifted) for absorption and lower (red-shifted) for emission, due to the nuclear reorganization energy associated with each transition (see Fig. 18.3.3). This result (termed "the Stokes shift") has remarkable practical consequences in the context of electrooptical and photovoltaic devices, as it means that driving the system "upwards" in energy from the thermal ground electronic state to the excited state requires more energy than can be gained by the corresponding "downwards" transition. This is attributed to the fact that in both cases, regaining a quasi-equilibrium thermal state of the nuclei after the electronic transition involves a nonradiative transition in which heat is emitted to the surroundings.

Notice that, in the absence of coupling to the surrounding low-frequency modes, the absorption and emission line shapes converge to a single delta peak, centered

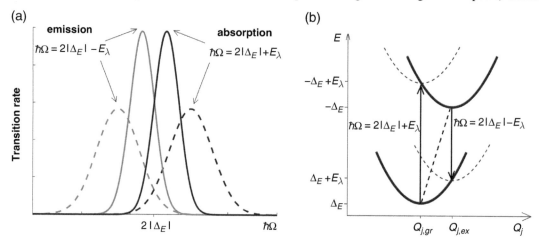

Figure 18.3.3 (a) The effect of nuclear reorganization on the absorption and emission rates for the field-driven spin-boson model, in the semiclassical limit. The rates are Gaussian functions of the field frequency, peaked at $\hbar\Omega = |2\Delta_E| + E_\lambda$ for absorption, and $\hbar\Omega = 2|\Delta_E| - E_\lambda$ for emission. The peaks are Stokes-shifted by the nuclear reorganization energy, E_λ, with respect to the "adiabatic" transition energy, $|2\Delta_E|$. The solid and dashed curves correspond to E_λ being equal to 10% and 30% of $|2\Delta_E|$, respectively. (b) Potential energy curves for a single low frequency mode at the ground and excited electronic states. The maximal absorption (emission) rate is obtained when the transition is "activation-less," namely, when the minimum of the $\hbar\Omega$-shifted ground (excited) state potential crosses the excited (ground) state potential. The "adiabatic" transition energy, $|2\Delta_E|$, is marked by the dashed diagonal line.

at the "adiabatic excitation energy," $\Omega = -2\Delta_E/\hbar$. This can be readily realized by taking the limit, $E_\lambda \to 0$ (Ex. 18.3.8), which also reproduces the perturbative calculation of the transition rate between pure initial and final states (see the discussion of periodic driving in Section 15.6, Eq. (15.6.28), and Ex. 15.6.10). "Pure electronic transitions" are characteristics of atoms in the gas phase, but rarely observed in nanoscale many-atom systems. In a condensed-phase environment (and even in polyatomic molecules in the gas phase), line broadening and the Stokes shift between the absorption and emission lines are commonly observed. Characteristic examples are the absorption and fluorescence spectra of molecular chromophores in solid matrices, lattices, and solutions, or exciton formation and relaxation in quantum dots and bulk materials. Notice that when the surrounding mode frequencies are much smaller than the absorption/emission transition rates, their averaged effect on the transition rate can be rightly treated as an ensemble average over a distribution of different static microscopic environments to which the electronic transition is coupled [18.5]. The Gaussian broadening of the absorption/emission lines due to coupling to low-frequency modes is therefore closely related to "inhomogeneous broadening." (See Chapter 19 for a discussion of inhomogeneous broadening in the limit of weak coupling to the bosonic environment.)

Exercise 18.3.8 *One of the representations of Dirac's delta is $\delta(x) = \lim_{\varepsilon \to +0} \sqrt{\frac{1}{4\pi\varepsilon}} e^{\frac{-x^2}{4\varepsilon}}$. Use it to show that in the limit of vanishing coupling to the nuclear modes, both the absorption and the emission rates are peaked at the "adiabatic" energy gap, $-2\Delta_E$, namely $\lim_{E_\lambda \to 0} k_{gr \to ex}(\Omega) = \frac{2\pi}{\hbar} \frac{|\mu_{gr,ex}|^2}{4} \delta(-2\Delta_E - \hbar\Omega)$, $\lim_{E_\lambda \to 0} k_{ex \to gr}(\Omega) = \frac{2\pi}{\hbar} \frac{|\mu_{gr,ex}|^2}{4} \delta(2\Delta_E + \hbar\Omega)$. Compare the result with the direct calculation of transition rate between pure states in Eq. (15.6.29).*

Vibronic Spectra

In many cases the electronic transition is coupled to "high-frequency" nuclear modes. This is a common situation in molecules, where electronic excitations affect chemical bonds. For example, electronic population transfer from bonding to antibonding molecular orbitals (see Section 14.3) would result in changes in bond distances, frequencies, dissociation energies, and so forth. The vibrational frequencies associated with chemical bonds near their stable configuration are typically much larger than the thermal energy (see Section 8.4), and the "low-frequency" (semiclassical) approximation just exercised does not apply to such vibrations. Yet, high-frequency modes are apparent in absorption and emission line shapes, and in many cases can be resolved experimentally, providing valuable information on the molecular structure and internuclear forces. The manifestation of high-frequency modes in absorption and emission spectral line shapes can be inferred directly from the general perturbative expression, Eq. (18.3.38), for the transition rates within the spin-boson model. For concreteness, let us consider a scenario in which the electronic transition is coupled to a dense set of low-frequency modes as well as to a single high-frequency mode. We shall denote the mode frequency as ω_0, and its displacement parameter (its effective coupling to the electronic transition) as Δ_0. Invoking the low-frequency approximation (Eq. (18.3.39)) only for the low-frequency modes, Eq. (18.3.38) obtains the form (Ex. 18.3.9)

$$k_{gr \to ex}^{(1)}(t) \cong \frac{|\mu_{gr,ex}|^2}{2\hbar^2} \, \text{Re} \int_0^t e^{\frac{-i\tau}{\hbar}(2\Delta_E + \hbar\Omega - E_\lambda)} e^{\frac{-E_\lambda k_B T \tau^2}{\hbar^2}}$$

$$e^{2i\Delta_0^2 \sin(\omega_0 \tau)} e^{-2\Delta_0^2(1-\cos(\omega_0 \tau))(1+2n(\omega_0))} d\tau$$

$$k_{ex \to gr}^{(1)}(t) \cong \frac{|\mu_{gr,ex}|^2}{2\hbar^2} \, \text{Re} \int_0^t e^{\frac{-i\tau}{\hbar}(-2\Delta_E - \hbar\Omega - E_\lambda)} e^{\frac{-E_\lambda k_B T \tau^2}{\hbar^2}}$$

$$e^{2i\Delta_0^2 \sin(\omega_0 \tau)} e^{-2\Delta_0^2(1-\cos(\omega_0 \tau))(1+2n(\omega_0))} d\tau.$$

(18.3.41)

For the high-frequency mode, the condition $\hbar\omega_0 \gg k_B T$ typically holds. Consequently, and for simplicity, we consider here the limit in which the thermal occupation number (Eq. (18.2.32)) can be approximated as zero, $n(\omega_0) \approx 0$, leading to

$$k_{gr \to ex}^{(1)}(t) \cong \frac{|\mu_{gr,ex}|^2}{2\hbar^2} e^{-2\Delta_0^2} \, \text{Re} \int_0^t e^{\frac{-i\tau}{\hbar}(2\Delta_E + \hbar\Omega - E_\lambda)} e^{\frac{-E_\lambda k_B T \tau^2}{\hbar^2}} e^{2\Delta_0^2 e^{i\omega_0 \tau}} d\tau$$

$$k_{ex \to gr}^{(1)}(t) \cong \frac{|\mu_{gr,ex}|^2}{2\hbar^2} e^{-2\Delta_0^2} \, \text{Re} \int_0^t e^{\frac{-i\tau}{\hbar}(-2\Delta_E - \hbar\Omega - E_\lambda)} e^{\frac{-E_\lambda k_B T \tau^2}{\hbar^2}} e^{2\Delta_0^2 e^{i\omega_0 \tau}} d\tau$$

(18.3.42)

Using the Taylor expansion, $e^{2\Delta_0^2 e^{i\omega_0 \tau}} = \sum_{n=0}^{\infty} \frac{(\sqrt{2}\Delta_0)^{2n}}{n!} e^{in\omega_0 \tau}$, and taking the upper limit in the time integral to infinity (in accordance with the validity of Fermi's golden rule), we obtain explicit expressions for the spectral line shapes (Ex. 18.3.10),

$$k_{gr \to ex}(\Omega) = \sum_{n=0}^{\infty} \frac{(\sqrt{2}\Delta_0)^{2n}}{n!} e^{-2\Delta_0^2} \frac{|\mu_{gr,ex}|^2}{4} \sqrt{\frac{T}{\hbar^2 k_B T E_\lambda}} e^{-\frac{(\hbar\omega_0 n + E_\lambda - 2\Delta_E - \hbar\Omega)^2}{4 k_B T E_\lambda}}$$

$$k_{ex \to gr}(\Omega) = \sum_{n=0}^{\infty} \frac{(\sqrt{2}\Delta_0)^{2n}}{n!} e^{-2\Delta_0^2} \frac{|\mu_{gr,ex}|^2}{4} \sqrt{\frac{\pi}{\hbar^2 k_B T E_\lambda}} e^{-\frac{(\hbar\omega_0 n + E_\lambda + 2\Delta_E + \hbar\Omega)^2}{4 k_B T E_\lambda}}.$$

(18.3.43)

The spectral lines are shown to be sums of Gaussian peaks, centered at $\hbar\Omega = -2\Delta_E + \hbar\omega_0 n + E_\lambda$ and $\hbar\Omega = -2\Delta_E - \hbar\omega_0 n - E_\lambda$, for absorption and emission, respectively. These peaks correspond to "vibronic transitions," in which the molecule gains (or loses) energy that equals the "adiabatic excitation energy," $-2\Delta_E$ (recall that $\Delta_E < 0$ by our convention), plus an integer number of vibration quanta, $\hbar\omega_0 n$ (see Fig. 18.3.4). Each vibronic peak is broadened and displaced by the reorganization energy, E_λ, attributed to the coupling to low-frequency modes. The peak heights are proportional to the square of the transition dipole and the field intensity, $|\mu_{gr,ex}|^2$ (see Eqs. (18.3.18, 18.3.20)), where the pre-factors, $(\sqrt{2}\Delta_0)^{2n} e^{-2\Delta_0^2}/n!$, depend on the number of vibration quanta involved in the transition. These pre-factors can be readily identified as the squares of Franck–Condon overlap integrals between the ground vibrational state at the initial electronic state and the nth vibrational state at the final electronic state (see Ex. 8.3.11 and Fig. 18.3.4). The ability to resolve individual vibrational line-shapes depends on the coupling of the vibronic transitions to low-frequency modes, and particularly on the ratio between the vibration quantum and the reorganization energy, $\hbar\omega_0/E_\lambda$. In a polar solvent, the reorganization energy is typically large due to efficient coupling to the polarizable medium, and the corresponding line broadening obscures the vibronic peaks. In the gas phase, however, and even in nonpolar solvents, the vibronic structure can be clearly resolved.

Exercise 18.3.9 (a) Show that the time integrals in Eq. (18.3.38) can be rewritten as
$$\int_0^t e^{\frac{\pm i\tau}{\hbar}(2\Delta_E+\hbar\Omega)} c_0(\tau) \prod_{j=1}^N c_j(\tau)d\tau,$$
where the single-mode correlation functions are defined as $c_j(\tau) = e^{2i\Delta_j^2 \sin(\omega_j\tau)} e^{-2\Delta_j^2(1-\cos(\omega_j\tau))(1+2n(\omega_j))}$. (b) Show that $c_0(\tau)$ can be rewritten as $c_0(\tau) = e^{-2\Delta_0^2(2n(\omega_0)+1)} e^{2\Delta_0^2[(n(\omega_0)+1)e^{i\omega_0\tau}+n(\omega_0)e^{-i\omega_0\tau}]}$. (c) Follow the low-frequency approximation, Eqs. (18.2.35–18.2.39), to show that $\prod_{j=1}^N c_j(\tau) = e^{\frac{i\tau}{\hbar}E_\lambda} e^{\frac{-E_\lambda k_B T\tau^2}{\hbar^2}}$, and derive Eq. (18.3.41).

(a)

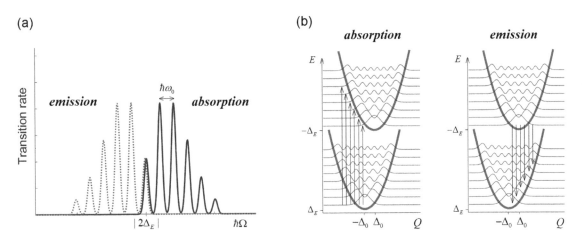

(b) absorption   emission

Figure 18.3.4 (a) Vibronic absorption and emission spectra calculated by Eq. (18.3.43) for the field-driven spin-boson model with a single high-frequency vibrational mode. Within the harmonic approximation, the absorption and emission peaks are symmetrically distributed with respect to the "adiabatic" transition energy, $|2\Delta_E|$. The different peaks (separated by the vibration quantum, $\hbar\omega_0 = 0.1|2\Delta_E|$) correspond to different changes in the vibration quantum number during radiation absorption/emission. The ambient temperature corresponds to $k_B T = 0.25\hbar\omega_0$, which implies that in the initial thermal states, only the vibrational ground state of the high-frequency mode is significantly populated. Each peak is broadened due to additional coupling of the electronic transition to multiple low-frequency modes, characterized by the reorganization energy $E_\lambda = 0.01|2\Delta_E|$. (b) Underlying potential energy curves and corresponding vibrational eigenstates probability densities for the high-frequency vibration. The vibronic transitions associated with the peaks in the spectra are marked by vertical arrows, where the Franck–Condon overlap integrals between the corresponding vibrational eigenstates determine the heights of the corresponding spectral peaks.

Exercise 18.3.10 *Introduce the Taylor expansion of $e^{2\Delta_0^2 e^{i\omega_0 \tau}}$ into Eq. (18.3.42) and then carry out the time integration to infinity. Notice that the time integrand is an even function of time, and use the identity, $\int\limits_{-\infty}^{\infty} dk\, e^{-zk^2} e^{ikx} = \sqrt{\frac{\pi}{z}} e^{\frac{-x^2}{4z}}$, to obtain Eq. (18.3.43).*

Exercise 18.3.11 *Using dimensionless position and momentum variables, $\hat{Q}$ and $\hat{P}$, a coherent state of a one-dimensional harmonic oscillator, $|\alpha\rangle$, is defined as $\langle Q|\alpha\rangle = (\frac{1}{\pi})^{1/4} e^{\frac{-(Q-\sqrt{2}\,\mathrm{Re}(\alpha))^2}{2}} e^{i\sqrt{2}\,\mathrm{Im}(\alpha)Q}$, where $\alpha \equiv \sqrt{\frac{1}{2}(Q_0 + iP_0)}$ (see Eqs. (15.4.11, 15.4.36)). The projections of the coherent state on the eigenstates of the harmonic oscillator Hamiltonian, $\frac{\hbar\omega}{2}(\hat{Q}^2 + \hat{P}^2)|\varphi_n\rangle = \hbar\omega(n + 1/2)|\varphi_n\rangle$, read $|\langle\varphi_n|\alpha\rangle|^2 = e^{-|\alpha|^2}\frac{|\alpha|^{2n}}{n!}$ (see Eq. (15.4.37)). Use this to show that the pre-factors multiplying the Gaussians in Eq. (18.3.43) are related to Franck–Condon overlap integrals, namely, given the harmonic oscillator ground state function, $\varphi_0(Q) = (\frac{1}{\pi})^{1/4} e^{\frac{-Q^2}{2}}$, show that*

$$\left| \int\limits_{-\infty}^{\infty} \varphi_n(Q + \Delta_0)\varphi_0(Q - \Delta_0)dQ \right|^2 = e^{-2\Delta_0^2} \frac{(\sqrt{2}\Delta_0)^{2n}}{n!}.$$

18.4 Förster Resonant Electronic-Energy (Exciton) Transfer (FRET)

A most important elementary process in nanoscale systems is electronic energy transfer between chromophores, often referred to as Fluorescence Resonance Energy Transfer (FRET) or Förster Resonance Energy Transfer, after Theodor Förster who explained the physics underlying this phenomenon [18.6]. In Section 18.3 we learned that electronic energy can be gained or lost by an isolated chromophore due to the interaction between its electric dipole and a time-periodic electric field. In a multichromophore system, however, the interaction between the electric dipoles of nearby chromophores can lead to direct energy transfer between them. Consider, for example, a chromophore in an excited electronic state (an energy "donor"). Instead of emitting its extra energy as electro-magnetic radiation (fluorescence), the energy can be transferred to a nearby chromophore (the energy "acceptor") via dipole–dipole interaction, resulting in an excitation of the latter, in a process termed FRET (see Fig. 18.4.1). When the interchromophore interaction is weak, this process can be readily associated with a rate constant, and when this rate is larger than the fluorescence rate, FRET dominates. In what follows we shall analyze the dependence of FRET on the parameters characterizing the chromophores in their microscopic environments. As we shall see, the rate of FRET increases with increasing spectral overlap between the emission line of the energy donor and the absorption line of the energy acceptor. If the donor and acceptor are of the same type, their emission and absorption lines are Stokes-shifted (see Section 18.3), and the overlap is relatively small. When the donor and acceptor are of a different type, the spectral overlap can become optimal, and the rate of FRET increases. The difference between the fluorescence of the originally excited chromophore (donor) and that of the acceptor enables us to measure the rate of FRET. Since

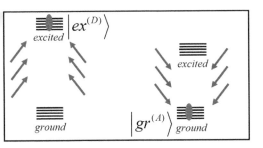
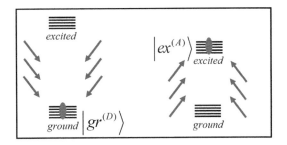

Figure 18.4.1 A schematic illustration of the states involved in FRET between two chromophores. Each chromophore is characterized by its ground and first excited electronic states, coupled to internal as well as external nuclear degrees of freedom in its local surroundings (marked symbolically by arrows). The left and right plots represent the two excited states of the bichromophoric system, where the excitation resides either on the donor (left) or on the acceptor (right).

this rate depends on the square of a dipole–dipole interaction matrix element (see what follows), it falls like R^{-6}, where R is the interchromophore distance. This dependence on R makes FRET a sensitive probe or interchromophore distances (as well as relative orientations) on the nanoscale.

The Model Hamiltonian

The zero-order Hamiltonian for the bi-chromophore system corresponds to two noninteracting chromophores, denoted arbitrarily as a donor (D) and an acceptor (A),

$$\hat{H}_0 = \hat{H}_0^{(D)} \otimes \hat{I}^{(A)} + \hat{I}^{(D)} \otimes \hat{H}_0^{(A)}. \tag{18.4.1}$$

Each chromophore Hamiltonian is projected onto the basis of its ground and first excited electronic states, which are orthonormal eigenstates of the respective electronic Hamiltonian within the Born–Oppenheimer approximation (see Eqs. (18.3.2, 18.3.3)),

$$\langle gr^{(D)}|gr^{(D)}\rangle = \langle ex^{(D)}|ex^{(D)}\rangle = 1 \quad ; \quad \langle gr^{(D)}|ex^{(D)}\rangle = 0$$
$$\langle gr^{(A)}|gr^{(A)}\rangle = \langle ex^{(A)}|ex^{(A)}\rangle = 1 \quad ; \quad \langle gr^{(A)}|ex^{(A)}\rangle = 0. \tag{18.4.2}$$

The single-chromophore Hamiltonians therefore read (see Eq. (18.3.7))

$$\hat{H}_0^{(A)} = \hat{H}_{gr,\mathbf{Q}_A}^{(A)}|gr^{(A)}\rangle\langle gr^{(A)}| + \hat{H}_{ex,\mathbf{Q}_A}^{(A)}|ex^{(A)}\rangle\langle ex^{(A)}|$$
$$\hat{H}_0^{(D)} = \hat{H}_{gr,\mathbf{Q}_D}^{(D)}|gr^{(D)}\rangle\langle gr^{(D)}| + \hat{H}_{ex,\mathbf{Q}_D}^{(D)}|ex^{(D)}\rangle\langle ex^{(D)}|. \tag{18.4.3}$$

Here, $\mathbf{Q}_D$ and $\mathbf{Q}_A$ stand for the (dimensionless) nuclear coordinates within the space of each chromophore (internal, or in the local surroundings). The corresponding nuclear Hamiltonians are modeled in terms of additive contributions of harmonic modes (Eqs.

(18.3.9–18.3.13)), where the nuclear equilibrium positions change upon a change in the electronic state of the relevant chromophore,

$$\hat{H}^{(A)}_{gr,\mathbf{Q}_A} = \Delta_E^{(A)} + \sum_{j_A} \frac{\hbar\omega_{j_A}}{2}(\hat{P}_{j_A}^2 + \hat{Q}_{j_A}^2) + \hbar\omega_{j_A}\Delta_{j_A}\hat{Q}_{j_A}$$

$$\hat{H}^{(A)}_{ex,\mathbf{Q}_A} = -\Delta_E^{(A)} + \sum_{j_A} \frac{\hbar\omega_{j_A}}{2}(\hat{P}_{j_A}^2 + \hat{Q}_{j_A}^2) - \hbar\omega_{j_A}\Delta_{j_A}\hat{Q}_{j_A}$$

$$\hat{H}^{(D)}_{gr,\mathbf{Q}_D} = \Delta_E^{(D)} + \sum_{j_D} \frac{\hbar\omega_{j_D}}{2}(\hat{P}_{j_D}^2 + \hat{Q}_{j_D}^2) + \hbar\omega_{j_D}\Delta_{j_D}\hat{Q}_{j_D}$$

$$\hat{H}^{(D)}_{ex,\mathbf{Q}_D} = -\Delta_E^{(D)} + \sum_{j_D} \frac{\hbar\omega_{j_D}}{2}(\hat{P}_{j_D}^2 + \hat{Q}_{j_D}^2) - \hbar\omega_{j_D}\Delta_{j_D}\hat{Q}_{j_D}. \tag{18.4.4}$$

The energy gap between the multidimensional potential energy surface minima at the ground and the excited states in each chromophore (the adiabatic excitation energy) is denoted as

$$2\Delta_E^{(A)} = \varepsilon_{gr}^{(A)} - \varepsilon_{ex}^{(A)},$$
$$2\Delta_E^{(D)} = \varepsilon_{gr}^{(D)} - \varepsilon_{ex}^{(D)} \tag{18.4.5}$$

where, by our convention, $\Delta_E^{(D)}, \Delta_E^{(A)} < 0$ (see Eq. (18.3.11)).

Focusing on electronic energy transfer between the two chromophores, we now restrict the discussion to the space of a single excitation within the bi-chromophore system. Namely, we further project the zero-order Hamiltonian onto the space in which one (and only one) chromophore is electronically excited (the single exciton manifold). The two relevant electronic states of the bi-chromophore system are products of single-chromophore states,

$$|D^*\rangle = |ex^{(D)}\rangle \otimes |gr^{(A)}\rangle \quad ; \quad |A^*\rangle = |gr^{(D)}\rangle \otimes |ex^{(A)}\rangle, \tag{18.4.6}$$

where $|D^*\rangle$ and $|A^*\rangle$ correspond to the exciton residing on the donor and the acceptor, respectively (see Fig. 18.4.1). The zero-order Hamiltonian in the single exciton manifold thus reads (Ex. 18.4.1)

$$\hat{H}_0 = \hat{H}_{D^*,\mathbf{Q}_D,\mathbf{Q}_A}|D^*\rangle\langle D^*| + \hat{H}_{A^*,\mathbf{Q}_D,\mathbf{Q}_A}|A^*\rangle\langle A^*| \tag{18.4.7}$$

$$\hat{H}_{D^*,\mathbf{Q}_D,\mathbf{Q}_A} = \hat{H}^{(D)}_{ex,\mathbf{Q}_D} + \hat{H}^{(A)}_{gr,\mathbf{Q}_A}; \quad \hat{H}_{A^*,\mathbf{Q}_D,\mathbf{Q}_A} = \hat{H}^{(D)}_{gr,\mathbf{Q}_D} + \hat{H}^{(A)}_{ex,\mathbf{Q}_A}. \tag{18.4.8}$$

Exercise 18.4.1 *The Hamiltonian of the bichromophoric system in the absence of interchromophore interaction ($\hat{H}_0$) is given by Eq. (18.4.1), where the single-chromophore Hamiltonians are given by Eq. (18.4.3). Calculate the matrix elements of $\hat{H}_0$ in the basis of the bi-chromophore states, $|D^*\rangle = |ex^{(D)}\rangle \otimes |gr^{(A)}\rangle$ and $|A^*\rangle = |gr^{(D)}\rangle \otimes |ex^{(A)}\rangle$, and derive Eq. (18.4.7).*

We now turn to modeling the interaction between the two chromophores. Again, a rigorous derivation based on quantum electrodynamics is beyond our scope. We restrict the discussion to a typical scenario in which the chromophore size (typically < 1 nm) is smaller than the interchromophore distance, R, and each chromophore is neutral. Expanding in powers of R the Coulomb interaction energy between the charge distributions at the two chromophores, the dominant term is the dipole–dipole interaction,

$$\hat{J} = \frac{\hat{\mathbf{d}}_D \cdot \hat{\mathbf{d}}_A}{R^3} - 3\frac{(\hat{\mathbf{d}}_D \cdot \mathbf{R}/R)(\hat{\mathbf{d}}_A \cdot \mathbf{R}/R)}{R^3}. \tag{18.4.9}$$

Here, $\hat{\mathbf{d}}_D$ and $\hat{\mathbf{d}}_A$ are the donor and acceptor dipole operators in the respective reference frames (see Eq. (18.3.17)), the vector connecting the centers of mass of the two chromophores is $\mathbf{R}$, where $R = |\mathbf{R}|$. The dipole–dipole interaction facilitates energy transfer between the donor and the acceptor. Notice that the interaction energy depends on the magnitude, as well as on the relative orientation of the two dipoles, and decays with the interchromophore distance as R^3. This long range nature of the interaction typically enables energy transfer at a substantial rate, even when the chromophores are a few nanometers apart. Indeed, electronic energy transfer is enabled over interchromophore distances in which charge transfer between the chromophores is excluded, owing to the exponential decay of the corresponding electronic tunneling matrix element (see Section 18.2).

Representing the dipole–dipole interaction in the basis of single exciton states, we obtain

$$\hat{V} = J_{D^*}(\mathbf{Q}_D, \mathbf{Q}_A)|D^*\rangle\langle D^*| + J_{A^*}(\mathbf{Q}_D, \mathbf{Q}_A)|A^*\rangle\langle A^*| + J_{D^*,A^*}|D^*\rangle\langle A^*| + J_{A^*,D^*}|A^*\rangle\langle D^*| \tag{18.4.10}$$

where $J_{D^*}(\mathbf{Q}_D, \mathbf{Q}_A) = \langle D^*|\hat{J}|D^*\rangle$, $J_{A^*}(\mathbf{Q}_D, \mathbf{Q}_A) = \langle A^*|\hat{J}|A^*\rangle$, $J_{D^*,A^*} = \langle D^*|\hat{J}|A^*\rangle$ and $J_{A^*,D^*} = \langle A^*|\hat{J}|D^*\rangle$. Notice that within the Born–Oppenheimer approximation (see Eqs. (18.3.20, 18.3.21)) the off-diagonal matrix elements of the dipole–dipole interaction do not depend on the nuclear coordinates (the Condon approximation).

The full Hamiltonian is the sum of the zero-order (Eqs. (18.4.4, 18.4.7, 18.4.8)) and the dipole–dipole interaction (Eq. (18.4.10)) terms, $\hat{H} = \hat{H}_0 + \hat{V}$. The zero-order Hamiltonian has the same structure as the corresponding Hamiltonian for charge transfer (18.2.14, 18.2.16)). The apparent difference is that in the case of energy transfer, the donor and acceptor states are defined in a bi-chromophore space, and accordingly the nuclear Hamiltonians are sums over modes from the two chromophores (Eq. (18.4.8)). The eigenstates of $\hat{H}_0$ are products of an electronic and a nuclear states, in the bi-chromophore system,

$$\hat{H}_0|D^*\rangle \otimes |\chi_{D^*,\mathbf{n}}\rangle = \varepsilon_{D^*,\mathbf{n}}|D^*\rangle \otimes |\chi_{D^*,\mathbf{n}}\rangle$$
$$\hat{H}_0|A^*\rangle \otimes |\chi_{A^*,\mathbf{m}}\rangle = \varepsilon_{A^*,\mathbf{m}}|A^*\rangle \otimes |\chi_{A^*,\mathbf{m}}\rangle. \tag{18.4.11}$$

The nuclear states are products of single-chromophore states,

$$\langle \mathbf{Q}_D, \mathbf{Q}_A|\chi_{D^*,\mathbf{n}}\rangle = \langle \mathbf{Q}_D|\chi_{ex,\mathbf{n}}\rangle\langle \mathbf{Q}_A|\chi_{gr,\mathbf{n}}\rangle$$
$$\langle \mathbf{Q}_D, \mathbf{Q}_A|\chi_{A^*,\mathbf{m}}\rangle = \langle \mathbf{Q}_D|\chi_{gr,\mathbf{m}}\rangle\langle \mathbf{Q}_A|\chi_{ex,\mathbf{m}}\rangle \tag{18.4.12}$$

which are products of displaced harmonic oscillator eigenfunctions,

$$\langle \mathbf{Q}_D | \chi_{ex,\mathbf{n}} \rangle = \varphi_{n_1}(Q_1 - \Delta_1) \cdot \varphi_{n_2}(Q_2 - \Delta_2) \cdot \varphi_{n_3}(Q_3 - \Delta_3) \cdots$$
$$\langle \mathbf{Q}_D | \chi_{gr,\mathbf{m}} \rangle = \varphi_{m_1}(Q_1 + \Delta_1) \cdot \varphi_{m_2}(Q_2 + \Delta_2) \cdot \varphi_{m_3}(Q_3 + \Delta_3) \cdots$$
$$\langle \mathbf{Q}_A | \chi_{ex,\mathbf{m}} \rangle = \varphi_{m_1}(Q_1 - \Delta_1) \cdot \varphi_{m_2}(Q_2 - \Delta_2) \cdot \varphi_{m_3}(Q_3 - \Delta_3) \cdots$$
$$\langle \mathbf{Q}_A | \chi_{gr,\mathbf{n}} \rangle = \varphi_{n_1}(Q_1 + \Delta_1) \cdot \varphi_{n_2}(Q_2 + \Delta_2) \cdot \varphi_{n_3}(Q_3 + \Delta_3) \cdots$$

(18.4.13)

The corresponding $\hat{H}_0$-eigenstates are

$$\varepsilon_{D^*,\mathbf{n}} = \Delta_E^{(A)} - \Delta_E^{(D)} - \sum_{j \in \{j_A, j_D\}} \frac{\hbar \omega_j}{2} \Delta_j^2 + \sum_j \hbar \omega_j \left(n_j + \frac{1}{2}\right)$$

$$\varepsilon_{A^*,\mathbf{m}} = \Delta_E^{(D)} - \Delta_E^{(A)} - \sum_{j \in \{j_A, j_D\}} \frac{\hbar \omega_j}{2} \Delta_j^2 + \sum_j \hbar \omega_j \left(m_j + \frac{1}{2}\right)$$

(18.4.14)

Thermal Energy Transfer Rates

We are interested in the rate of electronic energy transfer between the chromophores. As discussed in Chapter 17, perturbation theory is valid when the state-to-state coupling matrix elements are sufficiently small. In the present case, these matrix elements obtain the form $\langle \chi_{D^*,\mathbf{n}} | \chi_{A^*,\mathbf{m}} \rangle \langle D^* | \hat{J} | A^* \rangle$. Since $\langle D^* | \hat{J} | A^* \rangle \propto R^{-3}$, and since the multidimensional Franck–Condon factors, $\langle \chi_{D^*,\mathbf{n}} | \chi_{A^*,\mathbf{m}} \rangle$, are smaller than or equal to unity, perturbation theory becomes valid for sufficiently large interchromophore distances. Without loss of generality, we shall consider explicitly the transition from the donor to the acceptor (where similar considerations apply to the reverse process). The donor and acceptor manifolds of $\hat{H}_0$-eigenstates are associated with the projection operators (see Eq. (17.3.1)),

$$\hat{P}_{D^*} = |D^*\rangle\langle D^*| \quad ; \quad \hat{P}_{A^*} = |A^*\rangle\langle A^*|,$$

(18.4.15)

where $\hat{P}_{D^*}^2 = \hat{P}_{D^*}, \hat{P}_{D^*}\hat{P}_{A^*} = 0, \hat{P}_{A^*}^2 = \hat{P}_{A^*}$ and $[\hat{P}_{D^*}, \hat{H}_0] = [\hat{P}_{A^*}, \hat{H}_0] = 0$ (see Eqs. (17.3.2, 17.3.3)). The definition of a thermal energy transfer rate (see Section 17.3) assumes an initial state of the bi-chromophore system, in which all nuclei are at quasi equilibrium. The corresponding density operator reads (Eq. (17.3.10))

$$\hat{\rho}_{D^*}(0) = \frac{e^{-\hat{H}_0/(k_B T)} \hat{P}_{D^*}}{Z_{D^*}} \quad ; \quad Z_{D^*} = tr\{e^{-\hat{H}_0/(k_B T)} \hat{P}_{D^*}\}.$$

(18.4.16)

Notice that the thermal state of each nuclear mode is determined by the electronic state of the chromophore to which the mode is coupled. Implementing the general expression for the thermal rate to the present case, we obtain (Eq. (17.3.12))

$$k_{D^* \to A^*}^{(1)}(t) \cong \frac{2}{\hbar^2} \text{Re} \int_0^t tr\{\hat{\rho}_{D^*}(0)\hat{P}_{D^*}\hat{V}\hat{P}_{A^*} e^{\frac{i\hat{H}_0 \tau}{\hbar}} \hat{V} e^{\frac{-i\hat{H}_0 \tau}{\hbar}}\} d\tau.$$

(18.4.17)

Introducing explicitly the zero-order Hamiltonian (Eq. (18.4.7)) and the interaction (Eq. (18.4.10)), and taking the trace over the electronic degrees of freedom, the rate is expressed as a time integral over a nuclear correlation function (Ex. 18.4.2),

$$k_{D^* \to A^*}^{(1)}(t) = 2 \, \text{Re} \int_0^t c_{D^*, \mathbf{Q}_D, \mathbf{Q}_A}(\tau) d\tau$$

$$c_{D^*, \mathbf{Q}_D, \mathbf{Q}_A}(\tau) = \frac{|J_{D^*, A^*}|^2}{\hbar^2} tr_{\mathbf{Q}_D, \mathbf{Q}_A} \left\{ \frac{e^{-\hat{H}_{D^*, \mathbf{Q}_D, \mathbf{Q}_A}/(k_B T)}}{Z_{D^*}} e^{\frac{-i\hat{H}_{D^*, \mathbf{Q}_D, \mathbf{Q}_A}\tau}{\hbar}} e^{\frac{i\hat{H}_{A^*, \mathbf{Q}_D, \mathbf{Q}_A}\tau}{\hbar}} \right\}. \quad (18.4.18)$$

A formal expression for the "rate constant" of energy transfer can be obtained by evaluating the trace using a complete orthonormal system of multidimensional harmonic oscillator eigenfunctions. This yields (Ex. 18.4.3)

$$k_{D^* \to A^*}^{(1)}(t) = \frac{2|J_{D^*, A^*}|^2}{\hbar^2} \sum_{\mathbf{n}, \mathbf{m}} \frac{e^{-\varepsilon_{D^*, \mathbf{n}}/(k_B T)}}{Z_{D^*}} \, \text{Re} \int_0^t e^{\frac{-i(\varepsilon_{D^*, \mathbf{n}} - \varepsilon_{A^*, \mathbf{m}})\tau}{\hbar}} |\langle \chi_{D^*, \mathbf{n}} | \chi_{A^*, \mathbf{m}} \rangle|^2 d\tau.$$

$$(18.4.19)$$

In the limits of weak state-to-state coupling between an initial thermal state and a dense and wide manifold of final states, the integral over oscillating terms in Eq. (18.4.19) converges within a time that is much shorter than the characteristic time of energy transfer between the donor and the acceptor. (This is in close analogy to our detailed discussion of charge transfer (Section 18.2) and radiation absorption/emission (Section 18.3)). Therefore, the upper time-limit can be replaced by infinity, leading to Fermi's golden rule rate,

$$k_{D^* \to A^*} \cong \frac{2\pi}{\hbar} \sum_{\mathbf{n}, \mathbf{m}} \frac{e^{-\varepsilon_{D^*, \mathbf{n}}/(k_B T)}}{Z_{D^*}} |J_{D^*, A^*}|^2 |\langle \chi_{D^*, \mathbf{n}} | \chi_{A^*, \mathbf{m}} \rangle|^2 \delta(\varepsilon_{D^*, \mathbf{n}} - \varepsilon_{A^*, \mathbf{m}}). \quad (18.4.20)$$

A more informative expression is obtained by invoking the infinite time limit already in Eq. (18.4.18), and using the symmetry of the nuclear correlation function, $[c_{D^*, \mathbf{Q}_D, \mathbf{Q}_A}(\tau)]^* = c_{D^*, \mathbf{Q}_D, \mathbf{Q}_A}(-\tau)$ (Ex. 18.4.4), which leads to

$$k_{D^* \to A^*} = \int_{-\infty}^{\infty} c_{D^*, \mathbf{Q}_D, \mathbf{Q}_A}(\tau) d\tau. \quad (18.4.21)$$

We then notice that the nuclear function in Eq. (18.4.18) factorizes into a product of two independent dipole correlation functions (Eqs. (18.3.30–18.3.32)) of the two chromophores (see Ex. 18.4.5),

$$c_{D^*, \mathbf{Q}_D, \mathbf{Q}_A}(\tau) = \frac{16|J_{D^*, A^*}|^2 \hbar^2}{|\mu_{gr,ex}^{(D)}|^2 |\mu_{gr,ex}^{(A)}|^2} c_{ex}^{(D)}(\tau) \cdot c_g^{(A)}(\tau). \quad (18.4.22)$$

Substitution of this result in Eq. (18.4.21), one obtains the following expression for the energy transfer rate (see Ex. 18.4.6):

$$k_{D^* \to A^*} = \frac{8|J_{D^*, A^*}|^2 \hbar^2}{\pi |\mu_{gr,ex}^{(D)}|^2 |\mu_{gr,ex}^{(A)}|^2} \int_{-\infty}^{\infty} d\Omega \, k_{ex \to gr}^{(D)}(\Omega) k_{gr \to ex}^{(A)}(\Omega), \quad (18.4.23)$$

where $k_{ex \to gr}^{(D)}(\Omega)$ and $k_{gr \to ex}^{(A)}(\Omega)$ are, respectively, the rate expressions obtained in Eq. (18.3.37) for radiation emission from the donor, and radiation absorption by the acceptor. The rate of electronic energy transfer is shown to be proportional to the spectral overlap between the donor emission and the acceptor absorption spectral lines, namely an integral of the rate of emission by the donor multiplied by the rate of absorption by the acceptor, over all possible frequencies [18.6].

The Förster energy transfer rate is therefore maximized when the emission peak of the donor coincides with the absorption peak of the acceptor. Recalling that for any given chromophore the emission peak is (red-)shifted to lower frequencies with respect to the absorption peak (due to coupling of the electronic transitions to nuclear modes), the energy transfer rate is optimized when the donor and acceptor are different, where the acceptor's adiabatic excitation energy is smaller than that of the donor (see Fig. 18.4.2). As we can see, optimal energy transfer rate requires that the excited donor energy is not fully converted into the electronic excitation of the acceptor, where the difference is lost into heat production.

Exercise 18.4.2 *(This exercise is completely analogous to Ex. 18.2.4 for charge transfer): (a) Use the explicit expressions for the zero-order Hamiltonian (Eq. (18.4.7)) and show that $f(\hat{H}_0)|D^*\rangle = f(\hat{H}_{D,\mathbf{Q}_D,\mathbf{Q}_A})|D^*\rangle$ and $f(\hat{H}_0)|A^*\rangle = f(\hat{H}_{A^*,\mathbf{Q}_D,\mathbf{Q}_A})|A^*\rangle$, where $f(\hat{H}_0)$ is an analytic function of the respective operator. (b) Use the results of (a), the interaction operator (Eq. (18.4.10)), and the definitions of the initial and final ensembles (Eqs. (18.4.15, 18.4.16)) to derive Eq. (18.4.18) from Eq. (18.4.17). Recall that the trace over the full electronic and nuclear space can be expressed as $tr\{\hat{O}\} = tr_{\mathbf{Q}_D,\mathbf{Q}_A}\{\langle D^*|\hat{O}|D^*\rangle + \langle A^*|\hat{O}|A^*\rangle\}$.*

Exercise 18.4.3 *(This exercise is completely analogous to Ex. 18.2.4 for charge transfer). Derive Eq. (18.4.19) from Eq. (18.4.18) by evaluating the trace over the nuclear space using a complete set of eigenstates of the multidimensional Hamiltonian, $\hat{H}_{D^*,\mathbf{Q}_D,\mathbf{Q}_A}$, and an identity operator, expressed in terms of $\hat{H}_{A^*,\mathbf{Q}_D,\mathbf{Q}_A}$-eigenstates (Eqs. (18.4.11, 18.4.12)).*

Exercise 18.4.4 *The dipole–dipole correlation function is defined in Eq. (18.4.18). Show that $c^*_{D^*,\mathbf{Q}_D,\mathbf{Q}_A}(\tau) = c_{D^*,\mathbf{Q}_D,\mathbf{Q}_A}(-\tau)$. Use this result to derive Eq. (18.4.21) from Eq. (18.4.18), in the limit $t \to \infty$.*

Exercise 18.4.5 *The dipole–dipole correlation function, $c_{D^*,\mathbf{Q}_D,\mathbf{Q}_A}(\tau)$, is defined in Eq. (18.4.18). Use the decomposition of $\hat{H}_{D^*,\mathbf{Q}_D,\mathbf{Q}_A}$ and $\hat{H}_{A^*,\mathbf{Q}_D,\mathbf{Q}_A}$ in terms of "local" donor and acceptor modes (Eq. (18.4.8)), and the commutativity of donor-space and acceptor space operators, namely, $[\hat{H}^{(D)}_{gr,\mathbf{Q}_D},\hat{H}^{(A)}_{gr,\mathbf{Q}_A}] = [\hat{H}^{(D)}_{gr,\mathbf{Q}_D},\hat{H}^{(A)}_{ex,\mathbf{Q}_A}] = [\hat{H}^{(D)}_{ex,\mathbf{Q}_D},\hat{H}^{(A)}_{gr,\mathbf{Q}_A}] = [\hat{H}^{(D)}_{ex,\mathbf{Q}_D},\hat{H}^{(A)}_{ex,\mathbf{Q}_A}] = 0$, to express $c_{D^*,\mathbf{Q}_D,\mathbf{Q}_A}(\tau)$ in terms of the local dipole correlation functions, as defined in Eqs. (18.3.30–18.3.32).*

Exercise 18.4.6 *(a) Use the identity*

$$\int\limits_{-\infty}^{\infty} dt\, f(t)g(t) = \int\limits_{-\infty}^{\infty} d\Omega \frac{1}{\sqrt{2\pi}} \int\limits_{-\infty}^{\infty} dt\, e^{i\Omega t} f(t) \frac{1}{\sqrt{2\pi}} \int\limits_{-\infty}^{\infty} dt'\, e^{-i\Omega t'} g(t')$$

to express the time-integral over the nuclear correlation function in Eq. (18.4.22) as an integral over Ω. (b) Derive Eq. (18.4.23) by substitution of the result (a) in

Eq. (18.4.21) and identifying the donor emission rate and the acceptor absorption rate, as defined in Eq. (18.3.37).

It is instructive to express the energy transfer rate in terms of the spin-boson model parameters. Using the analogy between charge transfer (Eq. (18.2.26)) and electronic energy transfer (Eq. (18.4.18)), we obtain the following expression for the time-dependent energy transfer rate (the analogue of Eq. (18.2.33)):

$$k_{D^* \to A^*}^{(1)}(t) = \frac{2|J_{D^*,A^*}|^2}{\hbar^2} \text{Re} \int_0^t e^{\frac{-i\tau}{\hbar}(2\Delta_E^{(A)} - 2\Delta_E^{(D)})}$$

$$e^{2i \sum\limits_{j \in \{jD, jA\}} \Delta_j^2 \sin(\omega_j \tau)} \, e^{-2 \sum\limits_{j \in \{jD, jA\}} \Delta_j^2 (1 - \cos(\omega_j \tau))(1 + 2n(\omega_j))} \, d\tau. \qquad (18.4.24)$$

In the limits of low nuclear frequencies, $\frac{(\hbar \omega_j)^2}{k_B T E_\lambda} \ll 8\pi^2$, and weak dipole–dipole interaction $\frac{|J_{D^*,A^*}|^2}{2k_B T E_\lambda} \ll 1$ (see Eqs. (18.2.44, 18.2.45)), the transfer rate can be approximated as a constant (Fermi's golden rule), given by a particularly simple expression (Ex. 18.4.7),

$$k_{D^* \to A^*} = |J_{D^*,A^*}|^2 \sqrt{\frac{\pi}{\hbar^2 k_B T E_\lambda}} e^{\frac{-(E_\lambda - (2\Delta_E^{(A)} - 2\Delta_E^{(D)}))^2}{4k_B T E_\lambda}}. \qquad (18.4.25)$$

The result is perfectly analogous to the Marcus formula for charge transfer (Eq. (18.2.43)). In the case of energy transfer, the nuclear reorganization energy includes the contributions from both the donor and acceptor modes (internal as well as in their local surroundings), $E_\lambda \equiv \sum\limits_{j \in \{jD, jA\}} 2\Delta_j^2 \omega_j \hbar = \sum\limits_{j \in \{jD, jA\}} \frac{\hbar \omega_j}{2}(2\Delta_j)^2$, and the "driving force" is the difference between the adiabatic excitation energies of the donor and acceptor chromophores, $|2\Delta_E^{(D)}| - |2\Delta_E^{(A)}| = (\varepsilon_{ex}^{(D)} - \varepsilon_{gr}^{(D)}) - (\varepsilon_{ex}^{(A)} - \varepsilon_{gr}^{(A)})$ (recall that by our convention, $\Delta_E^{(D)}, \Delta_E^{(A)} < 0$; see Eq. (18.3.11)).

Notice that the Marcus-like formula for the Förster transfer rate (Eq. (18.4.25)) can be derived alternatively by using the low-frequency limit of the donor emission and acceptor absorption spectral lines (Eq. (18.3.40)) and by calculating the spectral overlap between them, according to Eq. (18.4.23) (see Ex. 18.4.8). Denoting the local chromophore reorganization energies $E_\lambda^{(D)} \equiv \sum\limits_{jD} 2\Delta_{jD}^2 \omega_{jD} \hbar$ and $E_\lambda^{(A)} \equiv \sum\limits_{jA} 2\Delta_{jA}^2 \omega_{jA} \hbar$, and the adiabatic excitation energies $|2\Delta_E^{(D)}|$ and $|2\Delta_E^{(A)}|$, the energy transfer rate is maximal when the overlap between the donor emission and acceptor absorption lines is optimal, namely, when $|2\Delta_E^{(D)}| - E_\lambda^{(D)} \cong |2\Delta_E^{(A)}| + E_\lambda^{(A)}$ (see Fig. 18.4.2).

Exercise 18.4.7 *The golden rule expression for the time-dependent charge transfer rate within the spin-boson model is given by Eq. (18.2.33). Invoking additional approximations, one obtains the semiclassical golden rule rate (Marcus formula), Eq. (18.2.43). Use the analogy between Eq. (18.2.33) and Eq. (18.4.24) to derive Eq. (18.4.25) for the electronic energy transfer rate, within the same set of approximations.*

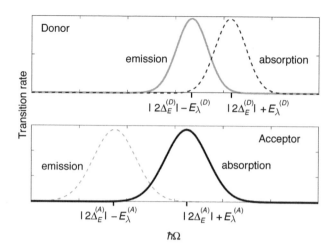

Illustration of the spectral lines of two chromophores. The donor emission and the acceptor absorption are presented as solid lines (the donor absorption and acceptor emission are plotted as dashed lines for completeness). The energy transfer rate is maximal for a pair of donor and acceptor that are different from each other, with maximal overlap between the donor emission and the acceptor absorption lines.

Exercise 18.4.8 *Use the semiclassical golden rule expressions for the absorption and emission spectral lines (Eq. (18.3.40)), with the local chromophore reorganization energies, $E_\lambda^{(D)} \equiv \sum_{j_D} 2\Delta_{j_D}^2 \omega_{j_D} \hbar$ and $E_\lambda^{(A)} \equiv \sum_{j_A} 2\Delta_{j_A}^2 \omega_{j_A} \hbar$, to derive the Marcus-like formula for the electronic energy transfer rate (Eq. (18.4.25)) as the spectral overlap integral, Eq. (18.4.23). Notice that the total reorganization energy includes all the modes of the donor and the acceptor chromophores, $E_\lambda = E_\lambda^{(D)} + E_\lambda^{(A)}$.*

Bibliography

[18.1] A. J. Leggett, S. Chakravarty, A. T. Dorsey, M. P. A. Fisher, A. Garg, and W. Zwerger, "Dynamics of the dissipative two-state system," Reviews of Modern Physics 59, 1 (1987).

[18.2] V. May and O. Kühn, "Charge and Energy Transfer Dynamics in Molecular Systems," 3rd ed. (Wiley, 2011).

[18.3] R. A. Marcus and N. Sutin, "Electron transfers in chemistry and biology," Biochimica et Biophysica Acta (BBA)-Reviews on Bioenergetics 811, 265 (1985).

[18.4] E. U. Condon, "Nuclear motions associated with electron transitions in diatomic molecules," Physical Review 32, 858 (1928).

[18.5] U. Peskin, Quantum mechanical averaging over fluctuating rates," Molecular Physics 110, 729 (2012).

[18.6] T. Förster, "Intermolecular energy migration and fluorescence," Annals of Physics 437, 55 (1948).

Open Quantum Systems

19.1 Exact Reduced System Dynamics

In many cases of interest, measurements performed on a large macroscopic system are aimed at some well-defined part of it. The relevant part of the full system is often referred to as "the reduced system." Imagine, for example, a distinguishable set of particles of interest embedded in a system of many other particles. The presence of interaction between the particles of interest and the other particles means that the system Hamiltonian is nonseparable, hence the particles of interest are generally entangled with the rest, and their exact dynamics is associated with the full (many-particle) system's wave function (in the case of a pure state) or a density matrix (in the general case). However, the measurement aims only at the reduced system, which, in quantum mechanical terms, means that the measurement operator is a trivial identity in the space of the other particles. Consequently, expectation values of system observables involve simple averaging with respect to the (accessible) states of the other particles. Such averaging has a remarkable effect on observables within the reduced system, and on phenomena such as exchange of energy between the reduced system and its environment (including relaxation, effective friction forces, and so forth), and de-coherence between the system Hamiltonian eigenstates, including disentanglement between particles. These phenomena are essential to the emergence of classical-like dynamics and irreversible processes in the system of interest.

Reduced dynamics can be of interest even for noninteracting particles when the measurement involves projection from the full single-particle Hilbert space into a part of it. For example, local measurements project onto a certain region in the particle's coordinate space. In such a case, the number of particles within the reduced system (the projected region of interest) does not need to be conserved, and the reduced system dynamics can exhibit a wealth of phenomena, such as transport of particles and/or energy, probability gain or loss, and resonant decay [7.3]. In Chapters 17 and 18 we already encountered examples of transfer of particles between two quantum states (e.g., a "donor" and an "acceptor") within a macroscopic system. These states were characterized by different subgroups of states projected out of the full system Hilbert space, where each subgroup corresponded to a reduced system.

Since reduced quantum systems can exchange particles and/or energy with the rest of "the universe," they are referred to as open quantum systems. The rest of the universe is referred to as their environment, or the surrounding. *Formally exact equations of motion for the open system dynamics can be readily derived, using appropriate projection operators*, as discussed in what follows. It is emphasized that these formal equations become cumbersome to implement, as they strictly involve the "dynamics of the universe." However, *in many cases, valid approximations lead to relatively simple closed equations for the reduced system dynamics. Most relevant in this context is the Markovian approximation, which essentially assumes a timescale separation between the (slow) environment-induced dynamics within the system, and the (fast) system-induced dynamics within its surroundings.* In this section we start from the exact expressions for the reduced dynamics, and the Markovian approximation is derived in the next sections.

Hilbert Space Projectors

Let us start by considering a generic system Hamiltonian, $\hat{H}(t)$, and two projection operators, $\hat{P}$, and $\hat{Q}$, whose sum composes the identity operator in the system's Hilbert space:

$$\hat{P} + \hat{Q} = \hat{I} \quad ; \quad \hat{P}^2 = \hat{P} \quad ; \quad \hat{Q}^2 = \hat{Q} \quad ; \quad \hat{P}\hat{Q} = 0. \tag{19.1.1}$$

We are interested in the projection of the exact wave function onto the $\hat{P}$-space, namely $\hat{P}\psi(t)$, where $\psi(t)$ is an exact solution of the Schrödinger equation,

$$\frac{\partial}{\partial t}|\psi(t)\rangle = -\frac{i}{\hbar}\hat{H}(t)|\psi(t)\rangle. \tag{19.1.2}$$

Substitution of the projection operators readily yields

$$\begin{aligned}
\frac{\partial}{\partial t}|\hat{P}\psi(t)\rangle &= -\frac{i}{\hbar}\hat{P}\hat{H}(t)|\hat{P}\psi(t)\rangle - \frac{i}{\hbar}\hat{P}\hat{H}(t)|\hat{Q}\psi(t)\rangle \\
\frac{\partial}{\partial t}|\hat{Q}\psi(t)\rangle &= -\frac{i}{\hbar}\hat{Q}\hat{H}(t)|\hat{P}\psi(t)\rangle - \frac{i}{\hbar}\hat{Q}\hat{H}(t)|\hat{Q}\psi(t)\rangle
\end{aligned}. \tag{19.1.3}$$

A formal solution to the second equation reads (see Ex. 19.1.1)

$$|\hat{Q}\psi(t)\rangle \equiv \hat{U}_Q(t,0)|\psi(0)\rangle - \frac{i}{\hbar}\int_0^t d\tau \hat{U}_Q(t,\tau)\hat{H}(\tau)|\hat{P}\psi(\tau)\rangle, \tag{19.1.4}$$

where a Q-space propagator, $\hat{U}_Q(t,\tau)$, is defined here by the differential equation

$$\frac{\partial}{\partial t}\hat{U}_Q(t,\tau) = -\frac{i}{\hbar}\hat{Q}\hat{H}(t)\hat{Q}\hat{U}_Q(t,\tau) \quad ; \quad \hat{U}_Q(t,t) \equiv \hat{Q}. \tag{19.1.5}$$

When the initial state is confined to the $\hat{P}$ space, we have $|\hat{Q}\psi(0)\rangle = 0$, and therefore,

$$\frac{\partial}{\partial t}|\hat{P}\psi(t)\rangle = -\frac{i}{\hbar}\hat{P}\hat{H}(t)|\hat{P}\psi(t)\rangle - \frac{1}{\hbar^2}\int_0^t d\tau \hat{P}\hat{H}(t)\hat{U}_Q(t,\tau)\hat{H}(\tau)|\hat{P}\psi(\tau)\rangle. \tag{19.1.6}$$

For a time-independent Hamiltonian, the time-evolution operator, $\hat{U}_Q(t,\tau)$, obtains the standard exponential form, $\hat{U}_Q(t,\tau) = \hat{Q}e^{\frac{-i(t-\tau)}{\hbar}\hat{Q}\hat{H}\hat{Q}}$ (see Eq. (15.2.11)), which leads to the explicit equation for the reduced space wave function, $|\hat{P}\psi(t)\rangle$:

$$\frac{\partial}{\partial t}|\hat{P}\psi(t)\rangle = -\frac{i}{\hbar}\hat{P}\hat{H}\hat{P}|\hat{P}\psi(t)\rangle - \frac{1}{\hbar^2}\int_0^t d\tau \hat{P}\hat{H}\hat{Q}e^{\frac{-i(t-\tau)}{\hbar}\hat{Q}\hat{H}\hat{Q}}\hat{Q}\hat{H}\hat{P}|\hat{P}\psi(\tau)\rangle. \qquad (19.1.7)$$

Transforming the $\hat{P}$-space wave function to the interaction picture representation (Eq. 15.3.7),

$$|\hat{P}\psi_I(t)\rangle \equiv e^{\frac{it}{\hbar}\hat{P}\hat{H}\hat{P}}|\hat{P}\psi(t)\rangle, \qquad (19.1.8)$$

the transformed equation reads

$$\frac{\partial}{\partial t}|\hat{P}\psi_I(t)\rangle = -\frac{1}{\hbar^2}\int_0^t d\tau e^{\frac{it}{\hbar}\hat{P}\hat{H}\hat{P}}\hat{P}\hat{H}\hat{Q}e^{\frac{-i(t-\tau)}{\hbar}\hat{Q}\hat{H}\hat{Q}}\hat{Q}\hat{H}\hat{P}e^{\frac{-i\tau}{\hbar}\hat{P}\hat{H}\hat{P}}|\hat{P}\psi_I(\tau)\rangle. \qquad (19.1.9)$$

As we can see, the projection of the full system wave function onto a part of the Hilbert space ($|\hat{P}\psi_I(t)\rangle$) is a solution of a homogeneous integrodifferential equation. When the initial state is not confined to the $\hat{P}$ space, the equation obtains an additional inhomogeneous term (Ex. 19.1.2),

$$\frac{\partial}{\partial t}|\hat{P}\psi_I(t)\rangle + \frac{1}{\hbar^2}\int_0^t d\tau e^{\frac{it}{\hbar}\hat{P}\hat{H}\hat{P}}\hat{P}\hat{H}\hat{Q}e^{\frac{-i(t-\tau)}{\hbar}\hat{Q}\hat{H}\hat{Q}}\hat{Q}\hat{H}\hat{P}e^{\frac{-i\tau}{\hbar}\hat{P}\hat{H}\hat{P}}|\hat{P}\psi_I(\tau)\rangle$$

$$= -\frac{i}{\hbar}e^{\frac{it}{\hbar}\hat{P}\hat{H}\hat{P}}\hat{P}\hat{H}\hat{Q}e^{\frac{-it}{\hbar}\hat{Q}\hat{H}\hat{Q}}|\hat{Q}\psi(0)\rangle. \qquad (19.1.10)$$

Eqs. (19.1.9, 19.1.10) are exact reformulations of the Schrödinger equation, but generally are not particularly useful. The need to follow the evolution of each past state ($|\hat{P}\psi_I(\tau)\rangle$; $0 \leq \tau \leq t$) under the $\hat{Q}$ space Hamiltonian at any time, t, is usually highly involved and does not have a particular advantage in comparison to solving the Schrödinger equation in the full space and projecting the final result onto the $\hat{P}$-space.

Exercise 19.1.1 *Use the identity $\frac{\partial}{\partial t}\int_0^t f(t,t')dt' = \int_0^t \frac{\partial}{\partial t}f(t,t')dt' + f(t,t)$ to show that the expression in Eq. (19.1.4) for the Q-space projection, $|\hat{Q}\psi(t)\rangle$, is indeed a solution to its defining equation, Eq. (19.1.3).*

Exercise 19.1.2 *Show that in the general case, where $|\hat{Q}\psi(0)\rangle \neq 0$, substitution of Eq. (19.1.4) in Eq. (19.1.3) results in an additional inhomogeneous term in the equation for the P-space projection. Show that for a time-independent Hamiltonian, the corresponding inhomogeneous equation in the interaction picture is Eq. (19.1.10).*

Liouville Space Projectors

The derivation of an equation for a reduced system dynamics in terms of projection operators is readily applicable also within Liouville's space. Recalling the Liouville–von Neumann equation (Eq. (16.4.3)) for the full density operator,

$$\frac{\partial}{\partial t}\hat{\rho}(t) = -\frac{i}{\hbar}\hat{L}(t)\hat{\rho}(t) \quad ; \quad \hat{L}(t)\hat{\rho}(t) = [\hat{H}(t), \hat{\rho}(t)], \tag{19.1.11}$$

and defining generic projection operators in Liouville's space, $\hat{P}_L$ and $\hat{Q}_L$,

$$\hat{P}_L + \hat{Q}_L = \hat{I}_L \; ; \; \hat{P}_L^2 = \hat{P}_L \; ; \; \hat{Q}_L^2 = \hat{Q}_L \; ; \; \hat{Q}_L\hat{P}_L = 0, \tag{19.1.12}$$

we are interested in following the dynamics of the reduced system density operator, defined as $\hat{\rho}_P(t) \equiv \hat{P}_L\hat{\rho}(t)$. Since Liouville's space is also a Hilbert space (see Section 16.3), an exact equation for $\hat{\rho}_P(t)$ is obtained by following the steps outlined in Eqs. (19.1.3–19.1.7). Indeed, replacing the system Hilbert space vector, $|\psi(t)\rangle$, by Liouville's space vector, $\hat{\rho}(t)$, the system Hamiltonian, $\hat{H}$, by the Liouville operator, $\hat{L}$, and the Hilbert space projectors by the Liouville space projectors, the exact result (for a time-independent Hamiltonian) reads (Ex. 19.1.3)

$$\frac{\partial}{\partial t}\hat{P}_L\hat{\rho}(t) = -\frac{i}{\hbar}\hat{P}_L\hat{L}\hat{P}_L\hat{\rho}(t) - \frac{1}{\hbar^2}\int_0^t d\tau \hat{P}_L\hat{L}\hat{Q}_L e^{\frac{-i\tau}{\hbar}\hat{Q}_L\hat{L}\hat{Q}_L}\hat{Q}_L\hat{L}\hat{P}_L\hat{\rho}(t-\tau), \tag{19.1.13}$$

which is analogous to Eq. (19.1.7) for a time-independent Hamiltonian. Transforming to an interaction picture,

$$\hat{\rho}_P^{(I)}(t) \equiv e^{\frac{it}{\hbar}\hat{P}_L\hat{L}\hat{P}_L}\hat{P}_L\hat{\rho}(t), \tag{19.1.14}$$

we readily obtain (Ex. 19.1.3)

$$\frac{\partial}{\partial t}\hat{\rho}_P^{(I)}(t) = -\frac{1}{\hbar^2}\int_0^t d\tau e^{\frac{it}{\hbar}\hat{P}_L\hat{L}\hat{P}_L}\hat{P}_L\hat{L}\hat{Q}_L e^{\frac{-i(t-\tau)}{\hbar}\hat{Q}_L\hat{L}\hat{Q}_L}\hat{Q}_L\hat{L}\hat{P}_L e^{\frac{-i\tau}{\hbar}\hat{P}_L\hat{L}\hat{P}_L}\hat{\rho}_P^{(I)}(\tau), \tag{19.1.15}$$

which is the Liouville space analogue of Eq. (19.1.9). The result is a homogeneous integrodifferential equation, where the time derivative of the reduced density operator $\hat{\rho}_P^{(I)}(t)$ depends on its past evolution starting at earlier times, $\hat{\rho}_P^{(I)}(\tau)$, which is generally too cumbersome to follow. Notice that when the initial state is not confined to the $\hat{P}_L$ space, the equation obtains an additional inhomogeneous term [19.1], by analogy to Eq. (19.1.10) (Ex. 19.1.3):

$$\frac{\partial}{\partial t}\hat{\rho}_P^{(I)}(t) + \frac{1}{\hbar^2}\int_0^t d\tau e^{\frac{it}{\hbar}\hat{P}_L\hat{L}\hat{P}_L}\hat{P}_L\hat{L}\hat{Q}_L e^{\frac{-i(t-\tau)}{\hbar}\hat{Q}_L\hat{L}\hat{Q}_L}\hat{Q}_L\hat{L}\hat{P}_L e^{\frac{-i\tau}{\hbar}\hat{P}_L\hat{L}\hat{P}_L}\hat{\rho}_P^{(I)}(\tau)$$

$$= -\frac{i}{\hbar}e^{\frac{it}{\hbar}\hat{P}_L\hat{L}\hat{P}_L}\hat{P}_L\hat{L}\hat{Q}_L e^{\frac{-it}{\hbar}\hat{Q}_L\hat{L}\hat{Q}_L}\hat{Q}_L\hat{\rho}(0). \tag{19.1.16}$$

Exercise 19.1.3 *(a) Start from the defining equations for the projected density operator, $\frac{\partial}{\partial t}\hat{P}_L\hat{\rho}(t) = -\frac{i}{\hbar}\hat{P}_L\hat{L}(t)\hat{P}_L\hat{\rho}(t) - \frac{i}{\hbar}\hat{P}_L\hat{L}(t)\hat{Q}_L\hat{\rho}(t)$; $\frac{\partial}{\partial t}\hat{Q}_L\hat{\rho}(t) = -\frac{i}{\hbar}\hat{Q}_L\hat{L}(t)\hat{Q}_L\hat{\rho}(t) - \frac{i}{\hbar}\hat{Q}_L\hat{L}(t)\hat{P}_L\hat{\rho}(t)$, and use the analogy to the derivation of Eq. (19.1.7), to derive Eq. (19.1.13) for time-independent Hamiltonians, when $\hat{Q}_L\hat{\rho}(0) = 0$. (b) Use Eq. (19.1.14) to derive Eq. (19.1.15) from Eq. (19.1.13). (c) Show that in the general case, where $\hat{Q}_L\hat{\rho}(0) \neq 0$, the inhomogeneous equation for the P-space projected density operator in the interaction picture is Eq. (19.1.16) (follow the analogy to Ex. 19.1.2).*

19.2 The Born–Markov Approximation in the System Hilbert Space

To follow exactly the system dynamics in a reduced Hilbert space, we must account continuously for the evolution of the reduced system states at past times, $\{|\hat{P}\psi_I(\tau)\rangle\}$, under the $\hat{Q}$-space Hamiltonian (see Eqs. (19.1.9, 19.1.10)). The equation of motion for the reduced system wave function can be simplified significantly, if "past states" are replaced by the "present state," namely $|\hat{P}\psi_I(\tau)\rangle$ is replaced by $|\hat{P}\psi_I(t)\rangle$, for any past time, $0 \leq \tau \leq t$. Implementing this in Eq. (19.1.9) (restricting hereafter to $|\hat{Q}\psi(0)\rangle = 0$) leads to the approximation

$$\frac{\partial}{\partial t}|\hat{P}\psi_I(t)\rangle \cong -\frac{1}{\hbar^2}\int_0^t d\tau\, e^{\frac{it}{\hbar}\hat{P}\hat{H}\hat{P}}\hat{P}\hat{H}\hat{Q}e^{\frac{-i(t-\tau)}{\hbar}\hat{Q}\hat{H}\hat{Q}}\hat{Q}\hat{H}\hat{P}e^{\frac{-i\tau}{\hbar}\hat{P}\hat{H}\hat{P}}|\hat{P}\psi_I(t)\rangle. \tag{19.2.1}$$

The "loss of memory" associated with this approximation characterizes Markov processes in which the future evolution of a system does depend on its past history. Notice that Eq. (19.2.1) still requires the backward time evolution of $|\hat{P}\psi_I(t)\rangle$ within the $\hat{Q}$-space, owing to the interaction between the $\hat{P}$ and $\hat{Q}$ subspaces, $\hat{Q}\hat{H}\hat{P}$. However, unlike in the exact expression, this operation is carried out independently, and only once per each time t, as in a reset process. As one can suspect, this simplicity is valid only when the coupling between the $\hat{P}$ and $\hat{Q}$ subspaces is sufficiently weak. Indeed, returning to the exact formulation (Eq. (19.1.9)) and using repeatedly the trivial relation between $|\hat{P}\psi_I(\tau)\rangle$ and $|\hat{P}\psi_I(t)\rangle$, namely $|\hat{P}\psi_I(\tau)\rangle = |\hat{P}\psi_I(t)\rangle - \int_\tau^t dt'\frac{\partial}{\partial t'}|\hat{P}\psi_I(t')\rangle$, *the approximation, Eq. (19.2.1), is shown to be the lowest-order term of an infinite series expansion of the exact equation in powers of the coupling operators, $\hat{Q}\hat{H}\hat{P}$ and $\hat{P}\hat{H}\hat{Q}$* (Ex. 19.2.1):

$$\frac{\partial}{\partial t}|\hat{P}\psi_I(t)\rangle = -\left[\int_0^t d\tau \hat{K}(t,\tau)\left[\hat{I} + \int_\tau^t d\tau'\int_0^{\tau'}d\tau''\hat{K}(\tau',\tau'')\right.\right.$$

$$\left.\left. \times \left[\hat{I} + \int_{\tau''}^t d\tau'''\int_0^{\tau'''}d\tau''''\hat{K}(\tau''',\tau'''')[\hat{I}+\cdots]\right]\right]\right]|\hat{P}\psi_I(t)\rangle$$

$$\hat{K}(t,\tau) \equiv \frac{1}{\hbar^2}e^{\frac{it}{\hbar}\hat{P}\hat{H}\hat{P}}\hat{P}\hat{H}\hat{Q}e^{\frac{-i(t-\tau)}{\hbar}\hat{Q}\hat{H}\hat{Q}}\hat{Q}\hat{H}\hat{P}e^{\frac{-i\tau}{\hbar}\hat{P}\hat{H}\hat{P}}. \tag{19.2.2}$$

The truncation of the infinite series at the first order in $\hat{K}(t,\tau)$ (namely, second order in the coupling between the $\hat{P}$ and $\hat{Q}$ subspaces) is reminiscent of the Born approximation in quantum scattering theory [7.1], and the first-order result (Eq. (19.2.1)) is often referred to as the Born–Markov approximation.

Exercise 19.2.1 *(a) Given the definition of the Kernel $\hat{K}(t,\tau)$ in Eq. (19.2.2), show that the exact equation for the projected state $|\hat{P}\psi_I(t)\rangle$, Eq. (19.1.9), reads $\frac{\partial}{\partial t}|\hat{P}\psi_I(t)\rangle =$*

$$-\int_0^t d\tau \hat{K}(t,\tau)|\hat{P}\psi_I(\tau)\rangle. \text{ (b) Derive the infinite series expansion in Eq. (19.2.2) by recur-}$$

sive application of the formal relation, $\hat{P}\psi_I(\tau) = \hat{P}\psi_I(t) - \int_\tau^t dt'\frac{\partial}{\partial t'}\hat{P}\psi_I(t')$.

Irreversible Dynamics and Exponential Decay

As an example of reduced Hilbert space dynamics, and for assessing the validity of the Markovian approximation within an exactly solvable model, let us consider the dynamics of a single state within a system of finite dimensions. When the state of interest is an eigenstate of some zero-order Hamiltonian, following its dynamics is closely related to following the probability amplitudes of transition to the other eigenstates. The latter was discussed extensively in Chapter 17, where we analyzed the conditions in which the total transition probability is associated with a well-defined rate constant, for example, $\frac{\partial}{\partial t}P_i(t)|_{t<<1/k_{i\to\{f\}}} \approx -k_{i\to\{f\}}$. Here we adopt a complementary view, by following the survival probability of a single state coupled to a manifold of other orthonormal states. As we shall see, when the Markovian approximation becomes valid, the survival probability follows an exponential decay in time, $\frac{\partial}{\partial t}P_i(t) = -k_{i\to\{f\}}P_i(t)$, at a rate which reproduces Fermi's Golden rule. Indeed, the result is consistent with Eq. (17.2.2), which identifies the decay rate constant with the initial time derivative of $P_i(t)$, and moreover, it implies that under certain conditions a quantum transition can become "irreversible" or "unidirectional" also in the long time-limit of the kinetics.

We start from the most general form of a discrete Hermitian model Hamiltonian,

$$\hat{H} \equiv \sum_{j=0}^{N} H_{j,j}|\varphi_j\rangle\langle\varphi_j| + \sum_{j>j'=0}^{N} \{H_{j,j'}|\varphi_j\rangle\langle\varphi_{j'}| + h.c.\}, \qquad (19.2.3)$$

where $\{|\varphi_j\rangle\}$, $j = 0,1,2,\ldots,N$, are orthonormal basis vectors, $\langle\varphi_{j'}|\varphi_j\rangle = \delta_{j,j'}$, corresponding to the eigenstates of some zero-order Hamiltonian (notice that $\{\hat{A}+h.c.\} \equiv \{\hat{A}+\hat{A}^\dagger\}$). We are interested in the time evolution under $\hat{H}$, of one of these basis states, denoted as $|\varphi_0\rangle$. For this purpose, let us introduce a projection operator into the space of the initial state $|\varphi_0\rangle$, $\hat{P} = |\varphi_0\rangle\langle\varphi_0|$, and a complementary projector, $\hat{Q} = \hat{I} - \hat{P} = \sum_{j=1}^{N} |\varphi_j\rangle\langle\varphi_j|$. We can then rewrite the full Hamiltonian as $\hat{H} = \hat{H}_0 + \hat{V}$, where $\hat{H}_0 \equiv \hat{P}\hat{H}\hat{P} + \hat{Q}\hat{H}\hat{Q}$ and $\hat{V} \equiv \{\hat{Q}\hat{H}\hat{P} + h.c.\}$. Denoting the eigenstates of this $\hat{H}_0$ as $\{|\chi_0\rangle, |\chi_1\rangle, |\chi_2\rangle, \ldots, |\chi_N\rangle\}$ and using them as an alternative basis, the full Hamiltonian is rewritten in the form (see Fig. 19.2.1 and Ex. 19.2.2),

$$\hat{H} = \hat{H}_0 + \hat{V}$$

$$\hat{H}_0 = \varepsilon_0|\chi_0\rangle\langle\chi_0| + \sum_{f=1}^{N} \varepsilon_f|\chi_f\rangle\langle\chi_f|$$

$$\hat{V} = \sum_{f=1}^{N} (V_{0,f}|\chi_0\rangle\langle\chi_f| + h.c.). \qquad (19.2.4)$$

Notice that $\hat{H}_0$ is in general different than the original zero-order Hamiltonian, whose eigenvectors are $\{|\varphi_0\rangle, |\varphi_1\rangle, \ldots, |\varphi_N\rangle\}$. However, the selected initial state, $|\varphi_0\rangle$ is, by construction, an eigenvector also of $\hat{H}_0$, namely

$$|\chi_0\rangle = |\varphi_0\rangle. \qquad (19.2.5)$$

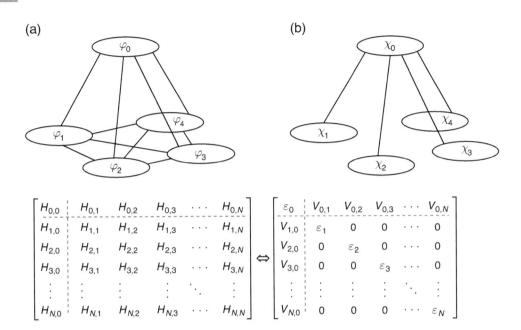

Figure 19.2.1 A schematic representation of the model Hamiltonian in an arbitrary orthonormal basis (a), and in a basis in which the zero-order Hamiltonian ($\hat{H}_0 \equiv \hat{P}\hat{H}\hat{P} + \hat{Q}\hat{H}\hat{Q}$) is diagonal (b). (For a concrete example, see Section 20.4, Fig. 20.4.1.)

The other eigenstates of $\hat{H}_0, \{|\chi_f\rangle\}$, $f = 1, 2, \ldots, N$ are the eigenstates of $\hat{Q}\hat{H}\hat{Q}$ (see Ex. 19.2.2), which are orthogonal to each other as well as to the initial state, $|\chi_0\rangle$.

Exercise 19.2.2 *Use the decomposition of the Hamiltonian in Eq. (19.2.3) in terms of the projection operators, $\hat{H} = \hat{P}\hat{H}\hat{P} + \hat{Q}\hat{H}\hat{Q} + \hat{P}\hat{H}\hat{Q} + \hat{Q}\hat{H}\hat{P}$, where $\hat{P} = |\varphi_0\rangle\langle\varphi_0|$, and $\hat{Q} = \hat{I} - \hat{P} = \sum_{j=1}^{N} |\varphi_j\rangle\langle\varphi_j|$. Denote the eigenstates of $\hat{P}\hat{H}\hat{P}$ and $\hat{Q}\hat{H}\hat{Q}$ as $|\chi_0\rangle$ and $\{|\chi_f\rangle\}$, respectively, to derive Eq. (19.2.4).*

The exact time evolution of the system under the full Hamiltonian, Eq. (19.2.4), is given by the Schrödinger equation, $i\hbar \frac{\partial}{\partial t}|\psi(t)\rangle = \hat{H}|\psi(t)\rangle$. It is convenient to transform to the interaction picture representation (Eqs. (15.3.7, 15.3.8)), where

$$|\psi_I(t)\rangle = e^{\frac{i\hat{H}_0 t}{\hbar}}|\psi(t)\rangle \quad ; \quad i\hbar\frac{\partial}{\partial t}|\psi_I(t)\rangle = e^{\frac{i\hat{H}_0 t}{\hbar}}\hat{V}e^{\frac{-i\hat{H}_0 t}{\hbar}}|\psi_I(t)\rangle. \tag{19.2.6}$$

We are interested in the reduced $\hat{P}$-space dynamics, which in the present example amounts to $|\chi_0\rangle\langle\chi_0|\psi_I(t)\rangle \equiv c_0(t)|\chi_0\rangle$. An exact equation for $\frac{\partial}{\partial t}c_0(t)$ can be obtained directly by implementing the general result, Eq. (19.1.9), for the Hamiltonian in Eq. (19.2.4), with the projection operators, $\hat{P} = |\chi_0\rangle\langle\chi_0|$ and $\hat{Q} = \sum_{f=1}^{N} |\chi_f\rangle\langle\chi_f|$ (see Ex. 19.2.3). Alternatively, we can use the complete basis of $\hat{H}_0$ eigenvectors to expand the state of the system at any time:

$$|\psi_I(t)\rangle \equiv c_0(t)|\chi_0\rangle + \sum_{f=1}^{N} c_f(t)|\chi_f\rangle. \tag{19.2.7}$$

Substitution of this expansion in Eq. (19.2.6) and projecting onto the different basis vectors yields (see Ex. 19.2.4)

$$\frac{\partial}{\partial t}c_0(t) = \frac{1}{i\hbar}\sum_{f=1}^{N} V_{0,f}e^{\frac{-i(\varepsilon_f-\varepsilon_0)t}{\hbar}}c_f(t) \quad ; \quad \frac{\partial}{\partial t}c_f(t) = \frac{1}{i\hbar}V_{0,f}^*e^{\frac{i(\varepsilon_f-\varepsilon_0)t}{\hbar}}c_0(t) \tag{19.2.8}$$

For the selected initial condition, $|\psi_I(0)\rangle \equiv |\chi_0\rangle$, we have, $c_0(0) = 1$, and $c_f(0) = 0$. Integrating the equation for $c_f(t)$, we obtain (Ex. 19.2.4)

$$\frac{\partial}{\partial t}c_0(t) = -\frac{1}{\hbar^2}\sum_{f=1}^{N}|V_{0,f}|^2\int_0^t d\tau e^{\frac{-i(\varepsilon_f-\varepsilon_0)(t-\tau)}{\hbar}}c_0(\tau). \tag{19.2.9}$$

Exercise 19.2.3 *Derive Eq. (19.2.9) by implementing Eq. (19.1.9) for the $\hat{P}$-space projection, $|\hat{P}\psi_I(t)\rangle$, with the Hamiltonian in Eq. (19.2.4), and the projection operators, $\hat{P} = |\chi_0\rangle\langle\chi_0|$ and $\hat{Q} = \sum_{f=1}^{N}|\chi_f\rangle\langle\chi_f|$.*

Exercise 19.2.4 *(a) Substitute the expansion of $|\psi_I(t)\rangle$ (Eq. (19.2.7)) in the Schrödinger equation (Eq. (19.2.6)), and project on the eigenstates of $\hat{H}_0$, to obtain the coupled equations for the expansion coefficients, Eq. (19.2.8). (b) Integrate Eq. (19.2.8) over time for the initial condition, $c_0(0) = 1$ and $\{c_f(0) = 0\}$, to obtain Eq. (19.2.9).*

The survival probability at time t is the probability of occupying the initial state (see Eq. (15.1.29)),

$$P_0(t) = |\langle\chi_0|\psi(t)\rangle|^2 = |\langle\chi_0|\psi_I(t)\rangle|^2 = |c_0(t)|^2. \tag{19.2.10}$$

The rate of change of the survival probability reads

$$\frac{\partial}{\partial t}P_0(t) = 2\,\text{Re}[c_0^*(t)\frac{\partial}{\partial t}c_0(t)], \tag{19.2.11}$$

which can be expressed exactly, using Eq. (19.2.9):

$$\frac{\partial}{\partial t}P_0(t) = -\frac{2}{\hbar^2}\text{Re}\left[\int_0^t d\tau\sum_{f=1}^{N}|V_{0,f}|^2e^{\frac{-i(\varepsilon_f-\varepsilon_0)(t-\tau)}{\hbar}}c_0^*(t)c_0(\tau)\right]. \tag{19.2.12}$$

As we can see, in the absence of coupling between the initial state and the other $\hat{H}_0$-eigenstates ($\{V_{0,f}\} = 0$), the survival probability remains unity at all times. This case corresponds to $|\chi_0\rangle$ being a stationary state of the system (an eigenvector of the full Hamiltonian). In any other case, however, $P_0(t)$ changes in time, where the rate of change at time t depends explicitly on the "history" of the survival probability amplitude, $c_0(\tau)$, at $0 \le \tau \le t$.

We now invoke a Markovian approximation (see Eq. (19.2.1) for the general case). Ignoring the "history," $c_0(\tau)$ is replaced by $c_0(t)$ under the time integral in Eq. (19.2.12), which yields a simple kinetic equation:

$$\frac{\partial}{\partial t}P_0(t) \cong -k_{0\to\{f\}}(t)P_0(t). \tag{19.2.13}$$

The time-dependent rate coefficient reads

$$k_{0\to\{f\}}(t) = \frac{2}{\hbar^2}\mathrm{Re}\left[\int_0^t d\tau\,\eta(t,\tau)\right] \quad ; \quad \eta(t,\tau) \equiv \sum_{f=1}^N |V_{0,f}|^2 e^{\frac{-i(\varepsilon_f-\varepsilon_0)(t-\tau)}{\hbar}}, \tag{19.2.14}$$

where $\eta(t,\tau)$ can be identified (Ex. 19.2.5) as the time correlation function of the interaction, $\eta(t,\tau) = tr\{e^{\frac{i\hat{H}_0(t-\tau)}{\hbar}}\hat{P}\hat{V}\hat{Q}e^{\frac{-i\hat{H}_0(t-\tau)}{\hbar}}\hat{Q}\hat{V}\hat{P}\}$.

Exercise 19.2.5   *Use the definition of the operators $\hat{H}_0$ and $\hat{V}$ in Eq. (19.2.4), and the projection operators, $\hat{P} = |\chi_0\rangle\langle\chi_0|, \hat{Q} = \sum_{f=1}^N |\chi_f\rangle\langle\chi_f|$, to show that the kernel, $\eta(t,\tau)$, as defined in Eq. (19.2.14) can be written as $tr\{e^{\frac{i\hat{H}_0(t-\tau)}{\hbar}}\hat{P}\hat{V}\hat{Q}e^{\frac{-i\hat{H}_0(t-\tau)}{\hbar}}\hat{Q}\hat{V}\hat{P}\}$.*

The validity of this Markovian approximation clearly depends on the properties of the exact time integral in Eq. (19.2.12). More precisely, the approximation is justified as long as $\eta(t,\tau)$ is strongly peaked around $\tau \approx t$ and decays to zero as a function of τ much faster than changes in the amplitude, $c_0(\tau)$. Moreover, a rapid decay of the integrand justifies replacement of the upper limit of the time integral by infinity, which implies that (except for an initial transient) a time-independent "rate constant" is obtained. Eq. (19.2.14) is then replaced by

$$k_{0\to\{f\}} \cong \frac{2}{\hbar^2}\mathrm{Re}\left[\int_0^\infty d\tau \sum_{f=1}^N |V_{0,f}|^2 e^{\frac{-i(\varepsilon_f-\varepsilon_0)\tau}{\hbar}}\right], \tag{19.2.15}$$

where the survival probability decays exponentially (see Eq. (19.2.13)),

$$\frac{\partial}{\partial t}P_0(t) \cong -k_{0\to\{f\}}P_0(t) \quad ; \quad P_0(t) \cong e^{-k_{0\to\{f\}}t}. \tag{19.2.16}$$

The expression for the decay rate constant (Eq. (19.2.15)) coincides with Fermi's golden rule result (see Eq. (17.2.7)) for the short-time rate of probability transfer from a pure state to an ensemble of final states (see Eq. (17.2.7)):

$$k_{0\to\{f\}} = \frac{2\pi}{\hbar} \sum_{f=1}^N |V_{0,f}|^2 \delta(\varepsilon_0 - \varepsilon_f). \tag{19.2.17}$$

This is naturally expected, since the short-time decay rate of the initial state population is identical to the rate of transition into all possible final states, as calculated by Fermi's golden rule (see Section 17.2). Importantly, the coincidence of these two approximated calculations suggests that the validity of the Markovian approximation (Eqs. (19.2.15–19.2.17)) is subject to the same limitations on the validity of Fermi's golden rule. The time integrand in the exact equation for the survival probability (Eq. (19.2.12)), is a multiplication of $\eta(t,\tau)$ by $c_0(\tau)c_0^*(t)$. The real part of similar functions to $\eta(t,\tau)$ was already analyzed in detail in Section 17.2. We recall that the oscillating contributions interfere constructively as $\tau \to t$; but in the weak coupling and wide band limits, dephasing at longer times (τ) leads to cancellation of contributions with opposite signs and to a decay toward zero. Notice that similar considerations apply here, although $\eta(t,\tau)$ in Eq. (19.2.12) is multiplied by the complex factor, $c_0(\tau)c_0^*(t)$. Roughly, as long as the amplitude, $c_0(\tau)$, deviates significantly from $c_0(t)$ only after $\text{Re}[\eta(t,\tau)]$ decays to zero, $c_0(\tau)c_0^*(t)$ can be replaced by the real-valued probability, $|c_0(t)|^2$, which justifies the result, Eq. (19.2.13).

We can readily verify that the Markovian approximation for the survival probability is consistent with a weak coupling limit by considering the exact expansion of $\frac{\partial}{\partial t}c_0(t)$ in powers of the interaction matrix elements $\{|V_{0,f}|^2\}$. Implementing the general formula, Eq. (19.2.2), for the model Hamiltonian defined by Eq. (19.2.4), with the projectors, $\hat{P}=|\chi_0\rangle\langle\chi_0|$ and $\hat{Q}=\sum\limits_{f=1}^{N}|\chi_f\rangle\langle\chi_f|$, the exact equation is reformulated as (Ex. 19.2.6)

$$
\frac{\partial}{\partial t}c_0(t) = -\left[\int_0^t d\tau\,\eta(t,\tau)\left[\hat{I}+\int_\tau^t d\tau'\int_0^{\tau'}d\tau''\eta(\tau',\tau'')\right.\right.
$$
$$
\left.\left.\times\left[\hat{I}+\int_{\tau''}^t d\tau'''\int_0^{\tau'''}d\tau''''\eta(\tau''',\tau'''')[\hat{I}+\cdots]\right]\right]\right]c_0(t). \tag{19.2.18}
$$

The Markovian approximation, Eqs. (19.2.13, 19.2.14), therefore corresponds to the first-order term in powers of the interaction correlation function, $\eta(t,\tau) = \sum_f|V_{0,f}|^2 e^{\frac{-i(t-\tau)}{\hbar}(\varepsilon_f-\varepsilon_0)}$, namely in powers of the squared coupling matrix elements (a Born–Markov approximation).

Exercise 19.2.6   *Derive Eq. (19.2.18) by implementing the general kernel formula (Eq. (19.2.2)) for the model Hamiltonian defined by Eq. (19.2.4), with the projectors, $\hat{P}=|\chi_0\rangle\langle\chi_0|$ and $\hat{Q}=\sum\limits_{f=1}^{N}|\chi_f\rangle\langle\chi_f|$.*

When the manifold of $\hat{H}_0$ eigenstates coupled to the initial state is sufficiently dense on the energy scale of the coupling matrix elements, the latter can be associated with a continuous function of the final state energy, $|V_{0,f}|^2 = \lambda_{0,f}^2(\varepsilon_f)$, and with a respective spectral density function (see Eq. 17.2.10), $J_{0,f}(\varepsilon) \equiv 2\pi\lambda_{0,f}^2(\varepsilon)\rho_f(\varepsilon)$. The interaction correlation function $\eta(t,\tau)$ is then related to the Fourier transform of the spectral density,

$$
\eta(t,\tau) = \frac{1}{2\pi}\int d\varepsilon\,J_{0,f}(\varepsilon)e^{\frac{-i(\varepsilon-\varepsilon_0)(t-\tau)}{\hbar}}. \tag{19.2.19}
$$

Invoking, for example, a finite and uniform band model, where $J_{0,f}(\varepsilon)\equiv J_b$ in the range $|\varepsilon| \le \varepsilon_b$, and zero otherwise (see Eq. (17.2.14)), the time-dependent derivative of the survival probability (Eq. (19.2.12)) reads (Ex. 19.2.7)

$$\frac{\partial}{\partial t}P_0(t) = \frac{-2J_b\varepsilon_b}{\pi\hbar^2}\mathrm{Re}\int_0^t d\tau e^{\frac{i\varepsilon_0\tau}{\hbar}}\frac{\sin(\varepsilon_b\tau/\hbar)}{\varepsilon_b\tau/\hbar}c_0^*(t)c_0(t-\tau). \tag{19.2.20}$$

As we can see, **when the bandwidth is taken to infinity, $\varepsilon_b \to \infty$, the conditions for validity of the Markovian approximation are automatically satisfied**. Indeed, performing the time integral, we obtain the Markovian result with no approximation in this infinite wide band limit (Ex. 19.2.7),

$$\frac{\partial}{\partial t}P_i(t) = -\frac{J_b}{\hbar}P_i(t), \tag{19.2.21}$$

where the decay rate is identified as the (constant) spectral density, divided by $\hbar$.

Exercise 19.2.7 *(a) In the continuous band limit, we have $\sum_{f=1}^{N}|V_{0,f}|^2 e^{\frac{-i(\varepsilon_f-\varepsilon_0)(t-\tau)}{\hbar}} \to$ $\frac{1}{2\pi}\int d\varepsilon J_{0,f}(\varepsilon)e^{\frac{-i(\varepsilon-\varepsilon_0)(t-\tau)}{\hbar}}$. Show that for $J_{0,f}(\varepsilon) \equiv J_b$ in the range $|\varepsilon| \leq \varepsilon_b$, and $J_{0,f}(\varepsilon) \equiv 0$ otherwise, the exact equation for the survival probability, Eq. (19.2.12), yields Eq. (19.2.20). (b) Show that in the limit $\varepsilon_b \to \infty$, the decay of the survival probability becomes exponential (Eq. (19.2.21)).*

Having established the conditions for an exponential decay of the initial survival probability, it is instructive to consider also the "fate" of the decaying state within the manifold of $\hat{H}_0$ eigenstates. Using Eq. (19.2.8) with the initial condition, $\{c_f(0) = 0\}$, $c_i(0) = 1$, the probability amplitude for populating the fth eigenstate ($|\chi_f\rangle$) reads

$$c_f(t) = \frac{V_{0,f}^*}{i\hbar}\int_0^t dt' e^{\frac{i(\varepsilon_f-\varepsilon_0)t'}{\hbar}}c_0(t'). \tag{19.2.22}$$

Assuming that the validity conditions for the Markovian approximation hold, the exact equation for $c_0(t)$ (Eq. (19.2.9)) can be approximated as

$$\frac{\partial}{\partial t}c_0(t) \cong -\int_0^\infty d\tau \sum_{f=1}^{N}\frac{|V_{0,f}|^2}{\hbar^2}e^{\frac{-i(\varepsilon_f-\varepsilon_0)\tau}{\hbar}}c_0(t). \tag{19.2.23}$$

Noticing that the real part of $\int_0^\infty d\tau \sum_{f=1}^{N}\frac{|V_{0,f}|^2}{\hbar^2}e^{\frac{-i(\varepsilon_f-\varepsilon_0)\tau}{\hbar}}$ equals half the exponential decay rate, $k_{0\to\{f\}}$ (see Eq. (19.2.15)), and denoting the corresponding imaginary part as $\mathrm{Im}\int_0^\infty d\tau \sum_{f=1}^{N}\frac{|V_{0,f}|^2}{\hbar^2}e^{\frac{-i(\varepsilon_f-\varepsilon_0)\tau}{\hbar}} \equiv \frac{\Delta}{\hbar}$, we obtain $\frac{\partial}{\partial t}c_0(t) \cong -\left(\frac{k_{0\to\{f\}}}{2} + i\frac{\Delta}{\hbar}\right)c_0(t)$, and therefore

$$c_0(t) \cong e^{\frac{-k_{0\to\{f\}}t}{2}}e^{-i\frac{\Delta t}{\hbar}}. \tag{19.2.24}$$

Substitution in Eq. (19.2.22) yields (Ex. 19.2.8)

$$|c_f(t)|^2 \xrightarrow[t \gg 1/k_{0\to\{f\}}]{} \frac{|V_{0,f}|^2}{(\varepsilon_f-(\varepsilon_0+\Delta))^2+(\hbar k_{0\to\{f\}})^2/4}. \tag{19.2.25}$$

As we can see, when the decay of the initial state is complete, $t \widetilde{>} 1/k_{0\to\{f\}}$, the other $\hat{H}_0$ eigenstates are populated according to their energy in a Lorenzian-like distribution (assuming weak dependence of the state-to-state coupling on the final state), whose width and center correspond, respectively, to the decay rate $(\Delta\varepsilon_f \sim \hbar k_{0\to\{f\}})$ and a displaced initial state energy, $\varepsilon_f \approx \varepsilon_0 + \Delta$. It is interesting to notice that the distribution of the population over a finite range of final state energies is in apparent contradiction with Fermi's golden rule rate expression, Eq. (19.2.17), which seems to impose strict energy conservation between the initial and final states. Indeed, the strict energy conservation and the finite lifetime of the initial population $(\tau_i = 1/k_{i\to\{f\}})$ seem to violate a "time–energy uncertainty" relation. This uncertainty is encoded, however, in the broadening of the final state distribution, Eq. (19.2.25), where $\Delta\varepsilon_f \tau_i \sim \hbar$. The effect of this broadening on the decay process itself is indeed missing in the Markovian approximation (in the second order in the coupling). It is accounted for in the higher orders of the expansion, Eq. (19.2.18).

Exercise 19.2.8 *Use Eqs. (19.2.22, 19.2.24) to show that* $|c_f(t)|^2 \cong \frac{|V_{0,f}|^2}{(\varepsilon_f - (\varepsilon_0 + \Delta))^2 + (\hbar k_{0\to\{f\}})^2/4}$
$(1 - 2e^{-k_{0\to\{f\}}t/2}\cos[(\varepsilon_f - (\varepsilon_0 + \Delta))t/\hbar] + e^{-k_{0\to\{f\}}t}).$

We end this section by noting that an exponential decay of the survival probability means that the decay process continues irreversibly. This is strictly true only in the absence of revivals of the interaction correlation function $(\eta(t,\tau))$, which characterize idealized models of decay into continuous wide bands. Nevertheless, exponential decay phenomena are commonly encountered in the real world in processes such as spontaneous emission of radiation (see Section 19.5), radioactive decay, atomic autoionization, and many more. A practical observation of exponential decay implies that revivals may occur only at times much longer than the relevant kinetics timescale, namely at $t \gg 1/k_{i\to\{f\}}$. This condition can be met in principle when the reduced system of interest (defined by the projector $\hat{P}$) is much "smaller" than the rest of the "universe," associated with the projector $\hat{Q}$ (in its phase space volume, in classical terms). In other cases, when the $\hat{P}$ and the $\hat{Q}$ spaces are of "comparable size," revivals become ubiquitous. Even in such cases, rapid relaxation of coherences within each space may still justify a Markovian approximation, which leads to kinetic equations for population transfer (master equations) instead of a unidirectional exponential decay.

19.3 The Born–Markov Approximation in Liouville Space

Reduced system dynamics is of interest in many cases. Consider the generic situation in which a "small" quantum system, typically associated with a finite number of particles and/or a finite number of degrees of freedom, interacts with a macroscopically large environment. In the context of nanoscale phenomena, such a system can be, for example, an atom, molecule, nanocrystal, or a defect (impurity), embedded in and interacting with a condensed phase environment, such as a solution, a solid crystal, a surface, and so on. Being interested in the small system, the detailed information with respect to the surroundings is in most cases not of prime interest and/or accessible to the measurement. Nevertheless, the interaction between the small system and its environment may have remarkable effects on the dynamics within the small system and hence on system observables. We are therefore often interested in the reduced system dynamics, in the presence of coupling to the environment. Since the latter is typically "thermodynamically large," it can be regarded as a reservoir (or a "bath") of energy and/or particles, characterized by its intensive (ensemble-averaged) properties, such as temperature, chemical potentials, and so on. Since the macroscopic environment is then associated with an incoherent ensemble, the most suitable framework for following the dynamics is Liouville's equation (see Section 16.4), where the state of the system in its environment is characterized by a density operator.

The exact dynamics of the reduced system of interest involves time evolution of past states of the reduced system under the complementary Hamiltonian (see Eq. (19.1.15)). Since the latter is typically associated with a macroscopically large number of degrees of freedom, it is practically intractable. Therefore, it is desirable to identify instead an effective Markovian Liouville equation, which corresponds to a time-local dynamical map within the space of reduced system density operators, namely $\frac{\partial}{\partial t}\hat{P}_L\hat{\rho}(t) = -\frac{i}{\hbar}\hat{L}_{eff}(t)\hat{P}_L\hat{\rho}(t)$. Such an equation indeed exists, where the effective (reduced space) Liouville operator obtains a generic Lindblad form, which guarantees that the dynamical map from $\hat{P}_L\hat{\rho}(0)$ to $\hat{P}_L\hat{\rho}(t)$ preserves nonnegativity of the system Hamiltonian eigenstate populations (complete positivity), and their accumulation to unity (trace preservation). The interested reader is referred to more specialized textbooks for the derivation of the general Lindblad form of the reduced Liouville operator ("the dissipator") [19.2]. Importantly, evaluating the exact reduced Liouville operator for a given Hamiltonian is a highly involved task, in general, unrealistic in most cases. Nevertheless, effective Markovian dissipators, cast into a generic Lindblad form, are often introduced phenomenologically to model reduced system dynamics. Moreover, as discussed in what follows, approximated Markovian dissipators can be rigorously derived from full Hamiltonian (Liouville space) dynamics by invoking several assumptions on the system–bath coupling. These approximations correspond essentially to scenarios in which there is a timescale separation between "fast" relaxation of system-induced perturbations to the bath, and "slow" bath-induced dynamics within the system.

Let us start from the exact equation for the dynamics of the reduced density operator (Eqs. (19.1.14, 19.1.15), where we shall restrict the discussion to $\hat{Q}_L\hat{\rho}(0) = 0$). Replacing the "past" reduced density operator $\hat{\rho}_P^{(I)}(\tau)$ by its "present" form, $\hat{\rho}_P^{(I)}(t)$, for any time, $0 \leq \tau \leq t$, we obtain a Born–Markov approximation:

$$\frac{\partial}{\partial t}\hat{\rho}_P^{(I)}(t) \cong -\frac{1}{\hbar^2}\int_0^t d\tau\, e^{\frac{it}{\hbar}\hat{P}_L\hat{L}\hat{P}_L}\hat{P}_L\hat{L}\hat{Q}_L e^{\frac{-i(t-\tau)}{\hbar}\hat{Q}_L\hat{L}\hat{Q}_L}\hat{Q}_L\hat{L}\hat{P}_L e^{\frac{-i\tau}{\hbar}\hat{P}_L\hat{L}\hat{P}_L}\hat{\rho}_P^{(I)}(t). \qquad (19.3.1)$$

We can readily verify that this approximation becomes formally valid in the limit of small coupling between the subspaces defined by the Liouville space projectors, $\hat{P}_L$ and $\hat{Q}_L$. Following the analogous derivation for Hilbert space projectors (see Ex. 19.2.1), the exact equation can be reformulated as

$$\frac{\partial}{\partial t}\hat{\rho}_P^{(I)}(t) = -\left[\int_0^t d\tau \hat{K}_L(t,\tau)\left[\hat{I}+\int_\tau^t d\tau'\int_0^{\tau'} d\tau'' \hat{K}_L(\tau',\tau'')\right.\right.$$

$$\times \left.\left.\left[\hat{I}+\int_{\tau''}^t d\tau'''\int_0^{\tau'''} d\tau'''' \hat{K}_L(\tau''',\tau'''')[\hat{I}+\cdots]\right]\right]\right]\hat{\rho}_P^{(I)}(t)$$

$$\hat{K}_L(t,\tau) \equiv \frac{1}{\hbar^2}e^{\frac{it}{\hbar}\hat{P}_L\hat{L}\hat{P}_L}\hat{P}_L\hat{L}\hat{Q}_L e^{\frac{-i(t-\tau)}{\hbar}\hat{Q}_L\hat{L}\hat{Q}_L}\hat{Q}_L\hat{L}\hat{P}_L e^{\frac{-i\tau}{\hbar}\hat{P}_L\hat{L}\hat{P}_L}, \qquad (19.3.2)$$

where *the approximation, Eq. (19.3.1), is readily identified as the lowest-order term of an infinite series expansion in powers of the interaction between the projected subspaces, $\hat{P}_L\hat{L}\hat{Q}_L$ and $\hat{Q}_L\hat{L}\hat{P}_L$*. The result, Eq. (19.3.1), is therefore often referred to as a Born–Markov approximation.

The System-Bath Hamiltonian and Nakajima–Zwanzig Projectors

We now distinguish the degrees of freedom of the system of interest from the rest by rewriting the full Hamiltonian in a generic form:

$$\hat{H} = \hat{H}_S + \hat{H}_B + \hat{H}_{SB} \quad ; \quad [\hat{H}_S,\hat{H}_B] = 0. \qquad (19.3.3)$$

Here, $\hat{H}_S$ and $\hat{H}_B$ denote the "system" and the "bath" Hamiltonians, respectively, where the system–bath coupling terms are included in $\hat{H}_{SB}$. Considering the decomposition of the system–bath Hamiltonian (Eq. (19.3.3)), the full Liouville operator can also be decomposed into additive contributions from the system, the bath, and their coupling:

$$\hat{L} = \hat{L}_S + \hat{L}_B + \hat{L}_{SB} \quad ; \quad [\hat{L}_S,\hat{L}_B] = 0, \qquad (19.3.4)$$

where

$$\hat{L}_S\hat{O} \equiv [\hat{H}_S,\hat{O}] \quad ; \quad \hat{L}_B\hat{O} \equiv [\hat{H}_B,\hat{O}] \quad ; \quad \hat{L}_{SB}\hat{O} \equiv [\hat{H}_{SB},\hat{O}]. \qquad (19.3.5)$$

While focusing on the reduced system dynamics and reduced system observables, in many cases it is justified to expect that the state of the environment remains essentially constant on the timescale of the bath-induced system dynamics. Notice that this does not imply that the bath is not influenced by the interaction with the system, but rather that system-induced perturbations to the bath relax much faster than the bath-induced dynamics within the system. As one can suspect, and as we shall indeed see in what follows, this simplified picture of timescale separation emerges when the coupling between the system and the bath is weak, and when the bath Hamiltonian is associated with a dense and wide band of eigenstates.

The idea of fast bath relaxation is naturally cast into useful projection operators in the Liouville space, as were introduced by Nakajima and Zwanzig. The projectors assume a reference stationary density operator, $\hat{\rho}_B$ (typically, the equilibrium density of the uncoupled reservoir (Section 16.5)), for the bath degrees of freedom, where

$$[\hat{\rho}_B, \hat{H}_B] = 0 \quad ; \quad tr_B\{\hat{\rho}_B\} = 1. \tag{19.3.6}$$

A projector from the full space into a reduced (product) space in which the system is disentangled from the bath is defined as

$$\hat{P}_L\hat{\rho}(t) \equiv \hat{\rho}_B tr_B\{\hat{\rho}(t)\} \equiv \hat{\rho}_B \hat{\rho}_S(t), \tag{19.3.7}$$

where the notation $tr_B\{\ldots\}$ corresponds to a partial trace (Ex. 19.3.1) over the bath degrees of freedom. Hence, the complementary projector, $\hat{Q}_L$, accounts for the system–bath correlations,

$$\hat{Q}_L\hat{\rho}(t) \equiv \hat{\rho}(t) - \hat{P}_L\hat{\rho}(t). \tag{19.3.8}$$

We can readily verify that these projectors satisfy the standard properties for two complementary projectors as defined in Eq. (19.1.12). Additionally (Ex. 19.3.2),

$$[\hat{P}_L, \hat{L}_S] = [\hat{Q}_L, \hat{L}_S] = [\hat{P}_L, \hat{L}_B] = [\hat{Q}_L, \hat{L}_B] = 0. \tag{19.3.9}$$

An important quantity of interest is the reduced system density,

$$\hat{\rho}_S(t) = tr_B\{\hat{\rho}(t)\}. \tag{19.3.10}$$

The trace over the space of the bath variables projects the full density operator onto the reduced system space. $\hat{\rho}_S(t)$ therefore contains the entire measurable information about the system, averaged over the bath degrees of freedom. Particularly, any system observable, $O_S(t)$ (see Eq. (16.2.4)), can be expressed in terms of the reduced density operator:

$$O_S(t) = tr\{\hat{O}_S\hat{\rho}(t)\} = tr_S\{tr_B\{\hat{O}_S\hat{\rho}(t)\}\} = tr_S\{\hat{O}_S tr_B\{\hat{\rho}(t)\}\} = tr_S\{\hat{O}_S\hat{\rho}_S(t)\}. \tag{19.3.11}$$

Exercise 19.3.1 *Let $\hat{A}$ be an operator in a Hilbert space that is a tensor product of two Hilbert subspaces, spanned by the orthonormal vector sets, $\{|s\rangle\}$ and $\{|b\rangle\}$ (without loss of generality, $|s\rangle$ and $|b\rangle$ can correspond to states of "a system" and "a bath," respectively). Expanding the operator in the product basis, $\{|s\rangle \otimes |b\rangle\}$, we have (see also Eq. (11.6.14)) $\hat{A} = \sum_{s,b,s',b'} A_{s,b,s',b'}|s\rangle\langle s'| \otimes |b\rangle\langle b'|$, where the elements, $A_{s,b,s',b'}$, are the matrix representation of $\hat{A}$. The partial traces of $\hat{A}$ with respect to each of the subspaces are defined as $tr_B\{\hat{A}\} \equiv \sum_{b''}\langle b''|\hat{A}|b''\rangle$ and $tr_S\{\hat{A}\} \equiv \sum_{s''}\langle s''|\hat{A}|s''\rangle$ (see Eq. (15.5.3) and Ex. 15.5.1 for the definition of a trace). (a) Show that the partial trace of $\hat{A}$ with respect to one of the subspaces is an operator in the other subspace, for example, $tr_B\{\hat{A}\} = \sum_{s,s'} \alpha_{s,s'}|s\rangle\langle s'|$, where $\alpha_{s,s'} = \sum_{b''} A_{s,b'',s',b''}$. (b) Let $\hat{S}$ and $\hat{B}$ be operators in the subspaces spanned by $\{|s\rangle\}$ and $\{|b\rangle\}$, respectively. Show that $tr_B\{\hat{S} \otimes \hat{B}\} = tr_B\{\hat{B} \otimes \hat{S}\} = tr_B\{\hat{B}\}\hat{S}$. (c) Let $\hat{S}$ be an operator in the subspace spanned by $\{|s\rangle\}$, and let $\hat{A}$ be an operator in the full (product) space. Show that $tr\{\hat{S}\hat{A}\} = tr_S\{\hat{S}tr_B\{\hat{A}\}\}$. (d) Let $\hat{B}$ be an operator in the subspace spanned by $\{|b\rangle\}$, and let $\hat{A}$ be an operator in the full (product) space. Show that $tr_B\{\hat{B}\hat{A}\} = tr_B\{\hat{A}\hat{B}\}$.*

Exercise 19.3.2   *(a) Show that $\hat{P}_L$ and $\hat{Q}_L$, as defined in Eqs. (19.3.7, 19.3.8), satisfy the relations $\hat{Q}_L^2 = \hat{Q}_L$, $\hat{P}_L^2 = \hat{P}_L$ and $\hat{Q}_L\hat{P}_L = \hat{P}_L\hat{Q}_L = 0$. (b) Show that $[\hat{P}_L, \hat{L}_S] = 0$ and $[\hat{P}_L, \hat{L}_B] = 0$.*

The Redfield Equation

It is convenient to expand the system–bath coupling as a sum over products of system and bath operators, denoted hereafter as $\{\hat{V}_1^{(S)}, \hat{V}_2^{(S)}, \ldots\}$ and $\{\hat{U}_1^{(B)}, \hat{U}_2^{(B)}, \ldots\}$, respectively:

$$\hat{H}_{SB} \equiv \sum_\alpha \hat{V}_\alpha^{(S)} \hat{U}_\alpha^{(B)}. \tag{19.3.12}$$

This expansion is rather general, as this form of $\hat{H}_{SB}$ corresponds to any analytic function in the system and bath operators. (Notice that each term in the sum need not be Hermitian, as long as the sum contains its Hermitian conjugate.) The NZ projection becomes especially useful, however, in the common case in which the bath coupling operators are orthogonal to the stationary bath density (in the sense of a Liouville's space inner product, Eq. (16.3.1)),

$$tr_B\{\hat{U}_\alpha^{(B)} \hat{\rho}_B\} = tr_B\{\hat{U}_\alpha^{(B)\dagger} \hat{\rho}_B\} = 0, \tag{19.3.13}$$

which means that observables associated with these operators vanish in the stationary state of the bath. In this case, the following identities hold for the projectors defned in Eqs. (19.3.7, 19.3.8) (see Ex. 19.3.3):

$$\hat{P}_L\hat{L}\hat{P}_L = \hat{P}_L\hat{L}_S\hat{P}_L \quad ; \quad \hat{P}_L\hat{L}\hat{Q}_L = \hat{P}_L\hat{L}_{SB} \quad ; \quad \hat{Q}_L\hat{L}\hat{P}_L = \hat{L}_{SB}\hat{P}_L. \tag{19.3.14}$$

Replacing $\hat{Q}_L\hat{L}\hat{P}_L$ by $\hat{L}_{SB}\hat{P}_L$ in the exact equation for the reduced system dynamics (Eq. (19.3.2)), the infinite-order expansion is explicitly expressed in terms of powers of $\hat{L}_{SB}$, and hence of the system–bath coupling operator, $\hat{H}_{SB}$. The Born–Markov approximation, Eq. (19.3.1), can then be identified as a weak coupling approximation, valid to second order in the system–bath interaction, $\hat{H}_{SB}$. Consistent with the neglect of higher orders in $\hat{H}_{SB}$, the $\hat{Q}_L$-space propagator in Eq. (19.3.1) can be approximated to the zeroth order in $\hat{L}_{SB}$:

$$e^{\frac{-i(t-\tau)}{\hbar}\hat{Q}_L\hat{L}\hat{Q}_L} \approx e^{\frac{-i(t-\tau)}{\hbar}\hat{Q}_L(\hat{L}_S+\hat{L}_B)\hat{Q}_L}. \tag{19.3.15}$$

Using Eqs. (19.3.9, 19.3.14, 19.3.15) as well as the equality, $\hat{P}_L e^{\alpha\hat{L}_B} = \hat{P}_L$ (see Ex. 19.3.4), the Born–Markov approximation (Eq. (19.3.1)) reads

$$\frac{\partial}{\partial t}\hat{\rho}_P^{(I)}(t) \cong -\frac{1}{\hbar^2}\int_0^t d\tau \hat{P}_L e^{\frac{it}{\hbar}(\hat{L}_S+\hat{L}_B)}\hat{L}_{SB}e^{\frac{-i(t-\tau)}{\hbar}(\hat{L}_S+\hat{L}_B)}\hat{L}_{SB}e^{\frac{-i\tau}{\hbar}(\hat{L}_S+\hat{L}_B)}\hat{\rho}_P^{(I)}(t), \tag{19.3.16}$$

or, recalling the definition of the interaction picture representation of the system–bath coupling, $e^{\frac{it}{\hbar}(\hat{L}_S+\hat{L}_B)}\hat{H}_{SB} = e^{\frac{it}{\hbar}(\hat{H}_S+\hat{H}_B)}\hat{H}_{SB}e^{\frac{-it}{\hbar}(\hat{H}_S+\hat{H}_B)} \equiv \hat{H}_{SB}^{(I)}(t)$, we obtain (Ex. 19.3.5)

$$\frac{\partial}{\partial t}\hat{\rho}_P^{(I)}(t) \cong -\frac{1}{\hbar^2}\hat{\rho}_B\int_0^t d\tau tr_B\{[\hat{H}_{SB}^{(I)}(t), [\hat{H}_{SB}^{(I)}(\tau), \hat{\rho}_P^{(I)}(t)]]\}. \tag{19.3.17}$$

Noticing that $\hat{\rho}_P^{(I)}(t) = e^{\frac{it}{\hbar}\hat{L}_S}\hat{P}_L\hat{\rho}(t) = \hat{\rho}_B e^{\frac{it}{\hbar}\hat{L}_S}\hat{\rho}_S(t)$ (see Eqs. (19.1.14, 19.3.7)), we obtain the equation of motion for the reduced system density operator, $\hat{\rho}_S(t)$ (Ex. 19.3.6):

$$\frac{\partial}{\partial t}\hat{\rho}_S(t) \cong -\frac{i}{\hbar}\hat{L}_S\hat{\rho}_S(t) - \frac{1}{\hbar^2}\int_0^t d\tau tr_B\{\hat{L}_{SB}e^{\frac{-i\tau}{\hbar}(\hat{L}_S+\hat{L}_B)}\hat{L}_{SB}e^{\frac{i\tau}{\hbar}(\hat{L}_S+\hat{L}_B)}\hat{\rho}_B\}\hat{\rho}_S(t), \quad (19.3.18)$$

or, equivalently (Ex. 19.3.7)

$$\frac{\partial}{\partial t}\hat{\rho}_S(t) \cong -\frac{i}{\hbar}[\hat{H}_S, \hat{\rho}_S(t)] - \frac{1}{\hbar^2}\int_0^t d\tau tr_B\left\{\left[\hat{H}_{SB}, \left[e^{\frac{-i\tau}{\hbar}(\hat{H}_S+\hat{H}_B)}\hat{H}_{SB}e^{\frac{i\tau}{\hbar}(\hat{H}_S+\hat{H}_B)}, \hat{\rho}_B\hat{\rho}_S(t)\right]\right]\right\}. \quad (19.3.19)$$

This result is known as the Redfield equation [19.3]. Notice that it can be derived directly by tracing the full Liouville equation over the bath degrees of freedom, keeping only terms up to second order in the system–bath coupling (see Ex. 19.3.8). Also notice that for initial states, in which the system and the bath are correlated, $\hat{Q}_L\hat{\rho}(0) \neq 0$, the exact equation obtains an additional inhomogeneous term (see Eq. (19.1.16)). Approximating this term up to second order in the system–bath coupling, the Redfield equation obtains a corresponding inhomogeneous term (see Ex. 19.3.9).

In the absence of initial system–bath correlation, the effect of the coupling of the system to the bath is introduced by a dissipator,

$$\hat{D}\hat{\rho}_S(t) \equiv -\frac{1}{\hbar^2}\int_0^t d\tau tr_B\{[\hat{H}_{SB}, [e^{\frac{-i\tau}{\hbar}(\hat{H}_S+\hat{H}_B)}\hat{H}_{SB}e^{\frac{i\tau}{\hbar}(\hat{H}_S+\hat{H}_B)}, \hat{\rho}_B\hat{\rho}_S(t)]]\}. \quad (19.3.20)$$

Using the expansion of the system–bath coupling in multiplications of system and bath operators ($\hat{H}_{SB} \equiv \sum_\alpha \hat{V}_\alpha^{(S)}\hat{U}_\alpha^{(B)}$; see Eq. (19.3.12)), the influence of the bath is captured in the correlation functions of the corresponding bath operators,

$$c_{\alpha,\alpha'}(\tau) \equiv tr_B\{\hat{U}_\alpha^{(B)}e^{\frac{-i\tau}{\hbar}\hat{H}_B}\hat{U}_{\alpha'}^{(B)}e^{\frac{i\tau}{\hbar}\hat{H}_B}\hat{\rho}_B\} \quad ; \quad \bar{c}_{\alpha,\alpha'}(\tau) = c_{\alpha,\alpha'}(-\tau), \quad (19.3.21)$$

where the dissipator obtains the form (see Ex. 19.3.10)

$$\hat{D}\hat{\rho}_S(t) = -\frac{1}{\hbar^2}\sum_{\alpha,\alpha'}\int_0^t d\tau\{c_{\alpha,\alpha'}(\tau)[\hat{V}_\alpha^{(S)}, e^{\frac{-i\tau}{\hbar}\hat{H}_S}\hat{V}_{\alpha'}^{(S)}e^{\frac{i\tau}{\hbar}\hat{H}_S}\hat{\rho}_S(t)]$$

$$+ \bar{c}_{\alpha',\alpha}(\tau)[\hat{\rho}_S(t)e^{\frac{-i\tau}{\hbar}\hat{H}_S}\hat{V}_{\alpha'}^{(S)}e^{\frac{i\tau}{\hbar}\hat{H}_S}, \hat{V}_\alpha^{(S)}]\}. \quad (19.3.22)$$

Introducing a complete orthonormal system of the reduced system Hamiltonian eigenstates, $\hat{H}_S|\varphi_n\rangle = E_n|\varphi_n\rangle$ (for simplicity, only a discrete spectrum is considered explicitly) and representing the system operators in this basis, $\rho_{n',n}(t) = \langle\varphi_{n'}|\hat{\rho}_S(t)|\varphi_n\rangle$ and $V_{n',n}^{(\alpha)} = \langle\varphi_{n'}|\hat{V}_\alpha^{(S)}|\varphi_n\rangle$, the Redfield equation (Eq. (19.3.19)) obtains the form

$$\frac{\partial}{\partial t}\rho_{n',n}(t) \cong -\frac{i}{\hbar}(E_{n'} - E_n)\rho_{n',n}(t) - \sum_{m',m}R_{n',n,m',m}(t)\rho_{m',m}(t), \quad (19.3.23)$$

where the dissipator is represented in terms of the Redfield tensor (super-operator; see Ex. 19.3.11),

$$
R_{n',n,m',m}(t) = \sum_{\alpha,\alpha'} \left\{ \sum_{l} [g_{l,m'}^{(\alpha,\alpha')}(t) V_{n',l}^{(\alpha)} V_{l,m'}^{(\alpha')} \delta_{m,n} + \overline{g}_{m,l}^{(\alpha',\alpha)}(t) V_{m,l}^{(\alpha')} V_{l,n}^{(\alpha)} \delta_{m',n'}] \right.
$$
$$
\left. - [g_{n',m'}^{(\alpha,\alpha')}(t) V_{n',m'}^{(\alpha')} V_{m,n}^{(\alpha)} + \overline{g}_{m,n}^{(\alpha',\alpha)}(t) V_{n',m'}^{(\alpha)} V_{m,n}^{(\alpha')}] \right\},
$$

(19.3.24)

and we introduced the Fourier integrals over the bath correlation functions,

$$
g_{n,n'}^{(\alpha,\alpha')}(t) \equiv \frac{1}{\hbar^2} \int_0^t c_{\alpha,\alpha'}(\tau) e^{\frac{-i\tau}{\hbar}(E_n-E_{n'})} d\tau \quad ; \quad \overline{g}_{n,n'}^{(\alpha,\alpha')}(t) \equiv \frac{1}{\hbar^2} \int_0^t \overline{c}_{\alpha,\alpha'}(\tau) e^{\frac{-i\tau}{\hbar}(E_n-E_{n'})} d\tau.
$$

(19.3.25)

Exercise 19.3.3   *For the system–bath operators as defined in Eqs. (19.3.3–19.3.6, 19.3.12–19.3.13), show that the projection operators $\hat{P}_L$ and $\hat{Q}_L$, as defined in Eqs. (19.3.7, 19.3.8), satisfy the relations in Eq. (19.3.14).*

Exercise 19.3.4   *(a) Use Eqs. (19.1.12, 19.3.9, 19.3.14, 19.3.15) in Eq. (19.3.1) to show that $\frac{\partial}{\partial t}\hat{\rho}_P^{(I)}(t) \cong -\frac{1}{\hbar^2} \int_0^t d\tau \hat{P}_L e^{\frac{it}{\hbar}\hat{L}_S} \hat{L}_{SB} e^{\frac{-i(t-\tau)}{\hbar}(\hat{L}_S+\hat{L}_B)} \hat{L}_{SB} \hat{P}_L e^{\frac{-i\tau}{\hbar}\hat{L}_S} \hat{\rho}_P^{(I)}(t)$. (b) Show that according to the definition of $\hat{P}_L$ (Eq. (19.3.7)), for any scalar α one has, $\hat{P}_L e^{\alpha\hat{L}_B} = \hat{P}_L$. Use this identity and the definition, $\hat{\rho}_P^{(I)}(t) \equiv e^{\frac{it}{\hbar}\hat{P}_L\hat{L}\hat{P}_L}\hat{P}_L\hat{\rho}(t)$ (Eq. (19.1.14)), to derive Eq. (19.3.16).*

Exercise 19.3.5   *(a) Use the identities for any scalar α, $e^{\alpha\hat{L}_S}\hat{\rho} \equiv e^{\alpha\hat{H}_S}\hat{\rho}e^{-\alpha\hat{H}_S}$, and $e^{\alpha\hat{L}_B}\hat{\rho} \equiv e^{\alpha\hat{H}_B}\hat{\rho}e^{-\alpha\hat{H}_B}$, to derive Eq. (19.3.17) from Eq. (19.3.16).*

Exercise 19.3.6   *Use the identities, $\hat{\rho}_P^{(I)}(t) = e^{\frac{it}{\hbar}\hat{L}_S}\hat{P}_L\hat{\rho}(t) = \hat{\rho}_B e^{\frac{it}{\hbar}\hat{L}_S}\hat{\rho}_S(t)$, and Eq. (19.3.16) to derive Eq. (19.3.18).*

Exercise 19.3.7   *Use the identity, $e^{\frac{it}{\hbar}(\hat{L}_S+\hat{L}_B)}\hat{A} = e^{\frac{it}{\hbar}(\hat{H}_S+\hat{H}_B)}\hat{A}e^{\frac{-it}{\hbar}(\hat{H}_S+\hat{H}_B)}$ to derive Eq. (19.3.19) from Eq. (19.3.18).*

Exercise 19.3.8   *Follow the alternative derivation given here to obtain the Redfield equation directly: (a) Transform the full Liouville equation, $\frac{\partial}{\partial t}\hat{\rho}(t) = \frac{-i}{\hbar}[\hat{H},\hat{\rho}(t)]$, with $\hat{H} = \hat{H}_S + \hat{H}_B + \hat{H}_{SB}$, into the interaction picture representation, $\frac{\partial}{\partial t}\hat{\rho}^{(I)}(t) = \frac{-i}{\hbar}[\hat{H}_{SB}^{(I)}(t),\hat{\rho}^{(I)}(t)]$, where $\hat{O}^{(I)}(t) \equiv e^{\frac{i}{\hbar}[\hat{H}_S+\hat{H}_B]t}\hat{O}(t)e^{\frac{-i}{\hbar}[\hat{H}_S+\hat{H}_B]t}$. (b) Integrate the equation over time and show that it can be rearranged exactly as $\frac{\partial}{\partial t}\hat{\rho}^{(I)}(t) = \frac{-i}{\hbar}[\hat{H}_{SB}^{(I)}(t),\hat{\rho}^{(I)}(0)] - \frac{1}{\hbar^2}\int_0^t dt'[\hat{H}_{SB}^{(I)}(t),[\hat{H}_{SB}^{(I)}(t'),\hat{\rho}^{(I)}(t)]] - \frac{i}{\hbar^3}\int_0^t dt'\int_{t'}^t dt''[\hat{H}_{SB}^{(I)}(t),[\hat{H}_{SB}^{(I)}(t'),[\hat{H}_{SB}^{(I)}(t''),\hat{\rho}^{(I)}(t'')]]]$. (c) Neglect the terms of third order and higher in the system–bath coupling and obtain an approximate Markovian equation for $\hat{\rho}^{(I)}(t)$, $\frac{\partial}{\partial t}\hat{\rho}^{(I)}(t) \cong \frac{-i}{\hbar}[\hat{H}_{SB}^{(I)}(t),\hat{\rho}^{(I)}(0)] - \frac{1}{\hbar^2}\int_0^t dt'[\hat{H}_{SB}^{(I)}(t),[\hat{H}_{SB}^{(I)}(t'),\hat{\rho}^{(I)}(t)]]$. (d) Defining the reduced system density operator*

as $\hat{\rho}_S^{(I)}(t) \equiv tr_B[\hat{\rho}^{(I)}(t)]$, and assuming an initial product density:, $\hat{\rho}^{(I)}(0) = \hat{\rho}(0) = \hat{\rho}_B \hat{\rho}_S(0)$, show that for $tr_B\{\hat{\rho}_B \hat{H}_{SB}\} = 0$ (Eqs. (19.3.12, 19.3.13)), we obtain $\frac{\partial}{\partial t}\hat{\rho}_S^{(I)}(t) =$
$-\frac{1}{\hbar^2} \int_0^t dt' tr_B[\hat{H}_{SB}^{(I)}(t), [\hat{H}_{SB}^{(I)}(t'), \hat{\rho}^{(I)}(t)]]$. (e) Without loss of accuracy to second order in the system–bath coupling, replace $\hat{\rho}^{(I)}(t) \cong \hat{\rho}_B \otimes \hat{\rho}_S^{(I)}(t)$ under the latter time integral and show that this leads to Eq. (19.3.17), and hence to the Redfield equation, Eq. (19.3.19).

Exercise 19.3.9   When the system and bath are initially correlated, $\hat{Q}_L \hat{\rho}(0) \neq 0$, the exact equation for the $\hat{P}_L$-space density operator obtains an inhomogeneous term (Eq. (19.1.16)). Expanding this term up to first-order in the system–bath coupling, show that Eq. (19.3.19) is generalized to $\frac{\partial}{\partial t}\hat{\rho}_S(t) \cong -\frac{i}{\hbar}[\hat{H}_S, \hat{\rho}_S(t)]$
$-\frac{1}{\hbar^2} \int_0^t d\tau tr_B\{[\hat{H}_{SB}, [e^{\frac{-i\tau}{\hbar}(\hat{H}_S + \hat{H}_B)} \hat{H}_{SB} e^{\frac{i\tau}{\hbar}(\hat{H}_S + \hat{H}_B)}, \hat{\rho}_B \hat{\rho}_S(t)]]\}$
$-\frac{i}{\hbar}\hat{\rho}_B tr_B\{[e^{\frac{it}{\hbar}(\hat{H}_S + \hat{H}_B)} \hat{H}_{SB} e^{\frac{-it}{\hbar}(\hat{H}_S + \hat{H}_B)}, \hat{\rho}(0)]\}$.

Exercise 19.3.10   Using the expansion of the system–bath coupling operator, $\hat{H}_{SB} \equiv \sum_\alpha \hat{V}_\alpha^{(S)} \hat{U}_\alpha^{(B)}$ (Eq. (19.3.12)), we obtain $e^{\frac{-i\tau}{\hbar}(\hat{H}_S + \hat{H}_B)} \hat{H}_{SB} e^{\frac{i\tau}{\hbar}(\hat{H}_S + \hat{H}_B)} \equiv \sum_\alpha \hat{V}_\alpha^{(S)}(\tau) \hat{U}_\alpha^{(B)}(\tau)$, where $\hat{V}_\alpha^{(S)}(\tau) \equiv e^{\frac{-i\tau}{\hbar}\hat{H}_S} \hat{V}_\alpha^{(S)} e^{\frac{i\tau}{\hbar}\hat{H}_S}$ and $\hat{U}_\alpha^{(B)}(\tau) \equiv e^{\frac{-i\tau}{\hbar}\hat{H}_B} \hat{U}_\alpha^{(B)} e^{\frac{i\tau}{\hbar}\hat{H}_B}$. Using these expressions, show that $\hat{D}\hat{\rho}_S(t)$, as defined in Eq. (19.3.20), can be expressed in terms of the bath coupling correlation functions $c_{\alpha,\alpha'}(\tau)$ and $\bar{c}_{\alpha,\alpha'}(\tau)$ (Eqs. (19.3.21, 19.3.22)).

Exercise 19.3.11   The Redfield equation for the reduced density operator (Eq. (19.3.19)) can be written as $\frac{\partial}{\partial t}\hat{\rho}_S(t) \cong -\frac{i}{\hbar}[\hat{H}_S, \hat{\rho}_S(t)] + \hat{D}\hat{\rho}_S(t)$, with $\hat{D}\hat{\rho}_S(t)$ defined according to Eq. (19.3.22). Defining $\rho_{n',n}(t) \equiv \langle \varphi_{n'}|\hat{\rho}_S(t)|\varphi_n\rangle$ and $V_{n',n}^{(\alpha)} \equiv \langle \varphi_{n'}|\hat{V}_\alpha^{(S)}|\varphi_n\rangle$, where $\{|\varphi_n\rangle\}$ are a complete orthonormal system of $\hat{H}_S$ eigenstates with the corresponding energies, $\{E_n\}$, derive Eq. (19.3.23), using the definitions in Eqs. (19.3.24, 19.3.25).

Notice that the time-dependent Redfield equation (Eq. (19.3.23)) is valid as long as the weak coupling assumption holds (namely, the equation is derived by replacing Eq. (19.3.2) by the Born–Markov approximation, Eq. (19.3.1)). The latter is limited to times that are short on the system–bath coupling period ($\hbar$ over the largest of the system–bath coupling matrix elements; see, e.g., Eq. (17.2.4)). In many cases of interest, however, these times are sufficiently long on the timescale of the reduced system dynamics, such that the latter is well approximated by the Redfield equation. Moreover, as discussed in what follows, the equation can be further simplified when the timescale for bath-induced dynamics in the system is well separated from the bath relaxation time (leading to the stationary Redfield equation) and from the periods of internal (bath-free) system dynamics (leading to the Pauli master equation).

19.4 The Stationary Born–Markov Approximation

Within the Born–Markov approximation, the dissipator (Eq. (19.3.24)) is time-dependent, owing to the Fourier integrals over the bath correlation functions (see

Eq. (19.3.25)). However, when the relaxation of these correlations is much faster than the bath-induced dynamics in the system, the upper time-limit in the Fourier integrals can be taken to infinity with no significant effect on the solution of the equation. In this case, the Redfield tensor (Eq. (19.3.24)) can be approximated as being time-independent:

$$R_{n',n,m',m}(t) \approx R_{n',n,m',m}^{(St)} = \sum_{\alpha,\alpha'} \{-[G_{n',m'}^{(\alpha,\alpha')} V_{n',m'}^{(\alpha')} V_{m,n}^{(\alpha)} + \overline{G}_{m,n}^{(\alpha',\alpha)} V_{n',m'}^{(\alpha)} V_{m,n}^{(\alpha')}]$$

$$+ \sum_{l} [G_{l,m'}^{(\alpha,\alpha')} V_{n',l}^{(\alpha)} V_{l,m'}^{(\alpha')} \delta_{m,n} + \overline{G}_{m,l}^{(\alpha',\alpha)} V_{m,l}^{(\alpha')} V_{l,n}^{(\alpha)} \delta_{m',n'}]\}, \qquad (19.4.1)$$

where the coefficients $G_{n,n'}^{(\alpha,\alpha')}$ and $\overline{G}_{n,n'}^{(\alpha,\alpha')}$ are the "half Fourier transforms" of the bath correlation functions,

$$G_{n,n'}^{(\alpha,\alpha')} \equiv \frac{1}{\hbar^2} \int_0^\infty c_{\alpha,\alpha'}(\tau) e^{\frac{-i\tau}{\hbar}(E_n - E_{n'})} d\tau \quad ; \quad \overline{G}_{n,n'}^{(\alpha,\alpha')} \equiv \frac{1}{\hbar^2} \int_0^\infty \overline{c}_{\alpha,\alpha'}(\tau) e^{\frac{-i\tau}{\hbar}(E_n - E_{n'})} d\tau.$$

$$(19.4.2)$$

The corresponding equation of motion for the reduced system dynamics under time-scale separation between the "fast" bath relaxation and the "slow" bath induced dynamics is sometimes referred to as the stationary Born–Markov (stationary Redfield) approximation [19.4],

$$\frac{\partial}{\partial t} \rho_{n',n}(t) = \frac{i}{\hbar}(E_n - E_{n'})\rho_{n',n}(t) - \sum_{m',m} R_{n',n,m',m}^{(St)} \rho_{m',m}(t). \qquad (19.4.3)$$

The conditions for the validity of this approximation can be readily identified by considering the explicit form of the bath correlation functions (Eq. (19.3.21)). Denoting the bath Hamiltonian eigenstates, $\hat{H}_B|b\rangle = E_b|b\rangle$, and recalling that $[\hat{\rho}_B, \hat{H}_B] = 0$, we have

$$c_{\alpha,\alpha'}(\tau) = \sum_{b,b'} \langle b|\hat{U}_\alpha^{(B)}|b'\rangle \langle b'|\hat{U}_{\alpha'}^{(B)}|b\rangle \rho_b e^{\frac{-i\tau}{\hbar}(E_{b'} - E_b)}$$

$$\overline{c}_{\alpha,\alpha'}(\tau) = \sum_{b,b'} \langle b|\hat{U}_\alpha^{(B)}|b'\rangle \langle b'|\hat{U}_{\alpha'}^{(B)}|b\rangle \rho_b e^{\frac{-i\tau}{\hbar}(E_b - E_{b'})}. \qquad (19.4.4)$$

These functions have the generic form of a sum of oscillating terms at the bath frequencies, $\{\omega_{b',b} = (E_{b'} - E_b)/\hbar\}$. Recalling that the relaxation time of a sum of oscillating terms decreases with increasing bandwidth (see the discussion of Fermi's golden rule in Section 17.2, and Fig. 17.2.2), the assumption of rapid decay of the bath correlation function, and hence the validity of the stationary Redfield approximation, become valid as the bandwidth increases. This implies that the matrix representations of the bath coupling operators $\{\langle b|\hat{U}_\alpha^{(B)}|b'\rangle\}$ are broad-banded, and/or that the distribution of the bath eigenstate populations, $\{\rho_b = \langle b|\hat{\rho}|b\rangle\}$, is sufficiently broad. The latter typically depends on the bath temperature, where the distribution width increases with increasing temperature (e.g., for a canonical ensemble, $\rho_b = e^{-E_b/(k_B T)}/Z$). Consequently, **the validity of the stationary Redfield approximation typically increases with increasing bath temperature.** (Notice that the validity of the time-dependent Redfield equation (Eqs. (19.3.23)) is not restricted by the bath temperature, as long as the weak coupling (Born–Markov) assumption holds, as we have discussed.)

Population Transfer and the Pauli Master Equation

Now let us recall that, in the absence of system–bath coupling, the system eigenstate populations are constant in time, and the dynamics is restricted to the oscillating coherences between pairs of system eigenstates (see Section 15.1). While this dynamics may be rich and interesting in its own right, here we wish to *focus on the bath-induced dynamics within the system under the influence of the system–bath coupling*. For this purpose, it is useful to transform Eq. (19.4.3) into the interaction picture representation (see Section 15.3), where in the absence of system–bath coupling the density operator is stationary, such that any system dynamics is solely attributed to the system–bath coupling. The transformed density matrix reads

$$\rho_{n',n}^{(I)}(t) = e^{i\omega_{n',n}t}\rho_{n',n}(t) \quad ; \quad \omega_{n',n} = \frac{E_{n'} - E_n}{\hbar}, \tag{19.4.5}$$

where

$$\frac{\partial}{\partial t}\rho_{n',n}^{(I)}(t) = -\sum_{m',m} R_{n',n,m',m}^{(I)}(t)\rho_{m',m}^{(I)}(t). \tag{19.4.6}$$

In this representation the coupling-free dynamics is encoded into the transformed dissipator (see Ex. 19.4.1):

$$R_{n',n,m',m}^{(I)}(t) = [R_{n',n,m',m}^{(St)}]e^{i(\omega_{n',n} - \omega_{m',m})t}. \tag{19.4.7}$$

The equation of motion for the reduced system density in the interaction picture (Eq. (19.4.6)) simplifies tremendously when the coupling-free system dynamics is much faster than the bath-induced dynamics. This scenario applies whenever the level spacings between (nondegenerate) eigenvalues of the bare system Hamiltonian are large in comparison to the system–bath coupling energy. *In these cases, coherences between the system eigenstates are oscillating rapidly (at frequencies $\{\omega_{n',n}\}$) on the timescale of the bath-induced dynamics*. Consequently, the corresponding elements of $R_{n',n,m',m}^{(I)}(t)$, average to near zero on the timescale of the bath-induced dynamics and can be approximately neglected with respect to its stationary elements. This is known as "the secular approximation," which is closely related to "the rotating wave approximation" (as encountered already in Section 18.3). To formally justify the resulting equation, let us assume first that the system coherences share a common oscillation period, $T_c = 2\pi/\omega$, such that $\omega_{n',n} \cong \omega \cdot l_{n',n}$, where $l_{n',n}$ is an integer. This assumption is not guaranteed to hold in general, but it is often fulfilled (exactly or approximately) for "small" (e.g., few-level) quantum systems (it is trivially correct for any two-level system). Second, let us assume that the common period T_c is still considerably short on the bath-induced dynamics timescale (while still being much longer than the relaxation time of the bath correlation functions (see Fig. 19.4.1), which should hold when the characteristic energy level spacings in the system are large in comparison to the bath spectral density (see Section 17.2)). Since changes in $\rho_{n',n}^{(I)}(t)$ within the period T_c are assumed negligible, we have $\frac{\partial}{\partial t}\rho^{(I)}_{n',n}(t) \approx \frac{1}{T_c}\int_{t-T_c/2}^{t+T_c/2} dt' \frac{\partial}{\partial t'}\rho^{(I)}_{n',n}(t')$ as well as

$\frac{1}{T_c}\int_{t-T_c/2}^{t+T_c/2} dt' \sum_{m',m} R^{(I)}_{n',n,m',m}(t')\rho^{(I)}_{m',m}(t') \cong \frac{1}{T_c}\int_{t-T_c/2}^{t+T_c/2} dt' \sum_{m',m} R^{(I)}_{n',n,m',m}(t')\rho^{(I)}_{m',m}(t)$. Using these approximations in Eqs. (19.4.6, 19.4.7), we obtain (Ex. 19.4.2)

$$\frac{\partial}{\partial t}\rho^{(I)}_{n',n}(t) \cong -\sum_{m',m} \overline{R}_{n',n,m',m}\rho^{(I)}_{m',m}(t), \tag{19.4.8}$$

$$\overline{R}_{n',n,m',m} = \frac{1}{T_c}\int_{t-T_c/2}^{t+T_c/2} dt' \sum_{m',m} R^{(I)}_{n',n,m',m}(t') = \delta_{l_{n',n},l_{m',m}} R^{(St)}_{n',n,m',m}. \tag{19.4.9}$$

Exercise 19.4.1 *Given Eq. (19.4.3) for the dynamics of the reduced system density matrix elements, $\frac{\partial}{\partial t}\rho_{n',n}(t) = \frac{i}{\hbar}(E_n - E_{n'})\rho_{n',n}(t) - \sum_{m',m} R^{(St)}_{n',n,m',m}\rho_{m',m}(t)$, and the transformation, $\rho^{(I)}_{n',n}(t) = e^{i\omega_{n',n}t}\rho_{n',n}(t)$ (Eq. (19.4.5)), derive Eq. (19.4.6) with the time-dependent tensor, $R^{(I)}_{n',n,m',m}(t)$, as defined in Eq. (19.4.7).*

Exercise 19.4.2 *Show that time averaging of Eq. (19.4.6) over the short oscillation period, T_c, leads to Eqs. (19.4.8, 19.4.9).*

The result is an effective equation for the reduced system density operator, $\hat{\rho}^{(I)}_S(t)$, where the bath-induced dynamics is captured in the time-independent super-operator $\overline{R}$ (Eq. (19.4.9)). It can be shown that this operator is of a generic Lindblad form (a generator of a dynamical semigroup), which assures that the sum as well as the nonnegativity of the system eigenstate populations are preserved over time. We direct the interested reader to complementary literature (e.g., [19.2]) with respect to the mathematical proofs of these properties. Here we address some practical consequences that can be readily deduced from Eqs. (19.4.8, 19.4.9).

First, we notice that since the off-diagonal elements of $\hat{\rho}^{(I)}_S(t)$ (coherences) between two nondegenerate eigenstates of the bath-free system Hamiltonian are associated with $l_{n',n} \neq 0$, the bath-induced changes in these coherences depend only on non-diagonal elements of $\hat{\rho}^{(I)}_S(t)$. This means that *the dynamics of coherences between nondegenerate eigenstates is decoupled from the dynamics of populations* (associated with $l_{n',n} = 0$).

Second, *in cases where the spectrum of the coupling-free system Hamiltonian is nondegenerate, bath-induced changes in the populations* (the diagonal matrix elements of $\hat{\rho}^{(I)}_S(t)$, corresponding to $l_{n',n} = 0$) *are independent of any coherences* (non-diagonal elements, which correspond to $l_{n',n} \neq 0$ in these cases). Denoting the populations of the system eigenstates as

$$P_n(t) = \rho_{n,n}(t) = \rho^{(I)}_{n,n}(t), \tag{19.4.10}$$

the corresponding equation for the bath-induced population changes reads (using Eqs. (19.4.8, 19.4.9))

$$\frac{\partial}{\partial t}P_n(t) \cong -\sum_m R^{(St)}_{n,n,m,m}P_m(t), \tag{19.4.11}$$

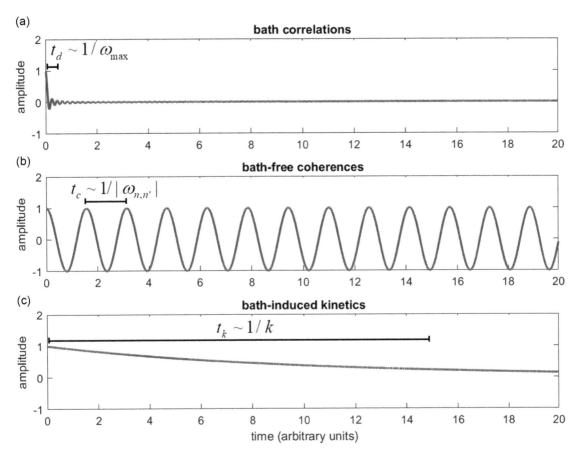

Figure 19.4.1 Illustrative representation of timescale separation underlying the weak coupling (Born–Markov) and the secular approximations. (a) The decay of bath correlation functions (Eq. (19.3.21)), determined by the bandwidth of the system bath coupling, $t_d \sim 1/\omega_{max}$. (b) Oscillating coherences at the bath-free system frequencies, $t_c \sim 1/|\omega_{n,n'}|$. (c) Bath-induced kinetics in the system (population transfer and/or decay of coherences), determined by the bath spectral density and temperature, $t_k \sim 1/k$. The characteristic timescales are indicated on each plot. The Born–Markov approximation is valid when the decay of bath correlations is much faster than both the bath-induced system dynamics, $t_d \ll f_k$ (the wide-band limit), and the bath-free system dynamics, $t_d \ll t_c$. The secular approximation becomes valid when, on top of these conditions, the bath-free system dynamics is much faster than the bath-induced system dynamics, $t_c \ll t_k$.

where (using Eq. (19.4.1))

$$R_{n,n,m,m}^{(St)} = \sum_{\alpha,\alpha'} \{-[G_{n,m}^{(\alpha,\alpha')} V_{n,m}^{(\alpha')} V_{m,n}^{(\alpha)} + \overline{G}_{m,n}^{(\alpha',\alpha)} V_{n,m}^{(\alpha)} V_{m,n}^{(\alpha')}]$$
$$+ \sum_l [G_{l,m}^{(\alpha,\alpha')} V_{n,l}^{(\alpha)} V_{l,m}^{(\alpha')} \delta_{m,n} + \overline{G}_{m,l}^{(\alpha',\alpha)} V_{m,l}^{(\alpha')} V_{l,n}^{(\alpha)} \delta_{m,n}]\} \tag{19.4.12}$$

Defining, $k_{m,n} \equiv \sum_{\alpha,\alpha'} \{G_{n,m}^{(\alpha,\alpha')} V_{n,m}^{(\alpha')} V_{m,n}^{(\alpha)} + \overline{G}_{m,n}^{(\alpha',\alpha)} V_{n,m}^{(\alpha)} V_{m,n}^{(\alpha')}\}$, and noticing that $k_{m,n} \geq 0$ (see the following), Eqs. (19.4.11, 19.4.12) obtain the form of a set kinetic equations for the state populations, often referred to as Pauli's master equation (Ex. 19.4.3),

$$\frac{\partial}{\partial t} P_n(t) = \sum_m k_{m \to n} P_m(t) - \sum_m k_{n \to m} P_n(t), \qquad (19.4.13)$$

where the state-to-state rates can be identified as

$$k_{m,n} = k_{m \to n} = \sum_{\alpha, \alpha'} \{ G_{n,m}^{(\alpha, \alpha')} V_{n,m}^{(\alpha')} V_{m,n}^{(\alpha)} + \overline{G}_{m,n}^{(\alpha', \alpha)} V_{n,m}^{(\alpha)} V_{m,n}^{(\alpha')} \}. \qquad (19.4.14)$$

As we can see, the time evolution of the reduced system populations is subject to probability conservation, namely (see Ex. 19.4.4)

$$\frac{\partial}{\partial t} \sum_n P_n(t) = 0. \qquad (19.4.15)$$

Reformulating the master equation in matrix form,

$$\frac{\partial}{\partial t} \mathbf{P}(t) = \mathbf{K} \mathbf{P}(t), \qquad (19.4.16)$$

the matrix elements of $\mathbf{K}$ read

$$[\mathbf{K}]_{n,m} = (1 - \delta_{m,n}) k_{m \to n} - \delta_{m,n} \sum_{n' \neq n} k_{n \to n'}. \qquad (19.4.17)$$

We can also see (Ex. 19.4.5) that the linear system of equations has a stationary ("steady state") solution, which we denote as $\mathbf{P}_S$:

$$\frac{\partial}{\partial t} \mathbf{P}_S = \mathbf{K} \mathbf{P}_S = 0. \qquad (19.4.18)$$

Exercise 19.4.3   *Substitute Eq. (19.4.12) in Eq. (19.4.11) to derive the Pauli master equation (Eq. (19.4.13)), with the population transition rates given in Eq. (19.4.14).*

Exercise 19.4.4   *Use the Pauli master equation (Eq. (19.4.13)) for $\frac{\partial}{\partial t} P_n(t)$, to show that $\frac{\partial}{\partial t} \sum_n P_n(t) = 0$. (Recall that each summation is over the entire spectrum of $\hat{H}_S$-eigenstates.)*

Exercise 19.4.5   *Use the structure of the matrix $\mathbf{K}$, as expressed in Eq. (19.4.17), to show that its rows are linearly dependent, and hence there exists a nontrivial solution, $\mathbf{P}$, to the homogeneous equation, $\mathbf{K} \mathbf{P} = \mathbf{0}$.*

We now turn to the explicit evaluation of the state-to-state population transfer rates in terms of the Hamiltonian parameters. The Hermiticity of the sum of system–bath coupling operators, $\hat{H}_{SB} \equiv \sum_{\alpha=1}^{N} \hat{V}_\alpha^{(S)} \hat{U}_\alpha^{(B)}$, means that the sum contains either Hermitian terms or sums of pairs of non-Hermitian terms and their conjugates. Consequently, we can show (Ex. 19.4.6) that $\sum\limits_{\alpha,\alpha'} \overline{G}_{m,n}^{(\alpha',\alpha)} V_{n,m}^{(\alpha)} V_{m,n}^{(\alpha')} = \left[\sum\limits_{\alpha,\alpha'} G_{n,m}^{(\alpha,\alpha')} V_{n,m}^{(\alpha')} V_{m,n}^{(\alpha)} \right]^*$, which means that the population transfer rates read

$$k_{m\to n} = \sum_{\alpha,\alpha'} 2\,\mathrm{Re}\{G_{n,m}^{(\alpha,\alpha')} V_{n,m}^{(\alpha')} V_{m,n}^{(\alpha)}\}, \tag{19.4.19}$$

or explicitly, using Eqs. (19.4.2, 19.4.4) and some algebra (see Ex. 19.4.7), we obtain

$$k_{m\to n} = 2\,\mathrm{Re}\frac{1}{\hbar^2} \int_0^\infty tr\{\hat{\rho}_B|\varphi_m\rangle\langle\varphi_m|\hat{H}_{SB}|\varphi_n\rangle\langle\varphi_n|e^{\frac{i\tau}{\hbar}(\hat{H}_B+\hat{H}_S)}\hat{H}_{SB}e^{\frac{-i\tau}{\hbar}(\hat{H}_B+\hat{H}_S)}\}d\tau. \tag{19.4.20}$$

Let us identify the uncoupled system and bath with a zero-order Hamiltonian defined as $\hat{H}_0 = \hat{H}_B + \hat{H}_S$, and the system–bath coupling with a perturbation operator, $\hat{V} = \hat{H}_{SB}$. We can now associate the initial and final states of the system with ensembles of $\hat{H}_0$-eigenstates, characterized by product states of the uncoupled bath eigenstates (defined by $\hat{H}_B|b\rangle = E_b|b\rangle$), and the respective eigenstates of the system Hamiltonian ($\hat{H}_S|\varphi_m\rangle = E_m|\varphi_m\rangle$). Defining projectors into these ensembles, $\hat{P}_{\{i\}} = |\varphi_m\rangle\langle\varphi_m| \otimes \sum\limits_b |b\rangle\langle b|$, and $\hat{P}_{\{f\}} = |\varphi_n\rangle\langle\varphi_n| \otimes \sum\limits_b |b\rangle\langle b|$, for the initial and final ensembles, respectively, and associating the initial state with the density operator, $\hat{\rho}_{\{i\}} = |\varphi_m\rangle\langle\varphi_m| \otimes \hat{\rho}_B$, Eq. (19.4.20) can be written as

$$k_{m\to n} = 2\,\mathrm{Re}\frac{1}{\hbar^2} \int_0^\infty tr\left\{\hat{\rho}_{\{i\}}\hat{P}_{\{i\}}\hat{V}\hat{P}_{\{f\}}e^{\frac{i\tau}{\hbar}\hat{H}_0}\hat{V}e^{\frac{-i\tau}{\hbar}\hat{H}_0}\right\}d\tau. \tag{19.4.21}$$

We can readily see that the population transfer rate is the infinite time limit of Eq. (17.3.12), derived in Chapter 17 for the transition rate between the two ensembles, which leads to Fermi's golden rule (Ex. 19.4.8). The rates or population transfer therefore obtain the standard FGR form,

$$k_{m\to n} = \frac{2\pi}{\hbar} \sum_{b,b'} \rho_b|\langle b'|\langle\varphi_n|\hat{V}|\varphi_m\rangle|b\rangle|^2 \delta(E_n - E_m - [E_b - E_{b'}]), \tag{19.4.22}$$

where $\hat{\rho}_B|b\rangle = \rho_b|b\rangle$. It is apparent from the result that the population transfer rates are nonnegative. This means that for a given distribution of initial state populations, not only does the sum of probabilities remain constant in time (Eq. (19.4.15)), but also each population remains nonnegative at all times (Ex. 19.4.9), namely

$$P_n(0) \geq 0 \Rightarrow P_n(t) \geq 0 \quad ; \quad t > 0. \tag{19.4.23}$$

Hence, the Pauli master equation (Eqs. (19.4.13, 19.4.16)) is consistent with the Lindblad requirements on the Markovian dynamical map from $\mathbf{P}(0)$ to $\mathbf{P}(t)$, which preserves nonnegativity of the system Hamiltonian eigenstate populations (complete positivity), and their accumulation to unity (trace preservation).

We can also see that the transition rate is subject to energy conservation, in the sense that the energy difference between the initial and final eigenstates of the system Hamiltonian ($E_n - E_m$) must be compensated for by similar transitions between the bath Hamiltonian eigenstates, associated with $E_b - E_{b'} = E_n - E_m$. In the case of coupling to a thermal bath at equilibrium, where $\rho_b = \frac{1}{Z}e^{-E_b/(k_B T)}$ (Eqs. (16.5.16–16.5.18)), the rates for bath-induced transitions in the forward and backward directions are shown to obey the detailed balance condition (see Ex. 19.4.10):

$$k_{m \to n} = k_{n \to m} e^{-(E_n - E_m)/(k_B T)}. \tag{19.4.24}$$

Using this condition in Eqs. (19.4.17, 19.4.18), we can readily verify the existence of a steady-state solution, in which the populations of the system Hamiltonian eigenstates are given by the Boltzmann distribution (see Ex. 19.4.11):

$$[\mathbf{P}_S]_n = \frac{1}{\sum_{n'} e^{-E_{n'}/(k_B T)}} e^{-E_n/(k_B T)}. \tag{19.4.25}$$

As we can see, under the assumptions of the Born–Markov and secular approxima- tions, *the steady state of the system (equilibrium in this case) "adopts" the constraint imposed by the coupling to the thermal bath, where the approximated reduced system density operator reads* $\hat{\rho}_S \cong e^{-\hat{H}_S/(k_B T)}/tr\{e^{-\hat{H}_S/(k_B T)}\}$.

To summarize the discussion of Pauli's master equation, we briefly review the con- ditions for its validity, as we have just outlined. First, the system–bath coupling energy must be sufficiently small, such that bath-induced dynamics in the system is fast on the timescale of coherent oscillations between the system and the bath (set by the inverse system–bath coupling energy). This justifies the time-dependent Born–Markov (Redfield) approximation. Then, for a wide and continuous bath spectral density, the relaxation of bath correlation functions should be sufficiently fast on the timescale of the bath-induced system dynamics, justifying the stationary Born–Markov (Redfield) approximation. Finally, when the bath-free system dynamics is much faster than the bath-induced dynamics (and yet slower than the relaxation of bath correlations) the secular (rotating wave) approximation is justified, leading to a probability-conserving quantum master equation. When the bath-free system Hamiltonian is nondegenerate, the kinetics of its eigenstate populations is decoupled from coherences between them, which justifies the Pauli master equation. Notice that in the discussion of population transfer between ensembles in Chapter 17, the presence of an initial "incoherent" state of a system was assumed, with coherences completely ignored. Here we see that the ignorance of coherences while following state populations can be rigorously justified when the assumptions underlying the Pauli master equation hold.

Exercise 19.4.6 *Show that for a Hermitian operator,* $\hat{H}_{SB} \equiv \sum\limits_{\alpha=1}^{N} \hat{V}_\alpha^{(S)} \hat{U}_\alpha^{(B)}$, *and using the definitions of* $G_{n,n'}^{(\alpha,\alpha')}$, $\overline{G}_{n,n'}^{(\alpha,\alpha')}$ *(Eq. (19.4.2)) and* $V_{n',n}^{(\alpha)} = \langle \varphi_{n'}|\hat{V}_\alpha^{(S)}|\varphi_n\rangle$, *we have*

$$\sum_{\alpha,\alpha'} \overline{G}_{m,n}^{(\alpha',\alpha)} V_{n,m}^{(\alpha)} V_{m,n}^{(\alpha')} = \left[\sum_{\alpha,\alpha'} G_{n,m}^{(\alpha,\alpha')} V_{n,m}^{(\alpha')} V_{m,n}^{(\alpha)} \right]^*, \text{ and therefore the population transition}$$

rates defined in Eq. (19.4.14) are real-valued (Eq. (19.4.19)).

Exercise 19.4.7 *Use Eqs. (19.4.2, 19.4.4) in Eq. (19.4.19) to derive Eq. (19.4.20).*

Exercise 19.4.8   *Use the definitions* $\hat{H}_0 = \hat{H}_B + \hat{H}_S$, $\hat{H}_B|b\rangle = E_b|b\rangle$, $\hat{\rho}_B|b\rangle = \rho_b|b\rangle$, $\hat{H}_S|\varphi_m\rangle = E_m|\varphi_m\rangle$, $\hat{P}_{\{i\}} = |\varphi_m\rangle\langle\varphi_m| \otimes \sum_b |b\rangle\langle b|$, $\hat{P}_{\{f\}} = |\varphi_n\rangle\langle\varphi_n| \otimes \sum_b |b\rangle\langle b|$, *and* $\hat{\rho}_{\{i\}} = |\varphi_m\rangle\langle\varphi_m| \otimes \hat{\rho}_B$ *to derive Eq. (19.4.22) from Eq. (19.4.21).*

Exercise 19.4.9   *Let us denote the probability of populating the nth eigenstate of $\hat{H}_S$ as $P_n(t)$. (a) Given $P_n(t) \geq 0$ for any n, and recalling that population transfer rates are non-negative ($k_{m \to n} \geq 0$), use Eq. (19.4.13) to show that when a certain probability vanishes, $P_n(t) = 0$, it means that $\frac{\partial}{\partial t}P_n(t) \geq 0$. (b) Use the result of (a) to show that if all the probabilities are nonnegative at $t = 0$ (namely $P_n(0) \geq 0$ for any n), they remain so at any later times, namely $P_n(t) \geq 0$ for $t > 0$.*

Exercise 19.4.10   *(a) For a canonical density operator, $\hat{\rho}_B = \frac{e^{-\hat{H}_B/(k_B T)}}{Z_B}$, use Eq. (19.4.22) to show that $k_{m \to n} = \frac{2\pi}{\hbar} \sum_{b,b'} \frac{e^{-E_b/(k_B T)}}{Z_B} |\langle b|\langle\varphi_m|\hat{V}|\varphi_n\rangle|b'\rangle|^2 \delta(E_m - E_n - [E_{b'} - E_b])$ and*

$k_{n \to m} = \frac{2\pi}{\hbar} \sum_{b,b'} \frac{e^{-E_{b'}/(k_B T)}}{Z_B} |\langle b|\langle\varphi_m|\hat{V}|\varphi_n\rangle|b'\rangle|^2 \delta(E_m - E_n - [E_{b'} - E_b])$.

(b) Replacing the discrete summations over the bath Hamiltonian eigenstates by energy integrals, where $|\langle b|\langle\varphi_m|\hat{V}|\varphi_n\rangle|b'\rangle|^2 \to \lambda(E_{b'}, E_b)$, and introducing the bath density of states, $\rho(E_b)$, show that $k_{m \to n} = \frac{2\pi}{\hbar} \int dE_b \int dE_{b'} \rho(E_{b'})\rho(E_b) \frac{e^{-E_b/(k_B T)}}{Z_B} \lambda(E_{b'}, E_b)\delta(E_m - E_n - E_{b'} + E_b)$ and $k_{n \to m} = \frac{2\pi}{\hbar} \int dE_b \int dE_{b'} \rho(E_{b'})\rho(E_b) \frac{e^{-E_{b'}/(k_B T)}}{Z_B} \lambda(E_{b'}, E_b)\delta(E_m - E_n - E_{b'} + E_b)$.

(c) Changing integration variables and defining the transition frequency, $\hbar\omega_{n,m} = E_n - E_m$, show that $k_{m \to n} = \frac{e^{\frac{-\hbar\omega_{n,m}}{2}/(k_B T)}}{Z_B} \frac{2\pi}{\hbar} \int dE\rho\left(E - \frac{\hbar\omega_{n,m}}{2}\right)\rho\left(E + \frac{\hbar\omega_{n,m}}{2}\right) \frac{e^{-E/(k_B T)}}{Z_B} \lambda(E - \frac{\hbar\omega_{n,m}}{2}, E + \frac{\hbar\omega_{n,m}}{2})$ and $k_{n \to m} = \frac{e^{\frac{\hbar\omega_{n,m}}{2}/(k_B T)}}{Z_B} \frac{2\pi}{\hbar} \int dE\rho\left(E - \frac{\hbar\omega_{n,m}}{2}\right)\rho\left(E + \frac{\hbar\omega_{n,m}}{2}\right) \frac{e^{-E/(k_B T)}}{Z_B} \lambda(E - \frac{\hbar\omega_{n,m}}{2}, E + \frac{\hbar\omega_{n,m}}{2})$, where $k_{m \to n} = k_{n \to m} e^{-(E_n - E_m)/(k_B T)}$.

Exercise 19.4.11   *Use the detailed balance condition, Eq. (19.4.24), to show that the Boltzmann probability distribution, $P_n(t) = const \cdot e^{-E_n/(k_B T)}$, is a stationary solution of the Pauli master equation (Eq. (19.4.13)).*

Pure Dephasing

Let us emphasize that population transfer between nondegenerate system Hamiltonian eigenstates vanishes if the system–bath coupling operators are all diagonal in the system eigenstate basis representation. This is apparent from the golden rule expression, Eq. (19.4.22), where $k_{m \to n} \propto |\langle b'|\langle\varphi_n|\hat{V}|\varphi_m\rangle|b\rangle|^2$. Recalling the explicit expansion of the coupling operator, $\hat{V} = \hat{H}_{SB} = \sum_\alpha \hat{V}_\alpha^{(S)}\hat{U}_\alpha^{(B)}$, population transfer depends on the existence of nonzero off-diagonal coupling matrix elements, $\hat{V}_{n,m}^{(\alpha)} = \langle\varphi_n|\hat{V}_\alpha^{(S)}|\varphi_m\rangle \neq 0$, which means that the coupling operators do not commute with the system Hamiltonian, $[\hat{H}_{SB}, \hat{H}_S] \neq 0$ (see Ex. 6.4.1).

In contrast, even when $[\hat{H}_{SB}, \hat{H}_S] = 0$, the system–bath coupling can still induce changes in the coherences between nondegenerate system Hamiltonian eigenstates.

This effect is often referred to as "pure dephasing." It can be readily verified by considering the reduced Liouville equation (Eqs. (19.4.8, 19.4.9)) under the constraint of "diagonal coupling":

$$\{V_{n,m}^{(\alpha)}\} = \{V_{n,n}^{(\alpha)}\delta_{n,m}\} \Rightarrow \langle\varphi_n|\hat{H}_{SB}|\varphi_m\rangle \propto \delta_{n,m}. \tag{19.4.26}$$

In this case, the Redfield tensor (Eq. (19.4.1)) obtains a diagonal form, $R_{n',n,m',m}^{(St)} \propto \delta_{m,n}\delta_{m',n'}$, (see Ex. 19.4.12), which means that

$$\frac{\partial}{\partial t}\rho_{n',n}^{(I)}(t) = -k_{n',n}\rho_{n',n}^{(I)}(t) \quad ; \quad n \neq n', \tag{19.4.27}$$

Each coherence is shown to change independently of other coherences and populations, where the bath-induced changes are introduced by

$$k_{n',n} = \sum_{\alpha,\alpha'}[G_{n',n'}^{(\alpha,\alpha')}V_{n',n'}^{(\alpha')} - \overline{G}_{n,n}^{(\alpha',\alpha)}V_{n,n}^{(\alpha')}](V_{n',n'}^{(\alpha)} - V_{n,n}^{(\alpha)}). \tag{19.4.28}$$

Using Eq. (19.4.5), we obtain in the Schrödinger picture

$$\frac{\partial}{\partial t}\rho_{n',n}(t) = -\{\text{Re}(k_{n',n}) + i[\omega_{n',n} + \text{Im}(k_{n',n})]\}\rho_{n',n}(t). \tag{19.4.29}$$

As we can see, the bath influence vanishes when $\langle\varphi_{n'}|\hat{H}_{SB}|\varphi_{n'}\rangle = \langle\varphi_n|\hat{H}_{SB}|\varphi_n\rangle$, namely, *when the diagonal bath coupling (Eq. (19.4.26)) is identical for two system Hamiltonian eigenstates, the coherence between them is invariant to this coupling*. Another necessary condition for the bath influence in this case is that the bath coupling correlation functions (see Eqs. (19.4.2, 19.4.4)) have a nonvanishing zero-frequency Fourier component, namely

$$G_{n,n}^{(\alpha,\alpha')} = \frac{1}{\hbar^2}\int_0^\infty c_{\alpha,\alpha'}(\tau)d\tau \neq 0 \quad ; \quad \overline{G}_{n,n}^{(\alpha,\alpha')} \equiv \frac{1}{\hbar^2}\int_0^\infty \overline{c}_{\alpha,\alpha'}(\tau)d\tau \neq 0. \tag{19.4.30}$$

Notice that the real part of $k_{n',n}$ is nonnegative. Using Eqs. (19.4.2, 19.4.4), we obtain (Ex. 19.4.13)

$$\text{Re}[k_{n',n}] = \frac{2\pi}{\hbar}\sum_{b,b'}\rho_b\frac{1}{2}|\langle b|\langle\varphi_{n'}|\hat{H}_{SB}|\varphi_{n'}\rangle|b'\rangle - \langle b|\langle\varphi_n|\hat{H}_{SB}|\varphi_n\rangle|b'\rangle|^2\delta(E_{b'} - E_b) \geq 0. \tag{19.4.31}$$

Therefore, when the coupling to the bath is diagonal (as defined by Eq. (19.4.26)), the magnitude of coherences between the system Hamiltonian eigenstates can only decrease during forward time evolution. This decay, which is decoupled from population transfer, is the "pure dephasing":

$$|\rho_{n',n}(t)| = e^{-\text{Re}[k_{n',n}]t}|\rho_{n',n}(0)|. \tag{19.4.32}$$

A remarkable consequence of pure dephasing is that in the infinite time limit the matrix representation of the reduced density matrix in the basis of $\hat{H}_S$ eigenstates approaches a diagonal form; namely, it becomes stationary regardless of its initial preparation.

Exercise 19.4.12 *(a) Show that under the constraint of "diagonal coupling" (Eq. (19.4.26)) the stationary Redfield tensor (Eq. (19.4.1)) obtains the form*
$R^{(St)}_{n',n,m',m} = k_{n',n}\delta_{m,n}\delta_{m',n'}$, where $k_{n',n} = \sum\limits_{\alpha,\alpha'} [G^{(\alpha,\alpha')}_{n',n'}V^{(\alpha')}_{n',n'} - \overline{G}^{(\alpha',\alpha)}_{n,n}V^{(\alpha')}_{n,n}](V^{(\alpha)}_{n',n'} - V^{(\alpha)}_{n,n})$
(Eq. (19.4.28)). (b) Use this result and Eqs. (19.4.8, 19.4.9) to derive Eq. (19.4.27). (c) Use Eq. (19.4.5) to derive Eq. (19.4.29).

Exercise 19.4.13 *Use Eqs. (19.4.2, 19.4.4) and the definition of $k_{n',n}$ (Eq. (19.4.28)) to show that* $\text{Re}[k_{n',n}] = \frac{1}{\hbar^2}\sum\limits_{b,b'}\int\limits_0^\infty \rho_b \cos[(E_{b'} - E_b)\tau/\hbar]|\langle b|\langle\varphi_{n'}|\hat{H}_{SB}|\varphi_{n'}\rangle|b'\rangle - \langle b|\langle\varphi_n|\hat{H}_{SB}|\varphi_n\rangle$
$|b'\rangle|^2 d\tau$ *and to derive Eq. (19.4.31).*

19.5 The Dissipative Qubit

As a concrete example, let us analyze in some detail the interaction of a two-level system (a TLS, or a qubit) with a bosonic environment, as captured in the spin-boson model. (Extensions to Fermionic environment will be addressed in the next chapter.) In Chapter 18 the corresponding TLS dynamics was analyzed in a local basis representation of the TLS (a "donor" state and an "acceptor" state), where the coupling to the bosonic environment was assumed diagonal in the local basis representation and "strong," whereas the off-diagonal coupling between the local states was regarded as a weak perturbation. In what follows, we consider a complementary scenario, where the off-diagonal coupling within the TLS is not necessarily weak, and the interaction between the TLS and the bosonic environment is diagonal in the system eigenstate representation and regarded as being in the weak-coupling limit.

The scenario of weak coupling of a few-level system (TLS in particular) to a bosonic bath is a generic one, with relevance to elementary processes encountered in atomic, molecular, optical, and material sciences on the nanoscale, as well as to quantum information processing, involving qubits in an inevitably dissipative environment. One example is spin relaxation in a condensed phase environment (in the harmonic approximation), which is the basis for magnetic resonance imaging techniques. Another example is the interactions of quantized dipoles in atoms, molecules, or nanoparticles with the quantum electromagnetic radiation field (photons), which underlies the phenomena of photon absorption and stimulated and spontaneous emission. (Notice that in Chapters 10, 13, and 14 we addressed single atoms or molecules as isolated quantum systems, but this holds only on timescales much shorter than the time for energy exchange with the electromagnetic radiation field.) Other examples include the coupling of spatially confined charge carriers or charge excitations to lattice vibrations (phonons) in solids, which underlies charge and energy transport in electronic and optoelectronic devices. These physical contexts are indeed very different, and a careful implementation of the spin-boson model to each specific system requires deep specific knowledge of the relevant context and consequently of the model parameters. *Here we give an introductory level understanding of the generic aspects and some common features of the reduced dynamics of a TLS (weakly) coupled to a bosonic bath.* The expert reader can identify some of the results with well-known concepts in different fields, such as Einstein coefficient in optics, spin relaxation times in nuclear magnetic imaging, or electron–phonon scattering rates in solid-state physics.

Focusing on the weak system–bath coupling limit, it is instructive to pre-diagonalize the "bath-free," two-level system Hamiltonian and to represent the spin-boson model (introduced in Chapter 18, Eq. (18.1.1)) in the basis of the TLS eigenstates, $|\varphi_g\rangle$ and $|\varphi_e\rangle$ for the ground and excited states, respectively. Without loss of generality, the average of the corresponding system eigenvalues is set to zero, such that the spin-boson model Hamiltonian obtains the form

$$\hat{H} = \hat{H}_S + \hat{H}_B + \hat{H}_{SB}$$

$$\hat{H}_S = \Delta_E \,\hat{\sigma}_z$$

$$\hat{H}_B = \sum_{j=1}^{N_\omega} \frac{\hbar\omega_j}{2}(\hat{P}_j^2 + \hat{Q}_j^2) = \sum_{j=1}^{N_\omega} \hbar\omega_j \left(\hat{b}_j^\dagger \hat{b}_j + \frac{1}{2} \right)$$

$$\hat{H}_{SB} = \hat{V}_S \hat{U}_B + \hat{V}_S^\dagger \hat{U}_B^\dagger \quad ; \quad \hat{U}_B = \sum_{j=1}^{N_\omega} \lambda_j \hat{b}_j. \tag{19.5.1}$$

Here $\hat{\sigma}_z = |\varphi_e\rangle\langle\varphi_e| - |\varphi_g\rangle\langle\varphi_g|$ (corresponding to the two-by-two Pauli matrix, σ_z (see Eq. (13.1.17))), $\Delta_E > 0$, $\hat{V}_S$ is any coupling operator in the space of the TLS, and $\hat{U}_B$ is the bath coupling operator, where $\hat{b}_j^\dagger$ and $\hat{b}_j$ are, respectively, creation and annihilation operators for a bath excitation at a frequency ω_j. Notice that the coupling is linear in the bosonic bath operators, which means that bath response is limited to the linear regime. This is consistent with the "weak coupling" limit, where the system–bath interaction can be formally expanded in a Taylor series, keeping only the linear terms. More specific justifications for the linear coupling arise in different contexts in different fields, for example, the harmonic (normal modes) approximation in disordered molecular systems, weak electron–phonon coupling in solids, or the dipole approximation in matter–radiation interaction.

We are interested in following the reduced TLS dynamics in its eigenstate representation under weak coupling to the bath. In the spirit of Nakajima and Zwanzig projection, it is natural to associate the initial state with an uncorrelated product of system and bath densities,

$$\hat{\rho}(0) = \hat{\rho}_B \otimes \hat{\rho}_S(0), \tag{19.5.2}$$

where the bath density operator is stationary. In the case of a thermal (canonical) ensemble,

$$\hat{\rho}_B = \frac{e^{-\hat{H}_B/(k_B T)}}{tr_B\{e^{-\hat{H}_B/(k_B T)}\}}. \tag{19.5.3}$$

As we can readily see, the bath density fulfils the conditions (Eq. (19.3.6)), $[\hat{\rho}_B, \hat{H}_B] = 0$, $tr_B\{\hat{\rho}_B\} = 1$, and the system bath coupling is consistent with the generic form of Eq. (19.3.12), $\hat{H}_{SB} \equiv \sum_\alpha \hat{V}_\alpha^{(S)} \hat{U}_\alpha^{(B)}$, where $tr_B\{\hat{U}_\alpha^{(B)}\hat{\rho}_B\} = tr_B\{\hat{U}_\alpha^{(B)\dagger}\hat{\rho}_B\} = 0$ (Eq. (19.3.13), see Ex. 19.5.1). Consequently, the reduced system density in the weak coupling limit follows the time-dependent Born–Markov (Redfield) equation (Eqs. (19.3.19, 19.3.20)),

$$\frac{\partial}{\partial t}\hat{\rho}_S(t) \cong -\frac{i}{\hbar}[\hat{H}_S,\hat{\rho}_S(t)]+\hat{D}\hat{\rho}_S(t), \tag{19.5.4}$$

where the dissipator reads in this case (see Eqs. (19.3.21, 19.3.22) and Ex. 19.5.2)

$$\hat{D}\hat{\rho}_S(t) = -\frac{1}{\hbar^2}\int_0^t d\tau \{c_e(\tau)[\hat{V}_S,e^{\frac{-i\tau}{\hbar}\hat{H}_S}\hat{V}_S^{\dagger}e^{\frac{i\tau}{\hbar}\hat{H}_S}\hat{\rho}_S(t)]+c_a(\tau)[\hat{V}_S^{\dagger},e^{\frac{-i\tau}{\hbar}\hat{H}_S}\hat{V}_S e^{\frac{i\tau}{\hbar}\hat{H}_S}\hat{\rho}_S(t)]+h.c.\}.$$

$$\tag{19.5.5}$$

The influence of the bosonic bath is shown to be captured in its correlation functions,

$$c_a(\tau) = tr_B\{\hat{U}_B^{\dagger}e^{\frac{-i\tau}{\hbar}\hat{H}_B}\hat{U}_B e^{\frac{i\tau}{\hbar}\hat{H}_B}\hat{\rho}_B\} = \sum_{j=1}^{N_\omega}|\lambda_j|^2 e^{i\tau\omega_j}n(\omega_j)$$

$$c_e(\tau) = tr_B\{\hat{U}_B e^{\frac{-i\tau}{\hbar}\hat{H}_B}\hat{U}_B^{\dagger}e^{\frac{i\tau}{\hbar}\hat{H}_B}\hat{\rho}_B\} = \sum_{j=1}^{N_\omega}|\lambda_j|^2 e^{-i\tau\omega_j}[1+n(\omega_j)], \tag{19.5.6}$$

where $n(\omega_j) = \frac{1}{e^{\hbar\omega_j/(k_B T)}-1}$ is the boson occupation number of the jth bath mode (see Ex. 18.2.7). Considering a continuous spectrum of bath frequencies, the correlation functions can be expressed in terms of the bath spectral density, $J(\hbar\omega) \equiv 2\pi\lambda^2(\hbar\omega)\rho(\hbar\omega)$, where the system–bath coupling function, $\lambda^2(\hbar\omega)$, and the bath density of states, $\rho(\hbar\omega)$, are related to the discrete frequency distribution as $\lambda^2(\hbar\omega_j) = |\lambda_j|^2$ and $\rho(\hbar\omega) = \sum_j \delta(\hbar\omega - \hbar\omega_j)$:

$$c_a(\tau) = \frac{\hbar}{2\pi}\int_0^\infty d\omega e^{i\tau\omega}J(\hbar\omega)n(\omega)$$

$$c_e(\tau) = \frac{\hbar}{2\pi}\int_0^\infty d\omega e^{-i\tau\omega}J(\hbar\omega)(n(\omega)+1) \tag{19.5.7}$$

When the multiplicity of the spectral density function and the boson occupation number $(J(\omega)n(\omega))$ is a sufficiently broad function of the bath frequencies (namely, in the wide band and high temperature limits), the relaxation time of the bath correlations can become faster than any significant change in the reduced density matrix. Consequently, the upper time limit in Eq. (19.5.5) can be taken to infinity (see a detailed analysis in Section 17.2 as well as Eqs. (19.4.1–19.4.3)), which yields the stationary Born–Markov approximation,

$$\hat{D}\hat{\rho}_S(t) \cong -\frac{1}{\hbar^2}\int_0^\infty d\tau \{c_e(\tau)[\hat{V}_S,e^{\frac{-i\tau}{\hbar}\hat{H}_S}\hat{V}_S^{\dagger}e^{\frac{i\tau}{\hbar}\hat{H}_S}\hat{\rho}_S(t)]+c_a(\tau)[\hat{V}_S^{\dagger},e^{\frac{-i\tau}{\hbar}\hat{H}_S}\hat{V}_S e^{\frac{i\tau}{\hbar}\hat{H}_S}\hat{\rho}_S(t)]+h.c.\}.$$

$$\tag{19.5.8}$$

Exercise 19.5.1 *(a) Use the commutation relation between the bosonic annihilation and creation operators (Eqs. (8.5.5–8.5.7)), $[\hat{b}_{j'},\hat{b}_j^{\dagger}] = \delta_{j,j'}$, to show that the traces over* the single-mode subspace, $tr_j\{\hat{b}_j f(\hat{b}_j^\dagger \hat{b}_j)\}$ and $tr_j\{\hat{b}_j^\dagger f(\hat{b}_j^\dagger \hat{b}_j)\}$, vanish for any analytic function $(f(\hat{A}) = \sum_{n=0}^\infty f_n \hat{A}^n)$. *(b) Given the definition of the bosonic bath Hamiltonian and coupling operators (Eq. (19.5.1) with $\hat{U}_B = \sum_{j=1}^{N_\omega} \lambda_j \hat{b}_j$), and the bath density operator (Eq. (19.5.3)), $\hat{\rho}_B = e^{-\hat{H}_B/(k_BT)}/tr_B\{e^{-\hat{H}_B/(k_BT)}\}$, use the result of (a) to show that $tr_B\{\hat{U}_B\hat{\rho}_B\} = tr_B\{\hat{U}_B^\dagger\hat{\rho}_B\} = 0$.*

Exercise 19.5.2 *(a) Use the explicit expressions (Eqs. (19.5.1, 19.5.3)), $\hat{U}_B = \sum_{j=1}^{N_\omega} \lambda_j \hat{b}_j$, $\hat{H}_B = \sum_{j=1}^{N_\omega} \hbar\omega_j(\hat{b}_j^\dagger \hat{b}_j + \frac{1}{2})$, $\hat{\rho}_B = \frac{e^{-\hat{H}_B/(k_BT)}}{tr_B\{e^{-\hat{H}_B/(k_BT)}\}}$, to show that the bath correlation functions, $c_e(\tau) = tr_B\{\hat{U}_B e^{\frac{-i\tau}{\hbar}\hat{H}_B}\hat{U}_B^\dagger e^{\frac{i\tau}{\hbar}\hat{H}_B}\hat{\rho}_B\}$ and $c_a(\tau) = tr_B\{\hat{U}_B^\dagger e^{\frac{-i\tau}{\hbar}\hat{H}_B}\hat{U}_B e^{\frac{i\tau}{\hbar}\hat{H}_B}\hat{\rho}_B\}$, read $c_e(\tau) = \sum_{j=1}^{N_\omega} |\lambda_j|^2 e^{-i\tau\omega_j}[1+n(\omega_j)]$ and $c_a(\tau) = \sum_{j=1}^{N_\omega} |\lambda_j|^2 e^{i\tau\omega_j} n(\omega_j)$.*

(b) For a general system–bath coupling operator, $\hat{H}_{SB} \equiv \sum_\alpha \hat{V}_\alpha^{(S)}\hat{U}_\alpha^{(B)}$ (Eq. (19.3.12)), the Redfield (Born–Markov) dissipator obtains the form of Eqs. (19.3.21, 19.3.22),

$$\hat{D}\hat{\rho}_S(t) = -\frac{1}{\hbar^2}\sum_{\alpha,\alpha'}\int_0^t d\tau\{c_{\alpha,\alpha'}(\tau)[\hat{V}_\alpha^{(S)}, e^{\frac{-i\tau}{\hbar}\hat{H}_S}\hat{V}_{\alpha'}^{(S)}e^{\frac{i\tau}{\hbar}\hat{H}_S}\hat{\rho}_S(t)] + \bar{c}_{\alpha',\alpha}(\tau)[\hat{\rho}_S(t)e^{\frac{-i\tau}{\hbar}\hat{H}_S}\hat{V}_{\alpha'}^{(S)}$$
$$e^{\frac{i\tau}{\hbar}\hat{H}_S},\hat{V}_\alpha^{(S)}]\},$$

where $c_{\alpha,\alpha'}(\tau) \equiv tr_B\{\hat{U}_\alpha^{(B)}e^{\frac{-i\tau}{\hbar}\hat{H}_B}\hat{U}_{\alpha'}^{(B)}e^{\frac{i\tau}{\hbar}\hat{H}_B}\hat{\rho}_B\}$ and $\bar{c}_{\alpha,\alpha'}(\tau) = c_{\alpha,\alpha'}(-\tau)$. Map the coupling operator defined in Eq. (19.5.1), $\hat{H}_{SB} \equiv \hat{V}_S\hat{U}_B + \hat{V}_S^\dagger\hat{U}_B^\dagger$, on this general form by identifying, $\hat{V}_S \equiv \hat{V}_1^{(S)}$, $\hat{U}_B \equiv \hat{U}_1^{(B)}$, $\hat{V}_S^\dagger \equiv \hat{V}_2^{(S)}$, $\hat{U}_B^\dagger \equiv \hat{U}_2^{(B)}$, to show that

$$\hat{D}\hat{\rho}_S(t) = -\frac{1}{\hbar^2}\int_0^t d\tau\{c_{1,2}(\tau)[\hat{V}_S, e^{\frac{-i\tau}{\hbar}\hat{H}_S}\hat{V}_S^\dagger e^{\frac{i\tau}{\hbar}\hat{H}_S}\hat{\rho}_S(t)] + c_{2,1}^*(\tau)[\hat{\rho}_S(t)e^{\frac{-i\tau}{\hbar}\hat{H}_S}\hat{V}_S^\dagger e^{\frac{i\tau}{\hbar}\hat{H}_S},\hat{V}_S]\}$$
$$-\frac{1}{\hbar^2}\int_0^t d\tau\{c_{2,1}(\tau)[\hat{V}_S^\dagger, e^{\frac{-i\tau}{\hbar}\hat{H}_S}\hat{V}_S e^{\frac{i\tau}{\hbar}\hat{H}_S}\hat{\rho}_S(t)] + c_{1,2}^*(\tau)[\hat{\rho}_S(t)e^{\frac{-i\tau}{\hbar}\hat{H}_S}\hat{V}_S e^{\frac{i\tau}{\hbar}\hat{H}_S},\hat{V}_S^\dagger]\}.$$

(c) Use the identities $c_{1,2}(\tau) = c_e(\tau)$ and $c_{2,1}(\tau) = c_a(\tau)$ to derive Eq. (19.5.5).

Recalling (Eq. (19.5.1)) of the bath-free TLS Hamiltonian eigenstates,

$$\hat{H}_S|\varphi_g\rangle = -\Delta_E|\varphi_g\rangle \quad ; \quad \hat{H}_S|\varphi_e\rangle = \Delta_E|\varphi_e\rangle. \tag{19.5.9}$$

The reduced system density operator can be represented as

$$\hat{\rho}_S(t) = \rho_{g,g}(t)|\varphi_g\rangle\langle\varphi_g| + \rho_{e,e}(t)|\varphi_e\rangle\langle\varphi_e| + \rho_{g,e}(t)|\varphi_g\rangle\langle\varphi_e| + \rho_{e,g}(t)|\varphi_e\rangle\langle\varphi_g|, \tag{19.5.10}$$

where $\rho_{g,g}(t) = \langle\varphi_g|\hat{\rho}_S(t)|\varphi_g\rangle$ and $\rho_{e,e}(t) = \langle\varphi_e|\hat{\rho}_S(t)|\varphi_e\rangle)$ are the ground and excited state populations, and the coherences between these states are $\rho_{e,g}(t) = \langle\varphi_e|\hat{\rho}_S(t)|\varphi_g\rangle$ and $\rho_{g,e}(t) = \langle\varphi_g|\hat{\rho}_S(t)|\varphi_e\rangle$.

The bath influence on the system depends on the nature of the coupling operators. **Let us consider first the case of off-diagonal coupling between the TLS and the bosonic bath,** setting

$$\hat{V}_S \equiv \mu|\varphi_e\rangle\langle\varphi_g| \quad ; \quad \hat{V}_S^\dagger = \mu^*|\varphi_g\rangle\langle\varphi_e|, \tag{19.5.11}$$

or, in terms of the Pauli spin operators, $\hat{\sigma}_+ = |\varphi_e\rangle\langle\varphi_g|$, and $\hat{\sigma}_- = |\varphi_g\rangle\langle\varphi_e|$,

$$\hat{H}_{SB} = \sum_{j=1}^{N_\omega} \mu\lambda_j\hat{b}_j\hat{\sigma}_+ + \mu^*\lambda_j^*\hat{b}_j^\dagger\hat{\sigma}_-. \tag{19.5.12}$$

This version of the spin-boson model is often used, for example, to describe the interaction of the quantum electromagnetic field with an electric dipole (corresponding to transition between two selected electronic states in an atom, molecule, impurity, and so forth), within the dipole and rotating wave approximations. (See Ex. 18.3.5 for the semiclassical version of the rotating wave approximation, and Ex. 19.5.6.)

Absorption, Stimulated Emission, and Spontaneous Emission

The dynamics of the different matrix elements of the reduced density operator within the stationary Born–Markov (Redfield) approximation can be obtained directly from Eqs. (19.5.4, 19.5.8). *For the populations, Pauli's master equation is readily obtained* (Ex. 19.5.3):

$$\frac{\partial}{\partial t}\rho_{g,g}(t) = k_{e\to g}^{em}\rho_{e,e}(t) - k_{g\to e}^{ab}\rho_{g,g}(t)$$
$$\frac{\partial}{\partial t}\rho_{e,e}(t) = k_{g\to e}^{ab}\rho_{g,g}(t) - k_{e\to g}^{em}\rho_{e,e}(t) \tag{19.5.13}$$

where the population transfer rates read

$$k_{e\to g}^{em} = 2|\mu|^2\mathrm{Re}\frac{1}{\hbar^2}\int_0^\infty d\tau c_e(\tau)e^{\frac{i\tau}{\hbar}2\Delta_E} = \frac{2\pi}{\hbar}\sum_{j=1}^{N_\omega}|\mu|^2|\lambda_j|^2[1+n(\omega_j)]\delta(\hbar\omega_j - 2\Delta_E)$$

$$k_{g\to e}^{ab} = 2|\mu|^2\mathrm{Re}\frac{1}{\hbar^2}\int_0^\infty d\tau c_a(\tau)e^{\frac{-i\tau}{\hbar}2\Delta_E} = \frac{2\pi}{\hbar}\sum_{j=1}^{N_\omega}|\mu|^2|\lambda_j|^2 n(\omega_j)\delta(\hbar\omega_j - 2\Delta_E), \tag{19.5.14}$$

or, in the case of a continuous bath spectral density,

$$k_{e\to g}^{em} = \frac{|\mu|^2}{\hbar}J(2\Delta_E)\left[n\left(\frac{2\Delta_E}{\hbar}\right) + 1\right]$$
$$k_{g\to e}^{ab} = \frac{|\mu|^2}{\hbar}J(2\Delta_E)n\left(\frac{2\Delta_E}{\hbar}\right). \tag{19.5.15}$$

As we can see, the rate of energy emission ($k_{e\to g}^{em}$) into the bosonic bath has two contributions:

$$k_{e\to g}^{em} = k_{e\to g}^{st} + k_{e\to g}^{se}. \tag{19.5.16}$$

The rate of stimulated emission ($k_{e\to g}^{st}$) depends on the thermal occupation number of the bath mode at the TLS transition frequency and is equal to the rate of energy absorption from the field,

$$k_{e\to g}^{st} = k_{g\to e}^{ab} = \frac{|\mu|^2}{\hbar}J(2\Delta_E)n\left(\frac{2\Delta_E}{\hbar}\right). \tag{19.5.17}$$

The spontaneous emission rate is independent on the bath temperature and corresponds to the interaction of the TLS with the vacuum (unoccupied) state of the bosonic field,

$$k_{e \to g}^{se} = \frac{|\mu|^2}{\hbar} J(2\Delta_E). \tag{19.5.18}$$

We can readily see that regardless of the initial state of the TLS, in the asymptotic time limit the relative populations of the ground and excited states are set by the bath temperature (see Ex. 17.3.7):

$$\frac{\rho_{e,e}(\infty)}{\rho_{g,g}(\infty)} = \frac{k_{g \to e}^{ab}}{k_{e \to g}^{em}} = e^{-2\Delta_E/(k_B T)}. \tag{19.5.19}$$

This result is a specific case of the detailed balance condition (Eq. (19.4.24)).

Exercise 19.5.3 *(a) Given Eqs. (19.5.4, 19.5.8), and defining the TLS eigenstate populations, $\rho_{g,g}(t) = \langle \varphi_g | \hat{\rho}_S(t) | \varphi_g \rangle$, and $\rho_{e,e}(t) = \langle \varphi_e | \hat{\rho}_S(t) | \varphi_e \rangle$, show that*
$$\frac{\partial}{\partial t} \rho_{g,g}(t) \cong - \frac{1}{\hbar^2} 2 \operatorname{Re} \int_0^\infty d\tau \{ c_e(\tau) \langle g | [\hat{V}_S, e^{\frac{-i\tau}{\hbar} \hat{H}_S} \hat{V}_S^\dagger e^{\frac{i\tau}{\hbar} \hat{H}_S} \hat{\rho}_S(t)] | g \rangle + c_a(\tau) \langle g | [\hat{V}_S^\dagger, e^{\frac{-i\tau}{\hbar} \hat{H}_S}$$
$$\hat{V}_S e^{\frac{i\tau}{\hbar} \hat{H}_S} \hat{\rho}_S(t)] | g \rangle \}.$$

(b) Introduce the identity operator in the space of the TLS, $\hat{I} = |\varphi_g\rangle\langle\varphi_g| + |\varphi_e\rangle\langle\varphi_e|$, to show that for off-diagonal TLS coupling operators, $\hat{V}_S \equiv \mu |\varphi_e\rangle\langle\varphi_g|$ and $\hat{V}_S^\dagger = \mu^* |\varphi_g\rangle\langle\varphi_e|$, this result reads $\frac{\partial}{\partial t} \rho_{g,g}(t) = k_{e \to g}^{em} \rho_{e,e}(t) - k_{g \to e}^{ab} \rho_{g,g}(t)$, where $k_{e \to g}^{em} = 2|\mu|^2 \operatorname{Re} \frac{1}{\hbar^2} \int_0^\infty d\tau c_e(\tau) e^{\frac{i\tau}{\hbar} 2\Delta_E}$ and $k_{g \to e}^{ab} = 2|\mu|^2 \operatorname{Re} \frac{1}{\hbar^2} \int_0^\infty d\tau c_a(\tau) e^{\frac{-i\tau}{\hbar} 2\Delta_E}$.

(c) Show similarly that $\frac{\partial}{\partial t} \rho_{e,e}(t) = k_{g \to e}^{ab} \rho_{g,g}(t) - k_{e \to g}^{em} \rho_{e,e}(t)$.

(d) Use the explicit expressions for the correlation functions (Eq. (19.5.6)) to derive Eq. (19.5.14).

(e) Use the expressions for the correlation functions for a continuous bath (Eq. (19.5.7)) to derive Eq. (19.5.15).

Coherence Transfer and the Bloch Equation

The off-diagonal matrix elements of the reduced TLS density operator correspond to coherences between the TLS eigenstates. Using the stationary Born–Markov (Redfield) approximation (Eqs. (19.5.4, 19.5.8)), *the time evolution of the coherences reads* (Ex. 19.5.4)

$$\frac{\partial}{\partial t} \rho_{e,g}(t) = \frac{-i}{\hbar} (2\Delta_E - \hbar\delta) \rho_{e,g}(t) - k^{dec} \rho_{e,g}(t)$$

$$\frac{\partial}{\partial t} \rho_{g,e}(t) = \frac{i}{\hbar} (2\Delta_E - \hbar\delta) \rho_{g,e}(t) - k^{dec} \rho_{g,e}(t). \tag{19.5.20}$$

As we can see, the Hermiticity of the density operator (Eq. (16.2.6)) is preserved in time, where $\rho_{g,e}(0) = \rho_{e,g}^*(0) \Rightarrow \rho_{g,e}(t) = \rho_{e,g}^*(t)$. In the absence of coupling to the bath, the coherences oscillate at the bath-free (field-free) TLS frequency, $2\Delta_E/\hbar$. *The off-diagonal coupling to the bosonic field is shown to induce a shift in the frequency of the coherence* *oscillations, as well as decay of the coherences in time.* The frequency shift corresponds to renormalization of the TLS energy gap by

$$\hbar\delta = \mathrm{Im}\frac{|\mu|^2}{2\pi}\int_0^\infty d\tau \int_0^\infty d\omega J(\hbar\omega)[2n(\omega)+1]e^{\frac{-i\tau}{\hbar}(2\Delta_E-\hbar\omega)}. \qquad (19.5.21)$$

Notice that this energy shift is present even at zero bath temperatures ($n(\omega) \to 0$), which reflects the effect of coupling between the TLS and the vacuum field (the Lamb shift [10.3][10.4]). *The decay of the coherences to zero (referred to as decoherence, or dephasing), $|\rho_{g,e}(t)| \propto e^{-k^{dec}t}$, is shown to be governed by a rate constant that is equal to the average of the energy absorption and emission rates* (Eqs. (19.5.14, 19.5.15)),

$$k^{dec} = \frac{1}{2}(k^{ab}_{g\to e} + k^{em}_{e\to g}). \qquad (19.5.22)$$

Notice that in the infinite time limit, the Born–Markov and secular approximations predict that the matrix representation of the reduced TLS density matrix becomes diagonal in the basis of the bath-free TLS eigenstates, where the state populations are Boltzmann-distributed. Hence, the system–bath coupling drives the system into an equilibrium-like stationary state, which reflects the bath temperature.

The TLS state populations and coherences are closely related to three components of the TLS dimensionless "spin" vector, defined as

$$\begin{aligned}
\langle\sigma_z(t)\rangle &\equiv tr\{\hat{\sigma}_z\hat{\rho}(t)\} \\
\langle\sigma_x(t)\rangle &\equiv tr\{\hat{\sigma}_x\hat{\rho}(t)\} \\
\langle\sigma_y(t)\rangle &\equiv tr\{\hat{\sigma}_y\hat{\rho}(t)\},
\end{aligned} \qquad (19.5.23)$$

where $\hat{\sigma}_x$, $\hat{\sigma}_y$, and $\hat{\sigma}_z$ are the Pauli spin operators corresponding to the Pauli matrices in the TLS eigenstate representation (Eq. (13.1.17)). Using Eqs. (19.5.13 and 19.5.20), we can readily obtain an equation of motion for the "spin" vector (Ex. 19.5.5),

$$\begin{aligned}
\frac{\partial}{\partial t}\langle\sigma_z(t)\rangle &= -(k^{ab}_{g\to e} + k^{em}_{e\to g})\langle\sigma_z(t)\rangle - k^{se}_{e\to g} \\
\frac{\partial}{\partial t}\langle\sigma_x(t)\rangle &= -k^{dec}\langle\sigma_x(t)\rangle - \frac{1}{\hbar}(2\Delta_E - \hbar\delta)\langle\sigma_y(t)\rangle \\
\frac{\partial}{\partial t}\langle\sigma_y(t)\rangle &= -k^{dec}\langle\sigma_y(t)\rangle + \frac{1}{\hbar}(2\Delta_E - \hbar\delta)\langle\sigma_x(t)\rangle.
\end{aligned} \qquad (19.5.24)$$

The result is well known as the Bloch equation in the context of magnetic resonance imaging. Indeed, associating the free TLS with a spin-half particle in a constant magnetic field along the z direction, namely $\Delta_E\hat{\sigma}_z = B_z\mu_B\hat{S}_z = \frac{B_z\mu_B\hbar}{2}\hat{\sigma}_z$, where the bosonic field represents lattice modes to which the spins are coupled, Eq. (19.5.24) describes the dynamics of the spin magnetization vector. The decay of the magnetization is characterized by the "spin-lattice relaxation" time, $T_1 = \frac{1}{k^{ab}_{g\to e}+k^{em}_{e\to g}}$, and the "spin-spin" relaxation time, $T_2 = 2T_1$, which identifies with the decoherence rate (Eq. (19.5.22)). Neglecting the TLS gap renormalization, $\hbar\delta << 2\Delta_E$, and defining the Larmor frequency, $\omega_0 = B_z\mu_B$, the Bloch equations are expressed as

$$\frac{\partial}{\partial t}\langle \sigma_z(t)\rangle = -\frac{1}{T_1}\langle \sigma_z(t)\rangle - k_{e\to g}^{se}$$

$$\frac{\partial}{\partial t}\langle \sigma_x(t)\rangle = -\frac{1}{T_2}\langle \sigma_x(t)\rangle - \omega_0\langle \sigma_y(t)\rangle \qquad (19.5.25)$$

$$\frac{\partial}{\partial t}\langle \sigma_y(t)\rangle = -\frac{1}{T_2}\langle \sigma_y(t)\rangle + \omega_0\langle \sigma_x(t)\rangle.$$

Exercise 19.5.4   *(a) The coherences between the TLS eigenstates are defined as $\rho_{g,e}(t) = \langle \varphi_g|\hat{\rho}_S(t)|\varphi_e\rangle$ and $\rho_{e,g}(t) = \langle \varphi_e|\hat{\rho}_S(t)|\varphi_g\rangle$. Introduce the identity operator in the space of the TLS, $\hat{I} = |\varphi_g\rangle\langle \varphi_g| + |\varphi_e\rangle\langle \varphi_e|$, into the stationary Redfield Equation (Eqs. (19.5.4, 19.5.8)), and show that for off-diagonal TLS coupling operators, $\hat{V}_S \equiv \mu|\varphi_e\rangle\langle \varphi_g|$ and $\hat{V}_S^\dagger = \mu^*|\varphi_g\rangle\langle \varphi_e|$, the coherences follow the equations $\frac{\partial \rho_{g,e}(t)}{\partial t} \cong \frac{i}{\hbar}2\Delta_E\rho_{g,e}(t) - \frac{|\mu|^2}{\hbar^2}\int_0^\infty d\tau\{[c_a(\tau) + c_e^*(\tau)]e^{\frac{-i\tau}{\hbar}2\Delta_E}\}\rho_{g,e}(t)$, and $\frac{\partial \rho_{e,g}(t)}{\partial t} \cong -\frac{i}{\hbar}2\Delta_E\rho_{e,g}(t) - \frac{|\mu|^2}{\hbar^2}\int_0^\infty d\tau[c_e(\tau) + c_a^*(\tau)]e^{\frac{i\tau}{\hbar}2\Delta_E}\rho_{e,g}(t)$. (b) Use the definition of the absorption and emission rates (Eq. (19.5.14)) and the definitions, $\delta \equiv \operatorname{Im}\frac{|\mu|^2}{\hbar^2}\int_0^\infty d\tau\{[c_a(\tau) + c_e^*(\tau)]e^{\frac{-i\tau}{\hbar}2\Delta_E}\}$ and, $k^{dec} \equiv \frac{1}{2}(k_{g\to e}^{ab} + k_{e\to g}^{em})$, to derive Eq. (19.5.20). (c) Use the explicit expressions for the correlation functions for a continuous bath (Eq. (19.5.7)) to derive Eq. (19.5.21).*

Exercise 19.5.5   *(a) Use Eq. (19.5.23), the Pauli spin matrices $\sigma_x = \begin{bmatrix} 0 & 1 \\ 1 & 0 \end{bmatrix}$, $\sigma_y = \begin{bmatrix} 0 & -i \\ i & 0 \end{bmatrix}$, $\sigma_z = \begin{bmatrix} 1 & 0 \\ 0 & -1 \end{bmatrix}$ and the TLS density matrix $\rho(t) = \begin{bmatrix} \rho_{e,e}(t) & \rho_{e,g}(t) \\ \rho_{g,e}(t) & \rho_{g,g}(t) \end{bmatrix}$ to show that $\langle \sigma_z(t)\rangle = \rho_{e,e}(t) - \rho_{g,g}(t)$, $\langle \sigma_x(t)\rangle = \rho_{g,e}(t) + \rho_{e,g}(t)$, $\langle \sigma_y(t)\rangle = -i\rho_{g,e}(t) + i\rho_{e,g}(t)$. (b) Use Eqs. (19.5.13, 19.5.20) for the time evolution of the density matrix elements to derive Eq. (19.5.24).*

Exercise 19.5.6   *The reduced dynamics of the TLS density operator (Eqs. (19.5.13, 19.5.20)), under the system–bath coupling, $\hat{H}_{SB} = \mu\hat{U}_B\hat{\sigma}_+ + \mu^*\hat{U}_B^\dagger\hat{\sigma}_-$, was derived from the stationary Redfield equation (Eqs. (19.5.4, 19.5.8)), allegedly with no farther approximations. This simplicity is attributed to the fact that the excitation ($\hat{\sigma}_+$) and de-excitation ($\hat{\sigma}_-$) TLS operators are coupled independently to $\hat{U}_B$ and to $\hat{U}_B^\dagger$, respectively. In a more general case, however, both $\hat{\sigma}_+$ and $\hat{\sigma}_-$ may couple to the same system–bath operators. As shown in what follows, the time evolution of the reduced density matrix is different in this case. Nevertheless, within the secular and rotating wave approximations, this difference is neglected and Eqs. (19.5.13, 19.5.20) are regained. Let us consider a system–bath coupling operator, $\hat{H}_{SB} = \mu(\hat{\sigma}_+ + \hat{\sigma}_-)(\hat{U}_B + \hat{U}_B^\dagger) \equiv \hat{V}_S(\hat{U}_B + \hat{U}_B^\dagger)$, where $\hat{V}_S = \mu(\hat{\sigma}_+ + \hat{\sigma}_-)$ is Hermitian and $\hat{U}_B = \sum_{j=1}^{N_\omega} \lambda_j\hat{b}_j$.*

(a) Show that in this case the Redfield dissipator (Eqs. (19.5.4, 19.5.8)) reads $\hat{D}\hat{\rho}_S(t) = -\frac{1}{\hbar^2}\int_0^\infty d\tau\{c(\tau)[\hat{V}_S, e^{\frac{-i\tau}{\hbar}\hat{H}_S}\hat{V}_S e^{\frac{i\tau}{\hbar}\hat{H}_S}\hat{\rho}_S(t)] + h.c.\}$, where $c(\tau) = c_e(\tau) + c_a(\tau)$.

(b) Defining $C \equiv \frac{|\mu|^2}{\hbar^2} \int\limits_0^\infty d\tau c(\tau) e^{\frac{-i\tau}{\hbar} 2\Delta_E}$ and $\overline{C} \equiv \frac{|\mu|^2}{\hbar^2} \int\limits_0^\infty d\tau c(\tau) e^{\frac{i\tau}{\hbar} 2\Delta_E}$, show that the corresponding time evolution of the reduced TLS density matrix is given by

$$\frac{\partial}{\partial t}\rho_{g,g}(t) = -[C+C^*]\rho_{g,g}(t) + [\overline{C}+\overline{C}^*]\rho_{e,e}(t)$$

$$\frac{\partial}{\partial t}\rho_{e,e}(t) = -[\overline{C}+\overline{C}^*]\rho_{e,e}(t) + [C+C^*]\rho_{g,g}(t)$$

$$\frac{\partial}{\partial t}\rho_{g,e}(t) = \frac{i}{\hbar}2\Delta_E\rho_{g,e}(t) - [C+\overline{C}^*]\rho_{g,e}(t) + [C^* +\overline{C}]\rho_{e,g}(t)$$

$$\frac{\partial}{\partial t}\rho_{e,g}(t) = \frac{-i}{\hbar}2\Delta_E\rho_{e,g}(t) - [C^* +\overline{C}]\rho_{e,g}(t) + [C+\overline{C}^*]\rho_{g,e}(t).$$

(c) Recalling the explicit form of the bosonic bath correlation functions (Eq. (19.5.6)), we obtain

$$C = \frac{|\mu|^2}{\hbar^2} \sum_{j=1}^{N_\omega} |\lambda_j|^2 \int\limits_0^\infty d\tau \left[e^{i\tau\omega_j} e^{\frac{-i\tau}{\hbar} 2\Delta_E} n(\omega_j) + e^{-i\tau\omega_j} e^{\frac{-i\tau}{\hbar} 2\Delta_E} [1+n(\omega_j)] \right]$$

$$\overline{C} = \frac{|\mu|^2}{\hbar^2} \sum_{j=1}^{N_\omega} |\lambda_j|^2 \int\limits_0^\infty d\tau \left[e^{i\tau\omega_j} e^{\frac{i\tau}{\hbar} 2\Delta_E} n(\omega_j) + e^{-i\tau\omega_j} e^{\frac{i\tau}{\hbar} 2\Delta_E} [1+n(\omega_j)] \right].$$

The rotating wave approximation implies that rapidly oscillating terms can be neglected next to slowly oscillating terms under the time integrals. Considering that both Δ_E and the bath frequencies are positive, show that this approximation means that $C \cong \frac{|\mu|^2}{\hbar^2} \int\limits_0^\infty d\tau e^{\frac{-i\tau}{\hbar} 2\Delta_E} c_a(\tau)$ and $\overline{C} \cong \frac{|\mu|^2}{\hbar^2} \int\limits_0^\infty d\tau e^{\frac{i\tau}{\hbar} 2\Delta_E} c_e(\tau)$.

(d) Recalling the definitions of the absorption and emission rates (Eq. (19.5.14)) and of $\delta \equiv \text{Im}\frac{|\mu|^2}{\hbar^2} \int\limits_0^\infty d\tau [c_a(\tau) + c_e^*(\tau)] e^{\frac{-i\tau}{\hbar} 2\Delta_E}$, show that $\text{Re}[\overline{C}] \cong k_{e\to g}^{em}/2$, $\text{Re}[C] \cong k_{g\to e}^{ab}/2$, and $\text{Im}[C+\overline{C}^*] \cong \delta$, and therefore

$$\frac{\partial}{\partial t}\rho_{g,g}(t) = -k_{g\to e}^{ab}\rho_{g,g}(t) + k_{e\to g}^{em}\rho_{e,e}(t)$$

$$\frac{\partial}{\partial t}\rho_{e,e}(t) = -k_{e\to g}^{em}\rho_{e,e}(t) + k_{g\to e}^{ab}\rho_{g,g}(t)$$

$$\frac{\partial}{\partial t}\rho_{g,e}(t) = \frac{i}{\hbar}2\Delta_E\rho_{g,e}(t) - \left[\frac{k_{g\to e}^{ab}+k_{e\to g}^{em}}{2}+i\delta\right]\rho_{g,e}(t) + \left[\frac{k_{g\to e}^{ab}+k_{e\to g}^{em}}{2}-i\delta\right]\rho_{e,g}(t)$$

$$\frac{\partial}{\partial t}\rho_{e,g}(t) = \frac{-i}{\hbar}2\Delta_E\rho_{e,g}(t) - \left[\frac{k_{g\to e}^{ab}+k_{e\to g}^{em}}{2}-i\delta\right]\rho_{e,g}(t) + \left[\frac{k_{g\to e}^{ab}+k_{e\to g}^{em}}{2}+i\delta\right]\rho_{g,e}(t).$$

(e) Transforming to the interaction picture representation, $\rho_{g,e}^I(t) = e^{-i2\Delta_E t/\hbar}\rho_{g,e}(t)$, $\rho_{e,g}^I(t) = e^{i2\Delta_E t/\hbar}\rho_{e,g}(t)$, the equations for the coherences obtain the form

$$\frac{\partial}{\partial t}\rho_{g,e}^I(t) = -\left[\frac{k_{g\to e}^{ab}+k_{e\to g}^{em}}{2}+i\delta\right]\rho_{g,e}^I(t) + \left[\frac{k_{g\to e}^{ab}+k_{e\to g}^{em}}{2}-i\delta\right]e^{-i4\Delta_E t/\hbar}\rho_{e,g}^I(t)$$

$$\frac{\partial}{\partial t}\rho_{e,g}^I(t) = -\left[\frac{k_{g\to e}^{ab}+k_{e\to g}^{em}}{2}-i\delta\right]\rho_{e,g}^I(t) + \left[\frac{k_{g\to e}^{ab}+k_{e\to g}^{em}}{2}+i\delta\right]e^{i4\Delta_E t/\hbar}\rho_{g,e}^I(t).$$

The secular approximation (see Eqs. (19.4.8, 19.4.9)) implies that the rapidly oscil-lating coefficients are negligible with respect to the stationary coefficients. Invoke this approximation and transform back the equations to the Schrodinger picture to regain the equations of motion for $\hat{H}_{SB} = \mu \hat{U}_B \hat{\sigma}_+ + \mu^* \hat{U}_B^\dagger \hat{\sigma}_-$ (Eqs. (19.5.13, 19.5.20)).

Diagonal Coupling and Pure Dephasing

We now focus on system–bath coupling that is diagonal in the system eigenstate basis. Physically, this type of coupling amounts to bath-induced variations in the TLS energy gap. As analyzed in the general case (see Section 19.4), *diagonal coupling cannot induce population transfer between the TLS eigenstates. Nevertheless, it is an important source of decoherence (dephasing) that is independent of the population transfer dynamics.* Without loss of generality, here the system–bath coupling is projected onto the excited state of the TLS, setting

$$\hat{V}_S \equiv \eta |\varphi_e\rangle\langle\varphi_e| = \eta \hat{\sigma}_+ \hat{\sigma}_-, \qquad (19.5.26)$$

where η is real-valued. Since $\hat{V}_s$ is Hermitian, the system–bath coupling (see Eq. (19.5.1)) becomes a single product of system and bath operators,

$$\hat{H}_{SB} = \hat{V}_S \hat{U}_B + \hat{V}_S^\dagger \hat{U}_B^\dagger = \hat{V}_S(\hat{U}_B + \hat{U}_B^\dagger) \quad ; \quad \hat{U}_B = \sum_{j=1}^{N_\omega} \lambda_j \hat{b}_j, \qquad (19.5.27)$$

and the dissipator obtains a simple form (compare to Eq. (19.5.5)),

$$\hat{D}\hat{\rho}_S(t) = -\frac{1}{\hbar^2} \int_0^t d\tau \{ [c_e(\tau) + c_a(\tau)][\hat{V}_S, e^{\frac{-i\tau}{\hbar}\hat{H}_s} \hat{V}_S e^{\frac{i\tau}{\hbar}\hat{H}_s} \hat{\rho}_S(t)] + h.c. \}, \qquad (19.5.28)$$

with the bath correlation functions as given in Eqs. (19.5.6, 19.5.7). Invoking the sta-tionary Born–Markov approximation (Eq. (19.5.8)), the time evolution of the state populations and coherences in the reduced TLS obtain the explicit form (Ex. 19.5.7),

$$\frac{\partial}{\partial t}\rho_{g,g}(t) = 0$$

$$\frac{\partial}{\partial t}\rho_{e,e}(t) = 0$$

$$\frac{\partial}{\partial t}\rho_{e,g}(t) = -[\mathrm{Re}\{k\} + \frac{i}{\hbar}(2\Delta_E + \hbar\,\mathrm{Im}\{k\})]\rho_{e,g}(t) \qquad (19.5.29)$$

$$\frac{\partial}{\partial t}\rho_{g,e}(t) = -[\mathrm{Re}\{k\} - \frac{i}{\hbar}(2\Delta_E + \hbar\,\mathrm{Im}\{k\})]\rho_{g,e}(t),$$

where (Ex. 19.5.8)

$$k = \frac{\eta^2}{2\pi\hbar} \int_0^\infty d\tau \int_0^\infty d\omega J(\hbar\omega)[\cos(\tau\omega)2n(\omega) + e^{-i\tau\omega}]. \qquad (19.5.30)$$

The diagonal coupling is shown to induce a real-valued shift to the TLS energy gap by
$\hbar \, \mathrm{Im}\{k\}$ *(and therefore to the oscillation frequency of the coherences), as well as a decay*
of the coherences to zero in the infinite time limit:

$$|\rho_{e,g}(t)| = |\rho_{e,g}(0)|e^{-k^{pd}t} \quad ; \quad k^{pd} = \mathrm{Re}\{k\} \geq 0. \qquad (19.5.31)$$

The (pure) dephasing rate, k^{pd} is shown to depend on the spectral density of the bosonic bath in the zero-frequency limit (see Ex. 19.5.8 and Eq. (19.4.30)). For any finite temperature, the rate is approximated as

$$k^{pd} \approx \lim_{\omega \to 0} \frac{k_B T \eta^2}{2\hbar} \frac{J(\hbar\omega)}{\hbar\omega} \qquad (19.5.32)$$

Notice that a finite spectral density at the zero-frequency limit corresponds to a "static noise" (or static disorder), where the diagonal coupling to the bath merely changes the TLS energy gap. Considering a low-frequency mode, $\hat{Q}_0 = (\hat{b}_0^\dagger + \hat{b}_0)/\sqrt{2}$, at a finite temperature, where the classical limit applies, Eqs. (19.5.26, 19.5.27) imply that the effective energy gap depends on the bath displacement, namely $2\Delta_E + \sqrt{2}\eta\lambda_0 Q_0$. The broadening of the distribution of static TLS energy gaps in an ensemble of otherwise identical quantum systems ("inhomogeneous broadening") can therefore be regarded as the physical origin of pure dephasing. In the context of magnetic resonance imaging, for example, inhomogeneity in the local magnetic field within a sample is a major source of dephasing, typically much faster than the decoherence due to population transfer between the TLS states (owing to off-diagonal system–bath coupling; see the preceding discussion). The effective decoherence rate (often referred to as $1/T_2^*$) is therefore much larger than $1/T_2$, as appears in Eq. (19.5.25).

Exercise 19.5.7   *Use Eqs. (19.5.26–19.5.28) and the definition, $k = \frac{\eta^2}{\hbar^2}\int_0^\infty d\tau \, [c_e(\tau) + c_a(\tau)]$, to derive Eq. (19.5.29) for the reduced density matrix elements.*

Exercise 19.5.8   *(a) Use the explicit expressions for the correlation functions in the case of a continuous boson bath (Eq. (19.5.7)) to show that $k = \frac{\eta^2}{\hbar^2}\int_0^\infty d\tau \, [c_e(\tau) + c_a(\tau)] = \frac{\eta^2}{2\pi\hbar}\int_0^\infty d\tau \int_0^\infty d\omega J(\hbar\omega)[\cos(\tau\omega)2n(\omega) + e^{-i\tau\omega}]$. (b) The bath-induced decay of the coherences is associated with the rate, $k^{pd} = \mathrm{Re}\{k\} \geq 0$ (Eq. (19.5.31)). Show that $k^{pd} = \lim_{\omega \to 0} \frac{\eta^2}{4\hbar} J(\hbar\omega)[2n(\omega) + 1]$, where for a finite temperature, $k^{pd} \approx \lim_{\omega \to 0} \frac{k_B T \eta^2}{2\hbar} \frac{J(\hbar\omega)}{\hbar\omega}$.*

Bibliography

[19.1]   C. Meier and D. J. Tannor, "Non-Markovian evolution of the density operator in the presence of strong laser fields," The Journal of Chemical Physics 111, 3365 (1999).

[19.2] H.-P. Breuer and F. Petruccione, "The Theory of Open Quantum Systems" (Oxford, 2002).

[19.3] A. G. Redfield, "The theory of relaxation processes," Advances in Magnetic and Optical Resonance 1, 1 (1965).

[19.4] D. Egorova, M. Thoss, W. Domcke, and H. Wang, "Modeling of ultrafast electron-transfer processes: Validity of multilevel Redfield theory," The Journal of Chemical Physics, 119, 2761 (2003).

Open Many-Fermion Systems

20.1 The Fock Space

In Chapters 13 and 14 we discussed in some detail the electronic structure of atoms, molecules, and periodic lattices, which are the "building blocks" of matter on the nanoscale. In these discussions the many-electron systems were closed, in the sense that the number of electrons was fixed. However, many of the relevant phenomena for nanoscience and technology involve transport of electrons between open systems, in which the number of electrons can repeatedly change. For example, when an atom or a molecule is adsorbed on a surface, it can exchange charge with the surface, leading to remarkable changes in electronic structure, nuclear configuration, chemical stability, and so forth. To fully characterize the dynamics associated with such charging/discharging processes, different charging states of the system of interest need to be accounted for simultaneously.

In this chapter we extend the treatment of many-electron systems to account for a variable number of electrons within the system, introducing the "second quantization" formulation. The state of a many-particle system is formulated in "Fock space," which is an extended Hilbert space, that relates to all possible particle numbers within the system. We shall focus here only on fermions, where the complementary discussion of second quantization for bosons, can be found elsewhere.

We start by recalling the generic form of the Hamiltonian for a system of N electrons. Considering, for example, an atom (Eq. (3.4.10)) or a molecule (Eq. (3.4.11)) with clamped nuclei, in the absence of external fields, the Hamiltonian (up to a trivial addition of a constant) can be cast as,

$$\hat{H}^{(N)} = \sum_{i=1}^{N} \hat{h}_i + \sum_{j>i=1}^{N} \hat{w}_{i,j}. \tag{20.1.1}$$

The first term is a sum over N single-particle Hamiltonians, accounting for each of the electrons, in the absence of the others. In terms of the electron coordinates, $\hat{h} \equiv \frac{-\hbar^2}{2m_e}\Delta_{\mathbf{r}} + V(\hat{\mathbf{r}})$, where $V(\hat{\mathbf{r}})$ is the "external single-particle potential," which is the Coulomb energy attributed to the electron owing to the spatial distribution of the nuclear charges. The second term is a sum over all electron-pair interactions. In terms of the electron coordinates, $\hat{w} \equiv \frac{Ke^2}{|\hat{\mathbf{r}}-\hat{\mathbf{r}}'|}$ is the Coulomb repulsion energy between the two electrons positioned at $\hat{\mathbf{r}}$ and $\hat{\mathbf{r}}'$. This term is universal, namely independent of any particular external potential.

Each single-particle state can be expanded in terms of a complete orthonormal system of state vectors, denoted as $\{|\Phi_k\rangle\}$. Here k is a set of quantum numbers, needed to uniquely define a single-particle state (e.g., its spatial orbital in the coordinates representation, as well as its spin state). The orthonormal set can be composed of, for example, the eigenstates of the single-particle Hamiltonian, $\hat{h}|\Phi_k\rangle = \varepsilon_k|\Phi_k\rangle$, or the eigenstates of the Fock operator (Eq. (13.3.18)). In general, the single-particle Hilbert space is spanned by an infinite set (see section 11.2), but for simplicity, and without loss of generality, we restrict the following discussion to finite dimensions, M, denoting the set of single-particle states (e.g., the number of orthonormal spin-orbitals) as $\{|\Phi_1\rangle,|\Phi_2\rangle,\ldots,|\Phi_M\rangle\}$, where

$$\langle\Phi_{k'}|\Phi_k\rangle = \delta_{k,k'}. \tag{20.1.2}$$

An N-particle vector space is a tensor product of N single-particle spaces (see Section 11.6). Consequently, in general, each N-particle state can be expanded as a linear combination of products of N single-particle states, where the number of linearly-independent products is M^N Recalling, however, that for N identical fermions a proper state must be antisymmetric with respect to any particle index permutation (see Section 13.3), the dimensions of the "physical" vector space for N electrons is much smaller. The antisymmetry property is guaranteed to hold when the product states are grouped into determinants (see Section 13.3). Each determinant is defined by an ordered selected set of N "populated" single-particle states out of the total number of M, namely a set of selected indexes, $l_1 < l_2 < \ldots < l_N$, where $\{l_j\} \in 1,2,\ldots M$ (see Eq. (13.3.2)):

$$\left\langle 1,2,\ldots,N \middle| \Psi^{(N)}_{\{l_1,l_2,\ldots,l_N\}} \right\rangle \equiv \frac{1}{\sqrt{N!}} \begin{vmatrix} \langle 1|\Phi_{l_1}\rangle & \langle 1|\Phi_{l_2}\rangle & \langle 1|\Phi_{l_3}\rangle & \cdots & \langle 1|\Phi_{l_N}\rangle \\ \langle 2|\Phi_{l_1}\rangle & \langle 2|\Phi_{l_2}\rangle & \langle 2|\Phi_{l_3}\rangle & \cdots & \langle 2|\Phi_{l_N}\rangle \\ \langle 3|\Phi_{l_1}\rangle & \langle 3|\Phi_{l_2}\rangle & \langle 3|\Phi_{l_3}\rangle & \cdots & \langle 3|\Phi_{l_N}\rangle \\ \vdots & \vdots & \vdots & \ddots & \vdots \\ \langle N|\Phi_{l_1}\rangle & \langle N|\Phi_{l_2}\rangle & \langle N|\Phi_{l_3}\rangle & \cdots & \langle N|\Phi_{l_N}\rangle \end{vmatrix}.$$

$$\tag{20.1.3}$$

Using the orthonormality of the single-particle states (Eq. (20.1.2)), we can readily verify that different N-particle determinants, corresponding to different selected sets $(\{l_1,l_2,\ldots,l_N\})$, are orthonormal:

$$\left\langle \Psi^{(N)}_{\{l_1,l_2,\ldots,l_N\}} \middle| \Psi^{(N)}_{\{l_1',l_2',\ldots,l_N'\}} \right\rangle = \delta\{l_1,l_2,\ldots,l_N\},\{l_1'l_2' \ldots,l_N'\}. \tag{20.1.4}$$

Since the number of different selected sets is $\frac{M!}{N!(M-N)!} = \begin{pmatrix} M \\ N \end{pmatrix}$, the corresponding determinants are a basis for an $\begin{pmatrix} M \\ N \end{pmatrix}$-dimensional vector space, $\{|\Psi^{(N)}\rangle\}$, where the proper states of an N-electron system are the span of $\{|\Psi^{(N)}\rangle\}$:

$$|\Psi^{(N)}\rangle = \sum_{\{l_1,l_2,\ldots l_N\}} a_{\{l_1,l_2,\ldots,l_N\}} |\Psi^{(N)}_{\{l_1,l_2,\ldots,l_N\}}\rangle. \tag{20.1.5}$$

Now let us notice that a formal representation of an N-particle state within the space of N' particles, with $N' \neq N$, corresponds to an improper (un-normalizable) state. Considering an extended vector space that is the direct sum of all the N-particle vector spaces, the inner product between vectors associated with different particle numbers within this space must therefore vanish:

$$\left\langle \Psi^{(N)}_{\{l_1,l_2,\dots,l_N\}} \middle| \Psi^{(N')}_{\{l_1',l_2'\dots,l_{N'}'\}} \right\rangle = \delta_{N,N'} \delta_{\{l_1,l_2,\dots,l_N\},\{l_1',l_2',\dots,l_{N'}'\}}. \tag{20.1.6}$$

Particularly, any N-particle state is orthogonal to the state of "no particles," corresponding to $N = 0$. This state is termed "the vacuum state," which we denote as $|\Psi^{(0)}\rangle$. **The extended vector space, which is the span of the extended basis,** $|\Psi^{(0)}\rangle, \{|\Psi^{(1)}\rangle\}, \{|\Psi^{(2)}\rangle\}, \dots, \{|\Psi^{(M-1)}\rangle\}, |\Psi^{(M)}\rangle$, **is termed the Fock space.** Each basis vector corresponds to a specific selection of single-particle states to be "occupied," out of the total number of M. Since in each basis vector, each single-particle state is either occupied or not (only two possibilities, in line with Pauli's exclusion principle, discussed in Section 13.3), the total number of possibilities of occupying the M single-particle states is 2^M, which is the dimension of the Fock space, $\sum\limits_{N=0}^{M} \binom{M}{N} = 2^M$. It is convenient to represent each basis vector in terms of a binary string, $|n_1,n_2,n_3,\dots,n_M\rangle$, corresponding to the "occupation number" of each of the M single-particle states, $\{|\Phi_1\rangle,|\Phi_2\rangle,\dots,|\Phi_M\rangle\}$, where "occupied" and "unoccupied" states are associated with $n_l = 1$ and $n_l = 0$, respectively:

$$\left| \Psi^{(N)}_{\{l_1,l_2,l_3,\dots,l_N\}} \right\rangle = |n_1,n_2,n_3,\dots,n_M\rangle \quad ; \quad n_l = \begin{cases} 1 & ; \quad l \in \{l_1,l_2,\dots,l_N\} \\ 0 & ; \quad otherwise \end{cases}. \tag{20.1.7}$$

(For example, if the total number of single-particle states is $M = 6$, and the indexes of the occupied states are $\{l_1,l_2,l_3\} = \{2,3,5\}$, the corresponding basis state would be denoted as $|\Psi^{(3)}_{\{2,3,5\}}\rangle = |0,1,1,0,1,0\rangle$). Using this notation, the representation of the vacuum state would be $|\Psi^{(0)}\rangle = |0,0,0,\dots,0\rangle$, the M basis states associated with a single particle in the system would be represented as

$$\{|\Psi^{(1)}\rangle\} = \begin{cases} |\Psi^{(1)}_{\{1\}}\rangle = |1,0,0,\dots,0\rangle \\ |\Psi^{(1)}_{\{2\}}\rangle = |0,1,0,\dots,0\rangle \\ |\Psi^{(1)}_{\{3\}}\rangle = |0,0,1,\dots,0\rangle \\ \vdots \\ |\Psi^{(1)}_{\{M\}}\rangle = |0,0,0,\dots,1\rangle \end{cases},$$

the $M(M-1)/2$ basis states associated with two particles in the system would be represented as

$$\{|\Psi^{(2)}\rangle\} = \left\{ \begin{array}{l} |\Psi^{(2)}_{\{1,2\}}\rangle = |1,1,0,\ldots,0\rangle \\ |\Psi^{(2)}_{\{2,3\}}\rangle = |0,1,1,\ldots,0\rangle \\ |\Psi^{(2)}_{\{1,3\}}\rangle = |1,0,1,\ldots,0\rangle \\ \quad\vdots \end{array} \right\},$$

and so forth, where the representation of the fully occupied state would be $|\Psi^{(M)}\rangle = |1,1,1,\ldots,1\rangle$. Recalling that (except for the vacuum state) each basis state is identified with a determinant (see Eq. (20.1.3)), the sign of each state depends on the order of the occupied single-particle state indexes, where for a fixed order, $l_1 < l_2 < \ldots < l_N$, $\{l_j\} \in 1, 2, \ldots, M$, the signs of all basis states are uniquely defined.

Exercise 20.1.1   *Write the binary strings for the 16 basis vectors defined by four single-particle states, $\{|\Phi_1\rangle, |\Phi_2\rangle, |\Phi_3\rangle, |\Phi_4\rangle\}$.*

Fock Space Operators

We now define operators in Fock space. Of particular importance are the operators that couple between the subspaces of different particle numbers. The electron creation operator in the lth single-particle state is denoted $\hat{a}_l^\dagger$ and can be defined by its operation on the vacuum state,

$$\hat{a}_l^\dagger |\Psi^{(0)}\rangle \equiv |\Psi^{(1)}_{\{l\}}\rangle. \tag{20.1.8}$$

The Hermitian conjugate of $\hat{a}_l^\dagger$ is denoted $\hat{a}_l$, and defined as usual (see Eq. (11.2.18)), $\langle\Psi^{(0)}|\hat{a}_l = \langle\Psi^{(1)}_{\{l\}}|$. Using the normalization conditions (Eq. (20.1.6)), we obtain (Ex. 20.1.2)

$$\hat{a}_l |\Psi^{(1)}_{\{l\}}\rangle = |\Psi^{(0)}\rangle, \tag{20.1.9}$$

which associates the operator $\hat{a}_l$ with electron annihilation at the lth single-particle state.

Exercise 20.1.2   *Use the normalization conditions (Eq. (20.1.6)), $\left\langle \Psi^{(1)}_{\{l\}} \middle| \Psi^{(1)}_{\{l\}} \right\rangle = 1$, and $\langle\Psi^{(0)}|\Psi^{(0)}\rangle = 1$, and the definition of the creation operator (Eq. (20.1.8)) to derive Eq. (20.1.9).*

Since an attempt to occupy the same single-particle state twice would correspond to a vanishing determinant (see Eq. (20.1.3)), we have $\hat{a}_l^\dagger |\Psi^{(1)}_{\{l\}}\rangle = (\hat{a}_l^\dagger)^2 |\Psi^{(0)}_{\{l\}}\rangle = 0$. Similarly, the number of particles in a given single-particle state cannot be smaller than zero, which means that $\hat{a}_l |\Psi^{(0)}\rangle$ must be orthogonal to any state in the Fock space, and therefore, $\hat{a}_l |\Psi^{(0)}\rangle = (\hat{a}_l)^2 |\Psi^{(1)}_{\{l\}}\rangle = 0$. These requirements can be summarized as

$$(\hat{a}_l^\dagger)^2 = (\hat{a}_l)^2 = 0. \tag{20.1.10}$$

We can readily verify that $\hat{a}_l$ and $\hat{a}_l^\dagger$ are non-Hermitian operators (Ex. 20.1.3). Their product, $\hat{a}_l^\dagger \hat{a}_l$, is Hermitian, where

$$\begin{aligned}
\hat{a}_l^\dagger \hat{a}_l |\Psi^{(0)}\rangle &= 0|\Psi^{(0)}\rangle \\
\hat{a}_l^\dagger \hat{a}_l |\Psi^{(1)}_{\{l\}}\rangle &= 1|\Psi^{(1)}_{\{l\}}\rangle
\end{aligned} \tag{20.1.11}$$

Both the unoccupied ($|\Psi^{(0)}\rangle$) and singly occupied ($|\Psi^{(1)}_{\{l\}}\rangle$) states are eigenstates of $\hat{a}_l^\dagger \hat{a}_l$, with eigenvalues corresponding to the number of electrons occupying the lth single-particle state, $N_l \in 0, 1$. The operator $\hat{a}_l^\dagger \hat{a}_l$ can therefore be identified as a physical observable corresponding to the electron number,

$$\hat{a}_l^\dagger \hat{a}_l \equiv \hat{N}_l. \tag{20.1.12}$$

Similarly,

$$\begin{aligned}
\hat{a}_l \hat{a}_l^\dagger |\Psi^{(0)}\rangle &= 1|\Psi^{(0)}\rangle \\
\hat{a}_l \, \hat{a}_l^\dagger |\Psi^{(1)}_{\{l\}}\rangle &= 0|\Psi^{(1)}_{\{l\}}\rangle
\end{aligned} \tag{20.1.13}$$

the eigenvalues of $\hat{a}_l \hat{a}_l^\dagger$ correspond to complementary numbers of "electron vacancies" or "holes" associated with the occupied and the empty states of the lth single-particle state. The operator $\hat{a}_l \hat{a}_l^\dagger$ can therefore be identified as an observable corresponding to the hole number,

$$\hat{a}_l \hat{a}_l^\dagger \equiv 1 - \hat{N}_l. \tag{20.1.14}$$

Combining Eqs. (20.1.12, 20.1.14), the operators $\hat{a}_l$ and $\hat{a}_l^\dagger$ are shown to satisfy an "anti-commutation" relation,

$$\{\hat{a}_l^\dagger, \hat{a}_l\} \equiv \hat{a}_l^\dagger \hat{a}_l + \hat{a}_l \hat{a}_l^\dagger = 1. \tag{20.1.15}$$

The proper two-electron basis states can be obtained by selecting two single-particle states out of the ordered list, $\{|\Phi_1\rangle, |\Phi_2\rangle, \ldots, |\Phi_M\rangle\}$. Selecting the states $|\Phi_{l'}\rangle$, and $|\Phi_{l'}\rangle$, where, $l < l'$, the corresponding two-particle state is the determinant, $|\Psi^{(2)}_{\{l,l'\}}\rangle$ (see Eq. (20.1.3)). This basis state is obtained from the vacuum by two successive operations of the relevant creation operators, namely,

$$|\Psi^{(2)}_{\{l,l'\}}\rangle \equiv \hat{a}_l^\dagger |\Psi^{(1)}_{\{l'\}}\rangle = \hat{a}_l^\dagger \hat{a}_{l'}^\dagger |\Psi^{(0)}\rangle, \tag{20.1.16}$$

and similarly (see Ex. 20.1.4),

$$|\Psi^{(0)}\rangle = \hat{a}_{l'} |\Psi^{(1)}_{\{l'\}}\rangle = \hat{a}_{l'} \hat{a}_l |\Psi^{(2)}_{\{l,l'\}}\rangle. \tag{20.1.17}$$

Recalling that exchanging two columns in a determinant is associated with a sign flip,

$$|\Psi^{(2)}_{\{l,l'\}}\rangle = -|\Psi^{(2)}_{\{l',l\}}\rangle, \tag{20.1.18}$$

it follows that changing the order of the two creation operators corresponds to a change of sign, which, together with Eq. (20.1.10), leads to the anti-commutation relations (see Ex. 20.1.5),

$$\{\hat{a}_l^\dagger, \hat{a}_{l'}^\dagger\} = 0 \quad ; \quad \{\hat{a}_l, \hat{a}_{l'}\} = 0. \tag{20.1.19}$$

Additionally, using Eqs. (20.1.16, 20.1.17), we obtain, $\left|\Psi_{\{l'\}}^{(1)}\right\rangle = -\hat{a}_l\left|\Psi_{\{l',l\}}^{(2)}\right\rangle =$
$-\hat{a}_l\hat{a}_{l'}^\dagger\hat{a}_l^\dagger|\Psi^{(0)}\rangle$, and $\left|\Psi_{\{l'\}}^{(1)}\right\rangle = \hat{a}_{l'}^\dagger|\Psi^{(0)}\rangle = \hat{a}_{l'}^\dagger\hat{a}_l\hat{a}_l^\dagger|\Psi^{(0)}\rangle$, and therefore $\hat{a}_{l'}^\dagger\hat{a}_l = -\hat{a}_l\hat{a}_{l'}^\dagger$.
Recalling Eq. (20.1.15), the result reads

$$\{\hat{a}_l, \hat{a}_{l'}^\dagger\} = \delta_{l,l'}. \tag{20.1.20}$$

Exercise 20.1.3 *Recalling the definition of a Hermitian operator (Eq. (11.2.20)), use
the matrix elements of the operators, $\hat{a}_l^\dagger$ and $\hat{a}_l$, between the states, $\left|\Psi^{(0)}\right\rangle$ and $\left|\Psi_{\{l\}}^{(1)}\right\rangle$, to
show that these operators are non-Hermitian.*

Exercise 20.1.4 *Use the normalization conditions (Eq. (20.1.6)), $\langle\Psi^{(0)}|\Psi^{(0)}\rangle = 1$,
$\left\langle\Psi_{\{l\}}^{(1)}\middle|\Psi_{\{l\}}^{(1)}\right\rangle = 1$, $\left\langle\Psi_{\{l,l'\}}^{(2)}\middle|\Psi_{\{l,l'\}}^{(2)}\right\rangle = 1$, and Eq. (20.1.16), to derive Eq. (20.1.17).*

Exercise 20.1.5 *Use Eqs. (20.1.16, 20.1.18) to derive Eq. (20.1.19).*

The canonical anti-commutation relations for the fermionic creation and annihilation operators (Eqs. (20.1.19, 20.1.20)) apply to all the single-particle states. Indeed, any many electron basis vector can be represented in terms of the set of operators, $\{\hat{a}_l^\dagger\}$, where, for an ordered sequence, $\{|\Phi_1\rangle, |\Phi_2\rangle, \ldots, |\Phi_M\rangle\}$, and for $l_1 < l_2 < \ldots < l_N$,

$$\left|\Psi_{\{l_1,l_2,\ldots,l_N\}}^{(N)}\right\rangle \equiv \hat{a}_{l_1}^\dagger\left|\Psi_{\{l_2,\ldots,l_N\}}\right\rangle = \hat{a}_{l_1}^\dagger\hat{a}_{l_2}^\dagger\hat{a}_{l_3}^\dagger\cdots\hat{a}_{l_N}^\dagger|\Psi^{(0)}\rangle. \tag{20.1.21}$$

Notice that since $|\Psi_{\{l',l'',\ldots\}}^{(N)}\rangle$ corresponds to a determinant (Eq. (20.1.3)), the operator $\hat{a}_l^\dagger|\Psi_{\{l',l'',\ldots\}}^{(N)}\rangle$ (with $l < l', l'', \ldots$) corresponds to an extended determinant in which a (first) row and a column are added. In binary string representation (Eq. (20.1.7)), Eq. (20.1.21) can also be written as

$$|n_1, n_2, n_3, \ldots, n_M\rangle = (\hat{a}_1^\dagger)^{n_1}(\hat{a}_2^\dagger)^{n_2}\cdots(\hat{a}_M^\dagger)^{n_M}|\Psi^{(0)}\rangle \quad ; \quad n_l = \begin{cases} 1 & ; \quad l \in \{l_1, l_2, \ldots, l_N\} \\ 0 & ; \quad \text{otherwise} \end{cases}, \tag{20.1.22}$$

where, $(\hat{a}_l^\dagger)^0 = 1$ and $(\hat{a}_l^\dagger)^1 = \hat{a}_l^\dagger$. Considering an N-electron system in a state, $|n_1, n_2, \ldots, n_M\rangle$, an electron creation in the kth single-particle state can therefore be represented as

$$\hat{a}_k^\dagger|n_1, n_2, \ldots, n_M\rangle = \begin{cases} 0 & ; \quad n_k = 1 \\ (-1)^{\sum_{j=1}^{k-1} n_j}|n_1, \ldots, n_k+1, \ldots, n_M\rangle & ; \quad n_k = 0 \end{cases}. \tag{20.1.23}$$

Notice that the state of the $N + 1$-electron system is a standard determinant (as defined by Eqs. (20.1.3, 20.1.7)), multiplied by an appropriate (Jordan–Wigner) phase, $(-1)^{\sum_{j=1}^{k-1} n_j}$ (see Ex. 20.1.6).

Similarly, an electron annihilation from the kth single-particle state is represented as

$$\hat{a}_k|n_1,n_2,\ldots,n_M\rangle = \begin{cases} 0 & ; \quad n_k = 0 \\ (-1)^{\sum_{j=1}^{k-1} n_j}|n_1,\ldots,n_k-1,\ldots,n_M\rangle & ; \quad n_k = 1 \end{cases}. \quad (20.1.24)$$

Notice that the occupation number of the kth single-particle state in the N-electron system is an eigenvalue of the number operator, $\hat{N}_k \equiv \hat{a}_k^\dagger \hat{a}_k$ (see Eq. (20.1.12)),

$$\hat{a}_k^\dagger \hat{a}_k|n_1,n_2,\ldots,n_M\rangle = n_k|n_1,n_2,\ldots,n_M\rangle. \quad (20.1.25)$$

Associating the total number of electrons in the state $|n_1,n_2,\ldots,n_M\rangle$ with the total number operator,

$$\hat{N} \equiv \sum_{k=1}^{M} \hat{N}_k = \sum_{k=1}^{M} \hat{a}_k^\dagger \hat{a}_k, \quad (20.1.26)$$

any basis vector is an eigenvector of $\hat{N}$,

$$\hat{N}|n_1,n_2,\ldots,n_M\rangle = N|n_1,n_2,\ldots,n_M\rangle \quad ; \quad N = \sum_{k=1}^{M} n_k, \quad (20.1.27)$$

where the total number of electrons is the corresponding eigenvalue.

Exercise 20.1.6   *Use the anti-commutation relations for the fermion operators (Eqs. (20.1.19, 20.1.20)), and Eqs. (20.1.21, 20.1.22) to derive Eqs. (20.1.23, 20.1.24).*

Matrix Representations

For one single-particle state ($M = 1$), the dimension of the Fock space is $2^1 = 2$, where the space is the span of two basis vectors corresponding to the occupied and unoccupied single-particle states, $|1\rangle$ and $|0\rangle$, respectively. Given a general state, $|\psi\rangle = \psi_1|1\rangle + \psi_0|0\rangle$, the corresponding vector representation reads (Eq. (11.2.7))

$$\psi = \begin{pmatrix} \psi_1 \\ \psi_0 \end{pmatrix} = \begin{pmatrix} \langle 1|\psi\rangle \\ \langle 0|\psi\rangle \end{pmatrix}, \quad (20.1.28a)$$

where the two basis vectors are unit vectors,

$$\mathbf{e}_1 = \begin{pmatrix} 1 \\ 0 \end{pmatrix} \quad ; \quad \mathbf{e}_0 = \begin{pmatrix} 0 \\ 1 \end{pmatrix}. \quad (20.1.28b)$$

The creation and annihilation operators for an electron are defined by the equations, $\hat{a}^\dagger|0\rangle = |1\rangle$, $\hat{a}^\dagger|1\rangle = 0$, and $\hat{a}|1\rangle = |0\rangle$, $\hat{a}^\dagger|1\rangle = 0$, respectively (Eqs. (20.1.8, 20.1.9)). Consequently, the matrix representations of these operators in the basis $\{|1\rangle \ |0\rangle\}$ read (Eqs. (11.2.11, 11.2.12))

$$\mathbf{a} = \begin{bmatrix} 0 & 0 \\ 1 & 0 \end{bmatrix} \quad ; \quad \mathbf{a}^\dagger = \begin{bmatrix} 0 & 1 \\ 0 & 0 \end{bmatrix}. \quad (20.1.29)$$

Similarly, the matrix representations of the electron and hole number operators, Eqs. (20.1.12, 20.1.14), read (Ex. 20.1.7)

$$\mathbf{a}^\dagger \mathbf{a} = \begin{bmatrix} 1 & 0 \\ 0 & 0 \end{bmatrix} \quad ; \quad \mathbf{a}\mathbf{a}^\dagger = \begin{bmatrix} 0 & 0 \\ 0 & 1 \end{bmatrix}. \tag{20.1.30}$$

We can readily verify that the matrix representations satisfy the anti-commutation relations, Eqs. (20.1.19, 20.1.20):

$$\{\mathbf{a}^\dagger, \mathbf{a}^\dagger\} = 0 \quad ; \quad \{\mathbf{a}, \mathbf{a}\} = 0 \quad ; \quad \{\mathbf{a}^\dagger, \mathbf{a}\} = \mathbf{I} \quad ; \quad \mathbf{I} = \begin{bmatrix} 1 & 0 \\ 0 & 1 \end{bmatrix} \tag{20.1.31}$$

Exercise 20.1.7 *Use the definitions $\hat{a}^\dagger|0\rangle = |1\rangle$, $\hat{a}^\dagger|1\rangle = 0$, $\hat{a}|1\rangle = |0\rangle$, and $\hat{a}^\dagger|1\rangle = 0$ to derive the (two-by-two) matrix representations of the operators, $\hat{a}^\dagger, \hat{a}, \hat{a}^\dagger\hat{a}$, and $\hat{a}\hat{a}^\dagger$ in the complete orthonormal basis $\{|0\rangle, |1\rangle\}$. Show that these matrices satisfy the anti-commutation relation, Eq. (20.1.31).*

For a system with many single-particle states ($M > 1$), the dimension of the Fock space is 2^M. This space is an ordered tensor product of M subspaces, corresponding to the the different single-particle states. Therefore, each basis vector of the Fock space is a tensor product of M basis vectors, each associated with one of the single-particle occupation states

$$|n_1, n_2, n_3, \ldots n_M\rangle = |n_1\rangle \otimes |n_2\rangle \otimes \cdots \otimes |n_M\rangle \quad ; \quad n_j \in \{0, 1\}. \tag{20.1.32}$$

For example, a basis vector, $|0, 1, \ldots, 1, 0\rangle$, corresponds to a tensor product,

$$|0, 1, \ldots, 1, 0\rangle \Leftrightarrow \overset{1}{\begin{pmatrix} 0 \\ 1 \end{pmatrix}} \otimes \overset{2}{\begin{pmatrix} 1 \\ 0 \end{pmatrix}} \otimes \cdots \otimes \overset{M-1}{\begin{pmatrix} 1 \\ 0 \end{pmatrix}} \otimes \overset{M}{\begin{pmatrix} 0 \\ 1 \end{pmatrix}}. \tag{20.1.33}$$

Using the vector representations, Eq. (20.1.28b), the basis vector is represented as a 2^M-dimensional unit vector,

$$\mathbf{e}_{0,1,\ldots,1,0} = \mathbf{e}_0 \otimes \mathbf{e}_1 \otimes \cdots \otimes \mathbf{e}_1 \otimes \mathbf{e}_0. \tag{20.1.34}$$

Electron creation and annihilation operators in the 2^M-dimensional Fock space are tensor products of operators in the M subspaces of the single-particle states. According to Eq. (20.1.23), the operator $\hat{a}_k^\dagger$, which creates an electron in the kth single-particle state, maintains the occupation numbers in all the other single-particle states, and multiplies the many-particle state by -1, for each occupied single-particle state whose index is smaller than k (the Jordan–Wigner phase). The matrix representation of $\hat{a}_k^\dagger$, denoted as $\mathbf{a}_k^\dagger$, must therefore satisfy the equation

$$\mathbf{a}_k^\dagger \cdot \mathbf{e}_{n_1, n_2, \ldots, n_k, \ldots, n_M} = \begin{cases} 0 & ; \quad n_k = 1 \\ (-1)^{\sum_{j=1}^{k-1} n_j} \mathbf{e}_{n_1, n_2, \ldots, 1, \ldots, n_M} & ; \quad n_k = 0 \end{cases}. \tag{20.1.35}$$

We can readily verify that the matrix $\mathbf{a}_k^\dagger$ reads

$$
\mathbf{a}_k^\dagger = \overset{1}{\begin{bmatrix} -1 & 0 \\ 0 & 1 \end{bmatrix}} \otimes \overset{2}{\begin{bmatrix} -1 & 0 \\ 0 & 1 \end{bmatrix}} \otimes \cdots \otimes \overset{k-1}{\begin{bmatrix} -1 & 0 \\ 0 & 1 \end{bmatrix}}
$$

$$
\otimes \overset{k}{\begin{bmatrix} 0 & 1 \\ 0 & 0 \end{bmatrix}} \otimes \overset{k-1}{\begin{bmatrix} 1 & 0 \\ 0 & 1 \end{bmatrix}} \otimes \cdots \otimes \overset{M-1}{\begin{bmatrix} 1 & 0 \\ 0 & 1 \end{bmatrix}} \otimes \overset{M}{\begin{bmatrix} 1 & 0 \\ 0 & 1 \end{bmatrix}}, \qquad (20.1.36)
$$

where, for any $j < k$, the identity matrix is replaced by $\begin{bmatrix} -1 & 0 \\ 0 & 1 \end{bmatrix} = -1 \begin{bmatrix} 1 & 0 \\ 0 & 0 \end{bmatrix} +$
$1 \begin{bmatrix} 0 & 0 \\ 0 & 1 \end{bmatrix}$. This means that the result accumulates a phase "1" if the jth state is empty, and a phase "-1" if the jth state is occupied.

Similarly, the matrix representation of the annihilation operator, $\hat{a}_k$, which complies with Eq. (20.1.24), reads

$$
\mathbf{a}_k = \overset{1}{\begin{bmatrix} -1 & 0 \\ 0 & 1 \end{bmatrix}} \otimes \overset{2}{\begin{bmatrix} -1 & 0 \\ 0 & 1 \end{bmatrix}} \otimes \cdots \otimes \overset{k-1}{\begin{bmatrix} -1 & 0 \\ 0 & 1 \end{bmatrix}}
$$

$$
\otimes \overset{k}{\begin{bmatrix} 0 & 0 \\ 1 & 0 \end{bmatrix}} \otimes \overset{k+1}{\begin{bmatrix} 1 & 0 \\ 0 & 1 \end{bmatrix}} \otimes \cdots \otimes \overset{M-1}{\begin{bmatrix} 1 & 0 \\ 0 & 1 \end{bmatrix}} \otimes \overset{M}{\begin{bmatrix} 1 & 0 \\ 0 & 1 \end{bmatrix}}. \qquad (20.1.37)
$$

We can verify (Ex. 20.1.8) that these matrix representations of the creation and annihilation operators satisfy the canonical anti-commutation relations, Eqs. (20.1.19, 20.1.20).

Exercise 20.1.8 *Show that the matrix representations of the creation and annihilation operators (Eqs. (20.1.36, 20.1.37)) satisfy the canonical anti-commutation relations for fermions, Eqs. (20.1.19, 20.1.20). (You can use the rules of tensor products multiplication, Eq. (11.6.21).)*

For concreteness, let us consider the Fock space for $M = 2$, which is the span of four (2^2) basis vectors, corresponding to the single-particle state occupation vectors, $|1,1\rangle$, $|1,0\rangle, |0,1\rangle, |0,0\rangle$,

$$
\mathbf{e}_{1,1} = \begin{pmatrix} 1 \\ 0 \end{pmatrix} \otimes \begin{pmatrix} 1 \\ 0 \end{pmatrix} = \begin{pmatrix} 1 \\ 0 \\ 0 \\ 0 \end{pmatrix}
$$

$$
\mathbf{e}_{1,0} = \begin{pmatrix} 1 \\ 0 \end{pmatrix} \otimes \begin{pmatrix} 0 \\ 1 \end{pmatrix} = \begin{pmatrix} 0 \\ 1 \\ 0 \\ 0 \end{pmatrix}
$$

$$\mathbf{e}_{0,1} = \begin{pmatrix} 0 \\ 1 \end{pmatrix} \otimes \begin{pmatrix} 1 \\ 0 \end{pmatrix} = \begin{pmatrix} 0 \\ 0 \\ 1 \\ 0 \end{pmatrix} \tag{20.1.38}$$

$$\mathbf{e}_{0,0} = \begin{pmatrix} 0 \\ 1 \end{pmatrix} \otimes \begin{pmatrix} 0 \\ 1 \end{pmatrix} = \begin{pmatrix} 0 \\ 0 \\ 0 \\ 1 \end{pmatrix}.$$

Using Eqs. (20.1.36, 20.1.37), the creation and annihilation operators are represented as

$$\hat{\mathbf{a}}_1^{\dagger} \equiv \begin{pmatrix} 0 & 1 \\ 0 & 0 \end{pmatrix} \otimes \begin{pmatrix} 1 & 0 \\ 0 & 1 \end{pmatrix} = \begin{pmatrix} 0 & 0 & 1 & 0 \\ 0 & 0 & 0 & 1 \\ 0 & 0 & 0 & 0 \\ 0 & 0 & 0 & 0 \end{pmatrix} \quad ;$$

$$\hat{\mathbf{a}}_1 \equiv \begin{pmatrix} 0 & 0 \\ 1 & 0 \end{pmatrix} \otimes \begin{pmatrix} 1 & 0 \\ 0 & 1 \end{pmatrix} = \begin{pmatrix} 0 & 0 & 0 & 0 \\ 0 & 0 & 0 & 0 \\ 1 & 0 & 0 & 0 \\ 0 & 1 & 0 & 0 \end{pmatrix}$$

$$\hat{\mathbf{a}}_2^{\dagger} \equiv \begin{pmatrix} -1 & 0 \\ 0 & 1 \end{pmatrix} \otimes \begin{pmatrix} 0 & 1 \\ 0 & 0 \end{pmatrix} = \begin{pmatrix} 0 & -1 & 0 & 0 \\ 0 & 0 & 0 & 0 \\ 0 & 0 & 0 & 1 \\ 0 & 0 & 0 & 0 \end{pmatrix} \quad ;$$

$$\hat{\mathbf{a}}_2 \equiv \begin{pmatrix} -1 & 0 \\ 0 & 1 \end{pmatrix} \otimes \begin{pmatrix} 0 & 0 \\ 1 & 0 \end{pmatrix} = \begin{pmatrix} 0 & 0 & 0 & 0 \\ -1 & 0 & 0 & 0 \\ 0 & 0 & 0 & 0 \\ 0 & 0 & 1 & 0 \end{pmatrix}. \tag{20.1.39}$$

It is easy to test explicitly (Ex. 20.1.9) that these matrix representations satisfy the canonical anti-commutation relations, Eqs. (20.1.19, 20.1.20), as well as Eqs. (20.1.23, 20.1.24).

Exercise 20.1.9 *Let us consider the Fock space corresponding to the two single-particle states, $\{|\Phi_1\rangle, |\Phi_2\rangle\}$, with the basis vectors, $|1,1\rangle, |1,0\rangle, |0,1\rangle, |0,0\rangle$. (a) Use Eqs. (20.1.23, 20.1.24) to show that*

$$\hat{a}_1^{\dagger}|0,0\rangle = |1,0\rangle \; ; \; \hat{a}_1^{\dagger}|0,1\rangle = |1,1\rangle \; ; \; \hat{a}_1^{\dagger}|1,0\rangle = \hat{a}_1^{\dagger}|1,1\rangle = 0$$

$$\hat{a}_1|1,0\rangle = |0,0\rangle \; ; \; \hat{a}_1|1,1\rangle = |0,1\rangle \; ; \; \hat{a}_1|0,0\rangle = \hat{a}_1|0,1\rangle = 0$$

$$\hat{a}_2^{\dagger}|0,0\rangle = |0,1\rangle \; ; \; \hat{a}_2^{\dagger}|1,0\rangle = -|1,1\rangle ; \hat{a}_2^{\dagger}|0,1\rangle = \hat{a}_2^{\dagger}|1,1\rangle = 0$$

$$\hat{a}_2|1,1\rangle = -|1,0\rangle ; \hat{a}_2|0,1\rangle = |0,0\rangle \; ; \; \hat{a}_2|1,0\rangle = \hat{a}_2|0,0\rangle = 0.$$

(b) Obtain the matrix representations of the creation and annihilation operators in the basis $\{|1,1\rangle, |1,0\rangle, |0,1\rangle, |0,0\rangle\}$, and compare the results to Eq. (20.1.39). (c) Check that the four matrices satisfy the anti-commutation relations for fermions, Eqs. (20.1.19, 20.1.20).

20.2 The Second Quantization Hamiltonian

We now turn to the representation of the Hamiltonian of a system of an indefinite number of electrons in the Fock space of M single-particle states. Since, by construction, the Fock space is a direct sum of N-particle subspaces, $|\Psi^{(0)}\rangle, \{|\Psi^{(1)}\rangle\}, \{|\Psi^{(2)}\rangle\}, \{|\Psi^{(M-1)}\rangle\}, |\Psi^{(M)}\rangle$, the projection of the Fock space Hamiltonian, $\hat{H}$, onto each N-particle subspace spanned by $(\{|\Psi^{(N)}\rangle\})$ must coincide with the corresponding N-particle Hamiltonian, $\hat{H}^{(N)}$ (Eq. (20.1.1)). Using a projection into the N-particle subspace, $\hat{P}_N = \sum_{\{l_1,l_2,l_3,\ldots,1_N\}} \left|\Psi^{(N)}_{\{l_1,l_2,l_3,\ldots,l_N\}}\right\rangle \left\langle\Psi^{(N)}_{\{l_1,l_2,l_3,\ldots,l_N\}}\right|$, this condition formally reads

$$\sum_{\{l_1,l_2,l_3,\ldots,l_N\},\{l_1',l_2',l_3',\ldots,l_N'\}} \left\langle\Psi^{(N)}_{\{l_1,l_2,l_3,\ldots,l_N\}}\left|\hat{H}^{(N)}\right|\Psi^{(N)}_{\{l_1',l_2',l_3',\ldots,l_N'\}}\right\rangle$$
$$\left|\Psi^{(N)}_{\{l_1,l_2,l_3,\ldots,l_N\}}\right\rangle\left\langle\Psi_{\{l_1',l_2',l_3',\ldots,l_N'\}}\right| = \hat{H}^{(N)}, \qquad (20.2.1)$$

or, considering any matrix element of $\hat{H}^{(N)}$,

$$\left\langle\Psi^{(N)}_{\{l_1',l_2',l_3',\ldots,l_N'\}}\left|\hat{H}\right|\Psi^{(N)}_{\{l_1,l_2,l_3,\ldots,l_N\}}\right\rangle = \left\langle\Psi^{(N)}_{\{l_1',l_2',l_3',\ldots,l_N'\}}\left|\hat{H}^{(N)}\right|\Psi^{(N)}_{\{l_1,l_2,l_3,\ldots,l_N\}}\right\rangle. \qquad (20.2.2)$$

We can verify that the Fock space Hamiltonian that complies with this identity for any fixed number of electrons ($N \leq M$) reads

$$\hat{H} = \sum_{i,j=1}^{M} h_{i,j}\hat{a}_i^\dagger\hat{a}_j + \frac{1}{2}\sum_{i,j,k,l=1}^{M} w_{i,j,k,l}\hat{a}_i^\dagger\hat{a}_j^\dagger\hat{a}_l\hat{a}_k, \qquad (20.2.3)$$

where the scalar parameters correspond to spin orbitals ($|\Phi_j\rangle \equiv |\varphi_j\rangle \otimes |\sigma_{m_{s,j}}\rangle$, see Eq. (13.3.16)),

$$h_{i,j} \equiv \langle\Phi_i|\hat{h}|\Phi_j\rangle = \delta_{m_{s,i},m_{s,j}}\int d\mathbf{r}\,\varphi_i^*(\mathbf{r})\left[\frac{-\hbar^2}{2m_e}\triangle_\mathbf{r} + V(\hat{\mathbf{r}})\right]\varphi_j(\mathbf{r})$$

$$w_{i,j,k,l} \equiv \langle\Phi_i|\otimes\langle\Phi_j|\hat{w}|\Phi_k\rangle\otimes|\Phi_l\rangle = \delta_{m_{s,i},m_{s,k}}\delta_{m_{s,j},m_{s,l}}Ke^2\int d\mathbf{r}\int d\mathbf{r}'\frac{\varphi_i^*(\mathbf{r})\varphi_j^*(\mathbf{r}')\varphi_k(\mathbf{r})\varphi_l(\mathbf{r}')}{|\mathbf{r}-\mathbf{r}'|}$$
$$(20.2.4)$$

Particularly, the diagonal matrix elements of $\hat{H}$ read (see Ex. 20.2.1)

$$\left\langle\Psi^{(N)}_{\{l_1,l_2,\ldots,l_N\}}\left|\hat{H}\right|\Psi^{(N)}_{\{l_1,l_2,\ldots,l_N\}}\right\rangle$$
$$= E_{\{l_1,l_2,\ldots,l_N\}} = \sum_{j\in\{l_1,l_2,\ldots,l_N\}=1}^{N} h_{j,j} + \frac{1}{2}\sum_{i,j\in\{l_1,l_2,\ldots,l_N\}=1}^{N}(w_{i,j,i,j} - w_{i,j,j,i}). \qquad (20.2.5)$$

As we can see, this result identifies with the expectation value of the N-electron Hamiltonian, with a single determinant state, as derived in Chapter 13 in the context of the Hartree–Fock approximation (see Eqs. (13.3.26–13.3.28) and Ex. 13.3.6). Notice that $w_{i,j,i,j}$ and $w_{i,j,j,i}$ are the Coulomb and exchange integrals, respectively.

Exercise 20.2.1 *(a) Using the binary string representation of a single determinant state (Eq. (20.1.7)), show that the expectation value of the second quantization Hamiltonian, Eqs. (20.2.3, 20.2.4), is given by Eq. (20.2.5). Compare the result to Ex. 13.3.6 for the energy expectation value of a single N-electron determinant. (b) Generalize the result of*

Ex. 13.3.6 for off-diagonal Hamiltonian matrix elements between different determinants,
$\left\langle \Psi^{(N)}_{\{l'_1,l'_2,l'_3,...,l'_N\}} \middle| \hat{H}^{(N)} \middle| \Psi^{(N)}_{\{l_1,l_2,l_3,...,l_N\}} \right\rangle$, *and show that the result coincides with the second*
quantization Hamiltonian matrix elements, $\left\langle \Psi^{(N)}_{\{l'_1,l'_2,...,l'_N\}} \middle| \hat{H} \middle| \Psi^{(N)}_{\{l_1,l_2,l_3,...,l_N\}} \right\rangle$.

In closed many-electron systems both the interaction $\hat{w}_{i,j}$ and the external potential (associated with $\hat{h}_i$) conserve the number of particles. Therefore, the Fock space
Hamiltonian must not couple between subspaces corresponding to different particle
numbers. To see that the Hamiltonian $\hat{H}$ in Eq. (20.2.3) complies with this requirement, we first notice that it commutes with the total electron number operator (see
Eq. (20.1.26) and Ex. 20.2.2),

$$[\hat{H}, \hat{N}] = 0. \tag{20.2.6}$$

Recalling that the basis vectors are eigenstates of the Hermitian number oper-
ator (Eq. (20.1.27)), we have $\left\langle \Psi^{(N')}_{\{l'_1,l'_2,...,l'_{N'}\}} \middle| \hat{H}\hat{N} \middle| \Psi^{(N)}_{\{l_1,l_2,...,l_N\}} \right\rangle = N \left\langle \Psi^{(N')}_{\{l'_1,l'_2,...,l'_{N'}\}} \middle| \right.$
$\left. \hat{H} \middle| \Psi^{(N)}_{\{l_1,l_2,...,l_N\}} \right\rangle$, and, $\left\langle \Psi^{(N')}_{\{l'_1,l'_2,...,l'_{N'}\}} \middle| \hat{N}\hat{H} \middle| \Psi^{(N)}_{\{l_1,l_2,...,l_N\}} \right\rangle = N' \left\langle \Psi^{(N')}_{\{l'_1,l'_2,...,l'_{N'}\}} \middle| \hat{H} \middle| \Psi^{(N)}_{\{l_1,l_2,...,l_N\}} \right\rangle$.
Eq. (20.2.6) therefore means that

$$\left\langle \Psi^{(N')}_{\{l'_1,l'_2...,l'_{N'}\}} \middle| \hat{H} \middle| \Psi^{(N)}_{\{l_1,l_2,...,l_N\}} \right\rangle = \left\langle \Psi^{(N')}_{\{l'_1,l'_2...,l'_{N'}\}} \middle| \hat{H}^{(N)} \middle| \Psi^{(N)}_{\{l_1,l_2,...,l_N\}} \right\rangle \delta_{N,N'}. \tag{20.2.7}$$

The conservation of particle number is therefore reflected in a block-diagonal structure
of the Fock space Hamiltonian.

Exercise 20.2.2 *(a) Use the anti-commutation relations for fermion creation and annihilation operators (Eqs. (20.1.19, 20.1.20)) and the definition of the electron number
operator (Eq. (20.1.26)) to show that* $[\hat{a}^\dagger_j, \hat{N}] = -\hat{a}^\dagger_j$, *and* $[\hat{a}_j, \hat{N}] = \hat{a}_j$. *(b) Use the
general operator identity,* $[\hat{A}\hat{B}, \hat{C}] = \hat{A}[\hat{B}, \hat{C}] + [\hat{A}, \hat{C}]\hat{B}$, *and the result of (a) to show that*
$[\hat{a}^\dagger_i \hat{a}_j, \hat{N}] = 0$. *(c) Use the results of (a) and (b) to show that the second quantization
Hamiltonian (Eq. (20.2.3)) commutes with the total electron number operator.*

Exercise 20.2.3 *Show that the second quantization Hamiltonian (Eq. (20.2.3)) for a
system of two orthonormal single-particle states ($|\Phi_1\rangle$ and $|\Phi_2\rangle$, selected as the eigen-
states of the single-particle Hamiltonian, $h_{i,j} = \varepsilon_j \delta_{i,j}$) reads $\hat{H} = \varepsilon_1 \hat{a}^\dagger_1 \hat{a}_1 + \varepsilon_2 \hat{a}^\dagger_2 \hat{a}_2 +
U \hat{a}^\dagger_1 \hat{a}_1 \hat{a}^\dagger_2 \hat{a}_2$, where $U = (w_{1,2,1,2} - w_{1,2,2,1})$.*

Notice that by selecting an infinite set of single-particle states, the Fock space
becomes a Hilbert space of infinite dimensions (see Section 11.2). Moreover, it is
often convenient to extend the Fock space by selecting a continuous, improper (see
Section 11.3) set of single-particle states. A commonly used choice associates the
single-particle states with "spin orbitals" (see Chapter 13), which are products of
eigenstates of the position and spin operators, $\{|\varphi_r\rangle \otimes |\sigma\rangle\}(\sigma \in (1/2, -1/2))$, with
the orthonormalization condition, $\langle \varphi_{r'}| \otimes \langle \sigma'| \cdot |\varphi_r\rangle \otimes |\sigma\rangle = \delta_{\sigma,\sigma'}\delta(\mathbf{r} - \mathbf{r}')$. Consider-
ing a generic orthonormal system of single electron states (spin orbitals) as discussed
above, $\{|\Phi_k\rangle\} \equiv \{|\varphi_k\rangle \otimes |\sigma_k\rangle\}$, and recalling the definition of a wave function in the
position representation, $\varphi_k(\mathbf{r}) = \langle \varphi_r|\varphi_k\rangle$ (Eq. (11.6.28)), we can use the expansion,

$|\varphi_{\mathbf{r}}\rangle \otimes |\sigma\rangle = \sum_{k=1}^{\infty} \delta_{\sigma_k,\sigma} \varphi_k^*(\mathbf{r})|\Phi_k\rangle$. Since an electron at a spin state $|\sigma\rangle$ in a position, $\mathbf{r}$, is represented as a linear combination of the generic set $\{|\Phi_k\rangle\}$, the corresponding electron creation operator is a linear combination of the creation operators within the set,

$$\hat{\psi}_{\mathbf{r},\sigma}^{\dagger} = \sum_{k=1}^{\infty} \delta_{\sigma_k,\sigma} \varphi_k^*(\mathbf{r}) \hat{a}_k^{\dagger}, \tag{20.2.8}$$

where it is emphasized that the index "k" corresponds to a specific spin-orbital. Similarly, the annihilation of an electron at spin state $|\sigma\rangle$ in a position, $\mathbf{r}$, is given by taking the Hermitian conjugate,

$$\hat{\psi}_{\mathbf{r},\sigma} = \sum_{k=1}^{\infty} \delta_{\sigma_k,\sigma} \varphi_k(\mathbf{r}) \hat{a}_k. \tag{20.2.9}$$

The operators $\hat{\psi}_{\mathbf{r},\sigma}^{\dagger}$ and $\hat{\psi}_{\mathbf{r},\sigma}$ are referred to as "field operators," which satisfy the following anti-commutation relations (Ex. 20.2.4):

$$\begin{aligned} \{\hat{\psi}_{\mathbf{r},\sigma}^{\dagger}, \hat{\psi}_{\mathbf{r}',\sigma'}\} &= \delta_{\sigma,\sigma'}\delta(\mathbf{r}-\mathbf{r}') \\ \{\hat{\psi}_{\mathbf{r},\sigma}^{\dagger}, \hat{\psi}_{\mathbf{r}',\sigma'}^{\dagger}\} &= 0 \\ \{\hat{\psi}_{\mathbf{r},\sigma}, \hat{\psi}_{\mathbf{r}',\sigma'}\} &= 0 \end{aligned} \tag{20.2.10}$$

The expression of the many-electron Hamiltonian in terms of the field operators reads (see Eq. (20.2.4) and Ex. 20.2.5)

$$\hat{H} = \sum_{\sigma} \int d\mathbf{r} \hat{\psi}_{\mathbf{r},\sigma}^{\dagger} \hat{h} \hat{\psi}_{\mathbf{r},\sigma} + \frac{1}{2} \sum_{\sigma,\sigma'} \int d\mathbf{r} \int d\mathbf{r}' \psi_{\mathbf{r},\sigma}^{\dagger} \hat{\psi}_{\mathbf{r}',\sigma'}^{\dagger} \hat{w} \hat{\psi}_{\mathbf{r}',\sigma'} \hat{\psi}_{\mathbf{r},\sigma}. \tag{20.2.11}$$

Exercise 20.2.4 *Use the anti-commutation relations for fermion creation and annihilation operators (Eqs. (20.1.19, 20.1.20)), and the definition of the field operators (Eqs. (20.2.8, 20.2.9)) to derive Eq. (20.2.10). Recall the formal definition of Dirac's delta in terms of a complete orthonormal set, Eq. (11.3.12).*

Exercise 20.2.5 *Accounting explicitly for the spin, σ_k, associated with each kth single particle state, the second quantization Hamiltonian (Eq. (20.2.3)) reads $\hat{H} = \sum_{k,k'=1}^{\infty} \delta_{\sigma_{k'},\sigma_k} h_{k,k'} \hat{a}_k^{\dagger} \hat{a}_{k'} + \frac{1}{2} \sum_{i,j,k,l=1}^{\infty} \delta_{\sigma_l,\sigma_i} \delta_{\sigma_k,\sigma_j} w_{i,j,k,l} \hat{a}_i^{\dagger} \hat{a}_j^{\dagger} \hat{a}_l \hat{a}_k$. Use the definitions of the field operators to derive this result from Eq. (20.2.11).*

20.3 Fermi–Dirac Distribution

Of special interest are open systems of many identical fermions, in the absence of interparticle interaction (a "Fermi gas"). This is a commonly used idealized (mean-field based) model that approximates to some extent single-electron observables in bulk metals. Therefore, it is useful for modeling an "electron reservoir." Here we focus on the equilibrium state for such a reservoir, and in the following sections we shall discuss exchange of charge carriers between a macroscopic fermion reservoir (e.g., an electrode surface) and a smaller system of interest coupled to it (e.g., an impurity such as an atom, molecule, or a quantum dot, adsorbed on that surface).

Let us choose an orthonormal set of single-particle states, as the eigenstates of the underlying singe-particle Hamiltonian, defined by the equation

$$\hat{h}|\Phi_k\rangle = \varepsilon_k|\Phi_k\rangle. \tag{20.3.1}$$

Each $|\Phi_k\rangle$ corresponds to some spatial wave function and a spin state. The corresponding fermion number operator reads in this case

$$\hat{N} = \sum_k \hat{a}_k^\dagger \hat{a}_k, \tag{20.3.2}$$

and, in the absence of interactions, the Fock space Hamiltonian reads (Eqs. (20.2.3, 20.2.4))

$$\hat{H} = \sum_k \varepsilon_k \hat{a}_k^\dagger \hat{a}_k. \tag{20.3.3}$$

In Chapter 16 we encountered the generic form of the equilibrium density operator for a grand canonical ensemble, namely for an open many-particle system, with fixed averages of energy and particle number per state (Eq. (16.5.25)). Using Eqs. (20.3.2, 20.3.3), the equilibrium density operator for a system of noninteracting fermions reads

$$\hat{\rho}^{(eq)} = \frac{e^{\frac{-1}{k_B T}\sum_k(\varepsilon_k-\mu)\hat{a}_k^\dagger \hat{a}_k}}{tr\{e^{\frac{-1}{k_B T}\sum_k(\varepsilon_k-\mu)\hat{a}_k^\dagger \hat{a}_k}\}}, \tag{20.3.4}$$

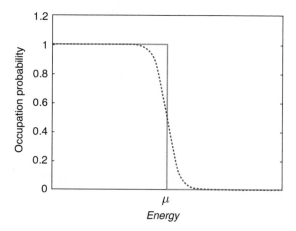

The Fermi–Dirac distribution function, representing the occupation probability per single-particle state in a system of noninteracting fermions at thermal equilibrium, as a function of energy. The sharp step marks the position of the chemical potential and corresponds to the zero-temperature limit. The distribution is broadened at a finite temperature, as shown by the dotted curve.

where k_B is Boltzmann's constant, T is the absolute temperature, and μ is the chemical potential. Of particular importance is the average occupation number of each single-particle state at equilibrium, defined as $N_l = tr\{\hat{N}_l \hat{\rho}^{(eq)}\} = tr\{\hat{a}_l^\dagger \hat{a}_l \hat{\rho}^{(eq)}\}$. Using the explicit expression for $\hat{\rho}^{(eq)}$, we obtain (see Ex. 20.3.1)

$$N_l = \frac{1}{1 + e^{(\varepsilon_l - \mu)/(k_B T)}}. \tag{20.3.5}$$

This result is known as the Fermi–Dirac distribution (see Fig. 20.3.1). It defines occupation at each single-particle state (e.g., each spin orbital), according to the state energy (ε_l), and the macroscopic parameters corresponding to the reservoir temperature (T) and its chemical potential (μ). At zero temperature, all the single-particle states whose energy is less than μ are fully occupied, whereas states at energy greater than μ are empty. (This reproduces Pauli's exclusion principle and the Aufbau principle, discussed in Section 13.3 in the context of the electronic ground state of N-electron systems.) At finite temperatures, the equilibrium distribution broadens, where there is a finite probability to find electrons at energies greater than the chemical potential, as well as electron vacancies (holes) at energies less than μ. Referring to the limit of a continuous (and nondegenerate) spectrum of the reservoir, the Fermi–Dirac distribution translates into continuous functions of the energy

$$f_e(\varepsilon) = \frac{1}{1 + e^{(\varepsilon - \mu)/(k_B T)}} \quad ; \quad f_h(\varepsilon) = 1 - f_e(\varepsilon), \tag{20.3.6}$$

where $f_e(\varepsilon)$ and $f_h(\varepsilon)$ are the probability densities of finding an electron or a hole, respectively, at the energy ε within the reservoir.

Exercise 20.3.1 *(a) Recalling that the trace of a tensor product of operators is a product of their traces, $tr\{\hat{A}_1 \otimes \hat{A}_2 \otimes \cdots \otimes \hat{A}_N\} = tr\{\hat{A}_1\} \cdot tr\{\hat{A}_2\} \dots tr\{\hat{A}_N\}$ (Ex. 15.5.1), use the commutativity of the number operators associated with the single-particle states to show that $tr\{\hat{N}_l \hat{\rho}^{(eq)}\} = tr_l\{\hat{a}_l^\dagger \hat{a}_l e^{\frac{-1}{k_B T}(\varepsilon_l - \mu)\hat{a}_l^\dagger \hat{a}_l}\}/tr_l\{e^{\frac{-1}{k_B T}(\varepsilon_l - \mu)\hat{a}_l^\dagger \hat{a}_l}\}$. (b) Calculate explicitly the trace in the subspace of the lth single-particle state to derive Eq. (20.3.5).*

20.4 Impurity Models

In many cases our focus of interest is a "small" system of fermions, which can exchange particles with a "large" fermion reservoir. A typical example is a molecule, adsorbed on the surface of a single crystal of bulk material. The interaction with the surface may lead to charging/discharging of the molecule, and consequently, to affecting its stable conformation, its binding energy to the surface, and even its chemical stability, inducing a chemical reaction. (This is the underlying principle of heterogeneous catalysis.) The system and the reservoir are different in many ways. For starters, while the system length scale is on the order of a nanometer, the bulk to which it is coupled may be macroscopically large. Additionally, the system is associated with a unique composition (e.g., specific atoms), whereas the bulk contains numerous replicas of identical unit cells. Finally, we are naturally interested in the effect of the reservoir on the small system observables, where the system needs to be modeled in some detail. The complementary effect of the system on the reservoir is often negligible, and in these cases, the fine details of the reservoir model are of less importance. The convenient partitioning of the entire system into a "small system of interest" and a reservoir is particularly instrumental when the coupling between these two subsystems is sufficiently weak, as discussed extensively in Chapter 19.

Focusing on exchange of particles between the system and the reservoir, it is natural to invoke some local basis for the single-particle Hilbert space, in which each basis state corresponds to a particle being either "inside" or "outside" the system. A commonly invoked approach relies on the tight-binding approximation in which the basis for the single-particle space is assumed to be both orthonormal and local (see Sections 14.4 and 14.5).

As a concrete example, let us consider an adsorbate on top of a bulk conductor (an "electrode"). A minimal tight binding model for the single-particle Hamiltonian [5.6] associates the adsorbate with a single terminal site, and the electrode with a semi-infinite uniform linear chain of sites (see Section 14.5 and Fig. 20.4.1). Invoking a basis of orthonormal localized states, $\langle \varphi_n | \varphi_{n'} \rangle = \delta_{n,n'}$, where $|\varphi_0\rangle$ and $\{|\varphi_1\rangle, |\varphi_2\rangle, |\varphi_3\rangle \ldots, |\varphi_{M_E}\rangle\}$ correspond to the adsorbate and conductor sites, respectively, the matrix representation of the single-particle Hamiltonian reads

$$\mathbf{h} = \begin{bmatrix} \varepsilon_0 & \gamma & & & 0 \\ \gamma & \mu & \beta & & \\ & \beta & \ddots & \ddots & \\ & & \ddots & \ddots & \beta \\ 0 & & & \beta & \mu \end{bmatrix}. \tag{20.4.1}$$

Here $\varepsilon_0 = \langle \varphi_0 | \hat{h} | \varphi_0 \rangle$ is the adsorbate on-site energy, $\gamma = \langle \varphi_0 | \hat{h} | \varphi_1 \rangle$ is the nearest-neighbor adsorbate–electrode coupling matrix element, $\beta = \langle \varphi_n | \hat{h} | \varphi_{n\pm1} \rangle$ is the coupling matrix element between neighboring electrode sites, and $\mu = \langle \varphi_n | \hat{h} | \varphi_n \rangle$ is the (constant) on-site energy at the electrode. (Notice that for a "half-filling model," in which the number of electrons present in the electrode is identical to the number of local sites, the on-site energy identifies with the chemical potential (see Ex. 20.4.1).)

The localized basis states enable us to conveniently divide the single-particle Hilbert space into the "adsorbate" and "electrode" subspaces, where the single-particle Hamiltonian, $\hat{h}$, is written as

$$\hat{h} \equiv \hat{h}_A + \hat{h}_E + \hat{h}_{A,E}. \tag{20.4.2}$$

Furthermore, it is convenient to transform into an orthonormal basis, in which $\hat{h}_A + \hat{h}_E$ is diagonal (see Fig. 20.4.1). Denoting this basis as $|\chi_0\rangle, |\chi_1\rangle, |\chi_2\rangle \ldots, |\chi_{M_E}\rangle$, where $\langle \chi_n | \chi_{n'} \rangle = \delta_{n,n'}$, the different parts of the Hamiltonian are expressed as

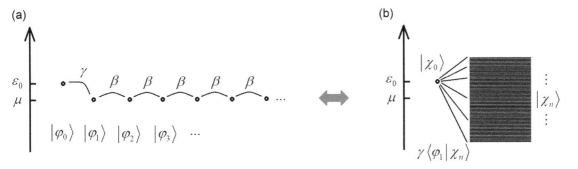

Figure 20.4.1   A schematic representation of a simple adsorbate–electrode model. The adsorbate is modeled as a single site at energy ε_0, coupled to the first site of a semi-infinite uniform chain (left plot), representing an "electrode." The chain is characterized by the on-site energy, μ, and the nearest-neighbor tunneling matrix element, β. (For a half filling model of noninteracting fermions in the electrode, the on-site energy coincides with the chemical potential.) The plot on the right represents the same model after block diagonalization of the electrode Hamiltonian, where the adsorbate is coupled to the band of delocalized electrode eigenstates, where the coupling is scaled by the projection of each delocalized eigenstate on the first ("surface") electrode site.

$$\hat{h}_E = \sum_{n,n'=1}^{M_E} \langle\varphi_n|\hat{h}|\varphi_{n'}\rangle |\varphi_n\rangle\langle\varphi_{n'}| = \sum_{n=1}^{M_E} \varepsilon_n |\chi_n\rangle\langle\chi_n|$$

$$\hat{h}_A = \langle\varphi_0|\hat{h}|\varphi_0\rangle |\varphi_0\rangle\langle\varphi_0| = \varepsilon_0 |\chi_0\rangle\langle\chi_0| \qquad (20.4.3)$$

$$\hat{h}_{A,E} = |\varphi_0\rangle\gamma\langle\varphi_1| + h.c. = \sum_{n=1}^{M_E} \gamma_n |\chi_0\rangle\langle\chi_n| + h.c. \quad ; \quad \gamma_n = \gamma\langle\varphi_1|\chi_n\rangle.$$

The eigenvalues and eigenvectors corresponding to the linear uniform chain model $(n = 1, 2, \ldots)$ are (see Eqs. (14.4.25, 14.4.26))

$$\varepsilon_n = \mu + 2\beta \cos\left(\frac{n\pi}{M_E + 1}\right) \quad ; \quad |\chi_n\rangle = \sum_{j=1}^{M_E} \sqrt{\frac{2}{M_E + 1}} \sin\left(\frac{n\pi j}{M_E + 1}\right) |\varphi_j\rangle, \qquad (20.4.4)$$

where each eigenvector is a linear combination of the localized conductor states. Notice that while these eigenvectors are delocalized over the electrode sites, the orthonormal basis $|\chi_0\rangle, |\chi_1\rangle, |\chi_2\rangle \ldots, |\chi_{M_E}\rangle$ is still "local" in the sense that a part of it ($|\chi_0\rangle$) is associated with the adsorbate site, and the other part ($|\chi_1\rangle, |\chi_2\rangle \ldots, |\chi_{M_E}\rangle$) is associated with the electrode (conductor) sites. (This locality reflects the underlying tight-binding model assumptions.)

Exercise 20.4.1   *The "electrode" part in the model depicted in Eq. (20.4.1) consists of a uniform linear tight-binding chain of M_E sites at the on-site energy, μ, with the nearest-neighbor coupling matrix elements, $\beta = -|\beta|$. The eigenvalues and eigenvectors of the corresponding model Hamiltonian were first introduced in Eqs. (14.4.25, 14.4.26) and are quoted in Eq. (20.4.4). Let us consider a "half-filling model" where the system is populated by noninteracting electrons whose number equals the number of electrode sites, M_E. (a) Considering the Pauli exclusion and the Aufbau principles (Chapter 13), show that for an even M_E, the energies of the highest occupied and lowest unoccupied eigenvectors of the chain Hamiltonian at zero temperature are, respectively,*

$\varepsilon = \mu + 2\beta \cos\left[\frac{\pi}{2}\left(1 - \frac{1}{M_E+1}\right)\right]$ and $\varepsilon = \mu + 2\beta \cos\left[\frac{\pi}{2}\left(1 + \frac{1}{M_E+1}\right)\right]$. (b) Show that for an infinite chain length, $M_E \to \infty$, these two energies coincide to the same value (the chemical potential of the many-electron system), which is equal to the on-site energy, μ. (c) Show that the energies of the highest occupied and lowest unoccupied eigenvectors of the chain Hamiltonian have the same value also for an odd M_E.

To account for many electrons in the adsorbate–electrode system, we can pair each spatial state with a spin state, $|\chi_n\rangle \otimes |\sigma\rangle$, $\sigma \in (1/2, -1/2)$, and construct a Fock space based on the orthonormal set of single-particle states. The many-electron Hamiltonian in this space reads (Eq. (20.2.3))

$$\hat{H} = \sum_{\sigma=-1/2}^{1/2} \sum_{i,j=0}^{M_E} h_{i,j} \hat{a}_{i,\sigma}^\dagger \hat{a}_{j,\sigma} + \frac{1}{2} \sum_{\sigma,\sigma'=-1/2}^{1/2} \sum_{i,j,k,l=0}^{M_E} w_{i,j,k,l} \hat{a}_{i,\sigma}^\dagger \hat{a}_{j,\sigma'}^\dagger \hat{a}_{l,\sigma'} \hat{a}_{k,\sigma}, \quad (20.4.5)$$

where, $h_{i,j} = \langle \chi_i | \hat{h} | \chi_j \rangle$ is the single-particle matrix element, and the electron-pair interactions are encoded in the scalars, $\{w_{i,j,k,l}\}$. It is customary to invoke a model of noninteracting fermions for the conductor, which means that the interaction is restricted to the adsorbate subspace; namely, $w_{i,j,k,l}$ is assumed to vanish unless $i = j = k = l = 0$. Denoting the "on-site" Coulomb repulsion at the adsorbate (often referred to as the Hubbard interaction parameter [20.1]) as

$$w_{0,0,0,0} = \langle \chi_0 | \otimes \langle \chi_0 | \hat{w} | \chi_0 \rangle \otimes | \chi_0 \rangle = Ke^2 \int d\mathbf{r} \int d\mathbf{r}' \frac{|\chi_0(\mathbf{r})|^2 |\chi_0(\mathbf{r}')|^2}{|\mathbf{r} - \mathbf{r}'|} \equiv U, \quad (20.4.6)$$

and using the explicit form of the single-particle Hamiltonian (Eqs. (20.4.2, 20.4.3)), the many-particle Hamiltonian, Eq. (20.4.5), reads (Ex. 20.4.2)

$$\hat{H} = \varepsilon_0 \hat{a}_{0,1/2}^\dagger \hat{a}_{0,1/2} + \varepsilon_0 \hat{a}_{0,-1/2}^\dagger \hat{a}_{0,-1/2} + U \hat{a}_{0,1/2}^\dagger \hat{a}_{0,1/2} \hat{a}_{0,-1/2}^\dagger \hat{a}_{0,-1/2}$$
$$+ \sum_{\sigma=-1/2}^{1/2} \sum_{i=1}^{M_E} [\varepsilon_i \hat{a}_{i,\sigma}^\dagger \hat{a}_{i,\sigma} + (\gamma_i \hat{a}_{0,\sigma}^\dagger \hat{a}_{i,\sigma} + h.c.)]. \quad (20.4.7)$$

Exercise 20.4.2 *Using the explicit form of the single-particle Hamiltonian, Eqs. (20.4.2, 20.4.3), in Eq. (20.4.5) and restricting the electron–electron interaction to the adsorbate space (Eq. (20.4.6)), derive Eq. (20.4.7).*

The many-particle adsorbate–electrode model (Eq. (20.4.7)) is identical in its structure to the Anderson impurity model [20.2] that was introduced for treating a magnetic impurity in a metal. Indeed, in both cases, a localized impurity state that can accommodate two electrons at opposite spin states is coupled to a reservoir of noninteracting electrons. This model can be readily generalized to account for impurities with many single-particle states by replacing the single-particle adsorbate Hamiltonian ($\hat{h}_A$ in Eqs. (20.4.2, 20.4.3)) with a tight binding model Hamiltonian for M_A-coupled adsorbate sites, that is,

$$\hat{h}_A \equiv \sum_{m_A, m'_A=1}^{M_A} h_{m_A, m'_A} |\varphi_{m_A}\rangle \langle \varphi_{m'_A}| = \sum_{m_A=1}^{M_A} \varepsilon_{m_A} |\chi_{m_A}\rangle \langle \chi_{m_A}|. \quad (20.4.8)$$

Invoking the electrode model, $\hat{h}_E$ (see Eq. (20.4.3)), and introducing nearest-neighbor coupling between the terminal site of the impurity (e.g., $|\varphi_{1_A}\rangle$) and the terminal site of the electrode (e.g., $|\varphi_{1_E}\rangle$), the generalized adsorbate–electrode model Hamiltonian reads

$$\hat{h} \equiv \hat{h}_A + \hat{h}_E + \hat{h}_{A,E}$$

$$\hat{h}_E = \sum_{m_E, m'_E = 1}^{M_E} \langle \varphi_{m_E} | \hat{h} | \varphi_{m'_E} \rangle | \varphi_{m_E} \rangle \langle \varphi_{m'_E} | = \sum_{m_E = 1}^{M_E} \varepsilon_{m_E} | \chi_{m_E} \rangle \langle \chi_{m_E} |. \qquad (20.4.9)$$

$$\hat{h}_{A,E} = |\varphi_{1_A}\rangle \gamma \langle \varphi_{1_E}| + h.c. = \sum_{m_E = 1}^{M_E} \sum_{m_A = 1}^{M_A} \gamma \langle \chi_{m_A} | \varphi_{1_A} \rangle \langle \varphi_{1_E} | \chi_{m_E} \rangle | \chi_{m_A} \rangle \langle \chi_{m_E} | + h.c.$$

Within the tight binding assumptions, the eigenstates of the uncoupled adsorbate and electrode Hamiltonians ($\{|\chi_{m_A}\rangle\}$ and $\{|\chi_{m_E}\rangle\}$, respectively, are an orthonormal basis for the spatial single-particle Hilbert space, $\langle \chi_m | \chi_{m'} \rangle = \delta_{m,m'}$. As in the previous section, we invoke a composite index to account also for the spin degree of freedom, $\{|\Phi_m\rangle\} = \{|\chi_m\rangle \otimes |\sigma_m\rangle\}$. Using the orthonormal basis of the single-particle states (spin orbitals), $\langle \Phi_m | \Phi_{m'} \rangle = \delta_{m,m'}$, the corresponding many-particle impurity model Hamiltonian obtains the generic form

$$\hat{H} = \hat{H}_A + \hat{H}_E + \hat{H}_{A,E}, \qquad (20.4.10)$$

$$\hat{H}_A = \sum_{m_A = 1}^{M_A} \varepsilon_{m_A} \hat{a}^\dagger_{m_A} \hat{a}_{m_A} + \sum_{i,j,k,l \in \{m_A\}} w_{i,j,k,l} \hat{a}^\dagger_i \hat{a}^\dagger_j \hat{a}_l \hat{a}_k$$

$$\hat{H}_E = \sum_{m_E = 1}^{M_E} \varepsilon_{m_E} \hat{a}^\dagger_{m_E} \hat{a}_{m_E} \qquad (20.4.11)$$

$$\hat{H}_{A,E} = \sum_{m_A = 1}^{M_A} \sum_{m_E = 1}^{M_E} v_{m_A} \gamma_{m_E} \hat{a}^\dagger_{m_A} \hat{a}_{m_E} + h.c.,$$

where we denoted $v_{m_A} = \langle \chi_{m_A} | \varphi_{1_A} \rangle$ and $\gamma_{m_E} = \gamma \langle \varphi_{1_E} | \chi_{m_E} \rangle$.

Notice that the many-particle Hamiltonian, $\hat{H}$, is naturally cast in a "system–bath" form (see Eq. (19.3.3)). The impurity ($\hat{H}_A$) corresponds to the "small system of interest," which can accommodate a finite number of electrons, whereas the Hamiltonian $\hat{H}_E$ corresponds to a "bath," or an electron reservoir, which, for $M_E \to \infty$, can accommodate an infinite number of electrons. The system–bath interaction operator, $\hat{H}_{A,E}$, corresponds to electron transitions between the system and the bath subspaces, namely, annihilating an electron in a state confined within the system and creating an electron at a reservoir state, or vice versa. Within the tight binding model, the system–bath interaction depends on the intersite coupling matrix elements at the impurity–reservoir interface (e.g., the parameters γ_{m_E} and v_{m_A} depend on the projections of the reservoir and impurity Hamiltonian eigenstates on the terminal sites). When these coupling matrix elements are significantly smaller than the intersite coupling matrix elements within each subspace, the system–bath interaction can be treated perturbatively. The reduced system dynamics, and particularly the exchange of electrons between the impurity and the reservoir can then be described within the Born–Markov approximation, as introduced in Chapter 19.

20.5 Charge Exchange with a Fermion Reservoir

Reduced System Dynamics

Let us start by identifying the adsorbate with the system, and the electrode with a bath, rewriting the generalized impurity Hamiltonian, Eqs. (20.4.10, 20.4.11), as a standard system–bath Hamiltonian (see Eqs. (19.3.3, 19.3.12)):

$$\hat{H} = \hat{H}_S + \hat{H}_B + \hat{H}_{SB} \quad ; \quad [\hat{H}_S, \hat{H}_B] = 0$$

$$\hat{H}_S = \sum_{m=1}^{M_A} \varepsilon_m \hat{a}_m^\dagger \hat{a}_m + \sum_{i,j,k,l \in \{m\}} w_{i,j,k,l} \hat{a}_i^\dagger \hat{a}_j^\dagger \hat{a}_l \hat{a}_k$$

$$\hat{H}_B = \sum_{k=1}^{M_E} \varepsilon_k \hat{a}_k^\dagger \hat{a}_k \qquad\qquad\qquad\qquad\qquad (20.5.1)$$

$$\hat{H}_{SB} = \hat{V}_S \hat{U}_B + \hat{V}_S^\dagger \hat{U}_B^\dagger \quad ; \quad \hat{V}_S \equiv \sum_{m=1}^{M_A} v_m \hat{a}_m^\dagger \quad ; \quad \hat{U}_B \equiv \sum_{k=1}^{M_E} \gamma_k \hat{a}_k$$

Our prime interest is in the system dynamics in the presence of its coupling to the bath. For this purpose, it is sufficient to follow the time evolution of the reduced system density operator under the full Hamiltonian. In the limit of weak coupling to the bath, the Nakajima and Zwanzig projection scheme and the Born–Markov approximation can provide a reliable description of the reduced dynamics, as detailed in Sections 19.3 and 19.4 for the generic system–bath Hamiltonian. We can therefore implement this general formulation to the adsorbate–electrode Hamiltonian, emphasizing some special characteristics of fermion reservoirs.

The initial density operator is taken to be a product state, where the system density operator depends on its transient preparation, and the state of the macroscopically large reservoir is assumed to be at near equilibrium, corresponding to a grand canonical ensemble (Eq. (20.3.4)), namely

$$\hat{\rho}(0) = \hat{\rho}_B \otimes \hat{\rho}_S(0) \quad ; \quad \hat{\rho}_B = \frac{e^{\frac{-1}{k_B T} \sum_k (\varepsilon_k - \mu) \hat{a}_k^\dagger \hat{a}_k}}{tr\{e^{\frac{-1}{k_B T} \sum_k (\varepsilon_k - \mu) \hat{a}_k^\dagger \hat{a}_k}\}}. \qquad (20.5.2)$$

As we can readily verify, the bath density fulfils the conditions, $tr_B\{\hat{\rho}_B\} = 1$ and $[\hat{\rho}_B, \hat{H}_B] = 0$ (Eq. (19.3.6)). Additionally, the system–bath coupling is of the generic form of a sum of products of system and bath operators, namely $\hat{H}_{SB} \equiv \sum_\alpha \hat{V}_\alpha^{(S)} \hat{U}_\alpha^{(B)}$ (Eq. (19.3.12)), where the bath coupling operators are orthogonal to the bath equilibrium density (Eq. (19.3.13)), $tr_B\{\hat{U}_B \hat{\rho}_B\} = tr_B\{\hat{U}_B^\dagger \hat{\rho}_B\} = 0$ (see Ex. 20.5.1). Therefore, in the weak system–bath coupling limit, the reduced system density, $\hat{\rho}_S(t) = tr_B\{\hat{\rho}(t)\}$, satisfies the general Born–Markov (Redfield) equation (Eqs. (19.3.19–19.3.22)), where the influence of the bath is captured in terms of its coupling correlation functions (Eq. (19.3.21)). For the present system–bath model Hamiltonian (Eq. (20.5.1)), the Redfield equation obtains the form (see Ex. 20.5.2)

$$
\frac{\partial}{\partial t}\hat{\rho}_S(t) \cong -\frac{i}{\hbar}[\hat{H}_S, \hat{\rho}_S(t)]
$$
$$
-\frac{1}{\hbar^2}\int_0^t d\tau \{c_e(\tau)\left[\hat{V}_S, e^{\frac{-i\tau}{\hbar}\hat{H}_S}\hat{V}_S^\dagger e^{\frac{i\tau}{\hbar}\hat{H}_S}\hat{\rho}_S(t)\right]
$$
$$
+c_a(\tau)[\hat{V}_S^\dagger, e^{\frac{-i\tau}{\hbar}\hat{H}_S}\hat{V}_S e^{\frac{i\tau}{\hbar}\hat{H}_S}\hat{\rho}_S(t)] + h.c.\}, \qquad (20.5.3)
$$

where $c_a(\tau)$ and $c_e(\tau)$ are the fermionic coupling correlation functions. These are related to the equilibrium Fermi distribution functions (Eq. (20.3.6)) for electrons and holes in the reservoir (see Ex. 20.5.2):

$$
\begin{aligned}
c_a(\tau) &= tr_B\{\hat{U}_B^\dagger e^{\frac{-i\tau}{\hbar}\hat{H}_B}\hat{U}_B e^{\frac{i\tau}{\hbar}\hat{H}_B}\hat{\rho}_B\} = \sum_{k=1}^{M_E}|\gamma_k|^2 e^{\frac{i\tau}{\hbar}\varepsilon_k}f_e(\varepsilon_k) \\
c_e(\tau) &= tr_B\{\hat{U}_B e^{\frac{-i\tau}{\hbar}\hat{H}_B}\hat{U}_B^\dagger e^{\frac{i\tau}{\hbar}\hat{H}_B}\hat{\rho}_B\} = \sum_{k=1}^{M_E}|\gamma_k|^2 e^{\frac{-i\tau}{\hbar}\varepsilon_k}f_h(\varepsilon_k)
\end{aligned} \qquad (20.5.4)
$$

Exercise 20.5.1 *(a) Use the anti-commutation relation between the fermionic annihilation and creation operators, $\{\hat{a}_l^\dagger, \hat{a}_{l'}^\dagger\} = 0$, $\{\hat{a}_l, \hat{a}_{l'}\} = 0$, $\{\hat{a}_l, \hat{a}_{l'}^\dagger\} = \delta_{l,l'}$, (Eqs. (20.1.19, 20.1.20)), to show that the traces over a single-orbital Fock space, $tr_k\{\hat{a}_k f(\hat{a}_k^\dagger \hat{a}_k)\}$ and $tr_k\{\hat{a}_k^\dagger f(\hat{a}_k^\dagger \hat{a}_k)\}$, vanish for any analytic function, $f(\hat{A}) = \sum_{n=0}^{\infty} f_n\hat{A}^n$. (b) Given the definition of the fermion bath Hamiltonian and coupling operators (Eq. (20.5.1) with $\hat{U}_B \equiv \sum_{k=1}^{M_E}\gamma_k\hat{a}_k$), and the bath density operator (Eq. (20.5.2)), use the result of (a) to show that $tr_B\{\hat{U}_B\hat{\rho}_B\} = tr_B\{\hat{U}_B^\dagger\hat{\rho}_B\} = 0$.*

Exercise 20.5.2 *(a) Use the explicit expressions, $\hat{U}_B \equiv \sum_{k=1}^{M_E}\gamma_k\hat{a}_k$, $\hat{H}_B = \sum_{k=1}^{M_E}\varepsilon_k\hat{a}_k^\dagger\hat{a}_k$, $\hat{\rho}_B = e^{\frac{-1}{k_BT}\sum_k(\varepsilon_k-\mu)\hat{a}_k^\dagger\hat{a}_k}/tr\left\{e^{\frac{-1}{k_BT}\sum_k(\varepsilon_k-\mu)\hat{a}_k^\dagger\hat{a}_k}\right\}$, and the fermionic anti-commutation relations to show that the bath correlation functions, $c_e(\tau) = tr_B\{\hat{U}_B e^{\frac{-i\tau}{\hbar}\hat{H}_B}\hat{U}_B^\dagger e^{\frac{i\tau}{\hbar}\hat{H}_B}\hat{\rho}_B\}$ and $c_a(\tau) = tr_B\{\hat{U}_B^\dagger e^{\frac{-i\tau}{\hbar}\hat{H}_B}\hat{U}_B e^{\frac{i\tau}{\hbar}\hat{H}_B}\hat{\rho}_B\}$, read $c_e(\tau) = \sum_{k=1}^{M_E}|\gamma_k|^2 e^{\frac{-i\tau}{\hbar}\varepsilon_k}\left[1 - \frac{1}{1+e^{\frac{1}{k_BT}(\varepsilon_k-\mu)}}\right]$ and $c_a(\tau) = \sum_{k=1}^{M_E}|\gamma_k|^2 e^{\frac{i\tau}{\hbar}\varepsilon_k}\frac{1}{1+e^{\frac{1}{k_BT}(\varepsilon_k-\mu)}}$.*

  (b) For a general system–bath coupling operator, $\hat{H}_{SB} \equiv \sum_\alpha \hat{V}_\alpha^{(S)}\hat{U}_\alpha^{(B)}$ (Eq. (19.3.12)), the Redfield (Born–Markov) dissipator obtains the form of Eqs. (19.3.21, 19.3.22),
$$
\hat{D}\hat{\rho}_S(t) = -\frac{1}{\hbar^2}\sum_{\alpha,\alpha'}\int_0^t d\tau \left\{c_{\alpha,\alpha'}(\tau)\left[\hat{V}_\alpha^{(S)}, e^{\frac{-i\tau}{\hbar}\hat{H}_S}\hat{V}_{\alpha'}^{(S)}e^{\frac{i\tau}{\hbar}\hat{H}_S}\hat{\rho}_S(t)\right] + \overline{c}_{\alpha',\alpha}(\tau)\left[\hat{\rho}_S(t)e^{\frac{-i\tau}{\hbar}\hat{H}_S}\hat{V}_{\alpha'}^{(S)}\right.\right.
$$
$$
\left.\left. e^{\frac{i\tau}{\hbar}\hat{H}_S}, \hat{V}_\alpha^{(S)}\right]\right\}, \text{ where } c_{\alpha,\alpha'}(\tau) \equiv tr_B\{\hat{U}_\alpha^{(B)}e^{\frac{-i\tau}{\hbar}\hat{H}_B}\hat{U}_{\alpha'}^{(B)}e^{\frac{i\tau}{\hbar}\hat{H}_B}\hat{\rho}_B\} \text{ and } \overline{c}_{\alpha,\alpha'}(\tau) = c_{\alpha,\alpha'}(-\tau).
$$

Map the coupling operator defined in Eq. (20.5.1), $\hat{H}_{SB} \equiv \hat{V}_S \hat{U}_B + \hat{V}_S^\dagger \hat{U}_B^\dagger$, on this general form by identifying $\hat{V}_S \equiv \hat{V}_1^{(S)}$, $\hat{U}_B \equiv \hat{U}_1^{(B)}$, $\hat{V}_S^\dagger \equiv \hat{V}_2^{(S)}$, $\hat{U}_B^\dagger \equiv \hat{U}_2^{(B)}$ to show that

$$\hat{D}\hat{\rho}_S(t) = -\frac{1}{\hbar^2} \int_0^t d\tau \left\{ c_{1,2}(\tau) \left[\hat{V}_S, e^{\frac{-i\tau}{\hbar}\hat{H}_S} \hat{V}_S^\dagger e^{\frac{i\tau}{\hbar}\hat{H}_S} \hat{\rho}_S(t) \right] + c_{2,1}^*(\tau) \left[\hat{\rho}_S(t) e^{\frac{-i\tau}{\hbar}\hat{H}_S} \hat{V}_S^\dagger e^{\frac{i\tau}{\hbar}\hat{H}_S}, \hat{V}_S \right] \right\}$$

$$- \frac{1}{\hbar^2} \int_0^t d\tau \left\{ c_{2,1}(\tau) \left[\hat{V}_S^\dagger, e^{\frac{-i\tau}{\hbar}\hat{H}_S} \hat{V}_S e^{\frac{i\tau}{\hbar}\hat{H}_S} \hat{\rho}_S(t) \right] + c_{1,2}^*(\tau) \left[\hat{\rho}_S(t) e^{\frac{-i\tau}{\hbar}\hat{H}_S} \hat{V}_S e^{\frac{i\tau}{\hbar}\hat{H}_S}, \hat{V}_S^\dagger \right] \right\}.$$

(c) Use the identifications, $c_{1,2}(\tau) = c_e(\tau)$, $c_{2,1}(\tau) = c_a(\tau)$, to derive Eq. (20.5.3).

For "macroscopically" large reservoirs, the spectrum of the single particle is dense; namely, the energy spacings between nearest levels are much smaller than the system–bath coupling matrix elements. It is instructive to replace the discrete model (e.g., Eq. (20.4.4)) by a continuous band (see Section 14.5). The summation over the system–bath coupling matrix elements squared $\{|\gamma_k|^2\}$ can then be replaced by an integral over a continuous function of the energy, $\gamma^2(\varepsilon)$, weighted by the electrode's density of states (see Section 5.5), where $|\gamma_k|^2 = \gamma^2(\varepsilon_k)$. Introducing the respective spectral density function (see also Eq. (17.2.10)),

$$J(\varepsilon) \equiv 2\pi\gamma^2(\varepsilon)\rho(\varepsilon), \tag{20.5.5}$$

the bath correlation functions are shown to be

$$\begin{aligned} c_a(\tau) &= \frac{1}{2\pi} \int d\varepsilon \, e^{i\tau\varepsilon/\hbar} J(\varepsilon) f_e(\varepsilon) \\ c_e(\tau) &= \frac{1}{2\pi} \int d\varepsilon \, e^{-i\tau\varepsilon/\hbar} J(\varepsilon) f_h(\varepsilon) \end{aligned}, \tag{20.5.6}$$

where the integral is over the electrode band energies. The discrete summations (Eq. (20.5.4)) are readily recovered for $\rho(\varepsilon) = \sum_k \delta(\varepsilon - \varepsilon_k)$.

As an example, let us consider the electrode model of a uniform linear chain with M_E sites (Eqs. (20.4.3, 20.4.4) or Eq. (20.4.11)), in which the coupling of the chain to the system is restricted to its terminal site via a tunneling matrix element, γ. In this case, we obtain $|\gamma_k|^2 = 2|\gamma|^2 \left[1 - \frac{(\varepsilon_k - \mu)^2}{4\beta^2} \right] / (M_E + 1)$, where for $M_E \gg 1$, the density of states can be approximated as, $\rho(\varepsilon_k) \approx (M_E + 1)/(\pi\sqrt{4\beta^2 - (\varepsilon_k - \mu)^2})$ (see Ex. 20.5.3). Consequently, in the limit of a semi-infinite linear chain ($M_E \to \infty$), the spectral density (Eq. (20.5.5)) corresponds to a semielliptical band model [20.3] (see Ex. 20.5.3),

$$J(\varepsilon) = \frac{|\gamma|^2}{\beta^2} \sqrt{4\beta^2 - (\varepsilon - \mu)^2}. \tag{20.5.7}$$

In the case of "half-filling" of the chain, the on-site single-particle energy, μ, identifies with the chemical potential of the many-electron system (Ex. 20.4.1), which means that the spectral density obtains its maximal value at the chemical potential.

Exercise 20.5.3 *Consider the adsorbate–electrode model, characterized by the single-particle Hamiltonian in Eqs. (20.4.2, 20.4.3). The eigenvalues and eigenvectors of the* *single-particle electrode Hamiltonian (corresponding to a linear uniform chain) are given by Eq. (20.4.4) (see also Eqs. (14.4.25, 14.4.26)) where the adsorbate–electrode coupling is restricted to the terminal electrode site and depends on the projections of the chain eigenvectors ($\{|\chi_k\rangle\}$) on the first electrode site, namely $\gamma_k = \gamma\langle\varphi_1|\chi_k\rangle$ (see Eq. (20.4.3)). Show that: (a) $|\gamma_k|^2 = 2|\gamma|^2 \left[1 - \frac{(\varepsilon_k-\mu)^2}{4\beta^2}\right]/(M_E + 1)$. (b) The density of states for the linear chain model reads $\rho(\varepsilon_k) \equiv \left.\frac{\partial n(\varepsilon)}{\partial\varepsilon}\right|_{\varepsilon=\varepsilon_k} \approx (M_E + 1)/(\pi\sqrt{4\beta^2 - (\varepsilon_k - \mu)^2})$. (c) The respective spectral density of the adsorbate–electrode interaction (Eq. (20.5.5)) is given by Eq. (20.5.7).*

The bath correlation functions have the generic form of sums, Eq. (20.5.4) (or integrals, Eq. (20.5.6)), over oscillating waves in time, at frequencies corresponding to the single-particle energy levels at the electrode band $\{(\varepsilon_k/\hbar)\}$, weighted by the pre-factors, $|\gamma_k|^2 f_{e/h}(\varepsilon_k)$. For a dense and wide frequency band the net result is a decaying function in time (see, e.g., the discussion of the generic Fermi's golden rule in Section 17.2, and Fig. 17.2.2), where the time decay becomes more rapid as the bandwidth increases. The effective bandwidth is determined by the width of the spectral density function, $J(\varepsilon)$ (or the distribution of the system–bath tunneling matrix elements, $\{|\gamma_k|^2\}$), as well as by the effective energy width of the relevant Fermi function ($f_e(\varepsilon)$ or $f_h(\varepsilon)$). The width of the spectral density increases as the fastest time period of the bath dynamics increases (e.g., for the semi-infinite linear chain model, Eq. (20.4.4), the bandwidth equals $|4\beta|$, which corresponds to the tunneling frequency between neighboring sites, $2|\beta|/\hbar$). The width of the Fermi distribution increases with increasing bath temperature (see Fig. 20.3.1). When the multiplicity, $J(\varepsilon)f_{e/h}(\varepsilon)$, is a sufficiently broad function of the energy, the relaxation time of the bath correlations can become faster than any significant change in the reduced density matrix according to Eq. (20.5.3). When this holds, the upper limit in the time integral in Eq. (20.5.3) can be taken to infinity, which yields the stationary Born–Markov approximation. (See a detailed analysis in Section 17.2 as well as Eqs. (19.4.1–19.4.3), and a similar consideration for a bosonic bath, Eqs. (19.5.6–19.5.8)):

$$
\begin{aligned}
\frac{\partial}{\partial t}\hat{\rho}_S(t) &\cong -\frac{i}{\hbar}[\hat{H}_S, \hat{\rho}_S(t)] \\
&- \frac{1}{\hbar^2}\int_0^\infty d\tau\{c_e(\tau)\left[\hat{V}_S, e^{\frac{-i\tau}{\hbar}\hat{H}_S}\hat{V}_S^\dagger e^{\frac{i\tau}{\hbar}\hat{H}_S}\hat{\rho}_S(t)\right] \\
&+ c_a(\tau)\left[\hat{V}_S^\dagger, e^{\frac{-i\tau}{\hbar}\hat{H}_S}\hat{V}_S e^{\frac{i\tau}{\hbar}\hat{H}_S}\hat{\rho}_S(t)\right] + h.c.\}.
\end{aligned}
\tag{20.5.8}
$$

The state of the adsorbate system is fully captured in the reduced density matrix in the basis of eigenstates of the reduced system Hamiltonian, defined as

$$
\hat{H}_S|\Psi_m\rangle = E_m|\Psi_m\rangle.
\tag{20.5.9}
$$

In the presence of electron–electron interaction, each eigenstate is a specific superposition of many-electron occupation vectors (determinants). Nevertheless, recalling that the Hamiltonian commutes with the electron number operator, the matrix representation of the Hamiltonian in Fock space is block diagonal (see Eqs. (20.2.6, 20.2.7)), and each eigenstate of $\hat{H}_S$ is also an eigenstate of the corresponding electron number operator,

$$\hat{N}_S|\Psi_m\rangle = N_m|\Psi_m\rangle. \tag{20.5.10}$$

Introducing the eigenstates of $\hat{H}_S$ as a complete orthonormal basis for the space of the reduced system, where $[\hat{O}_S]_{m,m'} = \langle\Psi_m|\hat{O}_S|\Psi_{m'}\rangle$, Eq. (20.5.8) can be written as

$$\frac{\partial}{\partial t}[\hat{\rho}_s(t)]_{n',n} \cong -\frac{i}{\hbar}(E_{n'}-E_n)[\hat{\rho}_s(t)]_{n',n} - \sum_{m,m'}R^{(St)}_{n',n,m',m}[\hat{\rho}_s(t)]_{m',m}, \tag{20.5.11}$$

where the stationary Redfield tensor (Eq. (19.4.1)) obtains the form (Ex. 20.5.4)

$$R^{(St)}_{n',n,m',m} = A_{n',n,m',m} + A^*_{n,n',m,m'}$$

$$A_{n',n,m',m} = \delta_{m,n}\sum_k \frac{1}{\hbar^2}\int_0^\infty d\tau e^{\frac{-i\tau}{\hbar}(E_k-E_{m'})}\{c_e(\tau)[\hat{V}_S]_{n',k}[\hat{V}_S^\dagger]_{k,m'} + c_a(\tau)[\hat{V}_S^\dagger]_{n',k}[\hat{V}_S]_{k,m'}\}$$

$$-\frac{1}{\hbar^2}\int_0^\infty d\tau e^{\frac{-i\tau}{\hbar}(E_{n'}-E_{m'})}\{c_e(\tau)[\hat{V}_S^\dagger]_{n',m'}[\hat{V}_S]_{m,n} + c_a(\tau)[\hat{V}_S]_{n',m'}[\hat{V}_S^\dagger]_{m,n}\}. \tag{20.5.12}$$

Exercise 20.5.4 *Show that the matrix representation of Eq. (20.5.8) in the basis of the system Hamiltonian eigenstates (Eq. (20.5.9)) is given by Eqs. (20.5.11, 20.5.12).*

Master Equation

When the energy level spacings within the spectrum of the adsorbate Hamiltonian ($\hat{H}_S$, in the absence of coupling to the electrode) are much larger than the spectral density of the adsorbate–electrode coupling, the bath-free dynamics is much faster than the bath-induced dynamics within the adsorbate. In this limit, and when the spectrum of $\hat{H}_S$ is nondegenerate, the secular approximation is valid (Eqs. (19.4.8–19.4.11)). The dynamics of the populations of $\hat{H}_S$-eigenstates, $P_n(t) \equiv [\hat{\rho}]_{n,n}(t)$, is then effectively decoupled from the dynamics of the coherences between them, which means that the equation for the population can be derived by setting to zero off-diagonal matrix elements of the reduced density matrix, $[\hat{\rho}]_{n',n}(t)$, in Eq. (20.5.11). Noticing that $R^{(St)}_{n,n,m,m} = 2\,\text{Re}[A_{n,n,m,m}]$, we obtain Pauli's master equation (Eq. (19.4.13)) for the eigenstate population kinetics (Ex. 20.5.5):

$$\frac{\partial}{\partial t}P_n(t) \cong \sum_{n'}k_{n'\to n}P_{n'}(t) - \sum_{n'}k_{n\to n'}P_n(t)$$

$$k_{n\to n'} \equiv \left|[\hat{V}_S]_{n,n'}\right|^2\frac{J(E_n-E_{n'})}{\hbar}f_h(E_n-E_{n'}) + \left|[\hat{V}_S]_{n',n}\right|^2\frac{J(E_{n'}-E_n)}{\hbar}f_e(E_{n'}-E_n). \tag{20.5.13}$$

As in the general case discussed in Section 19.4, the rate coefficients $\{k_{n \to n'}\}$ are shown to be nonnegative, and consequently, the nonnegativity of each state population (as well as the sum of state populations) is conserved by the master equation. Additionally, the structure of the rate equation guarantees the existence of a stationary (steady-state) solution, where $\frac{\partial}{\partial t} P_n(t) = 0$ (see Eqs. (19.4.13–19.4.23)).

Exercise 20.5.5   *Show that setting the coherences of the reduced density matrix to zero in Eqs. (20.5.11, 20.5.12) leads to Eq. (20.5.13).*

The physical processes underlying the state-to-state population transfer become apparent by considering the explicit form of the system bath coupling operators within the system (Eq. (20.5.1)), $\hat{V}_S \equiv \sum\limits_{m=1}^{M_A} v_m \hat{a}_m^\dagger$,

$$[\hat{V}_S]_{n,n'} = \langle \Psi_n | \hat{V}_S | \Psi_{n'} \rangle = \sum_{m=1}^{M_A} v_m \langle \Psi_n | \hat{a}_m^\dagger | \Psi_{n'} \rangle. \tag{20.5.14}$$

Since the operator $\hat{a}_m^\dagger$ creates an electron at the mth single-particle state in the system, and since eigenstates associated with different numbers of electrons in the system are orthogonal to each other (Eq. (20.5.10)), the coupling matrix element, $[\hat{V}_S]_{n,n'}$, vanishes unless the system in the state $|\Psi_{n'}\rangle$ contains one electron less than the system in the state $|\Psi_n\rangle$, namely $[\hat{V}_S]_{n,n'} \propto \delta_{N_{n'} N_n - 1}$. Consequently, the state-to-state rates can be expressed as

$$k_{n \to n'} = \begin{cases} \left| [\hat{V}_S]_{n,n'} \right|^2 \frac{J(E_n - E_{n'})}{\hbar} f_h(E_n - E_{n'}) & ; \ N_{n'} = N_n - 1 \\ \left| [\hat{V}_S]_{n',n} \right|^2 \frac{J(E_{n'} - E_n)}{\hbar} f_e(E_{n'} - E_n) & ; \ N_{n'} = N_n + 1 \\ 0 & ; \quad otherwise. \end{cases} \tag{20.5.15}$$

The nonzero transition rates in the master equation therefore correspond to either a single electron emission from the system to the reservoir, or a single electron absorption by the system from the reservoir. We can readily verify (see Ex. 20.5.6) that the rate expressions in the master equation coincide with the results of Fermi's golden rule for transitions between different manifolds of the uncoupled adsorbate–electrode Hamiltonian eigenstates. Notice that if electron–electron interactions within the system can be neglected ($\hat{H}_S \approx \sum_{m=1}^{M_A} \varepsilon_m \hat{a}_m^\dagger \hat{a}_m$ in Eq. (20.5.1)), each electron emission or absorption involves one of the single-particle (spin-orbital) states (see Ex. 20.5.7).

Exercise 20.5.6   *The adsorbate–electrode Hamiltonian can be conveniently split into zero order and perturbation Hamiltonians, $\hat{H} = \hat{H}_0 + \hat{V}$, where $\hat{H}_0 = \hat{H}_S + \hat{H}_B$ and $\hat{V} = \hat{H}_{SB}$ (see Eq. (20.5.1)). Using Eq. (20.5.9), the eigenvectors of $\hat{H}_0$ are defined by the equation,*
$$\hat{H}_0 |\Psi_n\rangle \otimes |n_1, n_2, \ldots, n_{M_E}\rangle = \left(E_n + \sum_{k=1}^{M_E} n_k \varepsilon_k \right) |\Psi_n\rangle \otimes |n_1, n_2, \ldots, n_{M_E}\rangle. \ A \ charge \ transfer$$
event between the electrode and the adsorbate can be formulated as a transition between incoherent ensembles of (orthogonal) $\hat{H}_0$ eigenstates, where the initial and final states are defined, respectively, by projection operators to the initial and final system Hamiltonian eigenstates, $\hat{P}_{\{i\}} = |\Psi_n\rangle\langle\Psi_n|$, and $\hat{P}_{\{f\}} = |\Psi_{n'}\rangle\langle\Psi_{n'}|$. In the weak molecule–electrode

coupling limit, the charge transfer rate can be evaluated by Fermi's golden rule (see Chapter 17), where the electrode is initially in a thermal state, $\hat{\rho}_{\{i\}}(0) = |\Psi_n\rangle\langle\Psi_n| \otimes \hat{\rho}_B$. *Starting from the generic perturbative rate expression (Eq. (17.3.12)),* $k^{(1)}_{\{i\}\to\{f\}}(t) \cong$

$$\frac{2}{\hbar^2}\text{Re}\int_0^t tr\{\hat{\rho}_{\{i\}}(0)\hat{P}_{\{i\}}\hat{V}\hat{P}_{\{f\}}e^{\frac{i\hat{H}_0\tau}{\hbar}}\hat{V}e^{\frac{-i\hat{H}_0\tau}{\hbar}}\}d\tau,$$ *and using Eq. (20.5.4) for the bath correlation functions, show that the transition rates appearing in the master equation are reproduced at the infinite time limit (under the fast bath relaxation assumption), namely* $k_{n\to n'} = \lim_{t\to\infty} k^{(1)}_{\{i\}\to\{f\}}(t)$.

Exercise 20.5.7 *In the absence of electron–electron interaction, the system Hamiltonian in Eq. (20.5.1) reads* $\hat{H}_S = \sum_{m=1}^{M_A} \varepsilon_m \hat{a}^\dagger_m \hat{a}_m$. *(a) Show that the eigenvectors of* $\hat{H}_S$ *in this case are the M_A-particle determinants, $\{|n_1, n_2, n_3, \ldots, n_{M_A}\rangle\}$ (defined in Eqs. (20.1.3, 20.1.7)), with the corresponding eigenvalues, $\left\{\sum_{m=1}^{M_A} \varepsilon_m n_m\right\}$, where $\{\varepsilon_m\}$ are the single-particle Hamiltonian eigenstates. (b) Denoting each eigenstate by the vector of occupation numbers, $|n_1, n_2, n_3, \ldots, n_{M_A}\rangle \Leftrightarrow \mathbf{n}$ and using the system–bath coupling ($\hat{V}_S$) in Eq. (20.5.1), show that the transition rate between nondegenerate $\hat{H}_S$ eigenvectors (Eq. (20.5.13)) reads in this case $k_{\mathbf{n}\to\mathbf{n}'} = \sum_{m=1}^{M_A}[\delta_{\mathbf{n}-\mathbf{n}',\mathbf{e}_m}k_{e,m} + \delta_{\mathbf{n}'-\mathbf{n},\mathbf{e}_m}k_{a,m}]$, where $\mathbf{e}_m$ is a unit vector of length M_A with elements $[\mathbf{e}_m]_{m'} = \delta_{m,m'}$. Here $k_{e,m} \equiv |v_m|^2 \frac{J(\varepsilon_m)}{\hbar} f_h(\varepsilon_m)$ and $k_{a,m} \equiv |v_m|^2 \frac{J(\varepsilon_m)}{\hbar} f_e(\varepsilon_m)$ are rates of single-electron emission to the reservoir, or absorption from the reservoir. Notice that the rate vanishes unless the two occupation vectors are identical except for a flip between zero and one at the mth entry, which corresponds to either absorption or emission of a single electron at the mth single-particle state.*

Notice that each transition rate is proportional to the spectral density ($J(E)$) at the charging (or discharging) energy, as well as to the Fermi distributions, namely to the availability of a hole (for emission) or an electron (for absorption) at the electrode, at the appropriate transitions. The dependence of the Fermi distribution function on the temperature and chemical potential imposes a detained balance relation on the forward and backward transitions between the same many-electron states. Let us consider the many-electron states, $|\Psi_{n'}\rangle$ and $|\Psi_n\rangle$, whose total electron occupation numbers differ by one, where $N_{n'} = N_n - 1$. Using Eq. (20.5.15), the ratio between the rates of electron emission ($k_{em} = k_{n\to n'}$) and electron absorption ($k_{ab} = k_{n'\to n}$) between these states reads (Ex. 20.5.8)

$$\frac{k_{em}}{k_{ab}} = \frac{k_{n\to n'}}{k_{n'\to n}} = e^{\frac{E_n - E_{n'} - \mu}{k_B T}}. \tag{20.5.16}$$

The Equilibrium State

Using the condition, Eq. (20.5.16), in the Pauli master equation (Eq. (20.5.13)), we can readily identify the existence of a stationary (steady-state) solution $\left(\frac{\partial}{\partial t}P_n(t) = 0\right)$, in which the populations of the system Hamiltonian eigenstates are given by the distribution (see Ex. 20.5.9),

$$P_n = \frac{e^{\frac{-(E_n - \mu N_n)}{k_B T}}}{\sum_{n'} e^{\frac{-(E'_n - \mu N'_n)}{k_B T}}}. \tag{20.5.17}$$

As we can see, under the assumptions of the Born–Markov and secular approxima-tions, *the steady state of the system (equilibrium in this case) "adopts" the constraint imposed on the (grand canonical) fermion reservoir, where the approximated reduced sys-tem density operator reads*, $\hat{\rho}_S \cong e^{\frac{-(\hat{H}_S - \mu \hat{N}_S)}{k_B T}} / tr \left\{ e^{\frac{-(\hat{H}_S - \mu \hat{N}_S)}{k_B T}} \right\}$ (compare to Eq. (19.4.25) for the canonical case).

Exercise 20.5.8 *Use the explicit rate expressions (Eq. (20.5.15)) to derive the ratio in Eq. (20.5.16).*

Exercise 20.5.9 *Recalling that each system eigenstate is associated with a well-defined number of electrons (see Eqs. (20.5.9, 20.5.10)) and restricting to $N_{n'} = N_n - 1$, Eq. (20.5.16) can be written as $\frac{k_{n \to n'}}{k_{n' \to n}} = e^{\frac{(E_n - \mu N_n) - (E_{n'} - \mu N_{n'})}{k_B T}}$. Use this to show that the probability distribution in Eq. (20.5.17) is a stationary solution to the master equation, Eq. (20.5.13).*

The Case of a Single-State Impurity

As a concrete simple example, let us return to the case of a single-state adsorbate, as depicted in the model Hamiltonian described in Eqs. (20.4.1–20.4.7). For extra sim-plicity, the electronic interaction on the adsorbate (impurity) site is assumed to be sufficiently weak (setting $U \to 0$) to allow the many-electron Hamiltonian (Eq. (20.4.7)) to become effectively separable in the two spin states. In this case, for either of the two spin states, the adsorbate–electrode Hamiltonian reads

$$\hat{H} = \varepsilon_0 \hat{a}_0^\dagger \hat{a}_0 + \sum_{k=1}^{M_E} \varepsilon_k \hat{a}_k^\dagger \hat{a}_k + \left[\sum_{k=1}^{M_E} \gamma_k \hat{a}_0^\dagger \hat{a}_k + h.c. \right], \tag{20.5.18}$$

where the system–bath Hamiltonian, Eq. (20.5.1), obtains the form $\hat{H}_S = \varepsilon_0 \hat{a}_0^\dagger \hat{a}_0$, $\hat{H}_B = \sum_{k=1}^{M_E} \varepsilon_k \hat{a}_k^\dagger \hat{a}_k$, and $\hat{H}_{SB} = \hat{V}_S \hat{U}_B + \hat{V}_S^\dagger \hat{U}_B^\dagger$, where $\hat{V}_S \equiv \hat{a}_0^\dagger$ and $\hat{U}_B \equiv \sum_{k=1}^{M_E} \gamma_k \hat{a}_k$. The space of the adsorbate states is spanned by the "empty" ($|\Psi_0\rangle = |0\rangle$) and "occupied" ($|\Psi_1\rangle = |1\rangle$) states of the single adsorbate orbital. Therefore, the adsorbate in this model corresponds to a two-level system (TLS). Using the identities,

$$\hat{H}_S |0\rangle = 0|0\rangle \; ; \; \hat{H}_S |1\rangle = \varepsilon_0 |1\rangle$$
$$\hat{V}_S |0\rangle = |1\rangle \; ; \; \hat{V}_S |1\rangle = 0, \tag{20.5.19}$$

the representations of the TLS operators in the occupation basis read

$$\hat{H}_S = \varepsilon_0 |1\rangle\langle 1| \; ; \; \hat{V}_S = |1\rangle\langle 0| \; ; \; \hat{V}_S^\dagger = |0\rangle\langle 1|. \tag{20.5.20}$$

In Section 19.5 we analyzed bath-induced dynamics in a TLS coupled to a boson bath, which exchanges energy with the system. Here, the analogous treatment corresponds to particles exchanged between the TLS and the fermion reservoir. The general derivation of the quantum master equation (Eqs. (20.5.1–20.5.13)) clearly applies also for the present model (Eqs. (20.5.19, 20.5.20)). Let us emphasize that since the two eigenstates of $\hat{H}_S$ are nondegenerate, and since the matrix representation of the coupling operator $\hat{V}_S$ in the basis of $\hat{H}_S$-eigenstates is off diagonal, the secular approximation is not needed, and Pauli's master equation is obtained directly from Eqs. (20.5.11, 20.5.12) with no additional assumptions (see Ex. 20.5.10).

Denoting the four elements of the density matrix in the basis of the adsorbate occupation numbers,

$$\rho_{0,0}(t) = \langle 0|\hat{\rho}_S(t)|0\rangle \; ; \; \rho_{1,1}(t) = \langle 1|\hat{\rho}_S(t)|1\rangle$$
$$\rho_{1,0}(t) = \langle 1|\hat{\rho}_S(t)|0\rangle \; ; \; \rho_{0,1}(t) = \langle 0|\hat{\rho}_S(t)|1\rangle, \tag{20.5.21}$$

the diagonal matrix elements, $\rho_{1,1}(t)$ and $\rho_{0,0}(t)$, correspond, respectively, to the occupied and unoccupied states of the adsorbate site. Within the stationary Born–Markov approximation, the population dynamics is decoupled from the coherences ($\rho_{0,1}(t)$ and $\rho_{1,0}(t)$), and *Pauli's master equation for the adsorbate population is obtained*,

$$\frac{\partial}{\partial t}\rho_{0,0}(t) = k^{em}_{1\to0}\rho_{1,1}(t) - k^{ab}_{0\to1}\rho_{0,0}(t)$$
$$\frac{\partial}{\partial t}\rho_{1,1}(t) = k^{ab}_{0\to1}\rho_{0,0}(t) - k^{em}_{1\to0}\rho_{1,1}(t). \tag{20.5.22}$$

The transfer rates $k^{ab}_{0\to1}$ and $k^{em}_{1\to0}$ correspond, respectively, to electron absorption and emission by the adsorbate. These rates are Fourier transforms of the bath correlation functions (Eq. (20.5.4)),

$$k^{em}_{1\to0} = 2\,\mathrm{Re}\frac{1}{\hbar^2}\int_0^\infty d\tau c_e(\tau)e^{\frac{i\tau}{\hbar}\varepsilon_0} = \frac{2\pi}{\hbar}\sum_{k=1}^{M_E}|\gamma_k|^2\delta(\varepsilon_k - \varepsilon_0)f_h(\varepsilon_k)$$
$$k^{ab}_{0\to1} = 2\,\mathrm{Re}\frac{1}{\hbar^2}\int_0^\infty d\tau c_a(\tau)e^{\frac{-i\tau}{\hbar}\varepsilon_0} = \frac{2\pi}{\hbar}\sum_{k=1}^{M_E}|\gamma_k|^2\delta(\varepsilon_k - \varepsilon_0)f_e(\varepsilon_k), \tag{20.5.23}$$

or, in the case of a continuous electrode spectral density (Eq. (20.5.6)),

$$k^{em}_{1\to0} = \frac{1}{\hbar}J(\varepsilon_0)f_h(\varepsilon_0)$$
$$k^{ab}_{0\to1} = \frac{1}{\hbar}J(\varepsilon_0)f_e(\varepsilon_0) \tag{20.5.24}$$

We can readily see that regardless of the initial population of the adsorbate, the asymptotic time limit of the ratio between the two occupation states reaches a constant value (see Ex. 20.5.11), where

$$\frac{\rho_{1,1}(\infty)}{\rho_{0,0}(\infty)} = \frac{k^{ab}_{0\to1}}{k^{em}_{1\to0}} = e^{\frac{-(\varepsilon_0-\mu)}{k_BT}}. \tag{20.5.25}$$

Therefore, at equilibrium the adsorbate population adopts the Fermi distribution:

$$\rho_{1,1}(\infty) = \frac{1}{1+e^{\frac{\varepsilon_0-\mu}{k_BT}}} = f_e(\varepsilon_0). \tag{20.5.26}$$

This is a manifestation of the general results (Eqs. (20.5.16, 20.5.17)) for the present model.

Exercise 20.5.10   *(a) Given Eq. (20.5.8), and defining the TLS eigenstate populations, $\rho_{0,0}(t) = \langle 0|\hat{\rho}_S(t)|0\rangle$, and $\rho_{1,1}(t) = \langle 1|\hat{\rho}_S(t)|1\rangle$, show that $\frac{\partial}{\partial t}\rho_{0,0}(t) \cong -\frac{1}{\hbar^2}2\,\mathrm{Re}\int_0^{\infty}d\tau$*

$$\{c_e(\tau)\langle 0|[\hat{V}_S, e^{\frac{-i\tau}{\hbar}\hat{H}_S}\hat{V}_S^{\dagger}e^{\frac{i\tau}{\hbar}\hat{H}_S}\hat{\rho}_S(t)]|0\rangle + c_a(\tau)\langle 0|[\hat{V}_S^{\dagger}, e^{\frac{-i\tau}{\hbar}\hat{H}_S}\hat{V}_S e^{\frac{i\tau}{\hbar}\hat{H}_S}\hat{\rho}_S(t)]|0\rangle\}.$$

(b) Introduce the identity operator in the Fock space of the adsorbate, $\hat{I} = |0\rangle\langle 0| + |1\rangle\langle 1|$, to show that for the off-diagonal TLS coupling operators (Eq. (20.5.20)), $\hat{V}_S = |1\rangle\langle 0|$ and $\hat{V}_S^{\dagger} = |0\rangle\langle 1|$, this result reads $\frac{\partial}{\partial t}\rho_{0,0}(t) = k_{1\to 0}^{em}\rho_{1,1}(t) - k_{0\to 1}^{ab}\rho_{0,0}(t)$, where $k_{1\to 0}^{em} = 2\,\mathrm{Re}\frac{1}{\hbar^2}\int_0^{\infty}d\tau c_e(\tau)e^{\frac{i\tau}{\hbar}\varepsilon_0}$ and $k_{0\to 1}^{ab} = 2\,\mathrm{Re}\frac{1}{\hbar^2}\int_0^{\infty}d\tau c_a(\tau)e^{\frac{-i\tau}{\hbar}\varepsilon_0}$ (c) Show similarly that $\frac{\partial}{\partial t}\rho_{1,1}(t) = k_{0\to 1}^{ab}\rho_{0,0}(t) - k_{1\to 0}^{em}\rho_{1,1}(t)$. (d) Use the explicit expressions for the correlation functions (Eq. (20.5.4)) to derive Eq. (20.5.23). (e) Use the expressions for the correlation functions for a continuous bath (Eq. (20.5.6)) to derive Eq. (20.5.24).

Exercise 20.5.11   *Use Eq. (20.5.24) and the definition of the Fermi–Dirac distribution function (Eq. (20.3.6)) to derive Eq. (20.5.25).*

20.6 Nonequilibrium Fermion Systems

So far, we discussed "small" quantum systems that are either isolated from their surroundings or weakly coupled to a large reservoir at a near equilibrium state. In a more general case, an open quantum system can be coupled to several reservoirs, each at its own near equilibrium state. In this case the system is driven into a nonequilibrium state. Nevertheless, when the coupling to different reservoirs is weak, the open system can reach a "steady state" at asymptotic times, analogous to an equilibrium state in the sense that system observables become time-independent. Yet, this state is different, since stationary fluxes (currents) of energy and/or particles can pass through the boundaries between the small system and the different reservoirs at a constant, nonzero rate. This scenario characterizes elementary processes on the nanoscale, such as charge, spin, and energy transport, which are microscopic events underlying macroscopic transport phenomena such as electrical conductivity and heat transport, and are at the heart of numerous technological applications, including nanoscale electronics and spintronics, thermoelectric power, and optoelectric devices.

As we saw in Section 20.5, within the realm of the Born–Markov and the secular approximations, coupling of a small fermionic system to a single reservoir of noninteracting fermions drives the system into an equilibrium distribution, in which the system eigenstate populations reflect the constraints imposed on the external reservoir. In this section we elaborate on the effect of coupling the small system of interest to several reservoirs, each maintained in a different near equilibrium state, at different temperatures and/or chemical potentials. As we shall see, the system can reach a nonequilibrium steady state, in which energy and/or particles continuously flow between the reservoirs (namely, in a directed manner) through the system. An example of charge transport through a nanoscale conductor (e.g., a single molecule) between two macroscopic leads under a bias potential will be analyzed in what follows.

An Impurity Coupled to Several Fermion Reservoirs

The formal treatment of an impurity coupled to several fermion reservoirs turns out to be a rather straightforward extension of the case of a single reservoir, at least when the system–bath coupling is restricted to the single-particle level (electron exchange between the impurity and reservoirs of noninteracting particles), and within the validity of the Born–Markov approximation. In this case, the effect of multiple reservoirs on the system evolution is shown to be a sum of contributions of each reservoir. The theoretical formulation can therefore rely to a large extent on the one derived in the previous section for a single reservoir.

We start by generalizing the adsorbate–electrode model Hamiltonian, Eq. (20.5.1), to the case of several reservoirs.

$$\hat{H} = \hat{H}_S + \hat{H}_B + \hat{H}_{SB} \quad ; \quad [\hat{H}_S, \hat{H}_B] = 0$$

$$\hat{H}_S = \sum_{m=1}^{M_A} \varepsilon_m \hat{a}_m^\dagger \hat{a}_m + \sum_{i,j,k,l \in \{m\}} w_{i,j,k,l} \hat{a}_i^\dagger \hat{a}_j^\dagger \hat{a}_l \hat{a}_k$$

$$\hat{H}_B = \sum_K \hat{H}_K \tag{20.6.1}$$

$$\hat{H}_{SB} = \sum_K (\hat{V}_{S,K} \hat{U}_{B,K} + \hat{V}_{S,K}^\dagger \hat{U}_{B,K}^\dagger).$$

The impurity Hamiltonian, $\hat{H}_S$, is unchanged, where the single reservoir model is replaced by a sum of reservoir Hamiltonians, $\hat{H}_K = \sum_{k_K} \varepsilon_{k_K} \hat{a}_{k_K}^\dagger \hat{a}_{k_K}$. Here $K = 1, 2, \ldots$ is the reservoir index, and the operator $\hat{a}_{k_K}^\dagger$ creates an electron in the k_Kth single-particle state at the Kth reservoir. The corresponding system–reservoir coupling operators read $\hat{U}_{B,K} \equiv \sum_{k_K} \gamma_{k_K} \hat{a}_{k_K}$ and $\hat{V}_{S,K} \equiv \sum_{m=1}^{M_A} v_{m,K} \hat{a}_m^\dagger$, where $\hat{a}_m^\dagger$ creates an electron in the mth single-particle state at the system, and $v_{m,K}$ is the projection of that mth state on a local coupling site to the Kth reservoir.

The full Fock space is a tensor product of the system and all the independent reservoir spaces. An initially uncorrelated density operator in this space therefore reads

$$\hat{\rho}(0) = \hat{\rho}_S(0) \otimes \hat{\rho}_B \quad ; \quad \hat{\rho}_B = \hat{\rho}_1 \otimes \hat{\rho}_2 \otimes \hat{\rho}_3 \ldots = \prod_K \hat{\rho}_K, \tag{20.6.2}$$

where each reservoir density operator ($\hat{\rho}_K$) obtains the equilibrium form for a grand canonical ensemble, characterized by its own temperature, T_K, and chemical potential, μ_K:

$$\hat{\rho}_K = e^{\frac{-1}{k_B T_K} \sum_{k_K} (\varepsilon_{k_K} - \mu_K) \hat{a}_{k_K}^\dagger \hat{a}_{k_K}} / Z_K \quad ; \quad Z_K = tr_K \{ e^{\frac{-1}{k_B T_K} \sum_{k_K} (\varepsilon_{k_K} - \mu_K) \hat{a}_{k_K}^\dagger \hat{a}_{k_K}} \}. \tag{20.6.3}$$

Our prime interest is the time evolution of the reduced system density operator, $\hat{\rho}_S(t) = tr_B\{\hat{\rho}(t)\}$. When the second-order Born–Markov approximation applies, this equation obtains the generic Redfield form of Eqs. (19.3.19–19.3.22). Introducing the system–bath coupling operator, $\hat{H}_{SB} = \sum_K (\hat{V}_{S,K} \hat{U}_{B,K} + \hat{V}_{S,K}^\dagger \hat{U}_{B,K}^\dagger)$, into these equations, the influence of the dissipator on the system dynamics is shown to be a sum over independent contributions from the different reservoirs (Ex. 20.6.1):

$$\frac{\partial}{\partial t}\hat{\rho}_S(t) \cong -\frac{i}{\hbar}[\hat{H}_S, \hat{\rho}_S(t)]$$

$$-\sum_K \frac{1}{\hbar^2}\int_0^t d\tau \{ c_{e,K}(\tau) \left[\hat{V}_{S,K}, e^{\frac{-i\tau}{\hbar}\hat{H}_S} \hat{V}_{S,K}^\dagger e^{\frac{i\tau}{\hbar}\hat{H}_S} \hat{\rho}_S(t) \right]$$

$$+ c_{a,K}(\tau)[\hat{V}_{S,K}^\dagger, \; e^{\frac{-i\tau}{\hbar}\hat{H}_S} \hat{V}_{S,K} e^{\frac{i\tau}{\hbar}\hat{H}_S} \hat{\rho}_S(t)] + h.c.\} \tag{20.6.4}$$

Each Kth reservoir is represented in this equation by its specific coupling operator in the system, $\hat{V}_{S,K}$, and the corresponding correlation functions, $c_{e,K}(\tau)$ and $c_{a,K}(\tau)$, which read (compare to Eq. (20.5.4) for the case of a single reservoir)

$$c_{e,K}(\tau) = tr_K\{\hat{U}_{B,K} e^{\frac{-i\tau}{\hbar}\hat{H}_K} \hat{U}_{B,K}^\dagger e^{\frac{i\tau}{\hbar}\hat{H}_K} \hat{\rho}_K\} = \sum_{k_K=1}^{M_E} |\gamma_{k_K}|^2 e^{\frac{-i\tau}{\hbar}\varepsilon_{k_K}} f_h(\varepsilon_{k_K})$$

$$c_{a,K}(\tau) = tr_K\{\hat{U}_{B,K}^\dagger e^{\frac{-i\tau}{\hbar}\hat{H}_K} \hat{U}_{B,K} e^{\frac{i\tau}{\hbar}\hat{H}_K} \hat{\rho}_K\} = \sum_{k_K=1}^{M_E} |\gamma_{k_K}|^2 e^{\frac{i\tau}{\hbar}\varepsilon_{k_K}} f_e(\varepsilon_{k_K}) \tag{20.6.5}$$

Or, associating each reservoir with a continuous spectral density, $J_K(\varepsilon)$ (see Eq. (20.5.6)),

$$c_{e,K}(\tau) = \frac{1}{2\pi}\int d\varepsilon e^{-i\tau\varepsilon/\hbar} J_K(\varepsilon) f_{h,K}(\varepsilon)$$

$$c_{a,K}(\tau) = \frac{1}{2\pi}\int d\varepsilon e^{i\tau\varepsilon/\hbar} J_K(\varepsilon) f_{e,K}(\varepsilon) \tag{20.6.6}$$

The additivity of the contributions from the different reservoirs is a seemingly straightforward generalization of Eq. (20.5.3). Indeed, it is derived from the additive contributions of the different reservoirs to the single-particle Hamiltonian, from the lack of direct inter-reservoir coupling (the inherent locality associated with the tight binding model assumptions), and from the restriction of the system–bath coupling to the single-particle level. Notice, however, that even within this structure of the system–bath coupling, the interaction of each reservoir with the system may induce inter-reservoir correlations. Such correlations are missed within the Born–Markov approximation, as they appear beyond the second order in the system–bath coupling operator.

Exercise 20.6.1   *For a generic system–bath coupling term, $\hat{H}_{SB} \equiv \sum_\alpha \hat{V}_\alpha^{(S)} \hat{U}_\alpha^{(B)}$, the dissipator in the Born–Markov approximation obtains the form shown in Eqs. (19.3.21, 19.3.22). To bring to this form the coupling operator, $\hat{H}_{SB} = \sum_K (\hat{V}_{S,K} \hat{U}_{B,K} + \hat{V}_{S,K}^\dagger \hat{U}_{B,K}^\dagger)$, let us define $\hat{H}_{SB} \equiv \sum_{\alpha \in K,n} \hat{V}_\alpha^{(S)} \hat{U}_\alpha^{(B)} = \sum_K \sum_{n=1}^2 \hat{V}_{K,n}^{(S)} \hat{U}_{K,n}^{(B)}$, where $\hat{V}_{K,1}^{(S)} = \hat{V}_{S,K}$, $\hat{V}_{K,2}^{(S)} =$*

$\hat{V}_{S,K}^{\dagger}$, $\hat{U}_{K,1}^{(B)} = \hat{U}_{B,K} = \sum_{k_K} \gamma_{k_K} \hat{a}_{k_K}$, $\hat{U}_{K,2}^{(B)} = \hat{U}_{B,K}^{\dagger} = \sum_{k_K} \gamma_{k_K}^{*} \hat{a}_{k_K}^{\dagger}$, *such that the dissipa-*

tor (Eq. (19.3.22)) obtains the form $\hat{D}\hat{\rho}_S(t) = -\frac{1}{\hbar^2} \sum_{K,K'} \sum_{n,n' \in 1,2} \int_0^t d\tau \{ c_{K,n,K',n'}(\tau) [\hat{V}_{K,n}^{(S)},$

$e^{\frac{-i\tau}{\hbar}\hat{H}_S} \hat{V}_{K',n'}^{(S)} e^{\frac{i\tau}{\hbar}\hat{H}_S} \hat{\rho}_S(t)] + \bar{c}_{K',n',K,n}(\tau) [\hat{\rho}_S(t) e^{\frac{-i\tau}{\hbar}\hat{H}_S} \hat{V}_{K',n'}^{(S)} e^{\frac{i\tau}{\hbar}\hat{H}_S}, \hat{V}_{K,n}^{(S)}] \}$, *with the bath correla-*

tion functions, $c_{K,n,K',n'}(\tau) = tr_B \{ \hat{U}_{K,n}^{(B)} e^{\frac{-i\tau}{\hbar}\hat{H}_B} \hat{U}_{K',n'}^{(B)} e^{\frac{i\tau}{\hbar}\hat{H}_B} \hat{\rho}_B \}$, $\bar{c}_{K,n,K',n'}(\tau) = c_{K,n,K',n'}(-\tau)$,

where $\hat{H}_B = \sum_K \hat{H}_K$, $\hat{\rho}_B = \prod_K \hat{\rho}_K$. *(a) Use the results of Ex. 20.5.1 to show that the correla-*

tion functions involving different reservoirs vanish, namely $c_{K,n,K',n'}(\tau) = \delta_{K,K'} c_{n,n'}^{(K)}(\tau)$,

where $c_{n,n'}^{(K)}(\tau) = tr_K \{ \hat{U}_{K,n}^{(B)} e^{\frac{-i\tau}{\hbar}\hat{H}_K} \hat{U}_{K,n'}^{(B)} e^{\frac{i\tau}{\hbar}\hat{H}_K} \hat{\rho}_K \}$. *(b) Using this result, show that the*
dissipator reads

$$\hat{D}\hat{\rho}_S(t) = -\frac{1}{\hbar^2} \sum_K \sum_{n,n' \in 1,2} \int_0^t d\tau \{ c_{n,n'}^{(K)}(\tau) [\hat{V}_{K,n}^{(S)}, e^{\frac{-i\tau}{\hbar}\hat{H}_S} \hat{V}_{K,n'}^{(S)} e^{\frac{i\tau}{\hbar}\hat{H}_S} \hat{\rho}_S(t)]$$
$$+ \bar{c}_{n',n}^{(K)}(\tau) [\hat{\rho}_S(t) e^{\frac{-i\tau}{\hbar}\hat{H}_S} \hat{V}_{K,n'}^{(S)} e^{\frac{i\tau}{\hbar}\hat{H}_S}, \hat{V}_{K,n}^{(S)}] \}.$$

(c) Apply the treatment in Ex. (20.5.2) to the dissipator in (b) to derive Eq. (20.6.4),
where the absorption and emission correlation functions are defined in Eq. (20.6.5).

Assuming that fast bath relaxation (the stationary Redfield approximation) and the secular approximation hold independently for each of the reservoirs, and using the additivity of the dissipators associated with the different reservoirs, we can readily generalize the derivation leading from Eq. (20.5.8) to Eq. (20.5.15) to multiple reservoirs. The kinetics of population transfer between any two eigenstates, $|\Psi_{n'}\rangle$ and $|\Psi_n\rangle$, of the many-particle system Hamiltonian ($\hat{H}_S$) is then given by the general master equation,

$$\frac{\partial}{\partial t} P_n(t) \cong \sum_{n'} \left[\sum_K k_{n' \to n}^{(K)} \right] P_{n'}(t) - \sum_{n'} \left[\sum_K k_{n \to n'}^{(K)} \right] P_n(t). \qquad (20.6.7)$$

Here $P_n(t)$ is the transient population of the nth eigenstate, and $k_{n \to n'}^{(K)}$ is the contribution of the Kth reservoir to the overall state-to-state transition rate, which depends on the specific reservoir spectral density, temperature, and chemical potential (via the respective Fermi distribution functions):

$$k_{n \to n'}^{(K)} = \left| [\hat{V}_{S,K}]_{n,n'} \right|^2 \frac{J_K(E_n - E_{n'})}{\hbar} f_{h,K}(E_n - E_{n'})$$
$$+ \left| [\hat{V}_{S,K}]_{n',n} \right|^2 \frac{J_K(E_{n'} - E_n)}{\hbar} f_{e,K}(E_{n'} - E_n). \qquad (20.6.8)$$

Recalling that within the master equation, bath-induced transition rates are restricted to an exchange of a single electron (Eqs. (20.5.14, 20.5.15)), we have

$$k_{n \to n'}^{(K)} = \begin{cases} \left| [\hat{V}_{S,K}]_{n,n'} \right|^2 \frac{J_K(E_n - E_{n'})}{\hbar} f_{h,K}(E_n - E_{n'}) & ; N_{n'} = N_n - 1 \\ \left| [\hat{V}_{S,K}]_{n',n} \right|^2 \frac{J_K(E_{n'} - E_n)}{\hbar} f_{e,K}(E_{n'} - E_n) & ; N_{n'} = N_n + 1 \\ 0 & ; otherwise \end{cases} \qquad (20.6.9)$$

Notice that the transition rates are nonnegative. Hence, by its structure (see Eqs. (19.4.13-19.4.18)) Eq. (20.6.7) has a stationary solution, $\frac{\partial}{\partial t}P_n(t) = 0$. Nevertheless, unlike in the case of coupling to a single reservoir (Eqs. (20.5.16, 20.5.17)), if the different reservoirs are associated with different chemical potentials and/or temperatures, the stationary solution will not reflect the equilibrium distribution in any of the reservoirs. Instead, the reduced system density will adopt an "intermediate" steady state that does not comply with the constraints imposed on any of the reservoirs. This state will involve a constant flow of charge and/or energy between the reservoirs through the system.

Charge Transport through Nanoscale Conductors

Molecular conductance junctions are devices in which a single molecule is connected to two macroscopic electrodes [20.4]. The coupling between the molecule and the electrodes can be via chemical bonds of the molecule to atoms at the contact interface (metals, typically), or remote, as, for example, in a scanning tunneling microscope, discussed in Section 6.1. When a bias potential is applied, charge flows through the molecular conductor. The dependence of the current on the bias voltage can reveal the detailed electronic structure of the molecule and the contacts. Additionally, electronically inelastic charging and discharging events at the electrode interfaces drive the molecule far from equilibrium, possibly changing its conformational state and/or chemical composition. These processes affect the mechanical stability of the junction, as well as its response to changes in the bias potential. The latter reveals phenomena such as nonlinear current–voltage relations, negative differential resistance, and kinetic hysteresis, which are of fundamental interest as well as of potential relevance to nanoelectronics applications.

From a formal point of view, a single molecule junction can be modeled as an impurity, coupled to two reservoirs, where the bias potential is associated with a difference between the chemical potentials at the two reservoirs. Inelastic processes and mechanical conformational changes in response to the bias are critical for the description of realistic single molecule junctions [20.5]. Here, for pedagogical purposes, we shall ignore the mechanical degrees of freedom, discussing a hypothetical system (corresponding to "clamping" the atomic nuclei) with a "purely electronic" model Hamiltonian, namely a specific realization of Eq. (20.6.1). Only two reservoirs will be considered, corresponding to "right" and "left" electrodes, with $K \in R, L$. The respective chemical potentials would be denoted μ_R and μ_L, where we shall allow for $\mu_R \neq \mu_L$. Without loss of generality, we shall consider here only cases in which the same temperature applies to the two reservoirs.

We are interested in charge flow into and out of the impurity. The averaged total number of electrons on the impurity at a given time is given by the expectation value of the electron number operator, $\hat{N}_S = \sum\limits_{m=1}^{M_A} \hat{a}_m^\dagger \hat{a}_m$, namely

$$N_S(t) = tr_S[\hat{\rho}_S(t)\hat{N}_S]. \qquad (20.6.10)$$

Recalling that the number operator is diagonal in the system Hamiltonian eigenstate representation (Eq. (20.5.10)), it is convenient to evaluate the trace in this basis, where only the eigenstate populations (diagonal elements of the reduced system density matrix) are needed, $P_n(t) = [\hat{\rho}_S(t)]_{n,n}$,

$$N_S(t) = \sum_n P_n(t) N_n. \tag{20.6.11}$$

Within the Born–Markov and secular approximations, the populations of the system Hamiltonian eigenstates follow the master equation, Eq. (20.6.7), which reads in this case

$$\frac{\partial}{\partial t} P_n(t) \cong \sum_{n'} k^{(R)}_{n' \to n} P_{n'}(t) - \sum_{n'} k^{(R)}_{n \to n'} P_n(t) + \sum_{n'} k^{(L)}_{n' \to n} P_{n'}(t) - \sum_{n'} k^{(L)}_{n \to n'} P_n(t). \tag{20.6.12}$$

The change in the averaged total electron number within the system therefore reads

$$\frac{\partial}{\partial t} N_S(t) = \sum_{n,n'} k^{(R)}_{n' \to n} P_{n'}(t) N_n - \sum_{n,n'} k^{(R)}_{n \to n'} P_n(t) N_n + \sum_{n,n'} k^{(L)}_{n' \to n} P_{n'}(t) N_n - \sum_{n,n'} k^{(L)}_{n \to n'} P_n(t) N_n. \tag{20.6.13}$$

Since the contributions from the two reservoirs to $\frac{\partial}{\partial t} N_S(t)$ are additive, it is instructive to identify the transient particle fluxes into the system from each of the reservoirs,

$$J_R(t) \equiv \sum_{n,n'} k^{(R)}_{n' \to n} P_{n'}(t) N_n - \sum_{n,n'} k^{(R)}_{n \to n'} P_n(t) N_n$$

$$J_L(t) \equiv \sum_{n,n'} k^{(L)}_{n' \to n} P_{n'}(t) N_n - \sum_{n,n'} k^{(L)}_{n \to n'} P_n(t) N_n, \tag{20.6.14}$$

where

$$\frac{\partial}{\partial t} N_S(t) = J_L(t) + J_R(t). \tag{20.6.15}$$

The structure of the master equation, Eq. (20.6.12), guarantees the existence of a stationary solution (Eqs. (19.4.15–19.4.18)), in which the eigenstate populations are fixed in time, $\{P_n^{(st)}\}$. Setting $\frac{\partial}{\partial t} P_n^{(st)} = 0$, these populations satisfy the equation:

$$\sum_{n'} k^{(R)}_{n' \to n} P_{n'}^{(st)} - \sum_{n'} k^{(R)}_{n \to n'} P_n^{(st)} + \sum_{n'} k^{(L)}_{n' \to n} P_{n'}^{(st)} - \sum_{n'} k^{(L)}_{n \to n'} P_n^{(st)} = 0. \tag{20.6.16}$$

Using the matrix notations of Eqs. (19.4.16, 19.4.17), this equation can be written as a homogeneous system of linear equations,

$$(\mathbf{K}_R + \mathbf{K}_L) \mathbf{P}^{(st)} = 0, \tag{20.6.17}$$

where

$$[\mathbf{K}_R]_{n,m} = (1 - \delta_{m,n}) k^{(R)}_{m \to n} - \delta_{m,n} \sum_{n' \neq n} k^{(R)}_{n \to n'}$$

$$[\mathbf{K}_L]_{n,m} = (1 - \delta_{m,n}) k^{(L)}_{m \to n} - \delta_{m,n} \sum_{n' \neq n} k^{(L)}_{n \to n'}. \tag{20.6.18}$$

Since the total averaged electron number in the system is also constant in time ($\frac{\partial}{\partial t} N_S(t) = 0$, by Eq. (20.6.11)), the net particle flux into the system vanishes. However, unlike in an equilibrium state (obtained by coupling the system to a single reservoir), the average fluxes at the two system–reservoir interfaces, $J_L^{(st)}$ and $J_R^{(st)}$, do not necessarily vanish. Instead, the only restriction imposed by the stationarity condition (Eq. (20.6.16)) reads

$$J_L^{(st)} = -J_R^{(st)}. \qquad (20.6.19)$$

This restriction implies that any increase in the number of particles coming from the left reservoir, is compensated by an equivalent decrease in the number of particles leaving to the right reservoir (and vice versa). This steady state is characterized by a constant charge current through the system, given by the probability flux, multiplied by the electron charge. Without loss of generality, defining the current direction from left to right, the steady-state current reads

$$I_{L\to R}^{(st)} = e \sum_{n,n'} k_{n'\to n}^{(L)} P_{n'}^{(st)} N_n - e \sum_{n,n'} k_{n\to n'}^{(L)} P_n^{(st)} N_n = e \sum_{n,n'} k_{n'\to n}^{(L)} P_{n'}^{(st)} (N_n - N_{n'}), \qquad (20.6.20)$$

or $I_{L\to R}^{(st)} = e \sum_n [\mathbf{K}_L \mathbf{P}^{(st)}]_n N_n.$

Current through a Single-State Impurity

The solution of the homogeneous equation, Eq. (20.6.17), to obtain the steady-state populations, $\{P_n^{(st)}\}$ is generally straightforward. It may become cumbersome, however, when the number of many-particle states within the system is large. It is instructive, therefore, to relate to an analytically solvable model, which captures some qualitative essence. For this purpose, we refer to the single-state impurity model discussed in Section 20.5. The model Hamiltonian, Eq. (20.5.18), is extended to account for two ("right" and "left") reservoirs,

$$\hat{H} = \varepsilon_0 \hat{a}_0^\dagger \hat{a}_0 + \sum_{k_L=1}^{M_L} \varepsilon_{k_L} \hat{a}_{k_L}^\dagger \hat{a}_{k_L} + \sum_{k_R=1}^{M_R} \varepsilon_{k_R} \hat{a}_{k_R}^\dagger \hat{a}_{k_R} + \left[\sum_{k_L=1}^{M_L} \gamma_{k_L} \hat{a}_0^\dagger \hat{a}_{k_L} + \sum_{k_R=1}^{M_R} \gamma_{k_R} \hat{a}_0^\dagger \hat{a}_{k_R} + h.c. \right].$$
$$(20.6.21)$$

The eigenstates of the system Hamiltonian in this case correspond to the unoccupied ($|0\rangle$) and occupied ($|1\rangle$) single-particle states. There are two transition rates between these states at each of the reservoirs, as illustrated in Fig. 20.6.1. Setting $n, n' \in (0,1)$, the master equation for the corresponding eigenstate populations, $P_0(t) = \rho_{0,0}(t)$ and $P_1(t) = \rho_{1,1}(t)$, reads (see Eq. (20.6.12))

$$\frac{\partial}{\partial t} P_0(t) = (k_{1\to 0}^{(R)} + k_{1\to 0}^{(L)}) P_1(t) - (k_{0\to 1}^{(R)} + k_{0\to 1}^{(L)}) P_0(t)$$
$$\frac{\partial}{\partial t} P_1(t) = (k_{0\to 1}^{(R)} + k_{0\to 1}^{(L)}) P_0(t) - (k_{1\to 0}^{(R)} + k_{1\to 0}^{(L)}) P_1(t), \qquad (20.6.22)$$

and the corresponding transient probability fluxes read (Eq. (20.6.14))

$$J_L(t) = e k_{0\to 1}^{(L)} P_0(t) - e k_{1\to 0}^{(L)} P_1(t)$$
$$J_R(t) = e k_{0\to 1}^{(R)} P_0(t) - e k_{1\to 0}^{(R)} P_1(t) \qquad (20.6.23)$$

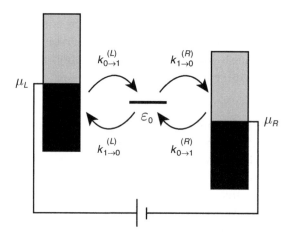

A schematic representation of the kinetics of charge transport through a single-state impurity. The electron reservoirs are characterized by the conductance bands, filled up to the respective chemical potentials. (Black and gray areas correspond to electronically filled and empty reservoir states at zero temperature.) The master equations are characterized by four rate constants, corresponding to electron absorption or emission at each electrode interface.

At steady state, the populations satisfy the stationarity condition, $\frac{\partial}{\partial t} P_0(t) = \frac{\partial}{\partial t} P_1(t) = 0$ (Ex. 20.6.2), which yields

$$P_0^{(st)} = \frac{k_{1\to0}^{(R)} + k_{1\to0}^{(L)}}{k_{1\to0}^{(R)} + k_{1\to0}^{(L)} + k_{0\to1}^{(R)} + k_{0\to1}^{(L)}} \; ; \; P_1^{(st)} = \frac{k_{0\to1}^{(R)} + k_{0\to1}^{(L)}}{k_{1\to0}^{(R)} + k_{1\to0}^{(L)} + k_{0\to1}^{(R)} + k_{0\to1}^{(L)}}, \qquad (20.6.24)$$

and the steady-state charge current therefore reads (Eq. (20.6.20))

$$I_{L\to R}^{(st)} = e k_{0\to1}^{(L)} P_0^{(st)} - e k_{1\to0}^{(L)} P_1^{(st)}. \qquad (20.6.25)$$

The total charge on the impurity at steady state, and the charge current through it depend on the microscopic model parameters, as well as on the constraints imposed on the reservoirs. To analyze these dependencies, we use the expression of the state-to-state transition rates at the Kth electrode in terms of the corresponding spectral densities and Fermi distribution functions (see Eq. (20.5.24)):

$$k_{1\to0}^{(K)} = \frac{1}{\hbar} J_K(\varepsilon_0) f_{h,K}(\varepsilon_0) \; ; \; k_{0\to1}^{(K)} = \frac{1}{\hbar} J_K(\varepsilon_0) f_{e,K}(\varepsilon_0). \qquad (20.6.26)$$

Eq. (20.6.24) therefore yields (Ex. 20.6.3)

$$P_1^{(st)} = \frac{J_R(\varepsilon_0)}{J_R(\varepsilon_0) + J_L(\varepsilon_0)} f_{e,R}(\varepsilon_0) + \frac{J_L(\varepsilon_0)}{J_R(\varepsilon_0) + J_L(\varepsilon_0)} f_{e,L}(\varepsilon_0). \qquad (20.6.27)$$

The result is an average of the two Fermi distribution functions at the two reservoirs, weighted by their respective spectral densities. Notice that when the two reservoirs have the same chemical potential and temperature, the system reaches an equilibrium state in which the impurity population reflects a single Fermi distribution, just as in the case of coupling to a single reservoir (see Eq. (20.5.26)). When the reservoirs differ in their chemical potentials, a nonequilibrium steady state is reached, in which the relative

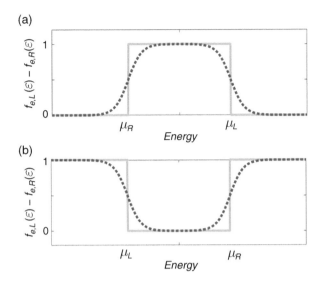

Figure 20.6.2 Fermi's conductance window for a finite bias potential between "left" and "right" reservoirs. Only charging energies that are "inside the conductance window" contribute to the left-to-right current, where the sign of the current is set by the sign of the window function, which is positive for $\mu_L > \mu_R$ (a) and negative for $\mu_R > \mu_L$ (b). The solid and dotted lines correspond to zero and finite temperatures, respectively.

dominance of each reservoir is determined according to its spectral density, namely, according to the relative system–bath coupling.

The steady-state current is obtained by using Eqs. (20.6.26, 20.6.27) in Eq. (20.6.25) (see Ex. 20.6.4):

$$I_{L \to R}^{(st)} = \frac{e}{\hbar} \frac{J_L(\varepsilon_0) J_R(\varepsilon_0)}{J_R(\varepsilon_0) + J_L(\varepsilon_0)} [f_{e,L}(\varepsilon_0) - f_{e,R}(\varepsilon_0)]. \tag{20.6.28}$$

The steady-state current is shown to be proportional to a "reduced spectral density," $\bar{J}(\varepsilon_0) = \frac{J_L(\varepsilon_0) J_R(\varepsilon_0)}{J_R(\varepsilon_0) + J_L(\varepsilon_0)}$, and to "Fermi's conductance window," which is the difference between the Fermi distributions at the two leads, $f_{e,L}(\varepsilon_0) - f_{e,R}(\varepsilon_0)$, at the impurity charging energy, ε_0 (see Fig. 20.6.2). Notice that when the spectral densities at the two reservoirs are the same, $J_L(\varepsilon_0) = J_R(\varepsilon_0) = J(\varepsilon_0)$, the reduced spectral density reads $\bar{J}(\varepsilon_0) = J(\varepsilon_0)/2$. When the spectral densities differ significantly, the current is nearly proportional to the smaller of the two, $\bar{J}(\varepsilon_0) \approx \min(J_L(\varepsilon_0), J_R(\varepsilon_0))$. Hence, the current is limited by the rate of charging/discharging at the electrode interface to which the coupling of the impurity is weaker. The Fermi conductance window can obtain, by definition, any value between -1 and 1, where positive and negative values of $f_{e,L}(\varepsilon_0) - f_{e,R}(\varepsilon_0)$ are associated with $\mu_L > \mu_R$ and $\mu_R > \mu_L$, respectively. The sign of the Fermi conductance window points to the direction of the net current at steady state, where currents in the "left-to-right" direction are positive, and vice versa. Notice that in any case, the current direction is from the higher to the lower chemical potential. However, when $\varepsilon_0 \ll \min\{\mu_L, \mu_R\}$ or $\varepsilon_0 \gg \max\{\mu_L, \mu_R\}$, the charging energy is said to be outside the "conductance window" and the current vanishes (see Fig. 20.6.3).

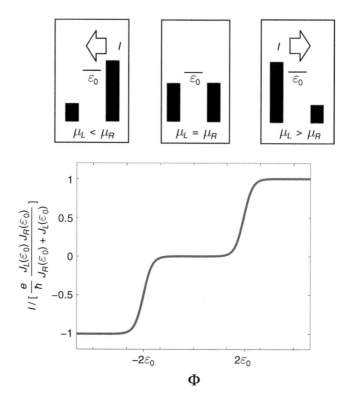

Figure 20.6.3 Current–voltage curve for the single-state impurity model. The Fermi energy at equilibrium is set to zero, and the impurity charging energy is taken to be positive, $\varepsilon_0 > 0$. The current is plotted in units of its maximal value, as a function of the bias potential between the two reservoirs, $\Phi \equiv (\mu_L - \mu_R)/e$, assuming a symmetric drop on the two contacts, $\mu_L = e\Phi/2 = -\mu_R$. The steps in the current correspond to the entrance of the charging energy into Fermi's conductance window, where the current direction is from the higher to the lower electrode chemical potential.

The current depends on the applied bias voltage between the left and right electrodes, $\Phi \equiv (\mu_L - \mu_R)/e$, as well as on the impurity charging energy, ε_0; namely, on the energy difference between the charged impurity and the Fermi energy of the electrodes at equilibrium. The current–voltage curves can therefore reveal the impurity charging energy, where for a symmetric voltage drop, namely $\mu_L = e\Phi/2 = -\mu_R$, a step in the current is revealed at $\Phi = 2\varepsilon_0$ (see Fig. 20.6.3). According to Eq. (20.6.28), the step-height equals $e\bar{J}(\varepsilon_0)/\hbar$, which depends on the coupling strength to the two reservoirs, and the step width increases with the temperature, which reflects the broadening as the Fermi conductance window as the reservoir's temperature increases.

Within the Born–Markov approximation, as the reservoirs' temperature goes to zero, so does the step-width in the current-voltage curve (Fig. 20.6.3). Moreover, as the coupling to the reservoirs (via the spectral densities) increases, the current seems to increase indefinitely. These two effects reflect an inaccuracy of the Born–Markov approximation, in which the state-to-state rate calculation is subject to strict energy conservation. We can readily see (for a detailed discussion, see Section 19.2 and Eq. (19.2.25)) that the distribution over final state energies per charging/discharging event is effectively Lorentzian-broadened (where the width corresponds to the transition rate, which is of the order of the spectral densities, divided by Plank's constant). This "intrinsic" (or "time-energy uncertainty") broadening implies that the steps in the current–voltage curve should have a finite width, and that with increasing spectral densities the current through the impurity cannot increase indefinitely, since the corresponding energy broadening of the impurity state energy may exceed the conductance window. These effects are accounted for in higher-order treatments of the system–bath coupling beyond the Born–Markov approximation [20.6] [20.7], including treatments based on quantum scattering theory [5.6], or Green's functions [20.8].

Exercise 20.6.2   *Use Eq. (20.6.22) for calculating the system eigenstate populations at steady state, $P_0^{(st)}$ and $P_1^{(st)}$, and show that the normalized populations $(P_0^{(st)} + P_1^{(st)} = 1)$ are given by Eq. (20.6.24).*

Exercise 20.6.3   *Use Eqs. (20.6.24, 20.6.26) to derive Eq. (20.6.27).*

Exercise 20.6.4   *Use Eqs. (20.6.26, 20.6.27) in Eq. (20.6.25) to derive Eq. (20.6.28).*

Transport through a Noninteracting Impurity

In the case of a multistate impurity, there are multiple charging/discharging energies, each corresponding to the addition/substruction of a single electron to the small many-particle system. Within the Born–Markov and secular approximations (assuming weak coupling to the reservoirs and non-degenerate impurity Hamiltonian), the contributions of the different charging/discharging events to the current are additive, resulting in a multistep current–voltage curve. In general, even in this case the nonequilibrium state in a multistate conductor is too cumbersome to follow analytically. Nevertheless, for a noninteracting system (when both electron–electron and vibronic interactions are negligible) the expressions for the nonequilibrium eigenstate populations and the steady-state current through the impurity are rather transparent and intuitive.

First, let us recall (see Ex. 20.5.7) that in the absence of interactions within the system (i.e., $\hat{H}_s = \sum_{m=1}^{M_A} \varepsilon_m \hat{a}_m^\dagger \hat{a}_m$ in Eq. (20.6.1)) the eigenvectors of the system Hamiltonian are determinants, denoted as $\{|n_1, n_2, n_3, \ldots, n_{M_A}\rangle\}$. Each determinant is fully characterized by the M_A-dimensional vector of occupation numbers, $\mathbf{n} = (n_1, n_2, n_3, \ldots, n_{M_A})$, where $n_m = 1$ and $n_m = 0$ correspond, respectively, to occupied and unoccupied single-particle states. The state-to-state transitions are restricted in this case to either absorption or emission of a single electron at the mth single-particle state, characterized by state specific absorption $(k_{a,m})$ or emission $(k_{e,m})$ rates. The transition rates between many-particle states therefore obtain the form (see Ex. 20.5.7),

$$k_{\mathbf{n}\to\mathbf{n}'} = \sum_{m=1}^{M_A} [\delta_{\mathbf{n}-\mathbf{n}',\mathbf{e}_m} k_{e,m} + \delta_{\mathbf{n}'-\mathbf{n},\mathbf{e}_m} k_{a,m}], \tag{20.6.29}$$

where $\mathbf{e}_m$ is a unit vector of length M_A, with elements $[\mathbf{e}_m]_{m'} = \delta_{m,m'}$. We can verify (Ex. 20.6.5) that in this case there is a steady-state solution $\left(\frac{\partial P_{\mathbf{n}}^{(st)}}{\partial t} = 0\right)$ to the master equation, Eq. (20.6.12), in which the population of the many-particle system eigenstates are products of occupation probabilities of the single-particle states, namely

$$P_{\mathbf{n}}^{(st)} = \prod_{m=1}^{M_A} (P_{m,[\mathbf{n}]_m}^{(st)}). \tag{20.6.30}$$

The probability attributed to the mth single particle state depends on whether this state is populated or not, within each $\mathbf{n}$th many-particle state; namely, whether $[\mathbf{n}]_m$ equals 'one' or 'zero,'

$$P_{m,1}^{(st)} = \frac{J_{R,m}(\varepsilon_m)}{J_{R,m}(\varepsilon_m) + J_{L,m}(\varepsilon_m)} f_{e,R}(\varepsilon_m) + \frac{J_{L,m}(\varepsilon_m)}{J_{R,m}(\varepsilon_m) + J_{L,m}(\varepsilon_m)} f_{e,L}(\varepsilon_m)$$
$$P_{m,0}^{(st)} = \frac{J_{R,m}(\varepsilon_m)}{J_{R,m}(\varepsilon_m) + J_{L,m}(\varepsilon_m)} f_{h,R}(\varepsilon_m) + \frac{J_{L,m}(\varepsilon_m)}{J_{R,m}(\varepsilon_m) + J_{L,m}(\varepsilon_m)} f_{h,L}(\varepsilon_m) \tag{20.6.31}$$

This result is a natural generalization of the result for a single-state impurity, Eq. (20.6.27), where the charging energy ε_0 is replaced by the respective single-particle state (orbital) energy, ε_m, and the spectral density at each reservoir is scaled by an m-dependent factor,

$$J_{K,m}(\varepsilon_m) \equiv |v_{m,K}|^2 J_K(\varepsilon_m). \tag{20.6.32}$$

Here, $|v_{m,K}|^2$ is the probability of occupying the (local) system site that is coupled to the Kth reservoir, at the mth single-particle state. According to Eq. (20.6.31), the steady-state probability of occupying the mth single-particle state by an electron/hole is a weighted average of the Fermi distributions of electrons/holes at the two reservoirs. The relative weights are determined independently per orbital, by its relative coupling to the two reservoirs.

Given the steady-state population of the eigenstates of the system Hamiltonian, an analytic expression for the steady-state current can also be derived (Ex. 20.6.6):

$$I_{L \to R}^{(st)} = \frac{e}{\hbar} \sum_{m=1}^{M_A} \frac{J_{R,m}(\varepsilon_m) J_{L,m}(\varepsilon_m)}{J_{R,m}(\varepsilon_m) + J_{L,m}(\varepsilon_m)} \left(f_{e,L}(\varepsilon_m) - f_{e,R}(\varepsilon_m) \right). \tag{20.6.33}$$

This result generalizes the expression derived for the single-state impurity (Eq. (20.6.28)) to the case of multiple-state impurity. As expected, the current through the different single-particle states is additive in this case. Each single-particle state (orbital) contributes independently to the current, where this contribution is proportional to a "reduced spectral density," $\bar{J}(\varepsilon_m) = \frac{J_{L,m}(\varepsilon_m) J_{R,m}(\varepsilon_m)}{J_{R,m}(\varepsilon_m) + J_{L,m}(\varepsilon_m)}$, and to the Fermi conductance window, evaluated at the single-particle state energy, $f_{e,L}(\varepsilon_m) - f_{e,R}(\varepsilon_m)$. This implies that the current–voltage plot (see Fig. 20.6.4) can reveal the single-particle energies of the impurity. Each such energy corresponds to a step, whose position is at a specific voltage $\Phi_m = 2\varepsilon_m$ (assuming a symmetric voltage drop on the contacts), and whose height depends on the coupling strength to the two reservoirs. An important aspect apparent from Eqs. (20.6.32, 20.6.33) is that the contribution of each single-particle state to the current increases with increasing projections of the corresponding orbital on the local sites to which the two electrodes are coupled ($|v_{m,L}|^2$ and $|v_{m,R}|^2$). In the context of molecular electronics, for example, this means that molecules with delocalized

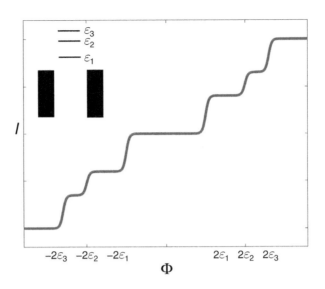

Figure 20.6.4 Current–voltage curve for a multistate noninteracting impurity model. The Fermi energy at equilibrium is set to zero, and the charging energies of the impurity single-particle states are taken to be positive, $\varepsilon_1, \varepsilon_2, \varepsilon_3, \ldots > 0$. The current is plotted as a function of the bias potential between the two reservoirs, $\Phi \equiv (\mu_L - \mu_R)/e$, assuming a symmetric drop on the two contacts, $\mu_L = e\Phi/2 = -\mu_R$. The steps in the current correspond to the entrance of the different orbital energies into Fermi's conductance window.

orbitals (which have simultaneous nonvanishing probability amplitudes at the contact sites to the two electrodes) are better conductors in a single molecule junction, which is indeed the case. Such molecules are typically associated with conjugated systems of π-electrons, as discussed in Section 14.4.

Notice that the current formula, Eqs. (20.6.32, 20.6.33), is accurate (apart from the missing broadening) only within the limitations of the Born–Markov and secular approximations; namely, when the single-particle states are well separated in energy (in the sense that the level spacings are large in comparison to the spectral densities) and the many-particle Hamiltonian is nondegenerate. When this is not the case, coherences between system eigenstates become important, and interference between different transport pathways leads to a rich behavior (e.g., [20.9]), which is beyond the expected step-like current–voltage curves predicted by the Born–Markov and secular approximations.

Exercise 20.6.5 *(a) Use the results of Ex. 20.5.7 to show that for a system of noninteracting fermions, the equation for the steady-state population of the $\mathbf{n}$th many-particle state $(P_{\mathbf{n}}^{(st)}$, Eq. (20.6.16)) reads $\sum_{m=1}^{M_A}[\delta_{\mathbf{n}'-\mathbf{n},e_m}k_{e,m} + \delta_{\mathbf{n}-\mathbf{n}',e_m}k_{a,m}]P_{\mathbf{n}'}^{(st)} - [\delta_{\mathbf{n}-\mathbf{n}',e_m}k_{e,m} + \delta_{\mathbf{n}'-\mathbf{n},e_m}k_{a,m}]P_{\mathbf{n}}^{(st)} = 0$, where $k_{e,m} = \frac{1}{\hbar}J_{R,m}(\varepsilon_m)f_{h,R}(\varepsilon_m) + \frac{1}{\hbar}J_{L,m}(\varepsilon_m)f_{h,L}(\varepsilon_m)$, $k_{a,m} = \frac{1}{\hbar}J_{R,m}(\varepsilon_m)f_{e,R}(\varepsilon_m) + \frac{1}{\hbar}J_{L,m}(\varepsilon_m)f_{e,L}(\varepsilon_m)$, and $J_{K,m}(\varepsilon) = |v_{m,K}|^2 J_K(\varepsilon)$. (b) Show that $P_{\mathbf{n}}^{(st)} = \prod_{m=1}^{M_A}(P_{m,[\mathbf{n}]_m}^{(st)})$, with $P_{m,[\mathbf{n}]_m}^{(st)} = \frac{k_{a,m}}{k_{e,m}+k_{a,m}}\delta_{[\mathbf{n}]_m,1} + \frac{k_{e,m}}{k_{e,m}+k_{a,m}}\delta_{[\mathbf{n}]_m,0}$, satisfies the equation for the steady-state population of the $\mathbf{n}$th many-particle state in (a). (c) Show that $P_{m,1}^{(st)} + P_{m,0}^{(st)} = $*

1, and therefore the sum over all the eigenstate populations reads $\sum_{\mathbf{n}} P_{\mathbf{n}}^{(st)} = 1$. *(d) Use the explicit expressions for the absorption and emission rates in (a) to derive Eq. (20.6.31) for the occupation probabilities of the single-particle states, at steady state.*

Exercise 20.6.6 *For a system of noninteracting fermions, the equation for the steady-state current (Eq. (20.6.20)) can be written as* $I_{L \to R}^{(st)} = e \sum_{\mathbf{n},\mathbf{n}'} k_{\mathbf{n}' \to \mathbf{n}}^{(L)} P_{\mathbf{n}'}^{(st)} [N_{\mathbf{n}} - N_{\mathbf{n}'}]$, *where* $\mathbf{n}$ *is an occupation vector,* $\mathbf{n} = (n_1, n_2, n_3, \ldots, n_{M_A})$. *(a) Using Eq. (20.6.29) for the state-to-state transition rates, and the results of Ex. 20.6.5 for the corresponding steady-state populations in this case,* $P_{\mathbf{n}}^{(st)} = \prod_{m=1}^{M_A} (P_{m,[\mathbf{n}]_m}^{(st)})$, *with* $P_{m,[\mathbf{n}]_m}^{(st)} = \frac{k_{a,m}}{k_{e,m} + k_{a,m}} \delta_{[\mathbf{n}]_m, 1} + \frac{k_{e,m}}{k_{e,m} + k_{a,m}} \delta_{[\mathbf{n}]_m, 0}$, *show that* $I_{L \to R}^{(st)} = e \sum_{m=1}^{M_A} [-P_{m,1}^{(st)} k_{e,m}^{(L)} + P_{m,0}^{(st)} k_{a,m}^{(L)}]$. *(b) Use the explicit expressions for the emission and absorption rates (see Ex. 20.6.5),* $k_{e,m} = \frac{1}{\hbar} J_{R,m}(\varepsilon_m) f_{h,R}(\varepsilon_m) + \frac{1}{\hbar} J_{L,m}(\varepsilon_m) f_{h,L}(\varepsilon_m)$, $k_{a,m} = \frac{1}{\hbar} J_{R,m}(\varepsilon_m) f_{e,R}(\varepsilon_m) + \frac{1}{\hbar} J_{L,m}(\varepsilon_m) f_{e,L}(\varepsilon_m)$, *to derive Eq. (20.6.33) for the steady-state current.*

Bibliography

[20.1] J. Hubbard, "Electron correlations in narrow energy bands," Proceedings of the Royal Society of London. Series A, Mathematical and Physical Sciences 277, 238 (1963).

[20.2] P. W. Anderson, "Localized magnetic states in metals," Physical Review 124, 41 (1961).

[20.3] M-C. Desjonqueres and D. Spanjaard, "Concepts in Surface Physics" (Springer, 1996).

[20.4] J. C. Cuevas and E. Scheer, "Molecular Electronics: An Introduction to Theory and Experiment" (World Scientific, 2nd Edition 2017).

[20.5] M. Galperin, M. A. Ratner and A. Nitzan. "Molecular transport junctions: Vibrational effects," Journal of Physics: Condensed Matter 19, 103201 (2007).

[20.6] L. Mühlbacher and E. Rabani, "Real-time path integral approach to nonequilibrium many-body quantum systems," Physical Review Letters 100, 176403 (2008).

[20.7] C. Schinabeck, A. Erpenbeck, R. Härtle and M. Thoss, "Hierarchical quantum master equation approach to electronic-vibrational coupling in nonequilibrium transport through nanosystems," Physical Review B 94, 201407 (2016).

[20.8] Y. Meir and N. S. Wingreen, "Landauer formula for the current through an interacting electron region," Physical Review Letters 68, 2512 (1992).

[20.9] R. Härtle, G. Cohen, D. R. Reichman and A. J. Millis, "Decoherence and lead-induced interdot coupling in nonequilibrium electron transport through interacting quantum dots: A hierarchical quantum master equation approach," Physical Review B 88, 235426 (2013).

Index